Plant Response to Stress

Functional Analysis in Mediterranean Ecosystems

NATO ASI Series

Advanced Science Institutes Series

A series presenting the results of activities sponsored by the NATO Science Committee, which aims at the dissemination of advanced scientific and technological knowledge, with a view to strengthening links between scientific communities.

The Series is published by an international board of publishers in conjunction with the NATO Scientific Affairs Division

A Life Sciences B Physics	Plenum Publishing Corporation London and New York
C Mathematical and Physical Sciences	D. Reidel Publishing Company Dordrecht, Boston, Lancaster and Tokyo
D Behavioural and Social Sciences E Applied Sciences	Martinus Nijhoff Publishers Boston, The Hague, Dordrecht and Lancaster
F Computer and Systems Sciences G Ecological Sciences H Cell Biology	Springer-Verlag Berlin Heidelberg New York London Paris Tokyo

Series G: Ecological Sciences Vol. 15

Plant Response to Stress

Functional Analysis in Mediterranean Ecosystems

Edited by

John D. Tenhunen

Systems Ecology Research Group, San Diego State University
San Diego, CA 92182, USA

Fernando M. Catarino

Department of Plant Biology, Faculty of Sciences
University of Lisbon, 1294 Lisbon, Portugal

Otto L. Lange

Lehrstuhl für Botanik II der Universität Würzburg
Mittlerer Dallenbergweg 64, 8700 Würzburg, FRG

Walter C. Oechel

Systems Ecology Research Group, San Diego State University
San Diego, CA 92182, USA

Springer-Verlag
Berlin Heidelberg New York London Paris Tokyo
Published in cooperation with NATO Scientific Affairs Division

Proceedings of the NATO Advanced Research Workshop on Plant Response to Stress – Functional Analysis in Mediterranean Ecosystems held at Hotel do Mar, Sesimbra, Portugal, October 13–18, 1985

ISBN 3-540-16082-5 Springer-Verlag Berlin Heidelberg New York
ISBN 0-387-16082-5 Springer-Verlag New York Berlin Heidelberg

Library of Congress Cataloging in Publication Data. NATO Advanced Research Workshop on Plant Response to Stress (1985 : Sezimbra, Portugal) Plant response to stress. (NATO ASI series. Series G, Ecological sciences ; vol. 15) "Proceedings of the NATO Advanced Research Workshop on Plant Response to Stress ... held at Hotel do Mar, Sesimbra, Portugal, October 13–18, 1985"—T.p. verso. 1. Botany—Mediterranean Region—Ecology—Congresses. 2. Plants, Effect of stress on—Mediterranean Region—Congresses. I. Tenhunen, John D., 1946-. II. North Atlantic Treaty Organization. III. Title. IV. Series: NATO ASI series. Series G, Ecological sciences ; vol. 15. QK314.5N38 1985 581.5'09182'2
87-23333
ISBN 0-387-16082-5

Printed in Germany

Printing: Druckhaus Beltz, Hemsbach; Bookbinding: J. Schäffer GmbH & Co. KG, Grünstadt
2131/3140-543210

PREFACE

This book is dedicated to international cooperation, understanding and peace. It is the end result of several years of cooperative work between scientists of three countries: the United States, Germany, and Portugal. The work presented, however, draws from a much broader base, hopefully achieving the objective of NATO Advanced Research Workshops, which have been established to allow and stimulate the exchange of new ideas and the synthesis of information by scientists of NATO countries.

The tasks of the workshop were several; to review established methodologies that have provided insight into ecosystem function and adaptations of plants in mediterranean climate zones; to examine new methodologies that have recently been applied in ecological studies and have provided new types of information; to summarize recent studies in mediterranean regions of plant water relations, photosynthesis and production, mineral nutrition, plant growth and development, and response to fire; to stimulate in particular an exchange of information among scientists of European Mediterranean countries; and to discuss means by which all of these objectives might be even more effectively achieved in the future through cooperative international research efforts. This variety of themes is clearly evident in the layout of the book.

Held in Sesimbra, Portugal in October of 1985, the workshop took place in a setting dominated by the National Park Serra da Arrábida. This setting was appropriate, having provided one of the research sites for our field studies, allowing field trips during which ecological field methods were demonstrated, and representing the need for long-term management of natural areas. Understanding resource use within these natural mediterranean ecosystems is difficult, but as pointed out by Professor Naveh, management of these ecosystems to provide the greatest benefits to all interest groups is a tremendous challenge. Several round-table discussions provided opportunities to discuss the relationship of ecosystem research to the solution of local environmental problems. The attendance, interest, and participation of a number of dignitaries and local funding agency personnel in these discussions was very much appreciated.
The success of this workshop as well as the program of studies carried out near Lisbon depended on outstanding efforts by innumerable persons. Recognizing the impossibility of listing all contributors, I would like to thank my colleagues and co-editors, the Catarino family, the Würzburg research team, the students of the Faculty of Sciences in Lisbon, J. S. Pereira and M. Chaves, the H. Walz family, A. Hoffman, M. Caldwell, P.C. Harley, O. Correia, M. David, O. Schulz-

Kampfhenkel and A. Pircher, the L. P. Pires family, and the staff of the Hotel do Mar in Sesimbra, Portugal for their generous support, assistance, and encouragement.

As it has contributed to the success of this effort, I would like to thank my wife Laura for her willingness to encounter both the elations and setbacks of an international family life during the period of our activites in Germany and Portugal. I also appreciate the dedicated work of Ms. Cathy Wermuth, who provided the technical services required in this desktop publishing project.

John D. Tenhunen

Contents

1. Methods and Experimental Strategies for Assessing Mediterranean Plant Function and Response to Stress

2. Plant Water Relations and Response to Water Stress

3. Photosynthesis and Primary Production under Mediterranean Climate Conditions

4. Carbon Cycling and Plant Response to Limited Nutrient Supply

5. Mediterranean Plant Growth, Structure, and Development

6. Fire Effects on Mediterranean Plant Populations

7. The Response to Stress at the Landscape Scale

8. Future Directions

1. Methods and Experimental Strategies for Assessing Mediterranean Plant Function and Response to Stress

Gas exchange methods for characterizing the impact of stress on leaves

O.L. Lange, P.C. Harley* , W. Beyschlag and J.D. Tenhunen*

Lehrstuhl für Botanik II, Universität Würzburg, Mittlerer Dallenbergweg 64, 8700 Würzburg, West Germany

INTRODUCTION

Mediterranean climates are characterized by moist, mild winters and hot, dry summers, thus exposing the leaves of evergreen sclerophyllous mediterranean vegetation to both of these weather regimes. Production and growth take place primarily during favorable conditions in the fall and spring, whereas metabolic activity becomes limited in most species as soil moisture stress and high atmospheric evaporative demand, coupled with high radiation loads, develop during the summer. While cold stress may occur during the winter that is harmful to cold-sensitive species, summer drought seems in general to play the major role in limiting growth of evergreen sclerophylls. They are usually replaced by rain-green malacophyllous deciduous vegetation as climatic conditions become drier towards the desert belts of the world.

Although extreme stress conditions during the mediterranean summer may lead to irreversible damage in the form of die-back due to drying or direct heat injury to foliage, evergreen leaves are adapted to mediterranean climates and normally survive the summer drought. Nevertheless, water and heat stress may cause severe but reversible changes in leaf metabolism affecting photosynthetic CO_2-exchange. Summer drought and heat stress not only decrease the absolute level of photosynthetic primary production (in the extreme case to zero), but also influence diurnal response patterns of CO_2-uptake.

The present chapter elaborates ways in which gas exchange measurements may be used to study and analyze the impact of environmental stresses on leaves. It will deal not only with the technical aspects of instrumentation design and operation, but also with conceptual experimental strategies, focusing on applications involving mediterranean sclerophyll vegetation.

INSTRUMENTATION FOR FIELD GAS EXCHANGE MEASUREMENTS

Enormous progress has been made over the last several years in developing instrumentation for gas-exchange measurements under field conditions, particularly with respect to portability. Many types of apparatus now available are

*Systems Ecology Research Group, San Diego State University, San Diego, California, USA

NATO ASI Series, Vol. G 15
Plant Response to Stress. Edited by J. D. Tenhunen et al.

capable of monitoring both CO_2 uptake and transpirational H_2O loss, as well as relevant environmental variables. In each of these systems, intact plant tissue is typically enclosed in a cuvette and CO_2 fluxes are measured using an infrared gas analyzer. The details of construction for each system vary considerably, however, with respect to the means for measurement of H_2O flux and the degree of control of environmental conditions within the cuvette. These depend on the specific purpose for which they were designed.

The choice of a gas exchange measuring system is influenced by requirements imposed due to the plant material under study, the nature of the experimental goals, and the need for portability. Size and geometry of the plant parts under investigation determine the size and shape of the cuvette, which determines in part requirements to achieve environmental conditioning in a particular climate. At one extreme, there are large cuvettes enclosing entire plants or even stands of crop plants (Reicosky and Peters 1977). Caldwell *et al.* (1983) used large cuvettes (400 l volume) that enclosed entire individuals of *Artemisia tridentata.* The interior of the cuvette requires temperature conditioning and rapid venting to avoid overheating, drawing down of CO_2 and increasing humidity. Long-term investigations of carbon balance and water relations can be carried out using experimental apparatus that include entire plants and, in addition, aspects of plant canopy function can be verified. Portability is not a design consideration for large cuvettes.

At the other extreme, small, easily portable porometer-type cuvettes are available, in which a single leaf or only a few square centimeters of one side of a leaf blade is exposed to the cuvette interior (e.g., Griffiths and Jarvis 1981; Field, Berry and Mooney 1982; Schulze *et al.* 1982; Williams, Gurner and Austin 1982). Large cuvettes generally dictate a long enclosure time of plant samples, on the order of days, due to slow system response time. With smaller cuvettes, a single measurement of net photosynthesis and transpiration may be obtained in less than a minute, and experimental designs may contain large numbers of replicates. Small cuvettes may also facilitate measurements in restricted locations, such as the inside of a shrub or tree canopy.

Actual measurements of gas exchange are often made under steady-state conditions. In many devices, the CO_2 and H_2O differentials across the cuvette are measured, when this signal is no longer found to change for a given set of conditions. In such instruments, the final CO_2 and H_2O levels are slightly different from those occurring naturally, depending on sample activity. Alternatively, systems have been designed which compensate for the amount of H_2O added to the gas stream by transpiration and the amount of CO_2 removed in photosynthesis with measured additions of dry air and 1% CO_2, respectively (Bingham *et al.* 1980; Field *et al.* 1982). In contrast to these two types of steady-state measurements, a system designed by LiCor (Lincoln, NE, USA) monitors the depletion of CO_2 and increase in H_2O under transient conditions immediately after the sample is incuded in the cuvette (Wells 1986). In this case, the measurements are made rapidly enough that changes in CO_2 and H_2O concentrations in the cuvette are assumed to have little effect on measured photosynthesis and transpiration rates.

Field measurements of plant gas exchange may be carried out with different scientific goals in mind, necessitating contrasting methodologies. The goal may be to study the behavior of the plant under ambient environmental conditions, as these conditions vary throughout the year. Thus, the conditions within the cuvette should as closely as possible approximate the environment experienced in the habitat with respect to irradiance, temperature and humidity. One approach has been to use an unclimatized cuvette, but to complete the measurements so rapidly that changes in the cuvette environment after enclosure are minimized. Holdtheide, Huber and Stocker(1936) conducted early experimentation of this type. In a later study, Lange (1962) designed a 'Klapp-Küvette' which automatically enclosed a leaf for one minute every five minutes. These early efforts were handicapped by low sensitivity and lack of stability of the gas analyzers then available. Recently, a new generation of analyzers has facilitated reliable and rapid measurements of CO_2 exchange and H_2O loss in small, portable unclimatized gas exchange cuvettes (see Lange *et al.* 1984; Wells 1986).

An alternative means of measuring in a close to natural environment is to use a cuvette system with electronic control that ensures that conditions inside the cuvette approximate those outside. Such systems are inevitably less portable and more complex in nature. Bosian (1955) was among the first to design such instrumentation which was improved upon by Koch and Walz (1967). Today, cuvettes of this type are often utilized in conjunction with fully equipped mobile field laboratories (Koch, Lange and Schulze 1971; Mooney *et al.* 1971; Beyschlag 1984; Tenhunen *et al.* 1984).

A second aim of ecophsyiological gas exchange investigations in the field may be to analyze plant functional response to specific environmental variables. Such information should allow one to interpret plant performance in terms of physiological mechanisms and to determine parameters necessary for modeling plant gas exchange. Control of temperature, humidity, incident irradiance, and CO_2 and O_2 partial pressure within the cuvette may be desirable depending on the goals of the study. During experimentation, a single environmental variable can be manipulated while the others are held constant and interactions between two or more variables may be investigated. Fully conditioned large plant chambers as well as small leaf cuvettes are now available for studies of this kind.

During our own investigations of gas exchange of mediterranean sclerophylls in Portugal, it was our aim to observe responses under natural conditions, but also to more fully analyze photosynthetic function in the leaf mesophyll. We used large conditioned chambers (ca. 5000 cm^3 volume) in which shoots with one to several leaves were enclosed for time periods of about one week (Beyschlag 1984; Tenhunen *et al.* 1984). Gas exchange and environmental data were recorded continuously at four-minute intervals over long periods of time, allowing generation of a large data base suitable for model verifications and statistical evaluation. The diurnal courses of leaf gas exchange, stomatal conductance, and environmental factors depicted in Figs. 1 through 3 illustrate the type of information obtained. One disadvantage of such measurements is that information is obtained for only a single shoot, and shoot to shoot variation has to be ignored (see methods described by Lawrence that overcome this difficulty - this volume). In order to increase sample size and provide more flexibility in experimental design, we used a portable 'minicuvette system' (Lange and Tenhunen 1984) and a portable 'H_2O/CO_2-porometer' in parallel with the large climatized cuvettes for rapid spot measurements (Lange and Tenhunen 1985). Conditioned plant chambers in combination with small cuvettes provide an ideal balance in studies that attempt to develop a picture of gas exchange behavior in response to natural conditions over the long-term as well as to detail responses of leaves on different plants and at different layers in plant canopies to individual environmental factors. It seems likely that with the increase in sophistication of electronic controls and miniaturization of component that is expected in the next few years, a single measurement system may soon be available that is suitable for both tasks.

CO_2-EXCHANGE OF MEDITERRANEAN SCLEROPHYLLS IN NATURE

Beyschlag (1984; see also Beyschlag *et al.*, Tenhunen *et al.* and Harley *et al.* this volume) used techniques such as those discussed above to investigate the CO_2-exchange of *Arbutus unedo* leaves throughout the year under ambient conditions in a natural evergreen scrub vegetation in Portugal (Research Station Quinta São Pedro, Sobreda). The results are summarized in Figs. 1 through 3, which depict diurnal courses of net photosynthesis (NP), transpirational water loss (Tr), leaf diffusive conductance (G), leaf water potential (Ψ), incident photosynthetically active radiation (PAR), and leaf temperature (T_L) for horizontally-oriented leaves at different times of the year. In the early April (Fig. 1), the leaves exhibit dome-shaped diurnal courses of photosynthesis, attaining relatively high rates of CO_2-uptake despite low irradiance. Water loss is small due to low temperatures and low leaf-to-air water vapor deficits. These results are similar to those obtained during the winter period and represent the general case during periods when leaves are not water stressed through low soil water availability and/or atmospheric stress. During the second half of April, maximum net photosynthesis rates occur early in the morning, and rates decline during the remainder of the day; the same trend is apparent with leaf conductance. As the season progresses and both atmospheric and soil moisture stress increase, such

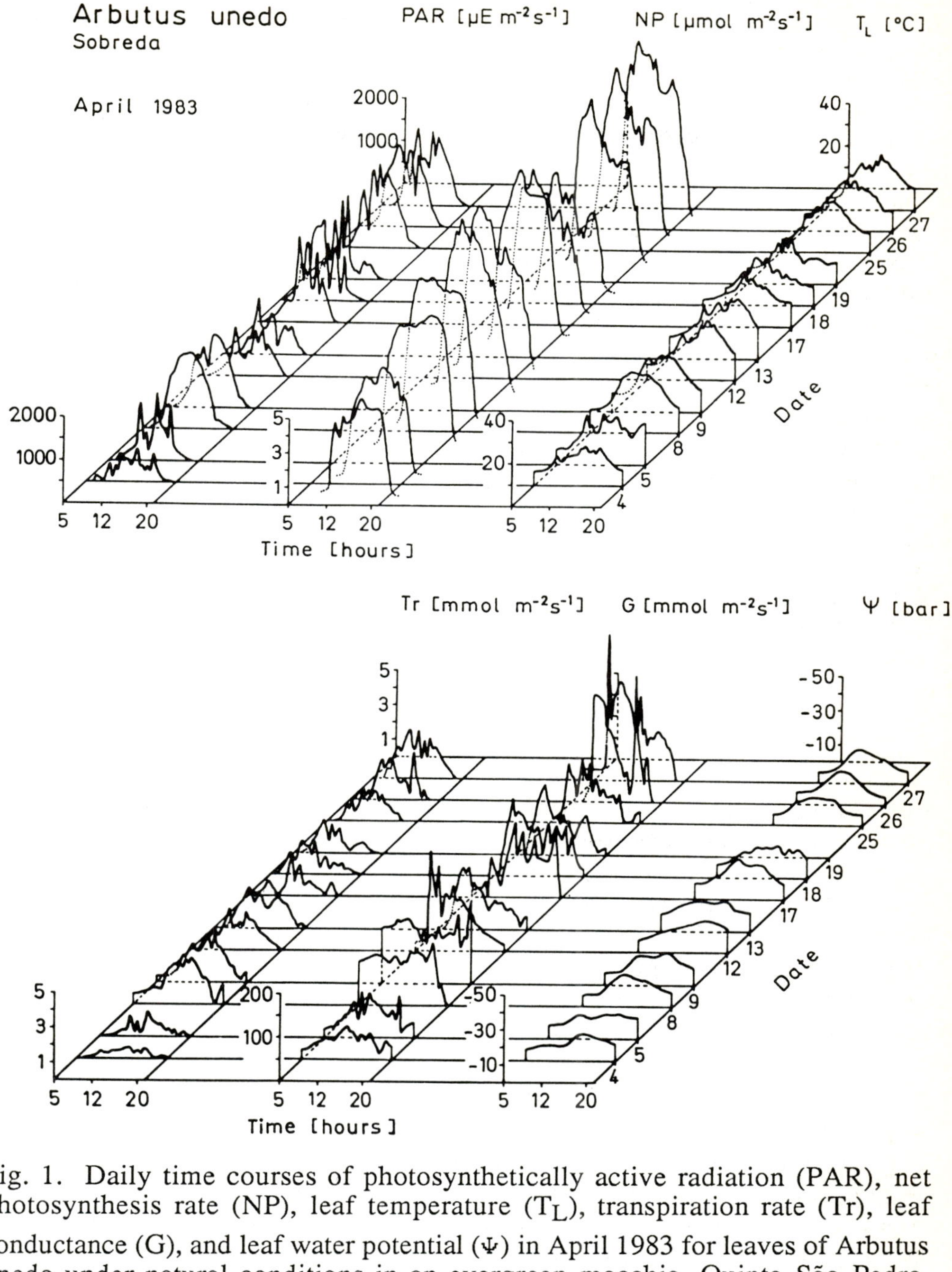

Fig. 1. Daily time courses of photosynthetically active radiation (PAR), net photosynthesis rate (NP), leaf temperature (T_L), transpiration rate (Tr), leaf conductance (G), and leaf water potential (Ψ) in April 1983 for leaves of Arbutus unedo under natural conditions in an evergreen macchia, Quinta São Pedro, Sobreda, Portugal (After Beyschlag 1984)

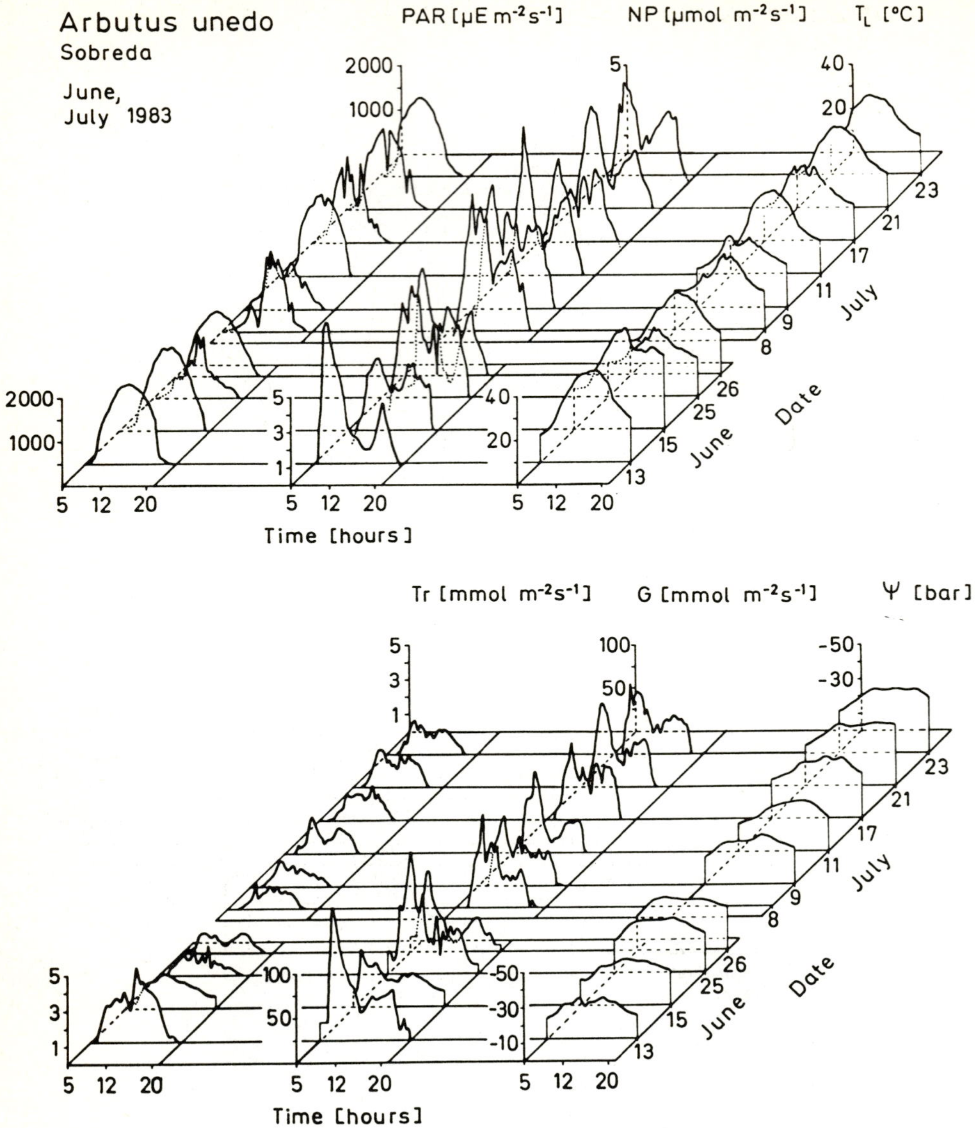

Fig. 2. Daily time courses of photosynthetically active radiation (PAR), net photosynthesis rate (NP), leaf temperature (T_L), transpiration rate (Tr), leaf conductance (G), and leaf water potential (Ψ) in June and July 1983 for leaves of Arbutus unedo under natural conditions in an evergreen macchia, Quinta São Pedro, Sobreda, Portugal (After Beyschlag 1984)

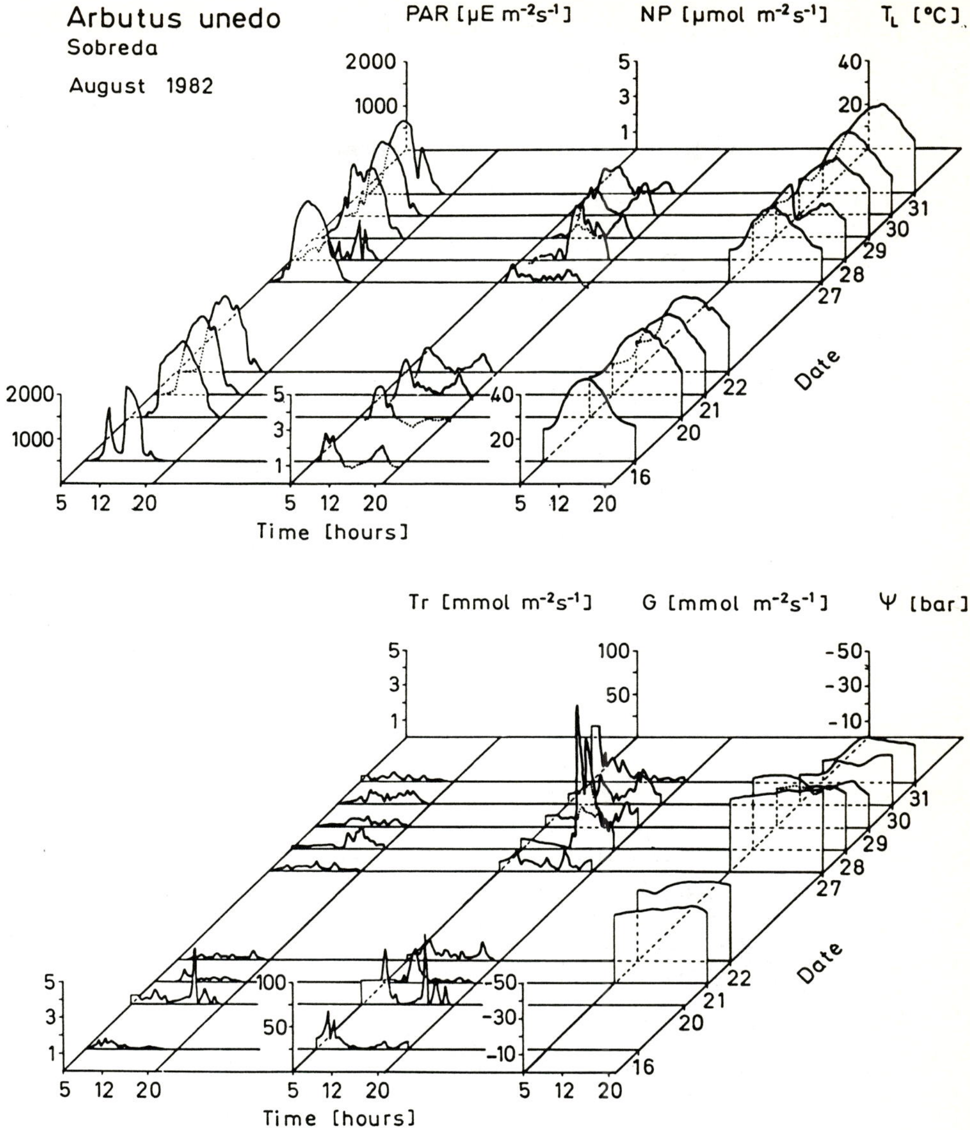

Fig. 3. Daily time courses of photosynthetically active radiation (PAR), net photosynthesis rate (NP), leaf temperature (T_L), transpiration rate (Tr), leaf conductance (G), and leaf water potential (Ψ) in August 1983 for leaves of Arbutus unedo under natural conditions in an evergreen macchia, Quinta São Pedro, Sobreda, Portugal (After Beyschlag 1984)

diurnal courses for CO_2 uptake are replaced by two-peaked curves with a strong midday depression of gas exchange as shown for June and July (Fig. 2). During extremely hot, dry conditions in August, characterized by low leaf water potentials, net photosynthesis is almost completely suppressed during midday, with low rates of CO_2 uptake occurring in the morning and afternoon (Fig. 3). Thus the period of October through May is characterized by low water stress and rates are high, but from June through September, maximal CO_2-assimilation declines markedly with decreasing leaf water potentials. Immediately following the first rain storms in the fall, water potentials recovered somewhat, but photosynthesis rates remained low (data not shown), indicating lack of immediate recovery from the effects of summer drought. Somewhat different behavior was found for the co-occurring species *Quercus coccifera* and *Quercus suber* as

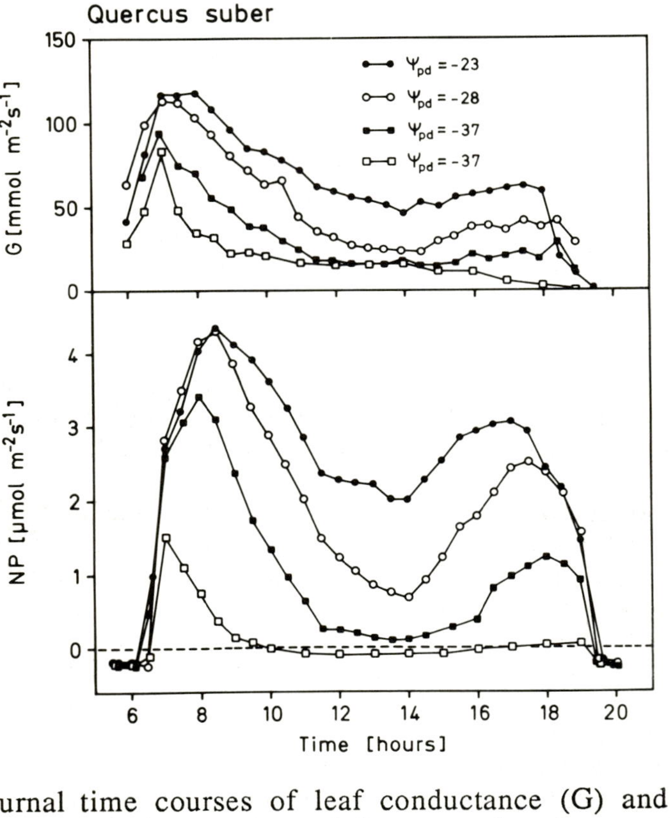

Fig. 4. Diurnal time courses of leaf conductance (G) and rates of net photosynthesis (NP) in leaves of potted *Quercus suber* plants on the third (solid circles), fourth (open circles), fifth (solid squares), and sixth (open squares) day of a soil drying cycle under "standard Portugal summer day" atmospheric conditions (see text) simulated in a growth chamber. Predawn leaf water potential (Ψ_{PD}) indicated in the top panel. (After J. Gebel, Diplom thesis, University of Würzburg)

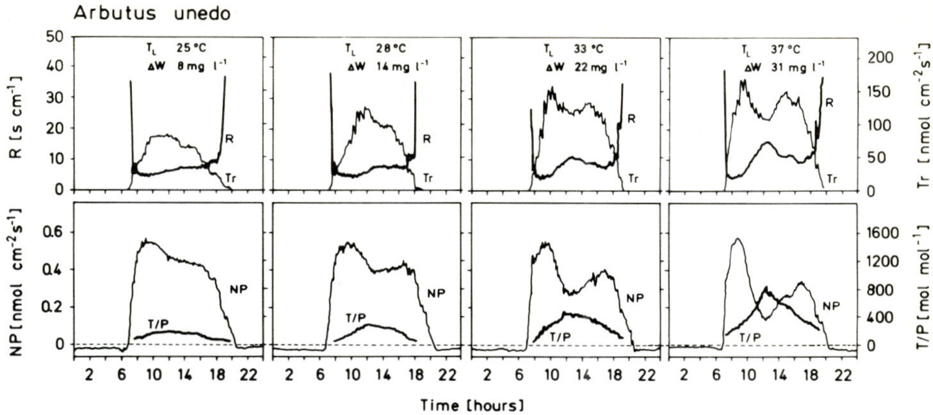

Fig. 5. Diurnal time courses of leaf resistance (R), rates of net photosynthesis (NP), rates of transpiration (Tr), and the ratio Tr/NP (T/P) with potted, well watered plants of *Arbutus unedo* on four consecutive days. Plants were subjected to simulated natural time courses of temperature, air humidity, and light (see text) according to the "standard Portugal summer day". Air temperatures (T_L) and leaf-to-air water vapor pressure differences (ΔW) were increased each day to the maximal value shown in the body of the figure. (After Tenhunen et al. 1981)

reported by Tenhunen *et al.* in this volume. In those species, a greater sensitivity to lower temperatures is found in winter and they exhibit a strong potential to increase NP and G immediately in response to rain in the fall.

The results reported above were obtained using a gas exchange system which maintained conditions in the cuvette similar to those experienced by nearby shoots. In order to elucidate the relative effects of soil vs. atmospheric stress, the enclosed shoot was subjected to diurnal courses of environmental conditions which simulated a typical summer day in the natural habitat, but avoided the inevitable daily variations in environmental conditions which occur in nature (Figs. 4 and 5 - 'Portugal standard summer day conditions'; see Tenhunen, Lange and Braun 1981; Lange, Tenhunen and Braun 1982). The results shown in Fig. 4 describe diurnal courses of net photosynthesis and leaf conductance obtained when a potted plant of the cork oak, *Quercus suber*, was subjected to progressive soil drying, leading to a reduction of predawn water potential from -23 to less than -37 bar. With decreasing predawn water potential, the diurnal pattern of CO_2-assimilation changed progressively from one characterized by a small midday depression to one characterized by a single, small morning peak in NP and G. Since atmospheric conditions remained identical throughout the experiment, these results clearly demonstrate the involvement of soil water stress both in altering the daily pattern of leaf gas exchange and in reducing absolute rates of CO_2 uptake.

A contrasting experiment using *Arbutus unedo* (Fig. 5), demonstrated that atmospheric water stress also exerts an independent effect on leaf activity. In a growth chamber, a continuously well-watered plant was subjected to simulated diurnal weather conditions in which maximum daily temperature increased from 23° C on the first day to 36° C on the last, with concomitant increases in leaf-to-air water vapor pressure differences. Under mild atmospheric stress, net photosynthetic rate was high and almost constant for most of the day. With increasing stress, leaf conductance decreased progressively during midday, and midday depression of photosynthesis and transpiration became more and more pronounced. Such behavior seems to be characteristic of most mediterranean-type sclerophyllous shrubs exposed to summer drought stress conditions, and the responses of a variety of species studied in Portugal are described in detail in this volume (Beyschlag *et al.*; Harley *et al.*; Meister *et al.*; Tenhunen, Beyschlag *et al.*; Tenhunen, Harley *et al.*; see also Correia *et al.*). Lawrence (this volume) discusses similar responses observed with leaves of matorral shrubs in Chile.

STRESS-INDUCED LIMITATION OF GAS EXCHANGE

Use of CO_2 Response Curves to Characterize the Photosynthetic Apparatus

The observed changes in leaf gas exchange behavior in response to stressful mediterranean summer conditions result due to interactive regulation by two distinct processes, diffusion of CO_2 gas to the sites of photosynthesis through the stomata on one hand, and metabolic fixation of CO_2 in the Calvin cycle utilizing trapped light energy on the other. To better understand plant adaptations to water stress, one of our goals has been to distinguish between limitations on production due to each of these processes as well as to determine whether changes in mesophyll photosynthetic capacity reflect stress induced damage and species sensitivity to water stress conditions.

From measurements of transpiration and leaf-to-air water vapor pressure gradients, it is relatively easy to determine leaf diffusive conductance which serves as a quantitative measure of stomatal aperture (see Sesták, Catsky and Jarvis 1971). It is more difficult, however, to characterize the instantaneous photosynthetic capacity of the mesophyll tissue. In order to do so, it is necessary to eliminate from consideration the variable diffusional CO_2 limitation imposed by the stomata. One has to measure 'behind the stomata', so to speak, and thereby relate rates of CO_2 assimilation to the partial pressure of CO_2 in the intercellular air space, so-called internal CO_2 or C_i. The latter can be calculated from

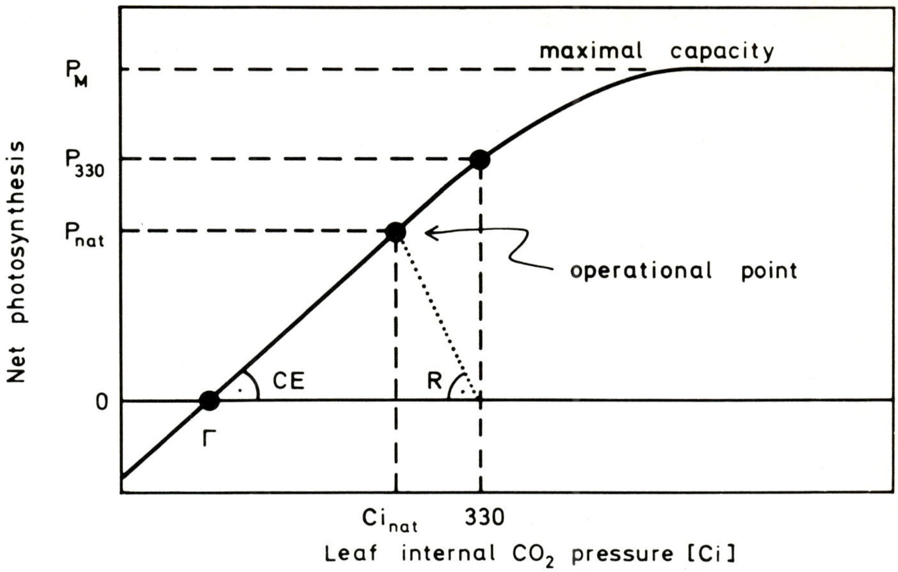

Fig. 6. Schematic presentation of the dependency of net photosynthesis rate (P) of a leaf on internal CO_2 partial pressure (C_i). See text for details

measured rates of stomatal conductance and net photosynthesis (see von Caemmerer and Farquhar 1981; Sharkey *et al.* 1982). In this context, the most reasonable standard for comparing and characterizing the status of the photosynthetic apparatus on the basis of gas exchange, is a CO_2 response curve of an intact leaf such as that in Fig. 6.

The response of NP follows saturation-type kinetics with respect to C_i. The maximal photosynthetic capacity (P_M) at saturating CO_2 concentration and saturating light intensity is thought to be determined by the maximal rate of regeneration of RuBP, the CO_2 acceptor. The initial slope of the response curve, designated carboxylation efficiency (CE), is thought to be determined by the kinetics of the carboxylating enzyme systems (Farquhar and von Caemmerer 1982). The CO_2 compensation point (Γ) is determined both by enzyme kinetics and the magnitude of mitochondrial respiration in the light. These three parameters may be utilized as indicators of the photosynthetic potential of a leaf, differ between species, and may change markedly as the result of stress (see further discussion of net photosynthetic potential in Tenhunen *et al.* - this volume). In addition, quantum yield (on an absorbed quanta basis) as determined from the initial slope of light responses of photosynthesis serves as another indicator of injurious effects of excessive irradiance, high or low temperatures, and low leaf water potentials on the photosynthetic apparatus (see Björkman

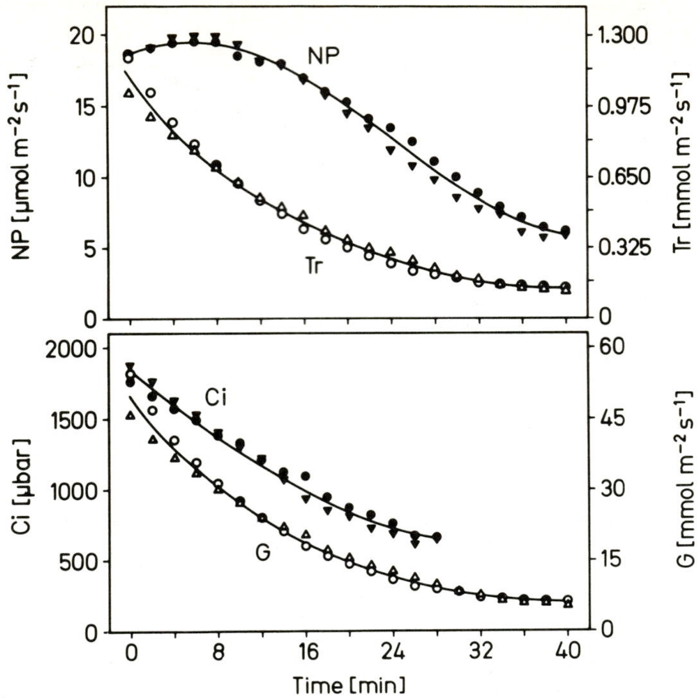

Fig. 7. During the summer drought period a leaf of *Arbutus unedo* was treated with CO_2-free air in the dark until stomata opened. Subsequently, it was subjected to saturating light for photosynthesis and high CO_2 partial pressure (beginning at time zero). Time courses are shown for the rate of net photosynthesis (NP), rate of transpiration (Tr), leaf conductance (G), and leaf internal CO_2 partial pressure (C_i). Measurements during two separate experiments with the same leaf are indicated by the triangles and circles

1981). Determination of CO_2 response curves in the field with Portuguese mediterranean type plants was accomplished using large gas exchange chambers and with the minicuvette system (see Tenhunen, Beyschlag *et al.*, Tenhunen, Harley *et al.* and Beyschlag *et al.* this volume). Light intensity, temperature and humidity were maintained constant, and external CO_2 concentration was varied stepwise to obtain the desired range of C_i values.

Under extreme drought conditions, when stomata close almost completely and leaf gas exchange is suppressed, a serious difficulty arises. In August, when predawn leaf water potentials of *Arbutus unedo* were below -40 bar, it was not possible to alter C_i by simply varying ambient CO_2. Stomata were so tightly closed that diffusion of CO_2 was severely restricted. Thus, in order to induce

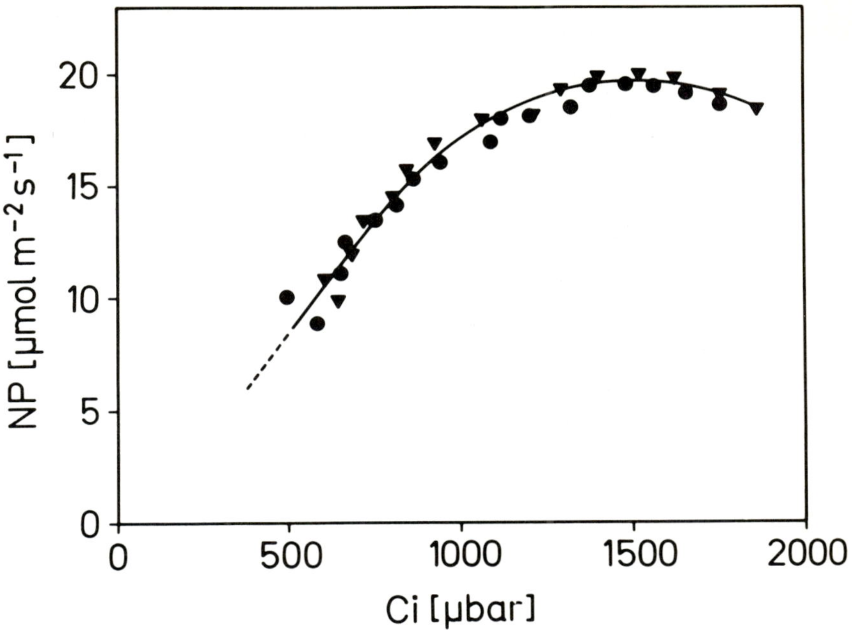

Fig. 8. Net photosynthesis rate (NP) of a leaf of *Arbutus unedo* plotted as a function of the calculated internal CO_2 partial pressure (C_i). Data were obtained in experiments of the type depicted in Fig. 7

stomatal opening, a leaf was enclosed in the dark inside the conditioned minicuvette which was flushed with CO_2-free air. After about 50 min, high stomatal conductances were obtained, and the leaf was immediately given a high ambient CO_2 concentration and saturating light (indicated as time zero in Fig. 7). Initially, C_i exceeded 2000 µbar, and rates of both CO_2 assimilation and transpiration were high. Subsequently, stomata began to close in response to high CO_2 levels and transpiration immediately decreased. Net photosynthesis, however, remained nearly constant for over 10 min, indicating that photosynthesis was CO_2-saturated. Only when C_i dropped below a value of approximately 1400 µbar did photosynthesis begin to decrease. Below 1400 µbar, concomitant changes in NP and C_i could be observed as long as transpiration readings were reliable and Ci could be calculated. Replicated experiments gave similar results, indicated by different symbols in Fig. 7. From such experiments CO_2 response curves were derived. Figure 8 shows one example in which data from two experiments are pooled. Such experiments demonstrate that the photosynthetic apparatus of a sclerophyll such as *Arbutus unedo* is still remarkably active, even when its stomata remain almost fully closed during the summer drought period and ambient CO_2 uptake rates are near zero.

Seasonal and Diurnal Changes in Characteristics of the Photosynthetic Apparatus

Figure 9 depicts CO_2 responses obtained for *Quercus suber* leaves in winter and early summer. Due to water and heat stress, the level of photosynthetic activity is substantially less during the summer measurement period. Both maximal photosynthetic capacity (P_M) and carboxylation efficiency (CE) are about half of the measured winter values (for more details, see Tenhunen, Beyschlag *et al* ., this volume). The temperature dependency of P_M also changes with increasing water stress as the season progresses, as discussed for *Quercus coccifera* in Tenhunen *et al.* (this volume). Photosynthetic capacity under the most extreme summer stress conditions is quite low. However, as with *Arbutus*, there is no total 'dormancy' of metabolic activity. More than 25 percent of the maximum photosynthetic capacity evident in winter remains present throughout the stressful summer period, even when CO_2 exchange under natural conditions is strongly suppressed.

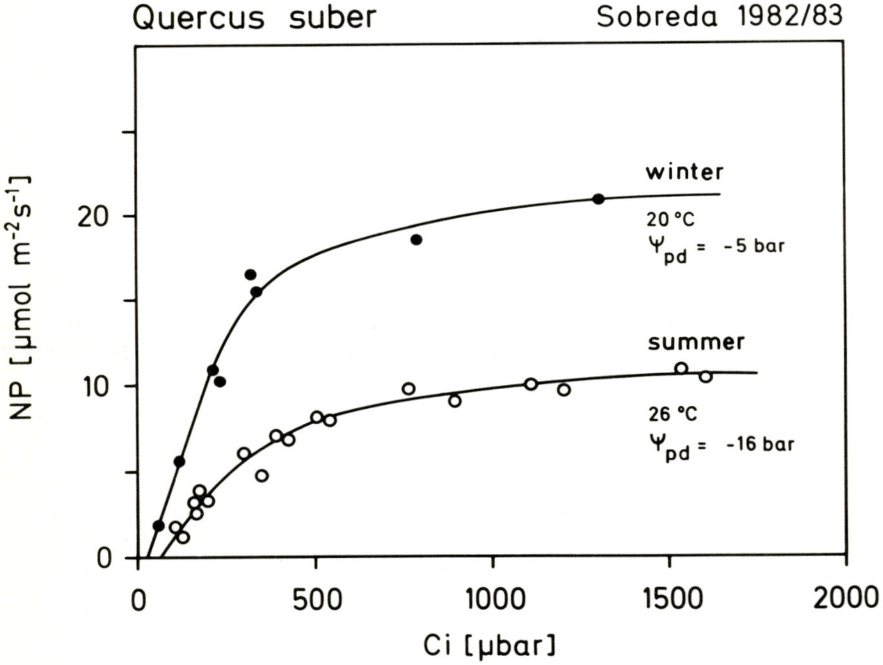

Fig. 9. Net photosynthesis rate of leaves of *Quercus suber* (saturating light) as a function of internal CO_2 partial pressure (C_i) observed during winter and summer with plants growing under natural conditions in a Portuguese macchia (Quinta São Pedro). Leaf temperature and predawn leaf water potential (Ψ_{PD}) as indicated

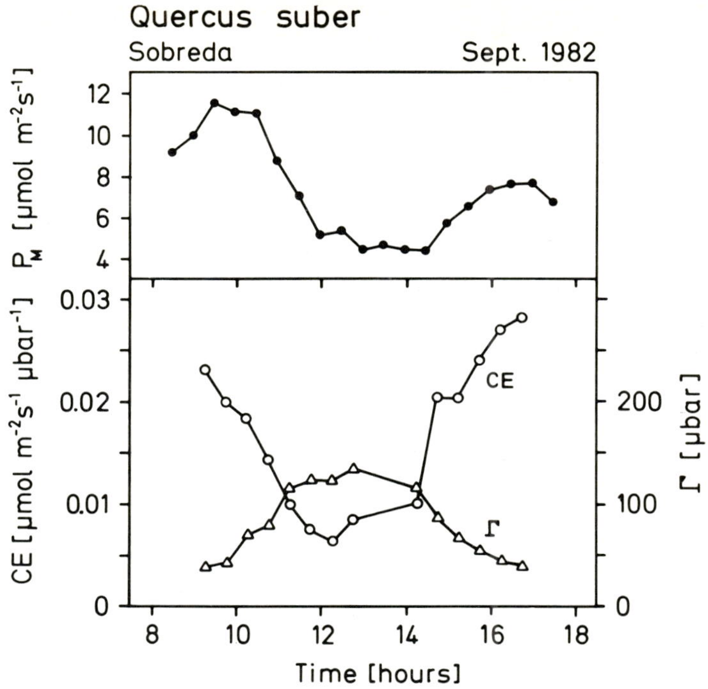

Fig. 10. Daily time courses of mesophyll photosynthetic capacity (P_M), carboxylation efficiency (CE), and leaf CO_2 compensation point (**Γ**) under "standard summer day" conditions for leaves of *Quercus suber* in the natural habitat in Portugal. See text for detailed explanation (After Tenhunen *et al.* 1984)

In addition to changing over the course of the seasons, activity of the photosynthetic apparatus also changes on a diurnal basis in response to increasing atmospheric stress during midday. Consequently, CO_2 assimilation is depressed during the hottest, driest period of the day, as discussed above. However, detailed analysis of this phenomenon is difficult, because repeated time-consuming determinations of CE and P_M during the course of a day are impractical. (The rapid development of miniaturized gas exchange systems, comparable otherwise to that used in this study, may help to alleviate this problem). We therefore applied the following methods (see Beyschlag 1984; Tenhunen *et al.* 1984). Using conditioned gas exchange chambers in the field, we imposed the artificial diurnal course of temperature, humidity and light, simulating conditions typical of mediterranean summer weather ('Portugal standard day' mentioned above). Thus, we were able to subject the leaves in the cuvette to identical, standardized conditions on several consecutive days. We then changed the ambient CO_2 concentration from day to day, resulting in a sequence of diurnal courses of net photosynthesis at 160, 350 and saturating (≈2500) μbar external CO_2 (see Beyschlag *et al.*; Tenhunen, Beyschlag *et al.*, both this volume). Thus, for each time of day, we have two points on the linear initial portion of the CO_2 response curve, defining CE and Γ, as well as a measure of P_M. Fig. 10 depicts the measured changes in CE, Γ and P_M obtained with *Q. suber* over the course of a 'Portugal standard day' in September. The stress-induced depression of the potential of the photosynthetic apparatus is obvious, as both P_M and CE are relatively high in the early morning and late afternoon but reach low values in the period around midday, while Γ shows the opposite pattern. This inhibition of biochemical photosynthetic activity during midday may be due to low ambient air humidity (see Lange *et al.* 1985), high leaf temperatures (see Weber *et al.* 1985), water stress and high irradiance (see Powles and Björkman 1982) acting alone or in combination, and may also involve a phytohormonal component (see Burschka *et al.* 1985).

Long-term seasonal patterns of drought and temperature stress and short-term environmental conditions interact with each other to determine the momentary efficiency of the photosynthetic apparatus under mediterranean climatic conditions. Tenhunen *et al.* (1985b) developed a hypothetical scheme of the seasonal pattern of CE in leaves of *Q. coccifera* in the morning when light saturation is first reached, and at midday. This hypothetical scheme is discussed and compared to measured diurnal courses of gas exchange in *Phillyrea angustifolia* by Harley *et al.* (this volume). An analysis based on shifts in CE allows a quantitative evaluation of both seasonal and diurnal stress impacts on the photosynthetic apparatus of plants. The physiological background of such changes in photosynthetic efficiency is discussed by Kaiser (this volume).

<u>Limitation of Potential Leaf Photosynthesis Due to Restricted Supply of CO_2</u>

As discussed above, photosynthetic performance under natural conditions is determined not only by the biochemical capacity of the photosynthetic machinery, but also by diffusional limitations restricting CO_2 supply. Partial pressure of CO_2 in the mesophyll tissue of the leaf is controlled by the external CO_2 partial pres-sure, the diffusion resistances through which CO_2 molecules must pass on their way from the atmosphere to the sites of carboxylation, and the biochemical acti-vity of the mesophyll tissue. Changes in stomatal aperture largely govern leaf conductance. Like the leaf mesophyll, stomata are sensitive to water shortage, low humidity and temperature stress.

The extent to which mesophyll carboxylation efficiency and stomatal conductance each contribute to the limitation of photosynthesis may be estimated from analyses of gas exchange measurements. The actual point at which a leaf is operating with respect to CO_2 ('operational point' in Fig. 6 resulting in a net

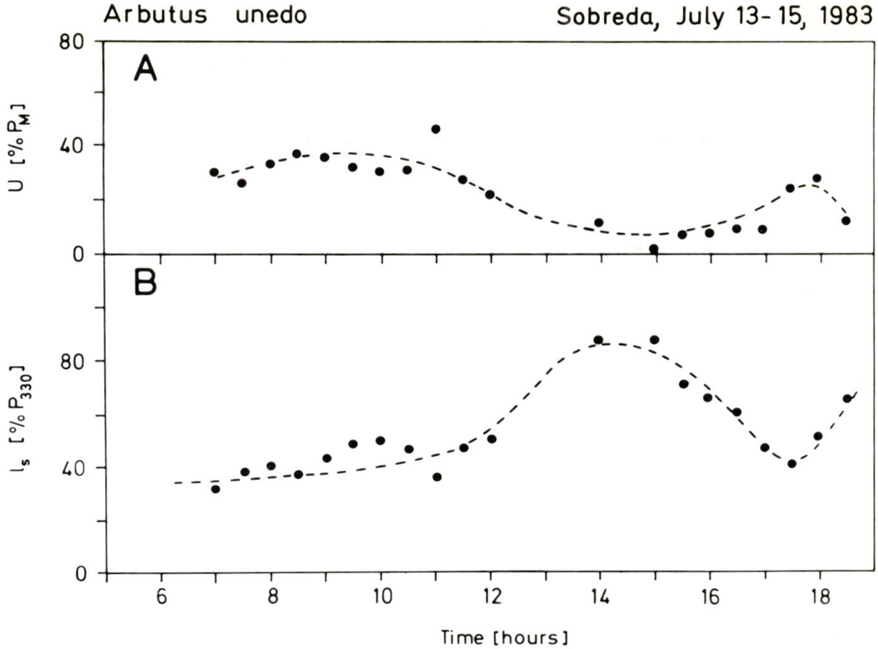

Fig. 11. Daily time courses of utilization (U, diagram A) of maximal photosynthetic capacity (P_M) and of stomatal limitation of net photosynthesis (l_s, diagram B - P_{330} indicates rate of net photosynthesis at internal CO_2 partial pressure of 330 µbar) for leaves of *Arbutus unedo* under natural conditions at Quinta São Pedro. Dates indicate three days of measurement of CO_2 responses, see text for details (After Beyschlag 1984)

photosynthesis rate P_{nat} at Ci_{nat}) may be determined graphically as the intersection of the photosynthesis versus C_i response curve and a line originating at ambient CO_2 partial pressure on the abscissa, C_{amb}, whose slope, R, is determined by leaf conductance (dotted line in Fig. 6, see Raschke 1979). Increasing leaf conductance will increase C_i and therefore photosynthesis, but even at infinite conductance C_i can never exceed C_{amb}. In order to assess the influence of stomata on gas exchange, we assume infinite conductance, resulting in a photosynthesis rate, P_{amb}. The difference between P_{amb} and P_{nat}, therefore, is due to stomatal limitations. Following Farquhar and Sharkey (1982), this stomatal limitation (l_s) may be expressed as follows: $100 \cdot (P_{amb}-P_{nat})/P_{amb}$. Increasing C_i will result in higher net photosynthesis rates until P_M is reached. The ratio P_{nat}:P_M (expressed as a percentage, U) is therefore an index of photosynthetic capacity utilization at any given time.

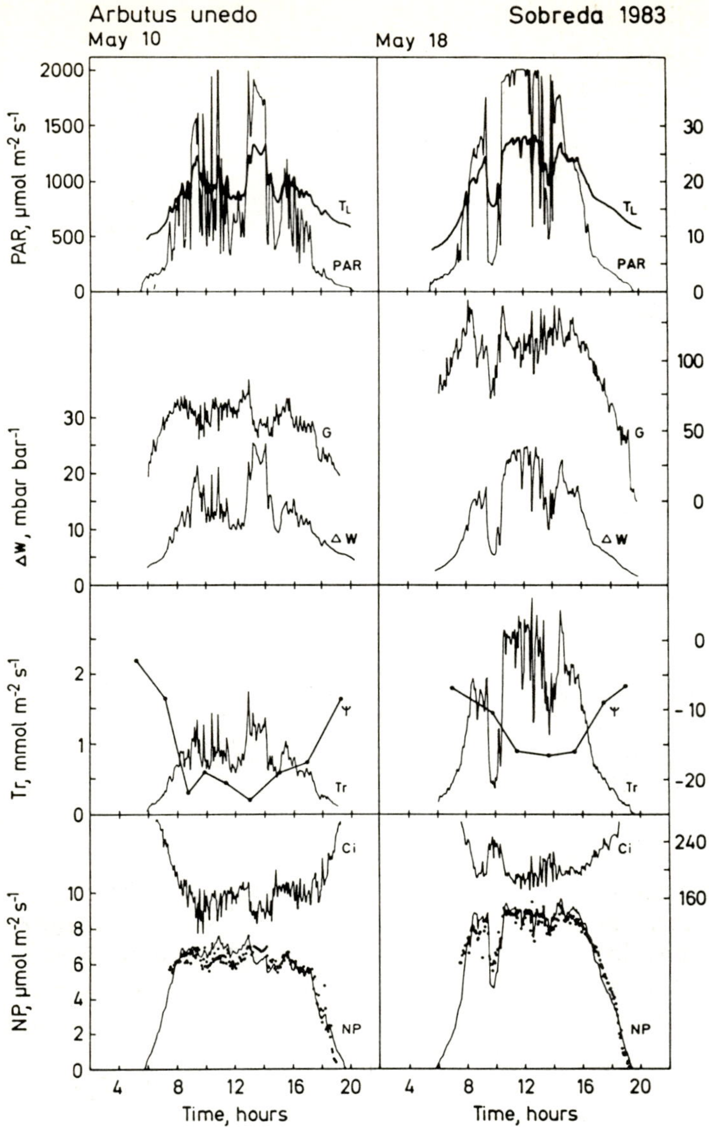

Fig. 12. Daily time courses of photosynthetically active radiation (PAR), leaf temperature (T_L), air-to-leaf water vapor pressure difference (ΔW), leaf conductance to water vapor (G), transpiration rate (Tr), leaf water potential (Ψ), net photosynthesis rate (NP), and calculated internal CO_2 partial pressure (C_i), measured on a shoot of *Arbutus unedo* under natural conditions at Quinta São Pedro, Sobreda, Portugal (After Harley *et al* . 1986). For net photosynthesis, the solid line represents measured data and symbols represent model predictions

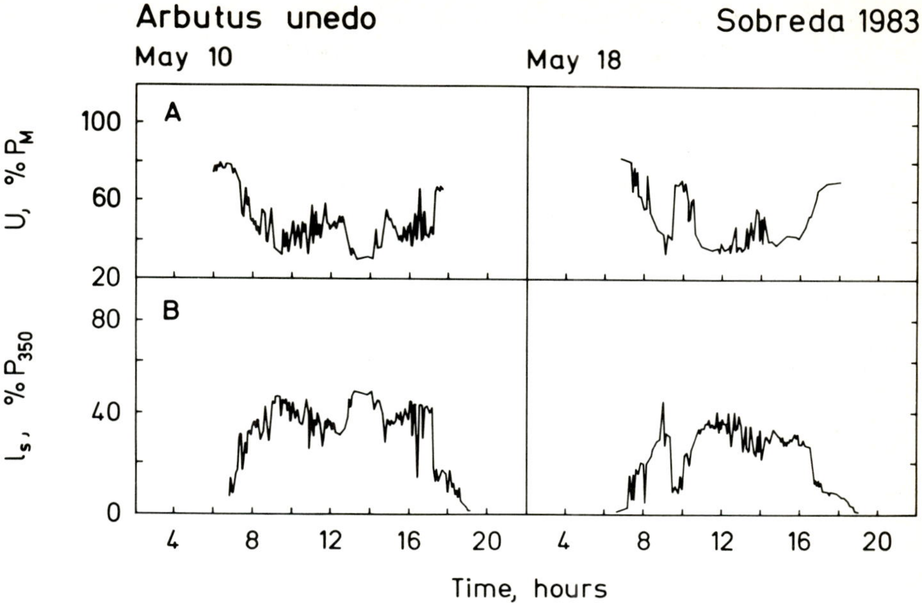

Fig. 13. Daily time courses of utilization (U) of maximal photosynthetic capacity (P_M) and of stomatal limitation of net photosynthesis (l_s) for leaves of *Arbutus unedo* under natural conditions at Quinta São Pedro. Results are for simulations on the two days depicted in Fig. 12 - see Fig. 11 and text for detailed explanation. (After Harley *et al.* 1986)

Both l_s and U, change over the course of a day, as leaf conductance and mesophyll photosynthetic capacity change. These limitations on net photosynthesis for fully illuminated, horizontally-oriented leaves of *Arbutus unedo* over the course of a mid-summer day in Portugal are shown in Fig. 11 (Beyschlag 1984). In the morning, about 40% of total photosynthetic capacity is utilized (panel A), but at midday this value drops below 20%. Thus, the leaf is operating far below the inflection point of the CO_2 response curve, suggested to be an optimal operational point by von Caemmerer and Farquhar (1981). The diurnal changes which occur are partly due to increased stomatal limitation (panel B) which reaches a value of over 80% during midday. Thus, limiting water loss is apparently of considerable importance to sclerophylls even at the expense of carbon gain, under the extreme conditions typical of mediterranean summers.

More general information on the changes in l_s and U may be derived by utilizing analytical models of photosynthesis, which are capable of simulating leaf performance under a wide variety of field conditions. Harley *et al.* (1986), for example, applied a biochemically based model to predict CO_2 assimilation rates for *Arbutus unedo* under springtime conditions (bottom panel in Fig.12). Environmental inputs driving the model were leaf temperature, incident PAR and stomatal conductance. The simulated diurnal patterns of U and l_s for the same two days are shown in Fig. 13. In the absence of water stress, predicted and measured l_s (bottom panel) fluctuated around 40% on May 10, indicating a 40% reduction in net photosynthesis imposed by the stomata, compared with values of 80% discussed above for water stress conditions (Fig. 11). On May 18, predicted and measured l_s values were slightly lower. Values for U decrease as low as 30-40% which is considerably higher than values obtained under stress conditions (Fig. 11). The ability to calculate l_s and U accurately with such models allows the possibility of estimating these limitations on photosynthesis under conditions other than those imposed during gas exchange measurements, for leaves oriented in different positions (see also Ehleringer and Comstock this volume) or for collections of leaves in canopies (Meister *et al.* this volume).

SUMMARY

Those of us engaged in trying to understand plant behavior under natural conditions have long been frustrated by the realization that the behavior of plants growing in the field is often impossible to duplicate under laboratory or greenhouse conditions. We have attempted to demonstrate through examples based on our work in Portugal, however, that the simultaneous development of new gas exchange techniques and new theoretical bases for interpreting gas exchange data make these exciting times for those of us interested both in describing plant photosynthetic behavior under field conditions and elucidating underlying causes of changes in this behavior with the development of environmental stress.

New theoretical ideas relating to leaf behavior leading to optimal water-use efficiency (Cowan 1982), or appropriate leaf strategies for optimal allocation of nitrogen (Mooney and Gulmon 1983; Field and Mooney 1986), for example, are guiding us in planning new experimental designs. Additionally, the development of analytical models of leaf photosynthesis, based almost entirely on known leaf physiology (Farquhar and von Caemmerer, 1982; Harley *et al.*, 1986), provide a potentially valuable tool for interpreting leaf behavior under field conditions. And finally, the emergence of extremely versatile and portable systems for measuring gas exchange are rapidly bringing to field studies the degree of accuracy and environmental control previously possible only under laboratory conditions.

REFERENCES

Beyschlag W (1984) Photosynthese und Wasserhaushalt von *Arbutus unedo* L. im Jahreslauf am Freilandstandort in Portugal. Gaswechselmessungen unter natürlichen Bedingungen und experimentelle Faktorenanalyse. Thesis University Würzburg

Björkman O (1981) Responses to different quantum flux densities. In: Lange OL, Nobel PS, Osmond CB, Ziegler H (eds), Encyclopedia of Plant Physiology, NS vol 12A: Physiological Plant Ecology I. Springer, Berlin-Heidelberg-New York-Tokyo, pp 57-107

Bosian G (1955) Über die Vollautomatisierung der CO_2-Assimilation bestimmung und zur Methodik des Küvettenklimas. Planta 45:470-492

Burschka C, Lange OL, Hartung W (1985) Effects of abscisic acid on stomatal conductance and photosynthesis in leaves of intact *Arbutus unedo* plants under natural conditions. Oecologia (Berl) 67:593-595

Caemmerer S von, Farquhar GD (1981) Some relationships between the biochemistry of photosynthesis and the gas exchange of leaves. Planta 153:376-387

Caldwell MM, Dean TJ, Nowak RS, Dzurec, RS, Richards JH (1983) Bunchgrass architecture, light interception, and water-use efficiency: assessment by fiber optic point quadrats and gas exchange. Oecologia (Berl) 59:178-184

Cowan, IR (1982) Regulation of water use in relation to carbon gain in higher plants. In: Lange OL, Nobel PS, Osmond CB, Ziegler H (eds), Encyclopedia of Plant Physiology, NS vol 12B: Physiological Plant Ecology II. Springer, Berlin-Heidelberg-New York-Tokyo, pp 589-613

Farquhar GD, von Caemmerer S (1982) Modelling of photosynthetic response to environment. In: Lange OL, Nobel PS, Osmond CB, Ziegler H (eds), Encyclopedia of Plant Physiology, NS vol 12B: Physiological Plant Ecology II. Springer, Berlin-Heidelberg-New York-Tokyo pp 549-587

Farquhar GD, Sharkey TD (1982) Stomatal conductance and photosynthesis. Annu Rev Plant Physiol 33:317-346

Field C, Berry J, Mooney HA (1982) A portable system for measuring carbon dioxide and water vapor exchange of leaves. Plant Cell Environment 5:179-186

Field C, Mooney HA (1986) The nitrogen-photosynthesis relationship in wild plants. In: Givnish T (ed), On the Economy of Plant Form and Function. Cambridge University Press, London (in press)

Griffiths JH, Jarvis PG (1981) A null balance carbon dioxide and water vapour porometer. J Exp Bot 32:1151-1168

Harley PC, Weber JA, Gates DM (1985) Interactive effects of light, leaf temperature, CO_2 and O_2 on photosynthesis in soybeans. Planta 165:249-263

Harley PC, Tenhunen JD, Lange OL (1986) Use of an analytical model to study limitations to net photosynthesis in *Arbutus unedo* under field conditions. Oecologia (Berl) 70:393-401

Holdheide W, Huber B, Stocker O (1936) Eine Feldmethode zur Bestimmung der momentanen Assimilationsgröße von Landpflanzen. Ber Deut Botan Ges 45:168-188

Koch W, Walz H (1967) Kleinklimaanlage zur Messung des pflanzlichen Gaswechsels. Naturwiss 54:321-322

Koch W, Lange OL, Schulze E.-D. (1971) Ecophysiological investigations on wild and cultivated plants in the Negev Desert. I. A mobile laboratory for measuring carbon dioxide and water vapour exchange. Oecologia (Berl) 8:296-309

Lange OL (1962) Eine "Klapp-Küvette" zur CO_2-Gaswechselregistrierung an Blättern von Freilandpflanzen mit dem URAS. Ber Deut Botan Ges 75:41-50

Lange OL, Tenhunen JD, Braun M (1982) Midday stomatal closure in mediterranean type sclerophylls under simulated habitat conditions in an environmental chamber. I. Comparison of the behaviour of various European mediterranean species. Flora 172:563-579

Lange OL, Kilian E, Meyer A, Tenhunen JD (1984) Measurement of lichen photosynthesis in the field with a portable steady-state CO_2-porometer. Lichenologist 16:1-9

Lange OL, Tenhunen JD (1984) A minicuvette system for measurement of CO_2 exchange and transpiration of plants under controlled conditions in field and laboratory. Heinz Walz Effeltrich pp 1-11

Lange OL, Tenhunen JD (1985) CO_2/H_2O porometer for the measurement of CO_2 gas exchange and transpiration of plants under natural conditions. Heinz Walz Effeltrich pp 1-13

Lange OL, Tenhunen JD, Beyschlag W (1985) Effects of humidity during diurnal courses on the CO_2- and light-saturated rate of net CO_2 uptake in the sclerophyllous leaves of *Arbutus unedo*. Oecologia (Berl) 67:301-304

Mooney HA, Dunn EL, Harrison AT, Morrow PA, Bartholomew B, Hays RL (1971) A mobile laboratory for gas exchange measurements. Photosynthetica 5:128-132

Mooney HA, Gulmon SL (1983) The determinants of plant productivity - Natural versus man-modified communities. Ecological Studies 44:146-158

Powles SB, Björkman O (1982) High light and water stress effects on photosynthesis in *Nerium oleander*. I. Photoinhibition of photosynthesis: Effect on chlorophyll fluorescence at 77 K in intact leaves and in chloroplast membranes. Carnegie Inst Wash Yearb 81:74-76

Raschke K (1979) Movements of stomata. In: Haupt W, Feinleib ME (eds), Encyclopedia of Plant Physiology, NS vol 7. Springer, Berlin-Heidelberg-New York, pp 383-441

Reicosky DC, Peters DB (1977) A portable chamber for rapid evapotranspiration measurements on field plots. Agronom Journ 69:729-732

Schulze E-D, Hall ED, Lange OL, Walz H (1982) A portable steady-state porometer for measuring the carbon dioxide and water vapour exchange of leaves under natural conditions. Oecologia (Berl) 53:141-145

Sesták Z, Catsky J, Jarvis, PG (eds) (1971) Plant photosynthetic production. Manual of methods. Junk Publishers, The Hague

Sharkey, TD, Imai K, Farquhar GD, Cowan IR (1982) A direct confirmation of the standard method of estimating intercellular partial pressure of CO_2. Plant Physiol 69:657-659

Tenhunen JD, Lange OL, Braun M (1981) Midday stomatal closure in mediterranean type sclerophylls under simulated habitat conditions in an environmental chamber. II. Effect of the complex of leaf temperature and air humidity on gas exchange of *Arbutus unedo* and *Quercus ilex*. Oecologia (Berl) 50:5-11

Tenhunen JD, Lange OL, Gebel J, Beyschlag W, Weber JA (1984) Changes in photosynthetic capacity, carboxylation efficiency, and CO_2 compensation point associated with midday stomatal closure and midday depression of net CO_2 exchange of leaves of *Quercus suber*. Planta 162:193-203

Tenhunen JD, Lange OL, Harley PC, Beyschlag W, Meyer A (1985) Limitations due to water stress on leaf net photosynthesis of *Quercus coccifera* in a Portuguese evergreen scrub. Oecologia (Berl) 67:23-30

Tenhunen JD, Meister HP, Caldwell MM, Lange OL (1984) Environmental constraints on productivity of the Mediterranean sclerophyll shrub *Quercus coccifera*. Proceedings of INTECOL workshop - Rates of Natural Primary Productivity and Agricultural Production. Options méditerranéennes (Instituto Agronomico Mediterraneo de Zaragoza) 84/1:33-53

Weber JD, Tenhunen JD, Lange OL (1985) Effects of temperature at constant air dew point on leaf carboxylation efficiency and CO_2 compensation point of diffferent leaf types. Planta 166:81-88

Wells J (1986) A portable photosynthesis system. In: Gensler W (ed), Advanced Agricultural Instrumentation. Martinus and Nijhoff, Publ., Dordrecht (in press)

Williams BA, Gurner PJ, Austin RB (1982) A new infrared gas analyser and portable photosynthesis meter. Photosynthesis Research 3:141-151

Rapid assessment of stress effects on plant leaves by chlorophyll fluorescence measurements

U. Schreiber and W. Bilger

Botanisches Institut der Universität Würzburg, Mittlerer Dallenbergweg 64, D-87 Würzburg, Federal Republic of Germany.

INTRODUCTION

Chlorophyll fluorescence serves as an intrinsic indicator of the photosynthetic reactions in the chloroplasts of green plants. The relationship between chlorophyll fluorescence and the mechanisms of photosynthesis have been the subject of a great number of investigations, since Kautsky discovered (Kautsky and Hirsch 1931) that fluorescence intensity in green leaves displays characteristic changes upon illumination (Kautsky effect - for reviews, see Papageorgiou 1975; Lavorel and Etienne 1977; Krause and Weis 1984). For almost half a century fluorescence has been mainly a tool for biophysically oriented researchers in studies of the primary processes of photosynthesis. Practical applications of the Kautsky effect in ecophysiological work had been limited by the availability of suitable instrumentation and by the complexity of the fluorescence information obtained *in vivo*. In recent years considerable efforts have been put into the development of field oriented fluorescence equipment and into the development of the methodology for analysing fluorescence data from intact leaves.

Recently a new measuring system has become available which permits the monitoring of fluorescence in full sunlight and the deconvolution of the complex fluorescence information by simple procedures (Schreiber *et al.* 1986). Hence, eventually chlorophyll fluorescence is prepared to step out of the biophysical dark laboratory into the daylight where "real" photosynthesis takes place. In the present contribution we wish to introduce the ecophysiologist to the principles of the chlorophyll fluorescence method in general and to the very recent developments in measuring techniques and data analysis in particular. Examples of practical applications are given for the cases of plant responses to heat, drought and light stress.

CHLOROPHYLL FLUORESCENCE AS AN INDICATOR OF PHOTOSYNTHESIS

Three main types of chlorophyll protein complexes are embedded into the thylakoid membranes of green plant chloroplasts: the light harvesting complex (LHC) (Chl a/Chl b protein complex), the photosystem I antenna complex (PS I)

NATO ASI Series, Vol. G15
Plant Response to Stress. Edited by J. D. Tenhunen et al.

and the photosystem II antenna complex (PS II). Proper functioning of the overall photosynthetic process requires balanced excitation of the two photosystems, brought about by appropriate association of the LHC with the two photosystems shown to be under regulatory control (Bennett *et al.* 1980). Chlorophyll fluorescence *in vivo* originates primarily from Chl $\underline{a}_{II}$ in PS II, in particular the so-called "variable fluorescence", i.e. the fluorescence increase observed upon illumination. Hence, changes in fluorescence yield should be regarded as primarily reflecting properties of excitation and energy conversion at PS II (see scheme of Fig. 1). However, due to the functional connection of PS II to the other components of the photosynthetic apparatus, indirectly fluorescence yield will be an indicator of the whole complex process. It depends on the specificity of the measuring technique, on the rationale of the experimental approach and the theoretical background for data analysis, to what extent such information on PS II may be traced back to information on overall photosynthesis and on specific aspects like photophosphorylation, water splitting or the carbon reductive cycle.

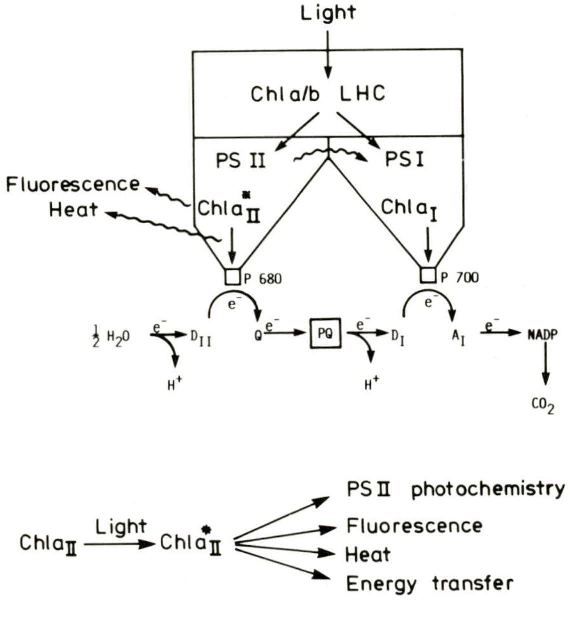

Fig. 1. Control of chlorophyll fluorescence by energy conversion at photosystem II reaction centers. Schematic arrangement of the major Chl protein antenna complexes with the reaction centers of the two photosystems (P 680 and P 700). The principal steps of energy transfer and primary electron transport reactions are indicated. Room temperature fluorescence originates mainly from PS II and is in competititon with other deexcitation processes

Assuming that *in vivo* fluorescence is an indicator of the excitation state density in PS II, fluorescence intensity is determined by the following factors: 1) The intensity of excitation light, 2) the fraction of excitation distributed to PS II, 3) the rate of excitation trapping and energy conversion at P 680, 4) the extent of excitation dissipation into heat, 5) the extent of excitation transfer ("spillover") from PS II to PS I.

THE KAUTSKY EFFECT

When a leaf is illuminated with continuous light following a period of darkness, fluorescence rises from a low level (F_O) via an intermediate level (I) to a peak level (P) and then decays through several phases via intermediate maxima (M_1, M_2, ...) to a value close to the original F_O-level (Kautsky effect). Fig. 2 shows a typical fluorescence induction curve of *Arbutus unedo* L., depicted with different time scales, as measured with a modulation fluorometer (see following section on measuring systems). The various induction transients reflect the onset of photosynthesis as it affects the excitation density within PS II. 1) the F_O-level, which is monitored by the weak modulated measuring beam in the preceding dark period, is a measure for the initial energy distribution to PS II and for the efficiency of excitation trapping at P 680. Under normal conditions, all centers are "open" at F_O, i.e. all primary acceptors, Q_A, are oxidized. 2) Upon onset of

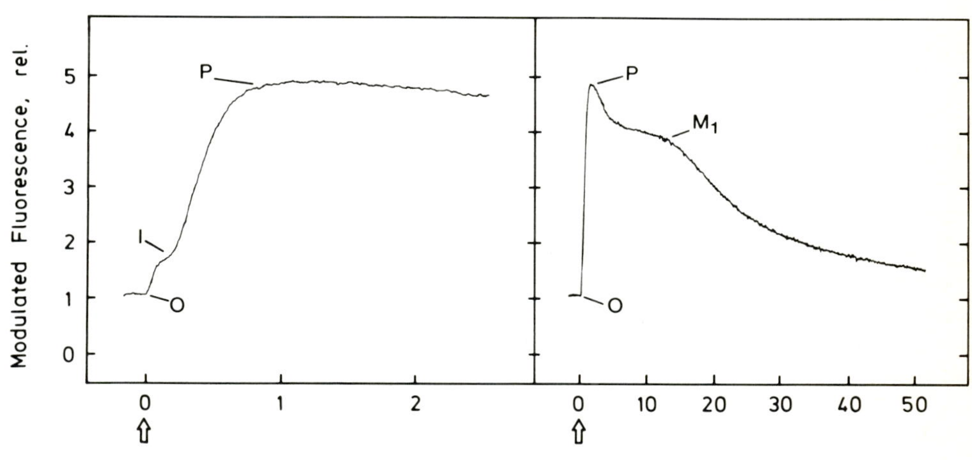

Fig. 2. Typical fluorescence induction kinetics (Kautsky effect). Time dependent changes of the fluorescence yield are shown at two different time scales. The characteristic fluorescence levels (O, I, P, M_1) are indicated. Measurement with modulation fluorometer (PAM 101 Chlorophyll Fluorometer). Object: Intact leaf of *Arbutus unedo*. Light intensity: 150 μmol m^{-2} s^{-1}. Temperature: 22° C

illumination, the first fluorescence rise (O - I) reflects reduction of Q_A which is in equilibrium with the secondary acceptor Q_B. Accumulation of Q_B^- must precede electron transport into the plastoquinone pool (PQ), as PQ accepts 2e$^-$ from $Q_B^=$ (Velthuys and Amesz 1974). Hence, the I-level corresponds to the equilibrium $Q_A^- Q_B \rightleftharpoons Q_A Q_B^-$ which may coexist with a fully oxidized PQ-pool. An extended step at the I-level may be considered to reflect a large PQ-pool. 3) The I - P rise parallels accumulation of $Q_A^- Q_B^=$, as the PQ-pool becomes reduced. This transient is typical for dark-adapted samples, because electron transport capacity at the PS I acceptor side deactivates in the dark. 4) The following decline of fluorescence reflects the light-activation of electron transport capacity at the PS I acceptor side, as well as the development of a number of overlapping "fluorescence quenching" processes, which will be discussed later in more detail (see section on Q-analysis).

The fluorescence induction kinetics described here and depicted in Fig. 2, represent the normal behaviour of a healthy leaf. Variation of extrinsic parameters, e.g. light intensity, temperature, gas composition, and preillumination, will cause characteristic changes in the induction pattern. For assessment of stress effects it is essential to control these extrinsic parameters during the actual fluorescence measurements, such that the response of the stressed leaf can be compared to the response of a control leaf under standard conditions. Stress-induced changes in the properties of the photosynthetic apparatus will be reflected in corresponding changes in the induction transients, analysis of which may yield information on the site and the extent of the stress-induced damage.

THE MEASURING SYSTEM

The conventional instrumentation for measuring chlorophyll fluorescence has been described in recent reviews (Schreiber 1983; Renger and Schreiber 1986) so that only the essential points will be summarized here. Some emphasis will be given to the description of a new type of modulation fluorometer (Schreiber *et al.* 1986) which appears to be particularly well suited for the needs of the ecophysiological researcher.

In any fluorescence measuring system, there has to be efficient separation of the fluorescence from the much stronger excitation light. This is achieved by the use of appropriate optical filters, e.g. short-pass filters ($\lambda < 660$ nm) for excitation and long-pass filters ($\lambda > 680$ nm) protecting the photodetector from stray excitation light while allowing passage of the longer wavelength fluorescence. In conventional fluorometers the excitation light at the same time serves as "actinic light" (light driving photosynthetic reactions). Hence, there have been considerable restrictions as to the spectral composition of actinic illumination. Furthermore, as conventional fluorometers can not distinguish between fluorescence and any other light reaching the photodetector, the leaf sample and the detector system has to be protected from ambient daylight. Therefore, conventional fluorometers have been of limited use for ecophysiological field work.

Recently, a pulse modulation chlorophyll fluorometer has been developed at our laboratory, which is capable of measuring fluorescence yield in full daylight, without being disturbed by large changes in unfiltered actinic illumination (Schreiber *et al.* 1986; Schreiber 1986). This instrument employs a novel measuring principle which is briefly described here. Fluorescence is excited repetitively by 1 μsec pulses of light from a light emitting diode (LED) passing through a short-pass filter (λ < 670 nm). The photodiode detector is protected by a long-pass filter (λ > 700 nm). A highly selective pulse amplification system ignores all signals except the fluorescence excited during the 1 μsec measuring pulses. Due to the large linearity range of the photodiode (up to 10^9) and to high amplifier selectivity, the measuring system tolerates extreme changes in actinic light intensity (up to several times the intensity of full sunlight) at weak

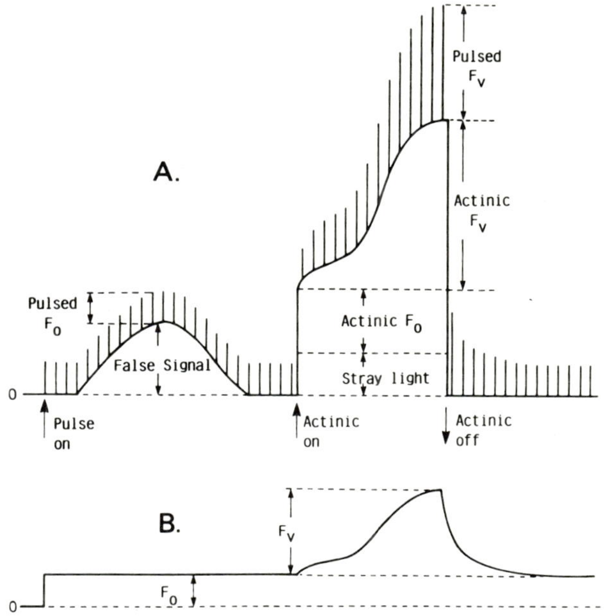

Fig. 3. Measuring principle of pulse modulation fluorometer. A. Unprocessed signal at photodiode. B. Processed signal at output of selective amplifier system. Brief pulses of measuring light are applied repetitively. Signal changes induced by actinic illumination and upon redarkening are displayed schematically. In practice, the non-pulsed signal may exceed the pulsed signal by more than a factor 10^6. Only the pulsed fluorescence is amplified and recorded as a continuous signal

measuring light intensities. Hence, with this type of modulation fluorometer, the experimenter has become free of any restrictions concerning intensity and spectral composition of illumination, and measurements can be carried out in daylight in the natural environment of the plant.

In practice, fluorescence measurements with the new type of modulation fluorometer are rather simple. When the pulsed measuring light is switched on, the recorded signal gives the fluorescence yield of the sample. Fluorescence yield is low when the leaf is dark-adapted (F_O-level), and it will display the characteristic changes of the Kautsky effect when actinic light is applied. Upon darkening, fluorescence yield relaxes via characteristic decay phases back to the F_O-level. The selective amplifier system assures that only the pulse modulated signal is recorded, but not stray signals or signals due to stray actinic light and the fluorescence excited by the actinic light. Fig. 3 shows in a schematic drawing the difference between the unprocessed signal at the detector, as it would be recorded by a conventional fluorometer, and the processed signal at the output of the selective pulse amplifier system.

For the ecophysiological researcher, not trained in instrument design and construction, commercial availability of suitable equipment is important. The pulse modulation fluorometer is available as a compact, integrated measuring system (PAM 101, 102, 103 Chlorophyll Fluorometer) with various accessories from H. Walz, Effeltrich, Germany. This system utilizes fiberoptics, can be battery operated and is suited for investigations in the field and in the laboratory.

Q-ANALYSIS BY THE SATURATION PULSE METHOD

When Kautsky postulated a close relationship between chlorophyll fluorescence yield and carbon dioxide assimilation (Kautsky and Hirsch 1931, 1934; Kautsky and Franck 1943), he encountered great scepticism among photosynthesis researchers, because the fluorescence response to CO_2-withdrawal was the opposite to that which was expected. Under conditions of steady state illumination, removal of CO_2 resulted in a decrease of fluorescence yield, rather than an increase which might be due to inhibition of photosynthetic energy conversion. It took more than 30 years for this discrepancy to be resolved. Starting with the work of Duysens and Sweers (1963) evidence has been accumulating that fluorescence quenching is composed of photochemical as well as non-photochemical components. Photochemical quenching has been called Q-quenching, since it is the primary acceptor of PS II, Q_A, in its oxidized state that determines the rate of excitation trapping at PS II (being in competition with fluorescence emission - see Fig. 1). There are several mechanisms by which fluorescence yield is quenched non-photochemically (for a review see Krause and Weis 1984). However, most of non-photochemical quenching has been found to be correlated with the "energization" of the thylakoid membrane, i.e. with the

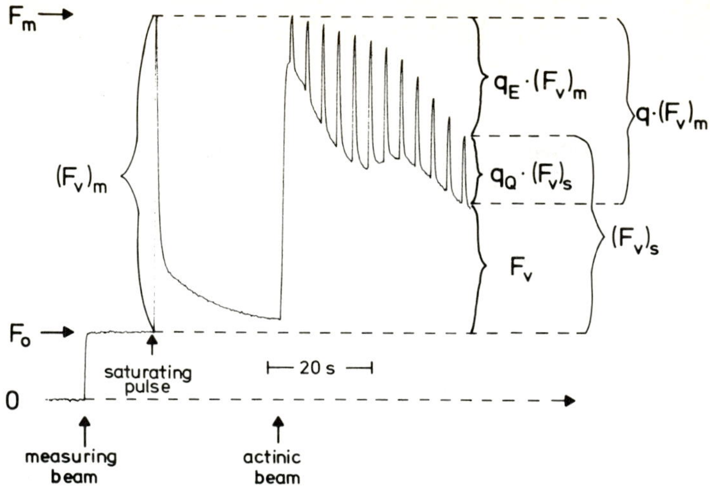

Definition of quenching coefficients

$$F_v = (F_v)_m - q \cdot (F_v)_m$$
$$= (F_v)_m - q_E \cdot (F_v)_m - q_Q \cdot (F_v)_s$$
$$= (F_v)_m - q_E \cdot (F_v)_m - q_Q \cdot [(F_v)_m - q_E \cdot (F_v)_m]$$

$$\frac{F_v}{(F_v)_m} = (1 - q_E) - q_Q \cdot (1 - q_E)$$
$$= (1 - q_E) \cdot (1 - q_Q) = 1 - q$$

$$F_v = (F_v)_s - q_Q \cdot (F_v)_s \qquad (F_v)_s = (F_v)_m - q_E \cdot (F_v)_m$$

$$\frac{F_v}{(F_v)_s} = 1 - q_Q \qquad \frac{(F_v)_s}{(F_v)_m} = 1 - q_E$$

$$q_Q = \frac{(F_v)_s - F_v}{(F_v)_s} \qquad q_E = \frac{(F_v)_m - (F_v)_s}{(F_v)_m}$$

Fig. 4. Definition of quenching coefficients and terminology of characteristic fluorescence parameters. F_O, fluorescence intensity of dark adapted sample with measuring beam of negligible actinic intensity. F_m, maximal fluorescence intensity obtained with dark adapted sample upon application of saturating light pulse. $(F_v)_m$, maximal variable fluorescence of dark adapted sample. F_v, variable fluorescence at any given time during induction. $(F_v)_s$, maximal variable fluorescence at any given time during induction observed with application of a saturation pulse light-driven formation of a proton gradient between the stroma and the thylakoid interior. Hence, non-photochemical quenching has been generally termed E-quenching.

In recent years, considerable progress has been made in differentiating between Q-quenching and E-quenching (Bradbury and Baker 1981; Krause *et al.* 1982; Quick and Horton 1984). Bradbury and Baker (1981) introduced the "light-doubling method" for applications with intact leaves, which has been further developed by the use of fluorescence modulation techniques (Dietz *et al.* 1985; Ögren and Baker 1985). With the development of the new pulse modulation fluorometer (Schreiber *et al.* 1986) which allows extremely high actinic intensities to be applied with low measuring light intensity, a very satisfactory procedure for differentiating between the different forms of quenching has become possible. At any instant, a short pulse (about 500 ms in length) of saturating white light (about 2000 W/m^2) may be applied to the sample to induce complete reduction of the PS II acceptor Q_A and, thus, to transiently remove all Q-quenching. Any remaining quenching is assumed to consist of E-quenching. In practice, repetitive application of such "saturation pulses" at desired intervals provides continuous information on the two main quenching components. Modern data acquisition systems and computer technologies provide the means for quick calculation of the corresponding quenching coefficients, which may be plotted simultaneously with the original fluorescence curves. The mathematical formalism for calculating the quenching coefficients q_Q and q_E is summarized in Fig. 4. The approach outlined above is based on the assumption that only variable fluorescence, i.e. fluorescence exceeding F_O, is affected by E-quenching. However, under certain conditions (as with particularly strong energization, e.g. after heat pretreatment) energy dependent quenching of F_O may occur and this must be taken into account (Bilger *et al.* in these proceedings; see also section on HEAT STRESS below). Furthermore, photoinhibitory treatment of leaves leads to a type of fluorescence quenching which in a formal sense is equivalent to E-quenching as defined above. This type of quenching may be distinguished from true E-quenching by its much slower relaxation kinetics upon darkening (see section on PHOTOINHIBITION below).

An example of the application of the saturation pulse method for differentiating between q_Q and q_E (q-analysis) is given in Fig. 5, which illustrates the response to CO_2-removal mentioned at the beginning of this section. Fig. 5A shows the original trace of modulated fluorescence for a light stimulated induction curve of a broad bean leaf and with saturation pulses given every 30 sec ("saturation pulse induction curve"). Upon removal of CO_2 there is a small decrease in F_v and a large decrease in $(F_v)_s$ suggesting a substantial increase in q_E. This is confirmed by the deconvoluted curves of q_Q and q_E in Fig. 5B. While CO_2-removal induced an increase of q_E, q_Q is decreased at the same time. The decrease in q_Q is smaller than might be expected. However, O_2 is also serving as an electron acceptor in

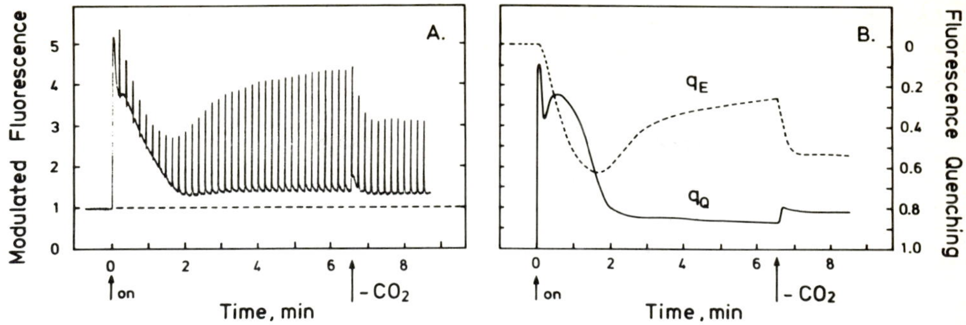

Fig. 5. Effect of CO_2-removal on fluorescence quenching. A. Saturation pulse induction curve. B. Time courses of quenching coefficients q_Q and q_E, obtained from the original trace (panel A). Note: The scale for fluorescence quenching has been reversed for better comparison with the original data. Where indicated by arrows, CO_2 was removed from the air stream by passage through saturated KOH solution. Experimental object: *Vicia faba* L. Light intensity: 200 µmol m^{-2} s^{-1}. Temperature: 22° C

photorespiration and in the Mehler reaction. When both CO_2 and O_2 are removed, q_Q decreases practically to zero (not in these figures, but see Dietz *et al.* 1985). The curves shown in Fig. 5 also demonstrate that additional information is obtained by analyzing data obtained at the onset of CO_2-fixation activity in the pre-darkened leaf for q_Q and q_E. During the first moments of illumination, before Calvin cycle enzymes have become activated, there is strong membrane energization (increase of q_E) and the level of reduction in the PQ-pool is high (low level of q_Q). As CO_2-fixation begins, ATP and NADPH are consumed, causing a relaxation in membrane energization (decrease of q_E) and reoxidation of PQ (increase of q_Q). Such information is not directly apparent from the induction curves without application of saturation pulses.

CORRELATION OF GAS EXCHANGE AND FLUORESCENCE DATA

In the past, leaf photosynthetic performance has been extensively studied via measurements of CO_2- and O_2-exchange. In this section, we wish to discuss the extent to which the data obtained by gas exchange and fluorescence techniques are correlated and complementary. At steady state and with continuous illumination, the overall rate of photosynthetic electron flow may be assumed to equal the rate

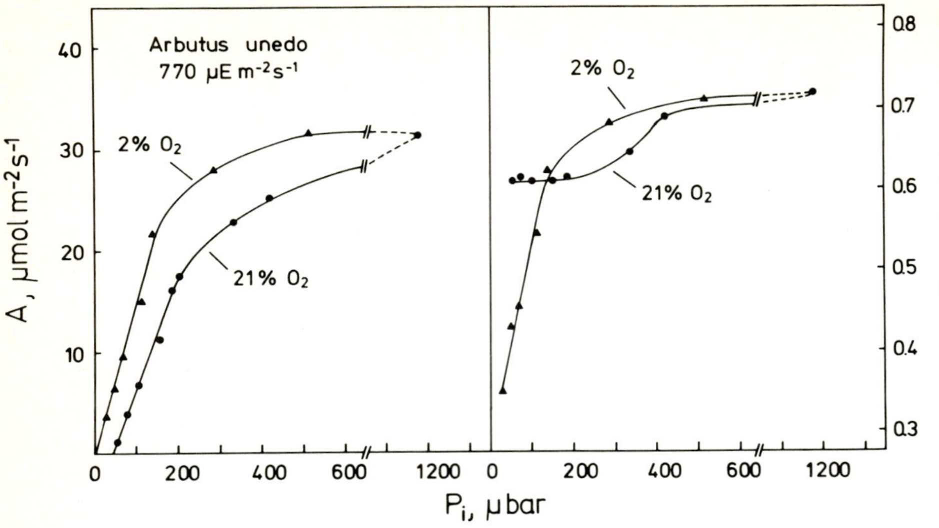

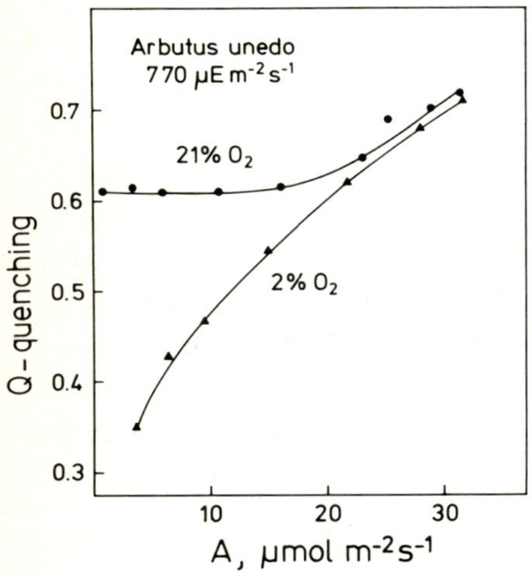

Fig. 6. Comparison of Q-quenching with net CO_2 assimilation rate, A, in *Arbutus unedo*. Dependence on the leaf internal CO_2 partial pressure at 21% and 2% O_2 concentration. Fluorescence and gas exchange were measured simultaneously, and internal CO_2 partial pressure, p_i, calculated as described elsewhere (Winter, Bilger and Schreiber in preparation). For every value of p_i, sufficient time was given for attainment of steady state. Light intensity: 770 μmol m^{-2} s^{-1}. Leaf temperature: 22° C. ΔW: 10 mbar bar^{-1}. The plot of Q-quenching vs. A (lower part of figure) was derived from the separate p_i dependencies (top part of figure)

of charge separation at PS II reaction centers. This is considered to be linearly related to photochemical fluorescence quenching (q_Q) (Lavorel and Etienne 1977). Therefore, a simple linear relationship between q_Q and assimilation rate, A, might be expected. This is only partially confirmed by the data presented in Fig. 6, which illustrates the CO_2-concentration dependencies of assimilation rate, A, and photochemical quenching, q_Q, at 21% and 2% O_2. There is a satisfactory one to one relationship between A and q_Q at 2% O_2. However in the presence of

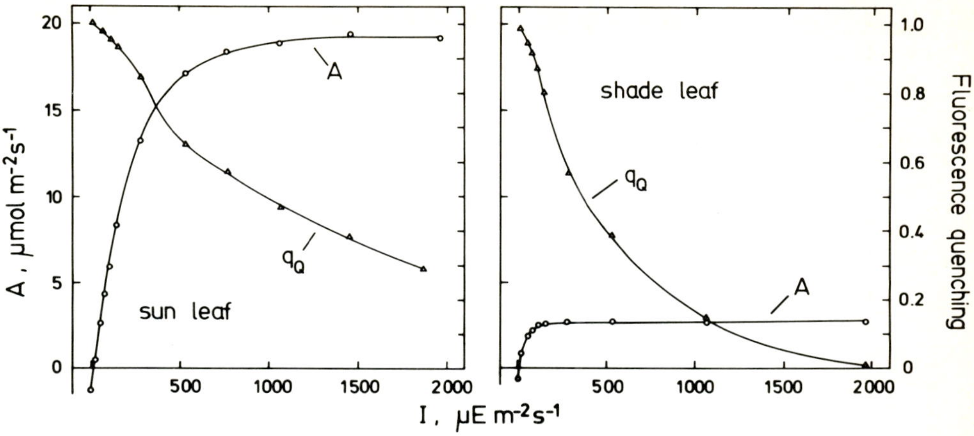

Fig. 7. Comparison of light saturation curves of Q-quenching and assimilation rate in sun and shade leaves of *Arbutus unedo*. Fluorescence and CO_2 fixation were measured simultaneously (see Fig. 6). At each light intensity, data were recorded after attainment of steady state conditions. Leaf temperature: 22° C. Gas composition: 350 µbar CO_2 and 21% O_2. ΔW: 10 mbar bar^{-1}. The sun leaf was from a plant grown outdoors under conditions of full sunlight and with high N_2-supply, while the shade plant was grown at approx 200 µmol m^{-2} s^{-1}

21% O_2 and with low CO_2-concentrations, decreases in A are not accompanied by a corresponding decrease in q_Q. Under these conditions molecular oxygen serves as an efficient electron acceptor to reoxidize the PQ-pool and to maintain high photochemical quenching. The independent effects on A and q_Q at low CO_2-concentration and at 21% O_2 are a measure of photorespiratory activity in the leaves. Combining gas exchange and fluorescence measurements provides information which is not readily available by either of the two methods alone.

In Fig. 7 the light intensity dependencies of q_Q and of CO_2-assimilation rate, A, are compared for sun and shade grown leaves of *Arbutus unedo*. Due to the different conditions of light adaptation (in addition the sun plant was also given high nitrogen supply) the two leaves sampled showed substantial differences in their light saturation characteristics. While assimilation rate was already saturated at 200 µmol m^{-2} s^{-1} in the shade leaf (A_{max}=2.6 µmol m^{-2} s^{-1}), the sun leaf was saturated only at 1000 µmol m^{-2} s^{-1} (A_{max} of 19 µmol m^{-2} s^{-1}). The differences in photosynthetic capacity are also reflected in the light intensity curves of q_Q. With increasing photon flux there is a decrease of q_Q, reflecting accumulation of electrons at the PS II acceptor side. This decrease is at first sigmoidal, then becomes close to logarithmic. In the sun leaf the sigmoidal phase is more extended than in the shade leaf. Furthermore, the rate, with which q_Q

approaches zero during the logarithmic phase is substantially lower in the sun leaf compared to the shade leaf.

While the differences in photosynthetic capacity are clearly reflected in the light curves of q_Q (Fig. 7), the theoretical basis for the relationship between assimilation rate and q_Q in such plots may be less obvious. For proper treatment of this point, the aspect of "limitation" of electron flow rate must be considered. For any light intensity, a hypothetical maximal rate of CO_2 uptake, A_m, may be assumed. This rate would be achieved if there were no limitation of electron flow. A_m may be compared to the observed rate, A. This leads to definition of a limitation coefficient, L:

$$A = A_m - A_m \cdot L \qquad (1) \qquad\qquad L = \frac{A_m - A}{A_m} \qquad (3)$$

$$(1 - L) = \frac{A}{A_m} \qquad (2) \qquad\qquad 1 > L > 0 \qquad (4)$$

With the given definition, the light intensity dependency of limitation, L, may be calculated from a light saturation curve of assimilation rate, A. This is shown for a theoretical light saturation curve based on the Smith equation in Fig. 8 (see

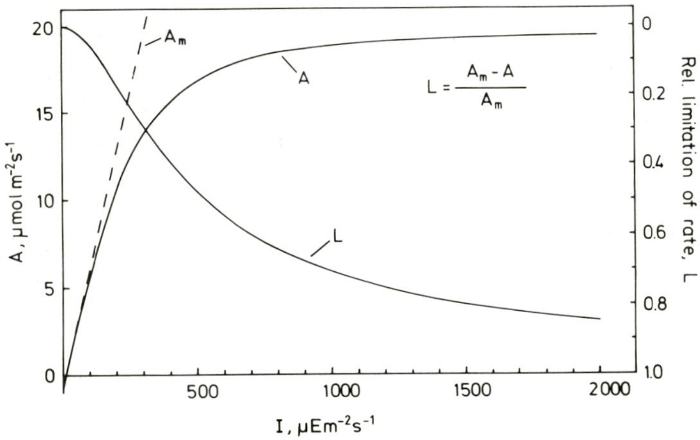

Fig. 8. Light intensity dependency of limitation of assimilation rate, L, calculated from a theoretical light saturation curve based on the Smith equation (see text). The following parameters derived from the experimental data of Fig. 7 are used in the equation: $\alpha = 0.0667$; $P_{ML} = 20.4$ µmol m^{-2} s^{-1}. Correction for dark respiration of 0.7 µmol m^{-2} s^{-1} was made. A_m corresponds to the rate $\alpha \cdot I$, with α representing the maximal quantum efficiency at low light intensity

Tenhunen *et al.* 1976). Plots for A and L are presented with the parameters chosen to match the experimental data of Fig. 7. A distinct similarity is found between the observed light saturation curve of q_Q (Fig. 7) and the theoretical light saturation curve of L. In both cases, a sigmoidal phase is followed by a close to logarithmic saturation characteristic. Quantitative correlation between L and Q-quenching should be good with the exception that electron flow to O_2 constitutes a limitation on CO_2- fixation rate that does not result in a decrease in q_Q (see Fig. 6).

LOCALIZATION OF LIMITING STEPS BY FLUORESCENCE INFORMATION

The question of "what limits photosynthesis" has occupied researchers for many years. In a recent review Sharkey (1985) divided limitations on the rate of photosynthesis into three general categories: (1) the supply or utilization of CO_2, (2) the supply or utilization of light, and (3) the supply or utilization of phosphate. This classification is based on information derived from gas exchange measurements which *a priori* assess the overall performance of the photosynthetic apparatus under steady state conditions. As was shown in the preceding section (see Fig. 6) fluorescence may also provide steady state information, allowing assessment of the relative overall rate. However, fluorescence may provide in addition more detailed information on specific sites of limitations, particularly those affecting "supply or utilization of light." The following examples illustrate how various treatments of intact leaves, all of which block overall CO_2 fixation, induce different fluorescence response patterns which are consistent with differing mechanisms of inhibition.

In Fig. 9 various fluorescence induction curves (quenching coefficients q_Q and q_E vs. time) are displayed. Different types of inhibitors were applied to the leaves via the upper epidermis, from which the cuticle was removed by gentle rubbing with carborundum powder. In the control leaf (panel A) the induction kinetics of q_Q and q_E initially show rapid reduction of the electron transport chain (decrease of q_Q to 0.4) followed by a slow reoxidation (increase of q_Q to 0.9). Membrane energization (initial increase of q_E to 0.6) relaxes again (decrease of q_E to 0.3) as the Calvin cycle and correlated utilization of ATP and decrease in membrane proton gradient set in (Schreiber *et al.* 1986). A leaf treated with the specific ATP-ase inhibitor Tentoxin (panel B) shows a rather slow and incomplete reoxidation of Q^-, (q_Q increases from 0.3 to 0.5). Instead of a relaxation in membrane energization, q_E rises to 0.65 in the steady state. Obviously, the lack of ATP-ase activity prevents Calvin cycle function and utilization of the membrane gradients. Since there is membrane energization and q-quenching indicates electron flow at low ATP-supply, this electron flow must depend on the

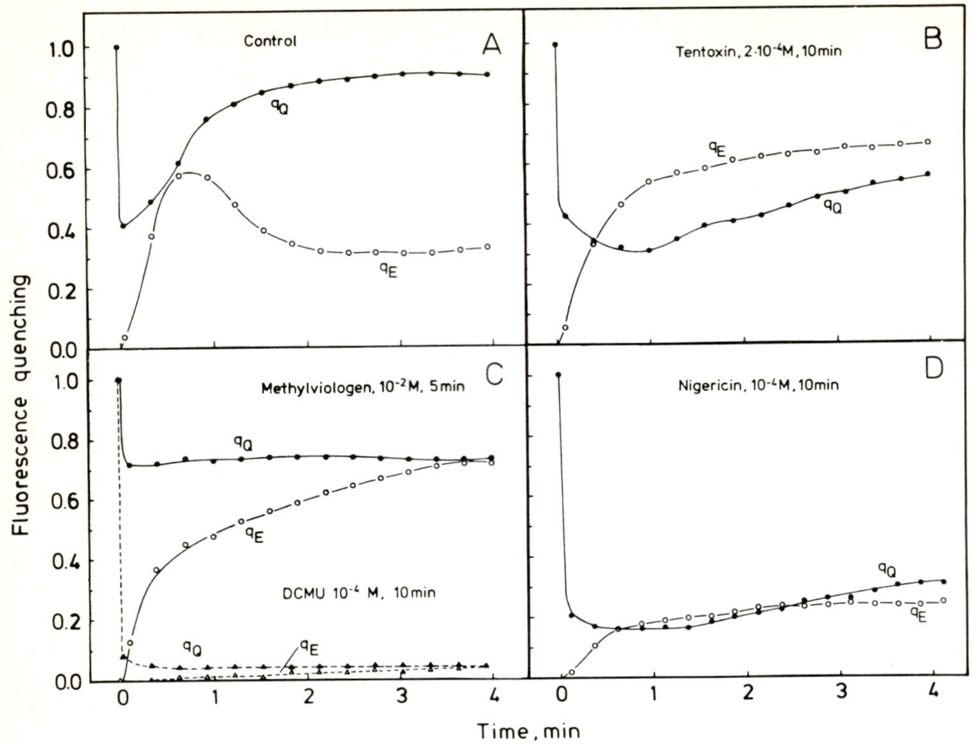

Fig. 9. Effect of various inhibitors on the induction kinetics of q_Q and q_E of intact spinach leaves. Quenching coefficients were determined from saturation pulse induction curves (see Figs. 4 and 5). The inhibitors were applied via the upper epidermis from which the cuticle was removed by gentle rubbing with carborundum powder. Concentration and incubation time of inhibitors as indicated in the figure. Light intensity: 200 μmol m^{-2} s^{-1}. Temperature: 22° C

availability of O_2 (not shown in the figures). In panel C, two substances are compared, which inhibit CO_2 fixation by different mechanisms. While DCMU (also known as the herbicide diuron) blocks electron transport at the PS II acceptor side, methylviologen (also known as the herbicide paraquat) serves as an efficient electron acceptor at PS I, thus preventing NADP reduction. Consistent with these different mechanisms of action, the induction kinetics of q_Q and q_E are very different. With methylviologen, q_Q rapidly decreases to 0.7, as compared to q_Q = 0.05 with DCMU and q_Q = 0.4 in the control. In contrast to the control, there are no further changes of q_Q with either herbicide after the initial change. Regarding membrane energization, q_E rises to a high value of 0.7 in the presence of methylviologen, while it stays very low (less than 0.05) in the presence of

DCMU. Leaf treatment with Nigericin leads to strong suppression of q_Q and q_E (Fig. 9, panel D). Nigericin is a protonophore, acting to dissipate the proton gradient across the thylakoid membrane. Obviously, the remaining membrane energization (q_E= 0.2) is not sufficient to supply the Calvin cycle with ATP.

These examples have shown how substances with specific and well-known mechanisms of action affect the induction pattern of q_Q and q_E. Knowledge of these changes in patterns may help in analysing the limitations of photosynthesis brought about by environmental stress. There may be only a few cases where stress causes damage which affects a single step in the photosynthetic chain of reactions. Nevertheless, fluorescence analysis is a rapid method. One therefore has the possibility of separating primary from secondary damage. In the following section examples are given for changes in fluorescence that occur in response to some of the major stress factors limiting photosynthetis of plants in their natural environments.

HEAT STRESS

Fluorescence methods have served as useful tools in the characterization of heat-induced changes of the photosynthetic apparatus (for review, see Krause and Weis 1984; Renger and Schreiber 1986). With the chlorophyll protein complexes embedded in the thylakoid membrane, chlorophyll fluorescence may be considered an intrinsic indicator for changes in membrane fluidity, stability and organization induced by heating. In the case of heat-stress and temperature-stress in general chlorophyll fluorescence serves as a sensitive indicator at two distinctly different levels. First, fluorescence changes may directly reflect structural changes within the membrane. Secondly, fluorescence induction kinetics and q-analysis will provide information on heat-induced changes in function of thylakoid membrane constituents and on overall photosynthetic performance.

The dark fluorescence yield, F_O, increases more than two-fold in the temperature region where heat-induced damage to the photosynthetic apparatus occurs (Schreiber and Berry 1977; Smillie and Nott 1979; Monson and Williams 1982). The critical temperature, T_c, where F_O begins to rise (with a constant heating rate of about 1° C/min) is correlated with the temperature at which photosynthetic activity begins to decline. Recordings of F_O(T) curves were found to provide a rapid means of determining heat tolerance of plants adapted to different climates. Recently, Bilger *et al.* (1984) compared T_c values from F_O(T) curves with T_{L5O} values (temperature leading to death of 50% surface area of leaves) derived from the conventional leaf necrosis tests. It was concluded that changes in F_O may be used satisfactorily to indicate the accumulation of heat experience by leaves which

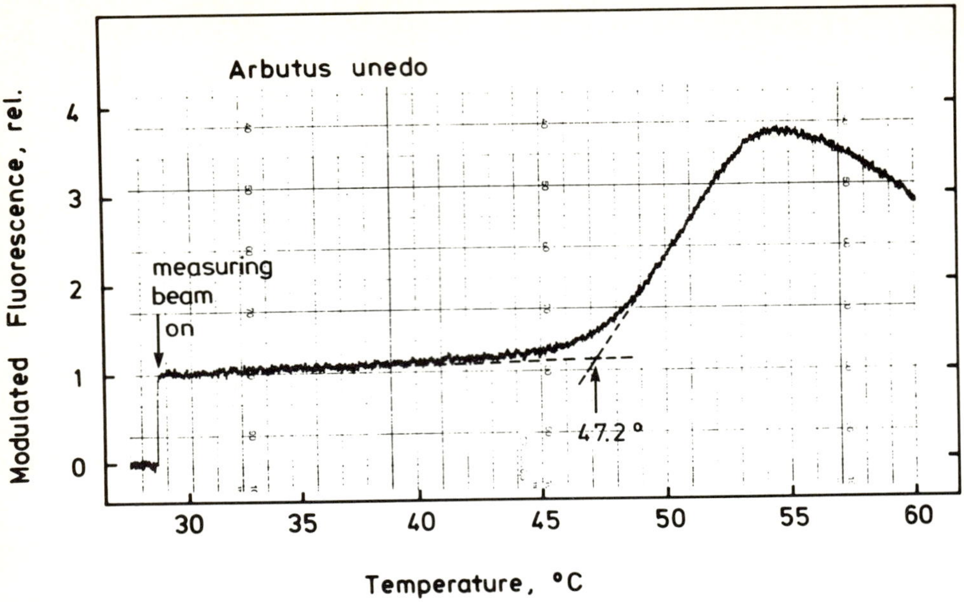

Fig. 10. $F_O(T)$ curve of *Arbutus unedo*. Modulated fluorescence was measured at F_O-conditions. Weak measuring light intensity: 1 W m^{-2}. Far-red background light (λ > 700 nm): 0.1 mW m^{-2}. Temperature was increased at a rate of about 1° C/min. The critical temperature for heat-induced damage at the membrane level (47.2° C) is determined as indicated

causes damage not only to the thylakoid membrane and eventually to the overall leaf tissue as well.

Until recently, continuous recording of the dark-fluorescence level, F_O, as required to obtain $F_O(T)$ curves, has been possible in only a few laboratories equipped with particularly sensitive fluorometers. The excitation light intensity has to be exceedingly low to avoid a light induced fluorescence increase, and detector sensitivity has to be correspondingly high. In most cases it was found useful to give far-red background light to reoxidize the PQ-pool. Otherwise this pool becomes reduced at elevated temperatures in the dark because dark respiration rate increases and decreases tissue O_2 content. This may be particularly disturbing in sclerophyllous leaves with high cuticular resistances. With the introduction of the pulse modulation fluorometer (see section on THE MEASURING SYSTEM), F_O-measurements at very low excitation levels and with far-red background illumination have become routinely accessible. Fig. 10 shows an $F_O(T)$ curve for *Arbutus unedo* as measured with the new modulation fluorometer.

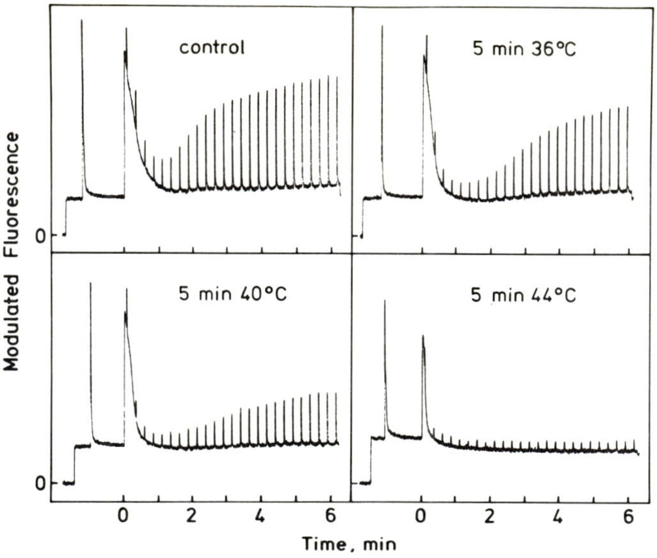

Fig. 11. Effect of heat-pretreatment on the saturation pulse induction kinetics of *Arbutus unedo*. Leaves were exposed for 5 min to the temperatures indicated in the dark. A recovery time of 5 minutes was allowed before conducting fluorescence measurements. Light intensity: 150 μmol m^{-2} s^{-1}. Temperature during fluorescence recording: 22° C

When evaluating $F_O(T)$ curve information, one must realize that the method applied monitors heat-induced changes of thylakoid membrane conformation at elevated temperatures. Upon recooling of the leaves, the increase of F_O is mostly reversible. Following heat-pretreatment, photosynthetic performance may be analysed at any control temperature by recording light-induced fluorescence changes and by q-analysis. Such analysis provides detailed information on the sites of primary heat-damage (see contribution in these proceedings by Bilger *et al.*). In Fig. 11 the effect of heat-pretreatment on the saturation pulse induction curves of *Arbutus unedo* is shown. It is apparent that $(F_v)_s$ (fluorescence measured in saturation pulses) is particularly sensitive to heat-pretreatment. In the control, a high level of $(F_v)_s$, associated with low q_E, indicates efficient energy utilization by Calvin cycle activity. With increasing heat-damage, q_E relaxation is progressively suppressed indicating an increased limitation of Calvin cycle activity. This limitation clearly precedes damage at the level of PS II, expressed by an increase in F_O and a decrease of $(F_v)_m$ (maximal variable fluorescence at $q_E = 0$, i.e. when dark-adapted - for more details see Bilger *et al.* in these proceedings).

The data presented in Fig. 11 display a phenomenon which we have ignored in the above analysis of fluorescence quenching: Following heat-pretreatment, energy dependent fluorescence quenching is very pronounced and fluorescence yield drops below the original dark-level, F_O. When actinic light is switched off in such a state of high energization, there is a rapid decrease of fluorescence to a dark-level, $F_O{}'$, followed by a slower increase to the original F_O-level. We have dealt with this phenomenon in a recent publication (Bilger and Schreiber 1987) and proposed that F_O-quenching (described by the quenching coefficient q_O) indicates a state 1-state 2 shift induced by LHC phosphorylation and consequent movement of LHC away from PS II towards PS I (for reviews on this subject, see Barber 1983; Krause and Weis 1984). As shown by Weis (1984, 1985), elevated temperatures enhance state 1-state 2 shifts in intact leaves. There is considerable physiological relevance to such state shifts, since they reflect the regulation of balanced energy distribution between the two photosystems. Recordings of F_O-quenching may provide an easy and rapid means of assessing such regulatory changes, without the need for spectrofluorometric analysis at 77K.

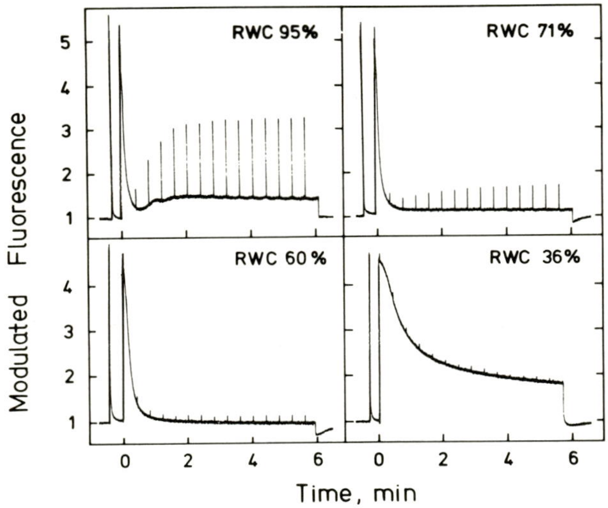

Fig. 12. Effect of water stress on the saturation pulse induction kinetics of *Arbutus unedo*. Leaves were dehydrated to the relative water contents indicated (RWC) in the dark. Light intensity: 125 µmol m^{-2} s^{-1}. Temperature: 20° C. CO_2-concentration: 5% in air

WATER STRESS

Water stress has been shown to cause substantial changes in the fluorescence induction pattern (Wiltens *et al.* 1978; Govindjee *et al.* 1981; Havaux and Lannoye 1983). In Fig. 12 the effect of dehydration on the saturation pulse fluorescence induction curves of *Arbutus unedo* is depicted. The leaves were dehydrated in darkness to avoid photoinhibition, and fluorescence measurements were carried out at high CO_2-concentration (5% in air) to minimize stomatal limitations. With decreasing relative water content the induction pattern is markedly affected. At 71% RWC the relaxation of E-quenching, reflecting onset of Calvin cycle activity in the control, is suppressed. With further dehydration, membrane energization first increases (60% RWC) and then decreases again (36% RWC). At the same time, there is a progressive decrease in photochemical quenching. As with heat-pretreatment (see Fig. 11), the energy-dependent quenching enhanced by dehydration does not only affect variable fluorescence but F_O as well. F_O-quenching, which can be determined upon turning-off actinic light, has to be taken into account in order to calculate values for Q-quenching and E-quenching.

In the experiments of Figs. 13 and 14, oxygen evolution rate and fluorescence quenching of *Arbutus unedo* leaves were measured simultaneously under steady state illumination at different relative water contents of the leaves. The measurements were carried out with a leaf disc oxygen electrode at high CO_2 concentration.

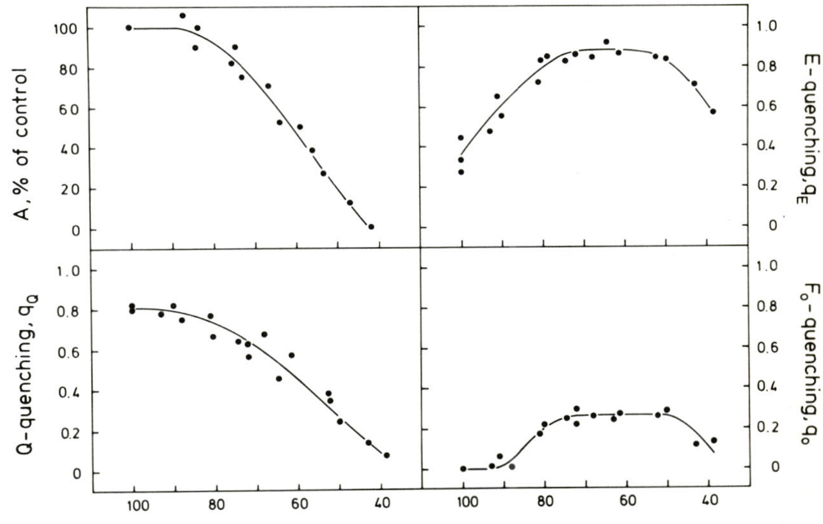

Fig. 13. Comparison of fluorescence quenching with assimilation rate as dependent on the relative water content in *Arbutus unedo* leaves. Assimilation rate, A, and fluorescence were measured simultaneously at steady state illumination in a leaf disc oxygen electrode (Hansatech LD 2 combined with PAM 101 Chlorophyll Fluorometer). To overcome stomatal closure induced by dessication, high CO_2-concentration was established by adding 0.1 N H_2SO_4 to an equal volume of 1 M $NaHCO_3$ in the leaf disc chamber

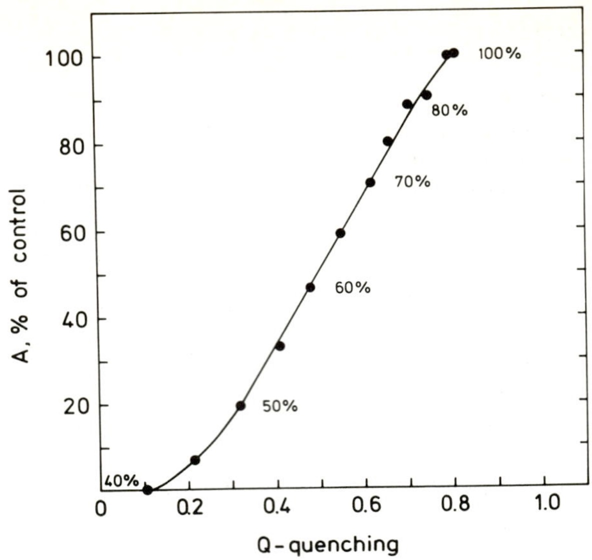

Fig. 14. Relationship of Q-quenching to assimilation rate at varying relative water contents in leaves of *Arbutus unedo*. Fluorescence and assimilation rate were measured simultaneously in a leaf disc electrode at high CO_2-concentration as described for Fig. 13. Relative water content is indicated in percent

Desiccation decreases both oxygen evolution and Q-quenching in an approximate one to one relationship, confirming that q_Q, as determined by the saturation pulse method, is a reliable measure of electron transport rate. Even at a very low relative water content of 36%, where electron transport is severely inhibited, variable fluorescence is almost unchanged (Fig. 12), suggesting that the PS II donor side and the primary reaction are not affected. The rapid light-induced development of q_E, even at 60% RWC (Fig. 12), suggests that electron transport reactions and consequent energization of the thylakoid membrane are not directly inhibited. Inhibition only occurs as the Calvin cycle fails to act as an electron acceptor. From these data it can not be decided whether the Calvin cycle enzymes or the ATP-ase is affected. In both cases enhanced membrane energization would be expected (see also effect of Tentoxin in Fig. 9). From results obtained by Kaiser (see Kaiser, this volume), an inhibition of Calvin cycle enzymes would appear most likely.

PHOTOINHIBITION

Light stress can cause an inhibition of photosynthesis, particularly when electron transport is already limited by other factors. It has been shown that chlorophyll fluorescence is a suitable indicator for assessment of photoinhibitory damage of the photosynthetic apparatus (Critchley and Smillie 1981; Powles and Björkman 1982; Kyle *et al.* 1984). In general, it has been reported that photoinhibition causes a decrease in variable fluorescence, F_V, suggesting primary damage at or close to the PS II reaction center. Measuring the light-induced fluorescence increase at liquid nitrogen temperature provides a reliable means of determining F_V. Using this method, the group of Björkman has extensively studied the ecological significance of photoinhibition when it occurs in combination with other stresses (for review see Björkman 1981; Powles 1984).

Using the modulated fluorescence method, F_V can be readily determined at any given temperature by application of a saturation pulse. However, as with the low temperature method, relaxation of energy dependent quenching must first have occurred. Fig. 15 shows the normal relaxation kinetics following brief non-photoinhibitory illumination. Upon darkening, $(F_V)_S$ (measured during a short saturation pulse) increases with biphasic kinetics, the second phase reaching the

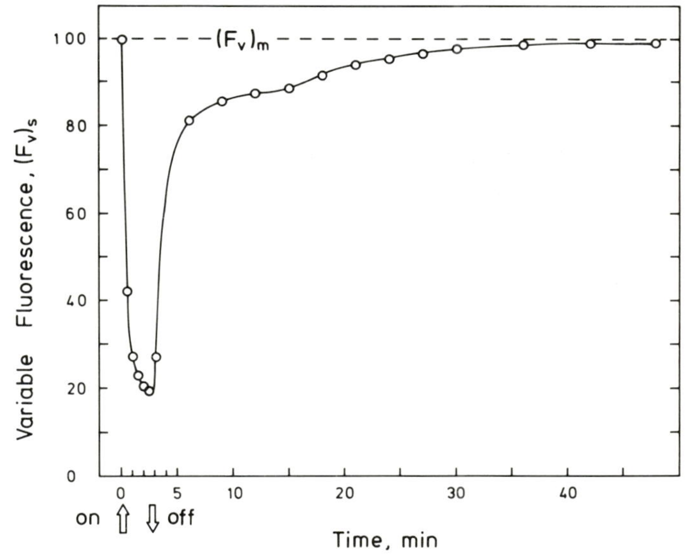

Fig. 15. Light induced decrease of variable fluorescence in *Arbutus unedo* by non-photochemical quenching and recovery of variable fluorescence upon darkening. Illumination with 500 μmol m^{-2} s^{-1} in air (see on/off arrows). Assessment of variable fluorescence by saturation pulses at a frequency which did not lead to accumulation of energy dependent quenching by the saturation pulses itself

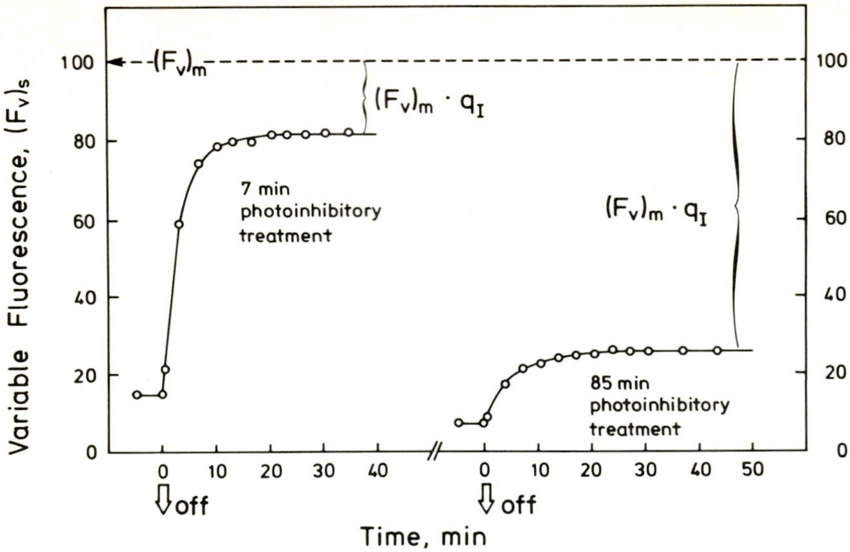

Fig. 16. Dark recovery of variable fluorescence following photoinhibitory treatment and determination of photoinhibitory quenching. Leaves of *Arbutus unedo* were illuminated for 7 min and 85 min with 1700 μmol m^{-2} s^{-1} white light in the absence of CO_2. Following photoinhibitory treatment dark recovery of variable fluorescence was measured as described for Fig. 15. Photoinhibitory quenching, $(F_v)_m q_I$, is apparent after full relaxation of energy dependent quenching. Temperature: 22° C

maximal level, $(F_v)_m$, within about 40 min. Following photoinhibitory treatment, even after prolonged dark treatment, full recovery of F_v is not observed (Fig. 16). In addition to q_Q and q_E, F_v is also affected by photoinhibitory quenching (characterized by the quenching coefficient q_I). Fig. 17 shows the development of photoinhibitory quenching with increasing exposure time of *Arbutus unedo* to strong illumination (1700 μmol m^{-2} s^{-1}). Two phases may be distinguished, confirming previous findings of Kyle *et al.* 1984; Krause *et al.* 1985; Samuelsson *et al.* 1985. There are indications that the termination of the first phase is caused by the onset of an effective repair process (Samuelsson *et al.* 1985). Possibly, the slowly recovering phase of F_v in Fig. 15 is one expression of such a process.

Kyle *et al.* (1984) proposed that under photoinhibitory conditions, reduced plastoquinone accumulates and is responsible for damage to the herbicide binding Q_B-protein, resulting in the inhibition of electron transport. We have shown that accumulation of electrons at the acceptor side of PS II will occur with increasing

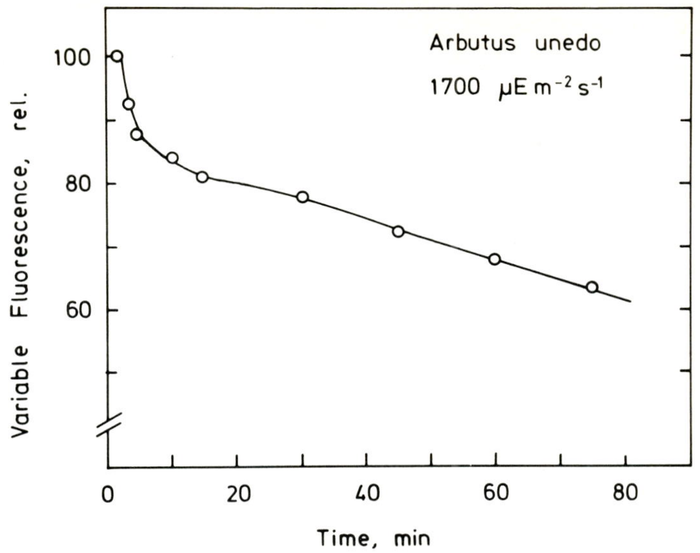

Fig. 17. Decrease in variable fluorescence as a function of duration of photoinhibitory treatment. Leaves of *Arbutus unedo* (grown at 300 μmol m^{-2} s^{-1}) were exposed for varying lengths of time to 1700 μmol m^{-2} s^{-1} white light. Following photoinhibitory treatment leaves were placed for 40 minutes in the dark in order that relaxation of energy dependent quenching occurred. Then variable fluorescence was assessed by application of a saturation pulse. Photoinhibitory treatment and fluorescence measurement were in air at 22° C

water stress in *Arbutus unedo* (Fig. 13), a condition which should favor photoinhibition. On the other hand low CO_2-concentrations did not lead to a highly reduced state of the electron transport chain (Fig. 6) because molecular oxygen still serves as an electron acceptor. The enhanced level of reduction of the electron transport chain, which we showed to occur upon dehydration, should be responsible for increased photoinhibition due to drought stress.

CONCLUSIONS

The results presented in the preceding sections have demonstrated the great potential of chlorophyll fluorescence in the analysis of photosynthetic response. There is a wide range of practical applications for chlorophyll fluorescence measurements in plant physiological and ecological work. The basic phenomenon of the "Kautsky effect" contains essential information on the state of the photosynthetic apparatus. The measurement principle is simple, non-destructive, and rapid. Some emphasis has been placed here on the new possibilities offered by a new type of modulation fluorometer which has become commercially available for field-oriented as well as laboratory investigations. In combination with the saturation pulse method the new system allows rapid and reliable determination of photochemical and non-photochemical quenching of fluorescence. Saturation pulse induction curves reveal information which complements that given by standard "Kautsky curves" and removes ambiguities in the interpretation of fluorescence changes. Applying the saturation pulse method (q-analysis), fluorescence measurements give information on photosynthetic performance even at steady state illumination. For quantitative determination of the photochemical quenching, q_Q, energy dependent quenching of dark fluorescence F_O, must also be taken into account.

Under appropriate conditions, there is good correlation between q_Q and photosynthetic electron transport rate as measured simultaneously via O_2-evolution or CO_2-fixation. Such correlation can be only expected when electron flow to O_2 as terminal acceptor is negligible (low O_2 or high CO_2), since for q_Q it is immaterial whether PS II acceptor reoxidation occurs via O_2 or via CO_2. With proper consideration of this aspect, fluorescence may also give insight into the function of the photorespiratory pathway. Generally speaking, fluorescence information may be considered complementary to that of gas exchange measurements. While the latter provides absolute overall rates, fluorescence gives relative rates and, in addition, information on the locations of specific steps limiting the overall process.

Practical applications of the chlorophyll fluorescence method were demonstrated by the assessment of photosynthesis limitations in *Arbutus unedo* induced by heat and water stress. The data presented show that the observed decrease in variable fluorescence caused by these stress factors is preceded by a decrease in photochemical quenching and an increase in energy dependent quenching. Hence, it is suggested that Calvin cycle reactions (or possibly the ATP-ase) are affected before any damage at the level of the primary reactions.

Light-stress (photoinhibition) was shown to cause a type of non-photochemical quenching (photoinhibitory quenching) which can be distinguished from energy dependent quenching by its irreversibility during extended dark periods. The saturation pulse method may offer a rather simple procedure for assessing photoinhibitory damage. A direct comparison of this method with the low temperature method (Powles and Björkman 1982) appears desirable.

In its natural environment, *Arbutus unedo* is frequently exposed to a combination of heat, drought and light stress. Its photosynthetic capacity is significantly reduced during the summer months (Beyschlag 1984). In future work, the application of the fluorescence method in field measurements should allow us to differentiate between the relative contributions of various stress factors to reductions in photosynthetic CO_2 uptake. Furthermore, it will be important to investigate by which means and to what extent plants are capable of acquiring resistance to stress factors in their natural habitat.

ACKNOWLEDGEMENTS

Many thanks are due to the Deutsche Forschungsgemeinschaft, to Ulrich Heber and Otto Lange for continuous support of our work. We are grateful to Klaus Winter and Werner Kaiser for stimulating discussion and for their help with gas exchange measurements. We also wish to thank Keith Rienits for correcting our English and Annette Weber for preparing the manuscript.

REFERENCES

Barber J (1983) Membrane conformational changes due to phosphorylation and the control of energy transfer in photosynthesis. Photobiochem Photobiophys 5:181-190

Bennett J, Steinback KE, Arntzen CJ (1980) Chloroplast phosphoproteins: Regulation of excitation energy transfer by phosphorylation of thylakoid membrane polypeptides. Proc Natl Acad Sci 77:5253-5257

Beyschlag W (1984) Photosynthese und Wasserhaushalt von *Arbutus unedo* L. im Jahreslauf am Freilandstandort in Portugal. Gaswechselmessungen unter natürlichen Bedingungen und experimentelle Faktorenanalyse. Dissertation, Würzburg

Bilger W, Schreiber U (1987) Energy dependent quenching of dark-level chlorophyll fluorescence in intact leaves. In: Govindjee *et al.* (eds), Excitation Energy and Electron Transfer in Photosynthesis, Photosynthesis Research Supp. Vol. Martinus Nijhoff/Dr. W. Junk Publishers, Dordrecht

Bilger W, Schreiber U, Lange OL (1984) Determination of leaf heat resistance: Comparative investigation of chlorophyll fluorescence changes and tissue necrosis methods. Oecologia 63:256-262

Björkman O (1981) Responses to different quantum flux densities. In: Lange OL, Nobel PS, Osmond CB, Ziegler H (eds) Encyclopedia of Plant Physiology, NS vol 12A. Springer, Berlin-Heidelberg-New York, pp 57-107

Björkman O, Powles SB (1984) Inhibition of photosynthetic reactions under water stress: Interaction with light level. Planta 161:490-504

Bradbury M, Baker NR (1981) Analysis of the slow phases of the *in vivo* chlorophyll fluorescence induction curve. Changes in redox state of photosystem II electron acceptors and fluorescence emission from photosystems I and II. Biochim Biophys Acta 635:542-551

Critchley C, Smillie RM (1981) Leaf chlorophyll fluorescence as an indicator of photoinhibition in *Cucumis sativus* L. Aust J Plant Physiol 8:133-141

Dietz KJ, Schreiber U, Heber U (1985) The relationship between the redox state of Q_A and photosynthesis in leaves at various carbon dioxide, oxygen and light regimes. Planta 166:219-226

Duysens LNM, Sweers HE (1963) Mechanism of two photochemical reactions in algae as studied by means of fluorescence. In: Studies on Microalgae and Photosynthetic Bacteria. University of Tokyo Press, Tokyo, pp 353-372

Govindjee, Downton WJS, Fork DC, Armond PA (1981) Chlorophyll a fluorescence transient as an indicator of water potential of leaves. Plant Sci Lett 20:191-194

Havaux M, Lannoye R (1983) Chlorophyll fluorescence induction: A sensitive indicator of water stress in maize plants. Irrig Sci 4:147-151

Kautsky H, Franck U (1943) Chlorophyllfluoreszenz und Kohlensäure-assimilation XI. Die Chlorophyllfluoreszenz von *Ulva lactuca* und ihre Abhängigkeit von Narkotika, Sauerstoff und Kohlendioxyd. Biochem Z 315:176-206

Kautsky H, Hirsch A (1931) Neue Versuche zur Kohlenstoffassimilation. Naturwissenschaften 19:964

Kautsky H, Hirsch A (1934) Das Fluoreszenzverhalten grüner Pflanzen. Biochem Z 274:422-434

Krause GH, Briantais JM, Vernotte C (1982) Photoinduced quenching of chlorophyll fluorescence in intact chloroplasts and algae. Resolution into two components. Biochim Biophys Acta 679:116-124

Krause GH, Köster S, Wong SC (1985) Photoinhibition of photosynthesis under anaerobic conditions studied with leaves and chloroplasts of *Spinacia oleracea* L. Planta 165:430-438

Krause GH, Weis E (1984) Chlorophyll fluorescence as a tool in plant physiology. II. Interpretation of fluorescence signals. Photosynth Res 5:139-157

Kyle DJ, Ohad I, Arntzen CJ (1984) Membrane protein damage and repair: Selective loss of a quinone-protein function in chloroplast membranes. Proc Natl Acad Sci USA 81:4070-4074

Lavorel J, Etienne AL (1977) *In vivo* chlorophyll fluorescence. In: Barber J (ed) Primary Processes of Photosynthesis. Elsevier, Amsterdam, pp 203-268

Monson RK, Williams GJ (1982) A correlation between photosynthetic temperature adaptation and seasonal phenology patterns in the shortgrass prairie. Oecologia 54:58-62

Ögren E, Baker NR (1985) Evaluation of a technique for the measurement of chlorophyll fluorescence from leaves exposed to continuous white light. Plant, Cell & Environment 8:539-547

Papageorgiou G (1975) Chlorophyll fluorescence: An intrinsic probe of photosynthesis. In: Govindjee (ed) Bioenergetics of Photosynthesis. Academic Press, New York, pp 319-371

Powles SB (1984) Photoinhibition of photosynthesis induced by visible light. Ann Rev Plant Physiol 35:15-44

Powles SB, Björkman O (1982) Photoinhibition of photosynthesis: Effect on chlorophyll fluorescence at 77K in intact leaves and in chloroplast membranes of *Nerium oleander*. Planta 156:97-107

Quick WP, Horton P (1984) Studies on the induction of chlorophyll fluorescence quenching by redox state and transthylakoid pH gradient. Proc R Soc Lond B 217:405-416

Renger G, Schreiber U (1986) Practical applications of fluorometric methods to algae and higher plant research. In: Govindjee, Amesz J and Fork DC (eds) Light Emission by Plants and Bacteria. Academic Press, New York, in press

Samuelsson G, Lönneborg A, Rosenquist E, Gustafsson P, Öquist G (1985) Photoinhibition and reactivation of photosynthesis in the cyanobacterium *Anacystis nidulans*. Plant Physiol 79:992-995

Schreiber U (1983) Chlorophyll fluorescence yield changes as a tool in plant physiology. I. The measuring system. Photosynth Res 4:361-373

Schreiber U (1986) Detection of rapid induction kinetics with a new type of high frequency modulated chlorophyll fluorometer. Photosynth Res 9:261-272

Schreiber U, Berry JA (1977) Heat-induced changes of chlorophyll fluorescence in intact leaves correlated with damage of the photosynthetic apparatus. Planta 136:233-238

Schreiber U, Schliwa U, Bilger W (1986) Continuous recording of photochemical and non-photochemical fluorescence quenching with a new type of modulation fluorometer. Photosynth Res 9

Sharkey TD (1985) Photosynthesis in intact leaves of C_3 plants: Physics, physiology and rate limitations. Bot Rev 51:53-105

Smillie RM, Nott R (1979) Heat injury in leaves of alpine, temperate and tropical plants. Aust J Plant Physiol 6:135-141

Tenhunen JD, Weber J, Yocum C, Gates DM (1976) Development of a photosynthesis model with an emphasis on ecological applications. II. Analysis of a data set describing the PM surface. Oecologia (Berl) 26:101-119

Velthuys BR, Amesz J (1974) Charge accumulation at the reducing side of system II of photosynthesis. Biochim Biophys Acta 333:85-94

Weis E (1984) Short term acclimation of spinach to high temperatures. Effect on chlorophyll fluorescence at 293 and 77K in intact leaves. Plant Physiol 74:402-407

Weis E (1985) Light- and temperature-induced changes in the distribution of excitation energy between photosystem I and photosystem II in spinach leaves. Biochim Biophys Acta 807:118-126

Wiltens J, Schreiber U, Vidaver W (1978) Chlorophyll fluorescence induction: An indicator of photosynthetic activity in marine algae undergoing desiccation. Can J Bot 56:2787-2794

Leaf absorptance and leaf angle: mechanisms for stress avoidance

J.R. Ehleringer and J. Comstock

Department of Biology, University of Utah, Salt Lake City, UT 84112 USA

INTRODUCTION

Leaf angle and leaf absorptance are two leaf parameters that link a leaf and its physical environment. These characters affect both the amount of solar radiation incident on a leaf and the fraction of that incident light which is actually absorbed. Differences in both leaf angle and leaf spectral characteristics have been known for some time for plants of divergent habitats, and it has been inferred that they confer some adaptive advantage to plants from arid habitats (Haberlandt 1884; Schimper 1903; Warming 1909). However, it is only relatively recently that the significance of these features to leaf temperature and water loss have been quantitatively evaluated (Raschke 1960; Gates 1963, 1980). Evaluating the interactions between energy balance, leaf temperature, transpiration and photosynthesis at the leaf level have provided a means for evaluating the significance of changes in leaf angle and/or absorptance to plant performance under different environmental conditions (Mooney 1972; Parkhurst and Loucks 1972; Cowan and Farquhar 1977; Ehleringer and Mooney 1978; Mooney and Gulmon 1979).

Changes in leaf angle and/or leaf absorptance affect leaf energy balance, leaf temperature, photosynthesis and transpiration through both direct and indirect interactions. The linkages between these parameters are shown in Fig. 1. The arrows indicate the direction in which the influence is exerted, and it can be seen that both leaf angle and leaf absorptance affect the amount of solar radiation absorbed by the leaf. As will be shown later, their impact is somewhat different with leaf angle influencing the diurnal/seasonal patterns of incident solar radiation and leaf absorptance reducing the total amount of solar radiation absorbed by the leaf irrespective of orientation effects. The amount of light absorbed will have a direct influence on both photosynthesis and on leaf temperature. Transpiration rate is influenced by leaf temperature which in turn is affected in a reciprocal fashion. Leaf conductance, determined anatomically by stomatal frequency and diameter, influences the exchange of both water vapor and carbon dioxide. Thus, leaf conductance directly influences photosynthesis, leaf temperature and transpiration. The consequences of changes in leaf angle and/or leaf absorptance on leaf performance can not be determined without knowing the other plant and environmental factors mentioned above, and the impact of angle/absorptance changes can either amplify or reduce photosynthesis and transpiration depending on the magnitude of the leaf conductance and water stress.

NATO ASI Series, Vol. G15
Plant Response to Stress. Edited by J. D. Tenhunen et al.

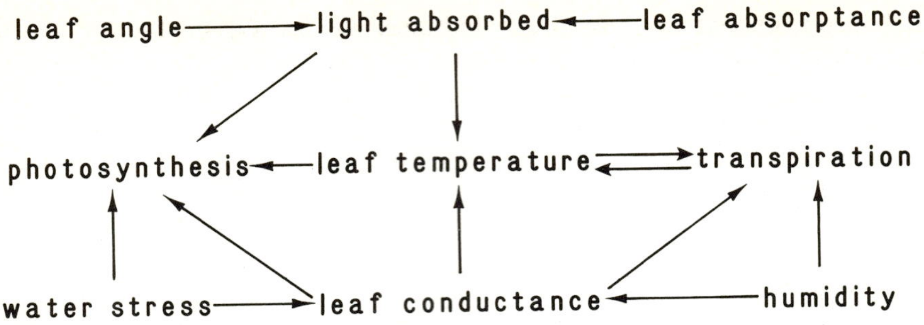

Fig. 1. Interactions between leaf angle, leaf absorptance, leaf metabolic parameters, and water stress

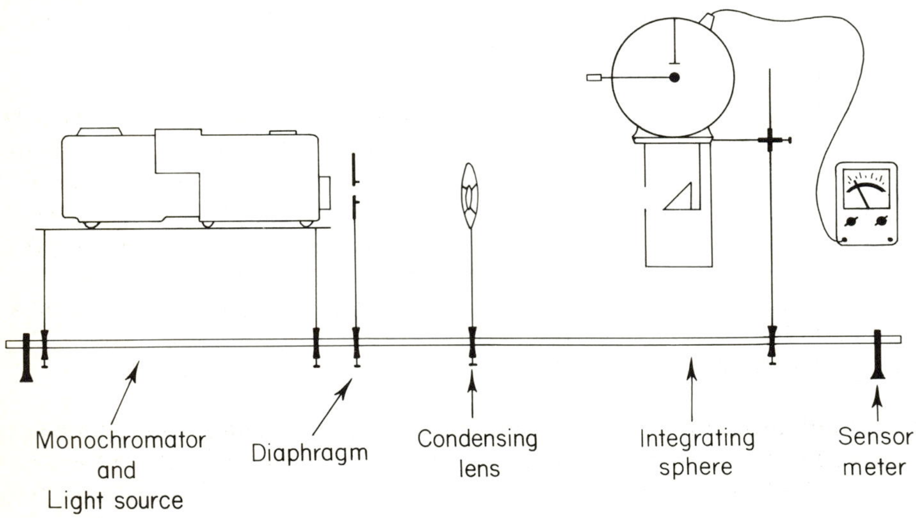

Fig. 2. Setup of an Ulbricht integrating sphere. Monochromatic light is supplied by a light source-monochromator and focused onto the sample inside the integrating sphere with a condensing lens. The base of the integrating sphere is covered with black velvet to reduce stray light. From Ehleringer (1981)

METHODOLOGY FOR LEAF ABSORPTANCE AND LEAF ANGLE MEASUREMENTS

Leaf absorptance is a measurement of the fraction of the incident photon or energy flux that is absorbed by the leaf. It can be measured at a single wavelength or over a specific waveband. When leaf absorptance are expressed for a particular waveband, the irradiance source (such as sunlight) must be specified.

Over the visible wavelengths (400-700 nm), the leaf absorptance to sunlight on either a photon or an energy flux basis are nearly identical and are frequently used interchangeably (Ehleringer and Björkman 1978, Gates 1980). Throughout this chapter, leaf absorptances unless otherwise mentioned have been measured on a photon basis. This is because of the rapidity and ease with which the leaf photon absorptance can be measured.

Leaf absorptances (a) are measured with an integrating sphere. There are two basic types: the Ulbricht sphere with the leaf sample inside the sphere during measurement (Fig. 2) and the Taylor sphere with the sample positioned in a port on the outside surface of the sphere during the measurement. The advantage of the Ulbricht sphere for leaf absorptance measurements is that only a single measurement need be made to obtain leaf absorptance. This is because both reflected and transmitted light are measured when the leaf is centrally located within the sphere. In contrast, separate measurements of both reflectance and transmittance are required by the Taylor sphere.

The leaf absorptance spectrum between 400-700 nm can differ greatly between plant species. In Fig. 3 are plotted leaf absorptance spectra for four species common in the hot desert habitat of the floor of Death Valley, California. Although leaves of all species exhibited a general trend of relative absorptance peaks between 400-500 nm and 600-700 nm, the absolute absorptance can differ depending both on leaf epidermal characteristics and on chlorophyll content. Lin and Ehleringer (1982) showed a strong correlation between leaf absorptance and total chlorophyll content with the greatest changes in leaf absorptance occurring during leaf maturation and senescence.

Although chlorophyll content changes may contribute to leaf absorptance values of leaves in Fig. 3 by affecting leaf transmittance (Table 1), the greatest factor affecting leaf absorptance is the leaf reflectance (Fig. 3). Epidermal modifications, such as the presence of waxes, trichomes, salt bladders, etc., create a reflective surface that shields the leaf interior from incident radiation, and thus reduces the leaf absorptance even though leaf chlorophyll may be high (e.g., *Tidestromia* in Fig. 3).

It is useful to have a single value which characterizes the absorptance for a particular leaf. This value represents the integral of a leaf absorptance spectrum with the spectral distribution of an irradiance source for specific wavebands. In this chapter, leaf absorptance will refer to the fraction of photons from sunlight at the earth's surface between 400-700 nm that are absorbed by the leaf. From correlation analyses with many plant species, Ehleringer (1981) showed that the leaf absorptance at 625 nm was a reliable indicator of the total leaf absorptance to sunlight between 400-700 nm. As can be seen from Table 1 and Fig. 3, there is a one to one correspondence between the leaf absorptance of each of the four species at 625 nm and the estimates of leaf absorptance between 400-700 nm.

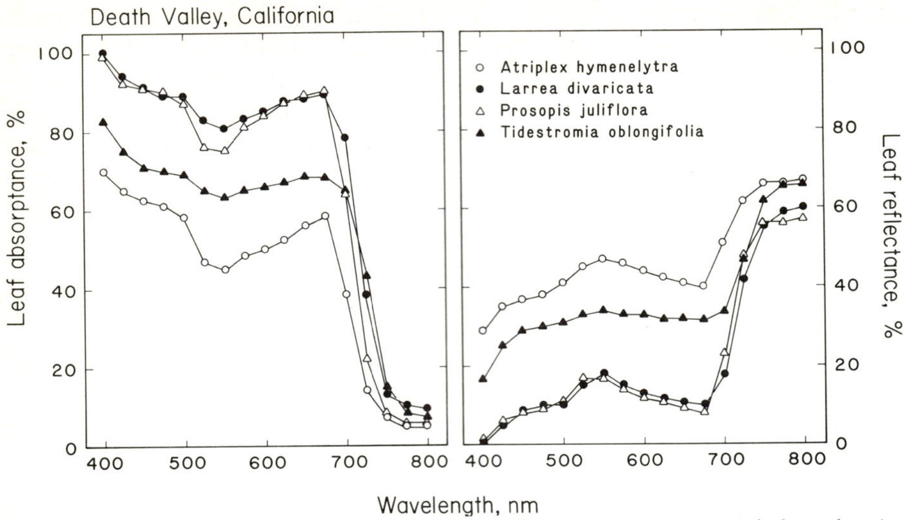

Fig. 3. Leaf absorptance (left plate) and leaf reflectance (right plate) as a function of wavelength between 400-700 nm for leaves of four common plants from the floor of Death Valley, California

Table 1. Leaf characteristics of four common perennials from a hot desert site on the floor of Death Valley, California.

Species	Total chlorophyll (mg m^{-2})	Chlorophyll a/b	Absorptance 400-700 nm (%)	Leaf angle (°)
Atriplex hymenelytra	158.9	3.10	52	67.5
Larrea divaricata	502.9	3.81	87	48.4
Prosopis juliflora	275.7	3.79	87	heliotropic
Tidestromia oblongifolia	332.4	3.78	67	----

While leaf absorptances between 400-700 nm ($a_{400-700}$) may be useful for calculating photon absorptance for photosynthetic estimates, it is the total energy absorptance between 400-3000 nm ($a_{400-3000}$) that is important for heat balance studies. Fortunately, there is a strong relationship between these two parameters. Ehleringer (1981) found that

$$a_{400-3000} = 0.73 \quad a_{400-700} - 11.9 \tag{1}$$

where both absorptances have units of percent.

Leaf angles are measured from the horizontal, using either an inclinometer or a protractor with plumb line. The leaf angle must be measured as the angle from the horizontal of a plane parallel to the leaf blade. This, of course, requires that the leaf azimuth be known. The leaf azimuth, which is the direction that the leaf blade plane is facing, is measured with reference to south. Thus, south has an azimuth value of 0°, east is -90°, west is $+90^\circ$, and north either $+180^\circ$ or -180°. Given the inclination (α_l) and azimuth (β_l) angles of a leaf and also the inclination (α_s) and azimuth (β_s) angles of the sun, the cosine of the angle of incidence (cos(i)) can be calculated as

$$\cos(i) = (\cos \alpha_l)(\sin \alpha_s) + (\sin \alpha_l)(\cos \alpha_s)(\cos (\beta_l - \beta_s)) \quad (2)$$

The cos(i) then becomes a measure of the fraction of the sun's direct rays that are incident on the leaf blade surface.

The distribution of azimuths within many plant canopies is often assumed to be even for model simulations (Ross 1975); however, known examples of strongly preferred leaf azimuths, such as in the compass plant *Lactuca serriola*, demonstrate that biased azimuthal distributions can have a significant impact on the solar radiation absorption patterns (Werk and Ehleringer 1984). Leaf azimuths for several crop species have been shown to be randomly distributed (Blad and Baker 1972), while in other cases there is a contagious or clumped component to leaf azimuthal distributions (Lemeur 1973; Lugg *et al.* 1981).

SPECIES-LEVEL PATTERNS ALONG ARIDITY GRADIENTS IN NORTH AND SOUTH AMERICA

In this section we describe examples of the types of morphological changes affecting leaf absorptance and leaf angle that are observed along aridity gradients in the arid, western portions of North and South America. This is done through both intrageneric comparisons of closely related species occurring in slightly different habitats and through comparisons of plant species occurring in geographically separated, but similar climatic zones.

Species Replacement Gradients

As noted previously, leaf pubescence is a character commonly associated with plants in arid habitats. Increases in leaf pubescence along a gradient of decreasing precipitation is not limited to a few species, but has been observed among many different families and genera (Ehleringer 1984). As one proceeds along an increasing aridity transect from conifer forest through mediterranean-climate sclerophylls to desert communities in southern California, intrageneric replacement series are common. No single species spans the entire transect, but there is a pattern such that on the wetter, more mesic sites species possess less pubescent leaves.

Moving to progressively drier portions of the transect, intrageneric replacements possess leaves with increasing pubescence. *Brickellia*, *Encelia*, *Eriogonum*, *Salvia*, and *Tetradymia* are examples of genera each of which contains a set of species exemplifying this aridity-pubescence replacement series.

As the degree of leaf pubescence increases, so does the leaf reflectance. *Salvia*, which occurs along an aridity gradient in southern California, can be used to illustrate this point. *S. mellifera* occurs on the wettest sites, and is first replaced by *S. leucophylla*, and then by *S. apiana* on progressively drier sites (Ehleringer 1984). Figure 4 shows the absorptance spectra for leaves from these three species. *S. mellifera* with a glabrate, green leaf has an absorptance spectrum between 400-700 nm that is typical for most green leaves. In the gray-leaved *S. leucophylla*, the absorptances at all wavelengths are reduced, but the absorptance spectrum is otherwise quite similar to the green-leaved *S. mellifera*. *S. apiana* leaves are white in appearance and clearly exhibit a much reduced absorptance at all wavelengths. The pubescence appears to be a blanket reflector, reflecting equally well at all wavelengths between 400-700 nm. A similar pattern has also been shown to occur in *Encelia*, which is another broad-leaved genus occurring along the same aridity gradients in southern California (Ehleringer and Björkman 1978).

Modifications in the leaf epidermis need not be restricted to increased pubescence. In other genera such as *Atriplex*, the epidermal modification is salt-containing bladders (Mooney *et al.* 1977), while in *Dudleya* and *Eucalyptus* waxes become more prevalent with aridity (Thomas and Barber 1974; Mulroy 1979).

Intrageneric studies of leaf angle changes with habitat are less common, but investigations with members of the evergreen-leaved genus *Arctostaphylos* indicate that leaf angles become steeper with habitat aridity. Shaver (1978) showed that species from more arid habitats have more vertically inclined leaves as well as having leaves with increased pubescence levels.

Coastal Scrub Vegetation

Leaf absorptances of the dominant perennial species in the coastal chaparral communities in southern California and in central Chile measured in the spring are quite similar (Tables 2 and 3). The leaf absorptances range between 75-84% and 74-87% in Californian and Chilean communities, respectively, whicle the average absorptances of leaves in each community are 82% and 81%. In neither of these two homologous-climate communities do there appear to be sharp differences in leaf spectral characteristics.

No data are available for the leaf angles of plants occurring in the coastal chaparral of these two areas.

Mediterranean-Climate Sclerophyllous Vegetation

In contrast to the coastal chaparral communities, the interior evergreen mediterranean-climate species of southern California and central Chile appear to have somewhat different leaf absorptances (Tables 4 and 5). The leaf absorptances ranged between 70-82% and 78-87% in Californian and Chilean evergreen chaparral communities, respectively. Correspondingly, the average leaf absorptance in the communities were 78% and 82%. The lower absorptances in the California chaparral may be an indication that this habitat is slightly drier, although the limited data preclude any general conclusion.

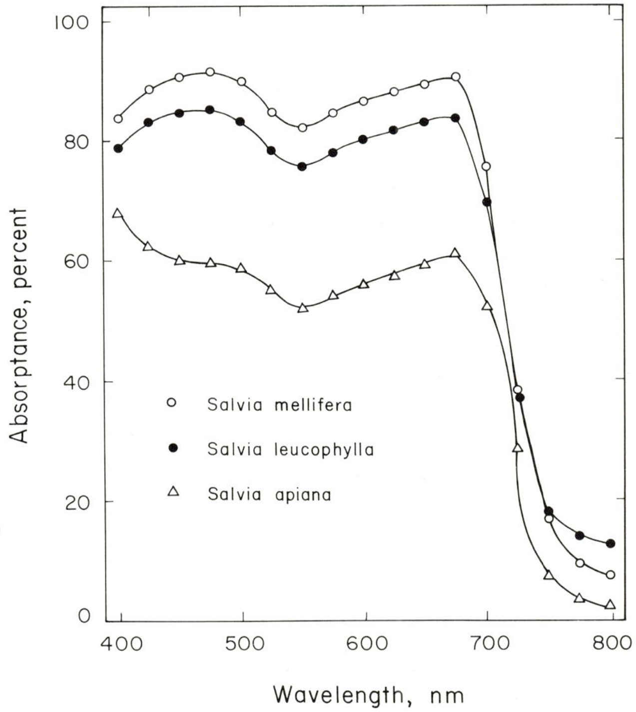

Fig. 4. Leaf absorptance between 400-800 nm for three *Salvia* species occurring along an aridity gradient in southern California. From Ehleringer (1984)

Table 2. Leaf absorptances (400-700 nm) of coastal chaparral species from Point Mugu, California. Data were collected during the spring and are means (n=5).

Species	Leaf absorptance (%)
Encelia californica	83.9
Eriogonum cinereum	83.2
Marrubium vulgare	81.8
Nicotiana glauca	83.2
Platanus racemosa	75.7
Rhus integrifolia	84.2
Salix lasiolepis	82.9
Salvia leucophylla	78.1
Salvia mellifera	83.7
Sambucus mexicana	83.3

Table 3. Leaf absorptances (400-700 nm) of coastal mattoral species from Pichidangi, Chili. Data were collected in the spring and are means (n=5).

Species	Leaf absorptance (%)
Baccharis concava	73.9
Cassia closiana	87.6
Eupatorium salviae	76.3
Fluorensia thurifera	80.4
Lepichinia chilensis	78.4
Lithraea caustica	87.2
Lobelia salicifolia	77.8
Lucuma valparidesea	86.0
Podanthus mitique	85.2
Proustia pungens	74.8

There are only a few observations of leaf angle distributions for different chaparral species. Lawrence (1975) compared leaf angle of two homologous species pairs in California and Chile (*Ceanothus greggii - Colliguaya odorifera* and *Adenostoma fasciculatum - Baccharis linearis*). There appeared to be little convergence in leaf angle among species pairs; leaf angles were 48° versus 72° in the *Ceanothus-Colliguaya* pair and 29° verus 63° in the *Adenostoma-Baccharis* pair. In a comparative study of *Arctostaphylos* species along the Sierra Nevada, Shaver (1978) found that as habitat aridity increased from the Central Valley

Table 4. Leaf absorptances (400-700 nm) of chaparral species from San Diego, California. Data were collected during the early summer and are means (n=3).

Species	Leaf absorptance (%)
Arctostaphylos patula	70.4
Ceanothus oliganthus	73.2
Cercocarpus ledifolius	79.5
Eriodyctyon californicum	79.6
Prunus ilicifolia	76.9
Quercus dumosa	81.8
Rhus laurina	82.1
Salvia mellifera	78.8

Table 5. Leaf absorptances (400-700 nm) of mattoral species from Santa Laura, Chile. Data were collected in the spring and are means (n=5).

Species	Leaf absorptance (%)
Colliguaya odorifera	81.7
Cryptocarya alba	83.6
Drimys winteri	78.7
Kageneckia oblonga	81.8
Lithraea caustica	87.2
Muehlenbeckia hastulata	80.8
Myrceugenella chequen	83.3
Quillaja saponaria	78.3
Sophora macrocarpa	85.4
Talguenia quiquinervis	80.4

floor to the mountain crests, there was also an increase in leaf angle of from 51° to 77°. More recently, Comstock and Mahall (1985) measured leaf angles of two coastal chaparral *Ceanothus* species. They found that during the spring growing season leaf angles averaged 60° and 64° for *C. crassifolius* and *C. megacarpus*, respectively. A larger comparison on California and Chile mediterranean-climate species was conducted by Parsons (1973). However, the qualitative nature of these observations only allow us to conclude that leaves of most species in both California and Chile were 45° or steeper. The absence of sufficient quantitative data on leaf angles of Californian and Chilean mediterranean climate plants prevents any firm evaluation of the extent to which this feature might be used to affect solar radiation absorption patterns during the summer stress periods.

Coastal Desert Vegetation

At latitudes more equatorial than the mediterranean-climate zone in both western portions of North and South America, there are pronounced coastal desert zones. Ehleringer *et al.* (1981) measured leaf absorptances of *Encelia* species occupying different positions along the precipitation gradient from coastal chaparral (300 mm annually) to desert (as low as 40 mm annually) in both continents. Leaf absorptances in *Encelia* ranged from 83% down to 45% over these gradients, and there was a remarkable similarity in the leaf absorptance vs. precipitation relationship along both transects.

Rundel *et al* . (1980) have measured both leaf angles and leaf absorptances of the dominant perennial species in the coastal Atacama Desert of northern Chile. In general, leaf absorptances were 75% or lower and leaf angles were greater than 50°. These leaf absorptance values are lower than have been measured for species in relatively wetter coastal chaparral communities (see above) and are more similar to the low leaf absorptances reported for leaves of species from interior desert communities (Ehleringer 1981, 1985).

COMMUNITY-LEVEL PATTERNS IN RESPONSE TO STRESS

California Mediterranean-Climate Vegetation

Mooney *et al.* (1974) evaluated leaf morphological characteristics of mediterranean-climate species in three communities along a precipitation gradient in northern Baja California. Leaves of most species were relatively small, mesophytic and often deciduous at the dry end of the transect and larger, evergreen and more sclerophytic at the wet end. The fraction of the species with vertically inclined leaves increased from 6% to 57% in going from the driest to the wettest site. Interestingly, there was no apparent change in the fraction of the species with "heavy leaf pubescence" from site to site along this transect. While steep leaf inclinations are often thought of as occurring in the more stressful environment, in this case it appears that plants at the drier sites maintain shallower leaf angles. Perhaps this was because drought-deciduous leaved species were a more significant component at the drier sites. In a similar comparative study of coastal sage and chaparral communities in southern California, Harrison *et al.* (1971) found that plants from the drier coastal sage habitat had leaves with greater pubescence (presumably lower leaf absorptance) than in plants from chaparral.

Wasatch Front Gradient

Along the Wasatch Front in Utah, growth is confined by cold winter temperatures and snow to a 4-6 month spring-summer period. Most of the precipitation falls during the winter and early spring periods with only 20% of the precipitation being received during the summer months. This means that plants are exposed to a summer drought period. Ehleringer (1985) measured leaf angles and leaf absorptances of the dominant species along an elevation gradient from the intermountain valleys to the upper meadows of the Wasatch Front. This corresponded to an annual precipitation cline of 350-1240 mm.

There were large changes in the leaf angles of plants along this transect (Fig. 5.) Leaf angles of shrubs and herbs decreased between the driest sites (saltbush community) and mountain brush communities (800 mm precipitation). At the higher annual precipitation level zones in the coniferous forests and alpine meadows, leaf angles were once again steeper than in the lower elevation mountain brush community. Two probable explanations for this are that 1) although the higher elevation plant communities received greater annual amounts of precipitation, most of it was as snow which melted and ran off before it could be utilized by the vegetation or 2) community leaf area indices were greatest in the mid-elevation sites and the competition for light favored lower leaf angles. Since all communities are exposed to the same summer drought period and the higher elevation plant communities have shorter growing seasons by virtue of colder spring temperatures, the plants in the uppermost elevation plant communities may spend more of the growing season exposed to decreased water availability than do the lower elevation mountain brush community.

While there were large changes in leaf angles, plants along this transect exhibited only minor, if any, changes in leaf absorptance (Fig. 5). The only components of the vegetation to show a decreased leaf absorptance were the shrubs at the two driest sites (saltbush and grassland communities). It is quite interesting in that both the Wasatch Front transect and the Baja California transect of Mooney *et al.* (1974), leaf angle appears to be the more variable character with aridity (although trends were opposite). Leaf absorptances appear to remain more constant, decreasing only in plants of the most arid habitats and then only after leaf angles have become steep. Related to this, Ehleringer (1981) demonstrated that in the most arid warm desert sites, leaf absorptances were on average lower than observed in the driest locations along the Wasatch Front and that shrubs were the only life form to have really low leaf absorptances (down to 30%).

STRESS REDUCTION BY DECREASING SOLAR RADIATION ABSORPTION

In aridland ecosystems, the predominant stress to which plants are subjected is water stress. This water stress can occur on a short-term basis, with diurnal

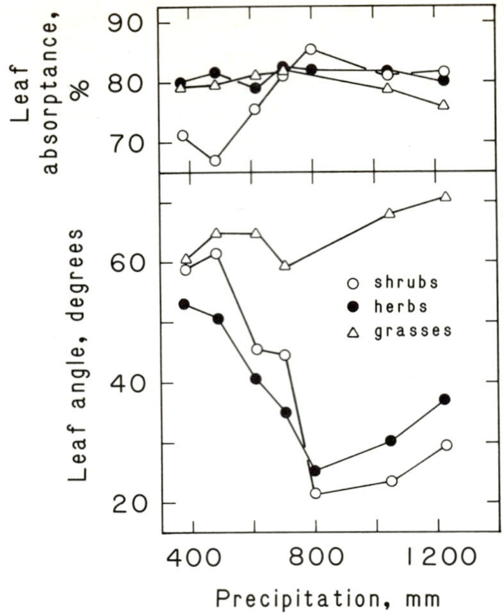

Fig. 5. Mean leaf absorptances and mean leaf angles for shrubs, herbs and grasses along an elevational transect in the Wasatch Front, Utah. From Ehleringer (1985)

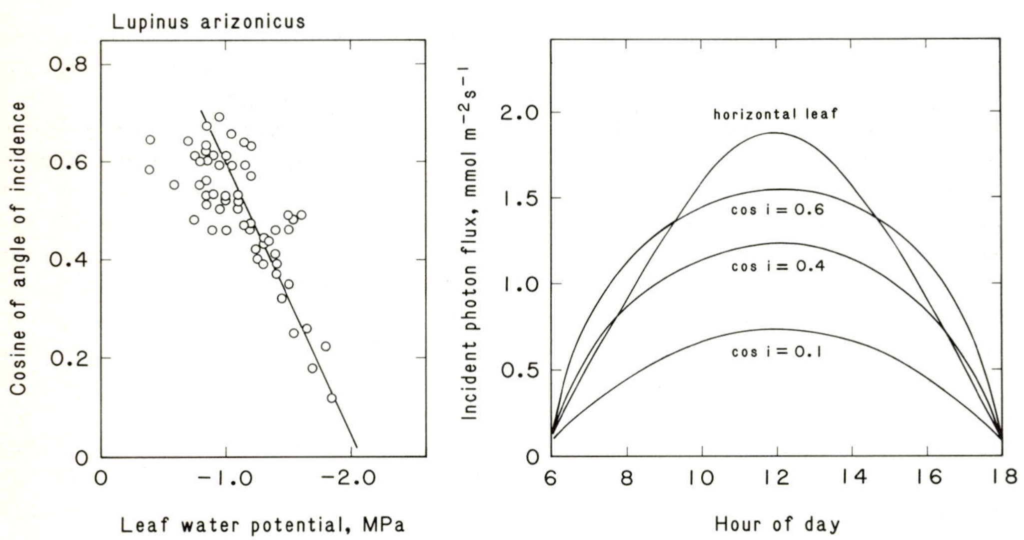

Fig. 6. The dependence of the cos(i) in leaves of *Lupinus arizonicus* on leaf water potential (left plate) and the calculated diurnal photon flux incident on leaves with different cos(i) values for April 1. Based on Forseth and Ehleringer (1980)

changes in the leaf to air humidity concentration gradient (Δw), or on a long term basis, through soil moisture depletion. In both cases, water stress results in a decreased leaf conductance, and since the carbon dioxide supply rate decreases with leaf conductance, a decreased need for photons to operate the photosynthetic light reactions (see Tenhunen *et al.* and Beyschlag *et al.* this volume). Photon flux levels beyond that required to saturate photosynthesis serve only to increase the energy load and heat the leaf; they may also be detrimental to the photosynthetic apparatus, resulting in photoinhibitory damage to the leaf (Björkman and Powles 1984; Ludlow and Björkman 1984). Increased leaf temperatures associated with high light absorption and low leaf conductances would serve only to increase the Δw gradient and therefore aggravate the water stress situation experienced by the plant. In situations where the water stress periods may be short or intermittent and in situations where there is still sufficient soil moisture to allow low rates of gas exchange, leaf retention and adjustment to water stress may result in positive rates of net carbon gain. It appears reasonable that as an adjustment to water stress, natural selection may have favored the evolution of features which result in a reduction of the amount of solar radiation absorbed by the leaf without going to the extreme of leaf abscision. We now review three types of responses induced by water stress in plants from arid lands which greatly influence the amount of solar radiation absorbed by the leaf.

"Steep Slopes" - the Leaf Angle Response

Reversible leaf angle changes are an active means for leaves to reduce the amount of absorbed solar radiation. Ehleringer and Forseth (1980) pointed out that leaves of many species move through the day following the path of the sun. These diaheliotropic movements result in a maximization of the solar radiation incident on the leaf (cos(i) = 1), and under well watered conditions, allow the leaf to photosynthesize continually at a high rate. However, under water stress conditions, leaves from many plant species also exhibit an active avoidance of the sun's direct rays. This paraheliotropic movement results in a decrease in the cosine of the angle of incidence. The value of cos(i) is dependent on the level of water stress (Fig. 6). The change from paraheliotropic to diaheliotropic movements is fully and rapidly reversible upon rewatering (Forseth and Ehleringer 1980).

In *Lupinus arizonicus*, there is a corresponding decrease in the amount of solar radiation incident on the leaf with a decrease in leaf conductance (Forseth and Ehleringer 1980). This pattern results in a balance between the supply of photons for the light reactions of photosynthesis with the carbon dioxide supply rate for the dark reactions. In such a way, stress or damage to the photosynthetic apparatus is minimized. Ludlow and Björkman have shown for *Macroptilium atropurpureum*, an Australian pasture plant with paraheliotropic leaf movements, that unfolding of the leaves under water stress does result in damage by excess light (photoinhibition).

Paraheliotropic leaf movements will also result in an increased leaf water use efficiency (ratio of photosynthesis to transpiration) under water stress (Forseth and Ehleringer 1983). The increase in water use efficiency necessarily results from a reduction in the energy load on the leaf and thus a reduction in the leaf temperature and Δw which would have occurred were leaves to have remained perpendicular to the sun's direct rays.

In many plant species, leaves are thought to remain stationary. However, leaf angles are not typically measured seasonally. Comstock and Mahall (1985) have measured leaf angles on two *Ceanothus* species in the California chaparral over an annual cycle from the wet, active growth season through the long, summer drought period. They observed that leaf angles did not remain fixed at a specific angle, but that leaf angles increased as leaf water potentials decreased (Fig. 7). Leaf angles of both *Ceanothus* were relatively steep at approximately 60° at the wettest part of the year. By August when the summer drought was greatest, leaf angles had increased to almost 85° in *C. crassifolium* and 75° in *C. megacarpus.*

The changes in leaf angle in *Ceanothus* leaves are completely reversible. This is an important feature since leaf life span is 14-15 months and leaf production is limited to the springtime (Comstock and Mahall 1985). Photosynthetic activity in mediterranean-climate sclerophylls such as *Ceanothus* requires a lengthy period of time for leaves to cover all production costs and to achieve a net profit in lifetime carbon gain (Miller and Stoner 1979). As such, the slow rate of net carbon gain may militate against irreversible leaf angle changes or morphological changes (such as leaf pubescence production discussed below), because these changes would be detrimental to productivity in the subsequent growing season.

"Folding Up" - The Concavity Response

A variation on the theme of leaf angle change is for leaves to fold about the midrib. Such a change in effect creates two leaf halves, both of which are steeply inclined. This form of leaf modification has received little attention since the early work of Bergen (1909), although it is clearly seen in mediterranean-climate plants such as *Rhus laurina.*

Data from several Sonoran Desert species indicate that leaf folding under extreme water stress conditions can effectively reduce the surface area exposed to the sun's direct rays by up to 73% (Table 6). The reduction in the amount of solar radiation absorbed may be greater than this as leaves which exhibit a leaf folding behavior often have a reduced leaf absorptance on their lower surface as compared to the upper surface. In some ways, this is similar to paraheliotropic leaf movements. However, leaf concavity results in a major reduction in light absorption only during midday; early in the morning and again late in the

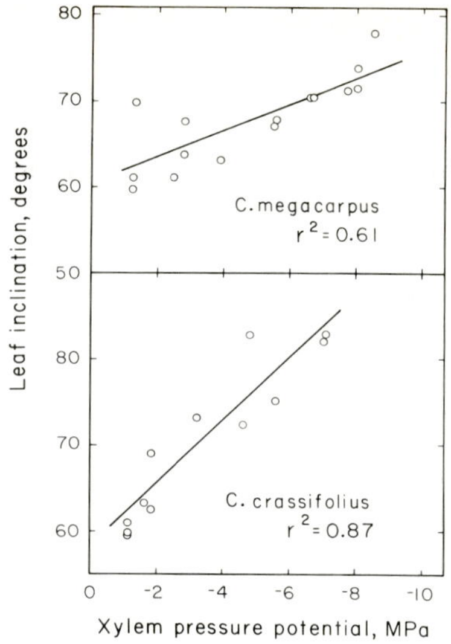

Fig. 7. Average leaf angle of *Ceanothus megacarpus* and *Ceanothus crassifolius* as a function of leaf water potential. From Comstock and Mahall (1985)

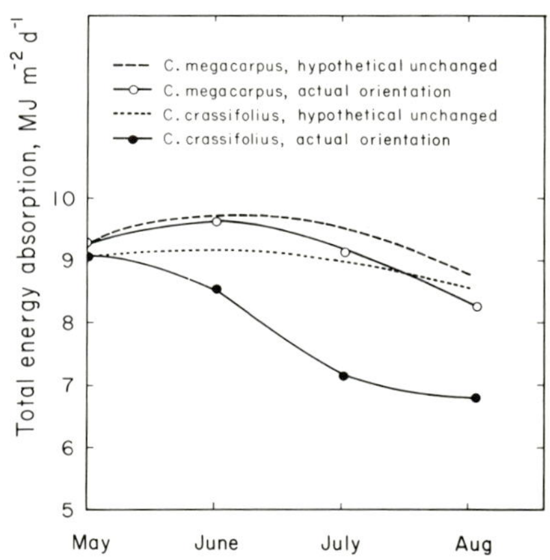

Fig. 8. Simulated values for average, total daily energy absorption using the measured leaf orientations and leaf absorptances for leaves of *Ceanothus megacarpus* and *Ceanothus crassifolius*. "Hypothetical unchanged" refers to the energy absorption that would have occurred had the leaf orientation remained unchanged from their values on May 15. From Comstock and Mahall (1985)

afternoon one half of the leaf is exposed to a high solar radiation load. Leaf concavity adjustments appear to be reversible, with unfolding occurring after water stress is relieved and soils are recharged with moisture.

"Changing Colors" - The Leaf Absorptance Response

In coping with prolonged summer water stress periods, a number of species produce leaves with different spectral characteristics for nondrought and drought periods. This is usually achieved by an epidermal modification that increases the leaf reflectance. From the gradient analyses described earlier, it appears that plants with increased leaf absorptances occur primarily at the driest ends of aridity gradients.

From a theoretical standpoint, we can ask what differences result from changing leaf angle versus changing leaf absorptance. The major differences between these two morphological changes are a potential change in the total amounts of solar radiation received and a change in the diurnal pattern of absorbed solar radiation. Consider three leaves with contrasting leaf absorptances and leaf angles. Leaf 1 will be typical of most mesophytic leaves with a leaf angle of 25° and a leaf absorptance of 85%; leaf 2 will have a leaf angle of 25°, but a leaf absorptance of 40%; leaf 3 will have a leaf angle of 70° and a leaf absorptance of 85%.

During the summer, leaf 1 will absorb the greatest amount of solar radiation, while leaves 2 and 3 will absorb similar daily totals of solar radiation (Fig. 9). The potential adaptive advantage then of a steep leaf angle (leaf 3) over a reduced leaf absorptance (leaf 2) is that there is essentially no cost associated with increasing leaf angle, whereas there is likely to be a substantial cost to produce the epidermal structures necessary to increase leaf reflectance. Thus, if plants retain leaves into the water stress periods, we might expect to see the geographic pattern observed in the Wasatch Front transect by Ehleringer (1985) (refer to Fig. 5). Another potential additional advantage to steep leaf angles is that greater amounts of solar radiation are absorbed early in the morning and again late in the afternoon where Δw gradients are likely to be lowest and most favorable to maximal leaf conductances.

Atriplex hymenelytra is an evergreen desert shrub that utilizes both steep leaf angle and a reduced leaf absorptance to reduce the absorbed heat load on the leaf. Leaf absorptances in *A. hymenelytra* are dependent on the extent to which salts in epidermal bladders have precipitated (Mooney *et al.* 1977). During the dry summer months, leaf absorptances are reduced to approximately 40%. Steep leaf angles combined with low leaf absorptances result in a lower leaf temperature than could be achieved by steep leaf angles alone, and certainly lower than were the leaf to have a high absorptance and a horizontal leaf angle (Fig. 10). Not only is water use reduced, thereby increasing the probability of surviving the drought period, but the reduced leaf temperatures result in an increased photosynthetic rate and an improved water use efficiency.

Leaf absorptance changes in *A. hymenelytra* are reversible, changing with changes in the water content of the epidermal bladders (Mooney *et al.* 1977). Since leaves on this plant are evergreen and produced only during the spring, this provides a mechanism for the plant to increase the net carbon contribution of the leaf by using the same leaf to adjust to changes in the water stress level over an entire annual cycle.

On the other hand, some desert plants, such as *Encelia farinosa*, produce a pubescent leaf covering to increase leaf reflectance. This change is irreversible; new leaves with different degrees of pubescence development and therefore different leaf absorptances are produced in response to changes in water stress (Fig. 11). Leaf production cycles may occur repeatedly in the same year depending on rainfall patterns. The advantage to plant performance is that leaves are finely adjusted to the degree of water stress in that habitat (Ehleringer and Mooney 1978). This results in enhanced carbon gain, reduced leaf temperatures and transpiration rates, and an improved water use efficiency. The disadvantage to the plant is, of course, the cost of frequently producing new leaves. *Encelia* species have deciduous leaves and these leaves can achieve extremely high net photosynthetic rates. It may be that since there is a significant carbon cost to producing leaf pubescence in *E. farinosa* (Ehleringer and Werk 1986), only plants with relatively high rates of carbon gain will use leaf pubescence as a protective means during water stress.

Table 6. Estimates of maximum leaf concavity for several arid land species in the Sonoran Desert, which appear capable of reversible leaf folding changes at the midrib in response to water stress (n=20).

Species	Apparent leaf width (mm)	Actual leaf width (mm)	Concavity (%)
Encelia californica	15.9	16.1	1
Encelia farinosa	10.7	13.4	20
Encelia radians	9.1	24.5	63
Jatropha cinerea	12.9	49.0	73

Another potential advantage to a reduced leaf absorptance is a reduction in the potential for photoinhibitory damage under water stress. Under water stress, photosynthesis in *E. farinosa* leaves becomes light saturated at progressively lower incident photon fluxes (Fig. 11). However, by reducing leaf absorptance the maximum absorbed photon flux is also reduced (indicated by arrows in Fig. 11), thereby alleviating photoinhibitory damage that occurs when plants under stress are exposed to irradiances above saturation (Björkman and Powles 1984; Ludlow and Björkman 1984). During the water-stress summer months, both low leaf absorptances and steep leaf angles should provide protection from

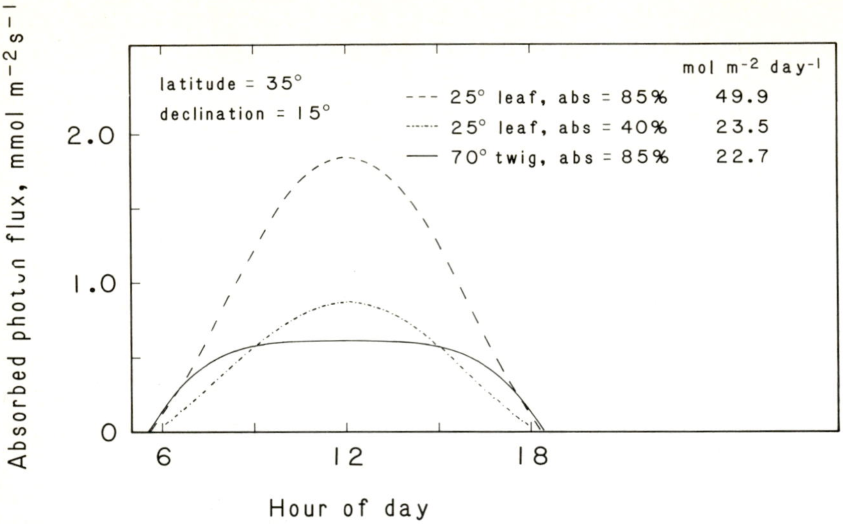

Fig. 9. The calculated diurnal photon flux absorbed by leaves with three contrasting leaf absorptances and leaf angles

Atriplex hymenelytra

	Leaf temperature (°C)	Transpiration (mmol m^{-2} s^{-1})	Photosynthesis (μmol m^{-2} s^{-1})	WUE (μmol / mmol)
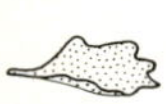	50	1.9	4	2.1
	47	1.7	7	4.1
	43	1.4	10	7.1

Fig. 10. Calculated relationships between leaf angle, leaf absorptance, photosynthesis, transpiration rate and water use efficiency for leaves of *Atriplex hymenelytra* during the summer. The air temperature was 40° and the leaf conductance to water vapor was 0.02 mol m^{-2} s^{-1}. Based on Mooney *et al.* (1977)

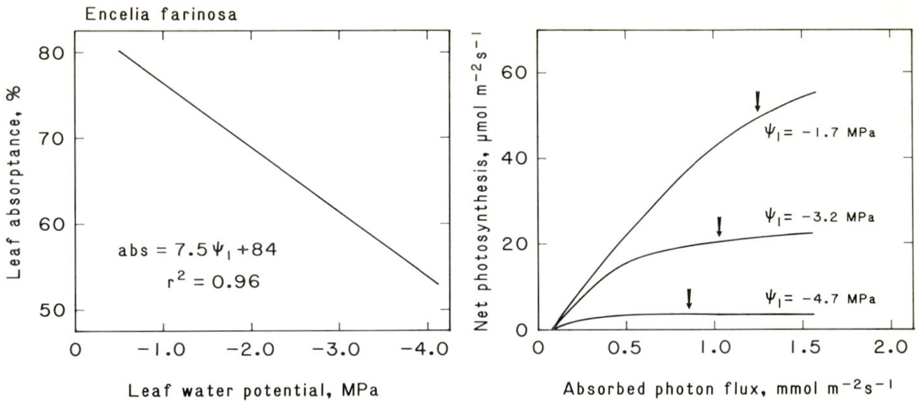

Fig. 11. The relationship between leaf absorptance and leaf water potential in leaves of *Encelia farinosa* (left plate) and the dependence of net photosynthesis on irradiance for leaves at three different water potentials in *Encelia farinosa* (right plate). The arrows indicate maximum midday absorbed photon fluxes given that leaves at different water potentials have different leaf absorptances. Based on Ehleringer (1982) and Ehleringer and Cook (1984)

photoinhibition. The advantage of a reduced leaf absorptance over a steep leaf angle in avoiding potential photoinhibitory damage lies in what happens if the drought period extends into the fall and early winter months. During the fall and early winter months, steeply inclined leaves will absorb much more solar radiation than will relatively horizontal leaves with reduced leaf absorptances (Ehleringer and Werk 1986). Thus, in habitats where the probability that water stress will be removed early in the fall is high (such as the chaparral), we might expect that steep leaf angles will be the common mechanism for reducing solar radiation absorption. In habitats such as coastal sage communities and warm deserts of North America, where drought periods are often longer, we might expect that a reduced leaf absorptance will be the more common mechanism for reducing solar radiation absorption. The data available for geographical trends in leaf absorptances and leaf angles support these predictions (Harrison *et al.* 1971; Mooney *et al.* 1974; Shaver 1978; Ehleringer 1981, 1985).

CONCLUSIONS

Considerable evidence exists showing that decreased absorptance and increased leaf angle function to protect leaf tissues from excessive irradiance under stressful conditions. The full implications of the use of leaf angle increases versus leaf absorptance decreases are not yet understood, but several points can be made from the existing data. Decreased leaf absorptance can provide a more complete protection from excess light levels than obtainable through orientation changes, but it is more costly and usually irreversible. In contrast, steep leaf angles can be reversible and, since they do not involve morphological changes, are thought to be less expensive. Steep leaf angles also shift the diurnal solar radiation absorption pattern away from midday to the morning and afternoon hours and thereby may result in greater leaf water use efficiencies. Sclerophylous shrubs of mediterranean-climate habitats tend to have steeply inclined leaves with high leaf absorptances. It is interesting to note that many of these shrubs also show pronounced midday stomatal closure even under moderate water stress levels (Tenhunen *et al.* 1980, 1981). The actual effect of steep leaf angles is also dependent on season, with total daily absorption being reduced in summer and less so if at all in winter.

Both leaf absorptance and leaf angle changes are common along aridity gradients. Leaf absorptance changes appear frequently only at the driest end of these transects, whereas leaf angles appear to change progressively along aridity transects. However, factors such as switches between deciduous and evergreen life forms complicate predictions.

Finally, changes in solar radiation absorption should reflect a balance between the irradiance requirements of photosynthesis and the negative effects of excessive energy loads. For those species which have been analyzed, it appears that plants have the ability to adjust this balance depending on the level of plant water stress.

REFERENCES

Bergen JY (1909) Concavity of leaves and illumination. Bot Gaz 48:459-461

Björkman O, Powles SB (1984) Inhibition of photosynthetic reactions under water stress: Interaction with light level. Planta 161:490-504

Blad BL, Baker DG (1972) Orientation and distribution of leaves within soybean canopies. Agron J 64:26-29

Comstock JP, Mahall BE (1985) Drought and changes in leaf orientation for two California chaparral shrubs: *Ceanothus megacarpus* and *Ceanothus crassifolius*. Oecologia (Berl) 65:531-535

Cowan IR, Farquhar GD (1977) Stomatal function in relation to leaf metabolism and environment. Symp Soc Exp Biol 31:471-505

Ehleringer JR (1981) Leaf absorptances of Mohave and Sonoran Desert plants. Oecologia (Berl) 49:366-370

Ehleringer JR (1982) The influence of water stress and temperature on leaf pubescence development in *Encelia farinosa*. Amer J Bot 69:670-675

Ehleringer JR (1984) Ecology and ecophysiology of leaf pubescence in North American desert plants. In: Rodriguez E, Healey PL, Mehta I (eds) Biology and Chemistry of Plant Trichomes. Plenum Press, New York, London, p 113

Ehleringer JR (1985) Changes in leaf characteristics in species along elevational gradients in the Wasatch Front, Utah. Amer J Bot

Ehleringer JR, Björkman O (1978) Pubescence and leaf spectral characteristics in a desert shrub, *Encelia farinosa*. Oecologia (Berl) 36:151-162

Ehleringer JR, Cook CS (1984) Photosynthesis in *Encelia farinosa* Gray in response to decreasing leaf water potential. Plant Physiol 75:688-693

Ehleringer JR, Forseth IN (1980) Solar tracking by plants. Science 210:1094-1098

Ehleringer JR, Mooney HA (1978) Leaf hairs: effects on physiological activity and adaptive value to a desert shrub. Oecologia (Berl) 37:183-200

Ehleringer JR, Mooney HA, Gulmon SL, Rundel PW (1981) Parallel evolution of leaf pubescence in *Encelia* in coastal deserts of North and South America. Oecologia (Berl) 49:38-41

Ehleringer JR, Werk KS (1986) Modifications of solar radiation absorption patterns and the cost-benefit implications at the leaf level. In: Givnish TJ (ed) On the Economy of Plant Form and Function. Cambridge University Press, London

Forseth IN, Ehleringer JR (1980) Solar tracking response to drought in a desert annual. Oecologia (Berl) 44:159-163

Gates DM (1963) Energy Exchange in the Biosphere. Harper and Row, New York

Gates DM (1980) Biophysical Ecology. Springer Verlag, Berlin-Heidelberg-New York

Haberlandt G (1884) Physiologische Pflanzenanatomie. Engelmann, Lehre

Harrison AT, Small E, Mooney HA (1971) Drought relationships and distribution of two mediterranean-climate California plant communities. Ecology 52:869-875

Lawrence WT (1975) A radiation model for chaparral canopies. Masters Thesis, San Diego State University

Lemeur R (1973) A method for simulating the direct solar radiation regime in sunflower, Jerusalem artichoke, corn and soybean canopies using actual stand structure data. Agric Meteorol 12:229-247

Lin ZF, Ehleringer JR (1982) Changes in spectral properties of leaves as related to chlorophyll and age in papaya. Photosynthetica 16:520-525

Ludlow MM, Björkman O (1984) Paraheliotropic leaf movement in Siratro as a protective mechanism against drought-induced damage to primary photosynthetic reactions: damage by excessive light and heat. Planta 161:505-518

Lugg DG, Youngman VE, Hinze G (1981) Leaf azimuthal orientation of sorghum in four row directions. Agron J 73:497-500

Miller, PC, Stoner WA (1979) Canopy structure and environmental interactions. In: Solbrig OT, Jain S, Johnson GB, Raven PH (eds) Topics in Plant Population Biology. Columbia University Press, New York, p 428

Mooney HA (1972) Carbon balance of plants. Ann Rev Ecol Syst 3:315-346

Mooney HA, Ehleringer JR, Bjorkman O (1977) The energy balance of leaves of the evergreen desert shrub *Atriplex hymenelytra*. Oecologia (Berl) 29:301-310

Mooney HA, Gulmon SL (1979) Environmental and evolutionary constraints on the photosynthetic characteristics of higher plants. In: Solbrig OT, Jain S,

Johnson GB, Raven PH (eds) Topics in Plant Population Biology. Columbia University Press, New York, p 316

Mooney HA, Gulmon SL, Parsons DJ, Harrison AT (1974) Morphological changes within the chaparral vegetation type as related to elevational gradients. Madroño 22:281-285

Mulroy TW (1979) Spectral properties of highly glaucous and non-glaucous leaves of a succulent rosette-plant. Oecologia (Berl) 38:349-357

Parkhurst D, Loucks O (1972) Optimal leaf size in relation to environment. J Ecol 60:505-537

Parsons DJ (1973) A comparative study of vegetation structure in the mediterranean scrub communities of California and Chile. PhD Thesis, Stanford University

Raschke K (1960) Heat transfer between the plant and the environment. Ann Rev Plant Physiol 11:111-126

Ross, J (1975) Radiative transfer in plant communities. In: Monteith JL (ed) Vegetation and the Atmosphere. Academic Press, London, New York, p 13

Rundel PW, Ehleringer JR, Mooney HA, Gulmon SL (1980) Patterns of drought response in leaf-succulent shrubs of the coastal Atacama Desert in northern Chile. Oecologia (Berl) 46:196-200

Shaver GS (1978) Leaf angle and light absorptance of *Arctostaphylos* species (Ericaceae) along elevational gradients. Madroño 25:133-138

Schimper, AFW (1903) Plant Geography upon a Physiological Basis. Clarendon Press, Oxford

Tenhunen JD, Lange OL, Braun M (1980) Midday stomatal closure in mediterranean type sclerophylls under simulated habitat conditions in an environmental chamber. II. Effect of the complex of leaf temperature and air humidity on gas exchange of *Arbutus unedo* and *Quercus ilex*. Oecologia (Berl) 50: 5-11

Tenhunen JD, Lange OL, Braun M, Meyer A, Lösch R, Pereira JS (1980) Midday stomatal closure in *Arbutus unedo* leaves in a natural macchia and under simulated habitat conditions in an environmental chamber. Oecologia (Berl) 147:365-367

Thomas DA, Barber HN (1974) Studies of leaf characteristics of a cline of *Eucalyptus urnigera* from Mount Wellington, Tasmania. II. Reflection, transmission, and absorption of radiation. Austr J Bot 22:701-707

Warming E (1909) Oecology of Plants: An Introduction to the Study of Plant Communities. Oxford University Press, London

Werk KS, Ehleringer JR (1984) Non-random leaf orientation in *Lactuca serriola* L. Plant Cell Environ 7:81-87

Methods for studying the mechanism of water stress effects on photosynthesis

W.M. Kaiser

Institut für Botanik und Pharmazeutische Biologie der Universität, Mittlerer Dallenbergweg 64, D-8700 Würzburg, FRG

INTRODUCTION

It is a triviality to state that any kind of stress will affect virtually all plant reactions, if only it is strong enough. But even mild stress of sufficient duration will induce a sequence of secondary events leading to changes in plant metabolism, structure and growth. These changes are far too complex to provide information on primary mechanisms of stress damage. A now well known phenomenon which illustrates the consequences of secondary stress damage is inhibition of the photosynthetic machinery due to exposure of plants to a combination of water stress and high light intensities. Photoinhibitory damage must not be considered as a primary water stress effect, since it is not observed at low light intensities (Powles 1984). In order to understand the manner in which photosynthetic processes are initially affected, water stress must be imposed experimentally in a well-defined manner and over a relatively short time interval. Unfortunately this means that experimental conditions will deviate considerably from what we call "natural." Therefore, the results obtained in this way must later be critically examined to establish their relevance under field conditions. In the following chapter, methods and results are described that outline our approach to determining the possible mechanisms by which water stress initially affects photosynthesis and plant metabolism.

MATERIALS AND METHODS

Spinacia oleracea (var. Monatol) was grown in a greenhouse with additional artificial illumination under a day/night cycle of 10/14 h and at a light intensity of 160 W/m^2 (HQi-lamps, 400 W, Schreder, Winterbach, FRG). For salt stress experiments, spinach was grown hydroponically in a modified Hoagland solution in a growth cabinet, but under otherwise similar conditions (Kaiser *et al.* 1983a). *Nerium oleander* and *Arbutus unedo* were grown in large pots in the greenhouse during winter, and outdoors in summer. Leaves were usually harvested in the morning and used immediately for the experiments. Preparation of leaf slices, as well as incubation conditions for osmotic stress experiments, have been described earlier (Kaiser 1982). Analysis of solutes in leaves and chloroplasts and procedures for measuring enzyme activities in an artificial stroma medium were recently published (Kaiser *et al.* 1986). Further experimental details are given in the legends of the figures and in the text.

NATO ASI Series, Vol. G15
Plant Response to Stress. Edited by J. D. Tenhunen et al.

RESULTS AND DISCUSSION

There is now good evidence that even mild water stress affects photosynthetic rates. The observed changes are due not only to increased stomatal resistance, but also to inhibition of metabolic processes in the mesophyll cell and chloroplast. Based on gas exchange measurements, the mesophyll resistance to intracellular CO_2 transport increases with water stress and the events behind that complex response remain to be detailed. Direct measurements of the mesophyll photosynthetic capacity of small leaves or leaf pieces are easily obtained by observing photosynthetic oxygen evolution polarographically in a commercially available oxygen electrode chamber. Stomatal limitations on photosynthesis are avoided by exposing materials to high CO_2 partial pressures (1 to 5% or greater - see below). These are produced with concentrated mixtures of carbonate/bicarbonate solutions in a very small, closed gas space. Such leaf oxygen electrode chambers are easily fitted with various optical instrumentation for carrying out simultaneous measurements of fluorescence or light scattering. In this manner, valuable information on the redox- or energy status of the photosynthetic machinery is obtained and this in turn facilitates localization of sites of inhibition (see Schreiber and Bilger, this volume). Water stress may be imposed on intact plants by withholding water, or in the case of detached leaves by exposure to a stream of dry air. Since the purpose here is to determine the manner in which photosynthetic processes are initially affected by water stress, the latter procedure seems better suited for a causal analysis of water stress damage. Rapid wilting should be carried out at light intensities significantly below the average light intensities experienced by the plants, in order to minimize photoinhibition.

For reasons which will become clear below, the relative water content (RWC) was usually used as a reference to describe leaf water status. A problem associated with the use of RWC is that water saturation may be somewhat loosely defined. In contrast to xerophytes, mesophytic leaves with extensible cell walls take up significant amounts of water from a water saturated chamber even when harvested from well-watered plants. Water-saturated spinach leaves can lose up to 30% of their water content before photosynthetic capacity begins to decline (Fig. 1A - *closed circles*), whereas photosynthetic rates of freshly harvested spinach leaves are often immediately affected by rather small water losses (Fig. 1A - *open circles*). A complete inhibition of photosynthesis usually occurs with a 70 to 80% water saturation deficit in spinach, depending on the initial water content.

When leaves from xerophytes were subjected to fast evaporative dehydration, photosynthetic oxygen evolution at 5% external CO_2 appeared to be more sensitively affected than in mesophytes. An example is given for *Arbutus unedo* in Fig. 2. However, when the ambient CO_2-concentration was close to 15%, the

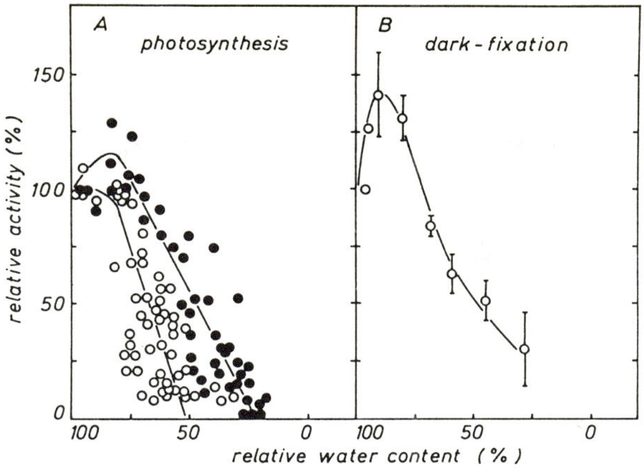

Fig. 1. Photosynthesis (A) and dark-CO_2-fixation (B) of spinach leaf discs as a function of the leaf relative water content (RWC). Both reactions were measured at 5% CO_2 in order to avoid as much as possible the stomatal limitation of CO_2 uptake. Photosynthesis was measured polarographically as oxygen evolution at saturating light intensity. 100%-rates were 250 to 350 µmoles CO_2 mg chlorophyll^{-1} h^{-1}. Each point represents a single measurement with an individual leaf disc (40 mm diameter) brought to the desired RWC by rapid wilting under a stream of air at room temperature and with dim room light. *Open circles* were obtained with freshly harvested leaves, closed circles with leaves which were saturated prior to the experiment by floating on water for about two hours. In (B), each point represents the mean from three independent experiments with two replicates each (+/- SD). Here, all leaves were water saturated prior to wilting. Incubation time for (B) was 15 min at 22° C. The bicarbonate solution was labelled with ^{14}C-bicarbonate to give a final specific activity of 37 kBq µmole^{-1}. After incubation, leaf discs were ground in liquid nitrogen and extracted for 2 hours in 4% perchloric acid. An aliquot (100 µl) of the extract was counted by liquid scintillation. Corrections were made for complete removal of non-fixed $^{14}CO_2$

sensitivity of *Arbutus* leaf discs to fast wilting was the same as for spinach. It is concluded that during severe water stress, stomatal resistance of xerophytic leaves can be so high that even 5% CO_2 is not enough to completely saturate the photosynthetic reactions (Fig. 2).

The small cuvettes used for measuring photosynthesis are also well suited for estimating the response of dark-CO_2-fixation to dehydration. For such experiments, the carbonate/bicarbonate solution has to be labelled with

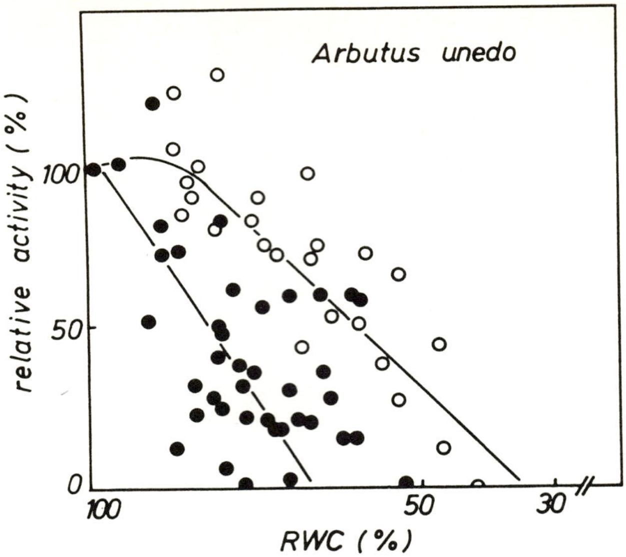

Fig. 2. Response of photosynthetic capacity of leaf discs from *Arbutus unedo* to rapid wilting. Photosynthesis was measured as oxygen evolution at very high CO_2 concentration. Closed circles represent values obtained at approximately 5% CO_2, open circles are values obtained at CO_2 concentrations close to 15%. Control rates were 70 μmoles mg chlorophyll^{-1} h^{-1} (corresponding to approximately 40 μmoles m^{-2} s^{-1}

appropriate amounts of $H^{14}CO_3$. Dark-CO_2-fixation via phosphoenolpyruvate carboxylase is interesting in the context of our analysis for two reasons: 1) the reaction takes place in the cytosol and therefore can give information about stress effects in that subcellular compartment and 2) it does not depend as directly as photosynthesis on membrane mediated reactions. As shown in Fig. 1B dark-CO_2-fixation is inhibited by rapid wilting to about the same extent as photosynthesis. From such experiments it becomes evident that a water saturation deficit of up to 30% has little effect on photosynthesis of mesophyll cells at saturating light intensities and CO_2 concentrations. Beyond that point, the inhibitory effects of water stress are not restricted only to membrane bound reactions like photosynthesis, but influence reactions occurring in the soluble phase as well (dark CO_2 fixation).

Another method of quickly imposing water stress in a controlled manner is to dehydrate tissue pieces, isolated organelles or cells in solutions containing

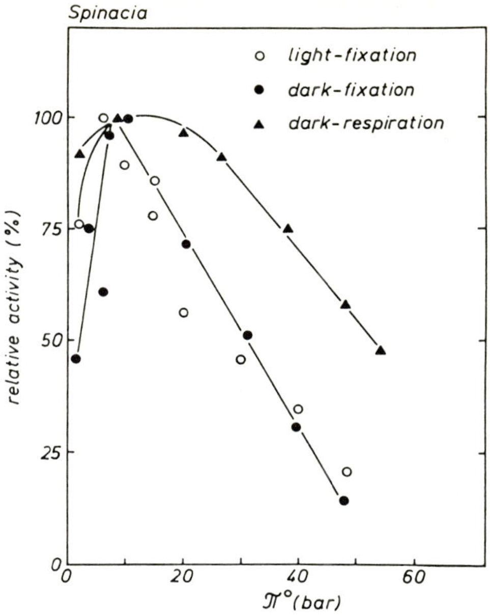

Fig. 3. Photosynthesis, dark-CO_2-fixation and respiration rates of spinach leaf slices in solution as a function of the external osmotic potential (sorbitol as osmoticum). Leaf slices were gently vacuum infiltrated prior to incubation with 10 mM radioactively labelled bicarbonate. For photosynthesis determinations, the intensity of white light (160 W/m^2) was saturating. Respiration was measured polarographically as oxygen uptake in a similar medium. From Kaiser *et al.* (1983b)

varying amounts of neutral osmotica such as sorbitol, sucrose or betaines. In the case of organelles or isolated protoplasts, this is the only possible way to control dehydration. With thin tissue slices immersed in solution, photosynthesis rates can be measured in the absence of stomatal limitations on gas exchange, if CO_2 is added as bicarbonate solution in sufficiently high concentrations to overcome the high diffusion resistance in the water phase. Tissue slices offer some advantage over isolated cells or protoplasts, since they may be studied even in an "overhydrated" state. Protoplasts suspended in isotonic or slightly hypertonic solutions are always in a state similar to that found when "moderate wilting" occurs. Furthermore, chloroplasts and protoplasts have been successfully isolated to date from only a few species, whereas leaf slices with substantial photosynthetic activity can easily be obtained from any plant leaf. Incubation media consist of a suitable buffer solution (pH 7 to 8) and 0.5 to 1 mM $CaCl_2$ to prevent solute leakage through cut sieve tubes or open plasmodesmata (Kaiser and Martinoia 1985).

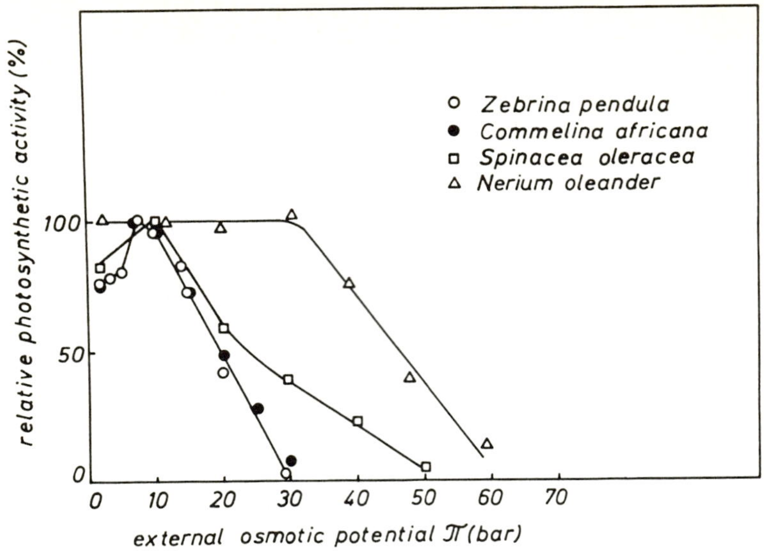

Fig. 4. Response to osmotic dehydration of photosynthesis of leaf slices from various hygro-, meso- and xerophytes (sorbitol as osmoticum). Conditions as in Fig. 3. From Kaiser (1982)

Fig. 3 shows that both photosynthetic capacity and rates of dark-CO_2-fixation of spinach leaf slices in solution are decreased by hypertonic stress (sorbitol as osmoticum). These effects due to osmotic dehydration may be compared to the results obtained with evaporative drying. Spinach leaf slices have an osmotic potential of the cell sap of about 400 to 450 mosmol kg^{-1}. A decrease of the relative water content of 50% should double this value to about 800 to 900 mosmol kg^{-1}, corresponding to about 20 bars. At an osmolality of the external sorbitol solution of about 20 bars, the photosynthesis rate of spinach leaf slices was inhibited by roughly 40% (Fig. 3, Fig. 4). A 50% decrease of the relative water content of fully water saturated leaves caused about the same degree of inhibition due to evaporative dehydration (Fig. 1A). Thus, at least for spinach, the water stress effect of each type of dehydration is similar. The main sources of error in such a comparison are possible permeation of cell membranes by the osmoticum and the unknown amount of extracellular water in leaves at various states of dehydration, which might comprise up to 30% of total leaf water content (Kyriakopoulos and Richter, 1981).

Based on changes in the relative water content during wilting, the metabolic reactions assayed in leaves from various plant species exhibited a similar sensitivity to water stress (Fig. 1, Fig. 2, see also Kaiser 1984). Osmotic dehydration, however, gave results which seemed completely contradictory at first sight: Photosynthesis rates in leaf slices from xerophytic plants were inhibited only at much higher external osmolalities than those from hygrophytes

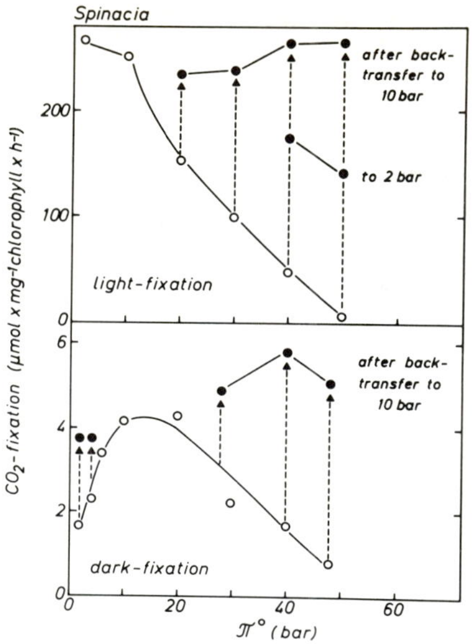

Fig. 5. Reversibility of osmotic dehydration effects on photosynthesis and dark-CO_2-fixation of spinach leaf slices. After incubating the slices in sorbitol media of different osmolalities for 15 min, the media were replaced by a medium containing either 0.33 M sorbitol (10 bar) or no sorbitol at all (2 bar). From Kaiser *et al.* (1983a)

and mesophytes (Fig. 4). In order to explain this discrepancy, it is helpful to have a working hypothesis as to the cause of the water stress induced inhibitions of metabolism mentioned above. When cells are dehydrated, internal solutes become concentrated. It is my opinion that this directly affects enzyme function in the water phase as well as on membranes.

An alternative mechanism that might explain impaired function is mechanical disruption. As a consequence of shrinkage, cells and organelles change their form and one might postulate also their surface area. But on the other hand, even strong shrinkage might occur without surface area changes, since even under severe dehydration, plasmolyzed protoplasts often remain connected to the cell walls by numerous filaments (which increase surface area). Chloroplasts in fully hydrated leaves are spherical or lens shaped. Isolated chloroplasts become flattened under osmotic dehydration (Heldt and Sauer 1971). Nevertheless, in these cases, shrinkage is thought to occur without changes in surface area and therefore with little mechanical stress. In the case of evaporative dehydration, protoplasts remain connected to the cell wall as long as the walls are flexible and change shape with water volume decrease or increase. Whether membrane rupture might be caused by sharp foldings due to extensive shrinkage, or by

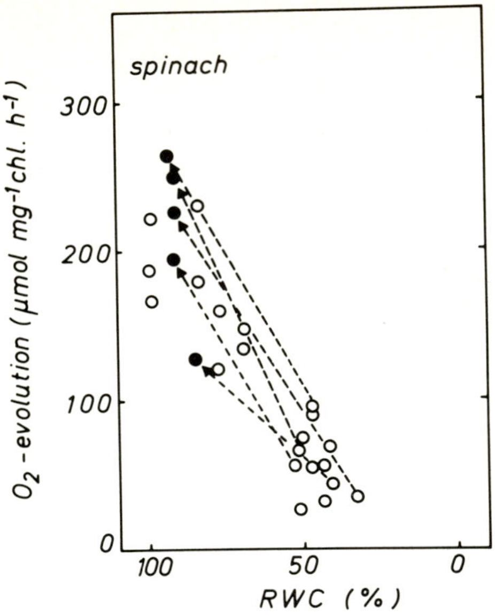

Fig. 6. Reversibility of evaporative dehydration effects. Photosynthesis of spinach leaf discs was measured in a leaf oxygen electrode chamber at high CO_2 partial pressure. After being wilted to the desired water deficit (RWC), photosynthesis was measured for about 10 min. Subsequently, leaf discs were rehydrated by floating on water for 60 min. After weighing, photosynthesis was measured again. *Open circles* give RWC and photosynthetic capacity after wilting, *closed circles* after rehydration

tearing the protoplast off the cell wall, effects of this sort are not reversible in a short period of time. However, the inhibition of enzymes *in vitro* by high salt concentrations has been shown to be quickly and completely reversible (Kaiser *et al.* 1981). Therefore, since water stress effects too are often readily reversible (Fig. 5, Fig. 6), whether obtained by osmotic or evaporative dehydration, it is proposed that the initial effects of moderate dehydration are mediated by increased internal solute concentrations. For a further analysis of possible solute effects during dehydration we will try to answer the following questions: 1) to what degree is the protoplast volume changed under osmotic stress? 2) what are the concentrations of major osmotically relevant solutes in chloroplasts? and 3) which of these solutes are potentially able to mediate impairment of photosynthesis under water stress?

Changes in leaf relative water content should essentially reflect changes in cell volumes. However, as mentioned above, some uncertainty is introduced due to the unknown amount of extracellular water and since different cell types within the tissue might lose more or less water if their initial turgor differed. In osmotic stress experiments, volume changes can be measured directly. A method of accomplishing this has been recently described (Kaiser 1982) and is briefly summarized here. Small leaf discs (5 to 6 mm diameter) are gently vacuum infiltrated with solutions containing ^3H_2O and ^{14}C-labelled sorbitol. After 30 to 60 min incubation on a shaker, discs are rapidly rinsed with unlabelled medium in order to remove label adhering to the surface. Subsequently, discs are extracted with perchloric acid. After counting radioactivity of both types present in an aliquot of the extract by liquid scintillation, we calculate three different volumes: a) total discs volume from the total amount of labelled water, which should be equally distributed throughout the tissue (this information can be checked by weighing the discs) b) the intercellular space volume from the total amount of labelled sorbitol, which is restricted to the water phase of cell walls, and c) the total protoplast volume as difference between (a) and (b). To obtain the volume sensitive to osmotic treatment, the total protoplast volume must be corrected for internal "solid space" (the so-called "non-osmotic space"), which is obtained by extrapolating the cell volume measured at various osmolalities to an infinite external osmotic potential (for details see Kaiser 1982).

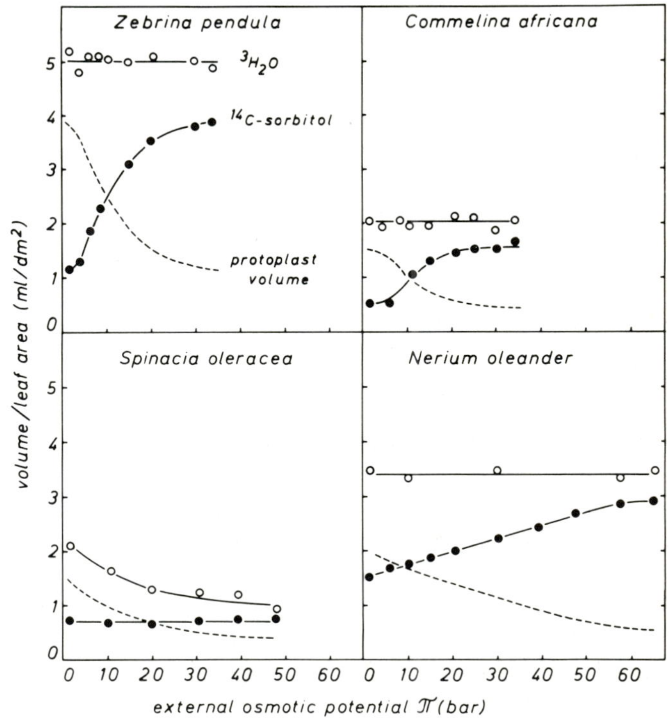

Fig. 7. Water space, sorbitol space and resulting total protoplast volume at different external osmolalities (sorbitol as osmoticum) in leaf discs from various plant species. From Kaiser (1982)

Using this method we have examined protoplast volume changes in leaf tissue from several species (Fig. 7). Expectedly, a given degree of protoplast shrinkage in the various species occurs in response to very different osmolalities in the external solution. The reason is obvious. The leaves examined have very different osmotic potentials of the cell sap in the fully hydrated state, e.g. 4 bars in *Zebrina* but 20 bars in leaves from *Nerium Oleander*. If turgor is zero, a 50% volume decrease is achieved by doubling the external osmolality. For *Zebrina*, this occurs between 4 to 8 bars but in *Nerium oleander* between 20 and 40 bars. These protoplast volume measurements enable us to compare the inhibition of photosynthesis by osmotic dehydration on the basis of relative volume changes. As shown in Fig. 8, leaves of various species now are found to respond quite similarly to dehydration. A certain volume decrease causes an inhibition of photosynthesis which is similar in all species tested, despite the fact that the osmolalities causing this volume decrease have to be very different.

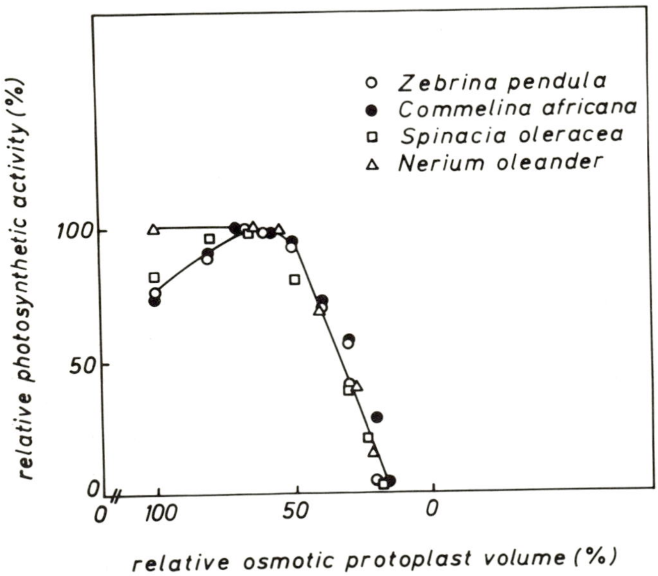

Fig. 8. Relative photosynthetic activity of leaf tissue from various plants plotted as a function of the relative osmotic protoplast volume (as percentage of the maximum rate and maximum volume). From Kaiser (1982)

Thus, it is not the osmolality or the water potential change as such that is inhibitory, but the concomitant volume change. How does this observation fit the above-mentioned "solute hypothesis"? There is no reason to assume that enzymes from different plant species respond differently to increased solute concentrations. At least for halophytes and non-halophytes this has been clearly established (Flowers *et al* . 1977). It is thought that different osmolalities of leaf cell sap are maintained in various species by production of different concentrations of so-called "compatible" solutes like sugars, amino acids or other compounds. On the other hand, the concentration of toxic (non-compatible) solutes should be similar in the metabolically active compartments of various species. The well-known high sugar concentration in leaf cell sap from xerophytes is in agreement with the above-mentioned hypotheses (Kaiser 1982).

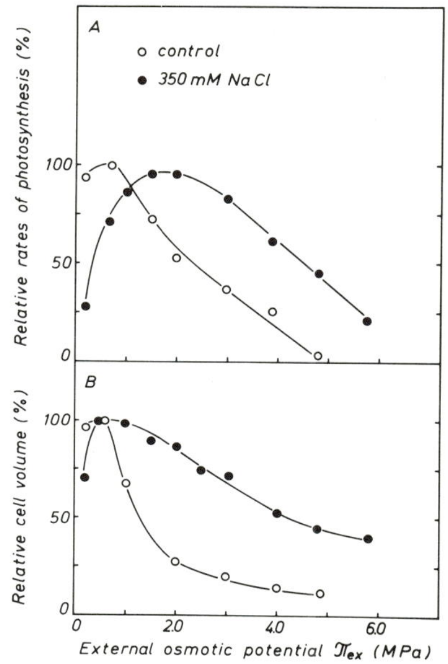

Fig. 9. Photosynthetic capacity and total osmotic protoplast volume in leaf tissue from spinach plants grown in the absence and presence of NaCl (350 mM), as a function of the osmolality of the suspending medium (Kaiser *et al.* 1983a). Note that at the lower osmolalities, photosynthesis of salt-grown plant leaf slices is strongly inhibited, since cells burst due to high turgor pressure

As a model system for testing these ideas, we can use plants grown in the absence or presence of high salt concentrations. Spinach is excellently suited for isolating intact chloroplasts and for analysing solutes therein. At the same time, it is moderately halotolerant. When spinach is grown hydroponically in solutions containing 350 mM NaCl, the cell sap osmolality is increased to about 900 mosmole kg^{-1} (Kaiser *et al.* 1983a). Accordingly, the response of leaf slices to increased external osmolalities for control plants and for salt-grown plants is as different as for hygrophytes or xerophytes (Fig. 9). When the response of photosynthesis to dehydration is compared in control- and salt-plants based on the relative volume change, both plant types show a similar sensitivity to dehydration (Fig. 10). From salt-grown spinach plants, intact chloroplasts have been obtained by a fast aqueous isolation technique (Kaiser *et al.* 1983a). An analysis of the solute content in leaves and chloroplasts revealed that leaves take up large amounts of salt. However, in chloroplasts, osmotic compensation appears to be

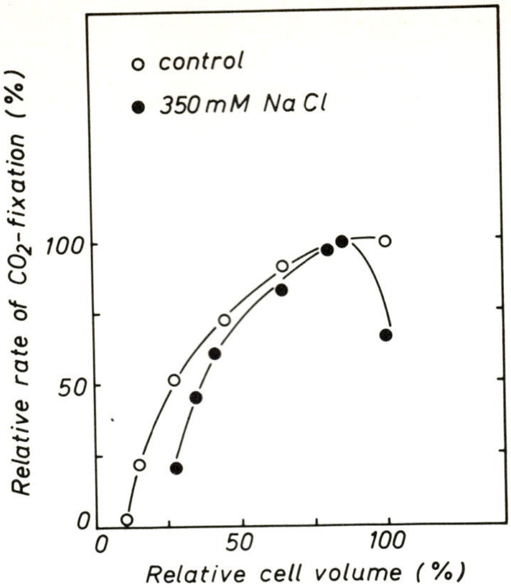

Fig. 10. Photosynthetic capacity of spinach leaf slices from plants grown in the absence (control) and presence of 350 mM NaCl, as a function of the total osmotic protoplast volume (relative values). Same experiment as in Fig. 9

achieved by accumulation of compatible organic solutes while the salt concentration constantly remains low (Kaiser *et al.* 1983a, Robinson *et al.* 1983). Although these observations are by no means proof that this strategy is adopted as part of the contrasting cellular regulation employed by mesophytes or xerophytes, they are certainly suggestive.

What are the potentially toxic solutes in metabolically active subcellular compartments? We have recently carried out a comprehensive analysis of the solute composition in leaves and chloroplasts from spinach (Kaiser *et al.* 1986), as well as from barley and peas (not shown). For many of the solutes, conventional analytical procedures are available. Inorganic anions, however, which were suspected to be the most likely candidates as noncompatible solutes, have rarely been studied. Based on several analytical procedures, it was found that in spinach chloroplasts, the concentration of major solutes is as follows: chloride 5-10 mM, nitrate 2-5 mM, phosphate 20-30 mM, sulfate 5-10 mM, potassium 150-180 mM, magnesium 15-20 mM, sugars 30-50 mM, amino acids 70-160 mM. Based on these data, an "artificial stroma medium" was developed containing all these solutes. By changing the overall concentration of this medium, the situation in chloroplasts during dehydration is mimicked, and the effect of the solutes on chloroplast enzymes or thylakoids can be examined. By changing the concentration of a single solute, we are able to distinguish between compatible and noncompatible solutes in the concentration range existing in chloroplasts.

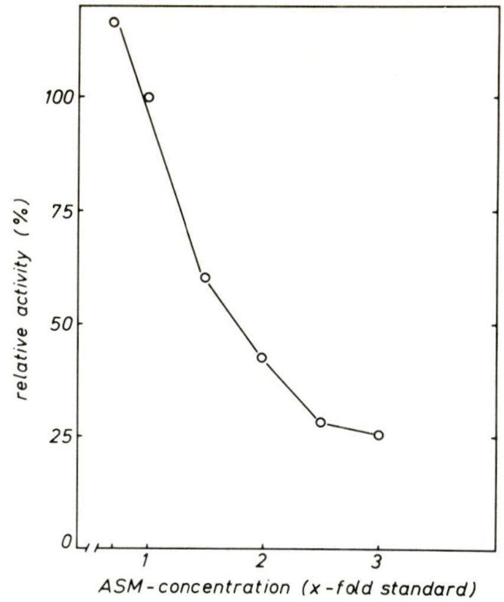

Fig. 11. Response of stroma enzymes to increased concentrations of an artificial stroma medium (ASM). Reaction measured was CO_2-fixation of a crude spinach leaf extract with ribose-5-phosphate and ATP as substrates, which includes phosphoribose isomerase (E.C. 5.1.3.4.), ribulose-5-phosphate kinase (E.C. 2.7.1.19) and ribulose-1,5-bisphosphate carboxylase/oxygenase (E.C. 4.1.39). From Kaiser *et al.* (1986)

An example of such an experiment is shown in Fig. 11. When the concentration of ASM is raised, enzyme activities are inhibited. A doubling of ASM, which should mimick a 50% water loss, usually caused a 50% reduction of enzyme activities. This inhibition was not restricted to chloroplast enzymes. Phosphoenolpyruvate carboxylase as the cytoplasmic key enzyme of dark-CO_2-fixation was inhibited as well (not shown), just as it was *in vivo*-in response to dehydration (Fig. 1B, Fig. 3). This might indicate that the concentration of incompatible solutes in the cytosol and in the chloroplast is similar. In addition, the solute effects are found to be nonspecific.

At the concentrations existing in chloroplasts, sugars and amino acids did not contribute to enzyme inhibition, nor did they prevent inhibition of reactions by other solutes. Only the divalent anions phosphate and sulfate appeared to be enzyme incompatible (Fig 12). At constantly low magnesium concentrations, amino acids were also inhibitory, but this was apparently due to chelation of magnesium, and did not occur when the concentration of amino acids and magnesium were increased in parallel (as in complete ASM).

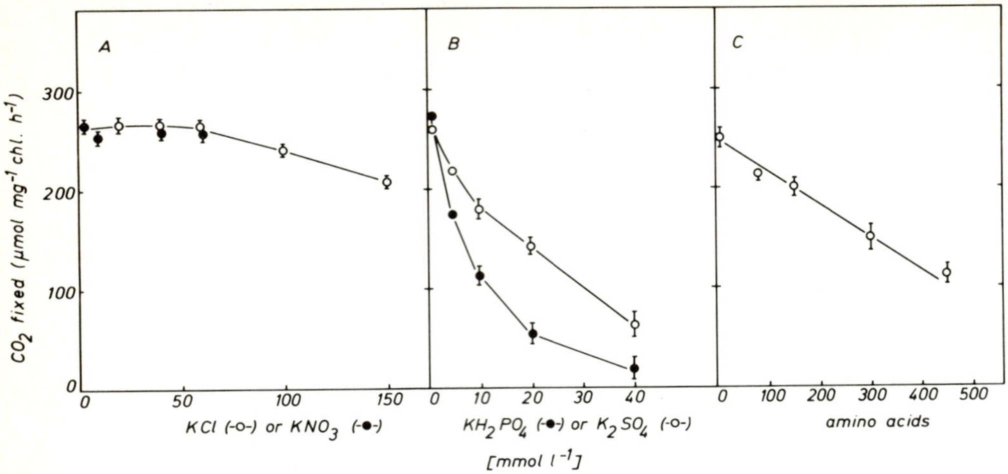

Fig. 12. Response of stroma enzymes to single compounds contained in an artificial stroma medium (ASM). Conditions as in Fig. 11, with the exception that the magnesium concentration was kept constant at 5 mM

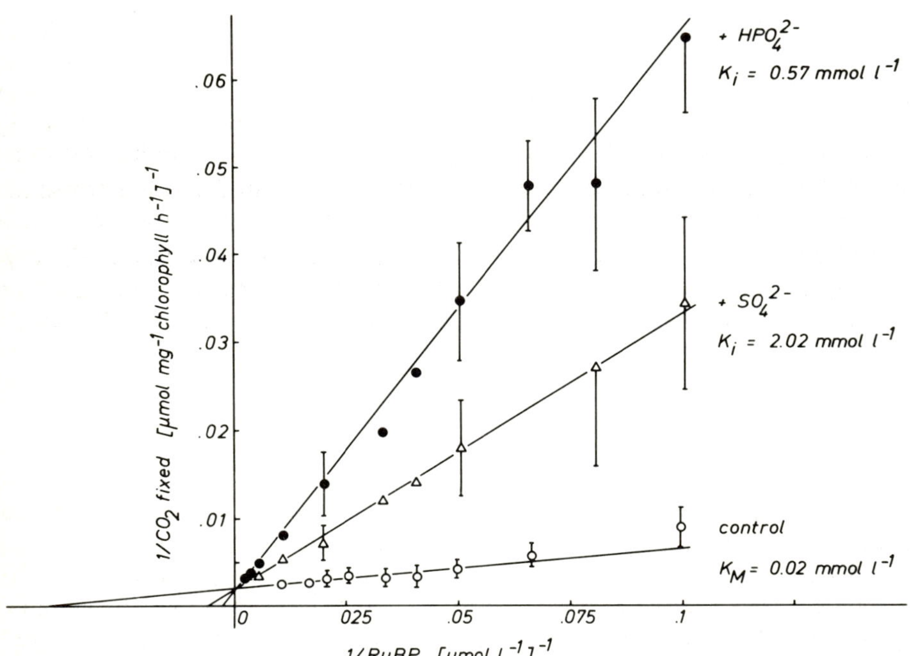

Fig. 13. Effect of sulfate and phosphate on the substrate affinity of ribulose-1,5-bisphosphate carboxylase. Bicarbonate concentration was 10 mM. From Kaiser *et al.* (1986)

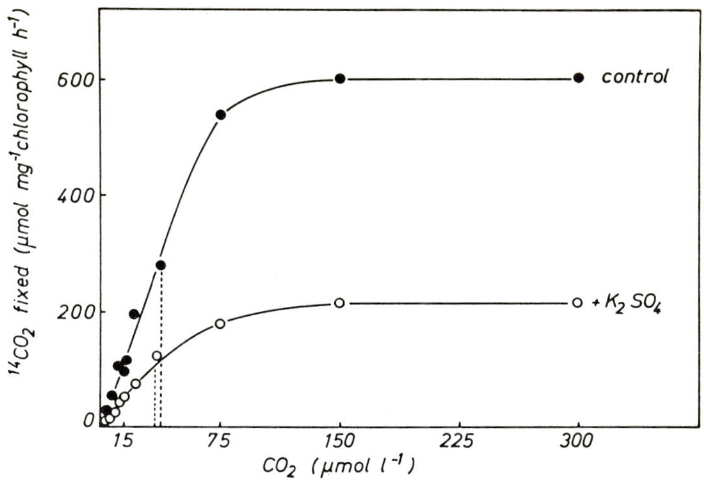

Fig. 14. Effect of sulfate (15 mM) on RuBP-carboxylase activity in a crude spinach leaf extract. RuBP-concentration was 0.5 mM. All solutions (pH 8) were flushed with CO_2-free air overnight prior to use. From Kaiser *et al.* (1986)

Ribulose-1,5-bisphosphate carboxylase as a single reaction was also inhibited by phosphate and sulfate with a similar sensitivity as the above-described sequence of three coupled Calvin cycle reactions. The inhibition of the carboxylase by phosphate and sulfate was competitive with respect to ribulose-1,5-bisphosphate, but not competitive with respect to CO_2 (Fig. 13, Fig. 14). The latter observation is of special interest, since it allows the prediction that under water stress *in vivo*, both the maximum capacity and the slope of the CO_2-response curve of leaves should be changed. This was in fact observed not only with intact plants (Jones and Fanjul 1983; see also Lange et al. this volume), but also with intact isolated chloroplasts or with leaf slices exposed to osmotic dehydration (Kaiser 1984a, 1984b).

CONCLUSIONS AND PERSPECTIVES

In our experiments, all plants tested showed a similar sensitivity to evaporative or osmotic dehydration, if the relative water content (RWC) or the relative osmotic cell volume was used as a reference. In the various species, quite different changes in water potential were responsible for producing a given (similar) degree of cell shrinkage (or change in RWC). This suggests a common mechanism by which damage occurs. Apparently, the physiological changes brought about by dehydration are quantitatively related to volume changes rather than to water potential changes. The widespread use of xylem water potential measurements as a reference for quantifying water or salt stress effects on plant metabolism may be questionable. Recently, this view has also been expressed by Sinclair and Ludlow (1985). Our results generally indicate that plants can lose about 20 to 30% of their saturation water content before inhibition of metabolic function occurs. Fifty percent inhibition occurs with approximately 50 to 60% water loss. At that stage, the inhibitory effects are usually reversible if the tissue is rapidly rehydrated. Reaction rates decrease to zero when 70 to 80% of cellular water is lost. Then, the deleterious effects are only partially reversible.

The rapidly reversible primary water stress effects may be mediated mainly by increased concentrations of inorganic anions like phosphate and sulfate. The response of enzymes *in vitro* to increased concentrations of an artificial stroma medium in principle matches the *in vivo* response of photosynthesis to dehydration. In such experiments with an artificial stroma medium, solute concentrations were varied, but the pH of the medium was kept constant. This is not necessarily the case *in vivo*. In osmotically stressed chloroplasts, light-mediated stroma alkalisation was less than in non-stressed chloroplasts (Berkowitz and Gibbs 1983). Though it has not yet been examined experimentally, the solute effects on enzyme kinetics should be intensified by a pH-shift toward more acidic values. It is, however, by no means clear whether the pH-changes that have been observed are a cause or a consequence of metabolic inhibition.

So far, solute effects have been examined only with water soluble stroma enzymes. Solute effects on peripheral membrane proteins are to be expected as well, and will be examined in the near future. In order to test the general validity of the solute hypothesis, much more information is needed on the solute composition and solute concentrations in plant cell organelles. The data presented here were obtained only from spinach. Unpublished results obtained with pea chloroplasts suggest that they have very similar concentrations of incompatible anions as does spinach. Data on chloroplasts from xerophytes or hygrophytes are completely lacking. Non-aqueous isolation techniques in combination with ion chromatography might allow a more complete analysis of chloroplasts from many other plant species in the future, thereby allowing a criticial examination of the solute hypothesis.

REFERENCES

Berkowitz GA, Gibbs M (1983) Reduced osmotic potential effects on photosynthesis. Identification of stromal acidification as a mediating factor. Plant Physiol 71:905-911

Flowers TJ, Troke PF, Yeo AR (1977) The mechanism of salt tolerance in halophytes. Annu Rev Plant Physiol 28:89-121

Heldt, HW, Sauer F (1971) The inner membrane of the chloroplast envelope as the site of specific metabolite transport. Biochim Biophys Acta 234:83-91

Jones HG, Fanjul L (1983) Effects of water stress on CO_2-exchange in apple. In: Marcelle R, Clijsters H, van Poucke M (eds), Effects of Stress on Photosynthesis. Martinus Nijhoff/ Dr. W Junk, The Hague, p 75

Kaiser WM (1982) Correlation between changes in photosynthetic activity and changes in total protoplast volume in leaf tissue from hygro-, meso- and xerophytes under osmotic stress. Planta 154:538-545

Kaiser WM, Weber H, Sauer M (1983a) Photosynthetic capacity, osmotic response and solute content of leaves and chloroplasts from *Spinacia oleracea* under salt stress. Z Pflanzenphysiol 113:15-27

Kaiser WM, Schwitulla M, Wirth E (1983b) Reactions in chloroplasts, cytoplasm and mitochondria of leaf slices under osmotic stress. Planta 158:302-308

Kaiser WM (1984a) Response of photosynthesis and of dark-CO_2-fixation to light, CO_2 and temperature in leaf slices under osmotic stress. J Exp Botany 35:1145-1155

Kaiser WM (1984b) Sites and mechanisms for the inhibition of photosynthesis by water stress. In: Sybesma C (ed), Advances in Photosynthesis Research Vol. 4, Martinus Nijhoff/ Dr. W Junk, The Hague, p 341

Kaiser WM, Martinoia E (1985) Absence of an apoplastic step in assimilate transport to the phloem? A comparison of assimilate efflux from leaf slices, mesophyll protoplasts and a unicellular green alga. J Plant Physiol 121:463-474

Kaiser WM, Schröppel-Meier G, Wirth E (1986) Enzyme activities in an artificial stroma medium: An experimental model for studying effects of dehydration on photosynthesis. Planta 167:292-299

Kyriakopoulos E, Richter H (1981) Pressure volume curves and drought injury. Physiol Plant 52:124-128

Powles SB (1984) Photoinhibition of photosynthesis induced by visible light. Annu Rev Plant Physiol 35:15-44

Sinclair TR, Ludlow MM (1985) Who taught plants thermodynamics? The unfulfilled potential of plant water potential. Aust J Plant Physiol 12:213-217

Robinson SP, Downton WJS, Millhouse JA (1983) Photosynthesis and ion content of leaves and isolated chloroplasts of salt stressed spinach. Plant Physiol 73:238-242

Coping with variability: Examples of tracer use in root function studies

M. M. Caldwell and D. M. Eissenstat

Department of Range Science and the Ecology Center, Utah State University, Logan, Utah 84322 USA

INTRODUCTION

Spatial and temporal variability is an inevitable element in most field ecological studies. Belowground, this temporal variability and spatial patchiness is also to be expected, even though it may be less immediately apparent. Thus, assessment of root system phenomena such as production or nutrient uptake in the heterogeneous soil environment is a problem requiring intensive sampling combined with some manner of coping with the variability. Improved resolution from increased sample size often reaches a point of diminishing gains and unreasonable costs. Furthermore, practical constraints often limit the number of samples that can be taken. Use of tracers, such as radioactive isotopes, provides numerous advantages for nondestructively tracking some belowground processes. Because this can be done nondestructively, the same plants or plots can be followed through time which reduces the variability in the determinations. This chapter will portray two examples of the use of tracers in root system function research and discuss these in relation to alternative approaches.

ROOT SYSTEM PRODUCTION

The assessment of annual root production, especially in communities of perennial vegetation, is a key component in understanding ecosystem primary production and nutrient cycling. In most systems, root system production is considerably greater than aboveground biomass production (Caldwell 1986). Assessing the magnitude of root production usually has been approached usually by three quite different techniques, one of which involves the use of tracers. These techniques will be discussed with an emphasis on the errors associated with each.

The most frequently used technique involves sequential belowground biomass determinations, most often by coring. This set of biomass samples through time usually exhibits periods of increasing and decreasing belowground biomass. These differences through time are considered to reflect the net result of new root growth and loss of roots in death and decay. During a period of increasing root biomass, there is very likely simultaneous herbivory and death of roots. Therefore, the net increase should be an underestimate of the true root production. (Apart from death, biomass lost in sloughing of root hairs, cortical cells, and rootcaps, exudation, etc. contribute to the underestimate of true root

NATO ASI Series, Vol. G15
Plant Response to Stress. Edited by J. D. Tenhunen et al.

productivity). The trends are seldom monotonic during the course of the growing season, but instead exhibit a series of peaks and troughs, especially if the sampling interval is reasonably short. An example of this is shown in Fig. 1. Different computational approaches have been taken with such data. These include simply taking the difference in maximum and minimum total root biomass during the year, or, alternatively, summing all positive biomass increments from this time series during the year. The latter approach will usually result in a larger estimate. If some estimates of the fraction of living roots from these samples can be made, these computations are sometimes performed on the live and dead root components in various formulations. These computational approaches are designed to minimize the underestimate of true net root production resulting from the loss of roots during periods of net increase in root biomass (Persson 1980, Hansson and Steen 1984).

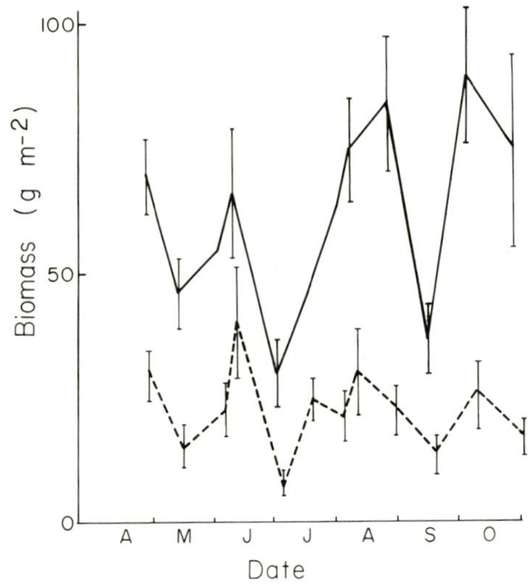

Fig. 1. Seasonal changes in live (dashed line) and dead (continuous line) root biomass as determined by sequential core sampling in a *Pinus sylvestris* plantation in Central Sweden. Adapted from Persson (1980)

For each root biomass sampling, there are associated errors due to the spatial distribution of roots and the sampling itself. The problem is to factor out these errors from the true temporal fluctuation in root biomass. Statistical approaches that estimate the magnitude of these errors have been developed and correction factors have been employed to adjust the calculated net production (Persson 1980, Hansson and Steen 1984). However, by either computation approach, whether taking the difference in minimum and maximum root biomass during the season, or summing all positive increments in root biomass, there is necessarily a bias involved because the troughs and peaks are being expressly selected out of the series of samples. Singh *et al.* (1984) have addressed this problem by using a model to simulate the "true" seasonal dynamics of root biomass for a grassland. Although the model took a growth and allocation simulation approach, it was "tuned" to actual root biomass data collected in an earlier study (Lauenroth and Whitman 1977). They compared the model results with root production calculated in various formulations from simulated sampling data they generated from the model with variances proportional to the mean and a coefficient of variation (0.32) actually encountered in the field sampling of Lauenroth and Whitman (1977) (Fig. 2). Depending on the mode of calculation and the frequency of collection of the simulated biomass data, net production relative to the model value was actually overestimated between 8 and 320% with possibilities of overestimates as high as 420% for a 10-yr period. For single years the overestimate could be much greater (860%). Statistical constraints could reduce this overestimation, but the errors were still daunting. Not surprisingly, when the summation of positive increments calculational approach was taken, the errors of overestimation increased with increasing sampling frequency because of the creation of artificial peaks and troughs. Their calculations also illustrated the limitations of simply increasing the number of samples.

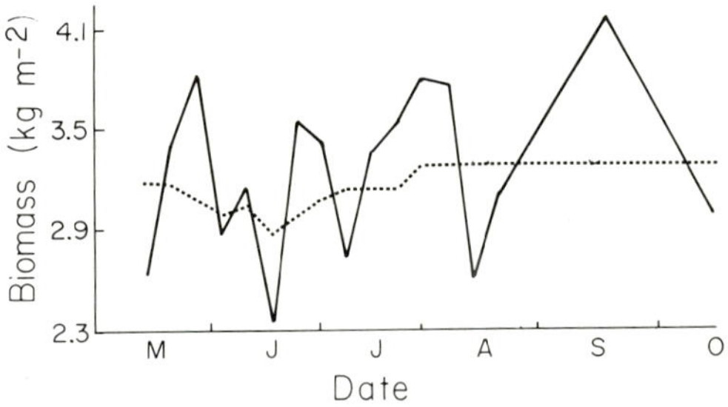

Fig. 2. The seasonal change in root biomass computed by a simulation model (dotted line) and the sequence of average root biomass from samples taken at approximately weekly intervals (continuous line). These sampling data were also simulated assuming random, normally distributed data with a coefficient of variation encountered in actual field sampling (see text). Adapted from Singh *et al.* (1984)

Additional procedures that can be used to reduce errors in the sequential harvesting technique are to use live roots, which increases the signal-to-noise ratio, and to use available information on root growth patterns in order to plan an efficient harvest schedule (Vogt *et al.* 1986).

The second frequently used technique is the ingrowth technique where root-free soil in mesh bags is placed in holes and subsequently removed at prescribed intervals. The root mass invading the mesh bags is used as an index of root productivity (Fabiao, Persson and Steen 1985, Persson 1979). The assumption plainly is that growth into the root-free soil in the mesh bag is the same as root production in the normal, undisturbed soil. This is an assumption that is difficult to test. If the soil becomes compacted in the process of boring the hole, root growth could be influenced (Shierlaw and Alston 1984). The disturbance of the soil in the process of removing the roots before the soil is placed in the mesh bags can cause changes in the mineral nutrient status of the soil (Runge 1983). This can, in turn, alter root morphology and patterns of root growth (Hackett 1968, Fitter 1982, Robinson and Rorison 1983). For example, one reaction to increased available nutrients is proliferation of root growth (St. John, Coleman and Reid 1983). Of course, other alterations of the soil environment such as loss of soil structure and reduced bulk density resulting from root extraction, could modify the pattern and quantity of root growth.

A third approach involves a tracer technique, namely, a determination of the change in $^{14}C/^{12}C$ in the structural carbon of the root system through time following a one-time $^{14}CO_2$-labeling of the plant shoot system, or what has been termed a ^{14}C-dilution technique (Caldwell and Camp 1974). Briefly, plants are labeled with ^{14}C by photosynthetic fixation of $^{14}CO_2$. Approximately one week following the labeling, presumably after the ^{14}C has been translocated throughout the plant, the structural carbon of the root tissues is sampled for the relative $^{14}C/^{12}C$ ratio. At a later point in time, usually at the end of the growing season, the root structural tissues are again sampled for the relative $^{14}C/^{12}C$ content. A turnover coefficient, i.e., the fractional renewal of the root system, is derived from the dilution of ^{14}C in the structural tissues of the root system through time due to the incorporation of new ^{12}C into the root structural carbon between the two sampling times (Fig. 3). The turnover coefficient can be multiplied by the root biomass determined at about the time of ^{14}C labeling to calculate the belowground production. This approach was first used to estimate belowground production in semiarid shrub steppe stands (Caldwell and Camp 1974). The calculated carbon investment in root production coincided with other components of the total carbon balance assessed independently for these shrub stands (Caldwell *et al.* 1977).

By using this ^{14}C-dilution approach, the need to assess root biomass at several times during the year is obviated; it only needs to be measured once at the time of labeling. Furthermore, separation of live and dead roots is not necessary. Avoided too are all the complications of peak-and-trough sampling in the sequential harvest approach. These errors involve missing peaks and troughs in the course of the sampling or inflating their importance as was discussed earlier. Also, the unaccounted death and loss of roots during periods of net biomass increase is unimportant in this approach because the ^{14}C-dilution approach

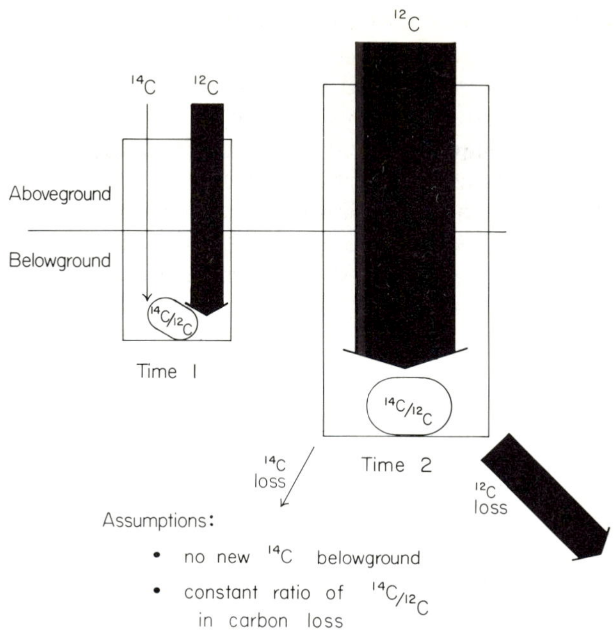

Fig. 3. Principal features of the ^{14}C-dilution technique for determination of belowground biomass turnover. The rectangle on the left represents carbon in the above- and belowground plant tissues at the time of labeling and the first determination of $^{14}C/^{12}C$ in structural tissues. The arrows represent incorporation of ^{14}C and ^{12}C at this time. The rectangle on the right represents carbon in the above- and belowground plant tissues and the incorporation of ^{12}C into the root structural tissues in the interval between Time 1 and Time 2. The $^{14}C/^{12}C$ ratio in structural tissues of the root system is again determined at Time 2. The assumptions are that no new ^{14}C is incorporated into belowground tissues after the sampling at Time 1 and that all losses from the root system structural tissues between Times 1 and 2 have a constant $^{14}C/^{12}C$ ratio in proportion to their representation in the belowground structural root tissues

accounts for the integrated root production over the entire interval between the times when the samples are taken.

Although the advantages of this ratio approach are clear, there are critical assumptions in the process which, if violated, can lead to serious errors. The first assumption is that when structural tissues of the root system are lost, i.e., when roots die and decay, the ^{14}C and ^{12}C are lost in the proportion in which they occur in the overall root system structural carbon pool. If this assumption holds, loss of root material between the two sampling times will not influence the determination of production due to dilution of the ^{14}C in the system. (The ratio analysis of $^{14}C/^{12}C$ is limited to the structural tissues, because the ^{14}C in these tissues will remain in the roots until they die and decay. Loss of ^{14}C from the labile carbon pool (e.g., starches and sugars) in respiration, translocation to the shoot system, or carbon exudates will not interfere with these estimates). Since ^{14}C in the structural tissues of the root system may tend to occur to a greater proportion in the roots that were younger and actively growing at the time of labeling and these roots would be less likely to die than older roots in the succeeding interval between $^{14}C/^{12}C$ samplings, there would be a tendency for the estimates of root production to be conservative (Caldwell and Camp 1974).

A second and more troublesome assumption is that after the time of the first sampling there will be no new ^{14}C contributed to the structural carbon of the root tissues. Following the assimilation of a single pulse of $^{14}CO_2$, the translocation and incorporation of ^{14}C-photosynthates into structural tissues throughout the plant is often considered to take place within a few days. Based on other independent ^{14}C-labeling studies, this appeared to be a reasonable assumption for the shrub species first used for this technique (Caldwell and Camp 1974) and, as mentioned earlier, the derived root production estimates corresponded well in the overall carbon budgets for these shrub stands (Caldwell *et al.* 1977).

Milchunas *et al.* (1985) have rigorously addressed this second assumption after their first attempts to use the ^{14}C-dilution approach resulted in very unreasonable turnover coefficients for the grass species they were investigating. In ^{14}C-budget studies followed through time for wheat and blue grama grass (*Bouteloua gracilis*), they found that an appreciable quantity of new ^{14}C was incorporated into the structural components of the root system over a 62-day period resulting in large negative errors (-50 to -118%) in their estimates of turnover. From a ^{14}C-balance study, they were able to devise correction factors for the amount of ^{14}C incorporated into the root systems following the first sampling. This reduced the errors in the turnover estimates to less than 2% in most cases. If it is not feasible to derive these correction factors, they suggested delaying the first $^{14}C/^{12}C$ sampling until a full year following the ^{14}C-labeling. This would mean that the second sampling would not take place until at least 18 months following the labeling.

With these refinements, Milchunas *et al.* (1985) suggest the $^{14}C/^{12}C$ ratio technique to have considerable promise in field studies. They compared the types of errors that can beset the sequential harvest and ^{14}C-dilution techniques for estimates of belowground production. Of the eight sources of error they identified to be involved in production estimates, seven were fully involved in the harvest technique while only four were involved in the ^{14}C-dilution approach and these were only partial errors, some of which would tend to cancel one another.

Although these three methods represent the most commonly used techniques for root production estimates, a couple of indirect approaches have been reported recently. One involves a total nitrogen budget analysis (Aber *et al.* 1985) and the other an estimate based on changes in root starch concentrations and soil temperatures (Marshall and Waring 1985). Both of these approaches depend on several assumptions that must be established for a particular ecosystem. Both may have promise in certain situations, but comparison with other techniques for assessing root production would still be recommended.

COMPETITION FOR PHOSPHORUS

Little progress has been made in the assessment of competition processes between higher plants under field conditions, and for good reason. Even with the use of tracers, such as radioisotopes, there are daunting problems in assessing which plant is acquiring more of the resource. Consider the following example: If one were to place ^{32}P-labeled phosphate as a plane of liquid between two neighboring grass species as shown in Fig. 4, an assessment of which plant takes up more of the label would involve a detailed sampling of tissues throughout both plants as well as an estimate of the biomass of each plant part for both species. This would be necessary because different species may distribute the label differently within the plant. Thus, one would need to know the total pool size of the label in the plants of the two species. Even if this were accomplished, the result would only portray the state of acquired label at one point in time.

By using two different isotope markers of phosphate (^{32}P and ^{33}P) and an indicator plant placed between the two competing grasses, as shown in Fig. 5, the ratio of phosphorus isotopes can be monitored in the indicator plant as an index of differential competitive effectiveness of the two grass species (Caldwell *et al.* 1985). Since one is only measuring isotope ratios in the single indicator plant, only small tissue samples of one plant part are needed since the two isotopes will be distributed within the plant in the same manner. Newly developing tissues, such as leaves, are sites of active deposition of phosphorus and useful types of tissue for sampling. Variability among tissue samples is to be expected as amounts of radioactive phosphate would depend on the sink-strength of the growing shoots and the extent of vascular connections between the sampled shoot and roots growing in the specific plane of isotope in the soil. An example of this is shown in Fig. 6 where the ratio of two phosphorus isotopes at different locations on an indicator plant, *Artemisia tridentata*, is shown. This diagrammatic top view of the shrub indicates variability, but no tendency for more isotope uptake on the side of the shrub corresponding to the location of that isotope in the soil.

Although several samples must be taken, only small tissue samples are needed for the isotope ratio determination. Therefore, samples can be taken at different times following the labeling providing information about differential uptake

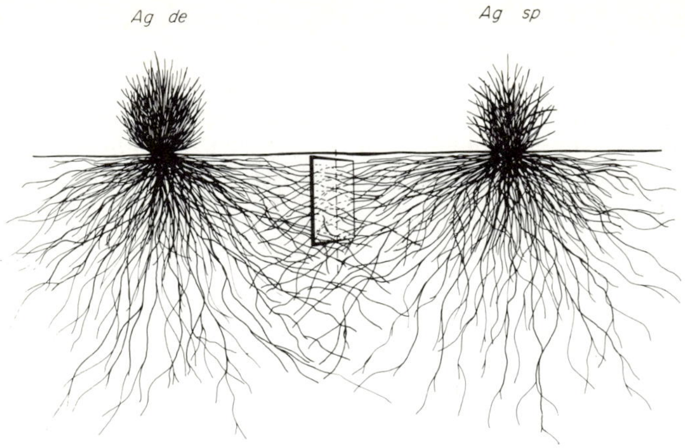

Fig. 4. An experimental scheme to test the differential acquisition of radioactive phosphate by two competing grass species, *Agropyron desertorum* (Ag de) and *Agropyron spicatum* (Ag sp). The isotope solution is placed in the soil between the two grasses through a row of small vertical holes

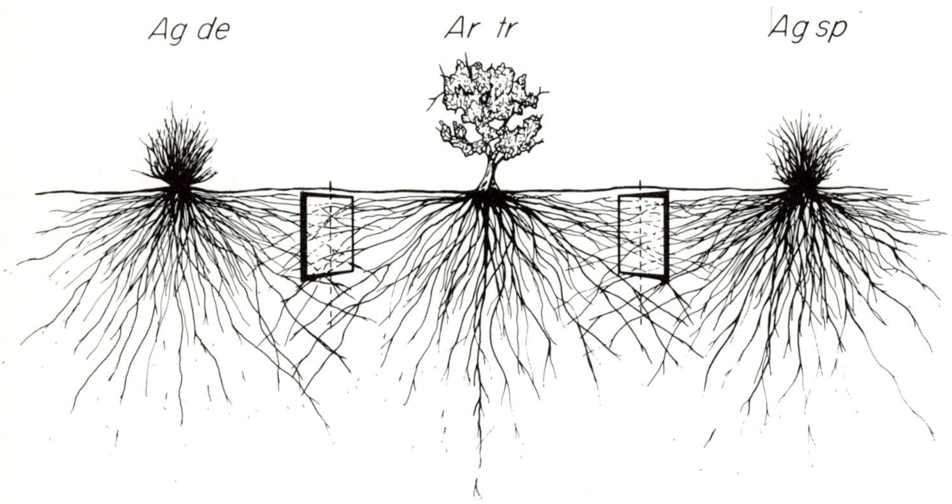

Fig. 5. An experimental scheme to test the competitiveness of two grass species, *Agropyron desertorum* and *Agropyron spicatum* by use of radioactive phosphate solutions placed in the soil between each of the grass species and indicator plant, *Artemisia tridentata*, (Ar tr). Different isotopes are used on either side of the indicator plant. See text for further details

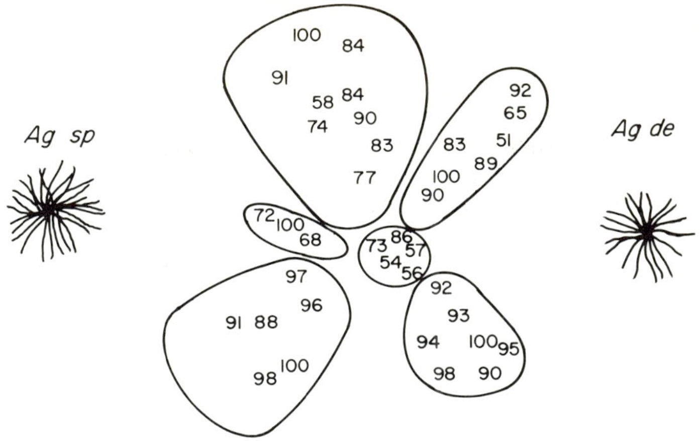

Fig. 6. The distribution of vegetative shoot samples collected at four different times from a single indicator *Artemisia tridentata* plant used in the experiment described in Fig. 5. The lobes represent different branches of the indicator plant as viewed from above. The numbers represent the percentage of the total radioactive phosphorus represented by the isotope placed between the indicator plant and the less competitive grass, *Agropyron spicatum*. Data were derived from a study reported in Caldwell *et al.* (1985)

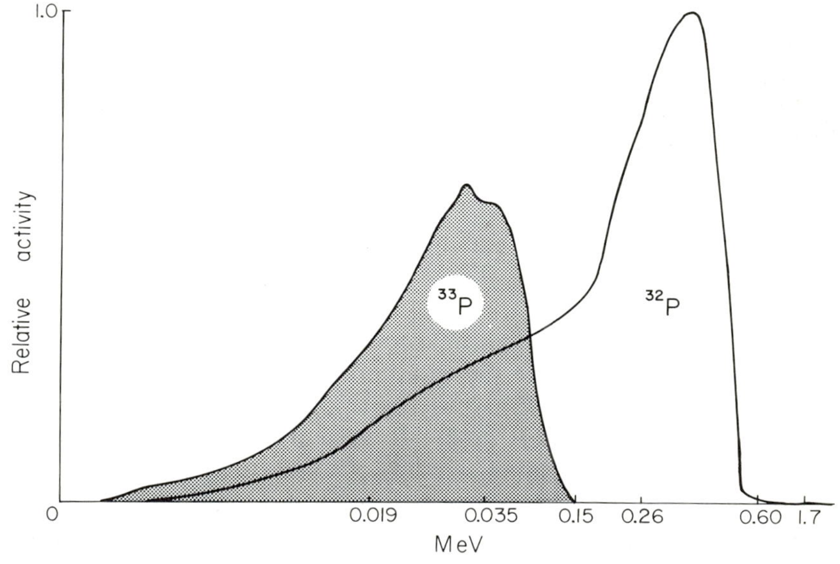

Fig. 7. Emission energy spectra (relative beta radiation at different energies) of the isotopes ^{32}P and ^{33}P, as determined by scintillation counting

through time. This is only limited by the half-life of the isotopes involved. For these phosphorus isotopes, extending an experiment beyond 60 days following labeling is not recommended.

Both phosphorus isotopes are beta radiation emitters, but the energy spectra do not overlap appreciably (Fig. 7) and it is possible to correct for this overlap. Thus, normal scintillation counting techniques are adequate for the double-isotope determinations.

In a single-isotope labeling experiment, the amount of phosphorus isotope is very important. The quantity of phosphorus acquisition can be highly variable among individual plants. Thus, when a single isotope is used, the variation within treatment means is usually very high. In a dual-labeling experiment, one compares differences within a plant, thus eliminating the inherent variability in phosphorus acquisition among plants. Table 1 illustrates the greater accuracy of using two phosphorus isotopes.

Table 1. Number of plant set replicates required to detect a given difference between two treatment means 80% of the time at $\alpha = 0.05$*.

	Difference between Means		
Type of Labeling	200%	50%	20%
Single	5	52	316
Dual	2	5	20

*Data for calculations came from an experiment comparing competitive effectiveness of *Agropyron spicatum* and *A. desertorum* (Caldwell *et al.* 1985).

The variance of ^{32}P and ^{33}P, and their respective ratios in the shoots of *Artemisia tridentata* were used to calculate the number of samples from the experiment reported in Caldwell *et al.* (1985); statistical procedures follow Snedecor and Cochran (1967). If one considers cost, ^{33}P is nearly 20 times more expensive than ^{32}P. However, since only small samples of shoot tissue are needed and many fewer experimental units, i.e., replicate plant sets, are required for a given level of accuracy, use of two isotopes should generally result in greatly reduced costs.

CONCLUSIONS

This chapter has discussed the merits and complications of two examples of the use of tracers in field belowground research. These take very different approaches in that the ^{14}C-dilution method is based on the change in the relative $^{14}C/^{12}C$ ratio through time as a means of determining the new ^{12}C being incorporated into the root system over time. On the other hand, the dual-isotope phosphorus experiments use the ratio of the two phosphorus isotopes in a single indicator plant as a measure of the competitive effectiveness of this plant with two different neighboring plants. These are but two of many uses of tracers in belowground system studies. Several others are covered in the chapter in this volume by Virginia.

ACKNOWLEDGEMENTS

Some of the information presented in this paper was derived from research supported by the National Science Foundation (BSR-8207171) and the Utah Agricultural Experiment Station. The drafting of figures by C. Warner is gratefully acknowledged.

REFERENCES

Aber JD, Melillo JM, Nadelhoffer KJ, McClaugherty CA, Pastor J (1985) Fine root turnover in forest ecosystems in relation to quantity and form of nitrogen availability: a comparison of two methods. Oecologia (Berl) 66:317-321

Caldwell MM, Camp LB (1974) Belowground productivity of two cool desert communities. Oecologia (Berl) 17:123-130

Caldwell MM, White RS, Moore RT, Camp LB (1977) Carbon balance, productivity and water use of cold-winter desert shrub communities dominated by C_3 and C_4 species. Oecologia (Berl) 29:275-300

Caldwell MM, Eissenstat DM, Richards JH, Allen MF (1985) Competition for phosphorus: Differential uptake from dual-isotope-labeled soil interspace between shrub and grass. Science 229:384-386

Caldwell MM (1986) Competition between root systems in natural communities. In: Gregory PJ, Lake JV, Rose D (eds.) Root Development and Function: Effects of the Physical Environment. Cambridge Univ. Press, in press

Fabio A, Persson HA, Steen E (1985) Growth dynamics of superficial roots in Portuguese plantations of *Eucalyptus globulus* Labill, studied with a mesh bag technique. Plant and Soil 83:233-242

Fitter AH (1982) Morphometric analysis of root systems: application of the technique and influence of soil fertility on root system development in two herbaceous species. Plant, Cell and Environment 5:312-322

Hackett C (1968) A study of the root system of barley. I. Effects of nutrition on two varieties. New Phytol 67:287-299

Hansson A, Steen E (1984) Methods of calculating root production and nitrogen uptake in an annual crop. Swedish J Agric Res 14:191-200

Lauenroth WK, Whitman WC (1977) Dynamics of dry matter production in a mixed-grass prairie in western North Dakota. Oecologia 27:339-351

Marshall JD, Waring RH (1985) Predicting fine root production and turnover by monitoring root starch and soil temperature. Can J For Res 15:791-800

Milchunas DG, Lauenroth WK, Singh JS, Cole CV, Hunt HW (1985) Root turnover and production by ^{14}C dilution: implications of carbon partitioning in plants. Plant and Soil 88:353-365

Persson H (1979) Fine-root production, mortality and decomposition in forest ecosystems. Vegetatio 41:101-109

Persson H (1980) Spatial distribution of fine-root growth, mortality and decomposition in a young Scots pine stand in Central Sweden. Oikos 34:77-87

Robinson D, Rorison IH (1983) A comparison of the responses of *Lolium perenne* L., *Holcus lanatus* L. and *Deschampsia flexuosa* (L.) Trin. to a localized supply of nitrogen. New Phytol 94:263-273

Runge M (1983) Physiology and ecology of nitrogen nutrition. In: Lange OL, Nobel PS, Osmond C, Ziegler H (eds) Encyclopedia of Plant Physiology, New Series, Vol 12C. Springer Verlag, Berlin-Heidelberg-New York, pp 163-200

Shierlaw J, Alston AM (1984) Effect of soil compaction on root growth and uptake of phosphorus. Plant and Soil 77:15-28

St John TV, Coleman DC, Reid CPP (1983) Growth and spatial distribution of nutrient-absorbing organs: selective exploitation of soil heterogeneity. Plant and Soil 71:487-493

Singh JS, Lauenroth WK, Hunt HW, Swift DM (1984) Bias and random errors in estimators of net root production: A simulation approach. Ecology 65:1760-1764

Snedecor GW, Cochran WG (1967) Statistical Methods. Iowa State Univ. Press, Ames

Vogt KA, Grier CC, Gower ST, Sprugel DG, Vogt DJ (1986) Overestimation of net root production: a real or imaginary problem? Ecology 67:577-579

Approaches for studying the function of deep root systems

R.A. Virginia and W.M. Jarrell*

Biology Department and Systems Ecology Research Group, San Diego State University, San Diego, California 92182 USA

INTRODUCTION

The distribution of roots in a soil column is such that root biomass and length generally decrease abruptly with depth (Böhm 1979). Most studies of roots have considered only those roots in the surface 0.5 to 1.0 m of soil. Although deeper roots have been noted and their significance discussed, few studies have examined the nature of these deep roots in any detail. The most frequently hypothesized function of these deep roots, relates to their ability to extract water which is available beyond the reach of other plants (Cannon 1911; Weaver 1926). However, little has been done to estimate the importance of deep roots in determining plant production or in maintaining the nutrient balance of the plant. The functional significance of N_2-fixation, root infection by mycorrhizal fungi and pathogens, and rhizosphere-related microflora and fauna of deep root systems have essentially been ignored..

Increasing interest in groundwater pollution problems has emphasized our need to understand the physiology and ecology of deep soils. Organic and inorganic solutes may be transformed by microflora or may be absorbed by roots. Processes in the vadose zone, between the bottom of the root zone and the groundwater (phreatic) surface, are of special concern. Low organic matter and the absence of roots might suggest very low microbial activity and little metabolic transformation of solutes. Recent studies, however, indicate that there may be a fairly diverse and active complex of organisms present in this deep soil layer (White *et al.* 1983; White *et al.* 1985). A few studies have been recently conducted which go beyond simply characterizing the biomass of roots at depth. For example, Rundel *et al.* (1982) describe results from field studies in which it would appear that deep roots (>4 m) are responsible for significant N_2-fixation in the subsaturated zone above the water table.

Deep soil studies have been inhibited by the sheer difficulties in obtaining useful samples from below 1 to 2 m. Great time and energy must be expended to obtain drill cores, to excavate the roots, and to obtain soil samples sufficiently large to characterize. Rooting density is often highly variable at depth, adding to the

---

*Dry Lands Research Institute and Department of Soil and Environmental Sciences, University of California, Riverside 92521 USA

NATO ASI Series, Vol. G15
Plant Response to Stress. Edited by J. D. Tenhunen et al.

difficulties of obtaining representative, reproducible samples. Fractured rock substrata are particularly difficult to excavate, and the samples recovered may be severely disturbed. Finally, deep soil organisms may be adversely affected by bringing them to the surface, through changes in temperature, humidity, and atmospheric composition (Dodge and Francis 1986).

Approaches seemingly suitable and promising for the study of deep root systems are summarized in this paper. We hope that we will stimulate more researchers to look below the soil surface, and to help assess the significance of this generally overlooked portion of the whole plant. While many books have been published since 1980 discussing root function and physiology, none have dealt extensively with methods for characterizing root systems themselves. Böhm's book (Böhm 1979) is probably the most complete monograph available which discusses techniques for the direct and indirect analysis of plant root systems. No recent update of this book has appeared.

ROOT PARAMETERS FOR STUDY

Where the direct observation or recovery of roots is involved, it is important to determine at the onset which parameters of the root system are to be measured. These parameters will vary, depending upon the objectives of the research. The most commonly assessed parameters include root mass, root length, root area or volume, root condition or health, and the degree of infection by symbionts or pathogens.

Root number alone can provide a good estimate of relative extent of rooting. This parameter may be utilized in studies employing all techniques of observation: excavating, coring, and rhizotron viewing. The number of roots exposed along the face of an excavation or on the surface of a soil core segment is easily counted. However, root number may not relate directly to any other root parameter except root length.

The determination of total root biomass is most important for studies concerned with belowground carbon allocation. However, since large woody roots usually dominate biomass totals but play little direct role in nutrient or water uptake, biomass alone tells us little about the relative activity of the root system. The difficulties associated with quantitative recovery of deep roots hampers the understanding of relative biomass allocation to root and shoot systems and the effects of plant stress on root/shoot partitioning.

Physiologically, root length and root surface area appear to be of greater significance in affecting plant function than root mass alone. Determination of these parameters has become experimentally more important recently, because of their inclusion in simulation models where rates of material uptake are limited by diffusion rates in soil and the distribution of fine root elements (e.g. Barber and

Bouldin 1984). The development of a line-intercept statistical technique to estimate root lengths has substantially simplified this otherwise tedious task (Newman 1966).

Root surface area may be a good index of the number of uptake sites on the root surface. Root surface area is relevant in studies concerning water uptake and the uptake of nutrients which reach the root primarily by mass flow, e.g. N and K in fertile soils. Root volume measurements can be related to the total absorbing surface area of the cortex and to the capacity of the root system to store nutrients (Russell 1977). Recent advances in image analysis may greatly shorten the time required to obtain estimates of root diameters as well as root lengths, thus allowing relatively rapid estimation of root total surface area and volume. McGowan *et al.* (1983) for example, have developed a technique to quantify live root length by first exposing the root sample to fluorescent dyes which preferentially stain live tissue, and by then examining a photograph of the sample with a image analysis system.

Since root infection by soil symbionts may have a large effect on the nutrient status of the plant, an assessment of root infection should be made when questions related to growth and nutrient accumulation are under investigation. Inspections of the root surface and cortex can indicate a great deal about the environment in which roots are growing. Nitrogen-fixing root nodules, both active and inactive, may appear on roots of legumes and actinorhizal plants. Mycorrhizal fungi, both vesicular-arbuscular and ecto-types, infect many plant roots (Mosse *et al.* 1981). The fraction of the total root length infected has become a common index of the significance of endomycorrhizae in the plant-soil system (Phillips and Hayman 1970; Ambler and Young 1977).

Presence of endoparasites, such as phytophagous nematodes, and of pathogenic microorganisms indicate the biological health of roots. These organisms affect root growth and function by increasing carbon loss due to exudation and by decreasing water and nutrient uptake by roots. Microscopic examination of roots can indicate the onset of senescence and cell damage due to stress or from grazing by herbivores. Discoloration of the cortex or change in length/width ratio (e.g. toxicity) can also lead to inferences about the relative health of roots.

METHODS OF ROOT STUDY

Root Excavation

The only feasible approach to examining and measuring the entire root system of a plant is direct excavation. Root system morphology, as characterized by depth and lateral extent, can be determined in excavation studies (Böhm 1979). Excavation may also be useful in studying overlap of roots between neighboring plants (Kummerow 1981). Root excavation is, however, very labor intensive and provides an observation of the root system at only one point in time. Therefore, excavations are of limited value when studying root function and root growth. Furthermore, excavation may not be an option where roots follow cracks in rocks or move through stony soil, unless heavy drilling equipment is available.

Large-scale root excavations have been conducted in mediterranean ecosystems, e.g. the Californian chaparral (Hellmers 1955; Kummerow *et al.* 1977) and the Chilean mattoral (Hoffmann and Kummerow 1978; Gilberto and Estay 1978). Water pressure has been used to expose roots of woody chaparral and matorral plants (Kummerow et al 1977). Air pressure may also be used to remove soil from around roots and has been applied in studies of forest ecosystems, cf. Böhm (1979). These techniques are useful for removing large volumes of soil. However, fine root recovery is difficult and fine textured clay soils are difficult to remove using water.

The depth and size class distribution of roots can be determined by excavation and by mapping the exposed roots. Deep trenches may be placed adjacent to the plant and the location of roots mapped or photographed against a grid system. Image analysis techniques can be used to process such photographs if the contrast in color between the roots and the substrate is sufficient. Root descriptions for many systems are available based on trenching techniques (Böhm 1979).

When information on root mass and the physiological condition of the roots is required, soil monoliths can be extracted from the ground and the roots separated by washing. A needle board may be driven into the monolith to retain roots in their natural position prior to removal of the soil. Using this technique, a three-dimensional view of the root system is obtained.

Soil Coring

Coring is one of the best methods for obtaining deep soil samples. Hollow tubes (Viehmeyer tube) may be driven into the soil and then extracted with jacks or pulleys (Viehmeyer 1929). Depths to 180 cm may be sampled in this manner. Heavy drilling equipment must be employed to reach depths below a few meters (Kelley *et al.* 1947). Depths below 30 m may be reached with commercially available drilling equipment. Such equipment has been used extensively to study groundwater and soil contamination but has been used much less by ecologists (McNabb and Mallard 1984). To extract an intact 10-cm-diameter core, a hole of approximately 16-cm diameter is drilled, and the required truck is large and heavy. Site disturbance may be substantial.

Continuous-flight auger systems can be used when quick access holes are required for deep placement of tracers or for insertion of neutron probe access tubes, root in-growth tubes or other monitoring devices. Undisturbed soil cores can be obtained with hollow-stem auger equipment that contains at least two sets of bits, usually an outer rotating bit which cuts the hole and a nonrotating inner coring bit. The coring bit slightly precedes the outer bit in order not to disturb the core. The outer bit cuts and/or removes the soil and this allows the inner bit to enter the soil with little application of pressure. Thus, sample compaction is minimized which might otherwise destroy the integrity of roots and kill soil fauna. The inner surface of continuous-flight auger systems may be flame sterilized if one is collecting samples to study soil biota or clean plastic liners can be inserted to contain the soil moved into the auger tube. These liners can be capped after obtaining the soil core and they facilitate handling and storage of cores after collection. Continuous hollow-stem augering equipment has been used by Virginia, Jarrell, Freckman and Whitford (in preparation) to study the deep soil ecology of the woody legume mesquite (*Prosopis glandulosa*). Soil cores have been collected to a depth of 15 m beneath mesquite trees growing along the edge of a playa on the Jornada Long-Term Ecological Research Site near Las Cruces, New Mexico.

Smaller more portable drilling devices are available for deep drilling. Osborne and Pelishek (1961) used a portable, rotary, three-way combination drill system for the installation of neutron probe access tubes to 10-m depths in southern California chaparral. The soil could be brought to the surface by wet or dry vacuum with this system.

An important limitation of all coring techniques is that individual samples represent a very limited volume of the rooted soil. If rooting densities are uniform in time and space, relatively few cores may be required to characterize the system. However, spatial and temporal variability in root distribution may be high, especially for irregularly spaced, taprooted species typical of many natural ecosystems. Small diameter cores will have a low probability of sampling enough of the deep root system to assure a representative picture.

Root Separation and Measurements

The degree of root recovery from soil samples depends on the extraction methods used. The high cost of obtaining root samples from deep soils means that efficient separation and measurement strategies must be employed, when processing the recovered soil. The most efficient methods involve floatation separation of the root material from the soil. Commercial root washers are available which greatly speed the rate at which root materials may be isolated (Smucker *et al.* 1982). Nevertheless, hand sorting may still be required to separate plant litter from the roots washed free of the soil.

Most investigators are interested in distinguishing between live and dead roots. This is usually a subjective decision based on the the color and physical appearance of the roots. Stains which react more extensively with living roots are also available (Joslin and Henderson 1984). It is often desirable to separate roots according to plant species, when these share the belowground resource. Differences in root color or morphology may allow this, based only on visual examination. Otherwise, interspecific biochemical or isotopic differences must be utilized. For example, roots of C_3 and C_4 plants will differ in stable carbon isotope abundance (Svejcar and Boutton 1985) and roots from N_2-fixing plants may have have a different nitrogen isotope ratio than non-fixing plants growing on the same site (Virginia and Delwiche 1982).

Root Observation Chambers

Rhizotrons permit direct observations of root growth along the face of a transparent barrier (Böhm 1979). Large rhizotrons are expensive to build, involve considerable soil disturbance during installation, and are not suitable for the study of roots below a few meters depth. Root periscopes or viewing tubes are a more attractive alternative for ecological studies of deep root systems. Transparent tubes of plastic are inserted into the soil to various depths. Miniature color video cameras have been developed which can be lowered into these tubes (called minirhizotrons) to view and record an image of roots which have contacted the surface (Upchurch and Ritchie 1983, 1984). To date minirhizotron studies have been conducted exclusively in agricultural settings where root distributions are uniform and rooting depths are shallow. Nonetheless, this technology provides the most feasible opportunity to directly study the phenology of deep roots.

ROOT SYSTEM FUNCTION

A number of indirect techniques have been developed which assess or infer root activity based on measurements of physical and biological properties of soils. These include observation of the seasonal pattern and depth distribution of soil water extraction, nutrient and isotopically labeled element accumulation in shoots, and recovery of soil biota typically associated with roots and rhizosphere processes.

Water Uptake: Soil Water Depletion

Water loss from the soil profile results from evaporation and root uptake. The temporal and spatial pattern of soil water depletion from the profile is therefore an indicator of root activity. Evaporation is significant near the soil surface (upper 1 m) but is not a significant pathway for water loss from greater depths. Thus, water depletion from subsurface soil layers can be attributed to root activity. Decreases in soil water content attributable to root activity can be monitored by coring and by gravimetric determination of soil water content in the recovered samples (Hanks and Ashcroft 1980; Gardner 1986). Changes in gravimetric water content and changes in root density are correlated (Böhm *et al.* 1977). Although sophisticated equipment is not required to determine soil water content, the difficulty in obtaining deep soil samples through time limits the utility of this approach. Furthermore, many cores may be required in heterogeneous soils to detect changes in soil water content

Since it is difficult to obtain deep soil samples, the most useful and efficient methods for assessing deep root activity are those where a measuring device is placed in the soil and soil water status is measured nondestructively throughout the growing season. Soil psychrometers, soil moisture resistance blocks, and the neutron probe meet these requirements. Soil water potential can be related to measurements of resistance of electrodes embedded in porous blocks made of gypsum or fiberglass or to the vapor pressure of water in soil determined using thermocouple psychrometers (Gardner 1986). The relationship between soil water potential and soil water content must be established if estimates of plant water use are desired. Arkley (1981) successfully used fiberglass moisture-temperature cells to study water extraction patterns to depths of 2.5 m by a mixed-coniferous forest in southern California. The sensors performed well for the entire 3.5-year study.

Thermocouple psychrometers measure the water potential of a soil by measuring the vapor pressure in equilibrium with the soil (Savage and Cass 1984). Psychrometers are very sensitive to thermal gradients within the psychrometer (Rawlins and Campbell 1986). This problem is less pronounced in deep soil applications, since thermal gradients are much smaller than those experienced in surface soil layers.

Soil volumetric water content can be measured by the neutron attenuation technique (Greacen 1981; Gardner 1986). A probe generating fast neutrons from an americium-beryllium source is lowered into the soil through an access tube, usually made of aluminium or PVC. Fast neutrons which collide with the soil water become thermalized (slow neutrons). The number of slow neutrons backscattered to a detector is proportional to volumetric soil water content. The advantage of this technique is that repeated measurements can be made on a relatively large soil volume. Rambal (1984) used this technique to study seasonal water extraction to 4.5 m in a *Quercus coccifera* evergreen scrub system for 7 consecutive years. It was necessary to develop a set of calibration curves for different volumes of rock in the soil and to adjust for variable rock distribution at the site. Nilsen *et al.* (1983) also used the neutron probe technique to examine water extraction by the desert phreatophyte mesquite to soil depths greater than 5 m in southern California.

There are a number of new technologies being developed which hold promise for studying roots and soil water. However, their application to field studies requires further refinement and testing. Proton (^{1}H) NMR imaging can be used to map the distribution of mobile water in plant roots (Bottomley *et al.* 1986). The NMR technique is nondestructive and uses nonionizing radiation. Other instruments that allow nondestructive soil moisture measurement and that might be applied to the study of deep soils are tensiometers for use in moist soil (Cassell and Klute 1986), time-domain reflectometers (Topp and Davis 1985), and fiber optic probes (Alessi and Prunty 1986).

Water Uptake: Source Separation by Isotopic Composition

Soil water may be available at depth in landscape positions where surface runoff water accumulates, in high infiltration soils, or where groundwater occurs (Chaney 1981). In these situations, plants often have a well developed surficial root system and a second deeper root system to utilize the subsurface moisture reserves (Rundel *et al* 1982). It is necessary to assess the relative contributions of surface soil moisture derived from precipitation and of deep soil water reserves to maintenance of plant production, if we are to understand the root function of plants with deep root systems.

The oxygen and hydrogen in plant xylem water is derived from soil water. Measurement of the natural abundance of ^{18}O and deuterium, D, isotopes in the transpiration stream of plants by stable isotope mass spectrometry may be used to quantify the relative importance of different sources of water used by plants. There is no significant isotopic fractionation during water uptake by roots. However, isotopic fractionation alters the ^{18}O and D isotopic abundances in precipitation and ground waters. Temperature variation causes the natural abundances of ^{18}O and D in precipitation to vary by season in temperate climates (Dansgaard 1964). Typically, precipitation during cooler winter months has a lower δD than summer precipitation. White *et al.* (1985) used differences in the δD of precipitation and soil water to determined the relative contribution of groundwater to the water budget of tree species growing in an eastern forest of the United States. Transpired water of trees growing in low-lying areas had an isotopic abundance near that of the available groundwater (δD of -60) during periods when the surface soil was dry. Shortly after rainfall, the δD in the xylem water shifted toward the precipitation isotope signature (δD of -20) indicating that the surface soil had become the major source of water to the plant.

Measurements of the $\delta^{18}O$ of xylem water may also distinguish between use of surface and deep water sources which have different oxygen isotopic ratios. Allison and Hughes (1983) determined in Australia the ^{18}O and D abundances to depths of 15 m in soil profiles below *Eucalyptus* shrub stands and below *Eucalyptus* shrub areas recently converted to wheat. Summer isotope ratios differed from the weighted annual precipitation average by -9 $\delta^{18}O$ and -40 δD. Isotopic fractionation resulted due to upward movement and evaporative loss of the lighter ^{16}O isotope. During winter at the cropped site, the isotope ratio of water in the surface soil increased indicating infiltration of precipitation with a higher δD and $\delta^{18}O$. However, there was little isotopic shift observed during winter in the surface soil at the native *Eucalyptus* site. This suggested that most of the rain had been intercepted by the shrub canopies and did not infiltrate into the soil.

Radioactive isotopes can also be used to assess water uptake by roots. Lewis and Burgy (1964) injected tritiated water into deep wells in a California oak woodland. Leaf water was extracted and analyzed for tritium activity using scintillation counting techniques. Plant uptake was detected indicating root activity to at least 38 m depth in this ecosystem.

Nutrient Tracers

Isotope enriched tracers and elements not usually abundant in soils may be applied to plants and/or soils to provide information about root function and depth distribution. The extent of root systems may be determined by labeling a plant and sampling the soil volume for the tracer. Alternatively, the soil may be labeled at different locations or depths and uptake of the label by the plant determined (Böhm 1979; Nye and Tinker 1977). Both types of experiments can help determine where active roots are located in the soil and when they are active. Placement of tracers in the soil may also be used to examine root competition for nutrients between adjacent plants (Caldwell *et al* 1985) and possible nutrient transfer from one plant to another via mycorrhizal fungal connections of root systems (Chiariello *et al.* 1982).

A number of tracers are available for the study of root function. Radioactive isotopes, especially ^{32}P and ^{33}P have been widely used (Nye and Tinker 1977, Caldwell and Eissenstat in this volume) as have certain stable isotopes such as ^{15}N. Elements normally scarce in most soils, such as lithium, boron, rubidium and strontium, have also been used as tracers (Böhm 1979; Fitter 1986). However, lithium and boron even in small quantities can affect plant metabolism.

Nitrogen Uptake

Low nitrogen availability is one of the most frequent limitations on plant productivity in natural ecosystems. Nitrogen compounds, enriched or depleted in the stable ^{15}N isotope, can be applied to the soil at various depths and distances from plants of interest. Small amounts of highly enriched ^{15}N can be applied to minimize the stimulation of plant growth. Incorporation of the isotopic label into aboveground tissues is evidence of root activity at the particular location or depth where applications were made.

This approach can be used to monitor N movement and uptake in natural and agricultural systems (see Hauk and Bystrom 1970). In many soils, the spatial and temporal variation in soil N distribution and in availability are high. This precludes uniformly labeling the entire rooting zone of a plant in the field. This heterogeneity, coupled with the high mobility of N (especially NO_3^-) in soil, complicates the interpretation of stable N isotope studies in plant-soil systems. Issues of non-uniformity of labeling and soil variability complicate all tracer studies (see Caldwell and Eissenstat in this volume).

Symbiotic Nitrogen Fixation

Some plants can utilize atmospheric N_2 by forming a symbiotic association with soil organisms. The soil bacteria *Rhizobium* and *Bradyrhizobium* infect roots of legumes, while nonlegumes are nodulated by *Frankia* actinomycetes. Methods to assess symbiotic N_2-fixation in the field have been reviewed (Bergersen 1980; Hauck and Weaver 1986; Rennie and Rennie 1983; Shearer and Kohl 1987; Virginia et al 1987). Marion (this volume) has reviewed the application of N isotope techniques to the study of nitrogen cycling in mediterannean ecosystems.

Traditional methods for assessing root nodule activity require a quantitatively sound nodule recovery and an assay for nitrogenase activity by the acetylene reduction method (Hardy *et al.* 1973). Nodules are exposed to an atmosphere containing 10% acetylene and the amount of ethylene evolved is determined by gas chromatography. Theoretically, one assumes that 1 mole of N_2 would be reduced for every 3 moles of acetylene reduced. However, significant deviations from this ratio have been noted and the value of this conversion factor should be determined by $^{15}N_2$ techniques for each system under study (Bergersen 1980).

The major limitation of the acteylene reduction assay for the study of deep root systems is the necessity to determine nodule mass per plant or per unit area. This is difficult even for shallow rooted herbaceous plants and is virtually impossible for deeply rooted woody plants, since symbiotic activity may occur at several meters depth (Virginia *et al.* 1986). Nodule activity is known to vary diurnally and seasonally making extrapolation of acetylene reduction rates to the quantity of N fixed by the root system difficult. The most promising method for assessing the contribution of N_2-fixation by deep roots involves measurements of the natural abundance of ^{15}N in soil and plant tissues to provide information on the N sources available to the plant. Some data indicate that natural ^{15}N abundance in plant tissues may be an indicator of rooting depth.

Most soils have an ^{15}N abundance which differs for that of atmospheric N_2 by values ranging from -5 to +15 $\delta^{15}N$ (per mil ^{15}N enrichment - Shearer *et al.* 1978). Deviations in the ^{15}N abundance of soil N from that of atmospheric N_2 result from isotopic fractionation during N transformations in soil which usually favor the movement and loss of the lighter ^{14}N isotope (Delwiche and Steyn 1970). The mechanism or mechanisms which result in depletion are not well known. Since soil N and the atmosphere often differ significantly in natural ^{15}N abundance, it is possible to determine from measurements of plant ^{15}N abundance the relative importance of these two sources of N to plants that are capable of symbiotic N_2 fixation.

Nitrogen fixing plants have been distinguished from non-fixing plants growing on the same site based on the natural ^{15}N abundance of plant tissues (Delwiche *et al.* 1979; Virginia and Delwiche 1982; Shearer *et al.* 1983). The following conditions and assumptions must be met for successful application of this method: 1) soil N must have a significantly different ^{15}N abundance than the atmosphere; 2) isotopic fractionation effects associated with N uptake by roots and N translocation within the plant cannot alter or mask ^{15}N differences between the fixing and non-fixing plants; and 3) suitable non-fixing control species must occur on the site. Since the ^{15}N abundance of available soil N is difficult to determine, the ^{15}N abundance of non-N_2-fixing control or reference plants is determined as a proxy for soil ^{15}N abundance. The reference plants integrate temporal and spatial variation in soil isotopic composition. Ideally, the fixing and reference plants should be of similar growth form and have comparable rooting structure (Virginia and Delwiche 1982).

The difference in the ^{15}N abundance between the N_2-fixing and non-N_2-fixing reference plants growing on the same site is directly proportional to the amount of N fixed by the symbiotic plant if there is no isotopic fractionation during N_2-fixation. However, a small but measurable, isotopic fractionation effect is often associated with the fixation of N_2. The fractional contribution of fixed N to the plant (FN_{dfa}, or "fixed N derived from the atmosphere") can be calculated from an isotope dilution expression, where $FN_{dfa} = (\delta^{15}N_0 - \delta^{15}N_t) / (\delta^{15}N_0 - \delta^{15}N_a)$; where $\delta^{15}N_0$ is the ^{15}N abundance of a suitable non-fixing control plant ; $\delta^{15}N_t$ is the ^{15}N abundance of the N_2-fixing plant; and $\delta^{15}N_a$ is the ^{15}N abundance of fixed N after it has been incorporated into plant tissue. The value of $\delta^{15}N_a$ can be obtained by growing the plant in a N-free medium.

The natural abundance method has been used to assess N_2-fixation by the deeply rooted desert woody legume mesquite (*Prosopis glandulosa*) in the California Sonoran Desert (Shearer *et al.* 1983). Nodulation occurs at depths greater than 4 m in this system (Virginia *et al.* 1986). Results from ^{15}N determinations indicated that most of N uptake by mesquite was derived from root symbiotic activity. The natural abundance method is the only feasible way to study the symbiotic activity of deep roots in such systems. The chief advantage of this method is that an analysis of easily collected aboveground tissues provides information on root system function. Perhaps the greatest advantage of the ^{15}N natural abundance method is that it provides a time-integrated measure of symbiotic activity by the plant.

Natural ^{15}N abundance may provide information on rooting depth. The ^{15}N abundance of non-fixing plants has been related to plant growth form. Plant foliar ^{15}N abundance was shown to decrease with a presumed increase in rooting depth (Virginia and Delwiche 1982) in a number of California ecosystems. Nitrogen isotope composition of the chaparral shrub chamise (*Adenostoma fasciculatum*) also seems to be related to rooting depth (Rundel personal communication). Chamise regenerates by both seed and resprouts after fire. The

$\delta^{15}N$ of chamise seedlings during the first year following fire in Sequoia National Park, California was the same as that of shallow rooted plants growing on the site. Chamise resprouts which had an established deep-root system had a $\delta^{15}N$ similar to that of other deeply rooted species. By the third year after the fire, chamise seedlings and resprouts had comparable ^{15}N abundance, indicating that deep rooting had been accomplished by the chamise seedlings (see Virginia *et al.* 1987).

Multiple Tracers

Frequently, due to changes in soil texture, soil water content, and soil nutrient availability within the soil profile, a number of distinct layers of absorbing roots may be distributed within the soil profile. In such instances, the utilization of multiple nutrient tracers may offer the best approach to answering questions concerning the location and timing of nutrient acquisition by the various root masses. Mutiple tracers may also be useful in the study of root competition between plants. Recently, Fitter (1986) applied $SrCl_2$, RbCl and LiCl at different depths to determine depth distribution of root activity as a function of time for several grassland species. Lipps *et al.* (1957) used ^{32}P and ^{90}Sr to examine root activity of a phreatophytic alfalfa stand in Nebraska with the groundwater at approximately 4 m depth.

ROOT ASSOCIATIONS

The importance of interactions between soil biota and plant roots in determining nutrient availability and uptake, root growth, and pathogen activity and disease has been noted by ecologists in several recent publications (Curl and Truelove 1986; Fitter *et al.* 1985; Mitchell and Nakas 1986). Soil organisms are dependent on plant litter and roots as a source of energy and nutrients for growth and reproduction. Some soil organisms, such as the mycorrhizal fungi are obligate mutualists and cannot complete their life cycle in the absence of host roots.

The abundance, depth distribution and species composition of the soil microflora and fauna may provide information about root distribution and relative root activity. Detailed studies examining relationships between roots and soil biota are lacking for mediterranean ecosystems (Di Castri and Vitali-Di Castri 1981). However, soil fauna might be expected at depth in mediterranean soils. Deep soils exist in mediterranean regions and the deep root systems of mediterranean shrubs and trees provide an input of organic matter to lower soil layers. In addition, the fractured bedrock of many mediterranean soils eases the penetration of soil organisms deep into the soil profile (Di Castri and Vitali-Di Castri 1981).

Rhizosphere Organisms

The distribution of many soil organisms is closely related to root distribution and root activity. Typically certain soil bacteria, fungi, and fauna are found in much higher numbers in the vicinity of roots (rhizosphere) than in bulk soil lacking roots (Curl and Truelove 1986). These rhizosphere organisms utilize organic materials and nutrients exuded from plant roots and from the breakdown of cell material sloughed off during root growth, such as root caps, epidermal cells, and root hairs. The abundance and diversity of rhizosphere organisms in a soil might provide an index of root mass and activity at a particular soil depth.

Fluorescent Pseudomonads

Investigators have noted a close relationship between roots and the rhizosphere bacteria *Pseudomonas fluorescens* and other species of this genus (Vancura 1980). These bacteria are found in high numbers near active roots. A number of other soil bacteria such as *Pseudomonas*, *Azospirillium*, *Arthrobacter*, *Rhizobium* and *Bradyrhizobium* exhibit a strong growth response in the presence of plant roots and root exudates (Curl and Truelove 1986). However, the numbers and nature of the bacteria inhabiting the soil rhizosphere depends on many variables. While generalizations about specific bacteria as definitive indicators of root activity are not possible, characterization of the depth distribution of the soil microbial community may provide presumptive evidence of root related processes.

Soil bacterial distribution under mesquite trees growing along the edge of a desert playa on the Jornada Long-Term Ecological Reseach site near Las Cruces, New Mexico were studied to depth of 15 m. The distribution of fluorescent pseudomonads could be related to root distribution in the upper 4 m for some but not all soil cores examined (Hemmingsen and Virginia unpublished data). Below this depth pseudomonads were lacking, even though some fine roots were recovered as deep as 12 m. At a nearby sand dune site also dominated by mesquite, fluorescent pseudomonads were not isolated from soil samples collected to 7 m depth.

Soil Fauna (Parasitic Nematodes)

Nematodes are among the dominant fauna in most soils including those in mediterranean ecosystems (Di Castri and Vitali-Di Castri 1981). Nematodes influence plant growth and root function by grazing on roots, soil microfauna and microflora (Freckman 1982). It is well known from agricultural studies that most plant parasitic nematodes exhibit host plant specificity (Webster 1972). Some species are restricted to a single host, while others such as *Melodygone incognita* have a broad host range. Recent studies of deep mesquite root systems in New Mexico have recovered plant parasitic nematode species as deep as 12 m (Freckman unpubl data). Investigators have found a relationship between free-living nematode population density and microbial biomass (Baath *et al.* 1981). Since microbial biomass and root activity are often related, nematode numbers may be a useful indicator of root biomass and activity. Such relationships, however, are largely speculative at this time. Studies attempting to elucidate the relationships among root parameters and the composition and population density of the soil faunal community are needed before this approach can be applied with confidence to the study of deep root systems.

Nitrogen-Fixing Symbionts

Plants capable of symbiotic N_2-fixation support specialized root structures termed nodules. The nodules contain the N_2-fixing endosymbiont. Legumes are nodulated by *Rhizobium* and *Bradyrhizobium* while non-legumes are nodulated by *Frankia* actinomyctes. Nodules can be identified by assaying for nitrogenase activity using the acetylene reduction assay or through morphological examination. The nodule can be sectioned, stained, and examined microscopically for the presence of infected cells (Dart 1975). The nodule symbiont may also be isolated, cultured and used to reinfect plants grown in the absence of nitrogen to confirm nodulation (Vincent 1970). Rhizobia can be isolated from surface-sterilized nodules by streaking crushed nodule contents on yeast extract mannitol agar plates. The isolation and culture of *Frankia* is more difficult (Burggraaf *et al.* 1981).

A high population density of rhizobia in a soil is indirect evidence of host plant root activity and nodulation at a particular soil depth. Estimation of the total number of rhizobia in a soil is difficult (Scott and Porter 1986). The most probable number (MPN) soil dilution technique is frequently used to estimate the population density of host nodulating rhizobia (*Rhizobium*, *Bradyrhizobium*) in a soil (Brockwell 1982; Vincent 1970). The host plant is inoculated with aliquots of a serial dilution series (10^{-1} to 10^{-6} is a typical range for dilutions) of the soil under study. The number of nodulating rhizobia in the soil sample can be calculated from the proportion of test plants forming nodules at each dilution. Scott and Porter (1986) make recommendations to improve the accuracy of the plant infection technique for counting rhizobia.

Virginia *et al.* (1986) proposed a modification of the MPN technique termed soil nodulation potential. They studied the population density of rhizobia from a mesquite (*Prosopis glandulosa*) stand growing near Harper's Well in the California Sonoran Desert. Soil samples were collected in 30 cm increments to a depth of 5-5.6 m. Seeds of the woody legume mesquite were grown in a 1:1 mix of field soil and sterile vermiculite. Plants were harvested after 12 weeks and the number of nodules on each seedling were counted to determine the nodulation potential of the soil sample. The nodulation potential of the soil samples and estimates of rhizobia numbers using the MPN technique for the same soil samples were correlated ($r=0.92$, $p<0.05$). The advantages of the nodulation potential approach are that fewer plants must be grown than with the MPN serial dilution methods and soil factors which influence nodulation in the field, other than symbiont population density (such as salinity, soil N content and soil P content), are allowed to exert an influence on nodulation. This approach should also be applicable to study of the non-legume *Frankia* association.

Mycorrhizal Fungi

The root nodule symbiosis is restricted to legumes and a number of woody non-legume species. In contrast, the roots of most plants are infected by mutualistic mycorrhizal fungi. Root infection by mycorrhizal fungi increases nutrient absorption by increasing the effective exploratory absorbing surface length of the root system (Mosse *et al.* 1981). Mycorrhizal fungi affect a number of plant processes. Mycorrhizal plants have been shown to have greater tolerance to a number of stress factors such as drought, saline soil, and root pathogens than non-mycorrhizal plants (Harley and Smith 1983). Since mycorrhizal fungi affect root and whole plant function, it is frequently necessary to assess the mycorrhizal status of plant roots in ecological studies.

Typically mycorrhizal infection decreases with increasing soil depth. However, mesquite roots recovered at 4-5 m depth in southern California were infected by vesicular-arbuscular mycorrhizae (Virginia *et al.* 1986). Mesquite roots recovered from 12 m soil depth in New Mexico were also mycorrhizal (Virginia and Jarrell unpubl. data). The root cortical cells of most vascular plants species can be infected with vesicular-arbuscular mycorrhizae (VAM) which are aseptate fungi of the Endogonaceae.

A summary of methods used in mycorrhizal research has been compiled by Schenck (1982). The objective of most ecological studies is to establish if the plant is mycorrhizal and to quantify the intensity of infection within the root system. VAM activity is usually studied by clearing and staining of roots (Phillips and Hayman 1970) and quantifying infection intensity. Mycorrhizal infection intensity can be determined by several approaches (Kormanik and McGraw 1982). The grid line intersect method can be used to provide information on total root length and the portion of root length colonized by VAM (Ambler and Young 1977). Roots are spread on a petri dish having a marked grid and are viewed with a dissecting scope. The presence or absence of mycorrhizal structures is noted, and the percentage of infected cross-points serves as an index of mycorrhizal infection intensity. Unfortunately, the technique does not account for the quantity of external hyphae, which is the factor which most closely related to the effectiveness of a given species (Hepper 1977).

As with rhizobia, the density of VAM mycorrhizal spores and propagules in a soil can be related to the presence of roots and root activity. Daniels and Skipper (1982) listed methods for extracting and quantifying VAM propagules from soil. The density of VAM propagules in soil can be determined by several methods. Automatic elutriator systems can be used to extract roots, VAM propagules and soil biota (e.g. nematodes) from the same soil sample (Byrd *et al.* 1976).

The density of effective VAM inoculum in soil can be determined from MPN soil dilution tests (Daniels and Skipper 1982). Field soil is mixed with sterile soil as the diluent at 10^{-3} to 10^{-5}. Seedlings of the desired test plant are grown for 6 weeks or more in the soil. Roots are then harvested and scored for the presence of VAM fungi. The MPN of viable inoculum can be determined from statistical tables.

Some plants such as trees are often infected with ectomycorrhizal fungi. These fungi form a mantle around the surface of the root and penetrate the root cortex intercellularly. Ectomycorrhizal infection usually produces a visible change in root morphology. The method selected to quantify the extent of ecotmycorrhizal fungi will depend on the objectives of the investigator (Grand and Harvey 1982). A common approach is to place root segments in subjective infection categories based on visual examination to determine the intensity of infection. Infection is expressed as the percent of short roots infected with mycorrhizal fungi. Alternatively, mycorrhizal tips can be counted. Relationships relating tip number to tip weight, surface area or volume can be derived to evaluate mycorrhizal biomass and other ecologically meaningful parameters.

SUMMARY AND FUTURE RESEARCH PRIORITIES

The potential importance of deep root systems to plant function has long been recognized. Until recently there have been few attempts to actually quantify the importance of deep roots in securing water and nutrients. The development of new methods and approaches which allow direct and indirect study of deep root function offers new and exciting possibilities. Modern drilling equipment can extract deep soil samples with minimal disturbance to roots and the soil biota. The application of stable isotope techniques using variations in natural abundance is especially exciting since analysis of easily collected aboveground tissues provides information on the function of deep root systems. Information on symbiotic nitrogen fixation and the identification of water sources used by the plant can be obtained with this approach. Although poorly described in natural systems, the structure, species composition and population densities of soil rhizosphere biota may provide information on the relative extent of root activity at a particular soil depth. The application of more than one of these approaches to a single system provides the best opportunity to better understand the functional importance of deep roots to plants.

ACKNOWLEDGEMENTS

The support of National Science Foundation Ecosystems Studies grants BSR-8216814 and BSR-8506807 is gratefully acknowledged.

REFERENCES

Alessi, RS, Prunty L (1986) Soil-water determination using fiber optics. Soil Sci Soc Am J 50:860-863

Allison GB, Hughes MW (1983) The use of natural tracers as indicators of soil-water movement in a temperate semi-arid region. J Hydrology 60:157-173

Ambler JR, Young JL (1977) Technique for determining root length infected by vesicular-arbuscular mycorrhizae. Soil Sci Soc Am J 41:551-556

Arkley RJ (1981) Soil moisture use by mixed conifer forest in a summer-dry climate. Soil Sci Soc Am J 45:423-427

Baath E, Lohm U, Lundgren B, Rosswall T, Soderstron B, Sohlenius B (1981) Impact of microbial feeding animals on total soil activity and nitrogen dynamics: A soil microcosm experiment. Oikos 37:257-264

Barber SA (1984) Soil Nutrient Bioavailability. John Wiley and Sons, New York, 398 p

Barber SA, Bouldin DR (1984) Roots, nutrient and water influx, and plant growth Agronomy Society of America Special Publication No. 4, 136 p

Bergersen FJ (ed) (1980) Methods for Evaluating Biological Nitrogen Fixation. John Wiley and Sons, New York, 702 p

Böhm (1979) Methods of Studying Root Systems. Springer-Verlag, Berlin-Heidelberg-New York, 188 p

Böhm W, Maduakor H, Taylor HM (1977) Comparison of five methods for characterizing soybean rooting density and development. Agron J 69:415-419

Bottomley PA, Rogers HH, Foster TH (1986) NMR imaging shows water distribution and transport in plant root systems *in situ*. Proc Nat Acad Sci USA 83:87-89

Brockwell J (1982) Plant-infection counts of rhizobia in soils. In: Vincent JM (ed) Nitrogen Fixation in Legumes. Academic Press, Sydney, p 41-58

Burggraaf AJP, Quispel A, Tak T, Valstar J (1981) Methods of isolation and cultivation of *Frankia* species from actinorhizas. Plant Soil 61:157-168

Byrd DW, Barker KR, Ferris H, Nusbaum CJ, Griffin WE, Small RH, Stone CA (1976) Two semi-automatic elutriators for extracting nematodes and certain fungi from soil. J Nematology 8:206-212

Caldwell MM, Eissenstat DM, Richards JH, Allen MF (1985) Competition for phosphorus: Differential uptake from dual-isotope-labeled soil interspaces between shrub and grass. Science 229:384-386

Cannon WA (1911) The Root Habits of Desert Plants. Carnegie Inst of Washington Publ No 131, Washington, DC, 96 p

Cassell DK, Klute, A.(1986) Water potential: tensiometry. In: Klute A (ed) Methods of Soil Analysis, Part 1 - Physical and Mineralogical Methods. Second Edition, Am Soc Agron, Madison, Wisconsin p. 563-596

Chaney WR (1981) Sources of water. In: Kozlowski TT (ed) Water Deficits and Plant Growth Vol. VI. Academic Press, New York, pp 1-47

Chiarello N, Hickman JC, Mooney HA (1982) Endomycorrhizal role for interspecific transfer of phosphorus in a community of annual plants. Science 217:941-943

Curl EA , Truelove B (1986) The Rhizosphere. Adv Series in Agric Sci 15, Springer-Verlag, Berlin-Heidelberg-New York, 288 p

Daniels BA, Skipper HD.(1982). Methods for the recovery and quantitative estimation of propagules from soil. In: Schenck NC (ed.). Methods and Principles of Mycorrhizal Research. The American Phytopathological Society, St. Paul, Minn., pp 29-35

Dansgaard W (1964) Stable isotopes in precipitation. Tellus 16:436-468

Dart PJ (1975) Legume root nodule initiation and development. In: Torrey JG and Clarkson DT (eds) The Development and Function of Plant Roots. Academic Press Inc., New York, pp 467-506

Delwiche CC, Steyn PL (1970) Nitrogen isotope fractionation in soils and microbial reactions. Environ Sci Technol 4:929-935

Delwiche CC, Zinke PJ, Johnson CM, Virginia RA (1979) Nitrogen isotope distribution as a presumptive indicator of nitrogen fixation. Bot Gaz 140:65-69

Di Castri F, Vitali-di Castri V (1981) Soil fauna of Mediterranean-climate regions. In: Di Castri F, Goodall DW, Specht RL (eds) Mediterranean-type Shrublands, Ecosystems of the World 11. Elsevier Sci Pub Co, Amsterdam, pp 445-478

Dodge CJ, Francis AJ.(1986) Anoxic collection and analysis of subsurface water samples. Research Highlights: Subsurface Transport Program, Brookhaven National Laboratory, U.S. Department of Energy, 18 p

Fitter AH (1986) Spatial and temporal patterns of root activity in a species-rich alluvial grassland. Oecologia (Berl) 69:594-599

Fitter AH, Atkinson D, Read DJ, Usher MB (1985) Ecological Interactions in Soil, Plants, Microbes and Animals. Brit Ecol Soc Spec Pub 4, Blackwell Sci Pub, Oxford, 451 p

Fox RL, Lipps RC (1964) A comparison of stable strontium and ^{32}P as tracers for estimating alfalfa root activity. Plant Soil 20:337-350

Freckman DW (ed) (1982) Nematodes in Soil Ecosystems. Univ of Texas Press, Austin, 206 p

Gardner WH (1986) Water content. In: Klute A (ed) Methods of Soil Analysis, Part 1 - Physical and Mineralogical Methods. Second Edition, Am Soc Agron, Madison Wisconsin, pp 493-544

Gilberto J, Estay H (1978) Seasonal water stress in some Chilean matorral shrubs. Bot Gaz 139:236-240

Grand LF, Harvey AE (1982) Quantitative measurement of ectomycorrhizae on plant roots. In: Schenck NC (ed) Methods and Principles of Mycorrhizal Research. The American Phytopathological Society, St. Paul, Minn, pp 157-164

Greacen EL (1981) Soil Water Assessment by the Neutron Method. CSIRO, Australia.

Hanks RJ, Ashcroft GL (1980) Applied Soil Physics. Soil Water and Temperature Applications. Springer-Verlag, Berlin-Heidelberg-New York, 159 p

Harley JL, Smith SE (1983) Mycorrhizal Symbiosis. Academic Press, New York, 483 p

Hardy RWF, Burns RC, Holsten RP (1973) Applications of the measurement of the acteylene-ethylene assay for measurement of nitrogen fixation. Soil Biol Biochem 5:47-81

Hauck RD, Bystrom M (1970) ^{15}N. A selected Bibliography for Agricultural Scientists. Iowa State Univ Press, Ames, 206 p

Hauck RD, Weaver RW (1986) Field Measurement of Dinitrogen Fixation and Denitrification. Soil Sci Soc Am Spec Publ No. 18, Madison, Wisconsin, 115 p

Hellmers H, Horton JS, Juhren G, O'Keefe J (1955) Root systems of some chaparral plants in southern California. Ecology 36:667-678

Hepper C (1977) A colorimetric method for estimating vesicular-arbuscular mycorrhizal infection in roots. Soil Biol Biochem 9:15-18

Hoffman A, Kummerow J (1978) Root studies in the Chilean matorral. Oecologia (Berl) 32:57-69

Joslin JD, Henderson GS (1984) The determination of percentages of living tissue in woody fine root samples using triphenyltetrazolium chloride. For Sci 30:965-970

Kelly OJ, Hardman JA, Jennings DS (1947) A soil-sampling machine for obtaining two-, three-, and four-inch diameter sores of undisturbed soil to a depth of six feet. Proc Soil Sci Soc Amer 12:85-87

Kormanik PP, A-C McGraw (1982) Quantification of vesicular-arbuscular mycorrhizae in plant roots. In: Schenck NC (ed) Methods and Principles of Mycorrhizal Research. The American Phytopathological Society, St. Paul, Minn, p. 37-45

Kummerow J (1981) Structure of roots and root systems. In: Di Castri F, Goodall DW, Specht RL (eds) Mediterranean-type Shrublands, Ecosystems of the World 11. Elsevier Sci Pub Co, Amsterdam, pp 269-288

Kummerow J, Krause D, Jow W (1977) Root systems of chaparral shrubs. Oecologia (Berl) 29:163-177

Lewis DC, Burgy RH (1964) The relationship between oak tree roots and groundwater in fractured rock as determined by tritium tracing. J Geophysical Res 69:2579-2588

Lipps RC, Fox RL, Koehler FE. (1957) Characterizing root activity of alfalfa by radioactive tracer techniques. Soil Science 84:195-204

McGowan M, Armstrong MJ, Corrie JA (1983) A rapid fluorescent-dye technique for measuring root length. Exp Agric 19:209-216

McNabb JF, Mallard GE (1984) Microbiological sampling in the assessment of groundwater pollution. In: Bitton G, Gerba CP (eds) Groundwater Pollution Microbiology. John Wiley & Sons, New York, pp 235-260

Mitchell MJ, Naksa JP(eds) (1986) Microfloral and Faunal Interactions in Natural and Agro-ecosystems. Developments in Biogeochemistry, Martinus Nijhoff/Dr W Junk Pub, Dordrecht, 505 p

Mosse B, Stribley DP, LeTacon F (1981) Ecology of mycorrhiza and mycorrhizal fungi. In: Alexander M (ed) Advances in Microbial Ecology Vol. 5. Plenum Press, New York, pp 137-210

Newman EI (1966) A method for estimating the total length of root in a sample. J Appl Ecol 3:139-145

Nilsen ET, Sharifi MR, Rundel PW, Jarrell WM, Virginia RA (1983) Diurnal and seasonal water relations of the desert phreatophyte *Prosopis glandulosa* (Honey Mesquite) in the Sonoran Desert of California. Ecology 64:1381-1393

Nye PH, Tinker PB (1977) Solute Movement in the Soil-Root System. Univ California Press, Berkeley, 342 p

Osborne JF, Pelishek RE (1961) Installing deep neutron tubes. Agric Engineering 42:611-612

Phillips JM, Hayman DS (1970) Improved procedure for clearing roots and staining parasitic and vesicular-arbuscular mycorrhizal fungi for rapid assessment of infection. Trans Brit Mycol Soc 55:158-161

Rambal S (1984) Water balance and pattern of root water uptake by a *Quercus coccifera* evergreen scrub. Oecologia (Berl) 62:18-25

Rawlins SL,Campbell GS. (1986) Water potential: thermocouple psychrometry. In: Klute A (ed) Methods of Soil Analysis, Part 1 - Physical and Mineralogical Methods. Second Edition,.Am Soc Agron, Madison, Wisconsin, pp 597-618

Rennie RJ, Rennie DA (1983) Techniques for quantifying N_2-fixation in association with nonlegumes under field conditions. Can J Microbiol 29:1022-1035

Rundel PW, Nilsen ET, Sharifi MR, Virginia RA, Jarrell WM, Kohl DH, Shearer GB (1982) Seasonal dynamics of nitrogen cycling for a *Prosopis* woodland in the Sonoran Desert. Plant Soil 67:343-353

Russell RS (1977) Plant Root Systems: Their Function and Interaction with the Soil. McGraw Hill, London, 298 p

Savage MJ, Cass A (1984) Measurement of water potential using in situ thermocouple hygrometers. Adv Agron 37:73-126

Schenck NC (ed). (1982) Methods and Principles of Mycorrhizal Research. The American Phytopathological Society, St. Paul, Minn, 244 p

Scott JM, Porter FE (1986) An analysis of the accuracy of a plant infection technique for counting rhizobia. Soil Biol Biochem 18:355-362

Shearer G, Kohl DH (1987) N_2-fixation in field settings: Estimations based on natural ^{15}N abundance. Aust J Plant Physiol (in press)

Shearer GB, Kohl DH, Chien SH (1978) The nitrogen-15 abundance in a wide variety of soils. Soil Sci Soc Am J 42:899-902

Shearer G, Kohl DH, Virginia RA, Bryan BA, Skeeters JL, Nilsen ET, Sharifi MR, Rundel PW, (1983) Estimates of N_2-fixation from variation in the natural abundance of ^{15}N in Sonoran Desert Ecosystems. Oecologia (Berl) 56:365-373

Slavik B (1974) Methods of Studying Plant Water Relations. Ecological Studies Series Vol 9, Springer-Verlag, Berlin-Heidelberg-New York, 449 p

Smucker AJM, McBurney, Srivastava AK (1982) Quantitative separation of roots from compacted soil profiles by the hydropneumatic elutriation system. Agron J 74:500-503

Svejcar TJ, Boutton TW (1985) The use of stable carbon isotopes analyses in root studies. Oecologia (Berl) 67:205-208

Topp GC, Davis JL (1985) Measurement of soil water content using time-domain reflectometry (TDR): A field evaluation. Soil Sci Soc Am J 49:19-24

Upchurch DR, Ritchie JR (1983) Root observations using a video recording system in mini-rhizotrons. Agron J 75:1009-1015

Upchurch DR, Ritchie JR (1984) Battery-operated color video camera for root observations in mini-rhizotrons. Agron J 76:1015-1017

Vancura V (1980) Fluorescent pseudomonads in the rhizosphere of plants and their relation to root exudates. Folia Microbiol 25:168-173

Viehmeyer FJ (1929) An improved soil-sampling tube. Soil Sci 27:147-152

Vincent JM (1970) A Manual for the Practical Study ot Root-nodule Bacteria. IBP Handbook No. 15. Blackwell Sci Publ, Oxford, 164 p

Virginia RA, Delwiche CC (1982) Natural ^{15}N abundance of presumed N_2-fixing and non-N_2-fixing plants form selected ecosystems. Oecologia (Berl) 54:317-325

Virginia RA, Jarrell WM, Rundel PW, Shearer G, Kohl DH (1987) The use of variation in the natural abundance of ^{15}N to assess symbiotic nitrogen fixation by woody plants. In: Rundel PW, Ehleringer JR, Nagy KA (eds) Stable Isotopes in Ecological Research. Ecological Studies Series, Springer-Verlag, Berlin-Heidelberg-New York, in press

Virginia RA, Jenkins MB, Jarrell WM (1986) Depth of root symbiont occurrence in soil. Biol Fert Soils 2:127-130

Weaver JE (1926). Root Development of Field Crops. McGraw-Hill, New York, 291 p

Weaver JE, Darland RW (1949). Soil-root relationship of certain native grasses in various soil types. Ecol. Monograph 19:303-338

Webster JM (1972) Economic Nematology. Academic Press, London, 563 p

White SWC, Cook ER, Lawrence JR, Broecker WS (1985) The D/H ratios of sap in trees: Implications for water source and tree ring D/H ratios. Geochim Cosmochimica Acta 49:237-246

White JT, Nickels JS,Parker JH, Findlay RH, Gehron MJ,Smith GA,Martz RF. (1985) Biochemical measures of the biomass, community structure and metabolic activity of the ground water biota. In: Ward CH (ed) Proceedings of the First International Conference of Ground Water Quality Research. Wiley-Interscience, New York

White DC, Smith GA, Gehron MJ, Parker JH, Findlay RH,Martz RF, Fredrickson HL. (1983) The ground water aquifer microbiota: biomass, community structure and nutritional status. Dev Ind Microbiol 24:201-211

Use of nitrogen-15 to assess terrestrial nitrogen cycling processes

G. M. Marion

Systems Ecology Research Group, San Diego State University, San Diego, California 92182 USA

INTRODUCTION

In a recent paper, Hauck (1982) estimated that over 3,000 articles have been written describing the use of ^{15}N in the agriculturally related sciences. In an earlier review, Hauck and Bystrom (1970) cited 1,000 articles dealing with ^{15}N in the agricultural sciences. This explosive increase in utilization of ^{15}N is mainly attributable to the greater availability of mass spectrometers and to the lower cost of ^{15}N-labeled materials. A recent development that should further increase utilization of ^{15}N is the development of commercial laboratories (e.g. Isotope Services Inc., Los Alamos, New Mexico, U.S.A.) that are capable of accurately analyzing labeled samples at moderate costs ($5 to $17 [U.S.A.] per sample).

The use of ^{15}N has won widespread acceptance as the most precise and, in general, accurate way of estimating nitrogen mobility within terrestrial ecosystems (Broadbent 1975; Edwards 1975; Hardy and Holsten 1977; Olson 1979; Hauck 1982; Vose *et al.* 1982). Although the radioactive isotope, ^{13}N, has been used for this purpose, the short half-life (ten minutes) limits its use to short-term experiments (Tiedje 1978).

The objectives of this paper are to:

(1) provide an overview of the ^{15}N methodology involved in
- (a) sample preparation and analysis,
- (b) natural abundance and applied tracer studies,
- (c) studies of plant and soil processes;

and

(2) evaluate the use of ^{15}N to elucidate nitrogen cycling processes in mediterranean-type ecosystems.

Selective studies demonstrating principles, methods, and applications will be discussed. Also, recent publications treating these subjects in more depth will be cited.

NATO ASI Series, Vol. G15
Plant Response to Stress. Edited by J. D. Tenhunen et al.

DEFINITIONS AND CALCULATIONS

The nomenclature used in natural abundance and applied tracer studies is somewhat different. For example, in natural abundance studies, two alternative ways of representing ^{15}N concentrations are widely used. The first is the abundance approach where

$$A = {}^{15}N/({}^{14}N + {}^{15}N) \qquad (1)$$

and the other is the ratio approach where

$$R = {}^{15}N/{}^{14}N. \qquad (2)$$

The abundance and ratio approach are related as follows:

$$A = R/(R+1) \qquad (3)$$

Another common way of representing isotopic ratios is the atom % ^{15}N which is defined by:

$$\text{Atom \% } {}^{15}N = 100\ A \qquad (4)$$

Frequently, authors present data as atom % ^{15}N excess which is the difference between a sample and a reference sample, usually the atmosphere. The ratio of ^{15}N to ^{14}N in atmospheric nitrogen is 1:272 which gives a natural ratio of 0.003676 which is equivalent to a natural abundance of 0.003663 or 0.3663 atom % ^{15}N (Hauck and Bremner 1976).

Because natural ^{15}N labeling is relatively minor, results are usually expressed as parts per thousand with use of the equations:

$$\delta^{15}N = 1000\ ((R(\text{sample}) - R(\text{standard}))/R(\text{standard})) \qquad (5)$$

or as

$$\delta^{15}N_a = 1000\ ((A(\text{sample}) - A(\text{standard}))/A(\text{standard})) \qquad (6)$$

The standard in Eqs. 5 and 6 is usually atmospheric nitrogen. Natural nitrogen isotopic ratios for terrestrial materials are generally with $\pm$ 50 $\delta^{15}N$ units (Wada *et al.* 1981). Even for the extreme case of 50 $\delta^{15}N$ units, Eqs. 5 and 6 differ by only 0.36 $\delta^{15}N$ units which is equivalent to 0.0001 atom % ^{15}N which is within the experimental error in measuring isotopic ratios. Therefore, Eqs. 5 and 6 can be considered to be equivalent at natural isotopic ratios.

For a simple chemical reaction such as:

substrate (s) -----> product (p) (7)

a useful relationship in fractionation studies is the isotopic fractionation factor ($\alpha_{p/s}$) which is defined by:

$$\alpha_{p/s} = R_{pi}/R_s \quad (8)$$

where R_{pi} is the isotopic ratio of an instantaneous increment of product and R_s is the isotopic ratio of the substrate (Mariotti *et al.* 1981). Some authors use $\beta (=1/\alpha)$ to represent the fractionation factor (Delwiche and Steyn 1970; Shearer *et al.* 1974). Another useful relationship in fractionation studies is the isotopic enrichment factor ($\epsilon_{p/s}$) which is defined by:

$$\epsilon_{p/s} = (\alpha_{p/s} - 1)1000 \quad (9)$$

(Mariotti *et al.* 1981). Positive isotopic enrichment factors imply an enrichment of ^{15}N in the product relative to the substrate, while negative values imply a depletion of ^{15}N.

In applied tracer (fertilizer) studies, the primary objective is usually to estimate the recovery of the applied label in various plant, soil, and water compartments. This can be done with either of the following equations for enriched ^{15}N samples.

$$X = (a - b)/(c - b) \quad (10)$$

or

$$\text{N Recovery (\%)} = 100\, P\, (a - b)/(F(c - b)) \quad (11)$$

where X = the fraction of the plant (or other) compartment originating from the added fertilizer, N Recovery (%) = the percent of the added fertilizer in the plant (or other) compartment, P = the total nitrogen in a compartment, F = the total fertilizer nitrogen, and a, b, and c are the atom % ^{15}N in the labeled compartment, unlabeled standard, and labeled fertilizer (respectively) (Edwards 1975; Hauck and Bremner 1976). For depleted ^{15}N samples, change the terms (a - b) and (c - b) in Eqs. 10 and 11 to (b - a) and (b -c), respectively. The overall accuracy of experimental techniques can be checked with zero-time (i.e. immediately after labeling) recoveries; zero-time recoveries should be 100 $\pm$ 5% (Buresh *et al.* 1982).

SAMPLE PREPARATION AND ANALYSIS

The most common procedure for analyzing samples for ^{15}N involves the following three steps: (1) conversion of sample nitrogen to ammonium by a Kjeldahl procedure, (2) oxidation of ammonium to nitrogen gas (N_2) by the sodium hypobromide reaction, and (3) determination of N_2 on a mass spectrometer (Fig. 1). Alternatives include the use of the Dumas furnace for direct conversion of sample nitrogen to N_2 gas and the use of an emission spectrometer for isotopic analyses.

The analytical determination of ^{15}N labeling in amino acid pools involves a two-step process: (1) separation of amino acids and (2) determination of the ^{15}N label. Methods which have been used to separate amino acids include: gas chromatography (Rhodes *et al.* 1981), thin-layer chromatography (Muhammad and Kumazawa 1976), and ion-exchange resin columns (Lewis and Stock 1978). The amino acids can be prepared for ^{15}N analysis by the conventional Kjeldahl digestion and reaction with sodium hypobromide (Lewis and Stock 1978), by combustion using the Dumas method (Muhammed and Kumazawa 1976), and by the direct determination of the N-heptafluorobutyryl isobutyl esters of amino acids in a mass spectrometer (Rhodes *et al.* 1981). Both mass and emission spectrometers have been used to determine ^{15}N concentrations. Since the amount of amino acid nitrogen is frequently low, the greater sensitivity of emission spectrometers to low levels of nitrogen gives them some advantage over mass spectrometers.

Accuracy and precision in isotopic analyses requires exacting attention to detail at all steps of the analytical procedure. Hauck (1982) lists and discusses 23 sources of error affecting nitrogen isotopic analyses (Table 1). Anyone contemplating ^{15}N analyses should be thoroughly familiar with these potential sources of errors. Three major sources of error in isotopic analyses are: (1) sampling, (2) cross contamination, particularly at the sample preparation and distillation steps, and (3) air leakage after sample oxidation (Fig. 1). To overcome the problems associated with subsampling, some workers (Ladd 1981; Buresh *et al.* 1982) recommend using the total plant or soil sample because sampling can be biased. Moore and Craswell (1976) found an atom % ^{15}N in enriched wheat plants for grain, chaff, stem, leaves, sterile tillers, and roots of 8.4 (0.1), 8.1 (0.1), 7.2 (0.3), 8.0 (0.2), 8.3 (0.2), and 7.1 (0.1), respectively. These values are the "a" values in Eqs. 10 and 11. To accurately assess the total fertilizer recovery, each plant compartment should be analyzed individually or the whole plant should be analyzed; it is less accurate to subsample a plant part and assume that it represents the whole plant. However in field experiments, it is often necessary to subsample. Methods for reducing cross-contamination at the distillation stage were discussed by Reeder *et al.* (1980) and Hauck (1982). A diffusion procedure can be used instead of the distillation procedure; the diffusion procedure is inherently less subject to cross-contamination than the distillation procedure and is more precise

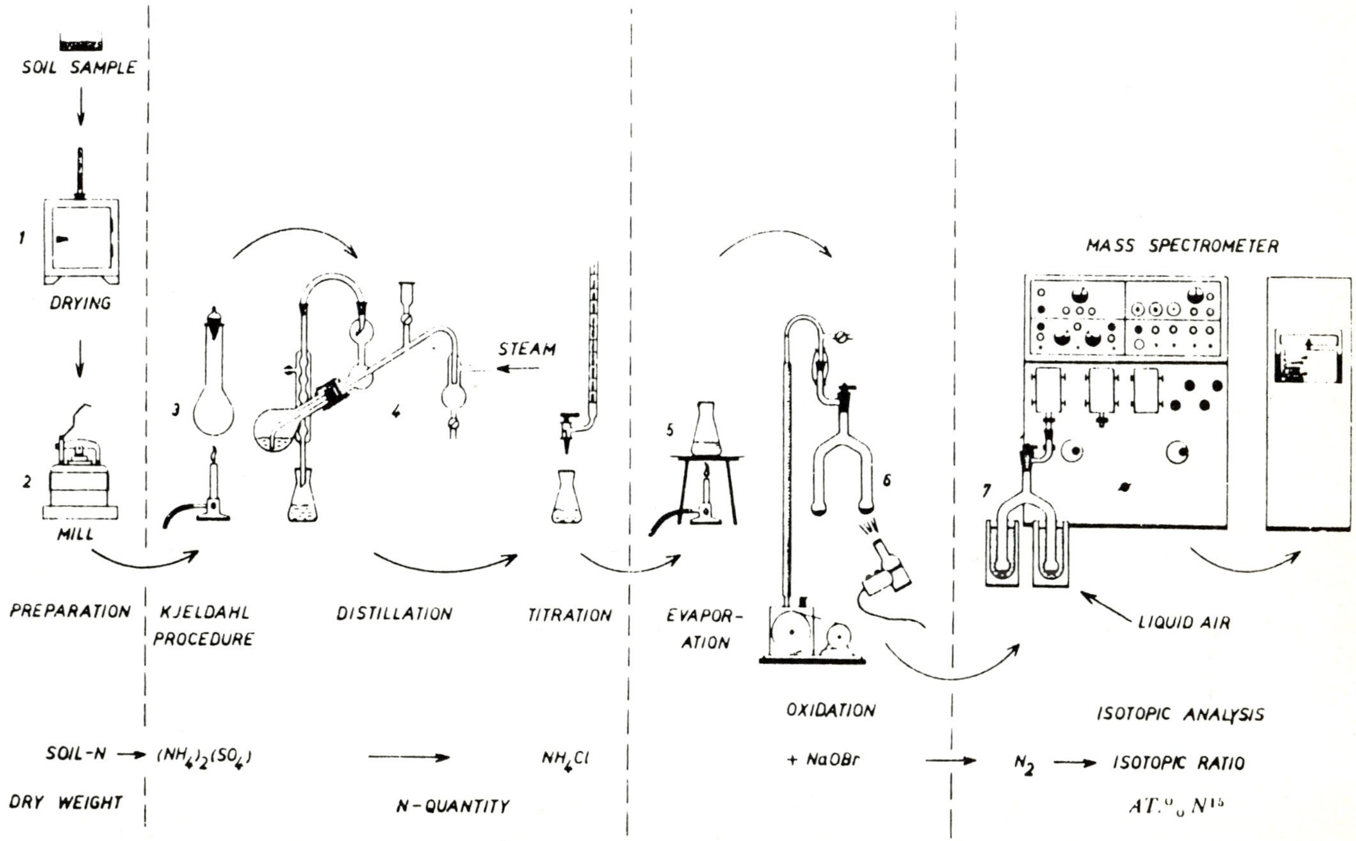

Fig. 1. Schematic diagram of the analytical methods discussed in the text. Reproduced with permission from Huser (1968)

Table 1. Summary of sources of error affecting N isotope-ratio analyses (Hauck 1982).

Sources of error (or operation leading to error)
Failure to recover NO_2-N during total N determination.
Failure to recover NO_3-N during total N determination.
Incomplete Kjeldahl digestion.
Nonsample N in Kjeldahl digest.
Incomplete liberation of NH_3 from solution.
Loss of NH_3 resulting from poor condenser efficiency.
Incomplete Devarda reduction of NO_3-N.
Cross-contamination during distillation.
Successive distillation of NH_3 from different N forms.
Contamination from nonsample N during NH_3 distillation.
Failure to acidulate distillate before concentration.
Excessive heating of NH_4Cl concentrates.
Sampling and sample preparation.
Contaminants in N_2 derived from NH_4-N.
Excess acid during NaOBr oxidation of NH_4-N.
Metal interference during NaOBr oxidation of NH_4-N.
NaOBr oxidation of amide-N to N_2.
Air leakage after N_2 formation.
Contaminants in N_2 mass spectrum.
Sample preparation for N isotope-ratio analysis by emission spectroscopy.
Analysis of N gas mixtures.
Nonrandom distribution of N isotopes.
Method of calculation.

(O'Deen and Porter 1979; Adamsen and Reeder 1983). Because the mass spectrometer is surrounded by an ocean of N_2 gas, the smallest leak after sample oxidation can lead to serious errors in atom % ^{15}N.

The machine of preference for ^{15}N analyses is the double collector, dual inlet mass spectrometer. The double collector allows both ^{28}N and ^{29}N to be analyzed simultaneously which eliminates several problems associated with a single collector machine. The dual inlet system allows the mass spectrometer to alternately analyze an unknown sample and a known reference sample. The mass spectrometer can determine a difference between two samples more accurately than it can an absolute ratio; if the reference is a sample of known isotopic composition, usually atmospheric N_2 gas, then the unknown isotopic ratio can be accurately determined. Atmospheric N_2 gas is a reliable standard for natural ^{15}N abundance measurements (Mariotti 1983). However, commercially bottled N_2 gas can vary significantly in ^{15}N concentrations from atmospheric N_2 (Mariotti 1983); but bottled N_2 gas can serve as secondary (or working) standard provided the tank is calibrated against a primary standard such as atmospheric N_2 gas. Preparation of the atmospheric N_2 standard is discussed by Mariotti (1983). The major advantage of the double collector, dual inlet mass spectrometer is its high sensitivity. With reasonable analytical care, a precision of 0.001 atom % ^{15}N or better is possible. With exacting attention, particularly to fractionation processes, a precision of 0.0002 atom % ^{15}N or better is possible which is necessary for natural abundance measurements. On the other hand, the single collector mass spectrometer has a precision of only 0.01 atom % ^{15}N (Hauck 1982). Major disadvantages of mass spectrometers are its high cost and large sample size (0.5 - 1.0 mg N). However, recent developments in inlet designs can increase mass spectrometer sensitivity to the μg level (Walker *et al.* 1975; McInteer and Montoya 1980). Emission spectrometers can routinely analyze much smaller sample sizes (1 - 10 μg N), but are less accurate and precise (0.01 atom % ^{15}N) at low enrichments (Hardy and Holsten 1977). For example, emission spectrometers can not be trusted below 0.2 atom % ^{15}N excess (Guiraud and Buscarlet 1975). For applied tracer studies at high enrichments, the single collector mass spectrometer, the double collector mass spectrometer, and the emission spectrometer are all adequate. The precision of the double collector, dual inlet mass spectrometer is necessary only in low enrichment, low depletion, and natural abundance studies.

For a fuller discussion of sample preparation and analysis, see the papers by: Bremner 1965; Bremner *et al.* 1966; Jansson 1968; Furst and Jonsson 1971; Proksch 1972; Hauck and Bremner 1976; Hardy and Holsten 1977; Buresh *et al.* 1982; Hauck 1982.

ENRICHED AND DEPLETED NITROGEN-15

Nitrogen materials with high levels of ^{15}N (enriched) and low levels of ^{15}N (depleted) relative to natural abundances can both be used in tracer experiments. The major advantage of depleted ^{15}N material is that it is available in kilogram amounts at lower cost per equivalent label than enriched ^{15}N material (Broadbent and Carlton 1978). However, materials depleted in ^{15}N are restricted to experiments where excessive dilution does not occur because the maximum depletion is only 0.366 atom % ^{15}N which is equivalent to an enriched ^{15}N of 0.732 atom % ^{15}N. For a machine with an analytical precision of 0.001 atom %

^{15}N, this would imply a maximum dilution of 366-fold before the sample is indistinguishable from the background. As a consequence, long-term experiments (e.g. several years) are best conducted with ^{15}N enriched labels. The only other problem associated with the analysis of ^{15}N depleted samples, other than those associated with enriched ^{15}N samples, is in the analysis of highly depleted samples (e.g. < 0.01 atom % ^{15}N) which can best be done by an indirect, isotopic dilution method (Hauck 1982).

ASSUMPTIONS, NATURAL NITROGEN-15 LEVELS, AND FRACTIONATIONS

Three fundamental assumptions underlie the use of applied tracers. These assumptions are: (1) complex elements (those containing two or more isotopes) in the natural state have a constant isotopic composition; (2) living systems can distinguish one isotope from another of the same element only with difficulty, if at all; and (3) the chemical identity of isotopes is maintained in biochemical systems (Hauck and Bremner 1976). These three assumptions are not entirely valid for all experimental conditions; but, they are generally valid for applied tracer studies (Hauck and Bremer 1976). Hauck and Bremner (1976) also list twelve other assumptions that should be kept in mind in the planning and conduct of tracer experiments. The potential ^{15}N user should review the latter paper.

These assumptions are not only important in applied tracer studies, but even more so in interpreting natural abundance levels. For example, plant and soil processes are known to fractionate ^{15}N (Fig. 2). In general, $\delta^{15}N$ values are positive, indicating a ^{15}N enrichment relative to the atmosphere (Table 2). Soils are typically enriched in ^{15}N with respect to either plants or the atmosphere (Tables 2 and 3). Nitrification, denitrification, and ammonia volatilization strongly fractionate ^{15}N (Fig. 2). These three processes are thought to be largely responsible for the enrichment of soil organic matter (Focht 1973; Shearer *et al.* 1974). Since most of these fractionations are caused by living organisms, such as microrganisms and plants, the fractionation factors are highly dependent on experimental conditions. For example, pH, moisture, oxygen level, temperature, medium, organisms, and energy sources are known to influence the fractionation factors (Cook *et al.* 1973; Edwards 1975; Chien *et al.* 1977; Mariotti *et al.* 1982; Bryan *et al.* 1983). Fractionation processes are critical to the interpretation of natural abundance measurements; examples of the role of fractionation processes on the interpretation of natural abundance measurements will be cited under the specific plant and soil processes.

A $\delta^{15}N$ of ten corresponds to an atom % ^{15}N excess of 0.0037 which is a trivial enrichment relative to those in most applied tracer studies. For applied tracer studies, one can generally assume that natural isotopic fractionations are negligible. However, fractionations can occasionally play a role even in applied tracer studies. For example, Edwards (1973) found statistically significant

Table 2. Summary of the mean values of $\delta^{15}N$, and standard deviations from the mean for various forms of nitrogenous compounds (Wada *et al.* 1975).

Nitrogenous compound	Mean $\delta^{15}N$	Standard deviation	Number of samples
Plant	0.31	2.8	23
Peat and coal	0.77	1.6	21
Soil	6.2	5.0	21
Animal	6.5	1.4	15
Crude oil	3.4	1.6	29
Oil brine	2.9	1.6	3
Mud	6.0	1.2	7
River org. N	8.9	6.4	4
Rain:			
NH_4	4.0	2.4	8
NO_3	1.7	3.4	9

Table 3. $\delta a^{15}N$ and N concentration of all plants sampled, N_2-fixing plants, and other plants presumed to be non-N_2-fixing. ***Indicates means of N_2-fixing plants are significantly different from the other plants by T-test, $p<0.001$ (adapted from Virginia and Delwiche 1982).

	$\delta_a{}^{15}N$		
Plants	Relative to atmosphere	Relative to total soil N	N concentration (% N)
All plants	2.2 ± 0.3	-4.9 ± 0.3	2.08 ± 0.17
N_2-fixing plants	0.8 ± 0.3***	-5.5 ± 0.3***	2.81 ± 0.17***
Other plants	2.7 ± 0.3	-3.8 ± 0.3	1.89 ± 0.16

differences in ^{15}N recoveries between depleted and enriched fertilizers in the nitrate compartment (Table 4); these significant differences may be due to the strong fractionation that occurs in the nitrification process (Fig. 2). Broadbent (1975), on the other hand, found much smaller differences between nitrogen recoveries in sudangrass by enriched and depleted nitrogen fertilizers (Table 5); in this case, none of the differences were statistically significant. The Moore and Craswell (1976) data cited previously demonstrated that internal plant fractionations can lead to sampling problems. These examples demonstrate that

Table 4. The average percentage contribution of applied aqua-ammonia fertilizer-N to extracted NO_3-N (Edwards 1973).

At % ^{15}N in fertilizer	% Contribution of fertilizer to NO_3-N 0.1 me rate	0.2 me rate	0.2 me + residue
0.0070	44.8±0.2	60.7±0.1	33.4±0.3
0.7306	47.8±0.3	63.7±0.3	37.4±0.3
2.9284	47.9±0.3	63.8±0.2	--
12.2037	47.6±0.2	63.9±0.2	37.8±0.2

Table 5. Percent uptake of labeled nitrogen from ^{15}N-depleted and ^{15}N-enriched fertilizers in 3 crops of Sudangrass (Broadbent 1975).

Fertilizer	Crop 1	2	3	Total
Depleted	69.3±1.4	7.7±0.7	2.9±0.5	79.9±1.3
Enriched	68.0±1.2	7.7±0.3	3.1±0.1	78.9±1.1

isotopic fractionations can play, at times, a significant role in quantifying nitrogen recoveries in applied tracer studies.

PLANT AND SOIL PROCESSES

Nitrogen Fixation

Since ^{15}N became available in the early 1940's, it has become the standard method against which other methods for estimating N_2 fixation are measured (Burris 1974; Hauck and Bremner 1976; Hardy and Holsten 1977; Vose *et al.* 1982). Hardy and Holsten (1977) have estimated that the ^{15}N method is about 100 times more sensitive than the difference method in estimating N_2 fixation. The direct method of estimating N_2 fixation requires reacting soil or nodules in the presence of a labeled N_2 atmosphere in a closed system; ^{15}N analysis of the soil (or nodules) and the closed atmosphere can be used to estimate N_2 fixation with the equation:

$$N_f = (a - b)/(c - b) \qquad (12)$$

where N_f = fraction of nitrogen fixed, a = atom % ^{15}N in soil (or nodules), b = atom % ^{15}N in standard atmosphere (0.3663), and c = atom % ^{15}N in labeled soil atmosphere (Vose *et al.* 1982). This method can be used to estimate both symbiotic and nonsymbiotic N_2 fixation. Silvester *et al.* (1985) have used this method to estimate symbiotic N_2 fixation in root nodules of five species in four genera of the Rhamnaceae in the matorral of central Chile. They found that N_2 fixation was associated with the apical portions of the nodules. They also compared the direct ^{15}N method to the acetylene reduction (C_2H_2) method and found that the N_2 fixation ratio of $C_2H_2/^{15}N$ was 3.22(0.29) which agreed well with the theoretical value of 3.00. The major advantage of the direct ^{15}N method is that it provides unequivocal proof of N_2 fixation. On the other hand, the required air-tight system is expensive, and the method only gives a short-term estimate of the N_2 fixation rate.

Nitrogen fixation for symbiotic associations can also be indirectly estimated by:

$$N_f = (c - a)/(c - b) \qquad (13)$$

where N_f = fraction of nitrogen fixed, a = atom % ^{15}N in fixing plant, b = atom % ^{15}N in standard atmosphere (0.3663), and c = atom % ^{15}N in non-fixing plants (Legg and Sloger 1976).

An alternative equation for estimating symbiotic N_2 fixation is:

$$N_f = (x - y)/(x - f) \qquad (14)$$

where x = $\delta^{15}N$ of leaves of non-fixer, y = $\delta^{15}N$ of leaves of fixer, f = $\delta^{15}N$ of leaves of fixer grown hydroponically without nitrogen (Shearer *et al.* 1983). Eqs. 13 and 14 can be used in both natural abundance and applied tracer studies. Eqs. 13 and 14 are equivalent only where f = 0.0 which implies that there is no fractionation in the N_2 fixation process. Since in general there is a small but significant fractionation in the N_2 fixation process (Fig. 2), Eq. 14 is preferable for estimating N_2 fixation in natural abundance studies. For example, using the data in Table 3 in Eq. 13, one estimates a N_2 fixation fraction of 0.70. Using the data in Table 3 in Eq. 14 with an "f" value of -1.3 (Shearer *et al.* 1983) leads to an estimate of 0.47 for the N_2 fixation fraction. Although the N_2 fixation fractionation is relatively small, it plays a major role in estimating fixation rates using the natural abundance method. On the other hand, in applied tracer studies, the magnitude of the tracer in the plants is usually so much larger than the value of "f" (Eq. 14) that Eqs. 13 and 14 can be considered to be equivalent in applied tracer studies.

In contrast to the direct method (Eq. 12), these indirect methods (Eqs. 13 and 14) give an integrated estimate of N_2 fixation over time. Clearly, for these indirect methods to work, $b \neq c$ or $x \neq f$. That is, there must be a clear difference in the isotopic composition of the potential nitrogen sources. The non-fixers acquire the bulk of their nitrogen from the decomposition of organic matter which is generally enriched in ^{15}N (Turner *et al.* 1983; Table 6). Fixers, on the other hand, have two potential nitrogen sources, the generally enriched soil organic matter and the unenriched atmosphere. Selecting a non-fixing reference plant is critical. Significantly different estimates of N fixation are possible depending on the choice of a reference plant (Rennie 1982; Mariotti *et al.* 1983; Chalk 1985). The reference plant should be chosen with the following criteria: (1) the non-fixer should have the same isotopic fractionation at absorption of mineral nitrogen as the fixer; and (2) the growth phases and extension of the root system should be as close as possible (Mariotti *et al.* 1983).

Table 6. $\delta_a{}^{15}N$ values at various soil depths (Rennie *et al.* 1976).

	Chernozemic dark brown			
	Total N		NO_3-N	
Depth (cm)	µg/g	$\delta_a{}^{15}N$	µg/g	$\delta_a{}^{15}N$
0-15	1,990	8.7	23	11.0
15-120	640	6.1	22	8.5
120-150	670	6.4	22	3.1
150-180	350	5.7	23	3.1
180-240	540	6.3	31	6.2
240-270	440	6.9	45	8.5

Virginia and Delwiche (1982) found that N_2 fixers had significantly lower ^{15}N abundances and higher total nitrogen concentrations than non-fixers (Table 3). They considered these two criteria to be presumptive evidence for N_2 fixation. Shearer *et al.* (1983) used natural abundances to estimate N_2 fixation by desert legumes (Table 7). Most of the sites showed significant N_2 fixation. On the other hand, Binkley *et al.* (1985) concluded that ^{15}N abundances did not provide a simple means for evaluating alder fixed nitrogen in a mixed coniferous forest. The major difficulty with the natural abundance approach is that the differences in natural ^{15}N labeling are small (approaching the limits of detection) and the potential fractionations in soils are many (Fig. 2). However, an appropriate reference plant can alleviate problems associated with fractionations. Although

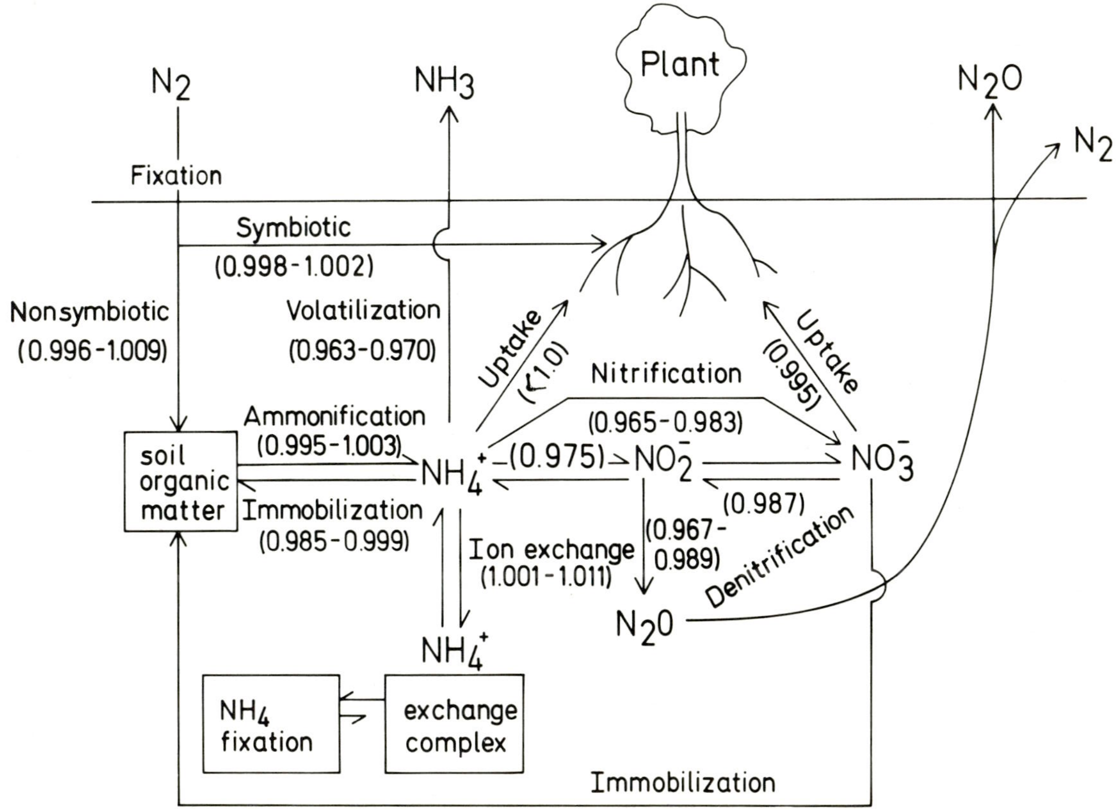

Fig. 2. Nitrogen isotopic fractionation factors (∝) for terrestrial plant and soil processes. Data sources for the fractionation factors were: Blackmer and Bremner (1977), Delwiche and Steyn (1970), Focht (1973), Freyer and Aly (1975), Ingerson (1953), Karamanos and Rennie (1978), Kohl and Shearer (1980), Mariotti *et al.* (1981, 1982, 1983), Rennie (1980), Urey (1947), Wada *et al.* (1975, 1981)

Table 7. Estimation of the fractional contribution of biologically-fixed N to *Prosopis* spp. and other desert legumes (Shearer *et al.* 1983).

Site	Legume	% N fixed mean S.E.
Harper's Well	*Prosopis* spp.	61 ± 7
Transect near Harper's Well	*Prosopis* spp.	43 ± 15
Nude Wash	All legumes	43 ± 10
	Prosopis spp.	65
Borrego Sink	*Prosopis* spp.	60 ± 5
Carrizo Badlands	*Prosopis* spp.	57 ± 6
Clark Dry Lake	*Prosopis* spp.	58 ± 5
Baja California, Mexico	All legumes	28 ± 11
(Near Catavina)	All Papilinoideae	70 ± 7
	Prosopis spp.	1 ± 5
	Acacia spp.	-12 ± 9

the natural abundance method for measuring N_2 fixation may be useful in some ecosystems, it may not be universally applicable.

To overcome the problem of a small differential in the isotopic composition between the atmospheric and soil nitrogen sources, workers have used labeled nitrogen fertilizers (Fried and Broeshart 1975; Vose *et al.* 1982), slow release labeled nitrogen fertilizers (Witty and Ritz 1984), and lightly enriched organic matter (Kohl and Shearer 1981; Ladd 1981) in order to increase the soil nitrogen isotopic composition. A problem associated with the use of high fertilizer levels is that fertilization decreases N_2 fixation rates. For example, N_2 fixation rates dropped from 44% to 4% as nitrogen addition ranged from 0 to 50 kg nitrogen (Table 8). It is also difficult to apply enriched N to a soil uniformly. Natural ^{15}N abundance methods have an advantage of greater homogeneity in soil N (Rennie 1982). Also the selection of a reference plant is more crucial with the enriched method than the natural abundance method. In particular, the reference plant should take up nitrogen from the soil over a given time interval in the same proportion as the fixing plant because the soil ^{15}N label is rapidly changing in enriched soils but markedly less so in unenriched soils. The use of lightly enriched organic matter in pot studies to screen plants for N_2 fixing ability is a particularly promising technique (Kohl and Shearer 1981; Ladd 1981).

Plant Nitrogen Uptake

Plant uptake of labeled nitrogen can be estimated with Eqs. 10 and 11. The ^{15}N method is generally accepted as the most accurate and precise method for

Table 8. The proportions of shoot N coming from fertilizer, soil, and biological fixation in soybean at 75 days. Calculations based on total N content, and also on ^{15}N content of the shoots (Ruschel *et al.* 1979).

Treatment	% N from fertilizer	% N from soil	% N_2 fixed
	Calculations based on total N		
0 kg N	0.0	46.46	53.54
25 kg N	37.79	32.77	29.44
50 kg N	36.96	30.00	33.04
Sig. diff. (P = 0.05)	4.15	8.23	10.06
	Calculations based on ^{15}N data		
0 kg N	0.0	56.00	44.00
25 kg N	10.31	62.32	27.37
50 kg N	23.67	72.09	4.24
Sig. diff. (P = 0.05)	3.29	7.67	11.03

estimating not only fertilizer recoveries but also N_2 fixation (Broadbent 1975; Edwards 1975; Olson 1979; Hauck 1982). An alternative method, the difference method, can estimate fertilizer recoveries or N_2 fixation by measuring the differences in nitrogen contents in plant and soil compartments between fertilized and unfertilized plots or between nitrogen fixers and non-fixers, respectively. Differences between the ^{15}N and difference methods can be large. For example, there was a 3.7-fold difference in apparent nitrogen recovery in a 25 kg nitrogen treatment and a 7.8-fold difference in nitrogen fixation in a 50 kg nitrogen treatment (Table 8). Hauck (1982) has estimated that the sensitivity of the ^{15}N method in estimating fertilizer recovery is 40-fold better than the difference method. Although ^{15}N methods are generally more accurate than alternative methods, this is not invariably so. ^{15}N methods rely on the validity of certain assumptions which are frequently hard to evaluate. Several authors have cautioned against assuming that ^{15}N methods are more accurate than alternative methods (Hauck and Bremner 1976; Chalk 1985).

Meints *et al.* (1975) have used unenriched nitrogen fertilizer to estimate fertilizer recoveries. This approach is similar to the use of natural abundances to estimate N_2 fixation in that it requires a significant difference in the isotopic nitrogen abundances between the plant nitrogen sources which are fertilizer nitrogen and soil organic matter. Nitrogen recoveries in plant material were underestimated in all six cases (Table 9). Also, the high errors associated with detecting small

Table 9. Estimation of fertilizer-derived N and standard deviation in plant materials (Meints *et al.* 1975).

Treatment	Fertilizer ^{15}N enrichment $\delta^{15}N$	Taylor's brome	Morrow plot 3 NA	Morrow plot 3 NB
		%		
First Harvest				
72 ppm N	0.1	22.6±6.0	72.4±14.8	42.3±16.9
73.5 ppm N	852.9	38.4±1.2	78.0±1.6	71.0±0.3
72.5 ppm N	1695.6	39.9±1.1	77.8±0.8	70.6±0.7
Second Harvest				
72 ppm N	0.1	21.0±14.2	33.0±10.6	17.5±24.2
73.5 ppm N	852.9	30.7±1.0	69.7±0.4	58.4±0.6
72.5 ppm N	1695.6	30.3±0.8	69.9±0.4	58.8±0.6

differences in isotopic composition are obvious in the standard deviations for the unenriched treatments. Meints *et al.* (1975) concluded that this method of using variation in the natural abundance of ^{15}N is qualitative or roughly quantitative. Karamanos and Rennie (1981) also used unenriched fertilizer to estimate recoveries; they concluded that this method is semi-quantitative. As was the case for N_2 fixation, the small difference in isotopic composition between fertilizer and soil nitrogen sources is the major source of error in estimating recoveries based on natural abundances.

Stock and Lewis (1984) have used tracer ^{15}N to estimate nitrate and ammonium uptake by an evergreen fynbos shrub species, *Protea repens*. They found that the nitrogen form did not cause any significant changes in the total nitrogen content of the plants, and no alteration in total nitrogen distribution patterns was evident. They concluded that *Protea repens* adapted to conditions prevalent in the latter stages of fynbos succession, not by showing an absolute requirement for either nitrogen form, but by an ability to absorb and assimilate small quantities of both nitrate and ammonium which satisfy the low nitrogen demand of these slow-growing, evergreen species. Stribley and Read (1974) found significant differences in the nitrogen isotopic composition of mycorrhizal and non-mycorrhizal *Vaccinium* plants; they speculated that mycorrhizal plants had access to other sources of nitrogen such as organic nitrogen which accounted for the difference in the nitrogen isotopic compositions.

Plant Metabolism

The assimilation of nitrogen into plant amino acids has been widely studied (Ivanko 1971; Muhammad and Kumazawa 1976; Lewis and Stock 1978).

Muhammad and Kumazawa (1976) found a higher rate of incorporation of ^{15}N into amino acids where ammonium was the nitrogen source than when nitrate was the source. They attributed this difference to the different pathways for ammonium and nitrate nitrogen incorporation into the amino acid pool and/or to lesser transport of nitrate from the source. Generally, amino acids that are formed primarily or secondarily in the Krebs cycle had greater incorporation of ^{15}N from both ammonium and nitrate feedings. These results corroborate earlier work by Ivanko (1971). Ivanko (1971) also found the nitrogen-starved plants had higher nitrogen uptake rates which was attributed to an increase of nitrogen acceptors during the starvation period.

Soil Ammonification, Immobilization, and Nitrification

Ammonification, immobilization, and nitrification are three processes along with plant uptake which most closely control the availability of nitrogen for plant growth in terrestrial ecosystems. Fig. 2 depicts autotrophic nitrification (i.e. ammonium---->nitrate). Recent work in a forest soil suggested that the potential for heterotrophic nitrification (i.e., organic matter----->nitrate), which is not depicted in Fig. 2, may be greater than the potential for autotrophic nitrification (Schimel *et al.* 1984). In these experiments, the ammonium pool was labeled with ^{15}N to demonstrate that nitrate production did not occur through autotrophic nitrification.

Because the ammonium and nitrate pools are relatively small and rapidly turning over, kinetic models are frequently used to characterize these pools and processes. Kirkham and Bartholomew (1954) developed a model for mineralization (ammonification) and immobilization of tracers in soil systems. The derived equations for mineralization (m) and immobilization (i) rates were:

$$m = (Mo - M) \log (HoM/HMo)/t \log(Mo/M) \tag{15}$$

$$i = (Mo - M) \log (Ho/H)/t \log(Mo/M) \tag{16}$$

where Mo is the mass of tracer plus non-tracer mineral atoms per unit mass of soil at time t = 0, Ho is the mass of tracer atoms per unit mass of soil at time t = 0, and M and H are the respective masses at time t = t. This model assumed no isotopic fractionation.

An alternative model, based on the mathematical analysis of isotopic dilution curves was used by Van Cleve and White (1980) to characterize tracer flux through forest-floor litter. The atom % ^{15}N excess in various soil pools is described with mathematical equations of the form:

$$E_t = A1 \exp(-g1t) + A2 \exp(-g2t) \tag{17}$$

where E_t = atom % ^{15}N excess in a pool at time t. With the derived A and g parameters, pool sizes, fluxes, and pool turnover times can be ascertained. This model assumes no fractionation and a steady state (i.e. pool sizes do not change). This model is more generally applicable than the previous model since it can be developed to handle any number of pools and fluxes. Interested workers are referred to the Van Cleve and White (1980) paper for an application of this technique to a soil system and to books by Rescigno and Segre (1966) and Shipley and Clark (1972) for more details on the theoretical foundations of the model.

Focht (1973) developed an explicit fractionation model based on first-order kinetics for the processes: ammonification, nitrification, and denitrification. This model considered ammonification of added organic matter, but did not consider ammonification and immobilization of nitrogen in soil organic matter. Shearer *et al.* (1974) developed an isotopic fractionation model based on first-order reaction kinetics which considered ammonification and immobilization of nitrogen in soil organic matter and nitrification. Both of these models have provided insights into the mechanisms controlling isotopic fractionation of nitrogen in soils. For example, the Focht model explained why nitrate originating from casein, ammonium, and nitrate are less enriched, respectively, in ^{15}N for equal nitrate concentration due to ammonium and nitrate being initially depleted in ^{15}N during ammonification and nitrification, respectively. The Shearer model predicted that the $^{15}N/^{14}N$ ratios of nitrate and ammonium depend in part upon the ratio of the rates of immobilization and ammonification.

Kohl *et al.* (1971) used natural variations in ^{15}N abundances of nitrate as an indicator of the relative roles of soil organic matter and fertilizer nitrogen as sources of pollutant nitrate. The assumptions in the Kohl *et al.* (1971) paper have been criticized (Hauck *et al.* 1972; Edwards 1975). However, other papers suggest that the natural abundance method can provide qualitative to semi-quantitative estimates of nitrate pollution provided the nitrate sources are sufficiently different in their ^{15}N concentrations and relevant fractionations are considered (Letolle and Mariotti 1974; Freyer and Aly 1975; Shearer *et al.* 1975).

Soil Denitrification

Soil denitrification refers to a sequence of steps leading to the reduction of nitrate with the formation of N_2O and N_2 gases. The steps are summarized by:

$$NO_3 \xrightarrow{k_1} NO_2 \xrightarrow{k_2} N_2O \xrightarrow{k_3} N_2 \tag{18}$$

A number of experimental approaches, including both ^{15}N and ^{13}N methods, have been used to study this process; see Tiedje (1978) for a general discussion of methodology. Because the end-products of denitrification are gases (N_2O and N_2), most direct experimental techniques involve closed systems. Acetylene inhibits N_2O reductase activity which results in the stoichiometric accumulation of N_2O gas (Knowles 1978; Tiedje 1978; Ryden *et al.* 1979). Experimental methods usually involve the addition of a labeled substrate (e.g. nitrate) followed by the measurement of the N_2O gas by gas chromatography and isotopic analysis by either a mass or emission spectrometer. Acetylene inhibits N_2O reductase and can also act as a substrate for the nitrogenase enzyme involved in N_2 fixation; denitrification and N_2 fixation can be measured simultaneously by monitoring N_2O and ethylene gas concentrations (Knowles 1978).

Under anaerobic conditions, denitrification can become the dominant process leading to the loss of nitrogen from soil systems. For example, Stanford *et al.* (1976) found greater incorporation of nitrate-^{15}N into gaseous products than into ammonium and organic matter; after 48 hours, none of the original nitrate-^{15}N was still present as nitrate. Results indicated that ammonium-^{15}N production preceded incorporation of ^{15}N into soil organic matter. The proportion of nitrate-^{15}N converted to ammonium-^{15}N and ^{15}N labeled organic nitrogen was higher with 2.5 mg than with 1 mg/g of added C. This finding may reflect the higher energy requirement for nitrate reduction to ammonium than for nitrate reduction to N_2O and N_2. Both assimilatory and dissimilatory nitrate reduction to ammonium are poorly understood soil processes; but, neither process appears to constitute a quantitatively important fate for nitrogen in most aerobic soils (Tiedje *et al.* 1981).

Mariotti *et al.* (1982) found the isotopic enrichment factor (Eq. 9) was a function of the rate of denitrification (k_2, Eq. 18):

$$\epsilon_{p/s} = 5.78 + 5.14 \ln (k_2) \tag{19}$$

The isotopic enrichment factors ranged from -11 (α = 0.989) to -33 (α = 0.967). Greatest enrichments were obtained for the lowest rate of denitrification under low temperatures and with undried soils. Lowest enrichments were found for high rates of denitrification with higher temperatures and additions of glucose to the soil.

Other Soil Processes

Ammonium fixation, ammonia volatilization, and ion exchange are three processes which can influence nitrogen cycling through terrestrial ecosystems (Fig. 2). Ammonium fixation refers to the process where ammonium is fixed in a non-exchangeable form by soil organic matter and 2:1 expanding lattice clay minerals; this process can be a significant sink for nitrogen. Olson (1979) has estimated that medium to fine textured soils containing 2:1 clays will contain in the order of 20 g/m^2 (30 cm depth) of fixed ammonium. In another paper, 76% of the recovered ^{15}N in soils after 40 months of field weathering was fixed ammonium (Kowalenko and Ross 1980). As this fixed ammonium is not isotopically exchangeable with labeled ammonium (Newman and Oliver 1966), it is generally inaccessible for plant use.

Because ammonia volatilization is controlled by physico-chemical factors, a conventional thermodynamic equilibrium relationship can be derived for the process. For the reaction:

$$^{14}NH_3(g) + {}^{15}NH_4(aq) \rightleftharpoons {}^{15}NH_3(g) + {}^{14}NH_4(aq) \qquad (20)$$

K = 0.967 (25° C, Urey 1947) and K = 0.970 (25° C, Ingerson 1953). This isotopic equilibrium constant also implies that the gas phase will be relatively depleted in ^{15}N and the solution phase will be relatively enriched in ^{15}N. Note that for an equilibrium reaction:

$$K = {}^{15}NH_3(g){}^{14}NH_4(aq)/{}^{14}NH_3(g){}^{15}NH_4(aq) = R(g)/R(aq) = \alpha \qquad (21)$$

The equilibrium constants given by Ingerson and Urey (respectively, 0.970 and 0.967) are within experimental error equal to the isotopic fractionation factor (α) reported by Wada *et al.* (1981) of 0.963. Ammonia volatilization losses increase with increasing pH. Reddy and Patrick (1979) found that greater losses of added ammonium- ^{15}N at higher pH values were apparently due to volatilization under oxidized conditions and due to volatilization and denitrification under reduced conditions.

The isotopic equilibrium reaction for ammonium exchange is given by:

$$^{14}NH_4(ex) + {}^{15}NH_4(aq) \rightleftharpoons {}^{14}NH_4(aq) + {}^{15}NH_4(ex) \qquad (22)$$

where

$$K(ex) = {}^{14}NH_4(aq)\ {}^{15}NH_4(ex)/{}^{15}NH_4(aq)\ {}^{14}NH_4(ex) = R(ex)/R(aq) = \alpha \qquad (23)$$

Values of K(ex) for exchange on potassium, calcium, and ammonium saturated soil clay colloids were 1.007, 1.011, and 1.003 (Karamanos and Rennie 1978). For exchange on kaolinite, K(ex) = 1.001 (Delwiche and Steyn 1970). Soil clay minerals preferentially adsorb ^{15}N leaving the soluble ammonium slightly depleted in ^{15}N.

NITROGEN-15 STUDIES IN MEDITERRANEAN-TYPE ECOSYSTEMS

^{15}N has been used as a tool in studying nitrogen cycling through mediterranean-type ecosystems (Table 10). Centers of ^{15}N studies include Stock and Lewis at the University of Cape Town in South Africa and Delwiche and Virginia in the U.S.A. (Table 10). Nitrogen fixation has been most widely studied; this may reflect the importance of this process in these periodically fire-disturbed ecosystems. The results of these studies were discussed at appropriate places in the text. In addition, there are two groups with expertise in both mediterranean-type ecosystems and ^{15}N methodology, but which have not used ^{15}N as a tracer in mediterranean-type ecosystems. These are the Montpellier (France) group which has used ^{15}N in agronomic studies (Fernandez and Warembourg 1983; Gandais-Riollet 1984; Bottner, 1985; Cortez *et al.* 1985); and the San Diego (U.S.A.) group which has used ^{15}N in tundra ecosystems (Marion *et al.* 1982; and several papers, in preparation by J. Kummerow, G. M. Marion, and W. C. Oechel).

Table 10. A summary of N-15 studies in mediterranean-type ecosystems.

Process	Location	Reference
Nitrogen fixation	Chile	Silvester *et al.* (1985)
Plant uptake and assimilation	South Africa	Lewis and Stock (1978)
Plant uptake and assimilation	South Africa	Stock and Lewis (1984)
Nitrogen cycling	South Africa	Stock (1985)
Nitrogen fixation	U.S.A.	Delwiche *et al.* (1979)
Nitrogen fixation	U.S.A.	Virginia and Delwiche (1982)

ACKNOWLEDGMENTS

I would like to thank the following individuals who responded to an inquiry concerning ^{15}N work by their groups: Drs. O. Balboa, F. J. Kruger, P. Lossaint, D. T. Mitchell, H. Mooney, J. Roy, P. W. Rundel, W. D. Stock, and R. A. Virginia. I would also like to thank Drs. J. Kummerow, K. Van Cleve, and R. A. Virginia for reviewing an earlier draft of this paper. I would also like to thank Drs. D.H. Kohl and G. Shearer for a critical review of a later version of this paper.

REFERENCES

Adamsen FJ, Reeder JD (1983) A comparison of diffusion and distillation methods for preparing samples for ^{15}N analysis. Soil Sci Soc Am J 47:612-613

Binkley D, Sollins P, McGill WB (1985) Natural abundance of nitrogen-15 as a tool for tracing alder-fixed nitrogen. Soil Sci Soc Am J 49:444-447

Blackmer AM, Bremner JM (1977) Nitrogen isotope discrimination in denitrification of nitrate in soils. Soil Biol Biochem 9: 73-77

Bottner P (1985) Response of microbial biomass to alternate wet and dry conditions in a soil preincubated with ^{14}C and ^{15}N labeled plant material. Soil Biol Biochem17:329-337

Bremner JM (1965) Isotope-ratio analysis of nitrogen in nitrogen-15 tracer investigations. In: Black CA (ed) Methods of Soil Analysis, Part 2. Am Soc Agron, Madison, Wisconsin, pp 1256-1286

Bremner JM, Cheng HH, Edwards AP (1966) Assumptions and errors in nitrogen-15 tracer research. In: The Use of Isotopes in Soil Organic Matter Studies, Pergamon Press, New York, pp 429-442

Broadbent FE (1975) Field measurements of N-utilization efficiency and nitrate movement in soils using N-depleted fertilizer. In: Proceedings of a Symposium on Isotope Ratios as Pollutant Source and Behavior Indicators, IAEA, Vienna, pp 373-382

Broadbent FE, Carlton AB (1978) Field trials with isotopically labeled nitrogen fertilizers. In: Nielsen DR, MacDonald JG (eds) Nitrogen in the Environment, Academic Press, New York, pp 1-42

Bryan BA, Shearer G, Skeeters JL, Kohl DH (1983) Variable expression of the nitrogen isotope effect associated with denitrification of nitrite. J Biol Chem 258:6813-6817

Buresh RJ, Austin ER, Craswell ET (1982) Analytical methods in nitrogen-15 research. Fert Res 3:37-62

Burris RH (1974) Biological nitrogen fixation, 1924-1974. Plant Physio 54:443-449

Chalk PM (1985) Estimation of N_2 fixation by isotope dilution. An appraisal of techniques involving ^{15}N enrichment and their application. Soil Biol Biochem 17:389-410

Chien SH, Shearer G, Kohl DH (1977) The nitrogen isotope effect associated with nitrate and nitrite loss from waterlogged soils. Soil Sci Soc Am J 41:63-69

Cook FD, Wellman RP, Krouse HR (1973) Nitrogen isotope fractionation in the nitrogen cycle. In: Ingerson E (ed) Proc Symp Hydrogeochem Biogeochem, Clark Company, Washington, DC pp 49-64

Cortez J, Billes G, Bottner P (1985) The isotopic composition of the soil microbial biomass and of adenine and guanine isolated from soil incubated with carbon-14 and nitrogen-15-labeled plant material: comparison with fumigation method. Soil Biol Biochem 17:773-778

Delwiche CC, Steyn PL (1970) Nitrogen isotope fractionation in soils and microbial reactions. Environ Sci Technol 4:929-935

Delwiche CC, Zinke PJ, Johnson CM, Virginia RA (1979) Nitrogen isotope distribution as a presumptive indicator of nitrogen fixation. Bot Gaz 140:S65-S69

Edwards, AP (1973) The measurement of the effects of isotopic discrimination where ^{15}N is used as a tracer in a complex system. In: Klein PD, Peterson SV (eds) Proceedings of the First International Conference on Stable Isotopes in Chemistry, Biology, and Medicine, NTIS, Springfield, VA, pp 246-251

Edwards AP (1975) Isotope effects in relation to the interpretation of nitrogen-15/nitrogen-14 ratios in tracer studies. In: Proceedings of a Symposium on Isotope Ratios as Pollutant Source and Behavior Indicators, IAEA, Vienna, pp 455-468

Fernandez MP, Warembourg FR (1983) Diffusion du ^{15}N dans le sol pendant la mesure de fixation biologique de l'azote. Plant Soil 73:431-434

Focht DD (1973) Isotope fractionation of nitrogen-15 and nitrogen-14 in microbiological nitrogen transformations. Theoretical model. J Environ Qual 2:247-252

Freyer HD, Aly AIM (1975) Nitrogen-15 studies on identifying fertilizer excess in environmental systems. In: Proceedings of a Symposium on Isotope Ratios as Pollutant Source and Behavior Indicators, IAEA, Vienna, pp 21-33

Fried M, Broeshart H (1975). An independent measurement of the amount of nitrogen fixed by a legume crop. Plant Soil 43:707-711

Furst P, Jonsson A (1971) Control and modification of methods for determination of ^{15}N in biological material. Acta Chem Scand 25:930-938

Gandais-Riollet N (1984) Effet rhizosphere sur l'activite microbienne vis a vis du carbone et de l'azote dans un sol enrichien debris vegetaux. Doctoral Dissertation, USTL, Montpellier, France

Guiraud G, Buscarlet LA (1975) Comparison of mass spectrometry and emission spectrometry for isotopic analysis of nitrogen. Int J Appl Radiat Isot 26:187-193

Hardy, RWF, Holsten RD (1977) Methods for measurement of dinitrogen fixation. In: Hardy RWF, Gibson AH (eds) A Treatise on Dinitrogen Fixation. Sect IV: Agronomy and Ecology, John Wiley and Sons, New York, pp 451-486

Hauck RD (1982) Nitrogen--Isotope-Ratio Analysis. In: Page AL, Miller RH, Keeney DR (eds) Methods of Soil Analysis. Part 2. (2nd Ed), Am Soc Agron, Madison, Wisconsin, pp 735-779

Hauck RD, Bartholomew WV, Bremner JM, Broadbent FE, Cheng HH, Edwards AP, Keeney DR, Legg JO, Olson SR, Porter LK (1972) Use of variations in natural nitrogen isotope abundance for environmental studies: a questionable approach. Sci 177:453-454

Hauck RD, Bremner JM (1976) Use of tracers for soil and fertilizer nitrogen research. Adv Agron 28:219-266

Hauck RD, Bystrom M (1970) ^{15}N. A Selected Bibliography for Agricultural Scientists. The Iowa State University Press, Ames, Iowa

Huser R (1968) Experiences with nitrogen-15 tracer techniques in estimating the microbial fixation of elementary nitrogen in the organic matter of forest soils. In: The Use of Isotopes in Soil Organic Matter Studies, Pergamon Press, Oxford, pp 457-469

Ingerson E (1953) Nonradiogenic isotopes in geology: a review. Bull Geol Soc Am 64:301-374

Ivanko S (1971) Metabolic pathways of nitrogen assimilation in plant tissue when ^{15}N is used as a tracer. In: Nitrogen-15 in Soil-Plant Studies, IAEA, Vienna, pp 119-156

Jansson SL (1968) Experimental techniques with nitrogen-15. In: The Use of Isotopes in Soil Organic Matter Studies. Pergamon Press, Oxford, pp 415-422

Karamanos RE, Rennie DA (1978) Nitrogen isotope fractionation during ammonium exchange reactions with soil clay. Can J Soil Sci 58:53-60

Karamanos RE, Rennie DA (1981) Changes and significance in natural nitrogen-15 abundance in residual nitrogen fertilizer studies. Can J Soil Sci 61:553-559

Kirkham D, Bartholomew WV (1954) Equations for following nutrient transformations in soil, utilizing tracer data. Soil Sci Soc Am Proc 18:33-34

Knowles R (1978) Common intermediates of nitrification and denitrification, and the metabolism of nitrous oxide. In: Schlessinger D (ed) Microbiology-1978. Am Soc Microbio, Washington, DC, pp 367-371

Kohl, DH, Shearer G (1980) Isotopic fractionation associated with symbiotic nitrogen fixation and uptake of nitrate ion by plants Plant Physiol 66:51-56

Kohl DH, Shearer G (1981) The use of soils lightly enriched in nitrogen-15 to screen for nitrogen-fixing activity. Plant Soil 60:487-489

Kohl DH, Shearer GB, Commoner B (1971) Fertilizer nitrogen: contributions to nitrate in surface water in a corn belt watershed. Sci 174:1331-1334

Kowalenko CG, Ross GJ (1980) Studies on the dynamics of recently clay-fixed ammonium ion using nitrogen-15. Can J Soil Sci 60:61-70

Ladd JN (1981) The use of nitrogen-15 in following organic matter turnover, with specific reference to rotation systems. Plant Soil 58:401-411

Legg JO, Sloger C (1976) A tracer method for determining symbiotic nitrogen fixation in field studies. In: Klein ER, Klein PD (eds) Proceedings of the Second International Conference on Stable Isotopes. NTIS, Springfield, VA, pp 661-666

Letolle R, Mariotti A (1974) Utilisation des variations naturelles d'abondance de l'azote-15 comme traceur en hygrogeologie. Premiers resultats. In: Proceedings of a Symposium on Isotope Techniques and Groundwater Hydrology, Unipub, New York, pp 209-220

Lewis OAM, Stock WD (1978) A preliminary study of the nitrogen nutritional status of members of the South Africa *Proteaceae*. J South Afr Bot 44:143-151

Marion GM, Miller PC, Kummerow J, Oechel WC (1982) Competition for nitrogen in a tussock tundra ecosystem. Plant Soil 66:317-327

Mariotti A (1983) Atmospheric nitrogen is a reliable standard for natural ^{15}N abundance measurements. Nature 303:685-687

Mariotti A, Germon JC, Hubert P, Kaiser P, Letolle R, Tardieux A, Tardieux P (1981) Experimental determination of nitrogen kinetic isotope fractionation: some principles; illustration for the denitrification and nitrification processes. Plant Soil 62:413-430

Mariotti A, Germon JC, Leclerc A (1982) Nitrogen isotope fractionation associated with the nitrite ion -----> nitrogen oxide (N_2O) step of denitrification in soils. Can J Soil Sci 62:227-241

Mariotti AJ, Mariotti F, Amarger N (1983) Use of natural nitrogen-15 abundance in the measurement of symbiotic fixation. In: Nuclear Technology for the Improvement of Pasture Management, IAEA, Vienna, pp 61-77

McInteer BB, Montoya JG (1980). Automation of a mass spectrometer for nitrogen isotope analysis. US Dept Energy Rep No LA-UR-80-245, Los Alamos, New Mexico

Meints VW, Shearer G, Kohl DH, Kurtz LT (1975) Comparison on unenriched versus nitrogen-15 enriched fertilizer as a tracer for nitrogen fertilizer uptake. Soil Sci 119:421-425

Moore AW, Craswell ET (1976) Non-uniformity of nitrogen-15 labeling in plant material. Comm Soil Sci Plant Anal 7:335-344

Muhammad S, Kumazawa K (1976). Use of emission spectrometry to trace ^{15}N-labeled ammonium and nitrate nitrogen in amino acids of rice panicle. In: Klein ER, Klein PD (eds) Proceedings of the Second International Conference on Stable Isotopes, NTIS, Springfield, VA, pp 674-682

Newman ACD, Oliver S (1966) Isotopic exchange of fixed ammonium. J Soil Sci 17:159-174

O'Deen WA, Porter LK (1979) Digestion tube diffusion and collection of ammonia for nitrogen-15 and total nitrogen determination. Anal Chem 51:586-589

Olson RA (1979) Isotope studies on soil and fertilizer nitrogen. In: Isotopes and Radiation in Research on Soil-Plant Relationships, IAEA, Vienna, pp 3-32

Proksch G (1972) Application of mass- and emission-spectrometry for $^{14}N/^{15}N$ ratio determination in biological material. In: Isotopes and Radiation in Soil-Plant Relationships Including Forestry, IAEA, Vienna, pp 217-225

Reddy CN, Patrick WH Jr (1979) Distribution of added labeled $(^{15}NH_4)_2SO_4$ in a flooded soil as influenced by redox potential and pH. In: Isotopes and Radiation in Research on Soil-Plant Relationships, IAEA, Vienna, pp 607-617

Reeder JD, O'Deen WA, Porter LK, and Lober RW (1980) A comparison of cross-contamination in distillation units used in total nitrogen and nitrogen-15 analyses. Soil Soc Am J 44:1262-1267

Rennie, DA (1980) Note on variations in nitrogen-15/nitrogen-14 ratios in soils and plants. In: Soil Nitrogen Fert. Pollut., Proc Rep Res Coord Meet, IAEA, Vienna, pp 237-239

Rennie DA, Paul EA, John SLE (1976) Natural nitrogen-15 abundance of soil and plant samples. Can J Soil Sci 56:43-50

Rennie RJ (1982) Quantifying dinitrogen (N_2) fixation in soybeans by ^{15}N isotope dilution: the question of the nonfixing control plant. Can J Bot 60:856-861

Rescigno A, Segre G (1966) Drug and Tracer Kinetics, Blaisdell Publ Co, Waltham, Massachusetts

Rhodes D, Myers AC, Jamieson G (1981) Gas chromatography mass spectrometry of N hepta fluorobutyryl iso butyl esters of amino-acids in the analysis of the kinetics of nitrogen-15 labeled ammonium assimilation in *Lemna-minor*. Plant Physiol 68:1197-1205

Ruschel AP, Vose PB, Victoria RL, Salati E (1979) Comparison of isotope techniques and non-nodulating isolines to study the effect of ammonium fertilization on dinitrogen fixation in soybean, *Glycine max*. Plant Soil 53:513-525

Ryden JC, Lund LJ, Letey J, Focht DD (1979) Direct measurement of denitrification loss from soils. Part 2 Development and application of field methods. Soil Sci Soc Am J 43:110-118

Schimel JP, Firestone MK, Killham KS (1984) Identification of heterotrophic nitrification in a Sierran forest soil. Appl Environ Microbiol 48:802-806

Shearer G, Duffy J, Kohl DH, Commoner B (1974) A steady-state model of isotopic fractionation accompanying nitrogen transformations in soil. Soil Sci Soc Am Proc 38:315-322

Shearer G, Kohl DH, Commoner B (1975) Use of variations in the natural abundance of ^{15}N to study sources, transformations and movement of nitrogen in a plant-soil-water system. In: Origin Fate Chem Residues Food Agric Fish, Proc Rep 2 Res Coord Meet, IAEA, Vienna, pp 77-104

Shearer G, Kohl DH, Virginia RA, Bryan BA, Skeeters JL, Nilsen ET, Sharifi MR, Rundel PW (1983) Estimates of N_2 fixation from variation in the natural abundance of ^{15}N in Sonoran Desert ecosystems. Oecologia (Berl) 56:365-373

Shipley RA, Clark RE (1972) Tracer Methods for *In Vivo* Kinetics, Academic Press, New York

Silvester WB, Balboa O, Martinez JA (1985) Nodulation and nitrogen fixation in members of the Rhamnaceae (*Colletia*, *Retanilla*, *Talguenea*, and *Trevoa*) growing in the Chilean matorral. Symbiosis 1:29-35

Stanford G, Legg JO, Staley TE (1976) Fate of ^{15}N labeled nitrate in soils under anaerobic conditions. In: Klein ER, Klein PD (eds) Proceedings of the Second International Conference on Stable Isotopes, NTIS, Springfield, VA, pp 667-673

Stock WD (1985) An investigation of nitrogen cycling processes in a coastal fynbos ecosystem in the South Western Cape Province, South Africa. PhD Dissertation, University of Cape Town, South Africa

Stock WD, Lewis OAM (1984) Uptake and assimilation of nitrate and ammonium by an evergreen fynbos shrub species *Protea repens Proteaceae*. New Phytol 97:261-268

Stribley DP, Read DJ (1974) The biology of mycorrhiza in the *Ericaceae*: IV. The effect of mycorrhizal infection on uptake of ^{15}N from labeled soil by *Vaccinium macrocarbon* Ait. New Phytol 73:1149-1155

Tiedje JM (1978) Denitrification in soil. In: Schlessinger D (ed) Microbiology-1978, Am Soc Microbiology, Washington DC, pp 362-366

Tiedje JM, Sorensen J, Change Y-YL (1981) Assimilatory and dissimilatory nitrate reduction perspectives and methodology for simultaneous measurement of several nitrogen cycle processes. In: Clark FE, Rosswall T (eds) Terrestrial Nitrogen Cycles: Processes, Ecosystem Strategies and Management Impacts, Ecol Bull (Stockholm) No 33

Turner GL, Bergersen FJ, and Tantala H (1983) Natural enrichment of ^{15}N during decomposition of plant material in soil. Soil Biol. Biochem 15:495-497

Urey HC (1947) Thermodynamic properties of isotope substances. J Chem Soc 562-581

Van Cleve K, White R (1980) Forest-floor nitrogen dynamics in a 60-year old paper birch ecosystem in interior Alaska. Plant Soil 54:359-381

Virginia, RA, Delwiche CC (1982) Natural ^{15}N abundance of presumed N_2-fixing and non-N_2-fixing plants from selected ecosystems. Oecologia (Berl) 54:317-325

Vose PD, Ruschel AP, Victoria RL, Saito SM, Matsui E (1982) Nitrogen-15 as a tool in biological nitrogen fixation research. In: Graham PH, Harris SC (eds) Biological Nitrogen Fixation in Tropical Agriculture, Cent Int Agric Trop, Cali, Columbia, p 575

Wada E, Kadonaga T, Matsuo S (1975) Nitrogen-15 abundance in nitrogen of naturally occurring substances and global assessment of denitrification from the isotopic viewpoint. Geochem J 9:139-148

Wada E, Shibata R, Torii T (1981) Nitrogen-15 abundance in Antarctica: origin of soil nitrogen and ecological implications. Nature 292:327-329.

Walker RL, Walton JR, Carter JA, Matthews DR (1975) Measurement system for total nitrogen and $^{15}N/^{14}N$ in plant tissue, soil, and water. In: Proceedings of a Symposium on Isotope Ratios as Pollutant Source and Behavioral Indicators, IAEA, Vienna, pp 429-438

Witty JF, Ritz K (1984) Slow-release nitrogen-15 fertilizer formulations to measure nitrogen fixation by isotope dilution. Soil Biol Biochem 16:657-661

Application of computer methods in the field to assess ecosystem function and response to stress

A. Cernusca

Institute of Botany, University of Innsbruck, Sternwartestr. 15, A-6020 Innsbruck, Austria

INTRODUCTION

Computer methods can be successfully applied in ecosystem research to measure data on functional processes in ecosystems and to analyze and synthesize results from large data sets through ecological modelling. In the study of ecosystems, the simultaneous measurement of more than 50 different factors is often necessary, whereby the individual measured data must be acquired and processed with a high degree of accuracy and frequency (Cernusca 1982a, 1985). Modern computerized measuring systems may function with over 250 sensors with no difficulty at all. Through the selection of a suitable AD-converter, both the measuring speed (10 measurements per second with microclimate measurements; up to 10,000 measurements per second with fluorescence investigations) and the digital resolution of the signal (8 bit = 256 to 4 1/2 digit which means a range of measurement of $\pm$20.000) can be adapted for each respective measurement task. A further advantage in using computers is that sensors with a non-linear calibration curve may be used. A linearization of the sensor response may be effected with a simple mathematical operation executed during data acquisition. Computer-based measuring systems can be quickly changed to accomplish very different measurement tasks by changing the software program, i.e. by changing the data acquisition and processing program.

With these measuring systems, complicated mathematical models of the processes under investigation may be utilized in conjunction with the acquired data to permit computation of derived variables and to automatically check the observations for errors. I am thinking here, for example, of the computation of CO_2-concentration in the interior of a leaf with the aid of mathematical models of diffusion processes or the computation of such complex variables as the Bowen-ratio (see Bowen 1926) or water use efficiency. With computerized data acquisition systems, these so-called complex factors are available in real time, and can be viewed on the display and stored on cassette for later detailed examination. Special test programs indicate possible measurement errors to the experimenter, and control programs allow, for example, the temperature and humidity in a photosynthesis measurement cuvette to be kept at values the same as sensed under naturally fluctuating external conditions. What is of decisive importance for planning and surveillance of experiments and in the long run for the successful completion of integrated research projects, however, is that with computerized

NATO ASI Series, Vol. G15
Plant Response to Stress. Edited by J. D. Tenhunen et al.

measuring systems the extensive data can be analyzed in detail and clearly displayed at the respective field site within hours.

Today, portable battery-run or solar powered data acquisition apparatus are being used increasingly in ecosystem research. Powerful micro-computers and recently even pocket-computers take over the automatic acquisition and processing of data (Cernusca 1982b, 1984, 1985). These compact portable units permit measurements even in remote areas and under extreme environmental conditions. The prerequisite for development of these apparatus has been the new CMOS-technology, which allows current-saving operation. In the following I would initially like to introduce one such pocket-computer measuring system called "MIKROMET" which I have developed, and then illustrate how it has been applied in ecosystem research and studies of stress ecology.

THE CONSTRUCTION OF MIKROMET

The central control unit of MIKROMET is encased in a splash-proof aluminum housing (Fig. 1). The housing contains an AD-converter, a pocket-computer (Sharp PC-1500) with a special interface for connection to the AD-converter, a 4-color printer/plotter, a cassette recorder, an automatic spooling device for the print-out chart and a storage battery. Up to 4 external multiplexers, each with 32 channels, can be connected via a control cable (up to 500 m long). Each multiplexer houses chopper stablized operational amplifiers and voltage regulators as well as various program-controlled measuring bridges for currents or voltages (range of ±20.000 mV), resistance thermometers and thermocouples (range of ±200.000 °C). Depending on the values recorded, external instruments can be controlled via digital I/O-ports. A serial interface permits data transfer to personal computers and mainframe computers. Programming in BASIC enables quick adaptation of measurement systems to specific tasks.

PROGRAMMING, OPERATION AND SURVEILLANCE

Modern data acquisition systems often work in an interactive mode, i.e. the user is aided by a menu of possible commands while operating the system and is made aware of input errors and errors during the course of measurement. Data acquisition programs for a number of ecological investigations have been developed for the MIKROMET. Using them requires no knowledge of programming whatsoever. The user is led through the program by means of questions appearing in the display of the computer. Consequently, the possibility of an operating error is excluded to a great extent. The full potential of such a computer-controlled data acquisition system can only be exploited, however, when the experimenter is able to adapt the program to a concrete task. Each data acquisition program consists of a BASIC-program which controls the measurement procedure and subsequent evaluation of measured data, and several

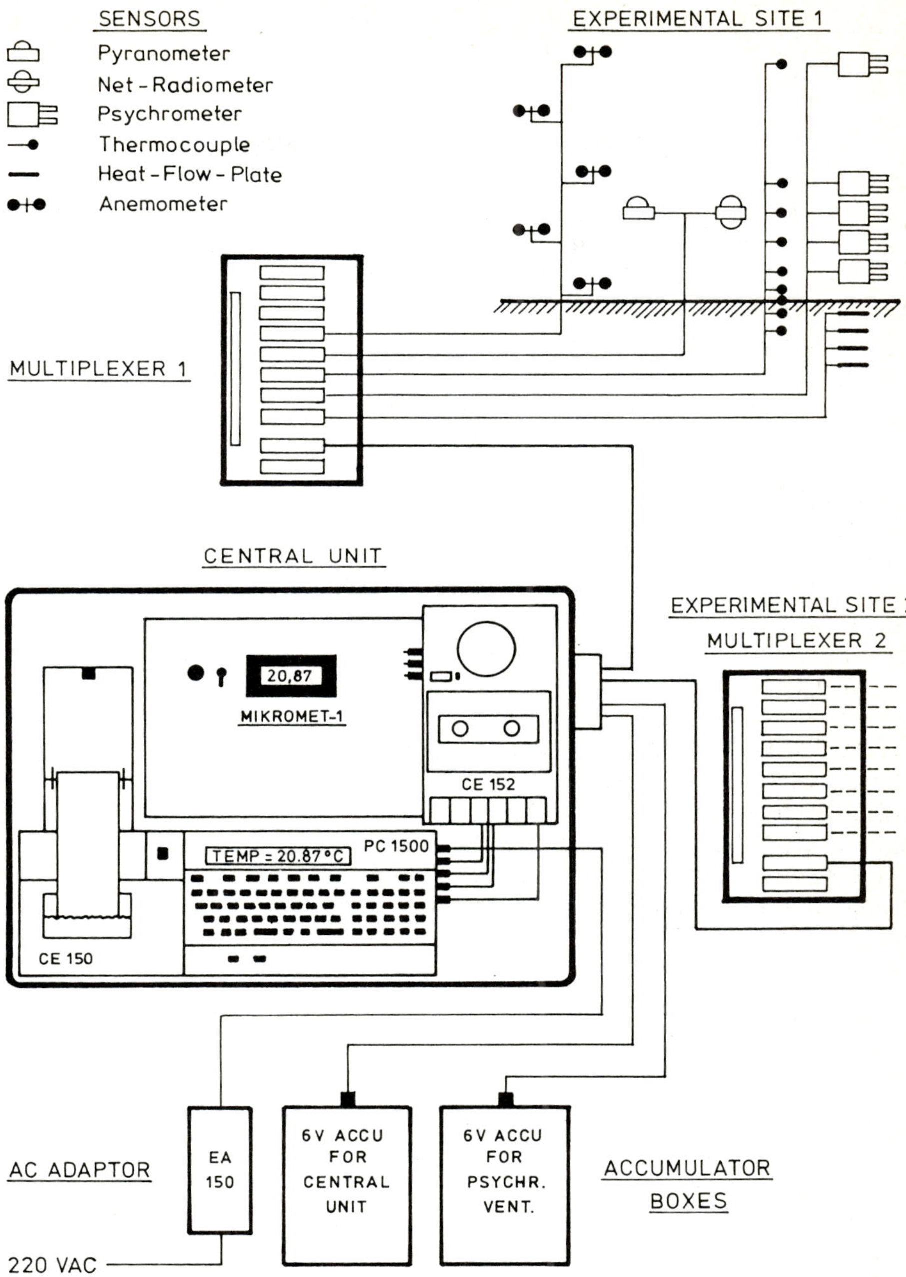

Fig. 1. Diagram of instrumentation of the MIKROMET for analyzing CO_2-exchange and energy regime at two experimental sites by means of the Bowen ratio method

machine programs which allow rapid control of the multiplexers and the transmission of data from the AD-converter to the computer. The following four machine program CALL-commands have been provided for the MIKROMET:

PORTON: This BASIC-command switches on the MIKROMET, i.e. the power supply for the peripheral modules (the AD-converter, measuring amplifier, channel selector, etc.) is activated.

PORTOFF: This command switches off the MIKROMET after completion of measurements.

CALL & C5, V: This command transmits the content of the variable V via the output-port as a control command. If, for example, the value 5 is allocated to variable V in the main program and this command is subsequently given, channel 5 is activated on the multiplexer.

CALL & D1, V: This command effects reading of the digits of the 4 1/2-digit AD-converter serially into the computer. The read-in of the measuring value is repeated according to the pre-set value of V. In this manner it is very easy to calculate a mean value, even during data acquisition.

A simple BASIC-program for automatic measuring of channels 0 - 15 of a multiplexer is illustrated in Fig. 2. At each sensor the AD-converter is read 5

```
 10: PORTON
 20: READINGS = 5
 30: FOR CHANNEL = 1 TO 16
 40: V = CHANNEL
 50: CALL &C5,V
 60: WAIT 0
 70: PRINT "channel;" V;
 80: V = READINGS
 90: CALL &D1, V
100: VALUE = B - A
110: PRINT VALUE
120: IF V=32 PRINT "ERROR"
130: M(CHANNEL) = VALUE
140: NEXT CHANNEL
150: PORTOFF
160: STOP
```

Fig. 2. BASIC-program for automatic measuring of channel 1 to 16 of a multiplexer times consecutively. The sensor and the value pertaining to it (the mean value of the 5 individual readings) are shown on the display. It is important for data acquisition that zero-point changes are automatically taken into account. In the same way amplitude checks for individual sensors can automatically be carried out and taken into account.

APPLICATIONS

The MIKROMET has already proven itself over long periods of field operation in difficult terrain (high-alpine, arctic) as the data acquisition and control unit for a computerized psychrometer, a null-balance porometer, and a portable fluorometer (Larcher and Cernusca 1985). It has been used for quantification of freezing processes in leaves, in analyses of CO_2-exchange with a fully climatized microcuvette, and for micrometeorological measurements designed to analyze of the energy balance, water regime and CO_2-exchange in grassland ecosystems (Cernusca *et al.* 1986). The parameters required for the description of matter and energy flows within ecosystems can be registered, analyzed and graphically represented in real time. Simultaneously, it is possible to recognize extraordinary events and to analyze these special cases causally.

Analysis of Ecosystem Energy Flows in the Central Caucasus by Means of MIKROMET

Comparative investigations of microclimate, CO_2-exchange and energy regime of a pasture grazed by sheep and an adjacent meadow (not grazed but mowed once a year) were carried out in the Central Caucasus using MIKROMET. Evapotranspiration, sensible heat convection and CO_2-flux between the atmosphere and the vegetated surface were calculated by means of the Bowen-ratio method, using gradients of temperature, water vapor and CO_2-concentration measured between the canopy and a height 120 cm above it. Soil heat flux was measured using heat flux plates and CO_2-release from the soil by means of the cuvette-method (open system). The streams of sample gas were controlled by MIKROMET. All results shown in Fig. 3 were plotted in real time.

Concerning plant community structure, the meadow differs from the pasture in a number of ways. In the pasture the height of the canopy decreases from 60 to 2 cm, phytomass and leaf area index (LAI) are only half that of the meadow. In the pasture, leaf inclination is very plagiophile in contrast to a predominantly erectophile leaf inclination in the meadow. Fifty percent of the phytomass in the pasture is concentrated in the lowermost 10 mm of the canopy. Therefore, it is inaccessible to the sheep. Concerning daily means of the energy budget, there are only slight differences between both communities on bright midsummer days. In the pasture the reduction of phytomass and LAI are largely compensated by an increased extinction of radiation and by increased values in leaf conductance. In

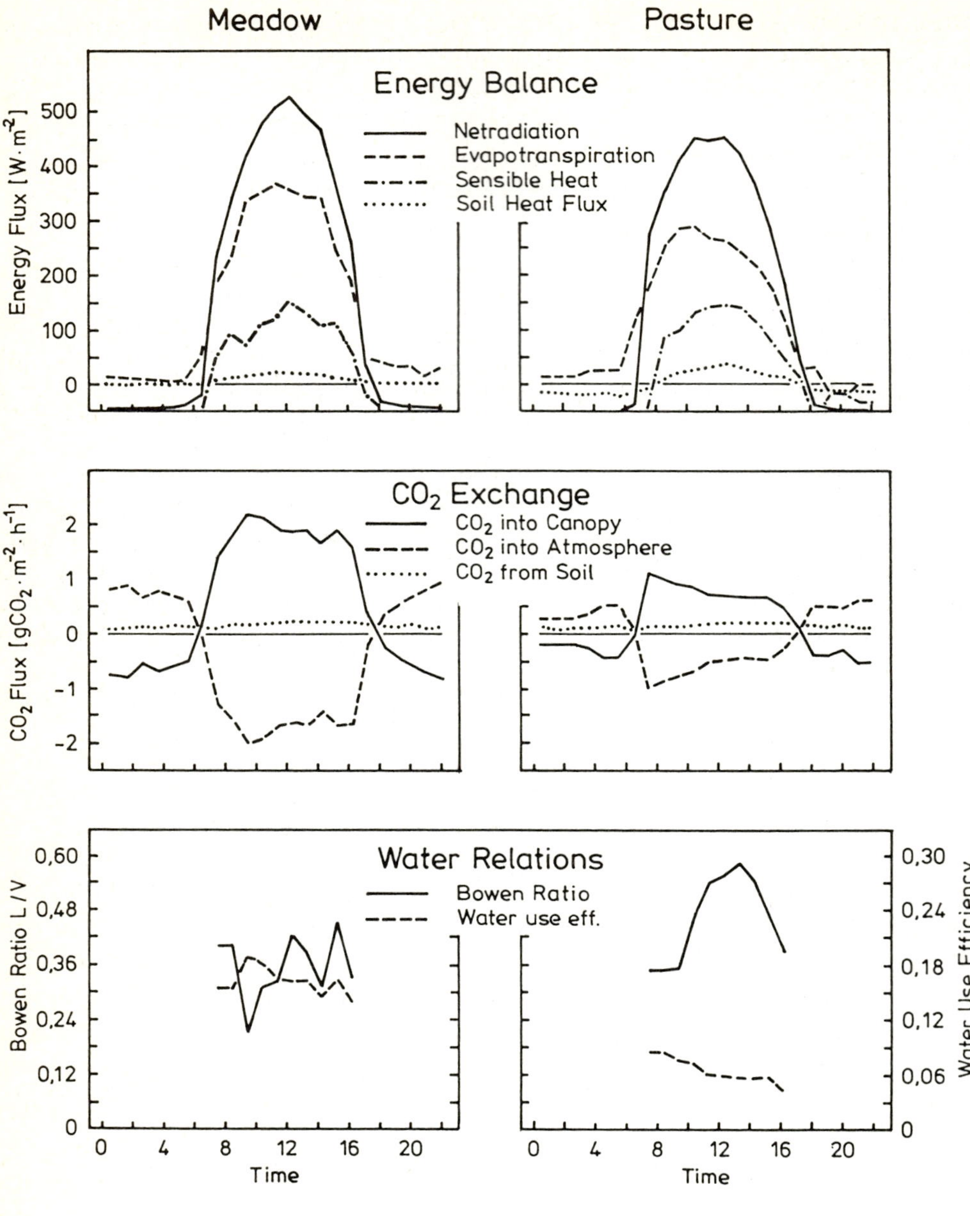

Fig. 3. Results obtained in real time from comparative investigations of CO_2-exchange and energy regime of a pasture grazed by sheep and an adjacent meadow (not grazed but mowed once a year) in the Central Caucasus using the MIKROMET

contrast to this, the reduced CO_2-exchange reflects a reduction of aboveground plant parts in the pasture. In the early afternoon, the increase of the Bowen-ratio indicates plant water stress due to limiting soil water conditions. The decrease in water use effiency indicates an additional limitation in photosynthesis by high temperatures. These results for the whole canopy agree well with results of single leaf experiments of CO_2-exchange (Cernusca *et al.* 1986, in preparation).

Computerfluorometry by Means of MIKROMET

Fluorescence measurements were a part of a project entitled "Ecological research on afroalpine flowering plants" (Prof. Beck, Univ. Bayreuth) on Mt. Kenya. Using the computerfluorometer, induction curves were registered automatically each hour throughout the night to study the effects of frost on various leaves of "giant rosette" plants. From the original recordings, one obtains the induction curves plotted on three different time scales (0 to 100 ms, 100 ms to 20 s and 0 to 240 s) along with calculated values of the maximum level P (peak of the fluorescence induction curve), the minimum level and the respective time taken to reach the different levels. Figure 4 shows typical reversible changes in fluorescence which result due to freezing, i.e. a higher P-level and a steeper rise to P, and increased inhibition of quenching from P to the terminal level (Bodner and Beck unpublished).

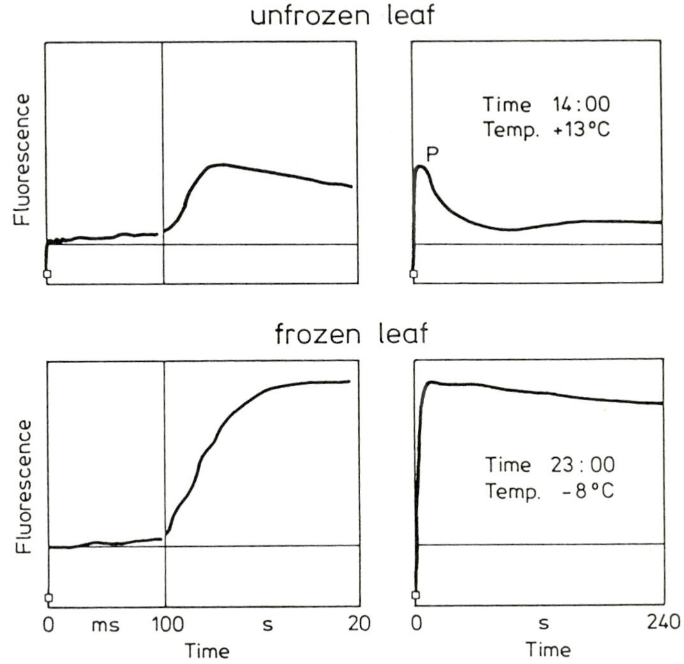

Fig. 4. Fluorescence induction curves on *Dendrosenecio brassica* automatically registered and analyzed using the computerfluorometer. The fluorescence shows typical changes due to freezing of the leaf

SUMMARY

In conclusion, I should like to draw attention to the fact that numerous ecological questions can be dealt with only through the application of automatic, computer-controlled measuring techniques such as those permitted by the MIKROMET system. This is especially the case when rapid data acquisition is required and when large amounts of data are to be collected (cf. Cernusca 1985, 1986).

REFERENCES

Bowen, IS (1926) The ratio of heat losses by conduction and evaporation from any water surface. Physiol Rev. 27:779-787

Cernusca A (1982a) Standardmeteorologie und Mikrometeorologie. In: Janetschek H (ed) Ökologische Feldmethoden. Verlag Ulmer, Stuttgart, pp 9-27

Cernusca A (1982b) Eingabe-/Ausgabe-Interface für Sharp Taschencomputer PC-1211. CHIP 6:148-150

Cernusca A (1984) MIKROMET - Automatische Datenerfassung, Auswertung und Steuerung über Mikrocomputer. Int Franz Sauberer Gedächtnissymposium, Wien, pp 69-71

Cernusca A (1985) Der Einsatz von Taschencomputern für programmgesteuerte Datenerfassung, Auswertung und Versuchsüberwachung. Verh Ges Ökologie (Bremen) 13:369-377

Cernusca A, Nachuzrisvili G, Abdaladse O, Gamkrelidse T, Sanadiradse G (1986) Messung des CO_2-Gaswechsels einer Mähwiese und einer Schafweide im Zentralkaukasus. Sitzungsber Österr Akademie Wiss, Math-nat Kl, in press

Larcher W, Cernusca A (1985) Mikrocomputergesteuerte mobile Anlage zum fluorometrischen Nachweis von Photosynthesestörungen. Sitzungsber Österr Akademie Wiss, Math-nat Kl, Abt I, 194:45-64

Studying long term community dynamics using image processing

D. Short and D. Short*

College of Sciences, San Diego State University, San Diego, CA 92182

INTRODUCTION

In most cases the longitudinal study of vegetation communities cannot be undertaken because of the extended time course needed for such an endeavor. However, in many parts of the world there are archives of aerial photographs extending 50 years into the past. With the appropriate image processing software and equipment, this historical sequence of photographs can be used to study community dynamics. This technique has recently been applied to the dynamics of oak woodland (*Quercus agrifolia*) in Southern California. This paper provides a description of the image processing system used in that study.

To recognize the potential of an image processing system, the formation of images must be understood. An excellent review of the theory of photographic imaging is provided by Heller (1970). Additional information can be found in Aldrich (1979) and in the American Society of Photogrammetry Manual (1975). When a scene is illuminated, the illuminating spectrum is modified in a characteristic manner depending on the physical properties of the objects in the scene forming the reflectance spectrum. In producing an image, the reflectance spectrum is sampled at each image point. For example, in black and white photography, the spectral response of the film at each wavelength is multiplied by the reflectance spectrum of the scene with the resulting curve being integrated and recorded as a grey level value on the film. In the case of color or color infrared film, the film consists of three separate layers each having a spectral response sensitive to a different portion of the spectrum. The use of filters can emphasize different portions of the spectral response curve of the film by attenuating certain portions.

Because black and white film records only a single grey value at each image point, it is impossible to adjust for differing levels of illumination. Consequently, it is very sensitive to changes in illumination caused by terrain, sun angle, and sky conditions. Feature detection is possible only under special circumstances. It is possible that several images of a single scene could be taken through different filters using black and white film. In the case of three images this would be equivalent to a single color image. By digitizing the color image through three different filters, three integrated samples of the reflectance spectrum at each image point in the picture can be obtained. By computing ratios of the three

*Department of Mathematics, Southwestern College, Chula Vista, CA 92010

NATO ASI Series, Vol. G15
Plant Response to Stress. Edited by J. D. Tenhunen et al.

different grey levels for each pixel, one obtains a measure of the color at that pixel. This method is not as sensitive to either terrain, sun angle, or sky conditions since the ratios are independent of total illumination. In this case, the dependence on these conditions occurs only when the shape of the illuminating spectrum changes and not just its magnitude. For vegetation identification and analysis, color infrared film is preferred since it is sensitive to longer wavelengths than standard color film. However, neither choice is ideal (Chiou 1984).

To take full advantage of the reflectance spectral information, an imaging spectrometer should be used. The current generation of imaging spectrometers normally covers the spectral range from 0.4 to 2.4 microns with a typical resolution of 0.01 to 0.02 microns. At each space coordinate, approximately 50 samples are taken of the reflectance spectrum over the range of the instrument. This allows a very accurate reconstruction of the reflectance spectrum curve at each point in the scene. For multispectral detection of features, imaging spectrometers provide the best data. However, most of the archived images are in black and white with a few recent series taken in color infrared and none with an imaging spectrometer. Thus, the identification of features is a major difficulty in working with the historical record of aerial photographs, since the data is dependent on the illumination of the scene.

DESCRIPTION OF THE IMAGE PROCESSING SYSTEM

The system accomplishes four major tasks; image digitization, scale correction, feature identification and data display. For photographs, a video camera digitization system is preferable. Such a system must allow the use of filters so that either a color or a color infrared photo may be split into its three component colors. Equipment with this capability is readily available.

The scale correction component is required to register all the photographs to a standard scale. Such software requires intensive computations and is frequently not available on smaller systems. This being the case, a detailed description of the algorithm is given. All such programs use various interpolation schemes to transform a given picture into another picture based on the correspondence between a given set of registration points. Examples of these points could be marked property corners, road or stream bed intersections, or other physical features which can be correlated between the given image and the scale corrected image. For scale correction, the minimum number of these registration points is three. In this case, a linear transformation between the two images can be constructed. With more registration points either higher order transformations or piecewise linear transformation are used.

Both of these can produce artifacts which for certain applications may prove undesirable. Thus, unless there is a specific reason intrinsic to the problem at hand such as the geometry associated with the camera system, either linear or piecewise linear should be used. The theoretical basis for defining a linear transformation given three pairs of registration points is based on barycentric coordinates. In this coordinate system each point is assigned a triple of numbers which represents the masses required at the registration points p, q and r in order for the given point X to be the center of mass. The transformation then assigns these masses to the corresponding registration points p', q' and r' in the new image and then calculates the resulting center of mass X'. This new center of mass X' is the image of the original point X under the transformation.

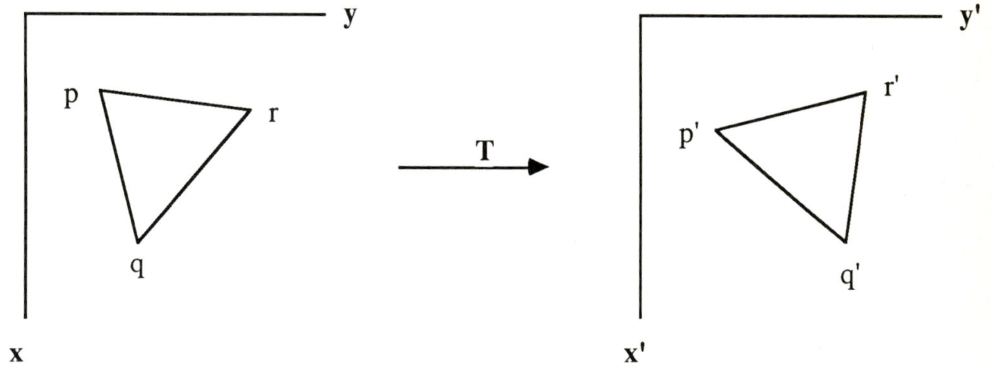

Fig. 1. Illustration of a transformation within the barycentric co-ordinate system

The formula describing the transformation T in terms of the coordinates of the registration points p,q,r and p',q',r' is given by:

$$\begin{bmatrix} x' \\ y' \\ 1 \end{bmatrix} = \begin{bmatrix} x'_{p'} & x'_{q'} & x'_{r'} \\ y'_{p'} & y'_{q'} & y'_{r'} \\ 1 & 1 & 1 \end{bmatrix} \begin{bmatrix} x_p & x_q & x_r \\ y_p & y_q & y_r \\ 1 & 1 & 1 \end{bmatrix}^{-1} \begin{bmatrix} x \\ y \\ 1 \end{bmatrix}$$

When such a linear transformation is defined and then implemented on a computer system, the question becomes: "What grey level should be assigned to each pixel in the new corrected picture?" Thus the pre-image of each point of the new picture must be identified, since the grey level of this point P is to be assigned to the new point P'. In most cases, this pre-image P does not coincide with a given pixel in the old picture, but lies between pixels.

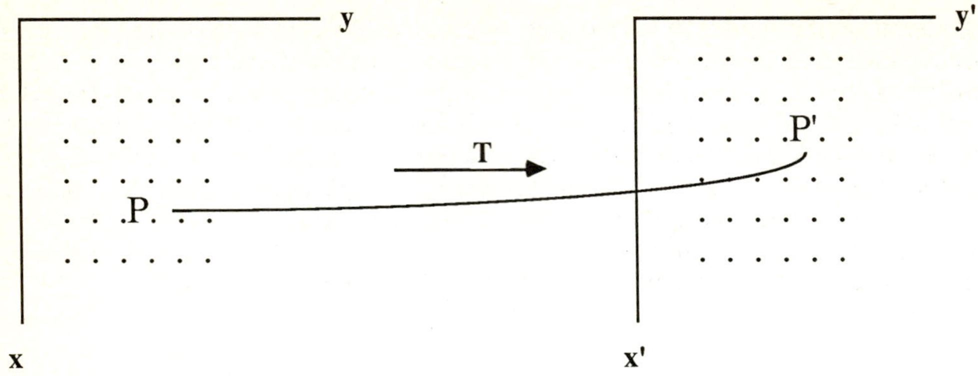

Fig. 2. A typical pixel pre-image

To determine the appropriate grey level, several methods are employed; 1) chopping which drops the fractional part of the pre-image's coordinates, 2) rounding which rounds the pre-image's coordinates, or 3) a weighted average of the adjacent grey values depending on the fractional portions in each coordinate. A danger in using the weighted average is that the resulting grey value is calculated and may not portray reality. In the case of both chopping and rounding an existing grey value is selected. In every case, the new image suffers a loss of resolution depending on the geometry of the transformation.

The third component of feature identification is very dependent on the scale and quality of the photographs to be used in the analysis. Both scale and quality can be directly related to image resolution. For example, a typical scale for an aerial photograph is 1:25000; this implies that 1.6 cm on the photo is equivalent to 400 meters on the ground. If the frame size for the image processing system is 400 pixels by 400 pixels, digitizing an area of 1.6 cm on a side will result in each pixel representing one square meter on the ground. Since the digitization followed by a scale correction will introduce a possible error of at least one pixel width, this implies a lower limit on the size of features to be isolated.

Suppose that a feature of reasonable size has been chosen, there are several methods for identifying this feature. In the case of color or color infrared, multi-spectral analysis can be used. This consists of forming three pictures of the scene representing three independent colors say red, green and blue. A region in the three dimensional space defined by the red, green and blue grey levels is associated with the feature to be detected. Usually this is done by designating a known portion of the feature to be identified. The system then uses the region defined by this sample as the detection criterion. Then all pixels satisfying the criteria of their red, green and blue grey levels being in this designated region will be identified with the feature.

If the feature has a unique color signature this method works very well. In many cases, vegetation identification by this method is complicated by shading and similar coloring of other types. In the case of black and white photographs, this multi-spectral method cannot be used since there is only one grey level associated with each pixel. The usual method is to associate a grey level interval with the desired feature. This is difficult because in most aerial photographs the illumination changes across the scene due to the terrain. Thus, an appropriate grey level in one area of the frame is not appropriate in another area. This effect is not as pronounced in the case of color or color infrared since the ratio of red to blue, blue to green, or green to red is not altered by intensity of the illumination.

To make the technique of thresholding of grey levels usable, the system must allow for masking a portion of a frame. Using masking software those areas of similar terrain aspect and hence illumination could be selected by the operator with thresholding of grey level values restricted to this designated region. Again, vegetation identification is complicated by shading and similar grey level values for other vegetation types. Another approach is also possible. If the desired feature is a boundary between two regions, the gradient of the grey level values can be calculated at each pixel forming a new picture which represents the change in grey level across the scene. A very high grey level in the gradient picture corresponds to a rapid change in grey levels in the original picture, while a low grey level corresponds to little change. A boundary between two regions in the original is denoted by a high grey level in the gradient picture. The thresholding technique with or without masking can be applied to the gradient picture to identify the boundary. For image processing purposes, the gradient of the grey level values denoted by g can be calculated at the (i,j) pixel by the simple approximation:

$$\mathrm{Grad}(g(i,j)) = |g(i,j) - g(i+1,j)| + |g(i,j) - g(i,j+1)|.$$

This formula represents a large savings in computer time over the usual definition involving the square root of the sum of squares. It can be implemented by subtracting the original picture grey level values from those shifted by one pixel. The final method employs a cursor controlled paint brush. With this method, the operator can designate pixels manually. In most cases, such a program is necessary to clean up the exceptional cases which the automatic techniques overlook.

The final system component is the data display software. A useful system must contain several utility programs for overlaying selected features on digitized photographs and changing the color of these overlays. For quantitative analysis, a program to obtain pixel counts by a specified color or grey level is needed. Since the photographs are all registered to a standard scale, a pixel count is equivalent to calculating the area of a selected feature. In addition, the capability to overlay a grid of a specified size is very useful. Typically, the output of the system would be produced on a plotter or on colored film of various formats.

The study of community dynamics requires the overlay of the selected community from different times with the differences highlighted.

COMMUNITY DYNAMICS OF *QUERCUS AGRIFOLIA* WOODLAND

In the application of image processing techniques to the study of community dynamics of *Quercus agrifolia* woodland, three sets of aerial photographs of the Comb's Camp area of the Sky Oaks Field Station were used. The project focused on the western 40 acres of this site within section 19. The first set was taken on July 1984 using color infrared film at a scale of 1:8500. At this scale, the 40-acre plot comprised approximately 25 cm^2 on the photograph which was of sufficient quality to provide a resolution of one meter. The second set was obtained from the USDA. These were taken in April 1980 with color infrared film at a scale of 1:25000. On these photographs, the same 40-acre plot comprised only 3 cm^2 with a resolution of approximately 2 meters. The final set was obtained from the National Archives. These photographs were taken during April 1939 on black and white film at a scale of 1:25000. Due to the large film grain size, the resolution is on the order of 2 to 3 meters.

The first step in registering these photographs to a common scale was to locate the section corners of section 19 on the ground. These locations were used to register the 1984 series with a scale of one meter per pixel. A quarter mile grid was superimposed on the corrected image dividing the section into 40-acre parcels. Attention was focused on the western 40-acre block containing the major stream bed of the Comb's Camp area. Since the man-made disturbances are of recent origin, they could not be used to register the 1939 photograph. However, by using rock outcrops which appear on all the photographs, registration was possible. The coordinates of a prominent point on three different rock outcrops were determined from the registered 1984 photograph. These were used to register both the 1980 and the 1939 photographs. In order to easily follow the evolution of each oak, computer generated blow-ups of two adjacent 100 meter on a side area in the creek bottom were obtained for all three years. These blow-ups were then used to make tracings and transparencies of the oak canopy in both 1984 and 1939. Finally, the two tracings were combined with the differences noted.

A study of these differences showed that the oaks in the creek bed are being slowly replaced by grassland, and new oaks are being established at the interface with the chaparral. This raises several interesting questions. Why is the woodland expanding away from the valley bottom into the chaparral? Why is the grassland expanding in the valley bottom? How does the past grazing and fire history affect these observations? To answer these questions, additional sites whose history is known are being studied by the techniques presented in this paper. In San Diego County, those sites located in the western portion are covered by the 1928 photographic series as the earliest and at about ten year intervals thereafter (Scheidlinger and Zedler 1980).

REFERENCES

Aldrich RC (1979) Remote sensing of wildland resources: A state-of-the-art review. USDA Forest Service General Technical Report RM-71. Fort Collins, Colo. p 56

American Society of Photogrammetry (1975) Manual of remote sensing. American Society of Photogrammetry, Virginia

Chiou WC (1984) Simple descriptors for contextual classification of hyperdimensional remotely sensed spectral data. Applied Optics 23:465-468

Heller RC (1970) Imaging with photographic sensors. In: Remote with special reference to agriculture and forestry. National Academy of Sciences, Washington, D.C. pp 35-72

Scheidlinger CR, Zedler PH (1980) Change in vegetative cover of oak stands in Southern San Diego County: 1928-1970. USDA Forest Service General Technical Report PSW-44:81-85

2. Plant Water Relations and Response to Water Stress

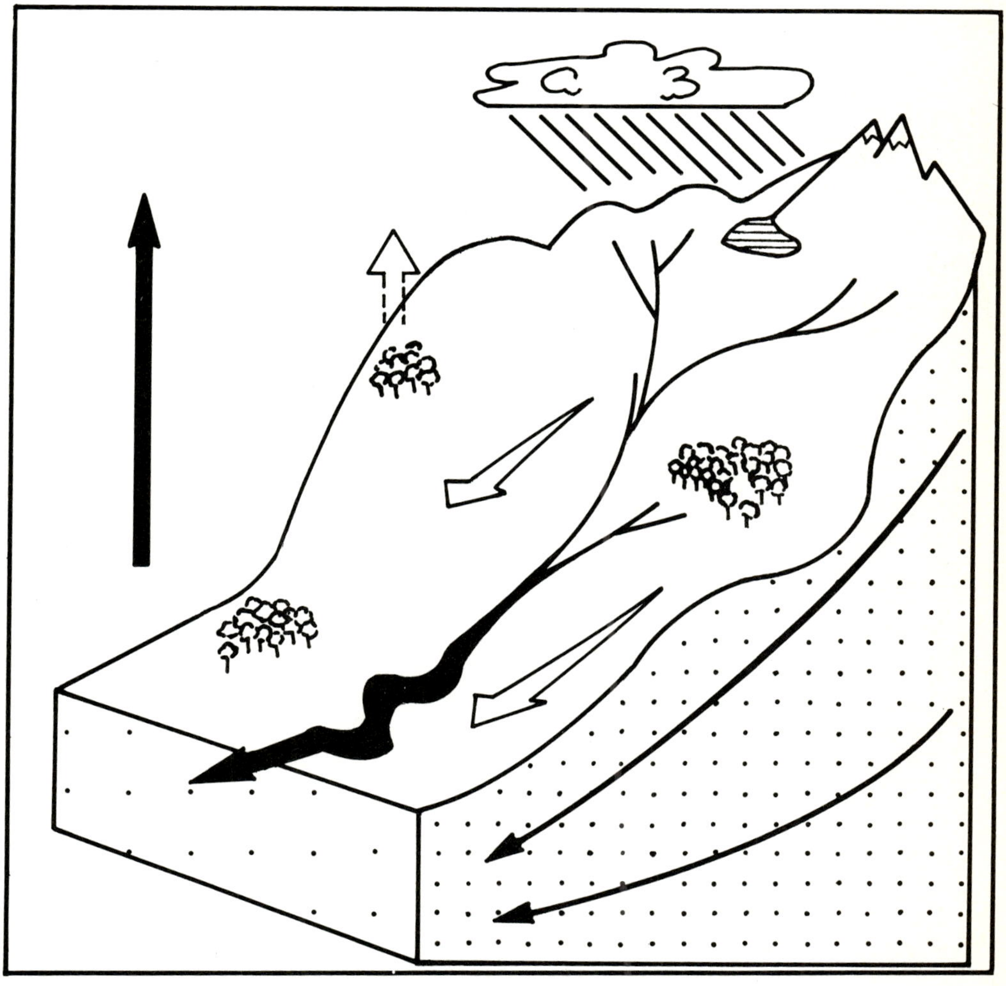

Response of two perennial grasses to water availability in different habitats related to successional change under Mediterranean climate conditions

J. Roy, E. Garnier and L. E. Jackson

Laboratoire de physiologie écologique, Centre Emberger, CNRS, BP 5051, 34033 Montpellier Cedex, France

INTRODUCTION

Water is the main factor limiting plant growth under mediterranean climate conditions. Relationships between drought duration and plant distribution, morphology, phenology, and physiology have been well established (Mooney and Dunn 1970; Mooney 1981; Miller 1982). The general pattern in vegetation changes (winter deciduous trees to evergreen trees to evergreen shrubs to drought deciduous shrubs and finally to succulents as drought duration increases) applies both for plants along spatial aridity gradients and for plants with roots which occupy different soil depths (Mooney 1982; Davies and Mooney 1985). Drought-evading herbaceous species are characteristically associated with the dominant woody species (Escarré *et al.* 1983; Naveh and Whittaker 1979) or may form communities of their own, such as grasslands (Heady 1977).

Among the drought-evading species, annual grasses have received most attention because of their importance in range management, especially in California (Gerakis *et al.* 1975; Hull and Muller 1976; Gulmon 1979; Ewing and Menke 1983a,b). Most studies on perennial grasses date back to the sixties, when English and Australian plant breeders looked for mediterranean ecotypes of cultivated plants to improve spring and autumn productivity of their native forage species (review in Roy 1982). Recent work deals with the relative ability of annual and perennial grasses to sustain drought (Jackson and Roy submitted), as well as ecotypic differentiation in relation to the aridity of the environment (Roy and Lumaret submitted). Ecotypic differentiation plays a major role in mediterranean ecosystems as far as grasses are concerned. In southern France, only 24% (in forest communities) and 35% (in grass communities) of the grass species are typically mediterranean compared to 60% and 79% respectively of the woody plants in the same communities. Mediterranean ecotypes of septentrional species and typical mediterranean grasses exhibit similar seasonal trends in water status and growth (Roy 1982).

However, even locally, climate is not homogeneous. It is modified by topography, soil characteristics and by plant cover. Here, we examine effects of the latter, namely to what extent a tree canopy modifies the water balance of the underlying herbaceous species, and how herbs (here perennial grasses) found preferentially in open stands or under a tree canopy differ in their response to

NATO ASI Series, Vol. G15
Plant Response to Stress. Edited by J. D. Tenhunen et al.

water availability. Such a study addresses questions related to the physiological ecology of plant succession. Comparison of responses of early and late successional plant species (herbs vs. trees) has shown a decrease in stomatal conductance (and transpiration rate) and an increase in the sensitivity of stomata to declining moisture levels as succession progresses (Bazzaz 1979). Only very few studies (e.g. Bacone *et al.* 1976), however, have refined this conclusion by studying species of a similar life form in different successional stages. Here we present data on water availability and related niche differentiation in two perennial grasses found in open stands (mid-successional stage) and under oak canopies (late successional stage).

MATERIALS AND METHOD

Dactylis glomerata L. ssp. *hispanica* and *Bromus erectus* Huds. are both caespitose perennial grasses with a broad geographical amplitude. Under the humid Mediterranean climate of southern France, these species occur as elements in communities which change as succession occurs and as abandoned vineyards revert to *Quercus pubescens* forests (Escarré *et al.* 1983). *D. glomera* is frequent in open swards but is progressively eliminated under dense oak canopies. On the contrary, *B. erectus* is the dominant species of the understory vegetation but is more rare in the open early stages of succession.

The field site was located near St. Martin de Londres, 25 km north of Montpellier. Vegetation in the sward, dominated by species in the *Labiaceae*, was typical of 15 to 20 year old abandoned vineyards. The adjacent woodlot was a *Quercus pubescens* coppice which was most recently cut 32 years ago. The vegetation was analysed in 240 squares (50x50 cm^2) in each plot. *D. glomerata* and *B. erectus* were present in 60 and 68% of the squares respectively in the sward and 17 and 96% of the squares in the woodlot. The percentage of soil covered by *D. glomerata* and *B. erectus* was 0.9 and 11% respectively in the sward and 0.3 and 63% in the woods.

Rainfall data (1981-82) are from the meteorological station of St. Martin de Londre located 1 km from the site. Throughfall was measured during autumn 1984. The gauges were cylindrical (12 cm diameter) and were laid on the soil. Leaf xylem water potential was measured with a Scholander pressure chamber (PMS Instruments, Corvallis, Oregon). All values indicated in the figures are the means of 3 measurements, except in experiments in which the kinetics of water potential recovery are reported where data values are for single observations.

To study root growth, 10 individuals of each species (5 from the sward and 5 from the woods) were first uniformly cultivated in a greenhouse for seven months to reduce phenotypic effects of their previous environmental exposure. In April, each plant was reduced to 3 tillers and transplanted to a glass-faced box, 100 cm deep by 50 cm wide by 15 cm thick. The boxes were tilted at a 10 degree angle with the glass downward. Plywood and polystyrene foam were placed on the glass surface to keep out light and reduce temperature fluctuations (see Bragg *et al.* 1983 for detailed information on the minirhizotron method). The soil was a mixture of 2/3 unleached red mediterranean soil and 1/3 horticultural compost. The whole profile was water saturated at the time of planting. No water was added afterward except as occurred due to occasional rainfalls. New roots appearing on the glass were recorded each two weeks for 2.5 months. Tracings of the roots were made on transparent paper with a pen of 2 mm width. Length of the roots was inferred from area measurements (Metraplan, Paris, France) made for each 10 cm horizontal soil stratum.

Biomass allocation to roots and aboveground organs was studied in plants grown from seed in monoculture in an experimental garden. Within each 1 m^2 plot, the spacing of plants between and along rows was 10 cm. Seeds were planted in mid-October 1982. Plants were well watered until 9 June. Rows of 4 plants of each species were harvested monthly. Roots were excavated from trenches dug below the sampled plants. The excavations were made below an area 40 cm long and 10 cm wide and extended to the bottom of the rooting zone. It was assumed that roots of those harvested plants that extended out of the 400 cm^2 sampled area were compensated for by presence of roots of neighboring plants in the harvest trenches. Spacing was planned to leave unsampled rows between the harvested plants. Dry weights of roots and aboveground organs were measured after drying 48 h at 80° C. Results for May, June and July are the mean of 4 plants watered as described above plus 3 plants whose watering was stopped at the end of April. Plants had a similar biomass allocation under the two treatments (Jackson and Roy 1986). Daily courses of water potential and stomatal conductance (diffusion porometer, Delta-T Device, Cambridge, England) were measured on May 25 and 27.

Growth of the two species was analysed along a soil water gradient. Individuals grown from seeds the previous year were reduced to four tillers and transplanted in April 1982 into 4 x 1 x 0.5 m^3 boxes filled with an unleached red mediterranean soil. The boxes were inclined and a constant water table was maintained (details in Roy 1985). Each box was divided by half along the length. One half was planted with *D. glomerata* and the other with *B. erectus* (density = 64 plants m^{-2}). One box had plants from the sward and another box plants from the woods. Results presented here are the mean obtained with both the sward and woods population since no ecotypic differentiation was found between those populations (Roy 1985). Half of a third box had a mixture of the two species from the sward. At mid-August, aerial parts of each individual were harvested and oven dried. For analysis, the gradient was divided in eight 50 x 50 cm^2 resource states, each with 16 plants (32 in the mixture). Results are the mean of the 13 heaviest plants, the 3 others being dead or very small as a result of unsuccessful transplant.

RESULTS

Effect of the Tree Canopy on the Water Balance of the Grasses

The seasonal course of the mid-afternoon leaf water potential of *Dactylis glomerata* and *Bromus erectus* is shown in Fig. 1. The development of the leaf canopy in mid-April improved the subsequent water balance of the woods populations. (The transmittance of the *Quercus pubescens* canopy was 0.75 during the winter and 0.35 after the development of the leaves.) As a consequence, green leaf biomass in July was 41 and 72% of the maximum spring green leaf biomass in the woods populations of *D. glomerata* and *B. erectus* respectively and only 0 and 10% in the sward populations (Garnier and Roy submitted).

The effect of the canopy on soil water status is complex. If the entire soil profile is considered, the soil water content at the end of the dry season is lower under the canopy compared to an open stand because of the higher evapotranspiration of the trees (Martson 1962; Fletcher and Lull 1963; Sartz 1972; Fardjah and Lemée 1980). However the tree canopy reduces the evapotranspiration of the understory herbaceous strata through a decrease in surface temperature and an increase in boundary layer resistance and air humidity. The result is an improved water status of the soil strata where herbs are rooted compared to an open stand (Fardjah and Lemée 1980). The spring predawn water potentials (Fig. 2)

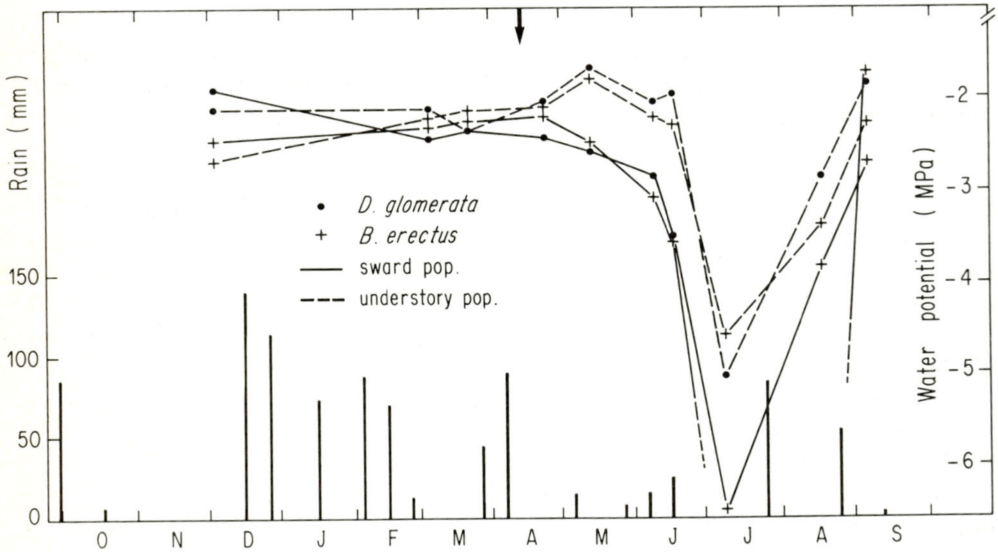

Fig. 1. Rainfall events and mid-afternoon leaf water potential of *D. glomerata* and *B. erectus* individuals in a sward and an adjacent woods during winter, spring, and summer. The arrow indicates the time of unfolding of the *Quercus pubescens* leaves

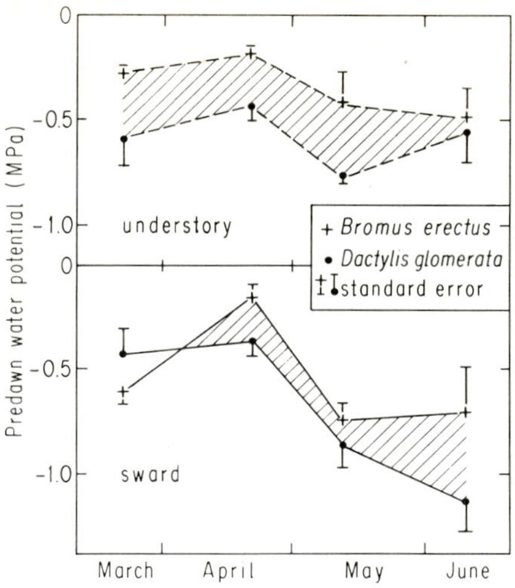

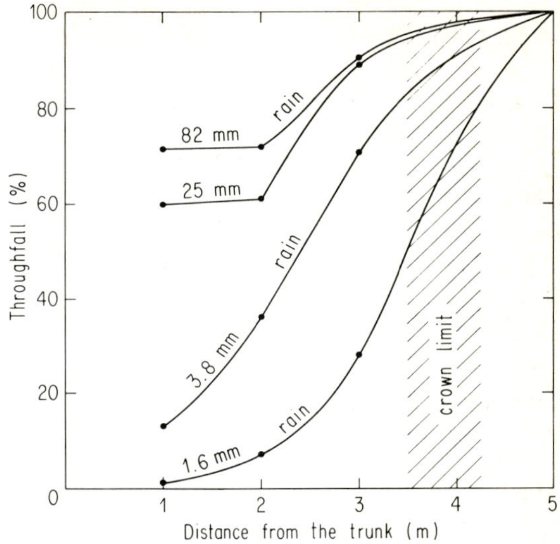

Fig. 2. Changes in predawn water potential of *D. glomerata* and *B. erectus* individuals in a sward and a woods during the spring and early summer

Fig. 3. Relative amount of water falling through a *Quercus pubescens* crown during rainfall events with different amounts of precipitations and at various distances from the trunk

confirm the more favorable water status of the woods individuals and indicate a substantial difference in soil moisture. The soil in the sward had a more negative water potential (by about 0.23 MPa in May and 0.43 in June) than the soil in the woods.

Canopy and litter rainfall interception may, however, prevent light summer rains from reaching the first soil layer. For leafy forests, canopy interception was estimated to be 10% and 20% of rainfall by Helvey and Patric (1965) and Larcher (1975) respectively. The percentage of water intercepted and lost is inversely related to the amount of precipitation (Fig. 3, Aussenac 1968; Larcher 1975). Under evergreen oak trees in Spain, Calabuig *et al.* (1980) found the percentage of rainfall interception to be 47% from June to August and 23% from September to November. Rain interception by litter averages 2 to 5% of the annual rainfall (Helvey and Patric 1965) or 1 to 2.9 mm of water per rain (Fardjah and Lemée 1980). At our field sites, the litter under the oak trees was 3 cm thick and was nonexistant in the open stand. However, the response of water potential to the mid-June rain (in particular *D. glomerata* - Fig. 1) indicates that despite interception, the effect of the canopy on the water status of the herbaceous layer is positive.

Daily Courses of Water Potential and Stomatal Conductance

The *in situ* daily courses of leaf water potential (Fig. 4) indicated that no substantial differences were experienced by the two species (see also Fig. 1). On June 10, *B. erectus* had a higher water potential than *D. glomerata* at dawn, but the reverse was true during the day when plants were transpiring. It suggests that either the soil-plant resistance to water flow is higher in *B. erectus* compared to *D. glomerata* or that transpiration rates are higher in *B. erectus*. The latter suggestion is supported by the results shown in Fig. 5. At a given water potential, the stomatal conductance of *B. erectus* is twice that of *D. glomerata*. Fornallaz (1982) also found similar water potential values in *D. glomerata* and *B. erectus* and a higher stomal conductance in the latter species. *B. erectus* was also found to have a greater area of transpiring surface. The green leaf area per plant was about the same in woods and swards individuals, but it was 80 cm^2 for *B. erectus* and 30 cm^2 for *D. glomerata* in May (Roy 1985).

A Field Assessment of Rooting Depth

The higher values of predawn water potential found in *B. erectus* compared to *D. glomerata* (Fig. 2) could be due either to a higher soil-root-stem conductance to water in that species or to the fact that its roots had access to soil zones with greater water availability. The former hypothesis was tested by studying the

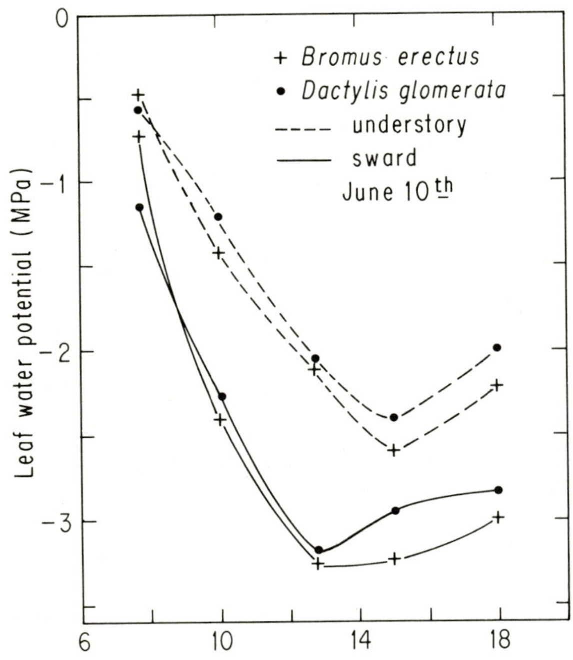

Fig. 4. Daily course of leaf water potential of *D. glomerata* and *B. erectus* individuals in a sward and a woods

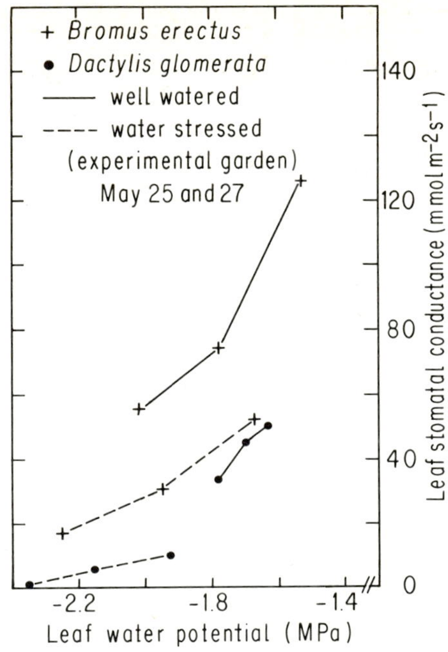

Fig. 5. Response of stomatal conductance to leaf water potential in *D. glomerata* and *B. erectus* plants grown under two water regimes

kinetics of water potential recovery after blocking transpiration. A species with a higher soil-root-stem conductance should come into equilibrium with the soil water potential faster than other species. In the middle of the afternoon, water potential of two adjacent *B. erectus* and *D. glomerata* individuals was measured. Then the plants were covered by a tank whose inner surfaces were lined with wet absorbing towels. Transpiration was blocked due to dark-induced stomatal closure and saturation of air water vapor. Water potential was subsequently measured at regular intervals. Figure 6 shows two of the 6 pairs of recovery responses that we obtained. Statistical analysis of the recovery kinetics was not possible, since the shape of the curve is a function of the initial water potential. Three experiments indicated a similar kinetic for the two species (as in Fig. 6) and 3 experiments indicated a slightly faster recovery in *D. glomerata* . Thus, the higher predawn water potential of *B. erectus* must result from better access to available water.

Since the predawn water potential was usually measured on adjacent *B. erectus* and *D. glomerata* individuals, distinctions in their rooting zones had to result due to differences in rooting depth, *D. glomerata* being shallower rooted. If this is true, the water potential of *D. glomerata* should recover faster than *B. erectus*

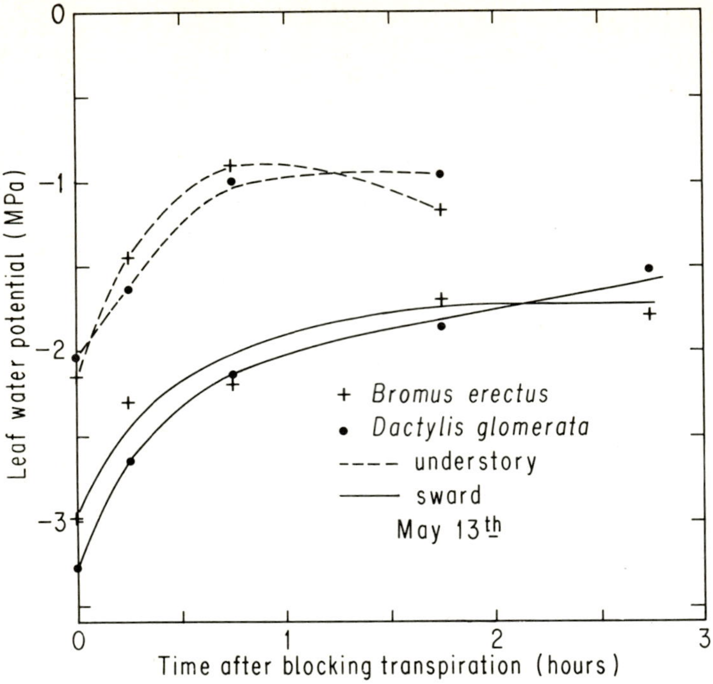

Fig. 6. Kinetics of water potential recovery after blocking transpiration of *D. glomerata* and *B. erectus* individuals in a sward and a woods

following soil surface watering since its active roots should be reached first by the descending moisture front. Figure 7 shows that this was actually the case. It was possible to measure the response of the plants to different amounts of watering (Fig. 7b). The initial and temporary decrease in water potential observed in *B. erectus* may reflect a lower transpiration rate due to the low water vapor pressure deficit of the air during watering. Although it was not carried out because the soil was too stony, it appears that the technique applied can also provide information on the actual depth of the active roots. The first step would be to determine the minimum watering required to get a complete recovery of the water potential (in our case, 10 mm was not enough, as indicated in Fig. 7b). The second step would be to determine the depth of the wetting front by digging holes with a borer.

Rooting Profiles and Biomass Allocation to Roots

The rooting profiles observed in the minirhizotron verified the conclusions of the field study: *B. erectus* had a deeper root system (Fig. 8). In addition to the vertical distribution of the root systems, the drawings reveal differences in root architecture. In the top layers of the soil, many lateral roots are horizontal in *D.*

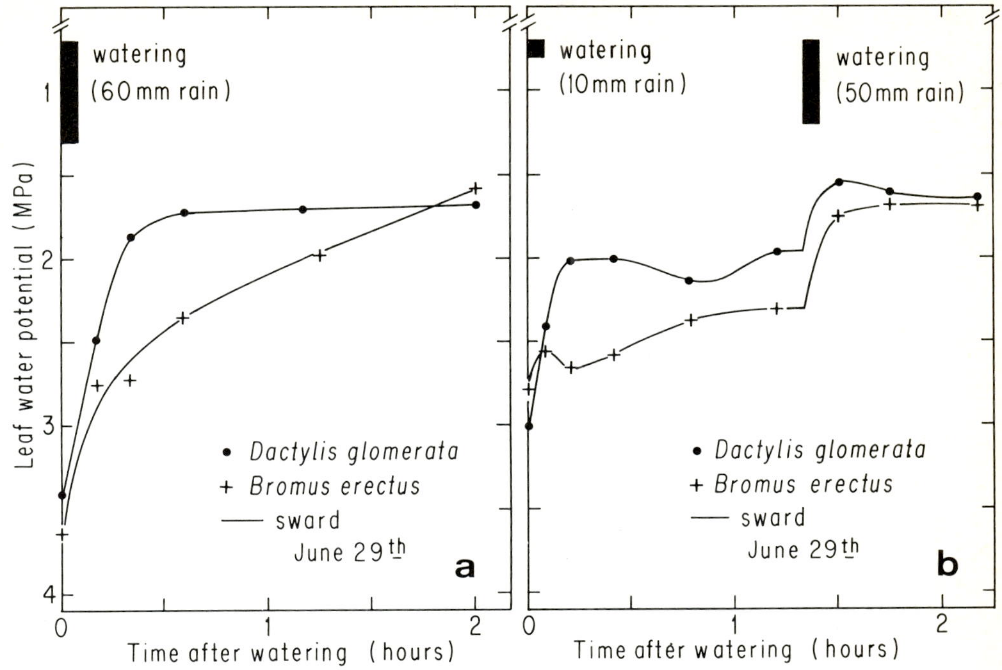

Fig. 7. Kinetics of water potential recovery of *D. glomerata* and *B. erectus* individuals following simulated rain events: a) a single 60 mm application and b) a 10 mm application followed by a 50 mm application

glomerata while they are vertical in *B. erectus*. As a result, the profile of root length shown in Fig. 8b does not reflect the profile for root absorption of materials. Most of the upper strata roots of *B. erectus* are only connections between the lower strata absorbing roots and the aerial part of the plant. *B. erectus* also had a greater total root length (21 m recorded on the glass compared to 14.7 m for *D. glomerata*; the ratio between species = 1.4).

Observed biomass distributions (Fig. 9) indicate a greater allocation to roots in *B. erectus* compared to *D. glomerata* after the winter period. Between May and July, the root biomass as a percentage of total plant biomass was maintained at a level approximately 50 to 60% higher in *B. erectus*. Since total plant biomass was similar in the two species, actual root biomass was higher in *B. erectus* (2.3 g in June compared to 1.55 g in *D. glomerata*). The corresponding root biomass ratio between species (1.55) is similar to the root length ratio between species observed in the minirhizotrons (1.4).

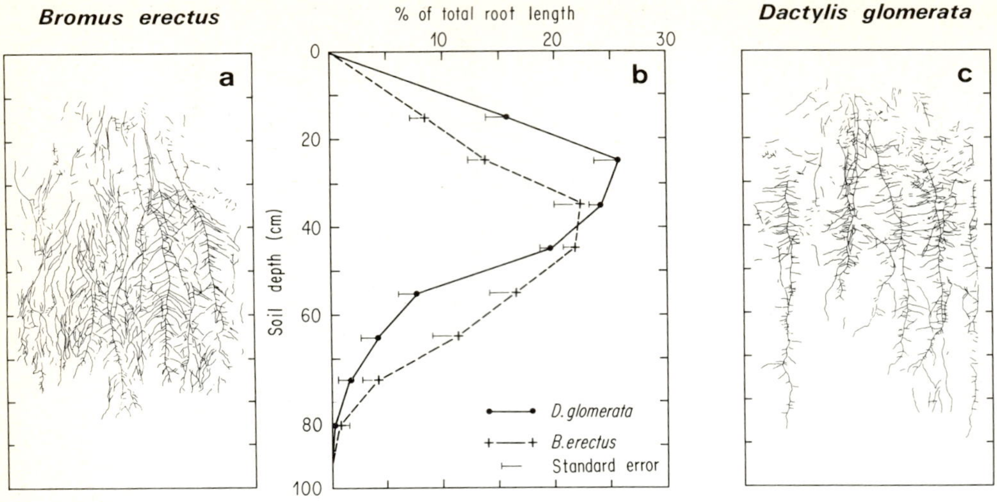

Fig. 8. Typical root configurations of *B. erectus* (a) and *D. glomerata* (c) as viewed in the minirhizotron and (b) the distribution of total root length of *D. glomerata* and *B. erectus* in nine soil horizons (10 cm thick layers)

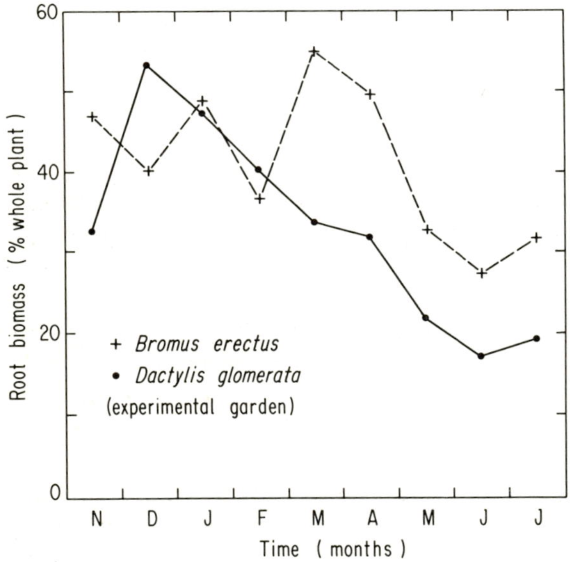

Fig. 9. Seasonal course of biomass allocation to roots in plants of *D. glomerata* and *B. erectus* in an experimental garden

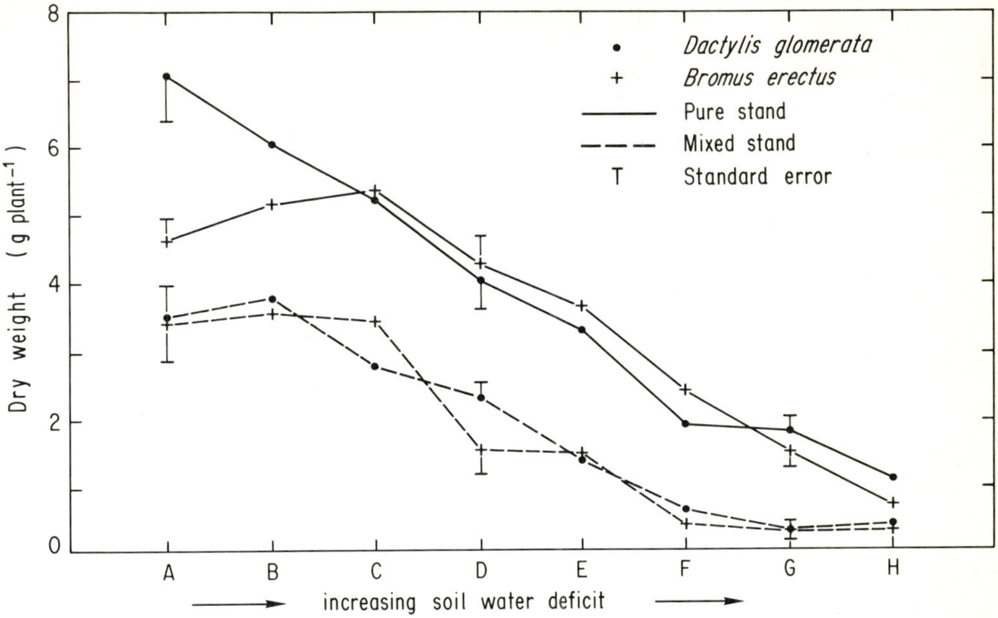

Fig. 10. Dry weight at harvest of aerial parts of *D. glomerata* and *B. erectus* plants cultivated along a soil water gradient in pure and mixed stands. Density in the mixed stand was twice the density of the pure stand

Growth in Response to Soil Water Deficits

Results presented in Fig. 10 indicate that growth of the two species was affected similarly by drought imposed in the inclined boxes. Although *D. glomerata* is more frequent in natural dry, open stands, it did not grow larger than *B. erectus* at the dry end of the gradient. In the pure stand, the biomass of *B. erectus* decreased in the wettest part of the gradient, probably because its deeper roots reached the water table and detrimental effects resulted from lack of oxygen (such a decrease occurred for both the woods and the sward populations, cf. data in Roy (1985), and similar results were obtained for the September regrowth). Both species had a similar biomass under wet conditions when grown in mixed stands, possibly because the limiting factor to growth may have been availability of soil mineral nutrients. At harvest, the aerial parts of the plants subjected to the three driest soil moisture levels were already dry. After the harvest of aerial parts, the soil water gradient deficit was removed by an abundant surface watering at the end of August. The number of green (i.e. alive) plants was counted one month later. Results (Fig. 11) indicate a higher drought survival in *D. glomerata*.

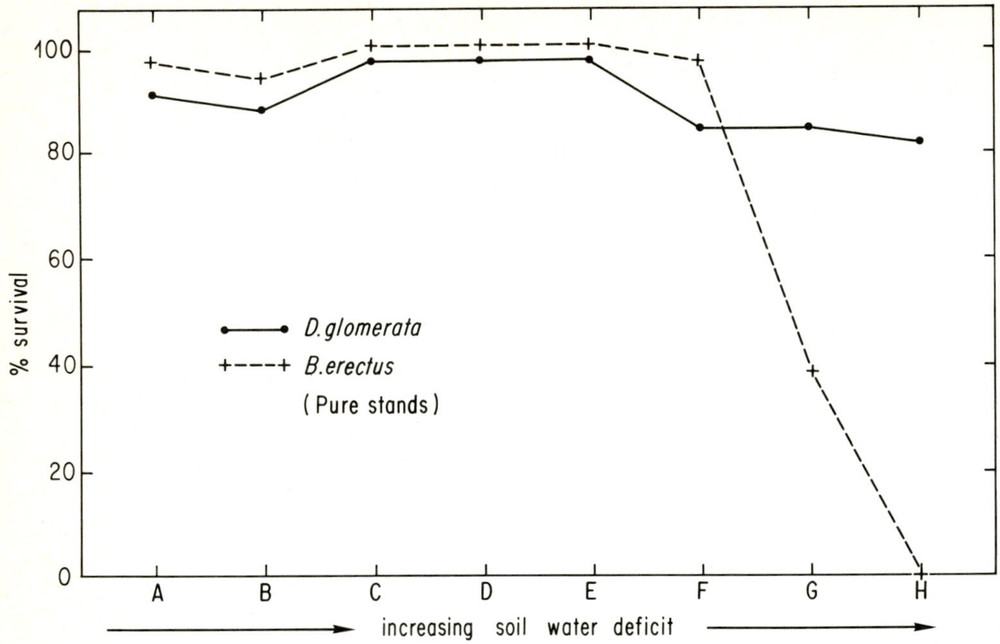

Fig. 11. Percent survival after rewatering of *D. glomerata* and *B. erectus* plants subjected to a gradient of drought imposed by a soil water gradient

DISCUSSION

Dactylis glomerata and *Bromus erectus* exhibit different capabilities and mechanisms for habitat exploitation. *B. erectus* has larger and deeper roots and has access to soil strata with greater water availability, but it also has a higher stomatal conductance and higher rates of transpiration compared to *D. glomerata*. Overall, the two species experience similar daily excursions in water potential most of the year. These species do not differ in their growth in response to water stress, but they do differ markedly in their dehydration tolerance. *D. glomerata* has a higher survival when exposed to severe and long periods of drought than *B. erectus*. Differences in the exploitation of various levels of the soil have been shown in other co-occurring species and were associated with differences in phenology (Parrish and Bazzaz 1976; Gulmon *et al.* 1983), small scale geographic distribution (Pickett and Bazzaz 1976) or intercompensatory mechanisms (Wieland and Bazzaz 1975) leading to niche differentiation in a single habitat.

The differences in water status and in capacity for water acquisition and use observed between these two species can explain in part their distributions. We have shown that plants in the sward experience a more severe drought than plants in the woods. There is no doubt that the higher dehydration tolerance of *D. glomerata* is an adaptive feature in the sward. The adaptive significance of differences in rooting depth is not as clear-cut. While the deeper rooted *B. erectus* can extract water from moist soil layers both in the sward and in the woods, the shallower rooted *D. glomerata* would experience unfavorable conditions in the dry sward if it did not compensate with low stomatal conductance and the ability to take advantage of the light summer rains (cf Fig. 7b). However, the fact that a high percentage of the water from those summer rains is intercepted by the tree canopy does not lead to a lower water potential of established *D. glomerata* plants in the woods (cf Fig. 1). Perhaps moisture availability from summer rains affects seedling establishment, which is very difficult in the woods. Of 54 tagged *B. erectus* seedlings, a census one year later showed that of the 7 seedlings still alive, 5 remained at the one tiller and two small leaves stage. All of the *D. glomerata* seedlings we were able to find and tag died.

The lower stomatal conductance observed in *D. glomerata* does not fit with the hypothesis, based on energy balance considerations, that plants in open environments would have higher stomatal conductances which allow potentially harmful high temperatures to be avoided (Bazzaz 1979). In our species, stomatal conductances seem to be related to the rooting depth. Whether a higher tolerance of photosynthesis to high temperatures may occur in *D. glomerata* has not yet been studied.

CONCLUSION

The two co-occurring perennial grasses *D. glomerata* and *B. erectus* utilize different strategies with regard to rooting depth and stomatal conductance to cope with the Mediterranean summer shortage of water. The differences in frequency of occurrence of those species at different successional stages after vineyards are abandoned is partly due to the higher dehydration tolerance of *D. glomerata*. Other mechanisms affecting their frequency are growth response to light intensity and the occurrence of safe sites for germination as related to seed size (Roy unpublished). The importance of water availability in determining the relative occurrence of the two species is confirmed by the fact that under a more arid Mediterranean climate (in the oak savannas of southern Spain), *B. erectus* does not grow and *D. glomerata* is found almost exclusively beneath the oak trees (Vacher 1984).

ACKNOWLEDGEMENTS

We thank A. Berger for comments on the manuscript and F. Jardon, A. Malvy, B. Orcel and L. Sonié for help with the measurements. Thanks are also due to the experimental field and drawing sections of the Centre Emberger. L.E. Jackson was supported by a NATO postdoctoral fellowship.

REFERENCES

Aussenac G (1970) Interception des précipitations par le couvert forestier. Ann Sci Forest 25:135-156

Bacone J, Bazzaz FA, Boggess WR (1976) Correlated photosynthetic responses and habitat factors of two successional tree species. Oecologia (Berl) 23:63-74

Bazzaz FA (1979) The physiological ecology of plant succession. Ann Rev Ecol Syst 10:351-371

Bragg PL, Govi G, Cannell RQ (1983) A comparison of methods, including angled and vertical minirhizotrons, for studying root growth and distribution in a spring oat crop. Plant Soil 73:435-440

Calabuig EL, Gago Gamallo ML, Gomez Gutierrez JM (1977) Influencia de la encina (*Quercus rotondifolia* LAM.) en la distribution del agua de lluvia. Anuario Cent Edaf Biol Apl Salamanca 4:143-159

Davis SD, Mooney HA (1985) Comparative water relations of adjacent California shrub and grassland communities. Oecologia (Berl) 66:522-529

Escarré J, Houssard C, Debussche M, Lepart J (1983) Evolution de la végétation et du sol après abandon cultural en région méditerranéenne: étude de succession dans les garrigues du Montpelliérais (France). Oecol Plant 4:221-239

Ewing AL, Menke JW (1983a) Reproductive potential of *Bromus mollis* and *Avena barbata* under drought conditions. Madroño 30:159-167

Ewing AL, Menke JW (1983b) Response of soft chest (*Bromus mollis*) and slender oat (*Avena barbata*) to simulated drought cycles. J Range Manage 36:415-418

Fardjah M, Lemée G (1980) Dynamique comparée de l'eau sous hêtraie et dans des coupes nues ou à Calamagrostis epigeios en forêt de Fontainebleau. Bull Ecol 11:11-31

Fletcher PW, Lull MW (1963) Soil moisture depletion by a hardwood forest during drought years. Soil Sci Soc Amer Proc 27:94-98

Fornallaz C (1982) Veränderungen des Wasserhaushaltes von *Bromus erectus* Huds. und *Dactylis glomerata* L. im Verlaufe einer Vegetationsperiode under feuchten und trockenen Bedingungen im Freiland. Diplomarbeit. Geobot Inst ETH, Stiftung Rubel, Zurich

Garnier E, Roy J (submitted) Modular and demographic analysis of plant leaf area in sward and wood populations of *Dactylis glomerata* L. and *Bromus erectus* Huds.

Gerakis PA, Guerrero FP, Williams WA (1975) Growth, water relations and nutrition of three grassland annuals as affected by drought. J Appl Ecol 12:125-135

Gulmon SL (1979) Competition and coexistence: three annual grass species. Amer Midl Nat 101:403-416

Gulmon SL, Chiariello NR, Mooney HA, Chu CC (1983) Phenology and resource use in three co-occurring grassland annuals. Oecologia (Berl) 58:33-42

Heady HF (1977) Valley grassland. In: Barbour MG, Major J (eds) Terrestrial Vegetation of California. John Wiley, New York, pp 491-514

Helvey JD, Patric JH (1965) Canopy and litter interception of rainfall by hardwoods of Eastern United States. Water Resource Res 1:193-206

Hull JC, Muller CH (1976) Response of California annual grassland species to variations in moisture and fertilization. J Range Manage 29:49-52

Jackson LE, Roy J (1986) Growth patterns of mediterranean annual and perennial grasses under simulated rainfall regimes of southern France and California. Oecol Plant 7:191-212

Larcher W (1975) Physiological ecology. Springer, Berlin-Heidelberg-New York, p 252

Marston RB (1962) Influence of vegetation cover on soil moisture in southeastern Ohio. Soil Sci Soc Amer Proc 26:605-608

Miller PC (1982) Environmental constraints to vegetation in mediterranean type ecosystems. Ecol Mediter 8:411-416

Mooney HA (1981) Primary production in mediterranean-climate regions. In: Di Castri F, Goodall DW, Specht RL (eds) Ecosystems of the World. II. Mediterranean-type Shrublands. Elsevier, Amsterdam, pp 249-255

Mooney HA (1982) Habitat, plant form and plant water relations in mediterranean-climate regions. Ecol Mediter 8:481-488

Mooney HA, Dunn EL (1970) Convergent evolution of mediterranean climate evergreen sclerophyll shrubs. Evolution 24:292-303

Naveh Z, Whittaker RH (1979) Structural and floristic diversity of shrublands and woodlands in Northern Israel and other Mediterranean areas. Vegetatio 41:171-190

Parrish JAD, Bazzaz FA (1976) Underground niche separation in successional plants. Ecology 57:1281-1288

Pickett STA, Bazzaz FA (1976) Divergence of two co-occurring successional annuals on a soil moisture gradient. Ecology 57:169-176

Roy J (1982) Intraspecific variation in the physiological characteristics of perennial grasses of the Mediterranean region. Ecol Mediter 8:435-448

Roy J (1985) Comparison of *Dactylis glomerata* and *Bromus erectus* populations from contrasted successional stages. In: Haeck J, Woldendorp JW (eds) Structure and Functioning of Plant Populations. II. Phenotypic and Genotypic Variation. North Holland Publishing Co, Amsterdam, pp 51-64

Roy J, Lumaret R (submitted) Associated clinal variation in leaf tissue water relations and enzyme polymorphism in *Dactylis glomerata* L. populations

Sartz RS (1972) Soil water depletion by a hardwood forest in southwestern Wisconsin. Soil Sci Soc Amer Proc 36:961-964

Vacher J (1984) Analyse phyto et agro-écologique des dehesas pastorales de la Sierra Norte (Andalousie occidentale, Espagne). CEPE/CNRS, Montpellier, p 195

Wieland NK, Bazzaz FA (1975) Physiological ecology of three codominant successional annuals. Ecology 56:681-688

Changes in aboveground structure and resistances to water uptake in *Quercus coccifera* along a rainfall gradient

S. Rambal and J. Leterme

C.E.P.E. - C.N.R.S., B.P. 5051, 34033 Montpellier Cédex, France

INTRODUCTION

We have assumed that changes in *Quercus coccifera* canopies along a rainfall gradient can be understood in terms of a system which optimizes net photosynthesis while minimizing water loss. This optimization process leads to perceptible differences in branch and leaf display patterns (Aber *et al.* 1982). The observed differences can be interpreted as drought-related adaptation mechanisms (Miller 1983) or due to modifications in the carbon allocation scheme. We further have assumed that stands on drier sites which undergo greater water stress should have low leaf area indices, while those on mesic sites should exhibit a tendency towards tall canopies with high leaf area indices. Our third assumption has been that changes in canopy structure will be accompanied by changes in rooting depth. In our case, the rooting depth is indirectly determined by measuring water uptake resistances, i.e. plant and soil resistances.

SITE DESCRIPTION AND METHODS

Observations were made at three sites located along an elevational transect in Languedoc (France), ranging from La Palme (near sea level) to Saint Martin de Londres (200 m above sea level). A fourth site was chosen near Murcia in southeastern Spain. These sites experience a wide range of precipitation (Table 1). All the locations are pure, 20 to 30 year old stands of *Quercus coccifera* growing either on hard or soft limestone.

Vertical profiles of the leaf and stem dry weight for 20-cm thick layers were determined to quantify canopy structure. These were determined by stratified clipping on five 1 m^{-2} samples. Sampling was undertaken during the winter period at the Saint Martin de Londres (SML), Montbazin - Gigean (MG) and La Palme (LP) sites. Another sampling was conducted during early summer in LP and Sierra de la Pila (SP - see Table 1). In late spring, after total defoliation by phytophagous insects, the MG site was burned by wildfire.

Field measurements of stand water relations were conducted between May and July 1983. Where possible, the analysis is based on two sampling days at each site during this period. Xylem water potential was obtained with a Scholander pressure chamber and transpiration rate by Huber's cut-leaf weighing technique.

NATO ASI Series, Vol. G15
Plant Response to Stress. Edited by J.D.Tenhunen et al.

Table 1. Characteristics of the four observation sites.

Sites	St. Martin de Londres (SML) Hérault-France	Montbazin-Gigean (MG) Hérault-France	La Palme (LP) Aude-France	Sierra de la Pila (SP) Murcia-Spain
Co-ordinates	43°49'N-3°47'E	43°30'N-3°40'E	43°N-3°02'E	38°10'N-1°10'W
Elevation (m)	200	120	30	900
Slope (%)	20	15	0-5	40-50
Aspect	SE	NW	ESE	NW
Mean annual precipitation (mm)	1188	737	595	301
Mean annual temperature (° C)	12.3	13.9	15.0	12.8

For steady state flow of water through the soil-plant-atmosphere continuum, the relation between the transpiration rate T, expressed in $m.s^{-1}$ and the changes in water potential between the soil Ψ and the xylem Ψ_x (MPa) can be written

$$T = \frac{\Psi - \Psi_x}{R_s + R_p} = \frac{\Delta\Psi}{R_s + R_p} \qquad (1)$$

where R_s and R_p (expressed in seconds) are the hydraulic resistance to flow in the bulk soil and root zone, called soil resistance, and the plant resistance along the xylem tissue pathway from the root to the leaf, respectively. If the moisture retention function of the soil is represented by

$$\Psi = \Psi_e (\Theta/\Theta_s)^{-\beta} \qquad (2)$$

where Ψ_e is the air entry water potential (MPa), Θ is the soil water content (cm^3 cm^{-3}), Θ_s is the saturated water content (cm^3 cm^{-3}), and β is an empirically determined constant, then the hydraulic conductivity, m s^{-1}, is given by (see Campbell 1974)

$$K = K_s (\Psi/\Psi_e)^{\frac{2}{\beta}+2} \qquad (3)$$

where K_s is the saturated hydraulic conductivity, m s^{-1}. Thus, the soil resistance is calculated from the soil water potential using the parametric form of the Gardner equation (Feedes and Rijtema 1972).

$$R_s = \frac{1}{BK} = \frac{1}{BK_s} \left(\frac{\Psi}{\Psi_e}\right)^{\frac{2}{\beta}+2} \qquad (4)$$

For a given time t_1,

$$R_t(t_1) = R_s(t_1) + R_p \qquad (5)$$

can be estimated from the slope of the curve relating T to $\Delta\Psi$. T and $\Delta\Psi$ values are measured during the early morning hours when the internally stored water contribute insignificantly to transpiraton. By repeating measurements at time t_2, we obtained a second equation

$$R_t(t_2) = R_s(t_2) + R_p \tag{6}$$

where R_p may be considered as a constant over time. These two linear equations can be easily solved. From Eq. 4, for soil water potentials Ψ_1 and Ψ_2 at times t_1 and t_2, we find

$$R_s(t_2) = R_s(t_1)\,(\Psi_2/\Psi_1)^{\frac{2}{\beta}+2} \tag{7}$$

Substituting for R_p from Eq. 6 into Eq. 5 and taking Eq. 7 into account, gives $R_s(t_1)$, $R_s(t_2)$ and R_p values.

RESULTS

The aboveground phytomass of stems decreased along the rainfall gradient from 2867 ± 303 (SD) g m^{-2} to 1058 ± 174 g m^{-2}. Differences in vertical profiles of stem and leaf biomass occur among stands (Fig. 1). To examine the possibility that continuous change may occur between these profiles and to characterize them, we calculated the arithmetic mean heights and the Fisher's skew coefficients. As mean annual precipitation decreases, mean height decreases. The skew coefficient decreases as well, moving from SML to LP (Table 2). For the SP site, it increases greatly to 0.44 due to an increase in the percentage of total stem phytomass in the first 20 cm above the ground.

Changes in foliage height profiles do not follow clear trends. During winter, the mean foliage heights for SML, MG and LP were 0.54, 0.63, and 0.35 m and the skew coefficients 0.17, -0.56 and 0.13 respectively. During summer, the mean heights for LP and SP were 0.28 and 0.36 m. The corresponding skew coefficients were 0.42 and -0.26. The 20-cm layer leaf dry weight was converted to leaf area index (LAI) using leaf specific weight. Leaf specific weights were 0.014 to 0.020 g cm^{-2} in SML, 0.012 to 0.022 in MG, 0.014 to 0.022 in LP and 0.017 to 0.022 in SP. The highest value corresponds to the highest layer, the lowest value is always associated with the lowest layer (see also Meister *et al.* in this volume). The leaf area indices obtained in winter were approximately 1.12 - 1.24 m^2 leaf m^{-2} ground. During the early summer period, LAI was 2.11 ± 0.14 (standard deviation) in LP and 1.48 ± 0.24 in SP. Rambal (1984) observed a LAI of 2.5 with a mean annual rainfall of 957 mm.

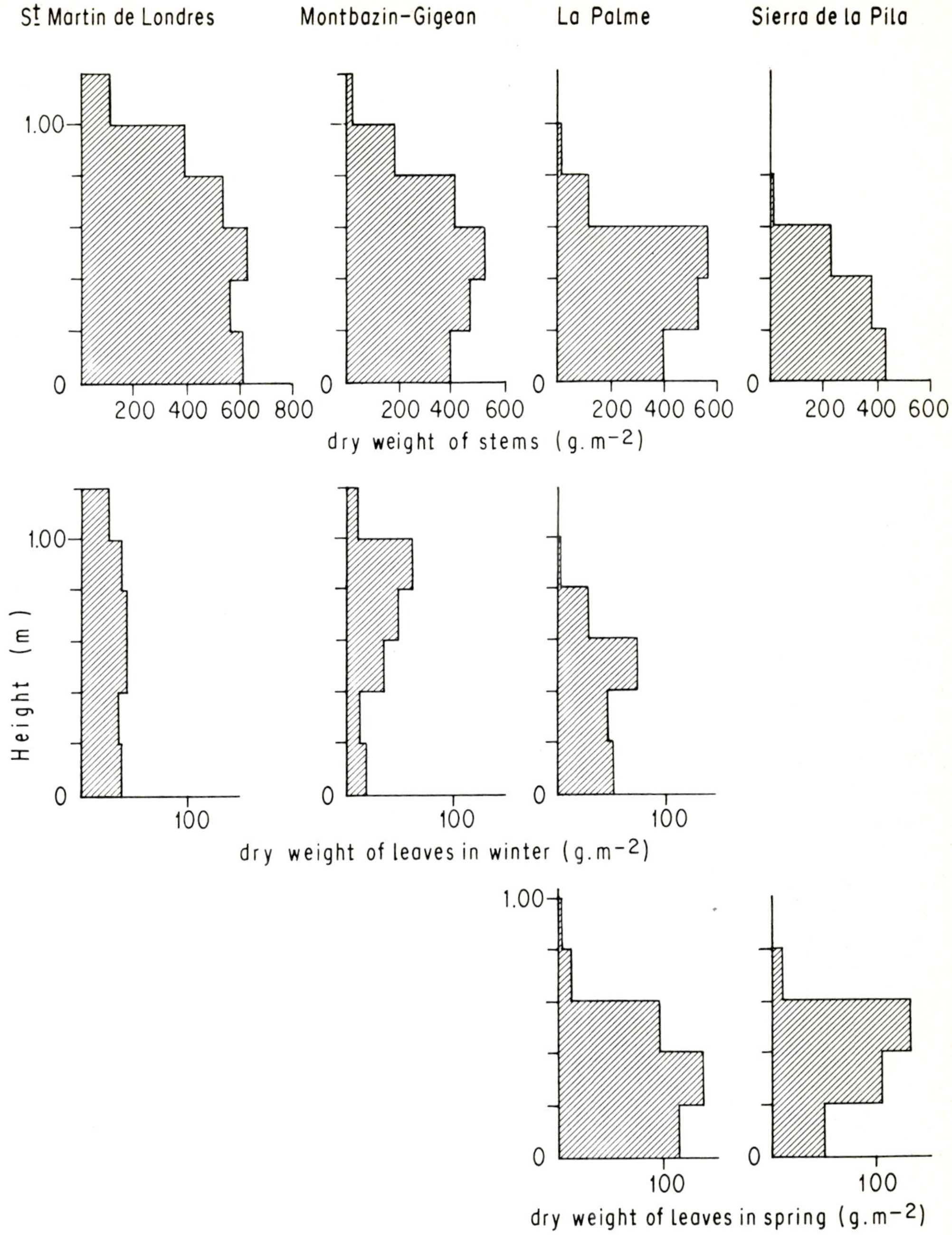

Fig. 1. Vertical distribution of stem and foliage biomass in *Q. coccifera* canopies at the SML, MG, LP and SP sites. The profiles are indicated for 20 cm canopy layers. All values are the mean of five 1 m^2 samples

Table 2. Structural characteristics of *Q. coccifera* canopies at the four sites respectively (see text).

Site	SML	MG	LP	SP
Total canopy height (m)	1.10-1.30	0.80-1.10	0.70-0.90	0.50-0.70
Stems:				
Total dry weight ($g\ m^{-2}$)	2867 ± 303	2016 ± 492	1657 ± 215	1058 ± 174
Arithmetic mean height (m)	0.49	0.46	0.35	0.26
Fischer's coefficient of skew	0.24	0.19	0.17	0.44
Leaves:				
Range of specific weight ($g\ cm^{-2}$)	0.014-0.020	0.012-0.022	0.014-0.022	0.017-0.022
Total winter dry weight ($g\ m^{-2}$)	215 ± 12	186 ± 40	199 ± 24	
Winter leaf area index ($m^2\ m^{-2}$)	1.24 ± 0.08	1.12 ± 0.24	1.16 ± 0.17	
Total summer dry weight ($g\ m^{-2}$)			349 ± 24	292 ± 33
Summer leaf area index ($m^2\ m^{-2}$)			2.11 ± 0.14	1.48 ± 0.24

Xylem water potential (Ψ_x) was rather high in LP on May 27 (-0.85 MPa) then decreased markedly in early summer to -1.45 MPa on July 8 (Fig. 2 lower sections) and -1.90 MPa on July 27. In SML, Ψ_x decreased less rapidly than in LP (-0.43 MPa on June 3 and -1.38 MPa on July 22 as in Fig. 2). The predawn potential in SP on June 24 was -1.33 MPa. The day upon which the predawn potential reached -1.33 MPa was estimated for each site by linear interpolation. This day was July 4 and July 20 in LP and SML respectively, suggesting a 26-day delay in reaching this stage of drought in SML as compared to SP and a 10-day delay between LP and SP. *Q. coccifera* in SML showed a seasonal variation of Ψ_x similar to that reported by Tenhunen *et al.* (1985) on the same species of oak near Lisbon, Portugal (see Tenhunen *et al.* in this volume).

The highest rates of transpirational water loss were 3.0 and 3.1 $\mu g\ H_2 0\ cm^{-2}\ s^{-1}$ observed on June 3 and July 8 in LP and SML respectively. These rates decreased to 1.7 and 1.8 on July 22 and 27 at the same sites. The highest measured rate in SP was 1.2 on June 24. All these values were generally lower than those obtained by Tenhunen *et al.* (1985). In order to solve the linear system of Eq. 5 and 6 for each site, the soil water potential was assumed to be equal to the dawn xylem potential minus -0.14 MPa, and the β parameter (Eq. 2) to be 4.06 (Rambal unpublished data). The latter value may be applied to a large range of clay loam soils (Gosh 1977). Assuming constant plant resistances, they were calculated as 1.63 and 1.32 x 10^9 s for SML and LP respectively (Table 3). These values are consistent with the R_p of 1.18 x 10^9 s previously estimated as appropriate for simulating the soil water balance of a *Quercus coccifera* stand (Rambal 1982). Assuming negligible stem and leaf resistance, the R_p decrease from SML to LP corresponds to a root length increase per unit area (Cowan 1975).

For days with approximately equal soil water potentials at the two sites, -1.24 and -1.27 MPa, the corresponding soil resistance ($R_{s)}$ were estimated respectively as 3.29 and 0.38 x 10^9 s (Table 3). If we considered the simplified model of root water uptake proposed by Feddes and Rijtema (1972), where the β parameter (Eq. 4) is in inverse ratio to the depth of the root system, the soil resistance difference may be interpreted as an increase in the depth of the root system from SML to LP. For the SP site, only one day of water relations measurements has been performed. Nevertheless, with Ψ = -1.19 MPa and assuming a similar plant resistance as for SML and LP, the calculated R_s has the same order of magnitude as found for SML. Therefore, the root system in SP appears to be a shallow root system. This hypothesis is partially confirmed by changes in the predawn xylem potential observed between 23 and 24 June. *Quercus coccifera* in SP was able to absorb enough water to significantly modify its water status until deeper wetting of the soil profile occured. The large response of predawn xylem potential to low level of precipitation (10 mm) would seem to indicate that more feeder roots are close to the soil surface at this site than at other sites.

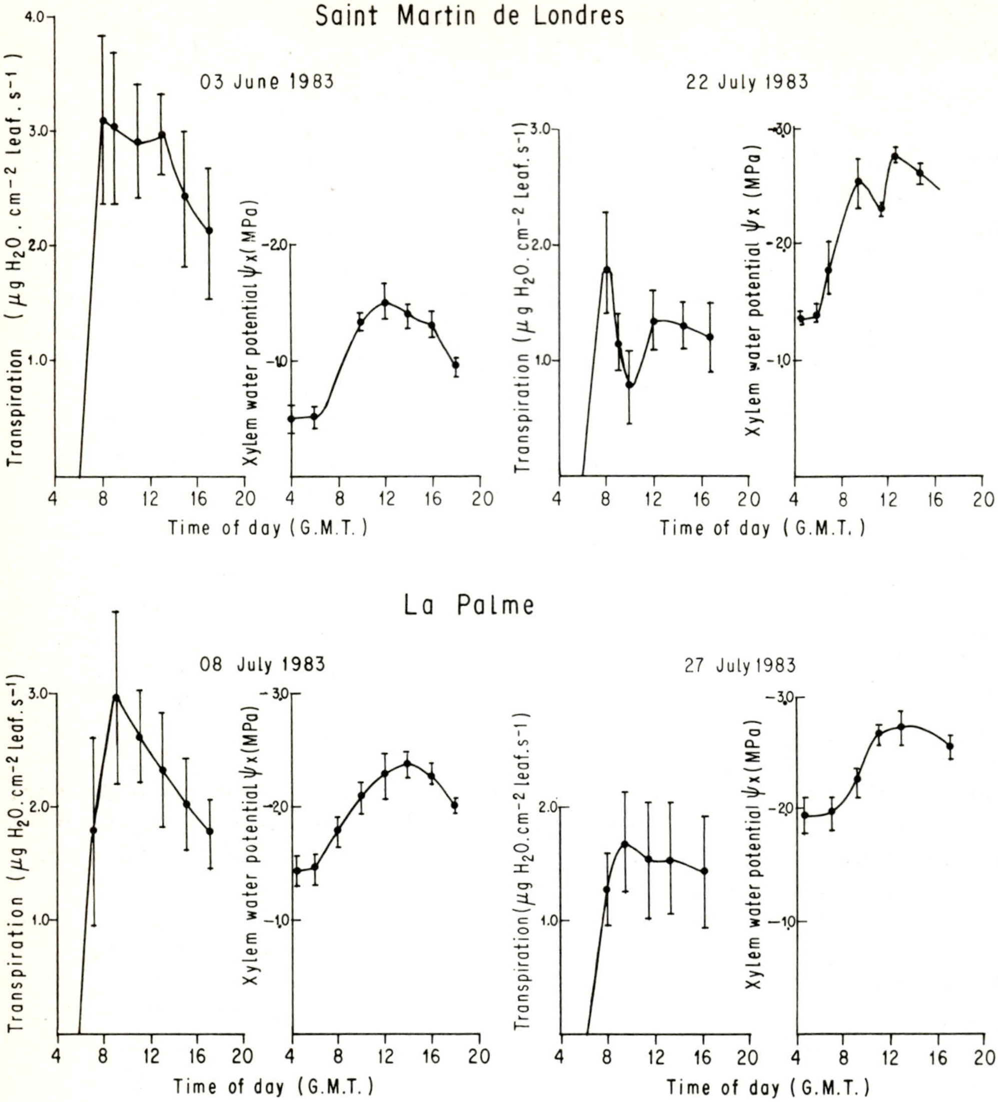

Fig. 2. Diurnal transpiration rate and xylem water potential in SML and LP for June and July 1983

Table 3. Elements for calculation of soil and plant resistances. The resistance values are expressed in 10^9 s and potential values in MPa.

Site	Date	Predawn Ψ_x	Ψ	R_t	R_s	R_p
SML	03 June	-0.49	-0.35	1.77	0.14	1.63
SML	22 July	-1.38	-1.24	4.92	3.29	1.63
LP	08 July	-1.41	-1.27	1.7	0.38	1.32
LP	27 July	-1.90	-1.76	2.17	0.85	1.32
SP	23 June	-1.95	--	--	--	--
SP	24 June	-1.33	-1.19	3.55	--	--

CONCLUSIONS

Sites with more mesic conditions have higher productivity which allows the maintenance of more supporting tissue, hence an increased shrub canopy height. The leaf area indices of *Quercus coccifera* canopies are similar to those in other mediterranean regions (Miller 1983). Decreases in total stem phytomass from mesic to xeric sites are accompanied by changes in canopy structure. Leaf area index decreases with precipitation decrease. This observation is consistent with results obtained by Grier and Running (1977) and Gholz (1982), who found a high correlation between LAI and site water balance. The results support the contention of Poole and Miller (1981) that the main response of shrubs to different precipitation regimes in the chaparral is to change leaf area index rather than physiological parameters.

Changes in leaf area index are associated with changes in rooting depth and modification of the carbon allocation scheme. Deep rooting patterns imply a greater carbon resource investment in the root system. Below some threshold value of annual precipitation, the plant cannot maintain this system. Then, *Quercus coccifera* seems to develop a shallow root system which is able to respond immediately to low levels of precipitation.

However, the ability to avoid damaging effects of drought must include mechanisms which enable the plants to remain productive when water resources are decreased. For example, Tenhunen *et al.* (1985, also in this volume) observed what appears to be a greater stability of the photosynthetic system of *Q. coccifera* compared to other sclerophyll species growing in the same site. Understanding of aboveground structure and rooting depth, and of their dependence on long-term water distribution must include a synoptic view of the interrelated effects of drought stress on water relations, CO_2-gas exchange and differential carbon allocation patterns.

ACKNOWLEDGEMENTS

The comments of John Tenhunen are gratefully acknowledged.

REFERENCES

Aber JD, Pastor J, Melillo JM (1982) Changes in forest canopy structure along a site quality gradient in southern Wisconsin. Am Midl Nat 108:256-265

Campbell GS (1974) A simple method for determining unsaturated conductivity from moisture retention data. Soil Sci 117:311-314

Cowan IR (1965) Transport of water in the soil-plant-atmosphere system. J Appl Ecol 2:221-239

Feddes RA, Rijtema PE (1972) Water withdrawal by plant roots. J Hydrol 17:33-59

Gholz HL (1982) Environmental limits on aboveground net primary production, leaf area, and biomass in vegetation zones of the Pacific Northwest. Ecol 63:469-481

Gosh RK (1977) Determination of unsaturated hydraulic conductivity from moisture retention function. Soil Sci 124:122-124

Grier CC, Running SW (1977) Leaf area of mature northwest coniferous forests: relation to site water balance. Ecol 58:893-899

Miller PC (1983) Canopy structure of mediterranean-type shrubs in relation to heat and moisture. In: Kruger FH, Mitchell DT, Jarvis JUM (eds) Mediterranean-type Ecosystem. Springer, Berlin-Heidelberg-New York, p 133

Poole DK, Miller DC (1981) The distribution of plant water stress and vegetation characteristics in southern California chaparral. Am Midl Nat 105:32-43

Rambal S (1982) Reflexions sur la prise en compte de la végétation et des caractéristiques hydrodynamiques du sol dans les modèles hydrologiques: le cas de bassins versants karstiques de la région nord Montpellièraine. Modélisation Mathématique et Simulation de Systèmes de l'Environnement. CNRS, Paris, p 85

Rambal S (1984) Water balance and pattern of root water uptake by a *Quercus coccifera* L. Evergreen scrub. Oecologia (Berl) 62:18-25

Tenhunen JD, Lange OL, Harley PC, Beyschlag W, Meyer A (1985) Limitations due to water stress on leaf net photosynthesis of *Quercus coccifera* in the Portuguese evergreen scrub. Oecologia (Berl) 67:23-30

Root to shoot communication of the effects of soil drying, flooding or increase salinity. A case for the involvement of plant growth regulators in a multiple chemical signal

W.J. Davies, P.G. Blackman, T.R. Lodge, Alexandra Rosa da Costa and Jane Metcalfe

Department of Biological Sciences, University of Lancaster, Bailrigg, Lancaster LA1 4YQ, UK

INTRODUCTION

The discovery of the first plant growth regulators in the 1930s arose out of enormous interest in the control of growth and development of plants. This interest continues today but there is still considerable uncertainty about the biosynthesis of growth regulators and their roles in plants. Nevertheless we are aware that environmental stresses can have significant effects on synthesis and metabolism. For example, very large increases in the endogenous content of abscisic acid (ABA) are promoted by increasing water deficit and by flooding. These observations, (reported in 1969 by Wright and co-workers at Wye College in the U.K.) probably more than any other single piece of work, stimulated interest in a link between stress, plant growth regulators and the growth and development of plants.

Gradually, it has become apparent that external applications of ABA can mimic many of the physiological and morphological effects of water deficit. It has become fashionable to think of ABA as the plant's 'stress hormone' and to attribute many responses to stress to changes in the endogenous concentration of this compound. Despite the interest in ABA and its effects, many workers still express doubts about the necessity of invoking a role for ABA in responses where turgor changes had previously been thought to be an adequate and appropriate mechanism to explain the response. Further doubts about a role for ABA come to light when apparently poor correlations are reported between, for example, the dynamics of the stomatal response to water deficit and the stress-stimulated increase in endogenous ABA (e.g. Beardsell and Cohen 1976).

The consensus of opinion today is that ABA may still have a central role to play in many of the plant's responses to stress but that other growth regulators and different components of the environment will interact with ABA to determine the final response. This paper will discuss the effects of ABA and highlight some of these important interactions. In addition, a role for plant growth regulators will be proposed, whereby chemical information generated by environmental stress may override any hydraulic effect on physiology, growth and development. We

NATO ASI Series, Vol. G15
Plant Response to Stress. Edited by J. D. Tenhunen et al.

propose that a combination of chemical and hydraulic influences may enable the plant to use water in an optimal fashion.

MIMICKING AND MODIFYING THE PLANT'S RESPONSES TO STRESS WITH PLANT GROWTH REGULATORS

Table 1 shows some of the changes in physiology, growth and development of the plant that can be imposed by withholding water from the soil. These effects have been highlighted because they can also be produced by applications of aqueous solutions of ABA to the plant.

Endogenous concentrations of ABA are often linearly related to leaf water potential below an often-distinct threshold value. The signal for production is not a decline in water potential but rather a decline in turgor (Pierce and Raschke 1980) or simply a shrinkage of cells (Hartung *et al.* 1983). There is some feeling that zero turgor may be the critical point of enhanced synthesis but many plants produce more ABA as soon as the turgor starts to fall or the cells start to shrink (Fig. 1). In addition to direct dehydration stress, ABA accumulation may be stimulated by increasing salinity (Walker and Dumbroff 1981), waterlogging (Mizrahi *et al.* 1970), high and low temperature stress (Chen *et al.* 1983), nutrient deficiency (Goldbach *et al.* 1974) and by pathogen attack (Ayres 1981). In many of these cases, however, the stimulus for enhanced ABA production will be a decline in leaf turgor.

It is easy to see from a consideration of this type of information and that contained in Table 1, why ABA has been considered to be an important stress hormone with a central role in modulating the plant's responses. Further, persuasive evidence for this contention is the 'wilty' habit of ABA-deficient mutants (e.g. Taylor and Rossall 1982) and the superior performance under drought of winter wheat plants selected for enhanced capacity to produce ABA (Quarrie and Lister 1983). Trials in plots which were sheltered from the rain showed that high- ABA selections out yielded low- ABA selections by 5 to 6%, both when fully irrigated and when given a post anthesis drought (Innes *et al.* 1984). These differences of 0.36 tonnes ha^{-1} were statistically significant. In addition, water use efficiency was higher in high-ABA selections. It would be unwise to generalise from one restricted data set but it seems likely that material of this type must provide useful information on a possible role for growth regulators in drought resistance.

Despite promising results of this kind there are increasing numbers of reports where close scrutiny of results reveals a poor correlation between changes in ABA content and stomatal conductance or water potential (e.g. Burschka *et al.* 1983). Often, as we have noted above, stomatal closure is seen to precede an increase in the endogenous concentration of ABA. In other cases, careful measurements of hormone concentrations and elucidation of mode of action seem

Table 1. Effects of water deficit that might be mimicked by applications of ABA and which may be important in drought responses.

Response to deficit	Consequence	Reference to work where ABA shown to have a similar effect
Decreased stomatal conductance	Reduced water loss? Increased WUE? Helps optimise water use?	Jones and Mansfield (1972) Mansfield (1976) Dubbe *et al.* (1978)
Inhibition of photosynthesis	Reduced growth, development and yield	Raschke (1982) Cornic and Miginiac (1983)
Enhanced water flux through roots	Increased water uptake may not occur when plant is transpiring rapidly Effects on ion concentration and membrane properties	Glinka (1980) Fiscus (1981) Davies *et al.* (1982)
Effects on assimilate distribution	May result in increased root/shoot ratio. Response very variable. Changes in root morphology can also result	Watts *et al.* (1981) Karmoker & Van Steveninck (1979)
Reduction in leaf growth	Reduced transpiring area. Smaller cells produced	Hall and McWha (1981) Quarrie and Jones (1977) Van Volkenburgh and Davies (1984)
Changes in leaf morphology e.g. thicker cuticle, more trichomes, reduced number of stomata	Reduced water loss	Quarrie (1985) Quarrie and Jones (1977)
Decreases in secondary tiller production (cereals)	More efficient use of assimilate. Decreased transpiring area	Quarrie (1982)
Earlier ear emergence and and flowering (cereals)	Shorter stems. Earlier grain filling period. More opportunity to extend grain filling period	Hall and McWha (1981) Quarrie (1982)
Pollen sterility	Reduces grain number and therefore yield	Morgan (1980) Saini and Aspinall (1982)
Effects on grain filling	Contradictory results	Dewdney and McWha (1979) King and Patrick (1982) Tietz and Dingkuhn (1981) Tietz *et al.* (1981)

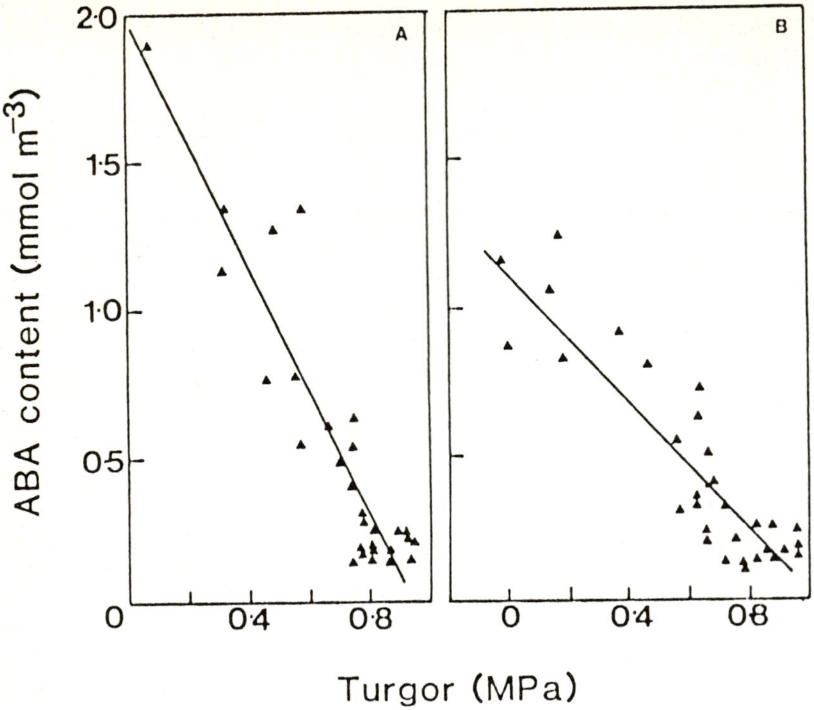

Fig. 1. The relationship between leaf ABA and leaf turgor for control (A) and drought-conditioned (B) plants of rice variety IR 20 during drought stress. Each point represents an individual leaf. Lines are fitted linear regressions. For (A) $y = 2.02 - 0.212x$, $r = 0.94$, $P < 0.001$. For (B) $y = 1.13 - 0.108x$, $r = 0.90$, $P < 0.001$. Both regression coefficient (α) and the intercept (β) differ significantly between A and B at $P < 0.001$. (Modified from Henson 1983)

to indicate that the effects of water deficit are not mediated through the change in concentration of ABA acting alone. For example, although application of ABA to growing leaves will reduce leaf growth rate, it is often found that the youngest, most rapidly growing leaves on the plant have the highest endogenous ABA concentrations. The mechanism of the ABA effect is apparently a reduction in the capacity of cell walls to respond to acidity (Van Volkenburgh and Davies 1983) but water deficit acts to reduce cell growth by reducing the capacity of cells to acidify the wall space (Van Volkenburgh and Boyer 1985).

The conclusion from this type of comparison must be that although increases in ABA concentration in leaves must have a central role in modulating the plant's response to stress, this is clearly too simplistic a view and other factors must be taken into account. Some of these factors are considered below.

A MORE DETAILED INTERPRETATION OF THE PLANT'S RESPONSES TO STRESS

Which Growth Regulators are Influenced by Stress and Where are They?

Table 2 summarises what information we have on the effects of water deficit on the so-called major plant growth regulators. The information is surprisingly sparse. This situation should be rectified in the next few years as immunoassays come into more common use and techniques like gc - ms become more widely available. We would also expect that increased importance might be attached to other classes of compounds (e.g. polyamines - Smith 1985). It is important that for all of these compounds we should have information not just on bulk shoot and root contents, but also information on subcellular location, sites of synthesis, rates of metabolism and identity of active and inactive metabolites.

Much of this information is available for ABA and our knowledge of sites of synthesis and storage allows us to put forward an explanation for the observation that stress-stimulated stomatal closure may precede the buildup of ABA in the leaf. Although ABA is synthesised in the cytoplasm of the leaf mesophyll cells, much ABA within the leaf of the well-watered plant is apparently contained within the chloroplasts. The high pH of chloroplasts makes them act as anion traps for ABA. The first response to water deficit may be a release of ABA to the cytoplasm and the apoplast and thus to sites of action, perhaps as a result of a reduction in stromal pH (Hartung *et al. 1981;* Cowan *et al.* 1982). This redistribution would be an effective, rapid response to developing water stress and would occur before ABA synthesis was enhanced. As yet, information of this quality is not available for the other plant growth regulators.

Interactions Between Different Growth Regulators and the Environment

Despite the fact that we have comparatively little information on the effects of water stress on the synthesis of growth regulators we do know that synthesis of individual regulators is influenced by other regulators. In turn, the activity of these will be affected by water deficit. Figure 2 shows the promotive effects of auxin and cytokinins on water-stress ethylene production via an effect on ACC synthesis. ABA can inhibit this process. In the water-stressed plant, an increase in ABA and a decrease in cytokinin and auxin concentrations would presumably offset the stimulation of ethylene synthesis.

This type of interaction is of interest in considerations of how endogenous concentrations of growth regulators may vary. Of more interest here are the interactions between different regulators and the environment in determining the physiological response to the degree of water stress that has developed. One of the reasons for the reportedly wide variety of effects of a single regulator on an individual process is a general failure to take account of interactions such as these. For example, the influence on stomata of variation in concentration of applied

Table 2. Effects of water deficits on endogenous concentrations of plant growth regulators.

Abscisic acid	Good evidence for increases in concentration as water potentials and turgors fall (see Quarrie 1985)
Auxin (IAA)	Scarce and conflicting evidence. Increase in IAA oxidase activity and reduced auxin transport with water deficit. (Darbyshire 1977; Hartung and Witt 1968; Davenport *et al.* 1977, 1980)
Gibberellins	Very little information. Endogenous GA activity may decline with water deficit (Aharoni *et al.* 1977; Aharoni and Richmond 1978; Hubick *et al.* 1985)
Cytokinins	Surprisingly little information. Deficit may reduce transport from roots. Itai and Vaadia 1965; Itai *et al.* 1968; Itai and Vaadia 1971b; Hubick *et al.* 1985)
Ethylene	Synthesis stimulated by water deficit (Yang and Pratt 1978; Apelbaum and Yang 1981)

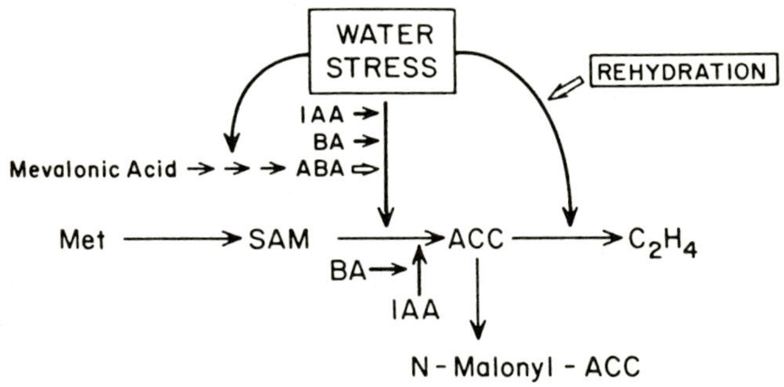

Fig. 2. Regulation of ethylene biosynthesis in water-stressed wheat leaves by plant growth regulators: black arrows = stimulation of the reaction; broad white arrows = inhibition of the reaction. (Modified from Yang *et al.* 1982)

and probably endogenous ABA will depend upon the potassium status of the plant (Radin 1984), endogenous auxin concentration (Snaith and Mansfield 1982) endogenous cytokinin concentration (Blackman and Davies 1983, 1984) and the CO_2 concentration (Fig. 3). It has recently become clear that inhibition of stomatal opening by ABA is also dependent upon the availability of calcium ions (De Silva *et al.* 1985). Age and drought history of the plant will also influence the ABA effect.

The interaction between ABA and cytokinin which is described above (Fig. 3) has now been demonstrated in several graminaceous species. Reports of cytokinin responses in broad leaved species are still few and far between (Jewer and Incoll 1981). It is interesting to note, however, that Radin *et al.* (1982) reported that kinetin could override the effect of ABA on cotton stomata but only in leaves that were deficient of nitrogen (Fig. 4). These results highlight the importance of possible interactions that can occur and how it may be unrealistic to expect simple correlations between endogenous concentrations of a single growth regulator and, for example, stomatal behaviour.

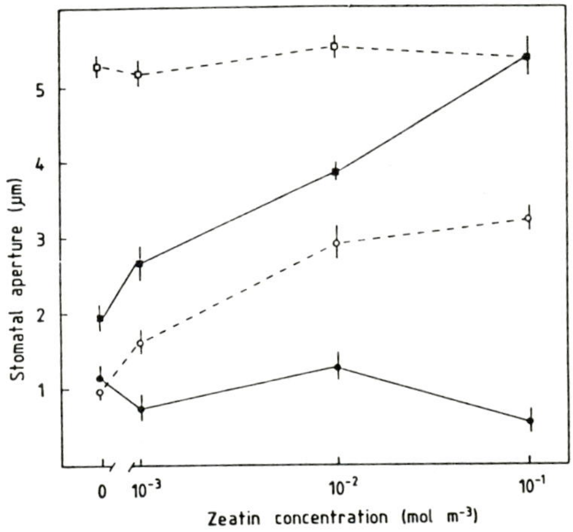

Fig. 3. Apertures of maize stomata from leaf pieces incubated on a range of zeatin concentrations with (——) or without (- - -) CO_2 and with (circles) or without (squares) ABA, (means of 60 observations + s.e.). Modified from Blackman and Davies 1984

WHY IS IT NECESSARY TO INVOKE THE EXISTENCE OF A CHEMICAL SIGNAL TO EXPLAIN THE EFFECTS OF WATER DEFICIT ON PLANTS?

We have considered above the evidence that plant growth regulators have potent effects on physiology and growth and that their synthesis and metabolism is markedly influenced by water deficit. Despite this type of evidence it is difficult in many studies to conclude unequivocally that a particular response shown by a

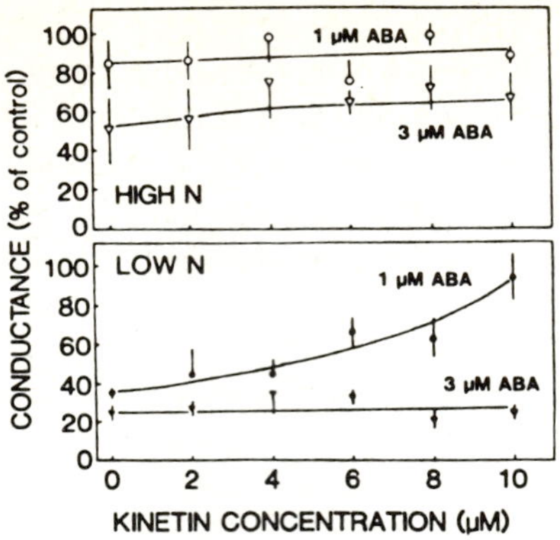

Fig. 4. Interactive effect of ABA and kinetin on conductance of cotton leaves. Plants were either well supplied with nitrogen (a) or nitrogen deficient (b). Values are means ± s.e. (Modified from Radin *et al.* 1982)

plant is produced by a growth regulator, rather than by a turgor change. Nevertheless, significant physiological changes accompanied by only small changes in plant water balance have led many workers to conclude that chemical regulation, perhaps via some signal moving from roots to shoots, may override any hydraulic influence in the shoots (e.g. Davies and Sharp 1981).

More recently, collection of more comprehensive data sets coupled with the use of some ingenious experimental techniques have provided firmer evidence for a chemical influence on shoot physiology that originates in the root and can act as an indicator of 'root stress'. For example, Turner *et al.* (1985) working with *Helianthus* plants dried down either by withholding water from the soil or by changing the vpd around the plant, failed to find any unique relationship between either the transpiration rate or leaf conductance and leaf water potential (Fig. 5). When leaf conductance was plotted as a function of water available in the soil, conductance decreased when about two-thirds of the water in the soil had been utilized in all experiments, irrespective of the vpd or the rate of transpiration (Fig. 6). This observation suggests strongly that soil and root water relations are controlling leaf gas exchange. Blackman and Davies (1985) have reached a similar conclusion using maize plants grown using a split root technique. When part of the root system was dried, stomata closed partially and remained closed, even though leaf turgors were comparable with those of well-watered plants (Fig. 7).

It is now apparent that an increase in soil salinity may exert much the same effect on the physiology and growth of plants. Termaat *et al.* (1985) have exposed wheat and barley plants to 100 millimolar NaCl, which reduced growth by approximately 20%. When sufficient pressure was applied to the roots of plants

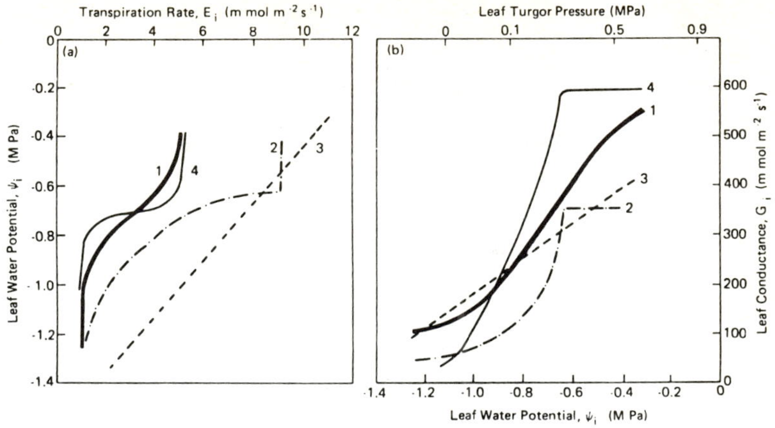

Fig. 5. Summary of the relationships between (a) rate of leaf transpiration and leaf water potential and (b) leaf water potential and leaf conductance for leaves in a cuvette as the extractable soil water content decreased: —, with a vapour pressure deficit of 10 Pa kPa^{-1} around the entire plant (Expt. 1); - • - •, with a vapour pressure deficit of 25 Pa kPa^{-1} around the entire plant (Expt. 2); - - -, for a vapour pressure deficit of 25 Pa kPa when the vapour pressure deficit around the leaf in the cuvette varied from 10 to 30 Pa kPa^{-1} while the remainder of the plant was maintained at a vapour pressure deficit of 10 Pa kPa^{-1} (Expt. 3); —, with the vapour pressure deficit around the plant varied from 10 to 30 pa kPa^{-1} while the leaf in the cuvette was maintained at a vapour pressure deficit of 10 Pa kPa^{-1} (Expt. 4). The numbers refer to the experiment number. (Modified from Turner *et al.* 1985)

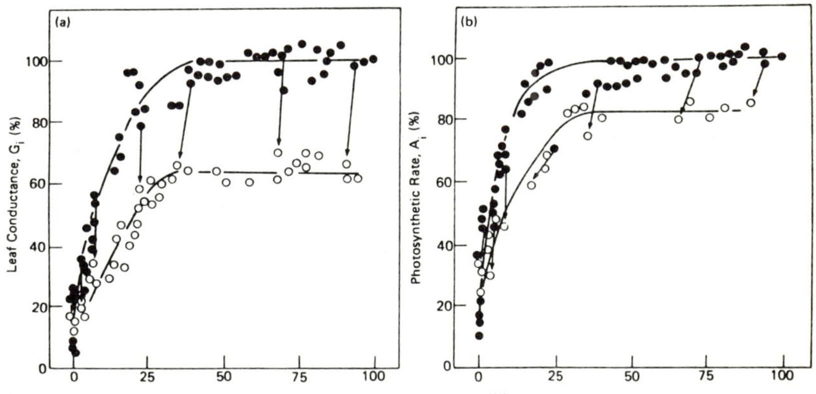

Fig. 6. The relationship between the percentage of extractable soil water and relative leaf conductance for leaves in an atmosphere with a vapour pressure deficit of (10) (•) and 25 (o) Pa kPa^{-1} in all experiments. The arrows give the changes in conductance when the vapour pressure deficit in the cuvette was varied from 10 to 25 Pa kPa^{-1}. (Modified from Turner *et al.* 1985)

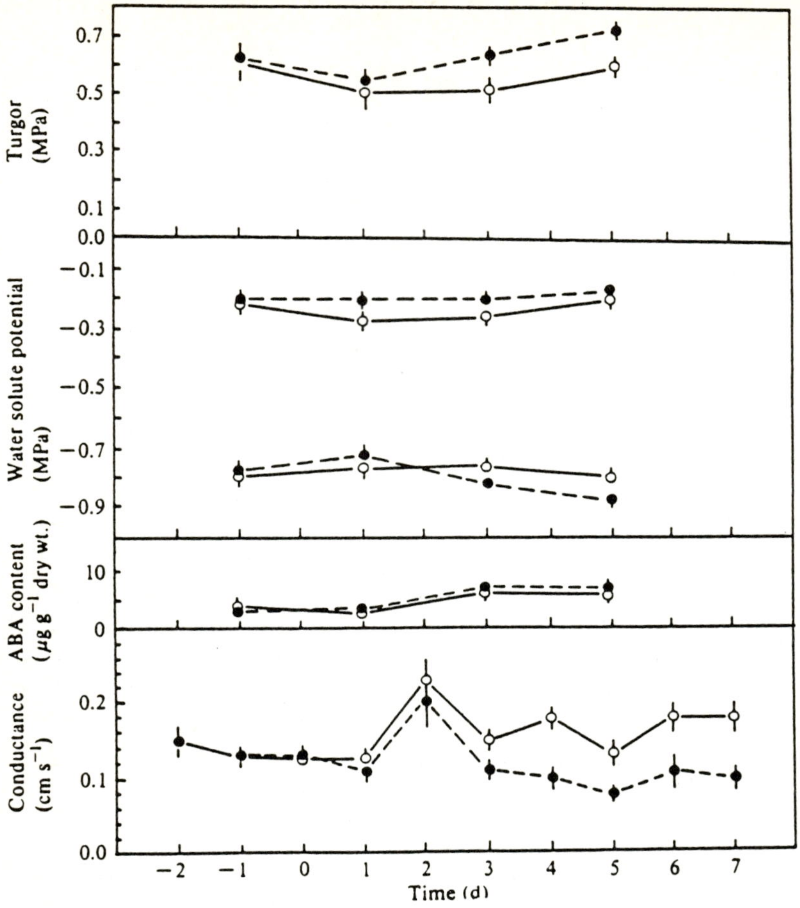

Fig. 7. Water relations, ABA content and stomatal behaviour of maize leaves in plants where water was withheld from part of the root which was split into two separate rooting containers. Control plants have both parts of the root system well-watered (o—o). Plants where part of the root system was dried (---) were last watered on day 0. Points are means of eight observations ± s.e. (Modified from Blackman and Davies 1985)

to offset the osmotic influence of the salt, there was no sustained influence of this treatment on growth rate (Fig. 8) or on transpiration rate. The theory is that the applied pressure increases turgor in the shoot and therefore turgor alone is not regulating growth and gas exchange of these plants. Specific effects of Na^+ and Cl^- have also been ruled out. The conclusion is that some chemical indicator of perturbation of the root environment is moving from the root to the shoot.

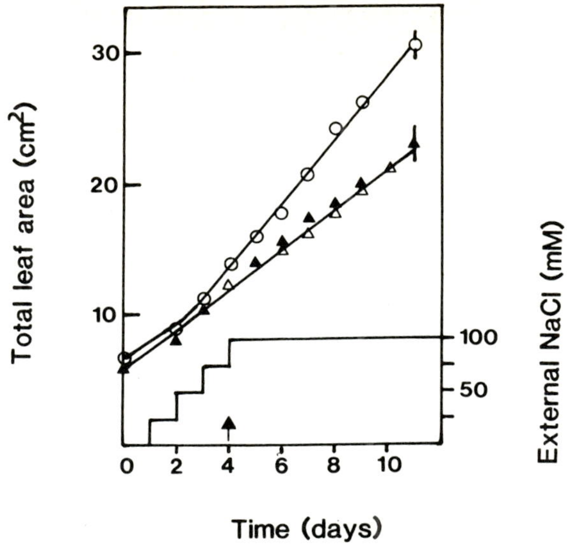

Fig. 8. Effect of NaCl and applied pressure on leaf expansion of wheat. The NaCl was increased in 25 mM increments daily to 100 mM, then a pressure of 0.48 MPa was applied to half the NaCl-treated plants (o). Arrow indicates where pressure first applied. Control plants (no NaCl) (open triangle) 100 mM NaCl; (closed triangle), 100 mM NaCl + pressure. Error bars indicate the average standard error for each group of plants (five plants/treatment). (Modified from Termaat *et al.* 1985)

A CHEMICAL INFLUENCE ON SHOOT PHYSIOLOGY DERIVED FROM THE ROOT

Effects on Gas Exchange

We have reviewed above the evidence for a chemical signal moving from roots to shoots to modify physiology and growth. Interestingly, Jones (1980) and Cowan (1982) have recently considered a role for a signal of this type to enable a plant to utilize water available to it in the soil in an optimal fashion. That is, the water available should be all used up (to maximise gas exchange) but at such a rate that the plant is able to complete its life cycle before this occurs. Using Jones' terminology, 'optimistic' or 'pessimistic' plants are likely to succumb to drought or use only a fraction of the water available to them. If the plant has evolved the capacity to use water in an optimal fashion, some measure of water availability (soil moisture status) must be available to allow it to regulate gas exchange accordingly. Cowan's more sophisticated analysis involves a similar mechanism in a plant which has evolved in an environment with a reasonably predictable frequency of rainfall. Optimal water use dictates that the plant measures the availability of water and thereby regulates gas exchange between rainfall episodes.

The Inadequacy of the Hydraulic Signal and the Independence of Root Tips

The type of hydraulic information available to shoots of plants growing in drying soil is illustrated with reference to Figure 9. Two maize plants are compared, one which has received water every day during the experimental period and another which has remained unwatered for 16 days. The unwatered plant is extracting most of its water from the bottom of the profile, while the the well-watered plant is removing most water from 20-30 cm below the soil surface. Despite this difference, both plants are extracting roughly the same amount of water from the soils (areas under the curves). This is the case even though one plant has not been watered for 16 days. Leaf turgors of unwatered plants are sustained throughout the experimental period (Sharp and Davies 1985). From these data it is difficult to conclude that a simple hydraulic signal is an adequate indicator to the shoots of a very significant decline in soil moisture status and water availability.

That additional information is available to shoots of such plants is illustrated by Figure 10. This figure shows the root distribution of the plants described in detail above. It is clear that after 16 days with no water added, plants have 75% of their root length in very dry soil. Is there any way that this information might be conveyed to the shoots? Our work on the water relations of maize roots has led us to the conclusion that maize root tips may be rather hydraulically isolated from the rest of the root system (Sharp and Davies 1979). It seems possible that in very dry soil, these tips may lose turgor, even though other parts of the plant have higher water potentials and turgors. Such a situation is illustrated in Fig. 11. In very dry soil, the net flow of water may be out of the root tip to the soil and root turgor may decline, even though other roots remain turgid and continue to take up water. We have some indication that root tip turgors do decline (Davies *et al.* 1986) and we might expect that root metabolism would be greatly influenced by such changes.

Growth Regulators as Signals

Two possibilities for chemical signals moving from dry root tips are either enhanced transport of some inhibiting influence (ABA?) or reduced transport of a positive influence required for 'normal' shoot growth and metabolism. ABA may be synthesised in roots in small quantities and this synthesis may be enhanced by water deficit (Walter *et al.* 1976) but we have found no evidence of enhanced ABA contents in maize plants showing reduced conductance when part of their root system has been dried (Fig. 7). A search for a promoter of opening of these stomata has revealed that low concentrations of both naturally-occurring and

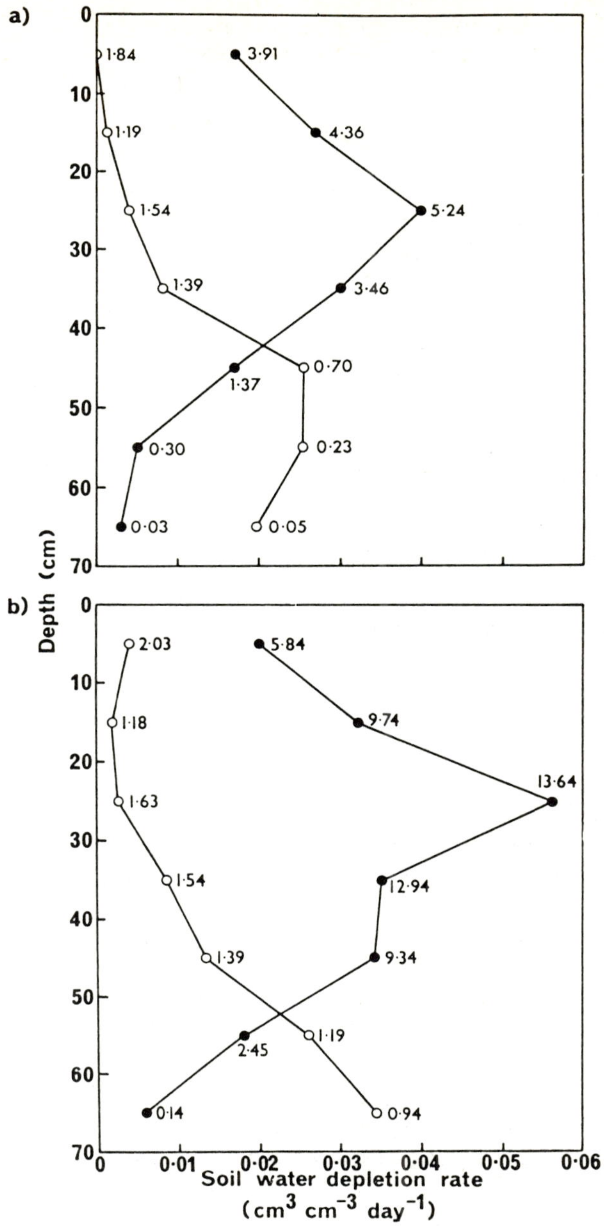

Fig. 9. The mean soil water depletion rate from consecutive 10 cm soil layers of tubes watered daily (•—•) and of tubes not watered after day 0 (o—o) during a) day 15 (well watered) and days 13 to 15 (unwatered), and b) day 18 (well watered) and days 16 to 18 (unwatered). The numbers against the points are the mean root length densities (cm cm^{-3}) in each soil on day 15 (a) and day 18 (b). (Modified from Sharp and Davies 1985)

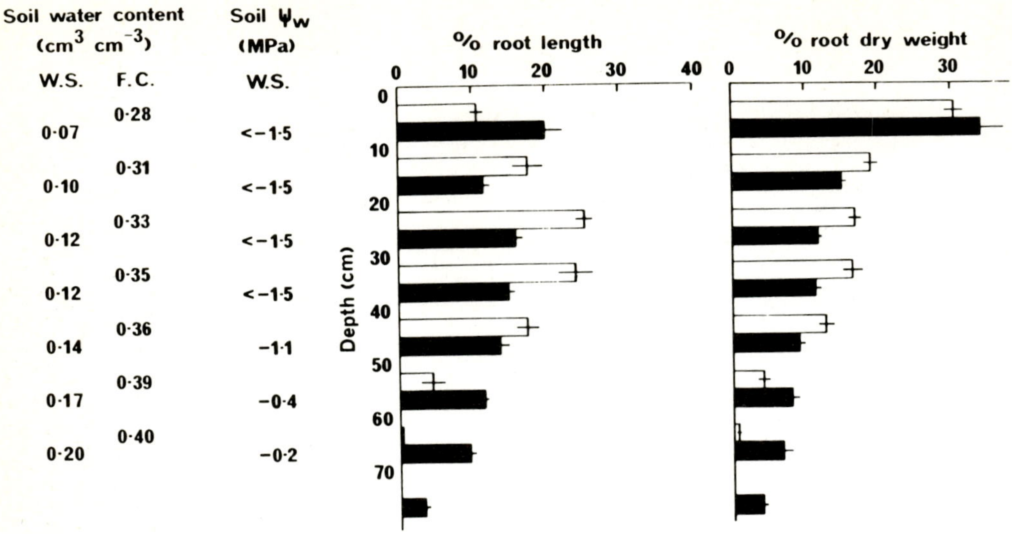

Fig. 10. The percentage of total root length and of total root dry weight in consecutive 10 cm soil layers after an 18 day period during which plants were either watered daily (open bars) or not watered after day 0 (shaded bars). Values are means of 5 determinations ± standard error. The mean soil water content and approximate bulk soil water potential of each layer are indicated. W.S., water stressed. F.C., field capacity. (Modified from Sharp and Davies 1985)

synthetic cytokinins reopen stomata to apertures comparable to those shown by well-watered plants (Fig. 12- Blackman and Davies 1985).

This result suggests the possibility that cytokinin supply from root tips may be reduced by soil drying (declining root turgor) and this may restrict stomatal opening and therefore water use by the plant. Total cytokinin supply would be an integrated indication of turgors of all the root tips and might act as a proportional controller of gas exchange as more and more root tips dry. It is possible that reduced cytokinin transport might also reduce growth of leaves (an additional means of reducing water use) (see Rosa da Costa *et al.* this volume), since a sustained cytokinin supply may be necessary to keep the walls of growing cells extensible.

If reduced cytokinin supply is to act as a uni-directional signal within the plant it is necessary that synthesis should be largely confined to roots. This may be the case (e.g. Van Standen and Davey 1979) but there is some controversy in this area (e.g. Salama and Wareing 1979). There is evidence that supply of cytokinin-like substances from roots is reduced by water deficit (Livne and Vaadia 1965). The

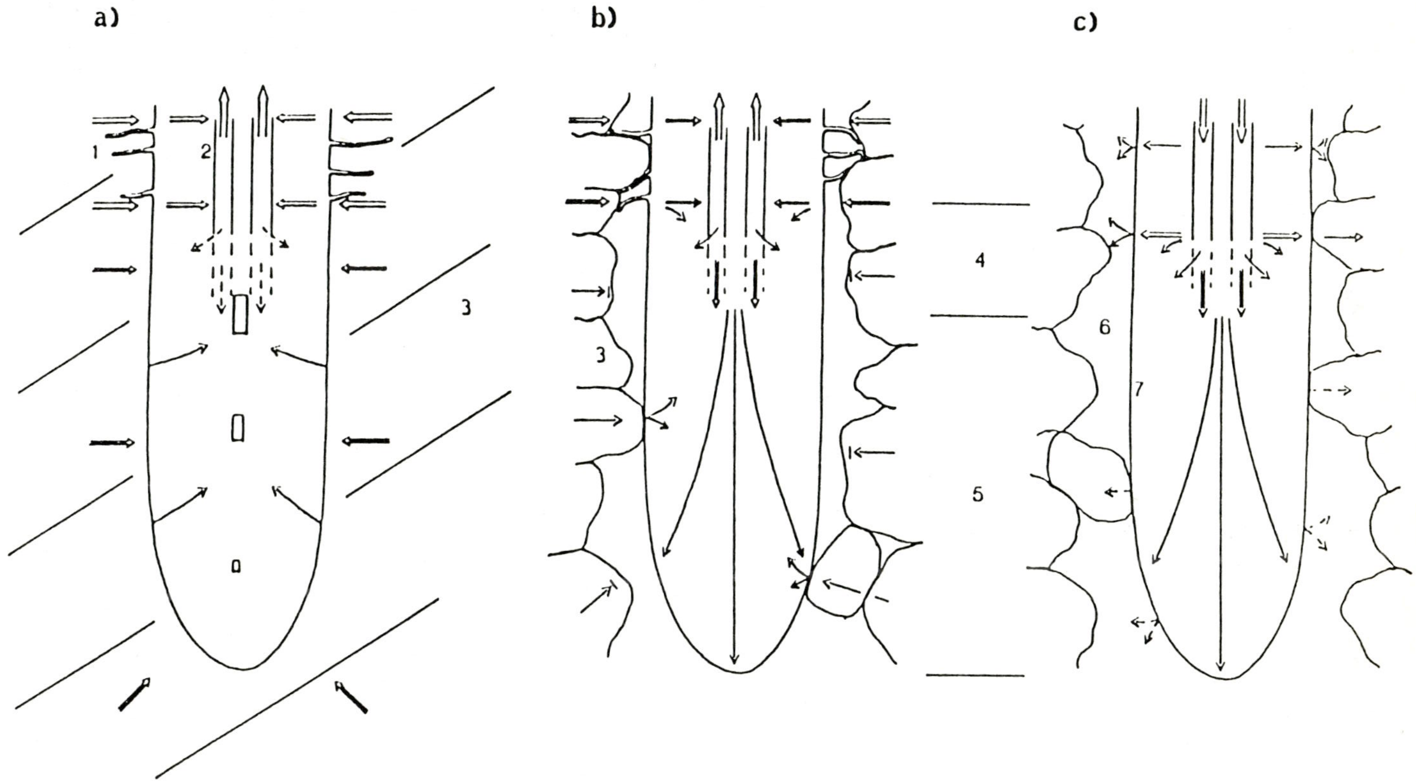

Fig. 11. A simplified diagram of the paths of water flow in the apex of a root growing in a) well watered soil, b) dry soil and c) very dry soil with a substantial vapour gap between the soil and the root. Broken arrows indicate possible paths of water flow dependent on the comparative water potentials of the root tissues and of the surrounding soil. Key: 1. Root hairs, 2. Xylem, 3. Soil, 4. Zone of xylem differentiation, 5. Non-differentiated zone, 6. Air filled pore, 7. Root surface not wetted by soil water. (Modified from Davies *et al.* 1986)

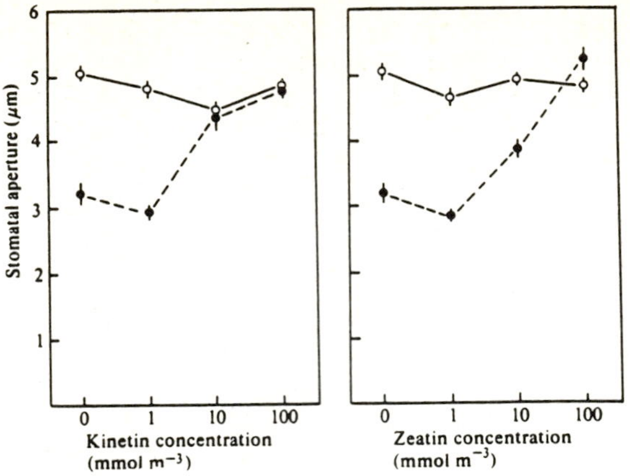

Fig. 12. Apertures of stomata of maize leaves incubated in the light in CO_2-free air on distilled water, kinetin or zeatin solutions. Leaves, which were taken from well-watered (o—o) or partly dried plants (•—•), were sampled 8 h into the light period on day 5 after water was first withheld from part of the root system of half of the plants. Points are means of 60 observations ± s.e. (Modified from Blackman and Davies 1985)

latter may result from the reversal of the flow of water in roots in very dry soil where root tips may be surrounded by a vapour gap (Fig. 11). Cytokinin synthesis is also reduced by flooding (Burrows and Carr 1969) and by increased salinity (Walker and Dumbroff 1981). Significantly, however, cytokinin applications did not enhance the growth and transpiration of the salt stressed wheat and barley plants that were discussed above (Termaat pers. comm.) and thus alternative chemical signals must exist.

CONCLUSIONS - A MULTIPLE CHEMICAL SIGNAL?

Although we have a good case for growth regulators acting independently of the hydraulic influence in plants as a chemical indication of root perturbation, it is unrealistic to expect that a single regulator will act in isolation. Rather, it seems likely that several components will interact to produce a physiological effect which cannot be explained as a result of a hydraulic change.

For example, we have described stomatal closure in turgid plants which are undergoing root drying. The same response can occur in flooded plants (Jackson and Kowalewska 1983). In both cases one might expect that the concentration of cytokinin moving from the roots would be reduced and this change alone may influence shoot physiology. In addition, however, both soil drying and flooding have been shown to result in nutrient deficiency in plants. If water deficit reduces the uptake of nitrogen and phosphorus (see Begg and Turner 1976; Shaner and Boyer 1976) and waterlogging reduces nitrogen uptake (e.g. Jackson and Kowalewska 1983) we might expect that nutrient deficiency would interact with the balance of both cytokinin and ABA in the plant to influence stomata and perhaps other aspects of shoot physiology. Reduced N and P status would increase the sensitivity of various processes to what may be a fairly constant ABA content (Radin *et al.* 1982). In essence, nutrient deficiency amplifies the effects of both growth regulators. Variation in the levels of these different components of what may be described as a multiple chemical signal could act to influence shoot physiology independently of any hydraulic influence moving from roots. The shoot might integrate all this information to optimise water use with respect to water availability. The multiple chemical signal is proposed as a sensitive link between shoot physiology and the plant's edaphic environment.

REFERENCES

Aharoni M, Richmond AE (1978) Endogenous gibberellin and and abscisic acid content as related to senescence of detached lettuce leaves. Plant Physiol 62:224-228

Aharoni M, Blumenfeld A, Richmond AE (1977) Hormonal activity in detached lettuce leaves as affected by leaf water content. Plant Physiol 59:1169-1173

Apelbaum A, Yang SF (1981) Biosynthesis of stress ethylene induced by water deficit. Plant Physiol 68:594-596

Ayres PG (1981) Responses of stomata to pathogenic micro-organisms. In: Jarvis PG and Mansfield TA (eds) Stomatal Physiology. Cambridge University Press, Cambridge, p 295

Beardsell MF, Cohen D (1975) Relationships between leaf water status, abscisic acid levels, and stomatal resistance in maize and sorghum. Plant Physiol 56:207-212

Begg JE, Turner NC (1976) Crop water deficits. In: Brady NC (ed) Advances in Agronomy, Vol. 28. Academic Press, New York, p 411

Blackman PG, Davies WJ (1983) The effects of cytokinins and ABA on stomatal behaviour of maize and *Commelina*. J Exp Bot 34:1619-1626

Blackman PG, Davies WJ (1984) Modification of the CO_2 responses of maize stomata by abscisic acid and by naturally-occurring and synthetic cytokinins. J Exp Bot 35:174-179

Blackman PG, Davies WJ (1985) Root to shoot communication in maize plants of the effects of soil drying. J Exp Bot 36:39-48

Burrows WJ, Carr DT (1969) Effect of flooding the root system of sunflower plants on the cytokinin content of the xylem sap. Physiol Plant 22:1105-1112

Burschka C, Tenhunen JD, Hartung W (1983) Diurnal variations in abscisic acid content and stomatal responses to applied abscisic acid in leaves of irrigated and nonirrigated *Arbutus unedo* plants under naturally fluctuating environmental conditions. Oecologia 58:128-131

Chen HH, Li PH, Brenner ML (1983) Involvement of abscisic acid in potato cold acclimation. Plant Physiol 71:362-365

Cornic G, Miginiac E (1983) Nonstomatal inhibition of net CO_2 uptake by (±) abscisic acid in *Pharbitis nil*. Plant Physiol 73:529-533

Cowan IR (1982) Regulation of water use in relation to carbon gain in higher plants. In: Lange OL, Nobel PS, Osmond CB, Ziegler H (eds) Water Relations and Carbon Assimilation Vol 12B Enc Plant Physiol NS Springer Verlag, Berlin, p 747

Cowan IR, Raven JA, Hartung W, Farquhar GD (1982) A possible role for abscisic acid in coupling stomatal conductance and photosynthetic carbon metabolism. Aust J Plant Physiol 9:489-498

Davenport TL, Morgan PW, Jordan WR (1977) Auxin transport as related to leaf abscission during water stress in cotton. Plant Physiol 59:554-557

Davenport TL, Morgan PW, Jordan WR (1980) Reduction of auxin transport capacity with age and internal water deficits in cotton petioles. Plant Physiol 65:1023-1025

Davies WJ, Sharp RE (1981) The root: a sensitive detector of a reduction in water availability? In: Kralovic J (ed) Mechanisms of Assimilate Distribution and Plant Growth Regulators. Slovak Soc of Agric, Ivanka pri Dunjani, Czechoslovakia, p 354

Davies WJ, Rodriguez JL, Fiscus EL (1982) Stomatal behaviour and water movement through roots of wheat plants treated with abscisic acid. Plant Cell and Env 5:485-493

Davies WJ, Metcalfe J, Lodge TA, Costa AR (1986) Plant growth substances and the regulation of growth under drought. Aust J Plant Physiol (in press)

Derbyshire B (1971) Changes in indoleacetic acid oxidase activity associated with plant water potential. Physiol Plant 25:80-84

De Silva DLR, Hetherington AM, Mansfield TA (1985) Synergism between calcium ions and abscisic acid in preventing stomatal opening. New Phytol (in press)

Dewdney SJ, McWha JA (1979) Abscisic acid and the movement of photosynthetic assimilates towards developing wheat grains. Z Pflanzenphysiol 92:183-186

Dubbe DR, Farquhar GD, Raschke K (1978) Effect of abscisic acid on the gain of the feedback loop involving carbon dioxide and stomata. Plant Physiol 62:413-417

Fiscus EL (1981) Effects of abscisic acid on the hydraulic conductance of and the total ion transport through *Phaseolus* root systems. Plant Physiol 68:169-174

Glinka Z (1980) Abscisic acid promotes both volume flow and ion release to the xylem in sunflower roots. Plant Physiol 65:537-540

Goldbach E, Goldbach H, Wagner H, Michael G (1975) Influence of N-deficiency on the abscisic acid content of sunflower plants. Physiol Plant 34:138-140

Hall HK, McWha JA (1981) Effects of abscisic acid on growth of wheat (*Triticum aestivum* L.) Ann Bot 47:427-433

Hartung W, Witt J (1968) Über den Einfluß der Bodenfeuchtigkeit auf den Wuchsstoffgehalt von *Anastaticahierochuntica* und *Helianthus annuus*. Flora (Jena) Abt. B 157:603-614

Hartung W, Heilmann B, Gimmler H (1981) Do chloroplasts play a role in abscisic acid synthesis? Plant Sci Lett 22:235-242

Hartung W, Kaiser WM, Burschka C (1983) Release of abscisic acid from leaf strips under osmotic stress. Z Pflanzenphysiol 112:131-138

Henson I (1983) Abscisic acid and water relations of rice (*Oryza sativa* L.): effects of drought conditioning on abscisic acid accumulation in the leaf and stomatal response. Ann Bot 52:247-255

Hubick KT, Taylor JS, Reid DM (1985) The effect of drought on levels of abscisic acid, cytokinins, gibberellins and ethylene in aeroponically-grown sunflower plants. Plant Growth Reg (in press)

Innes P, Blackwell RD, Quarrie SA (1984) Some effects of genetic variation in drought-induced abscisic acid accumulation on the yield and water use of spring wheat. J Agric Sci Camb 102:341-351

Itai C, Vaadia Y (1965) Kinetin-like activity in root exudate of water-stressed sunflower plants. Physiol Plant 18:941-944

Itai C, Vaadia Y (1971) Cytokinin activity in water stressed shoots. Plant Physiol 47:87-90

Itai C, Richmond A, Vaadia Y (1968) The role of root cytokinins during water and salinity stress. Israel J Bot 17:187-195

Jackson MB, Kowalewska AKB (1983) Positive and negative messages from roots induce foliar desiccation and stomatal closure in flooded pea plants. J Exp Bot 34:493-506

Jewer PC, Incoll LD (1981) Promotion of stomatal opening in detached epidermis of *Kalanchoe* by natural and synthetic cytokinins. Planta 153:317-318

Jones HG (1980) Interaction and integration of adaptive responses to water stress: the implications of an unpredictable environment. In: Turner NC, Kramer PJ (eds) Adaptation of plants to water and high temperature stress. Wiley and Sons, New York, p 436

Jones RJ, Mansfield TA (1972) Effects of abscisic acid and its esters on stomatal aperture and the transpiration ratio. Physiol Plant 26:321-327

Karmoker JL, Van Steveninck RFM (1979) The effect of abscisic acid on sugar levels in seedlings of *Phaseolus vulgaris* L. cv. Redland Pioneer. Planta 146:25-30

King RW, Patrick JW (1982) Control of assimilate movement in wheat. Is abscisic acid involved? Z Pflanzenphysiol 106:375-380

Mansfield TA (1976) Chemical control of stomatal movements. Phil Trans Roy Soc Ser B 284:471-482

Mizrahi Y, Blumenfeld A, Richmond AE (1972) The role of abscisic acid and salination in the adaptive response of plants to reduced root aeration. Plant Cell Physiol 13:15-21

Morgan JM (1980) Possible role of abscisic acid in reducing seed set in water stressed wheat plants. Nature 285:655-657

Pierce M, Raschke K (1980) Correlation between loss of turgor and accumulation of abscisic acid in detached leaves. Planta 148:174-182

Quarrie SA (1982) The role of abscisic acid in the control of spring wheat growth and development. In: Wareing PF (ed) Plant Growth Substances 1982. Academic Press, London, p 683

Quarrie SA (1984) Abscisic acid and drought resistance in crop plants. BPGRG News Bull 7:1-15

Quarrie SA, Jones HG (1977) Effects of abscisic acid and water stress on development and morphology of wheat. J Exp Bot 28:192-203

Quarrie SA, Lister PG (1983) Characterisation of spring wheat genotypes differing in drought-induced abscisic acid accumulation. In: Drought Stressed ABA Production. J Exp Bot 34:1260-1270

Radin JW (1984) Stomatal responses to water stress and to abscisic acid in phosphorus-deficient cotton plants. Plant Physiol 76:392-394

Radin JW, Parker LL, Guinn G (1982) Water relations of cotton plants under nitrogen deficiency. V Environmental control of abscisic acid accumulation and stomatal sensitivity to abscisic acid. Plant Physiol 70:1066-1070

Raschke K (1982) Involvement of abscisic acid in the regulation of gas exchange: evidence and inconsistencies. In: Wareing PF (ed) Plant Growth Substances 1982. Academic Press, London, p 683

Saini MS, Aspinall D (1982) Sterility in wheat (*Triticum aestivum* L.) induced by water deficit or high temperature: possible mediation by abscisic acid. Assist J Plant Physiol 9:529-537

Salama AM El-Da, Wareing PF (1979) Effects of mineral nutrition on endogenous cytokinins in plants of sunflower (*Helianthus annuus* L.). J. Exp Bot 30:971-981

Shaner DL, Boyer JS (1976) Nitrate reductase activity in maize leaves. II. Regulation of nitrate flux at low leaf water potential. Plant Physiol 58:505-509

Sharp RE, Davies WJ (1979) Solute regulation and growth by roots and shoots of water stressed maize plants. Planta 147:43-49

Sharp RE, Davies WJ (1985) Root growth and water uptake by maize plants in drying soil. J Exp Bot 36:1441-1456

Smith TA (1985) Polyamines. Ann Rev Plant Physiol 36:117-143

Snaith PJ, Mansfield TA (1982) Control of the CO_2 responses of stomata by indol-3-yl acetic acid and abscisic acid. J Exp Bot 33:360-365

Taylor IB, Rossall S (1982) The genetic relationship between the tomato mutants *flacca* and *lateral suppressor*, with reference to abscisic acid accumulation. Planta 154:1-6

Termaat A, Passioura JB, Munns R (1985) Shoot turgor does not limit shoot growth of NaCl-treated wheat and barley. Plant Physiol 77 (in press)

Tietz A, Dingkuhn M (1981) Regulation of assimilate transport in barley by the abscisic acid content of young caryopses. Z Pflanzenphysiol 104:475-479

Tietz A, Ludewig M, Dingkuhn M, Dorfling K (1981) Effect of abscisic acid on the transport of assimilates in barley. Planta 152:557-561

Turner NC, Schulze ED, Gollan T (1985) The responses to vapour pressure deficits and soil water content 11. In the mesophytic herbaceous species *Helianthus annuus*. Oecologia 65:348-355

Van Staden J, Davey JE (1979) The synthesis, transport and metabolism of endogenous cytokinins. Plant Cell and Env 2:93-106

Van Volkenburgh E, Davies WJ (1983) Inhibition of light stimulated leaf expansion by abscisic acid. J Exp Bot 34:835-845

Van Volkenburgh E, Boyer JS (1985) Inhibitory effects of water deficits on maize leaf elongation. Plant Physiol 77:190-194

Walker MA, Dumbroff EB (1981) Effects of salt stress on abscisic acid and cytokinin levels in tomato. Z Pflanzenphysiol 101:461-470

Walton DC, Harrison MA, Cote P (1976) The effects of water stress on abscisic acid levels and metabolism in roots of *Phaseolus vulgaris* L. and other plants. Planta 131:141-144

Watts S, Rodriguez JL, Evans SE, Davies WJ (1981) Root and shoot growth of plants treated with abscisic acid. Ann Bot 47:595-602

Wilson JA, Ogunkanmi AB, Mansfield TA (1978) Effects of external potassium supply on stomatal closure induced by abscisic acid. Plant Cell & Env 1:199-201

Yang SF, Pratt HK (1978) The physiology of ethylene in wounded plant tissues. In: Kahl G (ed) Biochemistry of wounded plant tissues. De Cruyter, Berlin, p 741

Yang SF, Hoffman HE, McKeon T, Riov J, Kao CH, Yung KH (1982) Mechanism and regulation of ethylene biosynthesis. In: Wareing PF (ed) Plant growth substances 1982. Academic Press, London, p 683

Stomatal conductance and water relations of shrubs growing at the chaparral-desert ecotone in California and Arizona

M.E. Blake-Jacobson

Systems Ecology Research Group, San Diego State University, San Diego, CA 92182 U.S.A.

INTRODUCTION

Water is one of the main factors limiting production in mediterranean-type ecosystems. The degree of stomatal opening and thereby leaf conductance largely determine rates of transpiration during drought periods and along with other factors control rates of carbon uptake (see Beyschlag *et al.,* Tenhunen *et al.* and other related papers in this volume) and water use efficiency. This study examined the water relations and control of stomatal conductance in several plant species subjected to drought, at each of three sites differing in amounts and seasonal distribution of precipitation and at intervals throughout all seasons. The data were gathered as part of a study of the water balance of shrubs growing at the chaparral-desert ecotones in Arizona and southern California, designed to examine potentials for anti-desertification (Roberts and Blake-Jacobson 1984).

MATERIALS AND METHODS

Three sites at chaparral-desert ecotones in Arizona and California were selected for study (Fig. 1). Three Bar in central Arizona, at the lower elevational edge of the Petran chaparral, was the wettest site. The Aravaipa site, at the southeastern distributional edge of Petran chaparral, was disturbed by grazing and man's activities. The California site, on the eastern edge of southern California chaparral adjoining the Sonoran desert, was the driest site. Mean annual precipitation was 434, 393 and 321 mm, respectively.

At each site, three chaparral and two desert shrub species were chosen for study. The same or similar species were studied at all sites. The species were widely distributed within their respective habitats (chaparral or desert) in addition to occurring at the ecotone. The three chaparral species were broadleaf evergreen sclerophyllous shrubs: *Rhus ovata* Wats., *Quercus turbinella* Greene ssp. *turbinella* and *Arctostaphylos* spp. Because no given *Arctostaphylos* species appeared at all three sites, we used *A. glauca* Lindl. in California and *A. pungens* H. B. K. in Arizona; both are shallow-rooted evergreens (Cable 1975; Thrower and Bradbury 1977). The desert species were all fine-leaved. One was the winter-deciduous *Acacia greggii* Gray., which was at every site. The second species was the evergreen *Larrea tridentata* (Sesse & Moc. ex DC.) Cov. at the

NATO ASI Series, Vol. G15
Plant Response to Stress. Edited by J.D. Tenhunen et al.

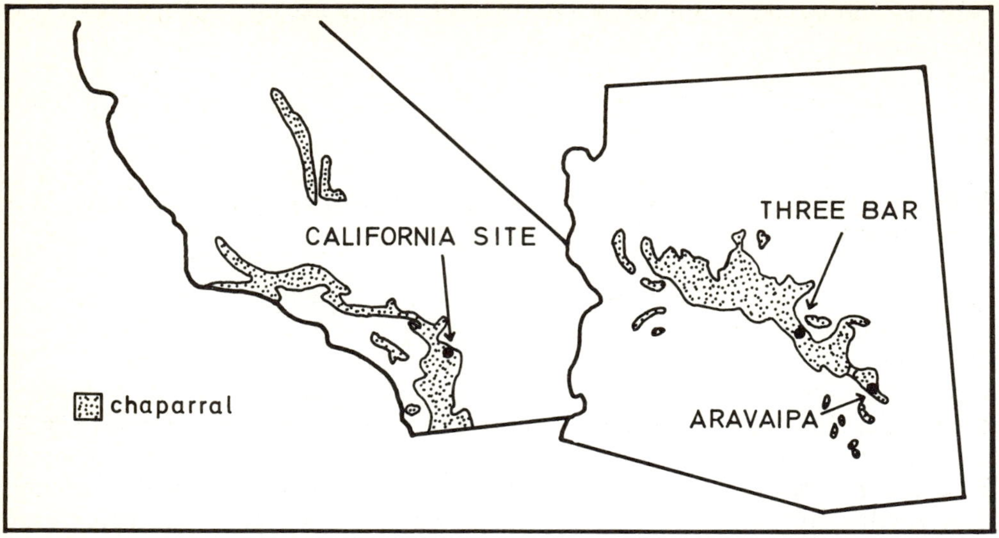

Fig. 1. Location of the three study sites in Southern California and in Arizona, U.S.A. Stipled areas indicate extensive chaparral vegetation. The area between the sites is mostly desert

California site and the winter-deciduous *Prosopis juliflora* (Schwartz) DC. var. *velutina* (Wooton) Sarg. at the two Arizona sites. Nomenclature follows Munz (1976), Kearney and Peebles (1960) and Benson and Darrow (1981). A brief overview of species characteristics is provided in Table 1.

Three individual shrubs were chosen from each of the five different species and were used throughout the study. Measurements at each site were replicated on two successive days at six-week intervals during the first study year and were repeated in three different seasons of the second year. There were 21 study days per site. The first study year ran from October 1981 to September 1982, and the second from October 1982 to November 1983.

Total solar radiation was measured with a Moll-Gorczynski type Kipp and Zonen solar radiometer. Wind speed was measured with a low stall speed generator type three-cup anemometer. Air temperature and relative humidity at 2 m were measured with a shaded thermistor and a Phys-Chemical relative humidity sensor housed in a ventilated weather shelter. Environmental factors were monitored at ten-second intervals throughout the study days, and half-hourly means of these observations were recorded with a Campbell Scientific Instruments CR21 Micrologger.

Table 1. Characteristics of the desert and chaparral species studied.

	Habitat	Leaf type	Rooting depth
Arctostaphylos pungens or *glauca*	chaparral	evergreen	shallow
Quercus turbinella	chaparral	evergreen	deep
Rhus ovata	chaparral	evergreen	deep
Acacia greggii	desert	winter deciduous	medium-deep
Prosopis juliflora var. *velutina*	desert	winter deciduous	deep
Larrea tridentata	desert	evergreen	shallow

Xylem pressure potentials were measured with a Scholander type pressure chamber (Scholander *et al.* 1965; Roberts *et al.* 1981). Moisture content in the upper soil profile was determined gravimetrically. Diurnal changes in stomatal conductance to water vapor, air temperature, and relative humidity within the shrub canopy were observed at approximately two-hour intervals from predawn to sunset, using a null-balance porometer and according to methods described by Beardsell *et al.* (1972).

RESULTS

Climate and Microclimate

Multi-year macroclimate data for precipitation and air temperature were obtained from stations nearby the study sites. Precipitation records were available for 31 years (1953 to 1983) for the Arizona sites but for only 8 years at the California site. Total rainfall increases from California to Aravaipa to Three Bar from 321 to 434 mm. Seasonal rainfall distribution varied among the sites. Three Bar and Aravaipa had 68% and 54%, respectively, falling in the winter months (October through April). The California site's winter rainfall was 78 or 69% of total rainfall, depending on whether data from 1983 were excluded or included in the averages. As mentioned above, data were available for only eight years here, and an exceptionally heavy rain storm in August 1983 skewed the average. For the two study years 1981-1983 the winter/summer distribution patterns were close to the average for all three sites.

Before the study began there was a very dry year. From October 1980 to October 1981 the Arizona sites received 75% of mean precipitation, while the California site received just over half of normal. The fall of 1981 was particularly dry at Aravaipa and California. The first study year, October 1981 to September 1982, was wetter at Three Bar (66% above the 434 mm mean) and slightly above the 393 mm normal at Aravaipa. The California site received less rain (23% below its mean of 372 mm) due to a dry winter. The second study year was very wet, being 113%, 65% and 100% above the means at the three sites, respectively. The two study years at both Arizona sites were among the 5 wettest years on record.

Temperature data, available for only 4 to 8 years, depending on site, show that plants at the California site are exposed to slightly more moderate temperatures than those at the Arizona sites. However, data for Aravaipa come from the Klondyke station which is 384 m lower in elevation and which would be warmer than the actual study site. Maximum monthly mean temperatures (usually the midday temperatures) show that Three Bar is warmer in spring and summer than the other two sites, but that winters are similar throughout. Temperature minima indicate that the California site is 1-2 degrees warmer than Three Bar, which is 2 degrees warmer than Klondyke; Klondyke is assumed to be warmer than Aravaipa.

Pairs of field days on which leaf conductances were measured had, in general, similar daily temperature maxima across all sites. Temperature minima on study days followed the pattern for monthly means, i.e. cooler at Three Bar and coolest at Aravaipa. Nighttime temperatures preceding study days occasionally fell below 5 degrees C; this occurred 7 times each at Aravaipa and Three Bar and once in California. Of these occasions, two at Aravaipa and four at Three Bar were minor, i.e. shorter than two hours in duration or only 1 degree below 5 degrees C.

Water Potentials

As would be expected, lower water potentials were observed in summer and in the dry fall of 1981, with recovery to higher water potentials in the wetter seasons (Figs. 2 through 6). The greater precipitation at Three Bar did not result in obvious differences in the types of patterns found for the annual course of water potentials of species common to the three sites. The lowest water potentials within each species were found at the California site. Observations during the second year of the study (crosses and squares in Figs. 2 through 6) are consistent with maintenance of higher water potentials due to the high level of precipitation that occurred.

The discussion which follows focuses on the extremes in water potentials encountered during the course of a day. Predawn potentials approximate the soil water potential experienced by the plant. They were measured within an hour of sunrise, the time of the least stress for plants. Predawn potentials are less negative (closer to zero) than the midday values. The term "midday" will be used to indicate the time when the lowest (most negative) values were measured, usually in the afternoon. Plants are most stressed at "midday". Delta potentials represent the difference within a day between predawn and midday values.

Of the chaparral shrubs studied, *Rhus ovata* had remarkably uniform potentials at all sites and throughout the study (Fig. 2). Its predawn water potential values ranged from -0.6 to -2.3 MPa, and midday values were consistently 1 to 1.5 MPa lower. On a given day, predawn and midday values varied little among individual plants, ranging +/- 0.25 MPa from their respective means. The data suggest an available water supply and little influence of seasonal drought on water potential. Thus, one might expect stomatal regulation in this species to be little affected by water potential.

The scrub oak, *Quercus turbinella*, displayed an annual pattern of water potentials (Fig. 3) typical of most chaparral species (Poole *et al.* 1981) and similar to those reported for *Quercus coccifera* and *Quercus suber* by Tenhunen *et al.* (this volume). *Quercus turbinella* had predawn water potentials ranging from -0.2 MPa in the wet seasons to -4.3 MPa in the dry seasons. At Three Bar, the wettest site, predawn potentials remained above -1.6 MPa on all study days. Water potentials measured in California had the greatest range, were lowest in the dry August of 1982, and were highest following the heavy August 1983 rains. Midday potentials decreased in spring faster than predawn values, thus delta potentials increased as the season progressed. At the height of the drought, delta potentials decreased depending on species and site, as some apparent lower limit in tolerable water potential was approached and almost complete stomatal closure occurred. Lowest midday potentials reached -4.33 MPa. The second study year was wetter, and all potentials were somewhat higher (especially higher after rain events) than during the equivalent seasons in the first study year.

The two *Arctostaphylos* spp. exhibited the greatest predawn water potential change with season, from -0.2 to -7.4 MPa (Fig. 4). The characteristics of this seasonal change were similar to that found with *Quercus turbinella* but more extreme. During the dry fall of 1981 *A. glauca* in California and *A. pungens* at Aravaipa had very low predawn values, down to -4.6 and -7.4 MPa, respectively. They recovered rapidly with onset of the wet season but again fell to well below -3 MPa during warm dry weather in July 1982. As with the oak, repeat measurements in the second year showed higher predawn and midday values during those seasons that were much wetter than corresponding seasons of the first year. Midday values were about 0.6 MPa below predawn values, from -0.8 to -7.4 MPa. At Three Bar, the delta potentials increased slightly as the season progressed from winter to summer. When predawn values fell to -4.5 MPa at

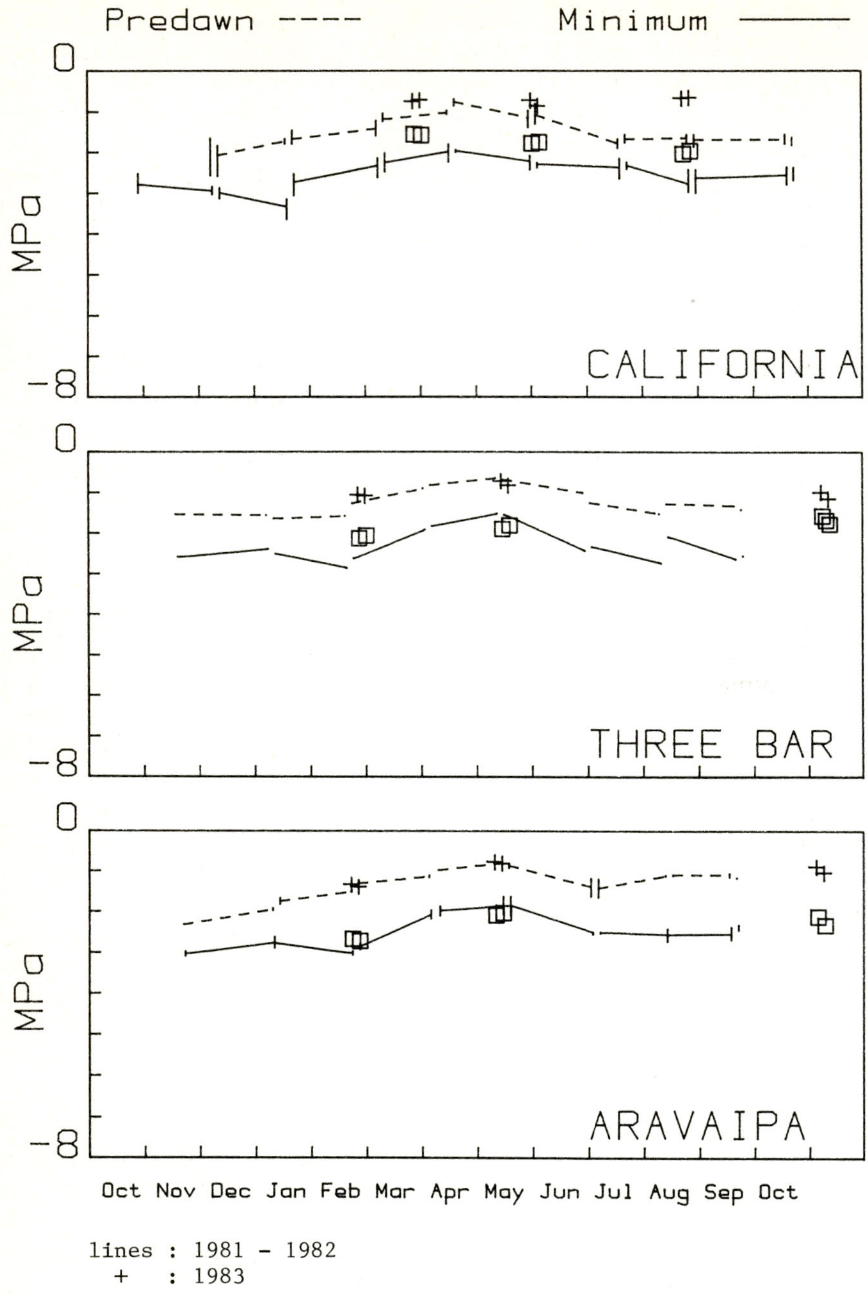

Fig. 2. Annual course of predawn water potential and minimum water potential observed during the day for *Rhus ovata*. Vertical bars indicate the maximum and minimum observed for the three replicate shrubs studied. Plus and square symbols indicate measurements carried out during the second year of the study

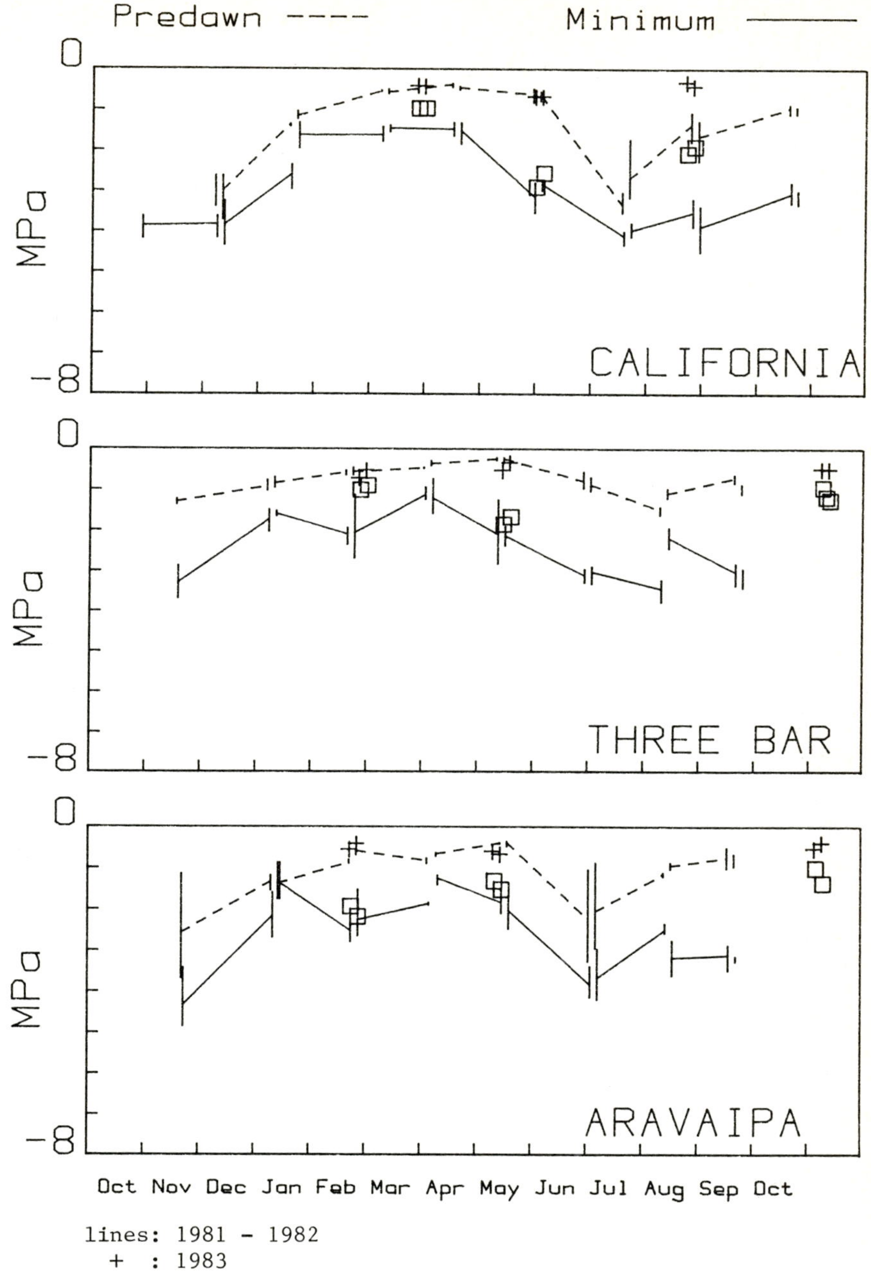

Fig. 3. Annual course as in Fig. 2 of predawn water potential and minimum water potential observed during the day for *Quercus turbinella*

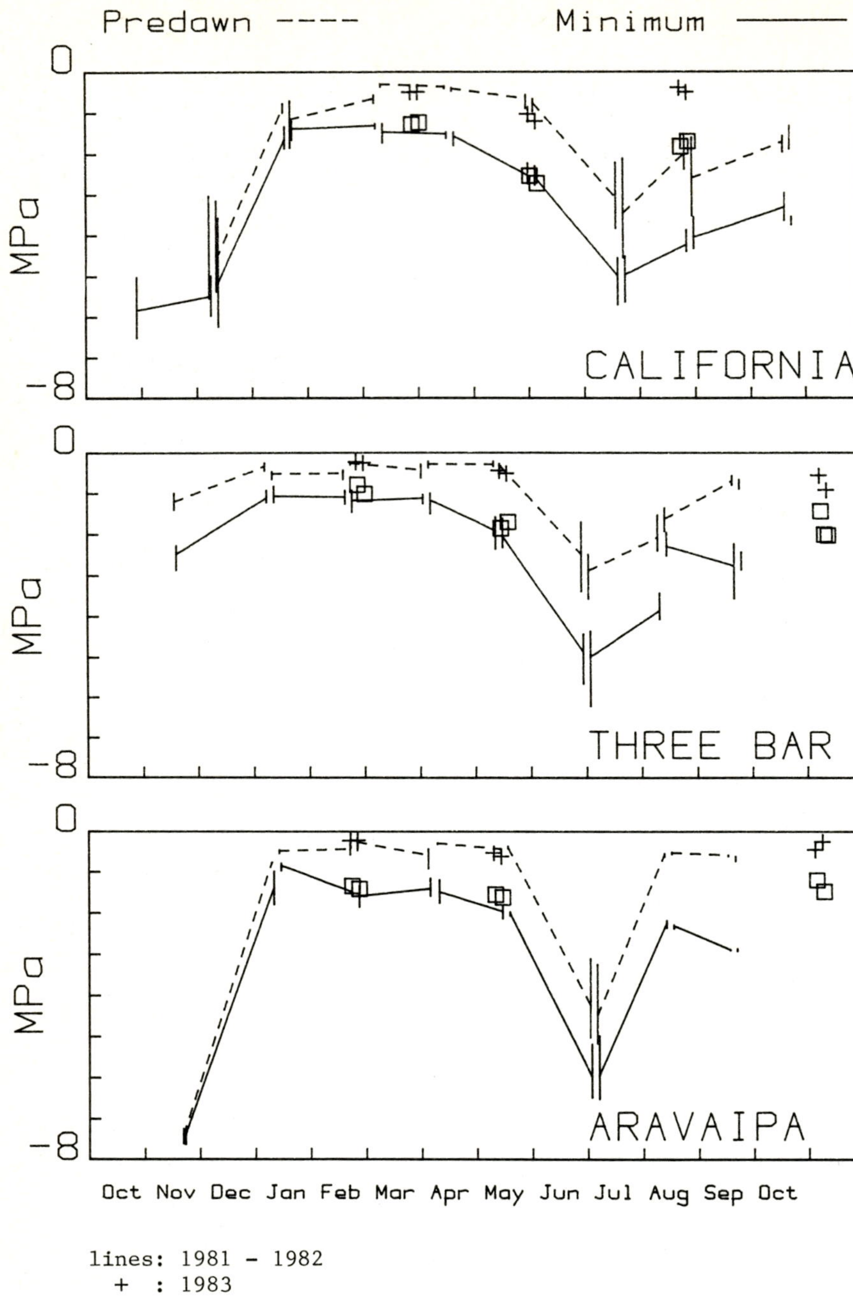

Fig. 4. Annual course as in Fig. 2 of predawn water potential and minimum water potential observed during the day for *Arctostaphylos glauca* at the California site and *Arctostaphylos pungens* at the Arizona sites

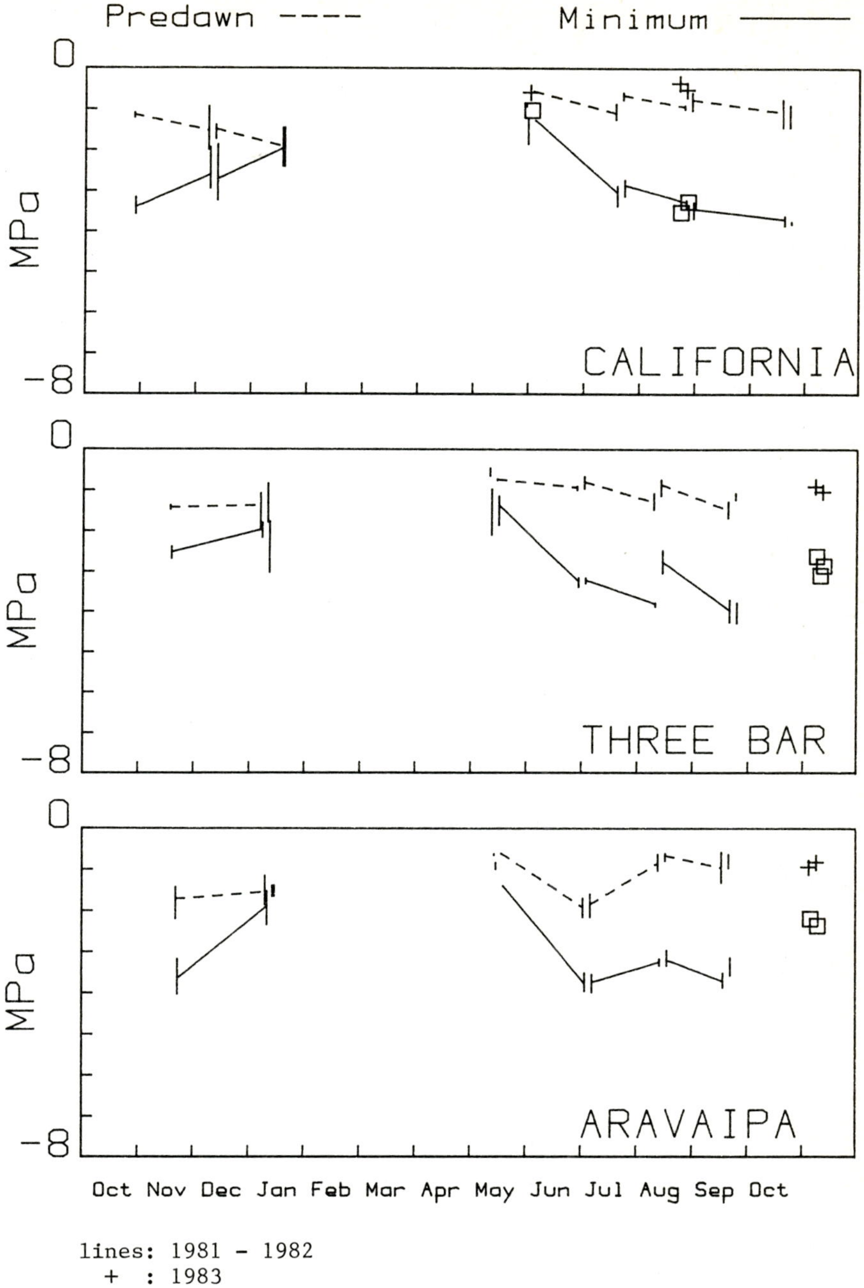

Fig. 5. Annual course as in Fig. 2 of predawn water potential and minimum water potential observed during the day for *Acacia greggii*

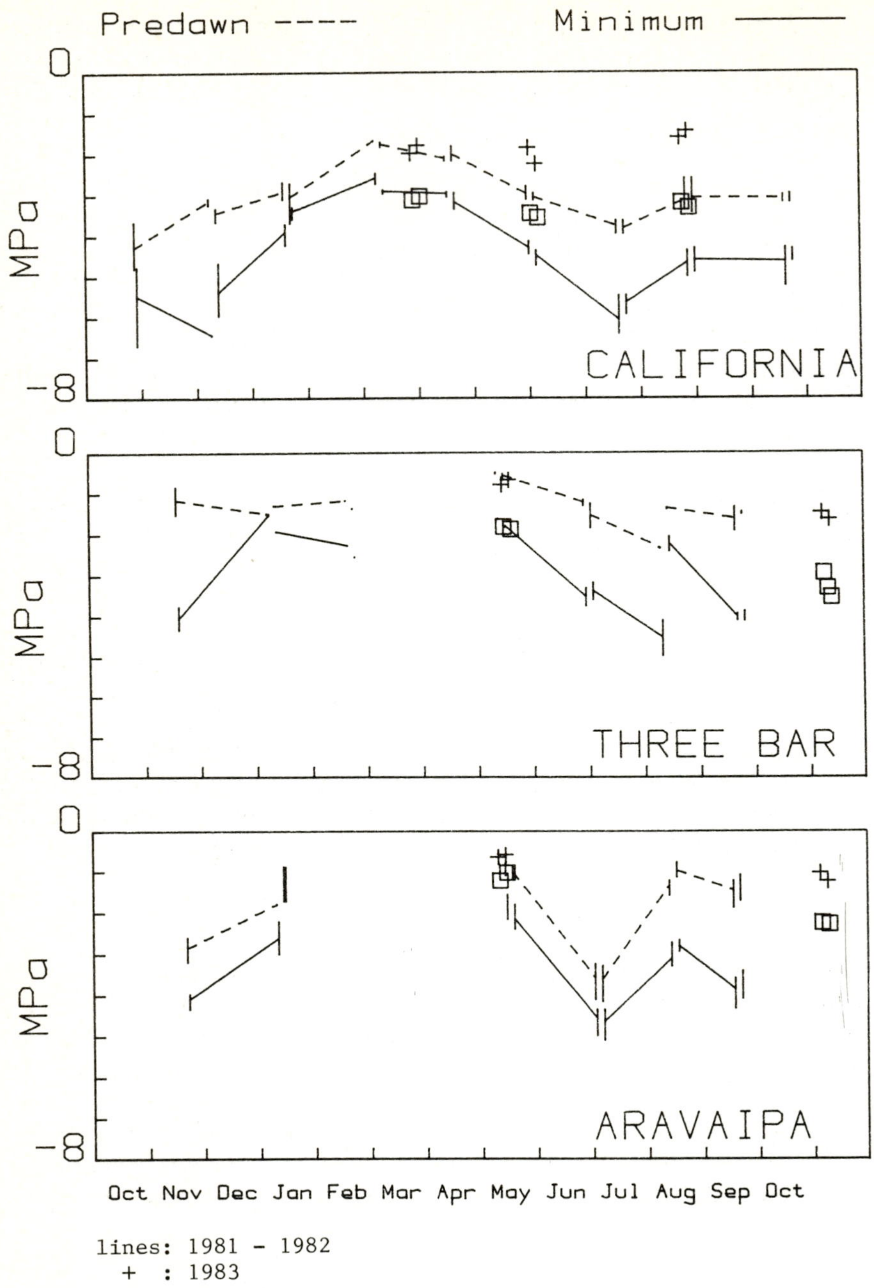

Fig. 6. Annual course as in Fig. 2 of predawn water potential and minimum water potential observed during the day for *Larrea tridentata* at the California site and *Prosopis juliflora* var. *velutina* at the Arizona sites

Aravaipa, delta values decreased. At very low predawn values (-8 MPa), deltas approached zero. Similar behavior to that found in *Arctostaphytlos* spp. was demonstrated for *Arbutus unedo*, a shallow-rooted shrub growing in Portugal (see Tenhunen *et al.* in this volume).

The winter-deciduous desert species *Acacia greggii* had predawn water potentials similar to those of *Rhus ovata* but values were slightly less negative (Fig. 5). Predawn values were between -0.4 and -1.9 MPa and midday values between -0.9 and -4.1 MPa. Compared to *Rhus*, midday water potentials decreased to lower values during the dry period (delta water potential was greater), indicating much freer use of available water. Compared to *Q. turbinella*, both predawn and midday water potentials of *A. greggii* were more negative in Arizona in November 1983. This is supportive of differential rooting in the two species. Moisture levels in the upper soil profile near both shrubs were the same.

The other desert species, *P. velutina* var. *juliflora* (winter-deciduous) and *Larrea tridentata* (evergreen) exhibited an annual pattern in water potential with characteristics similar to that of *Q. turbinella* (Fig. 6), although *Prosopis* at the Three Bar site responded similarly to *Acacia greggii*. Predawn potentials were 0.5 to 1 MPa more negative than those of *Q. turbinella* during long periods and ranged from -0.5 to -3.6 MPa. At the height of the drought, predawn potential did not decrease as much in *P. velutina* as in *Q. turbinella.* Midday water potential values of *P. velutina* ranged from -1.0 to -4.6 MPa. The least negative values were observed at Three Bar. Predawn water potentials of *Larrea tridentata* ranged from -1.6 to -3.8 MPa. Midday water potentials reached very low values from -2.9 to -6.4 MPa as found with *Arctostaphylos* but at no time did the delta water potentials decrease during drought.

Stomatal Conductance

During the extremely dry months in the fall of 1981, conductances of all species were low, especially at the California and Aravaipa sites (Figs. 7 through 11). Nevertheless, *Acacia greggii* at the California site was able to obtain adequate water supply to maintain conductance greater than 0.2 cm/s. The subsequent recovery of shrubs from drought during the winter months permitted increases in leaf conductance. The annual patterns in maximum conductance observed for the three chaparral species were strikingly similar. Although precipitation was greater during 1983, maximum conductances did not increase in any species above values found during the winter of 1981-1982.

Rhus ovata consistently had the lowest maximum conductances (on the average approximately 0.1 cm/s; Fig. 7). This trait apparently allows *Rhus* to maintain the high water potentials observed throughout the study and even during extended drought periods as in the fall of 1981. Interestingly, conductances as large as 0.2 cm/s were observed at Three Bar, the wettest site, during the spring of 1982 and at Aravaipa in late 1983.

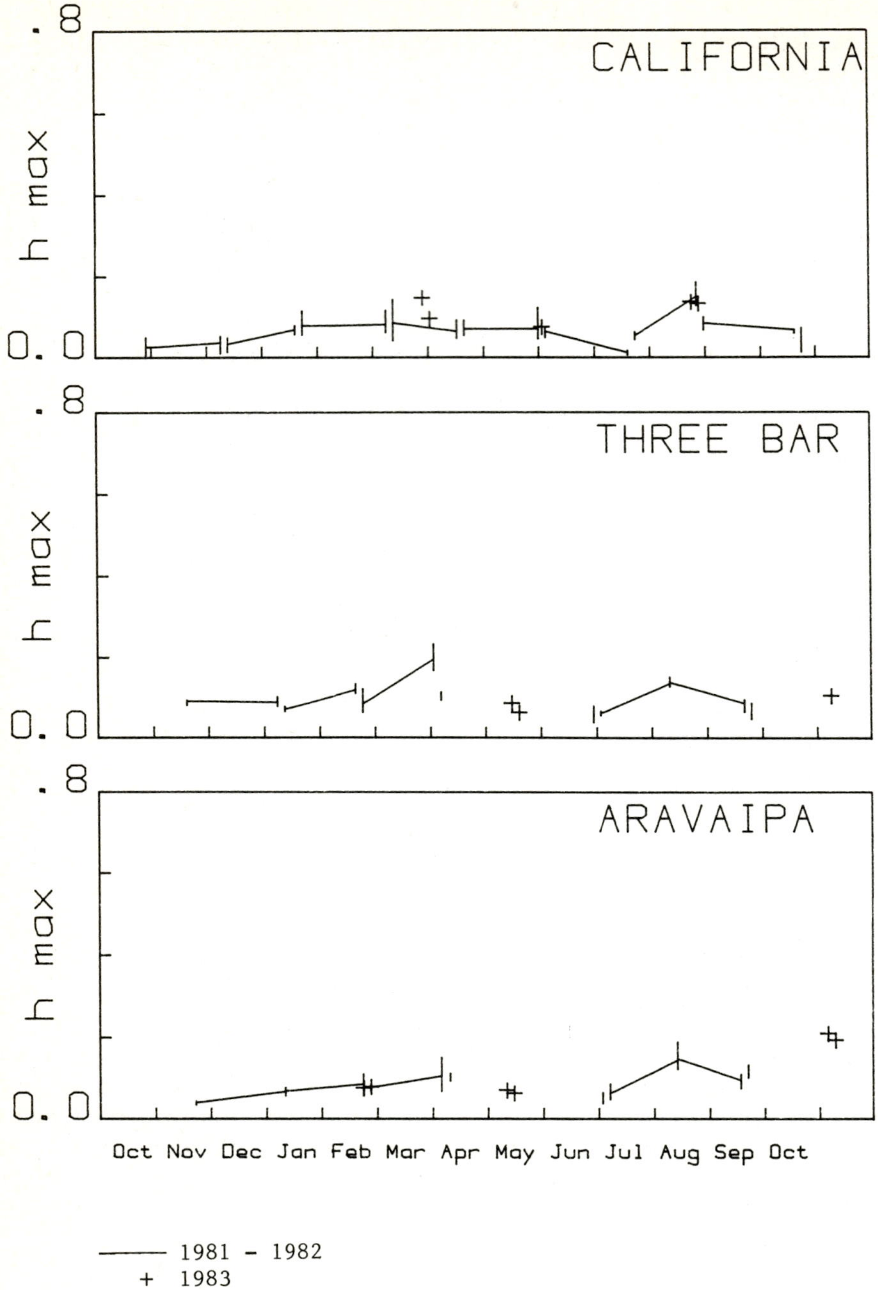

Fig. 7. Annual course of maximum leaf conductance observed during the day for *Rhus ovata*. Vertical bars indicate the maximum and minimum observed for the three replicate shrubs studied. Plus symbols indicate measurements carried out during the second year of the study

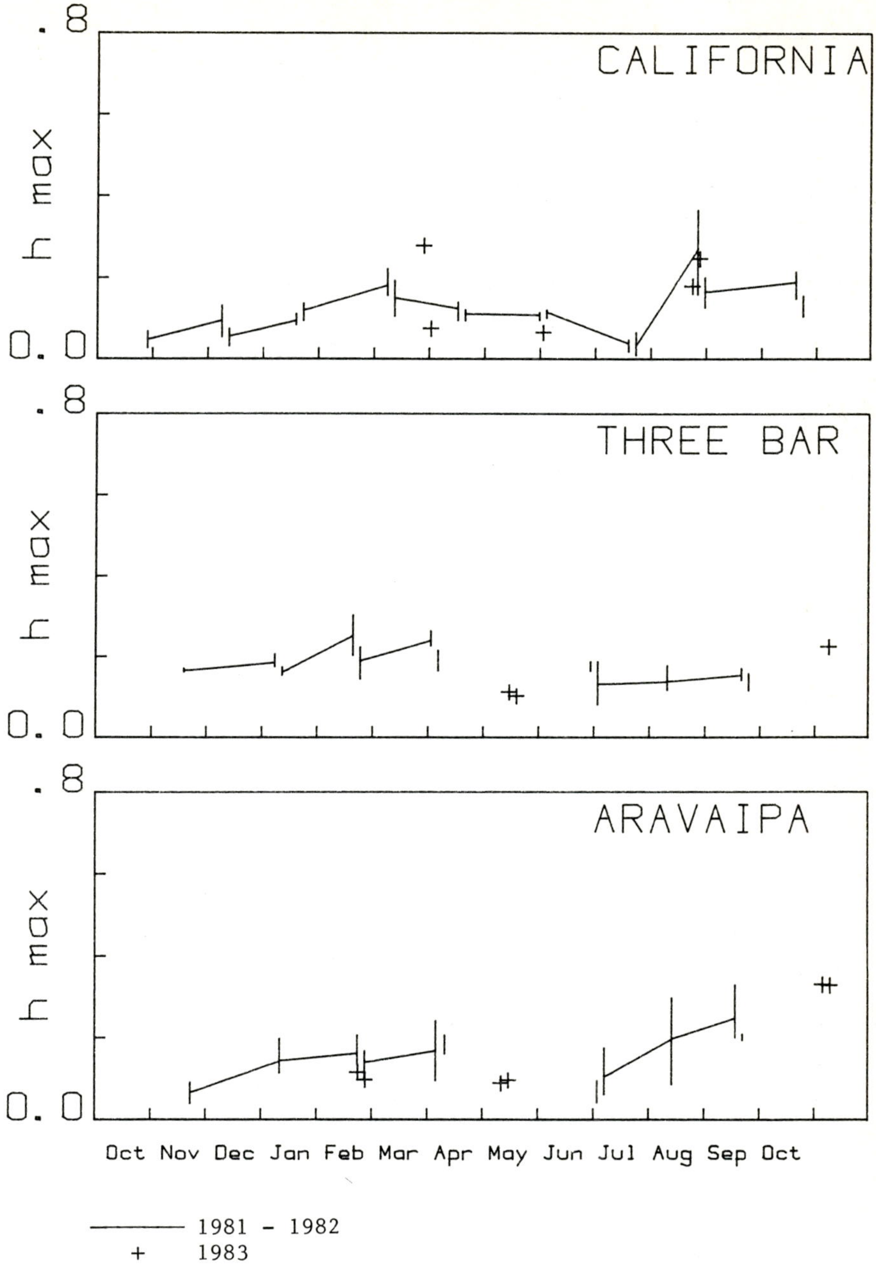

Fig. 8. Annual course as in Fig. 2 of maximum leaf conductance observed during the day for *Quercus turbinella*

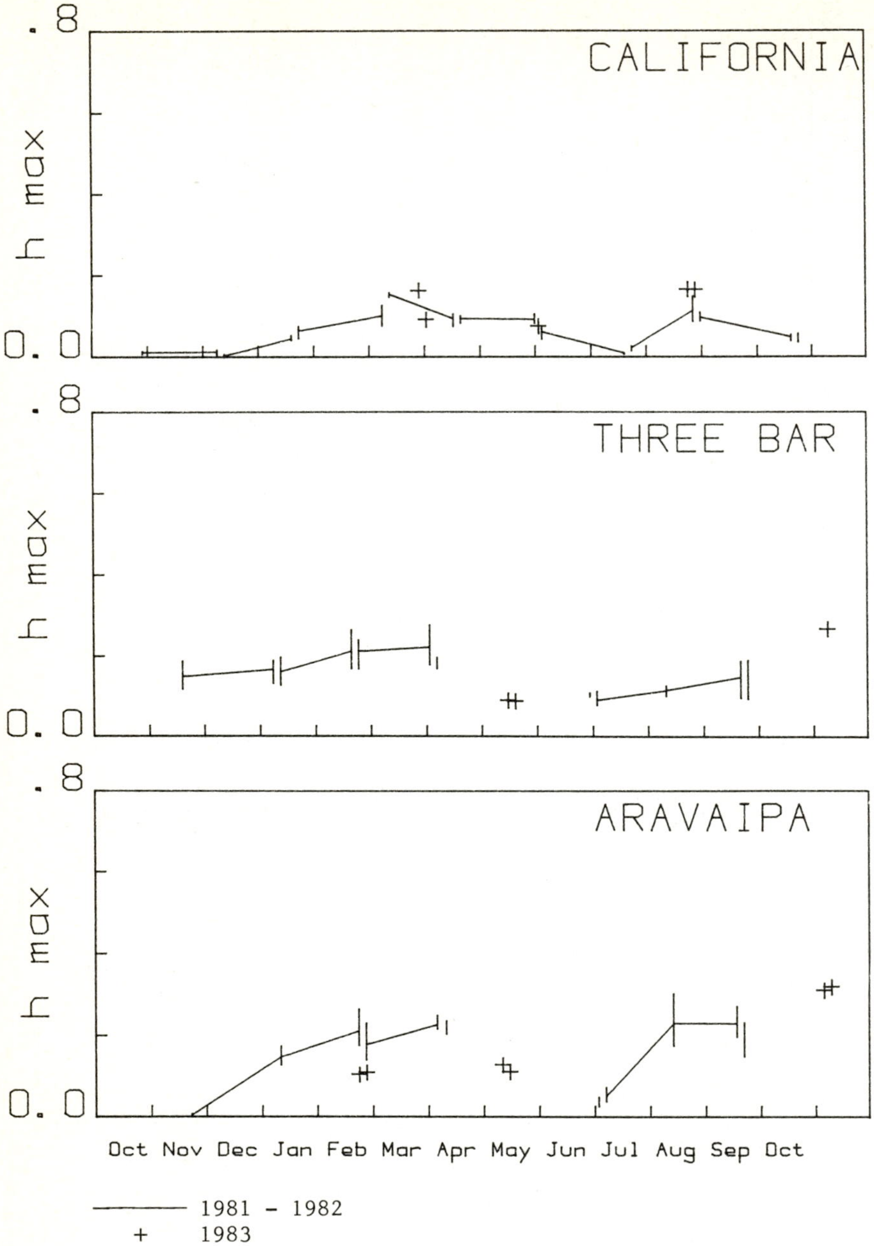

Fig. 9. Annual course as in Fig. 2 of maximum leaf conductance observed during the day for *Arctostaphylos glauca* at the California site and *Arctostaphylos pungens* at the Arizona sites

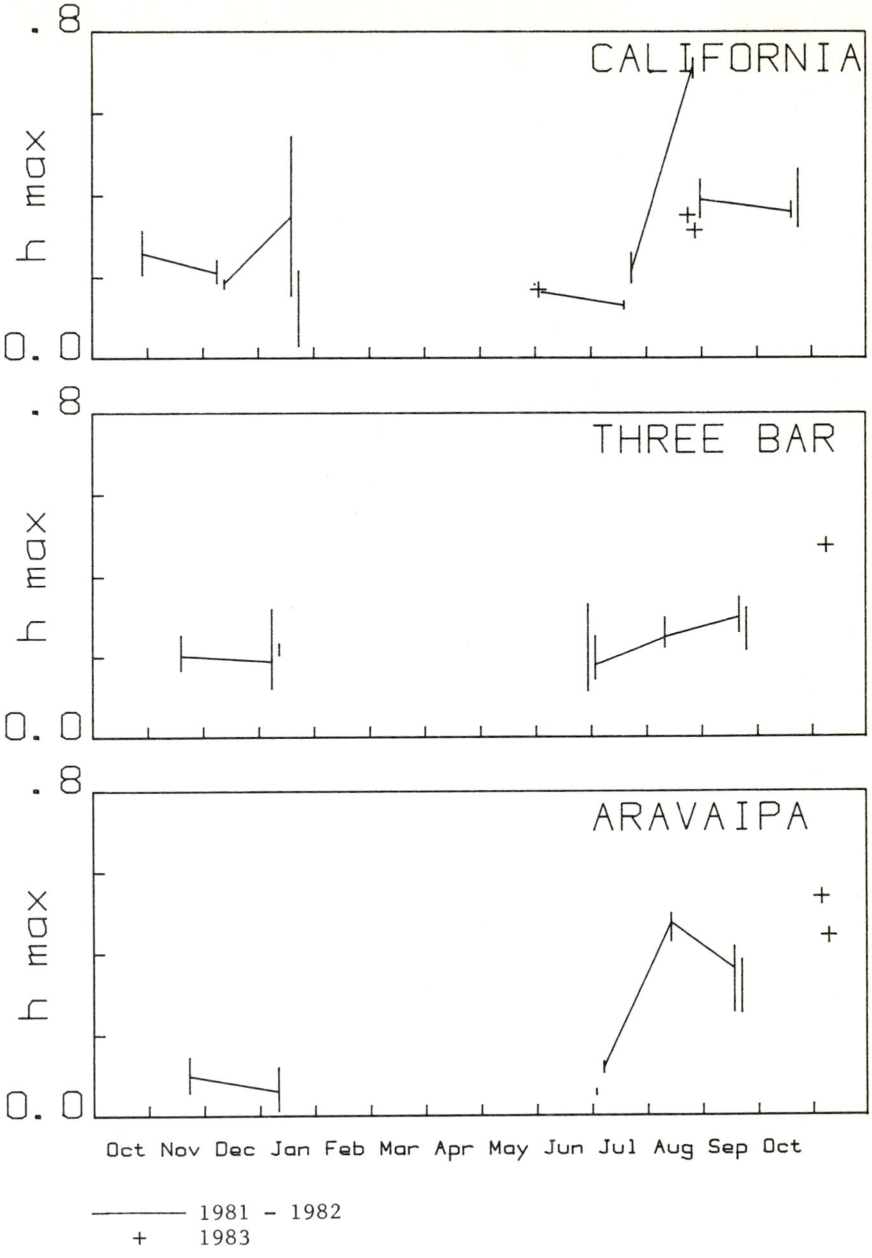

Fig. 10. Annual course as in Fig. 2 of maximum leaf conductance observed during the day for *Acacia greggii*

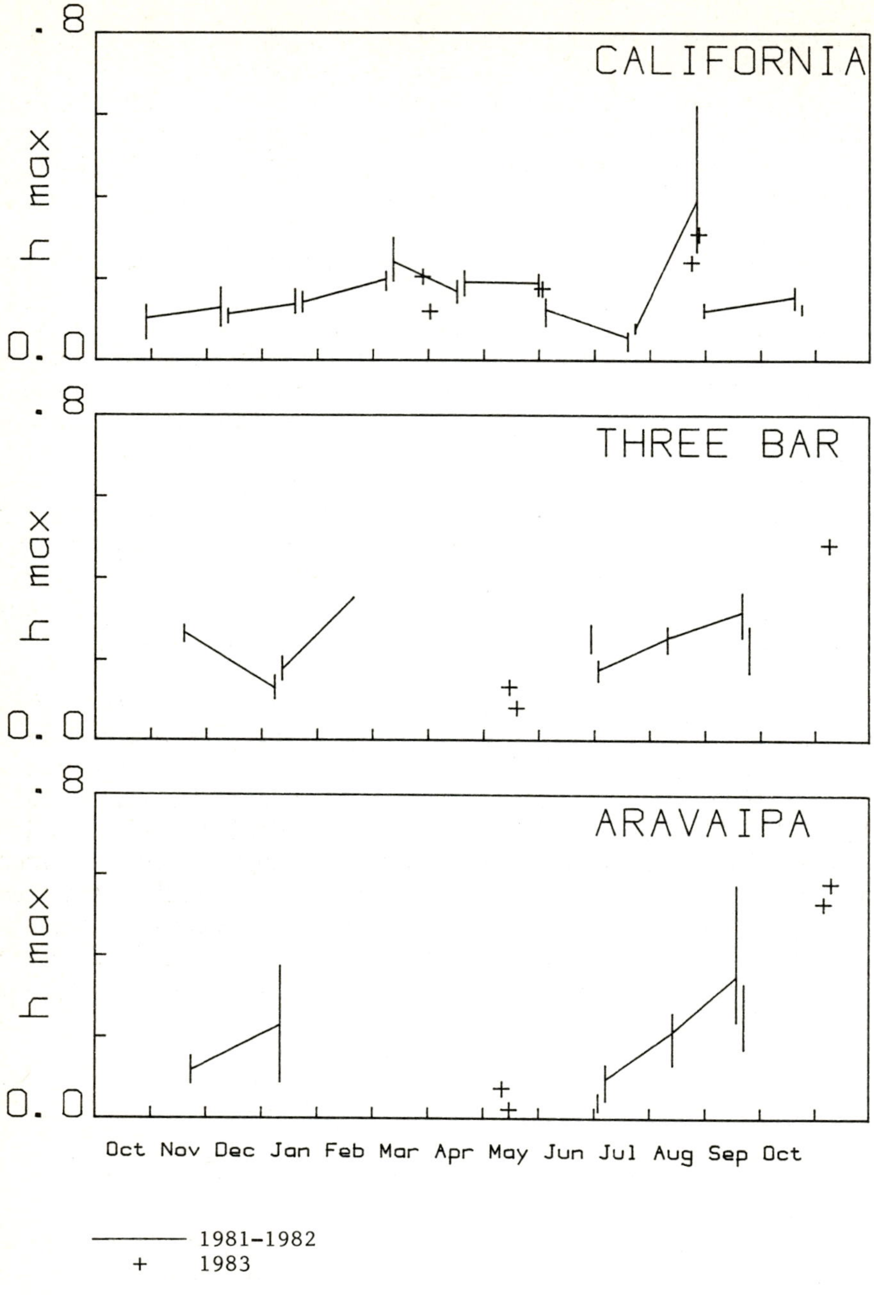

Fig.11. Annual course as in Fig. 2 of maximum leaf conductance observed during the day for *Larrea tridentata* at the California site and *Prosopis juliflora* var. *velutina* at the Arizona sites

Q. turbinella exhibited maximal conductances often more than twice as large as *Rhus ovata*. Conductance of *Q. turbinella* varied seasonally, was lowest in winter and in the dry summer, and was highest in spring and fall when soil moisture was high and temperatures were moderate (Fig. 8). In August, when temperatures were high, conductance was low on the first study day, but rains that afternoon caused higher conductances on the following day.

Conductances were very low in the *Arctostaphylos* spp. when water potentials were low, but recovered to higher levels (maximally 0.32 cm/s) when more soil moisture was available (Fig. 9). At the dry California site, conductance values were lower than those of *Quercus* and comparable to *Rhus*, except during the moist spring of 1982 when they increased to levels similar to that found for *Quercus*. At the wetter sites, conductance values were similar to those of *Quercus*.

The desert shrubs had higher conductances than the chaparral shrubs (Figs. 10 and 11). Maximal conductance was 0.39 cm/s in *L. tridentata* while *P. velutina* var. *juliflora* had higher conductances (up to 0.58 cm/s). *Acacia greggii* exhibited the highest conductances (0.72 cm/s). The desert deciduous shrubs are lavish water users. They are leafless, however, from January to April.

Midday closure of stomata of the chaparral shrubs was clearly suggested by data collected during the summer months. Similarly, indications of midday closure were found on several days of observations with *Prosopis* and *Larrea tridentata*. This agrees with results for desert shrubs reported by Nilsen *et al.* (1983) and Oechel *et al.* (1972). However, due to the large number of shrubs included in the study, consecutive samplings were completed only after relatively long intervals (approximately 2 to 3 hours), decreasing the ability to detect this phenomenon. Most often there was only a single maximum in leaf conductance. In winter, this maximum occurred near noontime. As air temperatures increased, the maximum was observed earlier in the morning followed by a continuous decline in conductance (cf. Lange *et al.* in this volume - Fig. 3).

DISCUSSION

These comparative observations demonstrate interesting differences in the abilities of shrub species growing side by side at the chaparral desert ecotone to exploit water resources and to maintain particular hydration states within the foliage during drought. The absolute conductance maxima observed for all species and sites are summarized along with the absolute minima in water potential in Fig. 12. Exposure of tissues to low water potentials increases in the chaparral shrubs from *Rhus ovata* to *Quercus turbinella* to *Arctostaphylos* spp. Similar differences between species and in minimum water potentials were observed between *Arctostaphylos glauca* and *Quercus dumosa* by Krause and

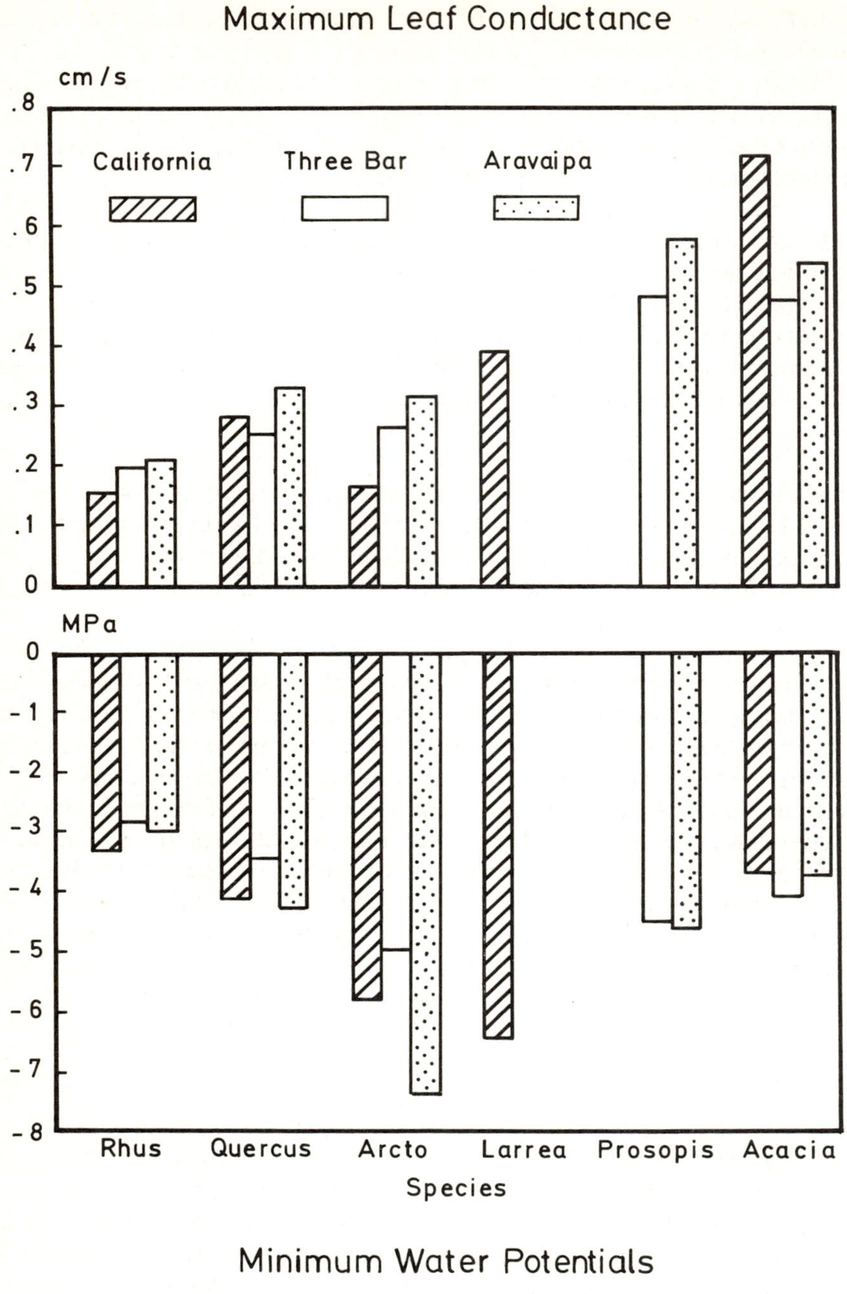

Fig. 12. Absolute leaf conductance maxima and water potential minima observed for all species and all sites

Kummerow (1977) and were maintained along an east-west transect of the Californian chaparral vegetation in *Rhus*, *Quercus*, and *Arctostaphylos* (Poole *et al.* 1981). Of the desert shrubs, *Acacia greggii* maintained the highest water potentials, although the water potentials which *Prosopis* experienced were somewhat similar. The more shallowly rooted *Larrea* experienced very low water potentials. In *Rhus*, *Quercus*, *Arctostaphylos*, and *Larrea*, increases in maximal conductance went hand in hand with more negative minimum water potentials. The deciduous desert shrubs exhibited the largest conductances, but nevertheless, maintained high water potentials.

Quercus turbinella and *Prosopis* decrease maximum conductances sharply when predawn water potentials fall to approximately -2.0 MPa. *Arctostaphylos* responds much sooner, i.e. near -1.0 MPa. *Acacia greggii*, *Larrea*, and *Rhus* exhibit a more continual decrease in conductance with more negative predawn water potentials. The desert shrubs in general exhibit greater maximum conductances and a broader range in conductance at all predawn water potentials. These conclusions are apparent from Fig. 13, in which the extremes in conductance based on all observations are bounded and plotted against predawn water potential on the days of measurement.

The patterns observed in the annual courses of water potential can be explained by differences in rooting depths as outlined in Table 1 and the differences in stomatal regulation observed. Both *Rhus ovata* and *Acacia greggii* are able to tap water reserves throughout the year and in these chaparral-desert ecotone areas maintain high tissue water potentials. *Rhus ovata*, however, accomplishes this in part through conservative water use. *Acacia greggii*, on the other hand, exhibits similar characteristics but in addition uses approximately 3 times as much water per unit LAI as *Rhus ovata*. Further observations have indicated that both species may maintain similarly high LAI (based on projected shrub area) at the study sites. Although *Acacia* is able to avoid effects of the drought at the sites included in this study, predawn water potentials were reported to decrease to as low as -4.4 MPa at other desert sites (Szarek and Woodhouse 1978).

The high tissue water potentials significantly affect other physiological functions in the canopies of these species, e.g. *Rhus ovata* is known to maintain cambial activity throughout the year (Avila *et al.* 1975). We might expect this to be the case for *Acacia greggii* as well, since cambial activity of other *Acacia* species responds sensitively to water availability (Fahn *et al.* 1968). We may postulate that photosynthetic processes in leaves of these species will likewise be influenced by the favorable water status of the tissues. One would expect little seasonal variation in photosynthetic capacities. The high conductance of *Acacia greggii* at our study sites is undoubtedly coupled with an equally high photosynthetic capacity (Szarek and Woodhouse 1978), as commonly found in desert shrubs and perennials during periods with adequate water (Ehleringer and Mooney 1983). Bunce *et al.* (1979) found photosynthesis rates of *Acacia greggii* to decrease only at leaf water potentials below -3.3 MPa. With respect to photosynthetic rate,

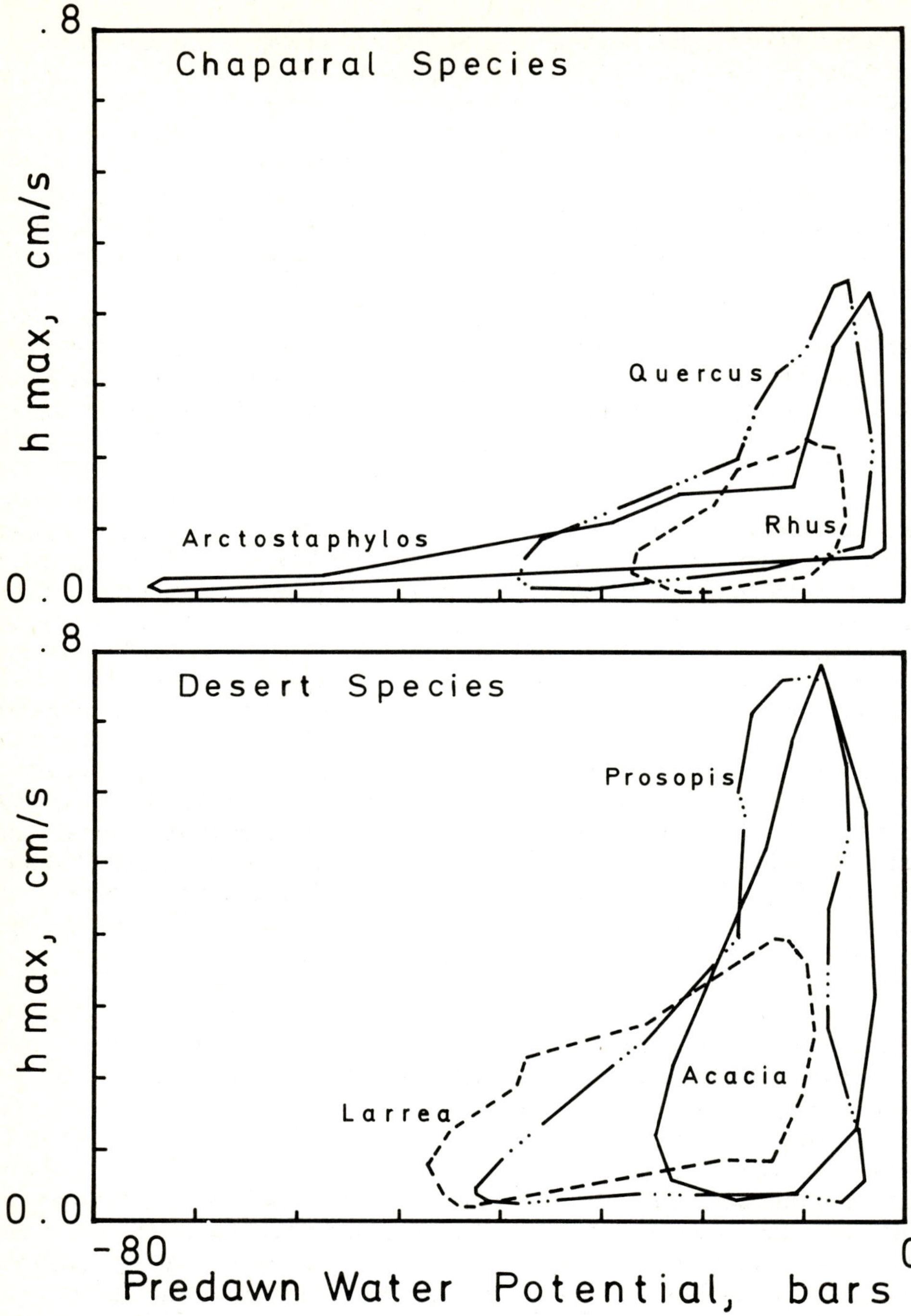

Fig.13. Boundaries of extremes in conductance for all species based on all observations plotted against predawn water potential on the days of measurement

Acacia greggii was more tolerant of low leaf water potentials than was *Larrea tridentata* (Bunce *et al.* 1979) The high photosynthetic rates of *Acacia* ensure effective water use efficiency and compensate in carbon gain for the 3 to 4 month leafless period during winter.

The response of *Prosopis* was comparable to that of *Acacia*, although water potentials at the Arizona sites decreased to more negative values. In general, the water potentials and conductances of *Prosopis* observed in these investigations were similar to those described for *Prosopis glandulosa* by Nilsen et al. (1983). As brought out in the latter study, *Prosopis* utilizes groundwater, maintains open stomata and transpires large amounts of water. Mooney *et al.* (1980) and Nilsen *et al.* (1983) have discussed the possibility that *Prosopis* moves groundwater into available storage in the upper soil layers during the night for use during the daytime period.

Quercus turbinella, *Arctostaphylos* spp., and *Larrea tridentata* have many characteristics in common. Leaves of all of these species are subjected to seasonal fluctuations in water availability. This is particularly apparent from the annual courses of predawn water potential illustrated. The shallow rooted *Arctostaphylos* is most severely influenced, with predawn water potentials decreasing to -7.4 MPa. The results obtained by Tenhunen *et al.*(this volume) with the similarly shallow-rooted *Arbutus unedo,* would lead us to expect that water shortages significantly decrease carbon fixation capacities of these species during the summer. *Quercus turbinella*, on the other hand, is deep-rooted enough and decreases conductance during the dry period in such a fashion that predawn water potential does not decrease below approximately -3.5 MPa. Data obtained with other *Quercus* species at a variety of mediterranean climate sites suggest that whole plant response of *Quercus* is such that this water potential limit is not exceeded (cf. Tenhunen *et al.* and Rambal and Leterme in this volume). As a result, little decrease occurs in the photosynthetic capacity of the leaf mesophyll, and carbon uptake can increase rapidly after rain.

Along with maximum conductances that are higher than those of the chaparral shrubs, *Larrea* is known to have a relatively high leaf photosynthetic capacity (Mooney *et al.* 1978). Predawn water potentials of *Larrea tridentata* decreased to approximately -4.5 MPa in the study of Mooney *et al.*. Even during the period of most extreme drought, conductance remained above 0.1 cm/s and calculated transpiration rates were substantial. Field studies by Odening *et al* . (1974) would suggest carbon uptake rates at -4.5 MPa approximately 25% of maximum and continued carbon uptake down to a leaf water potential of -8.0 MPa. While data obtained by Mooney *et al.* (1977) with potted plants indicated significant decreases in mesophyll photosynthetic capacity at -3.6 MPa, this has not been substantiated for plants growing in the field. Mooney *et al.* (1978) found that photosynthesis rates did not decline in field-grown plants at water potentials during measurement of -3.4 MPa. Thus, we cannot rule out the possibility that *Larrea* behaves under field conditions in a fashion similar to *Quercus*, regulating water use relative to water supply such that damage to the photosynthetic processes in the mesophyll is avoided. Oechel *et al.* (1972) called attention to the effectiveness of stomatal control of water loss in the maintenance of relatively high water potentials during drought. Syvertson *et al.* (1975) even suggest that under dry conditions more extreme than found in this study, water loss may be so effectively restricted and soil temperature gradients may be such that net water accumulation may occur in the plant tissues during the day, thus maintaining high photosynthetic capacities. Whether high mesophyll photosynthetic capacity is maintained or whether the effect of changes in water availability reduces mesophyll photosynthesis in a manner similar to that in *Arctostaphylos*, it is apparent that *Larrea* is suited to function well under conditions where daily minimum water potential decreases to as low as -6.4 MPa. Water supply is then still adequate to support open stomata and positive carbon uptake.

While the present observations are related only to stomatal control and changes in xylem water potentials, we may use these data along with observations of micrometeorological conditions within canopies of these species to estimate total water use in whole plants. Information on leaf area index and percent cover due to each species and evaporation from the ground surface will allow us to examine the ways in which different species influence the hydrology of chaparral-desert ecotone areas.

REFERENCES

Avila G, Aljaro ME, Araya S, Montenegro G, Kummerow J (1975) The seasonal cambium activity of Chilean and Californian shrubs. Amer J Bot 62:473-478

Beardsell MF, Jarvis PG, Davidson B (1972) A null-balance diffusion porometer suitable for use with leaves of many shapes. J of Applied Ecology 9:677-690

Benson L, Darrow RA (1981) Trees and shrubs of the southwestern deserts. University of Arizona Press, Tucson

Bunce JA, Chabot BF, Miller LN (1979) Role of annual leaf carbon balance in the distribution of plant species along an elevational gradient. Bot Gaz 140:288-294

Cable DR (1975) Range management in the chaparral type and its ecological basis: The status of our knowledge. Rocky Mountain Range and Experimental Station. Forest Service USDA Research Paper RM-202, October, 1975

Ehleringer J, Mooney HA (1983) Productivity of desert and mediterranean-climate plants. In: Lange OL, Nobel PS, Osmond CB, Zieger H (eds) Encyclopedia of Plant Physiology, NS vol 12D: Physiological Plant Ecology IV. Springer, Berlin-Heidelberg-New York-Tokyo, pp 205-231

Fahn A, Waisel Y, Benjamin L (1968) Cambial activity in *Acacia raddiana* Sari. Ann Bot 32:677-686

Kearney TH, Peebles RH (1951) Arizona Flora. Second Edition (with supplement 1960). University of California Press, Berkeley, Los Angeles, London

Krause D, Kummerow J (1977) Xeromorphic structure and soil moisture in the chaparral. Oecol Plant 12:133-148

Mooney HA, Björkman O, Collatz GJ (1977) Photosynthetic acclimation to temperature and water stress in the desert shrub, *Larrea divaricata*. Carn Inst Wash Yearb 75:410-413

Mooney HA, Björkman O, Collatz GJ (1978) Photosynthetic acclimation to temperature in the desert shrub, *Larrea divaricata*. Plant Physiol 61:406-410

Mooney HA, Gulmon SL, Rundel PW, Ehleringer J (1980) Further observations on the water relations of *Prosopis tamarugo* of the Northern Atacama Desert. Oecologia (Berl) 44:177-180

Munz PA (1974) A flora of southern California. University of California Press, Berkeley, Los Angeles, London

Nilsen ET, Sharifi MR, Rundel PW, Jarrell WM, Virginia RA (1983) Diurnal and seasonal water relations of the desert phreatophyte *Prosopis glandulosa* (Honey Mesquite) in the Sonoran Desert of California. Ecology 64:1381-1393

Odening W, Strain BR, Oechel WC (1974) The effects of decreasing water potential on net CO_2 exchange of intact desert shrubs. Ecology 55:1086-1095

Oechel W, Strain BR, Odening WR (1972) Photosynthetic rates of a desert shrub, *Larrea divaricata* Cav., under field conditions. Photosynthetica 6:183-188

Poole DK, Roberts SW, Miller PC (1981) Water utilization. In: Miller PC (ed) Resource Use by Chaparral and Matorral. A Comparison of Vegetation Function in Two Mediterranean Type Ecosystems. Springer-Verlag, New York, Heidelberg, Berlin, 455 pp

Roberts SW, Blake-Jacobson ME (1984) Final report to the U.S. Department of Agriculture, Science and Education Administration, covering research entitled "Plant characteristics to reverse desertification and increase production in semi-arid lands" (USDA Grant 59-2069-1-2-084-0)

Roberts SW, Miller PC, Valamanesh A (1981) Comparative field and water relations of four co-occurring chaparral shrub species. Oecologia (Berl) 48:360-363

Scholander PF, Hammel HT, Bradstreet ED, Hemmingsen EA (1965) Sap pressure in vascular plants. Science 148:339-346

Syvertsen JP, Cunningham GL, Feather TV (1975) Anomalous diurnal patterns of stem xylem water potentials in *Larrea tridentata*. Ecology 56:1423-1428

Szarek SR, Woodhouse RM (1978) Ecophysiological studies of Sonoran Desert plants. III. The daily course of photosynthesis for *Acacia greggii* and *Cercidium microphyllum*. Oecologia (Berl) 35:285-294

Thrower NJW, Bradbury DE (eds) (1977) Chile-California Mediterranean Scrub Atlas. Dowden, Hutchinson, and Ross, Inc., Stroudsburg, Pennsylvania

Regulation of water use by four species of *Cistus* in the scrub vegetation of the Serra da Arrábida, Portugal

O. Correia, F. Catarino, J.D. Tenhunen* and O.L. Lange**

Departamento Biologia Vegetal, Faculdade de Ciencias, Universidade de Lisboa, Lisboa, Portugal

INTRODUCTION

Plants of mediterranean ecosystems are subjected annually to prolonged periods of drought. The drought resistance of individual species occurring in these regions is conferred on them by both morphological and physiological adaptations which provide both drought avoidance and drought tolerance mechanisms (Levitt 1980). In parts of the Serra da Arrábida region, *Cistus salvifolius*, *Cistus monspeliensis*, *Cistus albidus*, and *Cistus ladanifer* are sympatric species, although in general, differences in habitat preference have been identified based on moisture availability. *Cistus salvifolius*, for example, is the most prevalent species on less exposed north facing mountain slopes. It occurs in a smaller growth form on drier south facing slopes together with the other three more thermophilic species. The present comparative study was undertaken to examine possible differences in drought resistance among these four *Cistus* species more closely. Over an annual cycle, the water relations and stomatal response of all four species were observed at monthly intervals in an attempt to identify differential responses to stress conditions.

MATERIALS AND METHODS

The study was carried out between October 1982 and October 1983 on a south facing slope of Jaspe Peak in the Serra da Arrábida region of central Portugal (38° 27' 34" N, 9° 0' 20" W, elevation 270 m). A detailed description of the geography, climate, and vegetation in the Serra da Arrábida was reported by Catarino *et al.* (1982). Summers are warm and dry, while winters are characterized by relatively high precipitation, averaging 650 mm per year.

The shrubs selected for study were less than one meter in height and occurred in a 20 to 30 year old stand of mixed sclerophyll scrub. Branch ends were inserted into a cuvette mounted on a steady-state diffusion porometer (Li-1600, Li-Cor,

*Systems Ecology Research Group, San Diego State University, San Diego, California 92182 USA

**Lehrstuhl für Botanik II, Universität Würzburg, Mittlerer Dallenbergweg 64, D-8700 Würzburg, West Germany

NATO ASI Series, Vol. G15
Plant Response to Stress. Edited by J. D. Tenhunen et al.

Lincoln, Nebraska; see Schulze *et al.* 1982 for modifications employed) and oriented such that light incident on the leaves was essentially the same as before insertion. Incident light intensity (PAR, μmol m^{-2} s^{-1}), leaf temperature (T_L, ^oC), water vapor mole-fraction difference between leaf air spaces and external air (ΔW, mbar bar^{-1}), transpiration rate under cuvette conditions (Tr, mmol m^{-2} s^{-1}), and leaf conductance for water vapor (G, mmol m^{-2} s^{-1}) were determined for two shoots of each species each hour (for further details on methods see Tenhunen *et al.* 1981). Transpiration rates are reported on a projected leaf area basis. Xylem water potential (Ψ) was determined every two hours from dawn to dusk with a pressure chamber of the Scholander type (Scholander *et al.* 1965) using terminal shoots.

RESULTS

The annual course in xylem water potential found at Jaspe with the four species of *Cistus* is shown in Fig. 1. The lowest value of water potential observed during diurnal course measurements in different seasons (Ψ_{MIN}) and the predawn value (Ψ_{PD}) are indicated. Apparent from the figure is that at the beginning of the study in October 1982, Ψ_{PD} had fully recovered from the 1982 summer drought to a level approximately maintained through the winter and following spring.

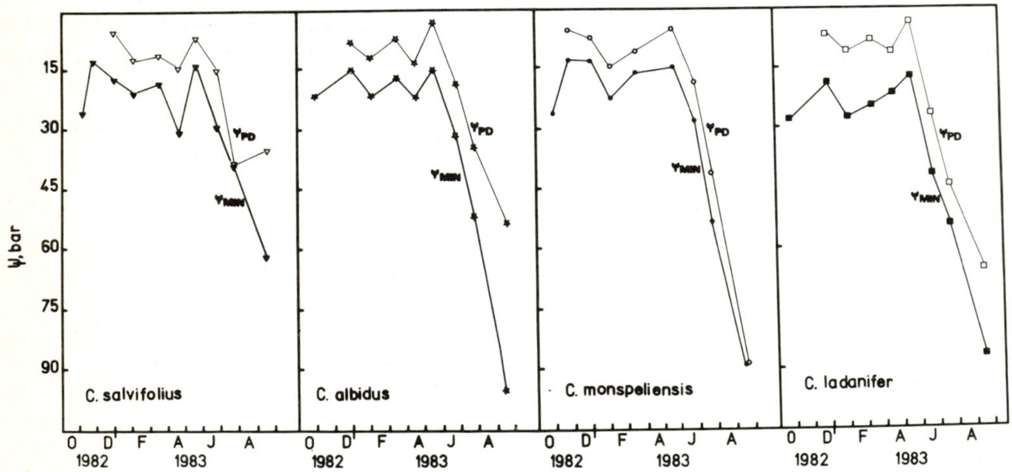

Fig. 1. Seasonal changes in terminal shoot predawn water potential (Ψ_{PD}) and the lowest water potential measured during the day (Ψ_{MIN}) for the species shown

Differences among species were not marked during this period of time and the average predawn water potential for all species was approximately -6 bar. Between October 1982 and the end of May 1983, diurnal decreases in water potential, i.e. the difference Ψ_{PD} - Ψ_{MIN}, remained approximately constant although differed for individual species. During the summer drought of 1983, water potentials decreased. Predawn water potential decreased least in *Cistus salvifolius* and *Cistus albidus*, suggesting that either water supply to these species during drought is superior or that water use is strongly reduced. In the case of *Cistus monspeliensis* and *Cistus ladanifer*, however, predawn water potential decreases during drought to very low levels. These latter species are, nevertheless, quite abundant and robust at the xeric study site. Despite a very low predawn water potential measured at the end of summer, large diurnal changes in water potential were observed with *Cistus ladanifer*.

Diurnal time courses for environmental factors, shoot water potential, leaf transpiration rate, and total leaf conductance are shown for six days between October 1982 and September 1983 in Figs. 2 through 7. The data shown indicate the average responses observed with two shoots of each of the four *Cistus* species. In general, dome shaped diurnal patterns were found for transpiration rate. Maximal rates of transpiration occurred near midday. During the winter, the diurnal course of conductance was generally similar to that of incident radiation (PAR). In addition,however, a tendency for conductance to decrease over the course of the day was apparent, i.e. conductances at the same light intensity and temperature were lower in the afternoon than in the morning. In the summer, conductance values were commonly highest in the early morning. The maximum value was observed between 8 and 9 o'clock, after which conductance steadily decreased. Decreases in conductance were correlated with increases in the water vapor mole-fraction difference between leaf and air (ΔW). In several instances, there was a suggestion that midday stomatal closure may occur in leaves of these shrub species (*C. ladanifer* and *C. monspeliensis* on October 26, 1982; *C albidus* on June 23, 1983) as has been shown for sclerophyll shrubs of the region (Tenhunen *et al.* 1981; Lösch *et al.* 1982; Lange *et al.* 1982).

Highest rates of transpiration were exhibited by *Cistus albidus* (approximately 6 mmol m^{-2} s^{-1} in May, see TR_{MAX} Fig. 8). Nevertheless, this species drastically reduced transpiration rate during summer drought. Despite a heavy pubescent covering on the leaves, the highest leaf conductance values were also measured with this species. *Cistus ladanifer* consistently exhibited the lowest transpiration rates (approximately 3.2 mmol m^{-2} s^{-1}) and had the lowest maximal conductance (G_{MAX} in Fig. 8). Leaf conductance values of *Cistus monspeliensis* and *C. salvifolius* were intermediate and approximately equal. In July, *Cistus monspeliensis* exhibited the highest rates of transpiration.

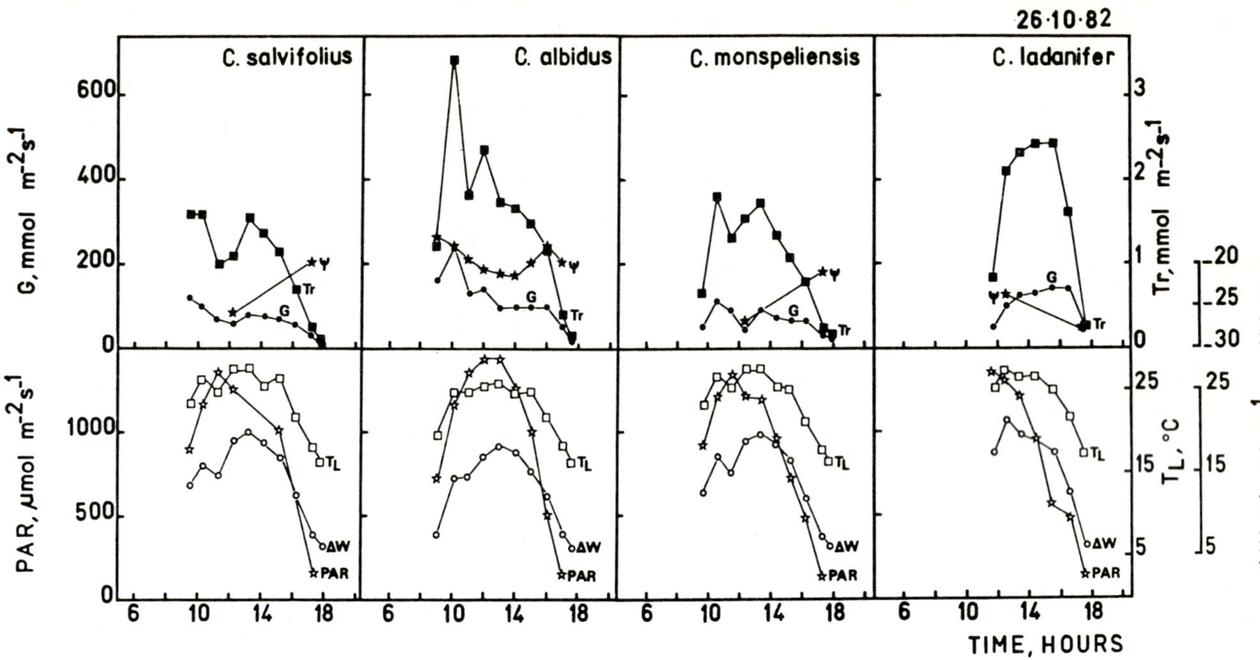

Fig. 2. Diurnal course of transpiration rate into the cuvette (Tr), total leaf conductance (G), shoot water potential (Ψ), incident photosynthetically active radiation (PAR), leaf temperature (T_L), and water vapor mole fraction difference between leaf and air (ΔW) for the *Cistus spp*. shown. Day, month, and year shown at the upper right hand corner of the figure

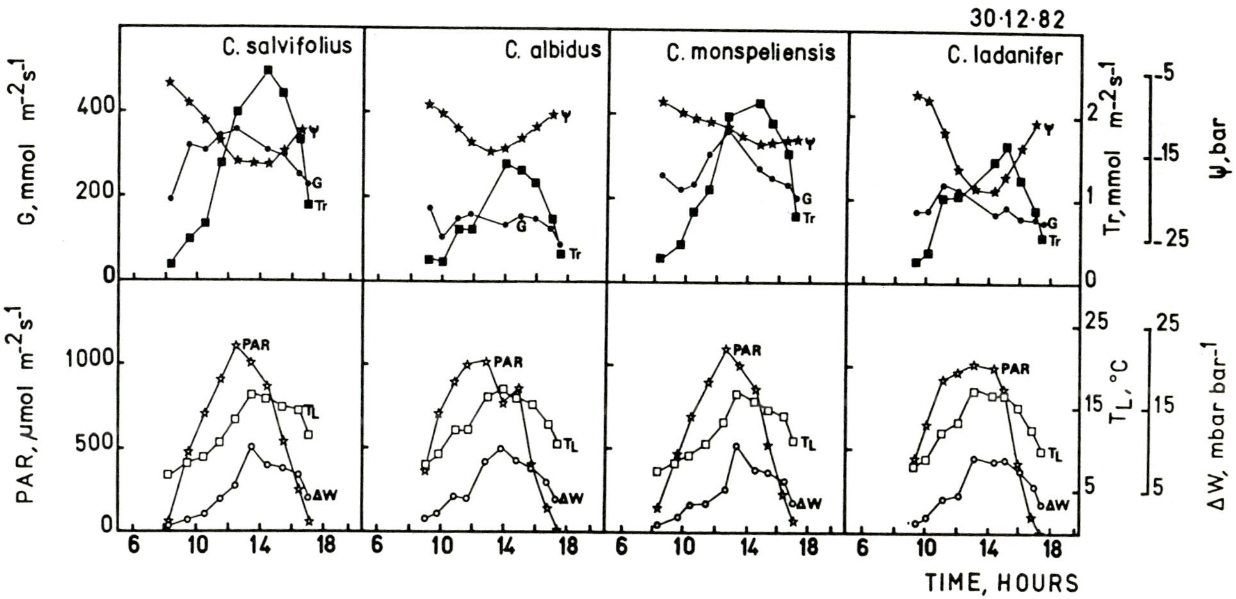

Fig. 3. Diurnal courses as in Fig. 2 for the *Cistus spp*. shown. Day, month, and year shown at the upper right hand corner of the figure

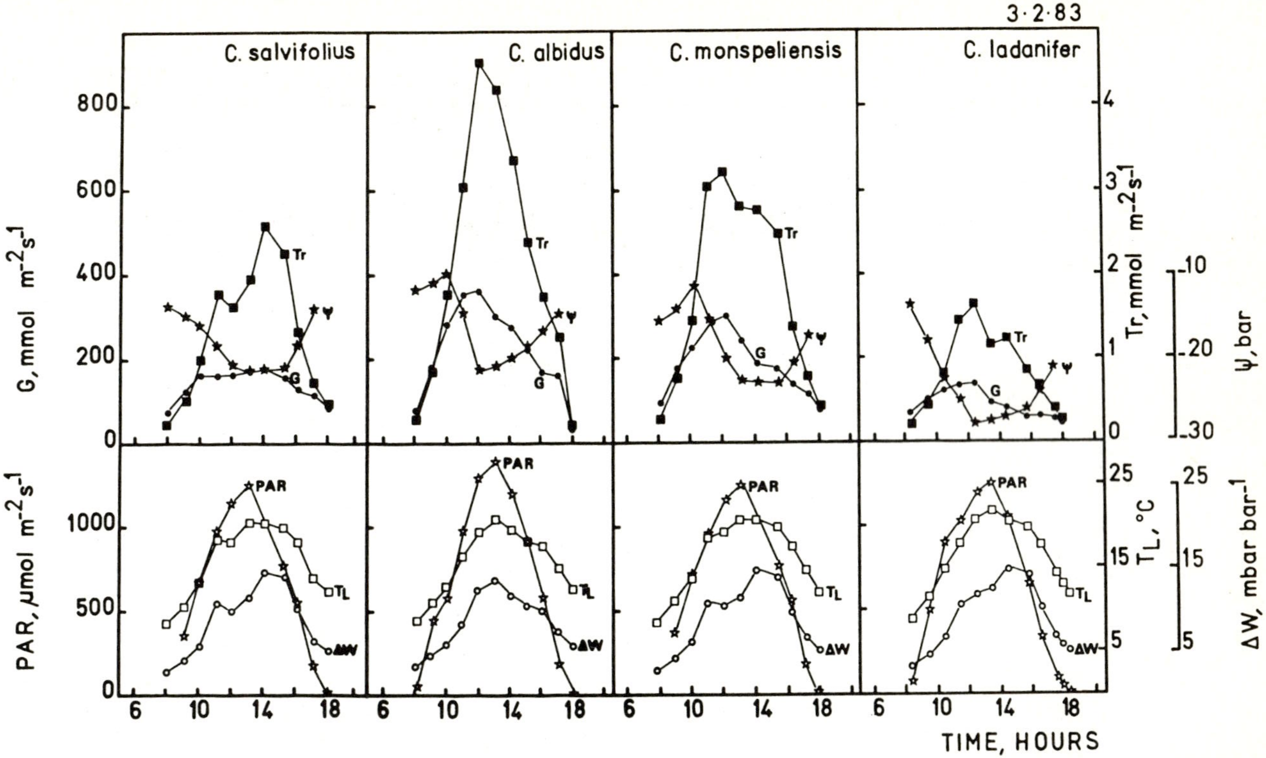

Fig. 4. Diurnal courses as in Fig. 2 for the *Cistus spp.* shown. Day, month, and year shown at the upper right hand corner of the figure

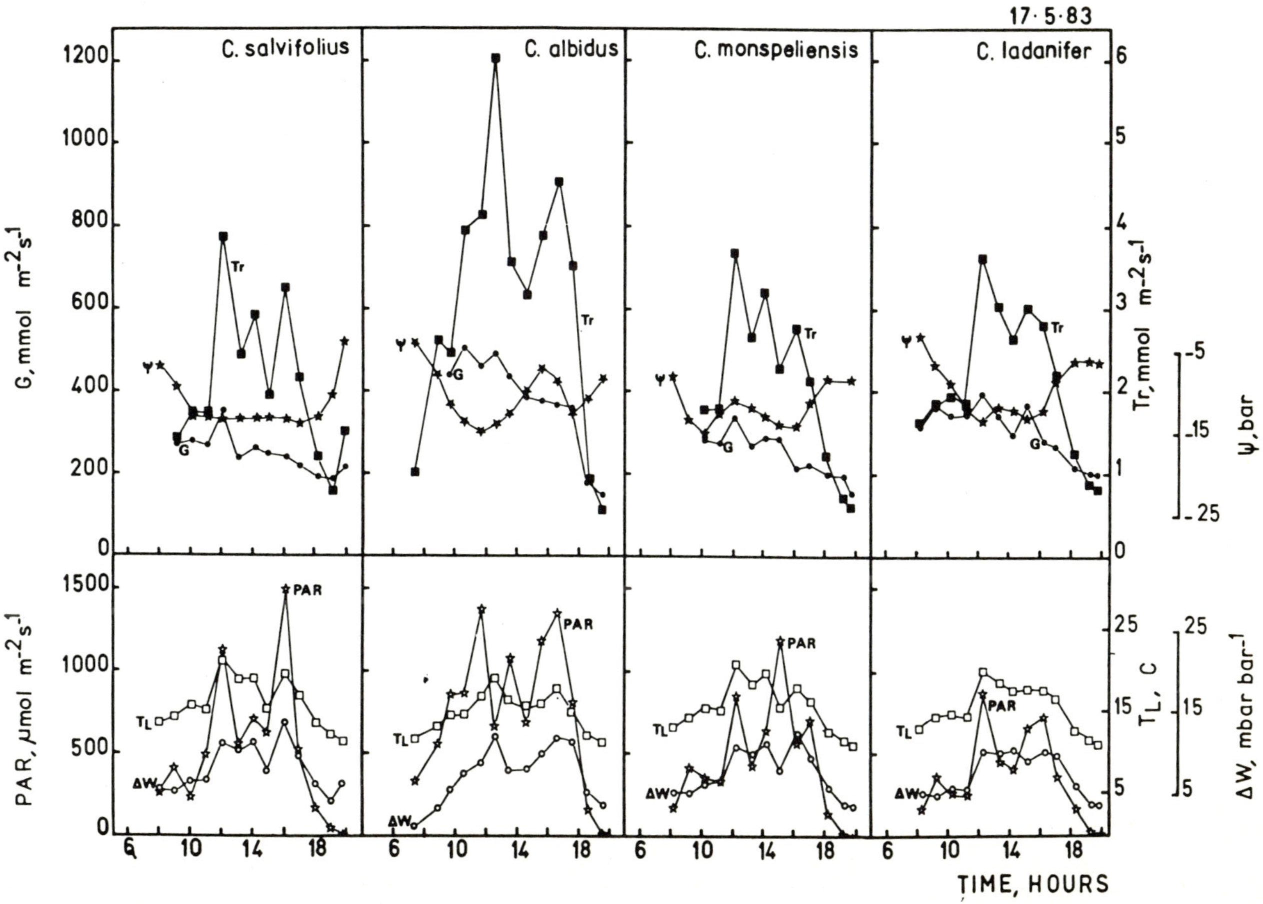

Fig. 5. Diurnal courses as in Fig. 2 for the *Cistus spp*. shown. Day, month, and year shown at the upper right hand corner of the figure

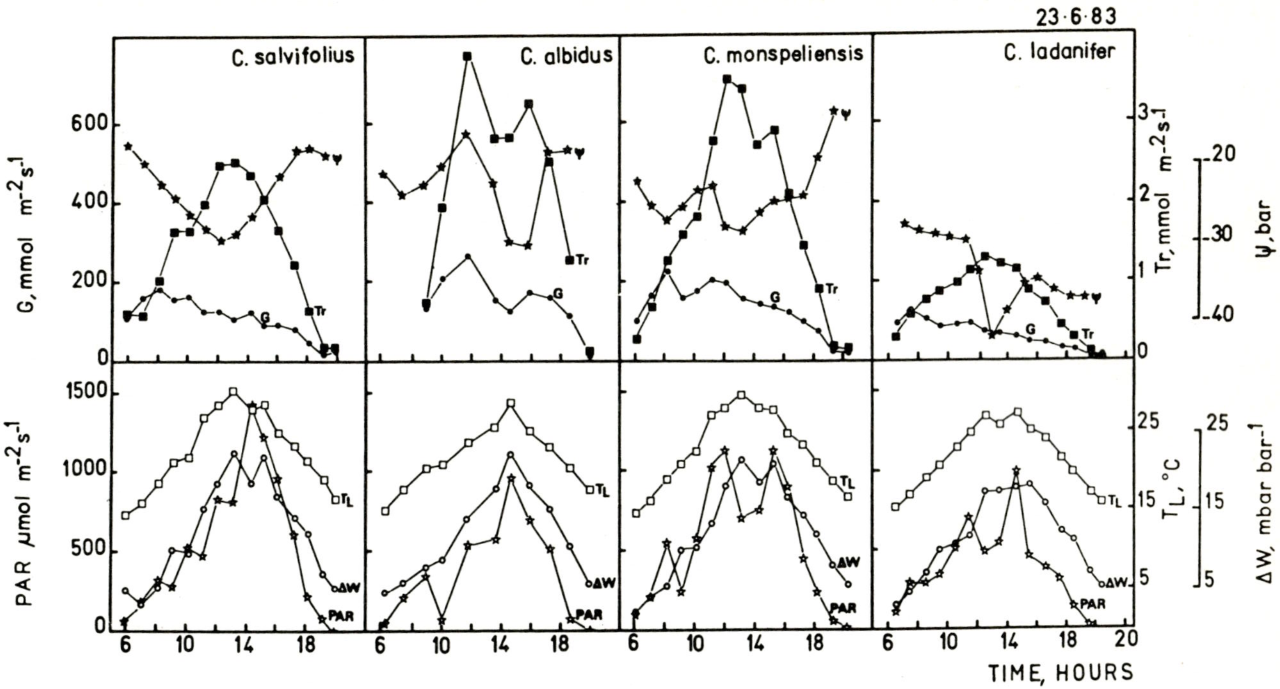

Fig. 6. Diurnal courses as in Fig. 2 for the *Cistus spp*. shown. Day, month, and year shown at the upper right hand corner of the figure

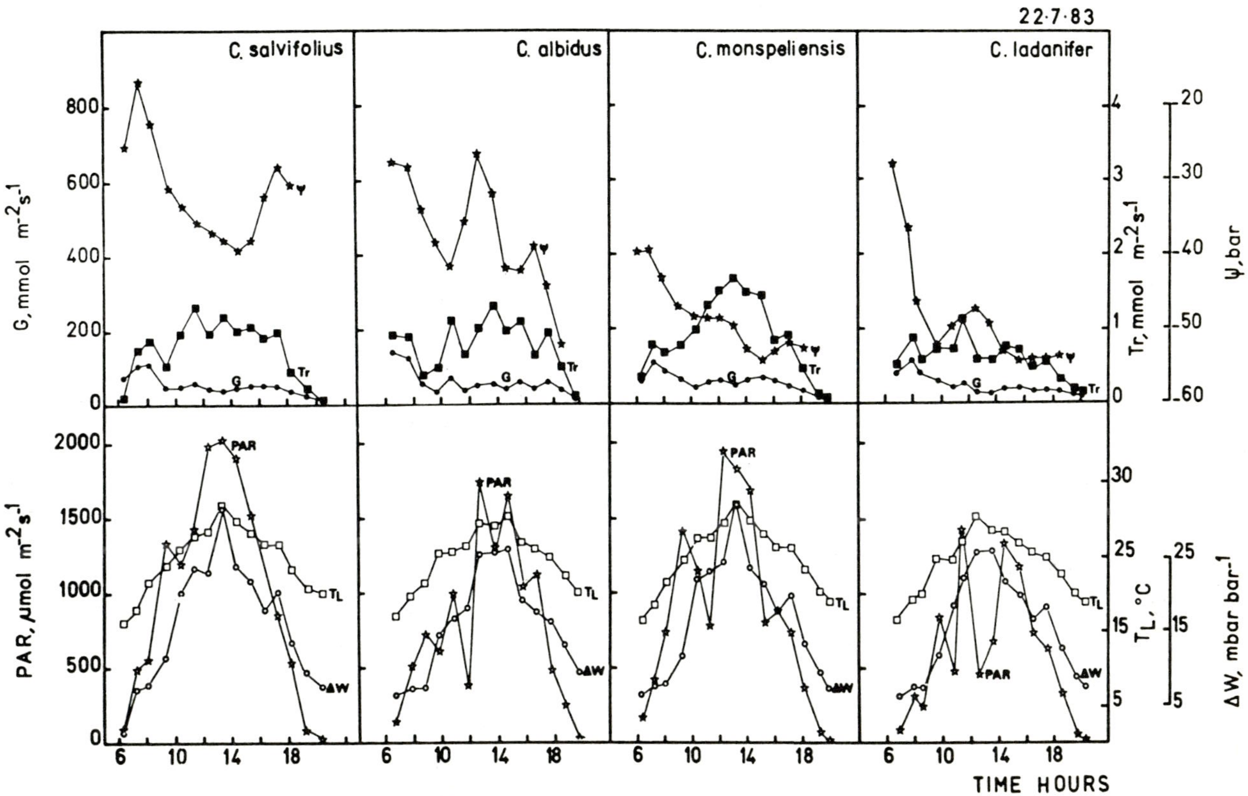

Fig. 7. Diurnal courses as in Fig. 2 for the *Cistus spp*. shown. Day, month, and year shown at the upper right hand corner of the figure

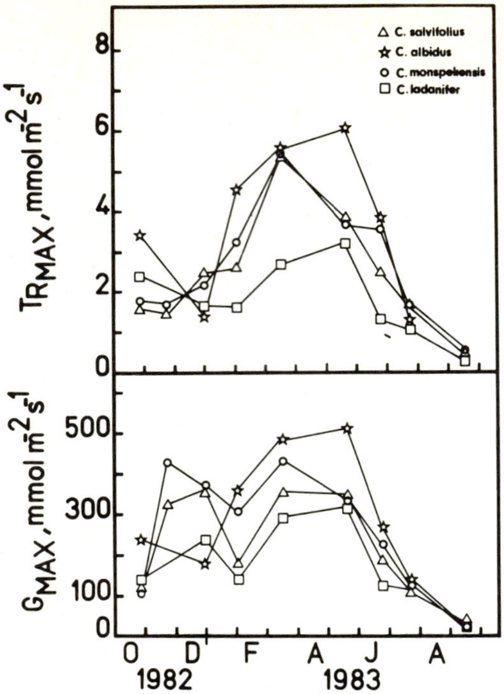

Fig. 8. Seasonal course of maximum transpiration rates (TR_{MAX}) and maximum leaf conductance (G_{MAX}) for the *Cistus* spp. shown

DISCUSSION

The results obtained in this study indicate a number of species specific differences that may be important in determining along with other factors the distributions of individual *Cistus* species in the mixed scrub vegetation of Serra da Arrábida. Most clearly, *Cistus ladanifer* exhibited a very distinct set of behavioral characteristics. The robust and large appearance of *C. ladanifer* shrubs at the study site might lead us to hypothesize that CO_2 uptake may continue to be positive under hot and dry conditions or at least that stress conditions do not immediately lead to detrimental effects at the biochemical level in leaf tissues during drought. The large leaf area maintained per plant is certainly advantageous when weather conditons improve in the fall. *Cistus ladanifer* appears to have the lowest maximum conductance of all species studied, a trait which has previously been found in evergreen Californian and Chilean shrub species to be associated with deep rooting (Poole *et al.* 1981). This must also be the case for *Cistus ladanifer*. Despite extremely negative predawn xylem water potentials during summer drought, a large change in water potential occurred over diurnal cycles suggesting substantial resistance in the water supply pathway as well as available water reserves.

In contrast to *C. ladanifer*, *Cistus salvifolius* appears to avoid low water potentials which occur in other species at the end of the dry season. It accomplishes this in a number of ways. First, drought conditions result in large decreases in maximal leaf conductance as in all other species. Water loss is further restricted by shedding of leaves. Other unpublished data have shown that as much as 80% of the winter leaf area is lost, and this in turn must increase the relative root area supplying remaining photosynthetic tissues. In addition, *Cistus salvifolius* grows at dry sites within the shelter of canopies provided by other shrub species. Studies in the Doñana National Park in Spain have shown that this obligate seeder (see discussion by Hilbert this volume) only becomes established and remains at those locations where shading occurs due to other shrubs. By these means of drought avoidance, *C. salvifolius* persists at the Jaspe site. We might hypothesize that its prevalance on north facing slopes indicates a relatively high potential photosynthetic capacity of leaves that is realized when water is available.

Present information allows a less satisfying interpretation of the behavior and strategies employed by *Cistus albidus* and *Cistus monspeliensis*. It appears that *Cistus albidus* is capable of high rates of gas exchange and that *Cistus monspeliensis* may have substantial water reserves available late in the dry season. Nevertheless, restrictions on their distribution and potential traits that might allow them to exploit particular niches must be sought in future studies. Perhaps further information on changes in leaf area index or on morphological structure, coupled with further studies of physiological behavior, will provide us with new insight.

REFERENCES

Catarino FM, Correia OA, Correia AI (1982) Structure and dynamics of Serra da Arrábida mediterranean vegetation. Ecologia Mediterranea VIII:203-222

Garcia Novo F (1977) The effects of fire on the vegetation of Doñana National Park (Spain). In: Mooney HA and Conrad CE (eds) Proceedings of the Symposium on the Environmental Consequences of Fire and Fuel Management,. pp 318-325

Lange OL, Tenhunen JD, Braun M (1982) Midday stomatal closure in Mediterannean type sclerophylls under simulated habitat conditions in an environmental chamber. Flora 172:563-579

Lösch R, Tenhunen JD, Pereira JS, Lange OL (1982) Diurnal courses of stomatal resistance and transpiration of wild and cultivated mediterranean perennials at the end of the summer dry season in Portugal. Flora 172:138-160

Levitt J (1980) Responses of plants to environmental stresses. Vol. II. Water, radiation, salt and other stresses. Academic Press, New York

Poole DK, Roberts SW, Miller PC (1977) Water utilization. In: Miller PC (ed), Resource Use by Chaparral and Matorral, Ecological Studies Vol 39. Springer, Berlin-Heidelberg-New York, pp 123-249

Scholander PF, Hammel HT, Bradstreet ED, Hemmingsen EA (1965) Sap pressure in vascular plants. Science 148:339-345

Schulze E-D, Hall AE, Lange OL, Walz H (1982) A portable steady-state porometer for measuring the carbon dioxide and water vapour exchanges of leaves under natural conditions. Oecologia (Berl) 53:141-145

Tenhunen JD, Lange OL, Pereira JS, Lösch R, Catarino F (1981) Midday stomatal closure in *Arbutus unedo* leaves: Measurement with a steady-state porometer in the Portuguese evergreen scrub. In: Margaris NS and Mooney HA(eds), Components of Productivity of Mediterranean-climate Regions, Basic and Applied Aspects. Dr. W. Junk, The Hague, pp. 61-69

Water use by *Quercus Ilex* L. in forests near Barcelona, Spain

M.P. Comín, A. Escarré*, C.A. Gracia, M.J. Lledó*, R. Rabella** , R. Savé** and J. Terradas**

Dept. d'Ecologia, Fac. Biologia, Universitat de Barcelona, Barcelona, Spain

INTRODUCTION

Quercus ilex evergreen-oak forests are often viewed as climax communities in large areas of the Iberian Peninsula influenced by a mediterranean type climate (see also discussion by Naveh in this volume). Nevertheless, there are genetic and floristic differences in *Q. ilex* populations and forests found throughout this region. In parts of northeast Spain under humid meso-mediterranean conditions, evergreen-oak belongs to the phytosociological association *Quercetum mediterraneo-montanum* (Bolós 1983) which is dominated by *Q. ilex* ssp. *ilex*. The water use by plants and hydrological balance in these ecosystems have been little studied, despite the importance of these forests for water yield, soil protection, and maintenance of environmental quality. We report here the results of studies at two experimental forest plots located in the Prades (l'Avic) and Montseny (La Castanya) mountain ranges. Data have been gathered to describe the water budget for a watershed on watertight rock. Other measurements have been focused on understanding the ecophysiology of *Q. ilex* and movement of water vertically through the forest canopy.

CLIMATE CONDITIONS AND WATER INPUTS

The watershed study plot at La Castanya is located in the Montseny Sierra 40 km north of Barcelona. The experimental plot has a north-northeast orientation and is located at the base of a steep slope. The substrate consists of Ordovicic schists. Climate conditions at this site are influenced by local topography. The annual rainfall averaged 862 mm during the period of our study. The summer is relatively dry, although rainfall is irregularly distributed throughout the year. Direct beam radiation does not reach the plot during winter months. Winter temperatures can be rather low and cold stress appears to be a major ecological factor shaping the vegetation. The average annual temperature is 9° C.

Mediterranean climate features are more pronounced at the l'Avic experimental area located 100 km southwest of Barcelona. Here the substrate consists of

*Dept. de Biología, Univ. Alicante, Ap. Correos 99, Alicante, Spain
**Dept. d'Ecologia, Fac. Ciències, Universitat Autònoma de Barcelona, Barcelona, Spain

NATO ASI Series, Vol. G15
Plant Response to Stress. Edited by J.D. Tenhunen et al.

Paleozoic schists. Rainfall is lower (621 mm) than in the Montseny area and clearly concentrated in spring and autumn. The watersheds are primarily northfacing, and freezing temperatures occur frequently in winter. The average annual temperature is higher (13° C)

We have measured precipitation, throughfall, stemflow, and interception in both areas. The data from Prades are more complete at the present time. There we used 50 collectors under the canopy for throughfall and 20 ring stemflow collectors on trees (*Quercus ilex, Arbutus unedo* and *Phillyrea media*). For 47 storms monitored during a two year period, the throughfall amounted to 68.7% of incoming precipitation, stemflow to 13.1% and interception to 18.2%. The throughfall percentage is similar to that found in La Castanya (Rodà 1983) and also in the Le Rouquet evergreen-oak forest by Loissant and Rapp (1971). This value is within the range measured for deciduous and coniferous temperate forests but below the average. Stemflow is probably higher in Prades than in other forests, due to canopy structure and forest density.

CONTROLS ON WATER USE WITHIN THE PLANT CANOPY

Vertical profiles of photosynthetically active radiation, leaf area index (LAI), and chlorophyll concentration were described by Gracia (1984), of stomatal conductance by Rabella *et al.* (1983) and response of leaves to desiccation by Escarré and Terradas (1983). From a tower in the forest, leaf and air temperature profiles and information on the water relations of leaves have been obtained during different seasons. From the direct measurement of radiation above the canopy and a model describing distribution of LAI within the canopy (Gracia 1984), we estimated the net radiation available in different canopy layers. Using these estimates along with measured leaf and air temperatures in each layer, we solved the energy balance of leaves in different levels (Fig. 1) and obtained estimates of transpirational water loss. An example of such results is illustrated in Fig. 2, where daily changes in relative transpiration rate of leaves in the upper, middle, and lower canopy are depicted for an early summer day.

The differences in contributions to the transpirational water loss by leaves in different layers are large (Fig. 2). Higher transpiration rates in sun leaves result in larger decreases in leaf water potentials during the day (Table 1). With branches previously water saturated and subjected to increasing temperatures in a controlled chamber, sun leaves exhibited a greater capacity than shade leaves to sustain transpiration above 30° C. Under natural conditions, the exposed sun leaves appear as well to reduce water loss by partial stomatal closure earlier than leaves of other levels (unpublished observations). This may be caused jointly by water deficits, high leaf temperature, and low humidity (see Tenhunen *et al.* in this volume). Leaves in the lowest part of canopy always have a very low conductance, and probably a very low assimilation rate.

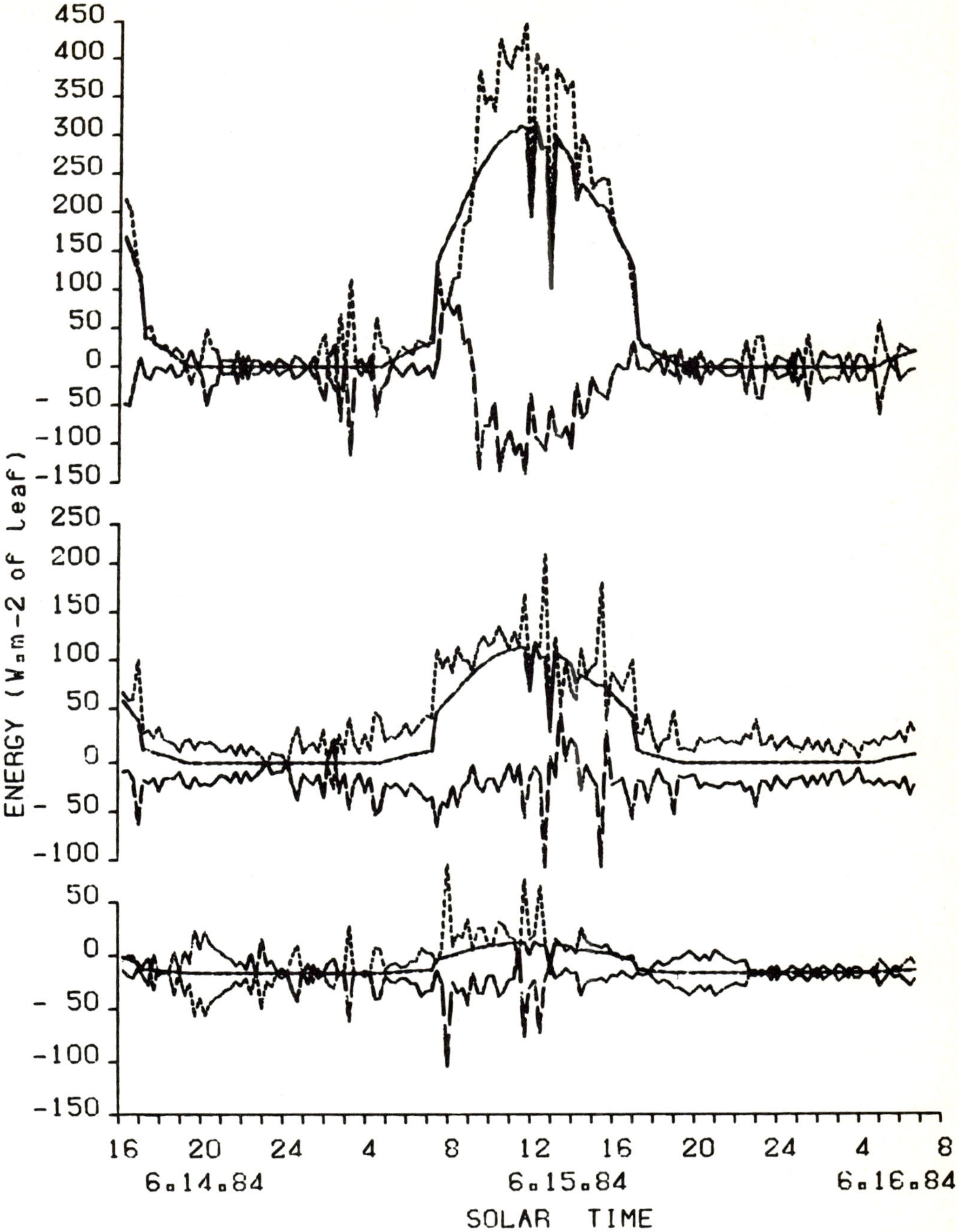

Fig. 1. Results of model analysis discussed in the text showing total net radiation absorbed by leaves (——), convective heat loss (------) and latent heat loss in transpiration (······) of leaves located at the upper, middle and lower part of the canopy of trees at the experimental station of La Castanya

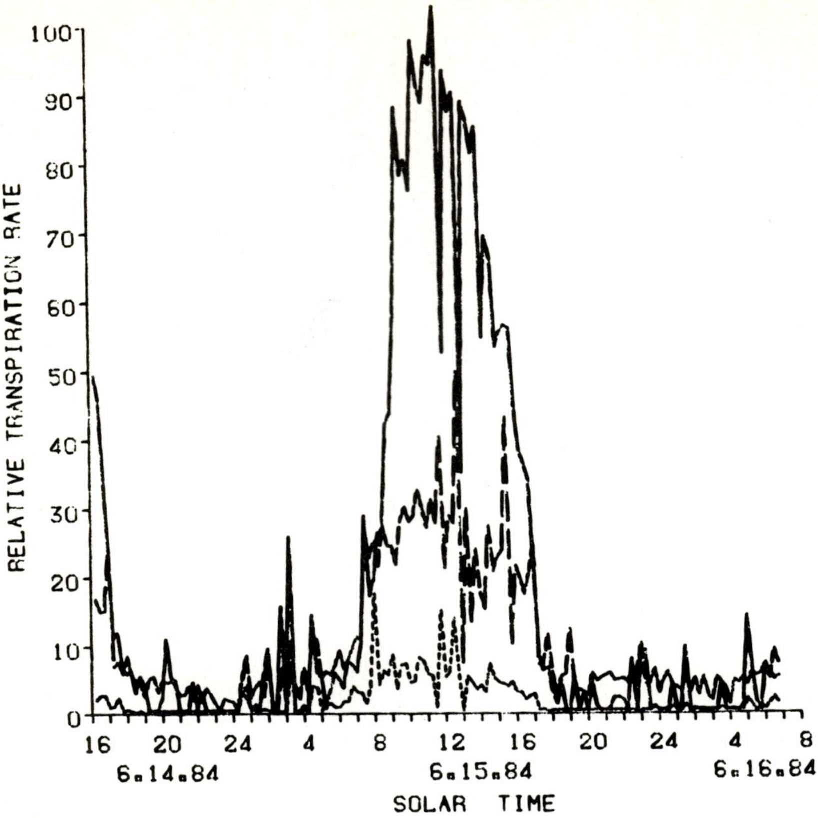

Fig. 2. Daily variation in the relative transpiration rate of *Quercus ilex* leaves located at the upper (——), middle (------) and lower (·····) part of the canopy as estimated from the model analysis described in the text. Transpiration rate of the uppermost leaves at noon is considered as 100%

Table 1. Maximum (Max) and minimum (Min) xylem water potential (bar) on days between July and October 1982 at four levels in the evergreen-oak canopy.

Height	10 m		8 m		6 m		4 m	
	Min	Max	Min	Max	Min	Max	Min	Max
Date								
06-07-82 (1)	-22.3	-17.6	-23.0	-15.2	-21.5	-15.3	-15.5	-11.5
11-08-82 (1)	-17.2	-6.6	-15.1	-6.0	-11.1	-4.4	-10.9	-3.8
12-08-82 (1)	-16.2	-4.0	-15.2	-3.4	-12.3	-3.3	-9.8	-2.8
10-09-82 (2)	-14.1	-3.8	-10.9	-3.0	-10.2	-3.2	-8.0	-2.4
22-09-82 (3)	-14.1	-4.1	-11.9	-4.0	-9.5	-3.5	-7.5	-2.7
14-10-82 (1)	-19.9	-8.8	-15.4	-7.3	-12.6	-7.7	-10.7	-6.0

(1): sunny day. (2): variable cloudiness (3): few clouds

Extreme water stress may be uncommon at the La Castanya site. Water potential did not decrease below -22.3 bars in the highest canopy level during the period of our observations, even though July 6, 1982 was considered to be the hottest day in a century in this region. Eckardt *et al.* (1977), Ausennac and Valette (1982), and Larcher (1960) similarly observed only moderately low water potentials in *Q. ilex* during summer drought. On the other hand, Tenhunen and Rambal found leaf conductances and net photosynthesis rates of *Quercus ilex* trees growing in plantations near Montpellier to be reduced at the end of the summer drought in September 1986 and to double under similar atmospheric conditions after the first fall rains. In these trees, stomata remained open for approximately 3 hours during the morning and carbon uptake was limited to this period (personal communication).

In La Castanya , cold stress is probably more important in determining the success of various species than is summer water stress. After a hard winter in 1985, many young sprouts of *Q. ilex* had fissures in their smooth, thin bark due to freezing. They did, nevertheless, leaf out during the following spring and summer. A high percentage of leaves exhibited symptoms of damage due to cold stress. Water potentials decreased in *Q. ilex* when exposure to cold temperatures occurred. On 15 February, with snowcover from 700 m to the summit and a temperature of -3° C at the highest station, water potentials varied from -5.3 ± 0.5 bars at 400 m to -28.0 ± 3.2 bars at 1140 m. Differences of 3 to 5 bars were found between the lowest and the highest stations on three days of 1983 (March 3, May 25, and September 21) under more moderate conditions. Again water potentials became increasingly negative with increasing altitude.

WATER BUDGET IN EVERGREEN-OAK WATERSHEDS

Water consumption by the forest can be estimated in different ways. The annual pattern of transpiration and evapotranspiration obtained with three different methods is depicted in Fig. 3. Yearly water budgets were determined for a 55.5 ha watershed of the Prades area. The forest biomass was 115 t ha^{-1}. Water inputs for the three years studied were 619.5, 526.2 and 585.7 mm. Corresponding water discharges were only 17.9, 81.6 and 78.4 mm (Fig. 4). Thus, the total evapotranspiration must have been 601.5 mm (97.1% of input), 444.5 mm (84.5%) and 504.9 mm (86.2%). For the entire three year period, evapotranspiration was 89.6% of the water input and only 10.4% drained as streamflow. One goal of our studies is to use the model discussed above to estimate total water use by the forest. The total water use estimate should then agree with measured watershed output.

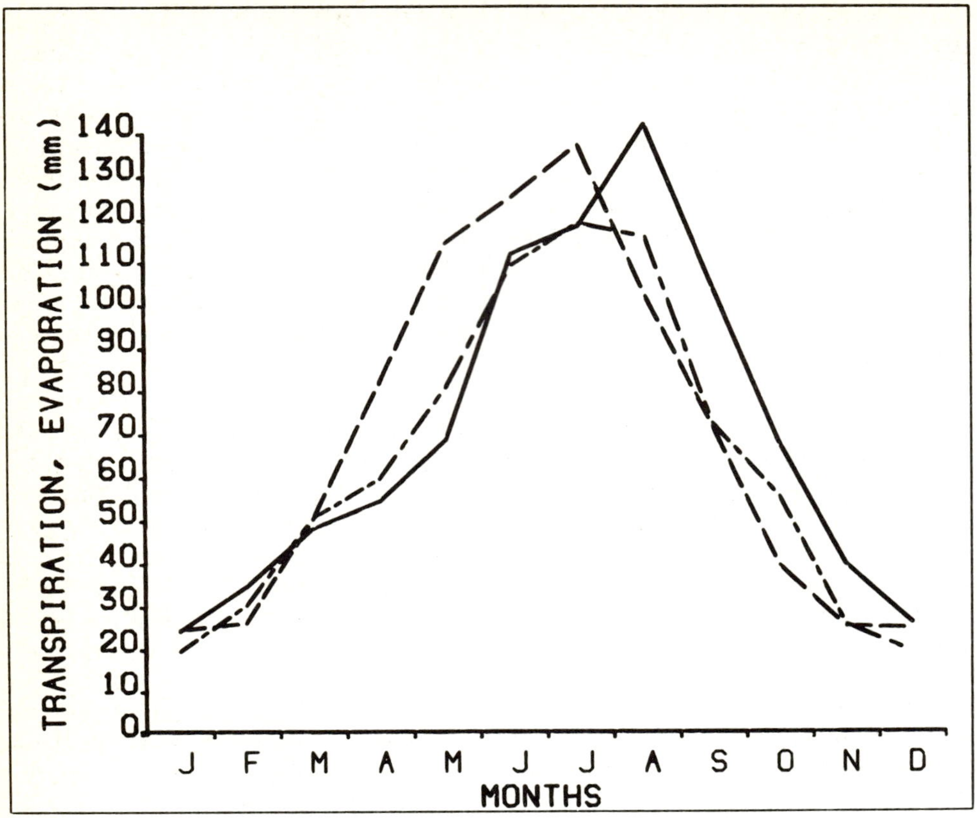

Fig. 3. Monthly variation in transpiration rate of the forest canopy based on water loss rates of individual leaves measured by the quick weighing method (– – –) and potential evapotranspiration evaluated using the methods of Turc (·····) and Thornthwaite (——) for La Castanya during 1980. Values for transpiration of individual leaves of different ages are multiplied by corresponding leaf biomass and day lengths to obtain a total monthly rate. In the calculation of the Turc estimate, real radiation measurements of Gracia (1985) were used

CONCLUSIONS

The water balance of evergreen-oak forests at the study sites has the following characteristics. Interception is low, and it reduces only slightly water input. Stemflow in comparison is relatively high. Regulation of water loss from the canopy is efficient and sun and shade leaf adaptations play an important role. Sun leaves appear to reduce water flow under dry conditions before the rest of the canopy. Leaves of lower part of the canopy always have low conductances and probably low assimilation rates. In the La Castanya plot, water stress was

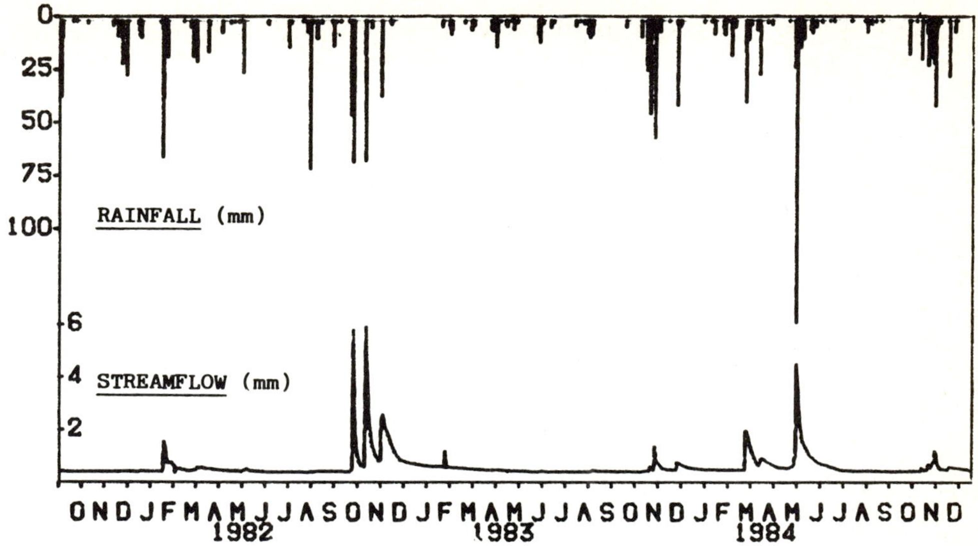

Fig. 4. Daily rainfall and discharge of the experimental watershed of l'Avic (Prades) during the study period

moderate and cold stress seems more important. In Prades, water input is lower and evapotranspiration higher, and it appears that both stresses are significant. Total water consumption in Prades is near 90% of actual input of water to the soil system (rainfall minus interception), so water yield is very low.

REFERENCES

Aussenac G, Valette JC (1982) Comportament hidrique de *Cedrus atlantica, Quercus ilex* et *Quercus pubescens* et de divers pins dans le Mont Ventoux. Ann Sci Forest 39, Vol 1:41-62

Bolos O (1983) La vegetació del Montseny. Diputació de Barcelona. Servei de Parcs Naturals

Eckardt FE, Berger A, Methy M, Heim G, Sauvezon R (1977) Interception de l'énergie rayonnante, échanges de CO_2, régime hydrique et production chez différents types de végétation sous climat méditerranéen. In: Moyse A (ed) Les processus de la production végétale primaire. Gauthier-Villars, Paris, pp 1-75

Gracia CA (1983) La clorofila en los encinares del Montseny. Tesis doctoral. Dep Ecologia, Univ Barcelona

Larcher W (1960) Transpiration and photosynthesis of detached leaves and shoots of *Quercus pubescens* and *Q. ilex* during desiccation under standard conditions. Bull Res Counc Israel 8:213-224

Loissant P, Rapp M (1971) Repartition de la matiere organique, productivite et cycles des elements mineraux dans écpsystèmes de climat mediterraneen. Proc Brussels Symp UNESCO, Paris. pp 597-617

Rabella R, Save R, Terradas J (1983) Pefiles verticales de conductancia estomática en *Quercus ilex* y *Fagus sylvatica*. Sem Reservas de la Biosfera. (Huelva)

Rodà F (1983) Biogeoquimica de les aigües de pluja i drenatje en ecosistemes forestals del Montseny. Tesis doctoral. Dep Ecologia, Univ Autónoma de Barcelona. Bellaterra

Soil drying and the resulting chemical and hydraulic effects on leaf growth

A. Rosa da Costa, J. Metcalfe, T. A. Lodge and W. J. Davies

Department of Biological Sciences, University of Lancaster, Bailrigg, Lancaster, LA1 4YQ UK

INTRODUCTION

There are now many reports in the literature that stomatal behaviour, leaf growth and other aspects of shoot physiology are influenced by changes in the edaphic environment in the vicinity of the plant root. Examples of such changes are soil drying and flooding with fresh or saline water. The general view is that these perturbations may lead to reduction in leaf turgor and may therefore influence physiology via a change in water relations. More recently, however, it has become clear that this is not always the case and that shoot physiology will show a response to soil drying (Davies *et al.* 1980; Davies and Sharp 1981), salinity (Termaat *et al.* 1985) and freshwater inundation (Jackson and Kowalewska 1983) even when leaf water relations are not perturbed.

One explanation for these observations is that changes in the edaphic environment cause the root to generate some chemical signal which may move to the shoot and override any hydraulic influence on growth or stomatal behaviour (Davies *et al.* 1986). Suggestions of this kind were made by Itai and Vaadia (1965) who demonstrated a water deficit-induced reduction in the supply of cytokinin-like substances moving from roots. To further investigate relationships of this kind Blackman and Davies (1985) utilized a split-root technique to dry part of the root system of maize plants while the other part was well supplied with water. This system ensured that leaf turgor remained high even when soil around part of the root system dried substantially. Despite high leaf turgor, stomatal conductances of half dried plants were reduced and stomata remained partially closed. This closure could be overcome by incubating leaf pieces in a medium containing cytokinin. The results from this experiment suggest that soil drying reduced the cytokinin supply from the roots of maize plants and this may cause restricted stomatal apertures and water use. Blackman and Davies (1983, 1984) had previously shown that external application of cytokinin had marked effects on stomatal behaviour of maize plants.

Jones (1980) and Cowan (1982) have argued that the passage of chemical information from the roots to the shoots may enable plants to utilize soil water in an optimal fashion with respect to time. In other words, all available soil water may be used, but at such a rate that it is not exhausted by the plant until it has completed its life cycle.

NATO ASI Series, Vol. G15
Plant Response to Stress. Edited by J. D. Tenhunen et al.

In this paper we report on an extension of our work with plants with split roots. If cytokinins are synthesized in root tips, cytokinin supply to shoots is likely to be a function of root tip water status which will determine root activity (see Davies *et al.* in this volume for further background information and supporting data). We have measured leaf growth, since there are many indications that leaf expansion rates may be slowed by withholding water, even when leaf turgors are sustained (Michelena and Boyer 1982). Restriction of the development of the transpiring area is another effective means whereby a root signal may modulate plant water use.

METHODS

Root systems of maize plants (John Innes Hybrid) were split between two pots containing J I No 2 potting compost, as described by Blackman and Davies (1985). Plants were 15 days from emergence when they were split and were then left in a high humidity environment to recover from the treatment for a further 14 days. After this time, water was withheld from half of the root system of half of the plants, while all other pots were watered to the drip point once each day.

Measurements of root tip water relations were made using a hygrometric technique (Sharp 1981) while shoot water relations were measured with the pressure chamber and dew point hygrometer. Roots were removed from the soil and carefully washed before root lengths were determined on an instrument similar to that described by Rowse and Phillips (1974). Leaf extension rates were logged using LVDTs connected to a BBC microcomputer. Hourly rates of leaf extension were measured throughout the experimental period.

RESULTS

Leaf Water Relations

Water was first withheld from half of the root system of some of the plants from day 0 and there was no evidence that at any time after this, leaf turgor was reduced compared to control plants (Fig. 1). There was therefore no hydraulic signal from the dry roots to perturb shoot physiology.

Leaf Growth

Leaves of both treated and control plants extended rapidly throughout the experimental period (Fig. 2). Daily mean growth rates during the light period (14 hourly measurements) of four individual plants are shown in Fig. 2 above. This graph does not show clearly that leaf growth rates are reduced in half watered plants, and is complicated by the rather different extension rates at the start of the experiment. We would expect a larger sample to indicate more

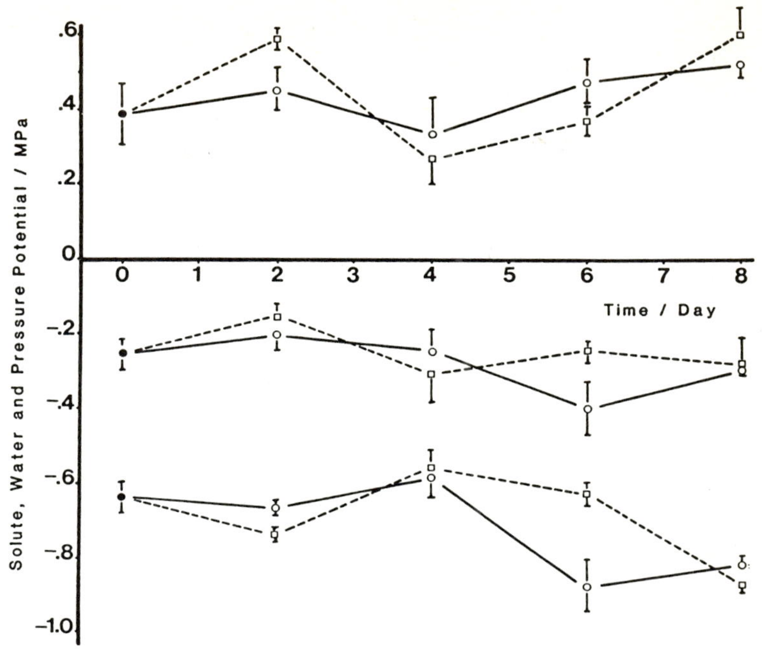

Fig. 1. Water (lower), solute (middle) and pressure (upper) potentials of the youngest expanding maize leaves taken from plants with roots split into two containers. Plants were either watered well throughout the experimental period (- - - -) or water was withheld from half of the root system from day 0 (—). Points are means of four measurements ± S.E.

clearly any changes occurring in leaf growth. There is some indication that late in the light period, leaf growth of half watered plants is reduced compared to growth of well-watered plants (Fig. 2 below).

Root Water Relations

There were no significant differences between the water relations of root tips from plants where the whole root system was well watered and water relations of root tips taken from the wet side of half watered plants (Fig. 3). Root tips in dry soil showed reduced water potentials within 2 days of water first being withheld from the soil. Root turgor was also reduced at this time but as the soil dried further, turgor was sustained as a result of solute accumulation in the root tips. Similar turgor regulation in maize roots has been reported by Sharp and Davies (1979).

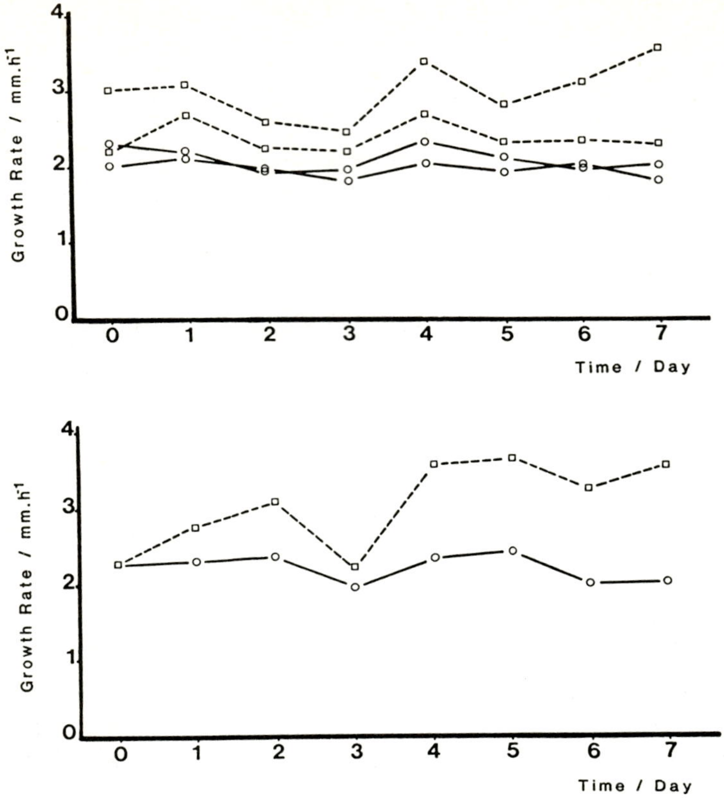

Fig. 2. Extension rates of young leaves of maize plants with split root systems. Plants were either watered well throughout the experimental period (- - - -) or water was withheld from half of the root system from day 0 (——):

above: Daily mean extension rates of two plants from each treatment

below: Mean extension rates during the seventh hour of each light period

Root Growth

Well-watered plants showed a reasonably steady root growth rate over the experimental period (Fig. 4) while growth of roots in the dry soil was restricted. The wet half of the half dried plants also showed restricted root growth until the end of the experimental period when there was a burst of root growth.

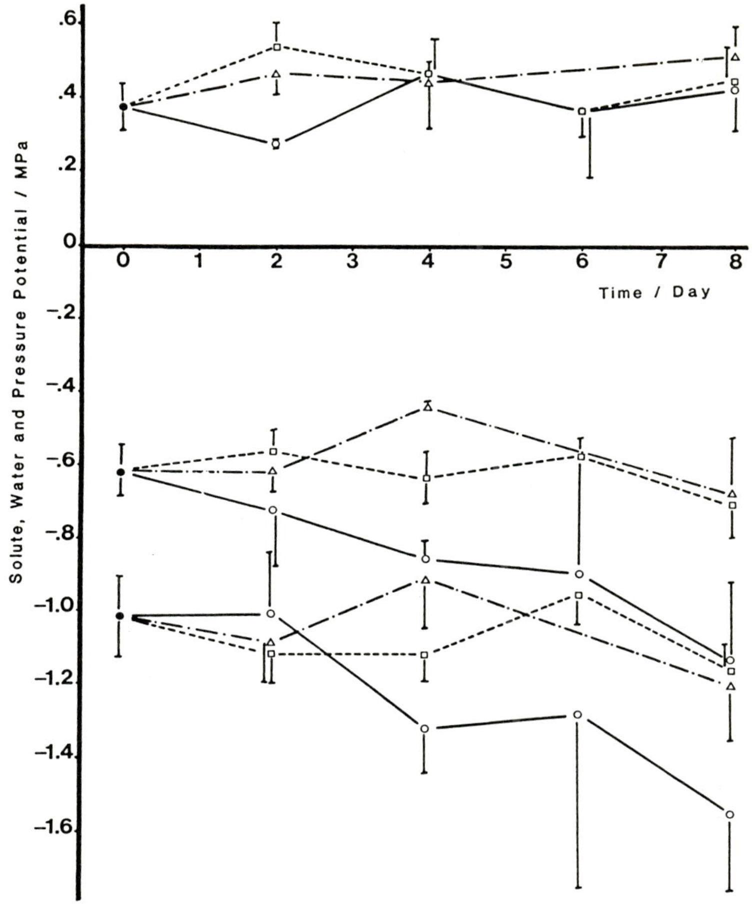

Fig. 3. Water relations as in Fig. 1 of root tips of maize plants which were grown with root systems split between two pots. Plants were either watered well (solid squares with dashes) or water was withheld from half of the root system from day 0. Data are shown for the well watered side of the half watered plants (Δ· - · - ·Δ) and for the drying side of the half watered plants (o——o)

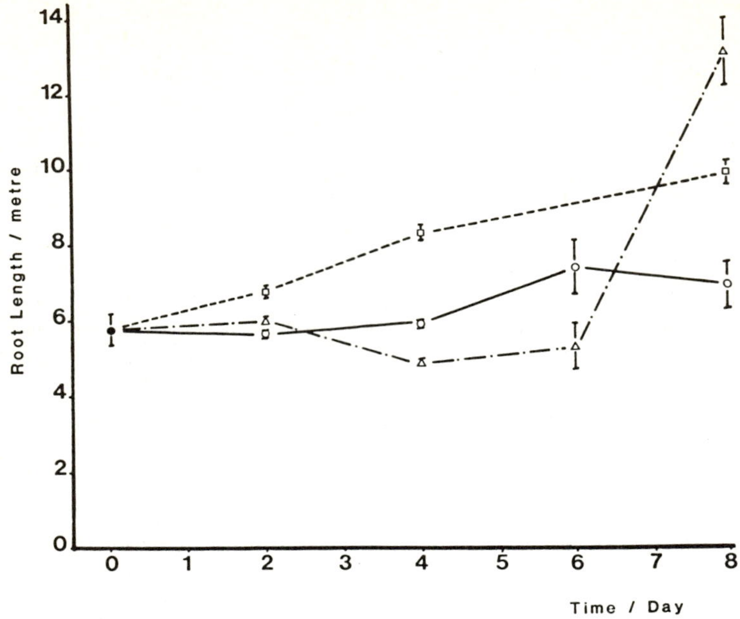

Fig. 4. Root lengths of maize plants grown with root systems split between two containers. Some plants were watered well throughout the experimental period while water was withheld from half the root system of other plants after day 0. Data are shown for well watered plants (total root system/2) (solid squares with dashes), for the well watered half of half watered plants (Δ· - · - ·Δ) and for the drying side of half watered plants (o——o). Points are means ± S.E.

DISCUSSION

Withholding water from part of a root system of a maize seedling has previously been shown to result in restricted stomatal apertures even though leaf turgors are not reduced compared to well-watered control plants (Blackman and Davies 1985). The explanation of this phenomenon may be a reduced supply of cytokinin moving from the roots in drying soil. In this experiment, reductions in leaf growth of the half-dried plants were not immediately obvious (Fig. 2). Nevertheless it does seem possible that late in the light period leaf growth of half-watered plants was reduced compared to growth rates shown by well-watered plants. (Fig. 2 below). This may be the result of a similar chemical signal moving from the roots, since leaf turgors of treated plants are not reduced (Fig. 1). We have not measured turgors of growing cells but it seems likely from the work of Michelena and Boyer (1982) that if turgors in the lamina are sustained then turgors in the growing zone will also remain high. Reduced leaf growth when turgor is high can presumably be explained by some change in cell wall properties. It is significant that cytokinin can enhance the extensibility of cell walls (see Cleland 1981). We might therefore argue that a reduced supply of cytokinin from roots would stiffen cell walls in the growing zone and reduce the growth of leaves even though turgor is high. We have shown that root growth in drying soil is greatly reduced (Fig. 4) and we might therefore expect a similar reduction in cytokinin production and cytokinin supply to the shoots. This may be due to reduced movement of xylem sap from the roots in dry soil.

Until variation in the concentrations of different cytokinins in plants can be determined, conclusions from experiments of this type must of necessity be speculative. The most significant observation made here is the suggestion that leaf growth late in the day may be reduced in plants with some roots in dry soil when compared to rates shown by plants with roots in wet soil. Blackman and Davies (1985) found similarly that stomatal behaviour of the two groups of plants did not differ until late in the light period. Their explanation was that this experimental system might allow plants to rehydrate completely at night and that restoration of root tip turgor might restore cytokinin synthesis and thus temporarily restore cytokinin balance in leaves. Concentrations would decline again as the next light period progressed and root tip turgors declined.

Although we have some evidence that root tip turgors do decline (Davies *et al.* 1986) in this experiment such decreases were not sustained beyond the first few days of the drying cycle. We might therefore expect cytokinin production to be sustained and thus if cytokinin concentrations in leaves are reduced these reductions must be explained by reduced transport from roots (see Davies *et al.* this volume). Other workers have found similar results (e.g. Hubick *et al.* 1985). Given this effect, it is difficult to explain the diurnal variation in cytokinin balance in leaves which is proposed above, at least without proposing diurnal variation in the rates of cytokinin synthesis and metabolism. There is no published evidence on this question but we are currently investigating diurnal variation in stomatal behaviour, leaf growth, root extension and root water relations of plants with roots in dry soil. This is all with a view to collecting more information on diurnal variation in chemical influences on stomata. If leaves are to optimise gas exchange on a diurnal basis it seems likely that this type of variation does exist.

CONCLUSIONS

The experiment reported here was an attempt to provide further evidence for a chemical messenger moving from roots in drying soil to shoots. This messenger could override the hydraulic influence on leaf growth and may therefore be another means of modulating gas exchange and optimizing water use with respect to time. There is some evidence that cytokinin supply from roots may be reduced by soil drying and that such a reduction may influence stomatal behaviour and leaf growth of maize plants.

ACKNOWLEDGEMENTS

We are pleased to acknowledge financial support for this work from Gulbenkian Foundation (ARC), the Agriculture and Food Research Council (JM) and the University of Lancaster Research Fund (TL).

REFERENCES

Blackman PG, Davies WJ (1983) The effects of cytokinins and ABA on stomatal behaviour of maize and *Commelina communis*. J Exp Bot 34:1619-1626

Blackman PG, Davies WJ (1984) Modification of the CO_2 responses of maize stomata by abscisic acid and by naturally occurring and synthetic cytokinins. J Exp Bot 35:174-179

Blackman PG, Davies WJ (1985) Root to shoot communication in maize plants of the effects of soil drying. J Exp Bot 36:39-48

Cleland RE (1981) Wall extensibility: hormones and wall extension. In: Tanner W and Loewus FA (1981) Encyclopedia of Plant Physiology, NS vol 13B, Plant Carbohydrates II. Springer, Berlin-Heidelberg-New York, pp 255-276

Cowan, IR (1982) Regulation of water use in relation to carbon gain in higher plants. In: Lange OL, Nobel PS, Osmond CB, Ziegler H (eds), Encyclopedia of Plant Physiology, NS vol 12B: Physiological Plant Ecology II. Springer, Berlin-Heidelberg-New York-Tokyo, pp 589-613

Davies WJ, Mansfield TA, Wellburn AR (1980) A role for abscisic acid in drought endurance and drought avoidance. In: Skoog F (ed) Plant Growth Substances 1979. Springer, Berlin-Heidelberg-New York, pp 242-253

Davies WJ, Sharp RE (1981) The root: a sensitive detector of a reduction in water availability? Kralovic J (ed) Mechanisms of Assimilate Distribution and Plant Growth Regulators. Slovak Society of Agriculture, Ivanka pri Dunaji, Czechoslovakia, pp 53-62

Davies WJ, Metcalfe J, Lodge TA, Costa AR (1986) Plant growth substances and the regulation of growth under drought. Aust J Plant Physiol 13:105-125

Hubick KT, Taylor JS, Reid DM (1985) The effect of drought on levels of abscisic acid, cytokinins, gibberellins and ethylene in aeroponically-grown sunflower plants. Plant Growth Reg 4:139-151

Itai C. Vaadia & (1965) Kinetin-like activity in root exudate of water-stressed sunflower plants. Physiol. Plant 18: 941-944

Jackson MB, Kowalewska AKB (1983) Positive and negative messages from roots induce foliar desiccation and stomatal closure in flooded pea plants. J Exp Bot 34: 493-506

Jones HG (1980) Interaction and integration of adaptive responses to water stress: the implications of an unpredictable environment. In: Turner MC and Kramer PJ (1980) Adaptation of plants to water and high temperature stress. Wiley and Sons, New York, pp 353-365

Michelena VA and Boyer JS (1982) Complete turgor maintenance at low water potentials in the elongating region of maize leaves. Plant Physiol 69: 1145-1149

Rowse HR, Phillips DA (1974) An instrument for estimating the total length of root in a sample. J Appl Ecol 11: 309-314

Sharp RE (1981) Mechanisms of turgor maintenance in *Zea mays* . Ph.D. Thesis, University of Lancaster, pp 315

Termaat A, Passioura JB, Munns R (1985) Shoot turgor does not limit shoot growth of NaC1-affected wheat and barley. Plant Physiol 77:869-872

3. Photosynthesis and Primary Production under Mediterranean Climate Conditions

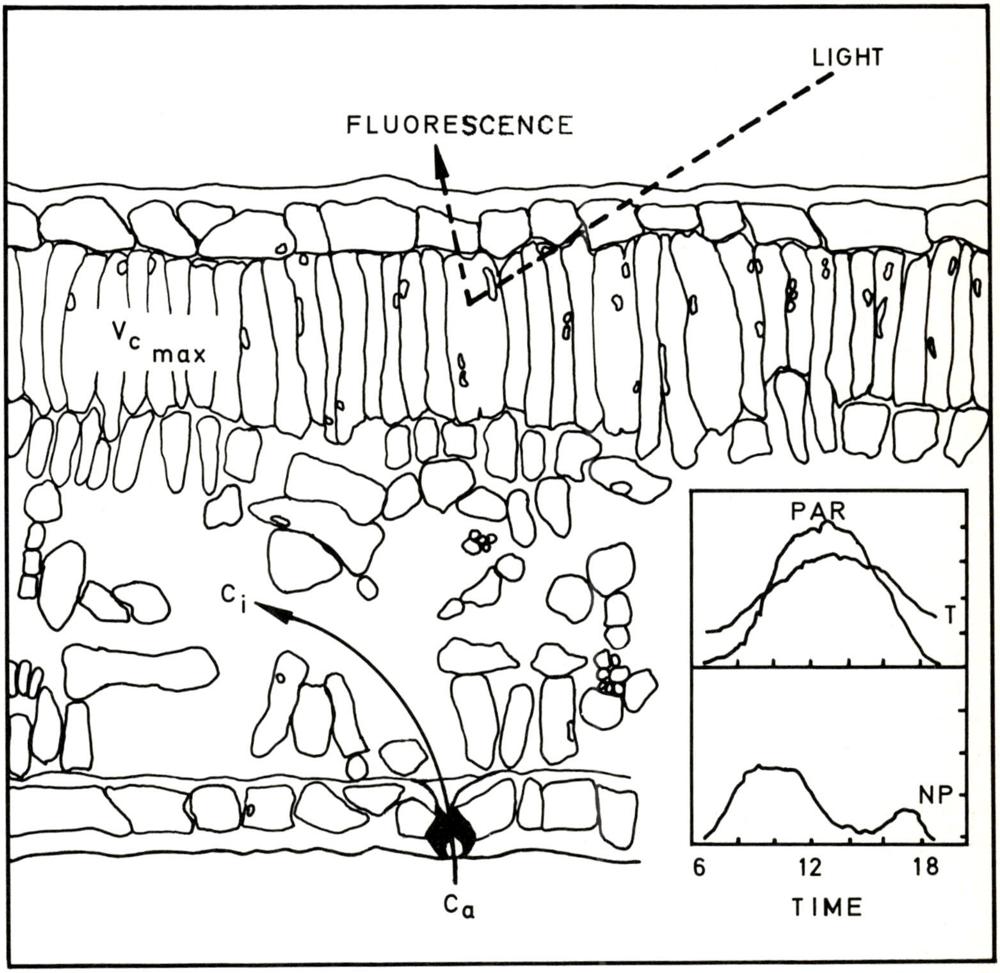

Gas exchange characteristics of representative species from the scrub vegetation of central Chile

W.T. Lawrence

Terrestrial Ecology Division, Center for Energy and Environment Research, University of Puerto Rico, G.P.O. Box 3682, San Juan, Puerto Rico 00936 USA

INTRODUCTION

In this chapter, I will review the history, scope, methods, and results of field-based plant ecophysiological research with matorral species from central Chile. An attempt will be made to relate photosynthesis and respiration responses of the native vegetation to seasonal and diurnal changes in site light, temperature and water availability characteristics. The results reported are used to evaluate our current knowledge of gas exchange ecophysiology of Chilean matorral species and to suggest where future research may be profitably directed.

HISTORY AND SCOPE OF INVESTIGATIONS

The history of gas exchange studies conducted in central Chile is relatively short. The first work commenced in late 1967 (Dunn 1970). This first investigation was funded as a collaborative effort between University of California (Los Angeles) and the University of Chile (Santiago) by the Ford Foundation. Later, International Biological Program (IBP) projects, funded by the National Science Foundation (NSF), compared southern California research sites with analogous sites in Chile. These studies gathered comparative data on gas exchange, production, plant water relations, and micrometeorology. Funding by the National Science Foundation also supported cooperative gas exchange work in central Chile, conducted by investigators from McGill University (Montreal, Canada), San Diego State University (San Diego, California), and the Universidad Católica de Chile (Santiago) from 1976 through 1980. In 1971 and 1972 gas exchange measurements were also made on Chilean scrub and coastal sage species planted in an experimental garden at Stanford University in Palo Alto, California (Gigon 1979). The bulk of the work mentioned above served as part of projects that emphasized a comparison of Chilean and Californian ecosystems and a test of the concept of convergence of ecosystem processes under similar climatic regimes (Cody and Mooney 1978; Miller 1981; Mooney and Dunn 1970a; Mooney *et al.* 1970). Ecophysiological data similar to that presented in this chapter, but for Californian sites, are found in many of the publications cited herein. Comparable studies from a research site in Portugal are reported by Tenhunen *et al.* (this volume).

NATO ASI Series, Vol. G15
Plant Response to Stress. Edited by J. D. Tenhunen et al.

Research Sites

The first published gas exchange measurements from matorral vegetation of central Chile were made near the La Rinconada Lo Cerda agricultural research station approximately 30 km east of Santiago in Quebrada La Plata (33° 33' S. 70° 30' W.) at 600 m elevation in the Andean foothills (Dunn 1970). Later work funded through the IBP and NSF was concentrated at Fundo Santa Laura (33° 44' S. 71° 00' W.) 30 km north northwest of Santiago on the eastern slopes of the coastal mountain range at 1000 m. At both sites man's impact is apparent, both through wood harvesting for firewood and charcoal, and grazing (Aschmann and Bahre 1977). This impact has produced a much more open aspect than in comparable sites in California. Thrower and Bradbury (1977) described in detail the Chilean and Californian sites used in these ecosystem level convergence investigations.

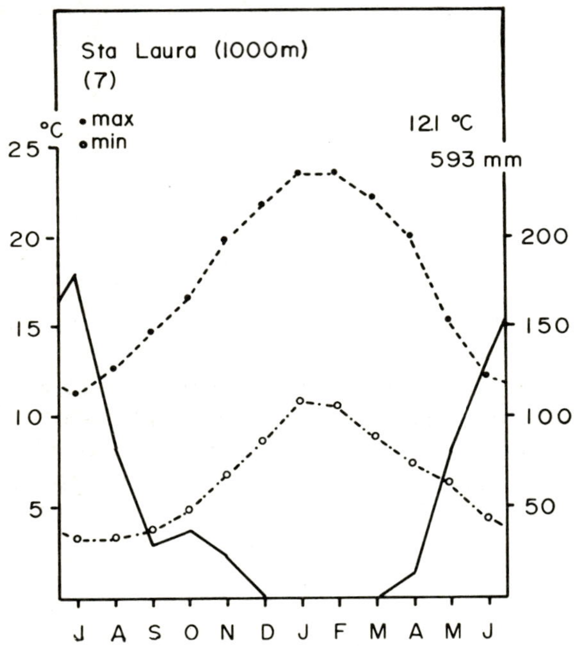

Fig. 1. Climate diagram for the Fundo Santa Laura, Chile site. Solid line is average monthly rainfall (mm), upper dotted line is monthly average maximum air temperature, and lower broken line is monthly average minimum air temperature (° C)

A climate diagram is presented for the Santa Laura site in Fig. 1. The mean annual temperature over a seven year measurement period was 12.1° C, the warmest summer month had on average max/mins of 23.6/10.9° C, the coolest winter month averaged max/mins of 11.3/3.3° C, and the seven year average precipitation was 593 mm (Miller and Hajek 1981). In central Chile the absolute maximum temperatures occur in summer (December through February) and reach 30 to 32° C, while the absolute winter minima are 2 to 4° C (July through

August). The absolute minimum temperatures are only 8 to 10° C in summer (Thrower and Bradbury 1977). The ocean (50 km to the west) moderates the climate of the Santa Laura site much more when compared with the Quebrada La Plata site of Dunn (1970) where the 1967 summer months had max/mins of 30.5/12.0° C and the winter, 14.0/0.0° C, with a 10 year average precipitation of 370 mm. The Santa Laura site which is closer to the coast is warmer in winter, cooler in summer, and has more precipitation on the average than the Andean foothill site. The more moderate coastal range environment has an important effect on plant phenology and growth. The average length of the growing season of four co-occurring evergreen shrubs was observed to be over 110 days at Santa Laura, and only 80 days in the Andean foothills (Montenegro *et al.* 1981).

The gas exchange work of Gigon (1979) was conducted in an experimental garden at Stanford, California approximately 35 km south of San Francisco at 20 m elevation. The average annual precipitation is 500 mm and the site is similar to the Santa Laura site in Chile with pronounced oceanic moderation of climatic extremes.

Species Studied

The plant species that have been examined in the three major investigative efforts from which gas exchange data are available, fall into several distinct growth form categories (Montenegro *et al.* 1981). These are listed in Table 1 along with sources of published gas exchange information relating to each of the species.

Table 1. Species list of Chilean matorral species used in various gas exchange studies. The species are keyed to the references of this paper as follows: (1) Dunn (1970), (2) Gigon (1979), and (3) Oechel and Lawrence (1981), Oechel *et al.* (1981), and Oechel and Lawrence (Unpublished Data). Species names follow the nomenclature of Muñoz-Pizarro (1966).

Species/Growth Form	Family	Investigators
Evergreen sclerophyllous shrubs:		
Colliguaya odorifera Mol.	Euphorbiaceae	(1,2,3)
Kageneckia oblonga R. and Pav.	Rosaceae	(1,2,3)
Lithraea caustica (Mol.) H. and A.	Anacardiaceae	(1,2,3)
Quillaja saponaria Mol.	Rosaceae	(1,3)
Summer deciduous half-shrubs:		
Satureja gilliesii (Grah.) Brig.	Labiatae	(3)
Lepechinia salviae (Lindl.) Epl.	Labiatae	(2)
Podanthus mitiqui Lindl.	Compositae	(2)
Summer deciduous thorny shrubs:		
Trevoa trinervis Miers	Rhamnaceae	(3)

Scope of Investigations

Dunn's work (1970) was carried out in two phases, a field study with *Colliguaya*, *Kageneckia*, *Lithraea*, and *Quillaja* at the Quebrada La Plata site in Chile, and a laboratory study, with *Kageneckia* and *Lithraea* only, conducted in California. The field work was undertaken from January through May 1968, thus, encompassing the transition to extreme drought in late summer and recovery of plants with cooler temperatures in early autumn (seasons are shifted by six calendar months in the southern hemisphere). To compensate for the unusually low precipitation during the previous winter and to experiment with the effects of water availability on gas exchange response, individual *Lithraea* and *Colliguaya* plants were irrigated weekly for three months. Tracking of photosynthesis under ambient conditions, observations of photosynthesis with saturating light, and measurements of dark respiration were carried out with the four species. Laboratory work, which was done with potted seedlings grown under 'spring' and 'summer' temperature conditions, concentrated on CO_2 and light response curves, and estimated photorespiratory fluxes, and stomatal and biochemical limitations on CO_2 fixation. The laboratory studies included several California chaparral species for comparison (Dunn 1970; Mooney and Dunn 1970b).

The ecophysiological investigations at Santa Laura (Lawrence and Oechel 1981; Oechel and Lawrence 1981; Oechel *et al.* 1981) were strictly field-based. The principal aim of the work was to construct light and temperature response surfaces of photosynthesis for the dominant species at the site. Detailed measurements of foliar gas exchange were made with ambient tracking and fixed temperatures cuvette techniques (Oechel and Lawrence 1979) using natural illumination. Gas exchange of distinct size classes of stems was also measured under ambient and controlled temperature regimes. Measurements were made at the time of maximum growth in 1976-1977 and 1979-1980 (December and January) and before the onset of drought. Most of the ecophysiological work was conducted with individuals of *Colliguaya*, *Lithraea*, *Satureja*, and *Trevoa*. Thus, evergreen, drought deciduous, and stem photosynthetic growth forms were included in the major effort. Due to the restricted number of species sampled during only two seasons with the controlled temperature approach, a wider survey of annual changes in maximum photosynthetic rates was carried out using a portable $^{14}CO_2$ labeling system (Martínez unpublished data). A modeling approach was later used to estimate annual carbon budgets for the four species mentioned above (Jacobson *et al.* 1981).

Only Gigon's study (1979) observed seasonal changes (from spring through autumn) of maximum photosynthesis rates, dark respiration, xylem pressure potential, and leaf stomatal and mesophyll resistance. The species used were at least two years old when measured, the evergreens being *Kageneckia*, *Lithraea*, and *Colliguaya*, and the drought deciduous *Lepechinia* and *Podanthus*. An irrigation treatment was also included. No attempt was made to collect data on the response to ambient changes in environmental conditions; monthly photosynthetic measurements were always made early in the morning under saturating artificial light at 25° C. This investigation also included comparisons with 'analogous' California chaparral species.

METHODS

Two gas exchange methodologies were used in these studies, infrared gas analysis for CO_2 and $^{14}CO_2$ labeling. In the former, the flux of carbon dioxide of the tissue under study is measured directly, while in the latter only total uptake of labeled carbon dioxide is determined in later laboratory analyses. Each method has its positive and negative features (see Sestak *et al.* 1971 for discussion). Infrared gas analyzer systems were used here to obtain temperature control during long-term ambient measurements and to allow manipulations of the light and temperature regime to which leaves were exposed. The $^{14}CO_2$ labeling systems were used to exploit their portability and potential for rapid sampling.

Fig. 2. Detailed view (left) of the thermoelectrically cooled cuvette described in Oechel and Lawrence (1979) and used by Oechel *et al.* in Chile. Placement of branch tip, stirring fan, and cooling fins are visible. General view (right) of a portion of the functioning six cuvette system. Tubing, power supplies, temperature controllers, cuvettes with support tripods, and miscellaneous cabling are seen in the foreground

Infrared Gas Analysis

Dunn (1970) and Gigon (1979) used temperature controlled cuvettes which traced their development back to Mooney *et al.* (1964, 1971) and Strain (1969). Oechel and Lawrence (1981), Oechel *et al.* (1981), and Lawrence and Oechel (1981) also employed temperature controlled cuvettes, but incorporated computer-aided control and data acquisition. Dunn (1970) and Gigon (1979) used single water cooled cuvettes, while the Oechel *et al.* (1981) system used six thermoelectrically cooled cuvettes simultaneously to achieve higher replication within a single species (Fig. 2). In both types of cuvettes the biological unit of gas exchange was a terminal stem, generally with multiple leaves, rather than a single leaf held perpendicular to the light source as is typical in laboratory experiments. This difference in experimental approach may complicate laboratory to field comparisons.

^{14}C Labeling Systems

Many samples were gathered at remote sites throughout the year using these techniques as described in Mustafa (1978) and Oechel and Mustafa (1979). After a 30-45 second labeling period in the cuvette, the labeled sample was collected or a fixed area was removed from the leaf with a hand operated punch. This sample was either refrigerated prior to drying, or killed in a chemical solution and then dried. Sample combustion released the labelled carbon from the material. The labelled carbon dioxide gas was trapped and radioactivity of the sample was determined in a liquid scintillation counter. Conversions from the activity of recovered ^{14}C to photosynthesis was made using the formulas reported in Tieszen *et al.* (1974).

DAILY COURSES OF GAS EXCHANGE

Dunn (1970) presents daily courses of gas exchange data for matorral shrubs. Data collected by Oechel and others (Oechel and Lawrence 1979) included an ambient tracking cuvette among the six in use at any time, but those data have not been published as independent data sets. Rather, the ambient data were included along with the other controlled temperature data in a summary of the response of CO_2 exchange to variation in light and temperature.

Dunn's paper (1970) reported mid-summer daily courses of photosynthesis for *Lithraea*, *Colliguaya*, *Quillaja*, and *Kageneckia* (Fig. 3). The effect of the extremely low precipitation prior to the study (172 mm versus a 60 year average of 370 mm) and its differential effects on the four species are seen in the figure. In addition to data presented here, gas exchange measurements were made throughout the summer (January through May) so that rough monthly and seasonal totals of net carbon fixed could be calculated. Leaf conductance values were not obtained in the field. Therefore, stomatal behavior can only be inferred.

Lithraea caustica

On February 29 (Fig. 3a), the xylem pressure potential for this species ranged from an early morning high of -1.1 MPa to a midday low of -3.4 MPa. A 30% reduction in photosynthesis below the maximum observed for that day occurred at the time of lowest xylem pressure potentials. Leaf temperature was also maximal during this period (see discussion by Tenhunen *et al.* in this volume concerning the possible role of high temperature and low humidity in stomatal control).

In January, early morning leaf water potentials was also near -1.0 MPa, but midday low did not reach -3.0 MPa. Air temperatures were also lower, never increasing above 25° C as it was a cloudy day. These more moderate conditions resulted in maximum photosynthetic rates of over 12 mg CO_2 dm^{-2} h^{-1}, while the February maximum (Fig. 3a) was only slightly over 7.2 mg CO_2 dm^{-2} h^{-1}. Thus, a 40% reduction occurred either due to the more severe drought conditions or increased stress due to high temperatures and high vapor pressure gradients between the leaf and air.

Colliguaya odorifera

This plant responded similarly in January to *Quillaja* on March 12. The early morning xylem potential was approximately -4.0 MPa which only became slightly more negative in the morning, and recovered to -3.0 MPa by 1200 hrs. It then declined to about -4.0 MPa again by late afternoon. As in *Quillaja,* maximum carbon gain was in the early morning when it reached 4.0 mg CO_2 dm^{-2} h^{-1}. Short periods of positive photosynthesis occurred at noon and in the late afternoon (Fig. 3b), although the maximum at these times was only about 1.0 mg CO_2 dm^{-2} h^{-1}. The leaves respired at nearly 2.0 mg CO_2 dm^{-2} h^{-1} at one point in the late morning. The results are similar to those reported by Tenhunen *et al.* with *Quercus suber* and *Quercus coccifera* (this volume). As in those species, the midday minimum in net photosynthesis occurred despite recovery in xylem water potential.

Quillaja saponaria

During the course of the day on March 12, xylem pressure potential was less than -3.2 MPa at 0600 hrs, dipped to a low of -4.5 MPa at 0700 hrs, and rose slightly during the day. The least negative leaf water potential in *Quillaja* was about the same as the most negative measured with *Lithraea* in February. However, the most negative values occurred in early morning with *Quillaja* rather than at midday. Positive carbon gain was limited to the very early morning, with

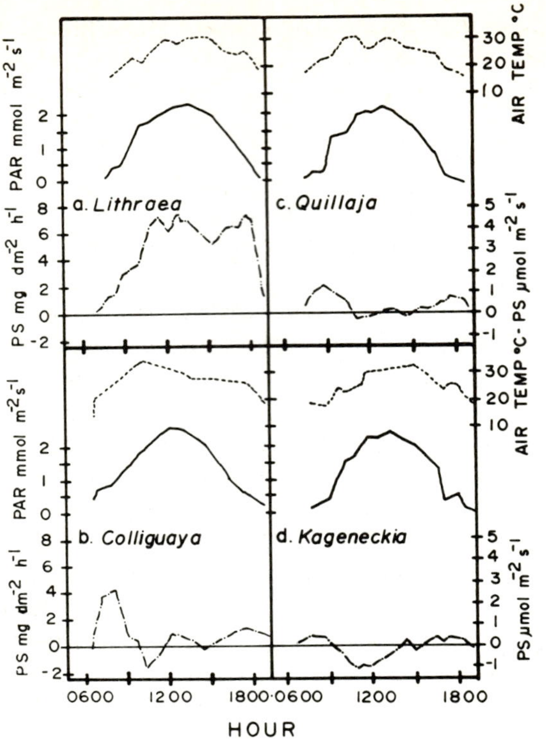

Fig. 3. Daily courses of air temperature (dotted line), PAR (solid line), and gas exchange (broken line) for *Lithraea caustica* (a), *Colliguaya odorifera* (b), *Quillaja saponaria* (c), and *Kageneckia oblonga* (d; from Dunn 1970). Measurements are for 29 February, 23 January, 12 March, and 29 February, 1968 respectively. Note dual scales for gas exchange in molar and mass units of carbon dioxide

maxima at 2.0 mg CO_2 dm^{-2} h^{-1}, and late afternoon with photosynthetic rates just reaching 1.5 mg CO_2 dm^{-2} h^{-1} (Fig. 3c).

Kageneckia oblonga

The measurements conducted with this species on February 29 indicated a strong midday depression of photosynthesis. Photosynthesis rate early and late in the day (Fig. 3d) did not exceed 1.0 mg CO_2 dm^{-2} h^{-1}. At midday leaves were respiring at more than 1.5 mg CO_2 dm^{-2} h^{-1}. Water potentials were approximately -3.5 MPa in the early morning, decreased to a midday low of more than -5.5 MPa, and recovered in late afternoon to about -2.7 MPa. The response was compatible with the interpretation that decreases in water potential itself might have decreased photosynthesis, but atmospheric stress (vapor pressure deficit, temperature) was similarly greatest during this period of the day. The effects of drought, at least in terms of reductions in carbon fixation, were most severe in this species.

TEMPERATURE AND LIGHT RESPONSE CURVES

The field investigations of Dunn (1970), Oechel and Lawrence (1981) and Oechel *et al.* (1981) devoted considerable attention to examing the effects of temperature on CO_2 gas exchange (humidity was not controlled in the cuvettes, although we can assume in general that the vapor pressure gradient between leaf and air was greatest when temperatures were highest). Lawrence and Oechel (1981) further reported the response of CO_2 gas exchange of both leaves and stems to temperature as observed in a 1979-1980 study. Light response curves were determined only by Oechel *et al.* (1981). Interpretation of the temperature responses observed by Dunn are complicated by the severe drought conditions under which he worked. Nevertheless, his measurements did include an aspect of seasonal change that is lacking in the November/December determinations of Oechel *et al.* The following discussion considers only the smoothed average response curves determined by statistical treatment of the data by the individual

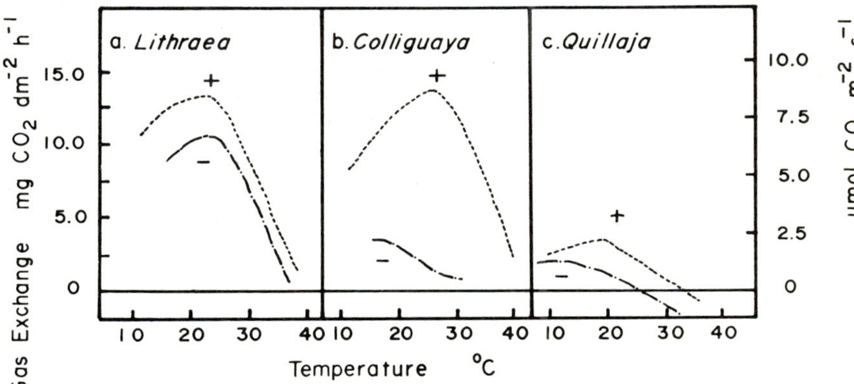

Fig. 4. Temperature response of photosynthesis for *Lithraea caustica* (a), *Colliguaya odorifera* (b), and *Quillaja saponaria* (c; from Dunn 1970). Upper curve is for irrigated plants (+), and lower curve for non-irrigated (-) in *Lithraea* and *Colliguaya* . In *Quillaja*, the upper curve is for mid-summer (January 30), and the lower is for early autumn (May 6). Note dual scales for gas exchange in molar and mass units of carbon dioxide. Actual data points are not drawn

authors. Dark respiration is referred to briefly in this section, but it is treated again in more detail in the section on MAXIMUM PHOTOSYNTHETIC RATES.

Lithraea caustica

Dunn (1970) and Oechel *et al.* (1981) found this species to be particularly insensitive to the effects of drought. Growth of *Lithraea* has been shown to be relatively independent of site water status (Montenegro *et al.* 1979a). The measured response curves indicated positive CO_2 uptake over a wide range of temperatures. In addition, Dunn (1970 - Fig. 4A) compared watered with unwatered shrubs and found relatively little difference in gas exchange rates between these treatments at saturating light levels when compared to other shrub species. The temperature optimum occurred at approximately 22.5° C in both treatments. The maximum photosynthesis rate of the watered plant was 13.1 mg CO_2 dm^{-2} h^{-1}, 25% higher than that of the unwatered plant. Increases in air temperatures above the optimum led to a rapid decrease in photosynthesis. In *Lithraea*, a 10° increase in temperature from the 22.5° optimum to 32.5° C resulted in a 56% decrease in photosynthesis. If extrapolated to 40° C, this would suggest an 86% decrease in rate at that temperature. At 10° C the rate was only 20% less than at the optimum.

The approach taken by Oechel and Lawrence (1979) provided detailed three dimensional surfaces describing the light and temperature dependencies of leaf gas exchange (Fig. 5). These response surfaces summarize results of continuous measurements obtained with six individuals over the course of a week. Light intensity was allowed to vary naturally and temperature was either held constant (using randomized 5° C daily temperature steps over the course of a week with a range in temperature between 5 and 40° C for each cuvette) or controlled to track ambient temperature. In contrast to results reported by Dunn (1970), the temperature response curves for *Lithraea* were similar at almost all light intensities and positive carbon fixation occurred at almost all light and temperature combinations. The differences in response reported by these authors must be due to seasonal differences and to less pronounced drought stress effects on photosynthesis at the Santa Laura site.

At saturating photosynthetically active radiation (PAR) levels above 1000 µmol m^{-2} s^{-1}, optimal photosynthetic rates of 8.3 mg CO_2 dm^{-2} h^{-1} occurred at temperatures between 20 and 25° C (Fig. 5). At 10° C photosynthesis was reduced about 34%, while increasing temperature to 30° resulted in only a 22% reduction. Nevertheless, at 40° C an 88% decrease in photosynthesis rate occurred. Carbon fixation is positive over all of the response surface except under conditions of of high temperatures (> 30° C) and low PAR (< 100 µmols m^{-2} s^{-1}). Later work undertaken by Lawrence and Oechel (1981) at Santa Laura examined temperature curves over 5 to 25° C (Table 2). A photosynthetic optimum of 13.7 mg CO_2 dm^{-2} h^{-1} occurred at 25° C, with a 12% decrease at 15°

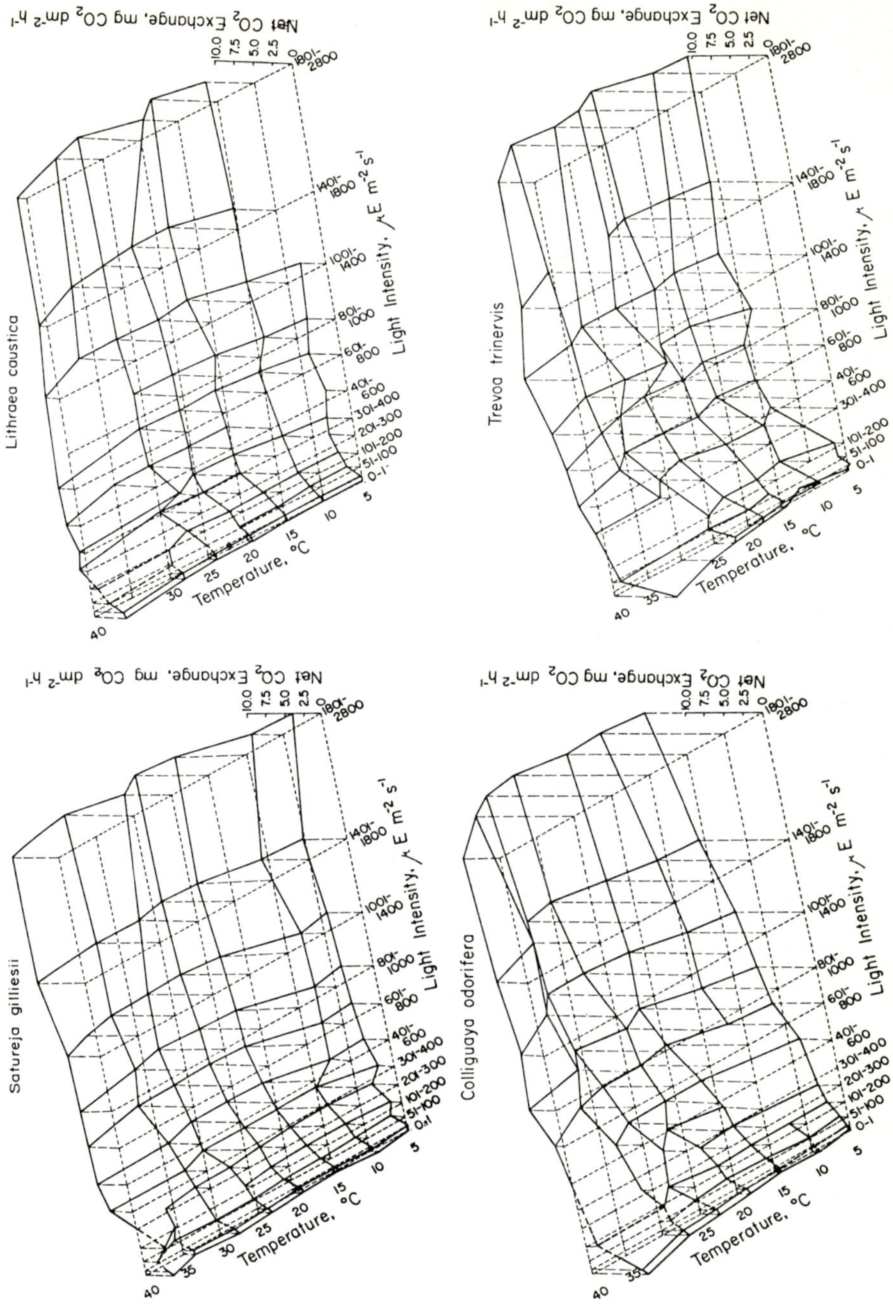

Fig. 5. Gas exchange response surfaces for *Lithraea caustica*, *Trevoa trinervis*, *Satureja gillesii*, and *Colliguaya odorifera* in November and December 1976. Adapted from Oechel *et al.* (1981)

Table 2. Temperature response of leaf and stem gas exchange from 1979-1980 summer season at Santa Laura (Lawrence and Oechel 1981). Dark respiration rates are under the column headed *Rd*, and the *Ps* column is that of gas exchange under saturating PAR. The *current* and *previous* rows are those for gas exchange of the last two years stem growth increments. Data are shown ± one standard error, units are mg CO_2 dm^{-2} h^{-1}.

Species	Tissue	5° Rd	5° Ps	15° Rd	15° Ps	25° Rd	25° Ps
Trevoa	leaves	-0.34 ± 0.02	14.53 ± 1.31	-0.30	18.45 ± 1.28	-0.88 ± 0.53	18.03 ± 1.84
	current	-0.20 ± 0.10	4.36 ± 1.11	-1.97 ± 0.22	7.07 ± 0.49	-4.44 ± 0.74	5.76 ± 1.64
	previous	-2.08 ± 0.28	1.52 ± 0.08	-4.43 ± 0.43	2.48 ± 0.29	-11.26 ± 1.12	-1.77 ± 0.36
Satureja	leaves	-0.60 ± 0.02	6.11 ± 0.41	-1.80 ± 0.10	10.97 ± 0.37	-4.28 ± 0.19	17.37 ± 0.17
	current	-0.75 ± 0.13	0.25 ± 0.05	-1.43 ± 0.11	0.34 ± 0.11	-2.53 ± 0.13	0.87 ± 0.09
	previous	-0.17 ± 0.03	-0.22 ± 0.04	-0.76 ± 0.06	-0.31 ± 0.05	-1.98 ± 0.10	-0.37 ± 0.11
Collig.	leaves	-0.71 ± 0.01	14.72 ± 2.69	-1.77 ± 0.12	14.76 ± 1.69	-5.50 ± 0.40	23.75 ± 2.50
	current	-0.43 ± 0.16	0.18 + 0.11	-1.40 ± 0.24	-0.59 ± 0.09	-4.35 ± 0.22	-2.62 ± 0.39
	previous	-0.52 ± 0.15	-0.51 + 0.35	-1.81 ± 0.16	-0.82 ± 0.53	-3.79 ± 0.18	-1.91 ± 0.39
Lithraea	leaves		3.32 ± 0.48	-0.29 ± 0.06	11.99 ± 1.38	-1.17 ± 0.08	13.74 ± 1.18
	current	-1.74 ± 0.22	-1.17 ± 0.46	-1.48 ± 0.32	-0.35 ± 0.35	-2.36 ± 0.41	1.17 ± 0.08
	previous	-0.86 ± 0.09	-0.51 ± 0.05	-1.22 ± 0.11	-1.20 ± 0.32	-2.90 ± 0.39	-2.42 ± 0.96

and a 75% decrease from optimum at 5° C. Leaf dark respiration rates increased four-fold from 0.29 to 1.17 mg CO_2 dm^{-2} h^{-1} between 15 to 25° C.

Trevoa trinervis

Trevoa trinervis is a small leaved, drought deciduous and stem photosynthetic species. Some suggestion has been made that nitrogen fixing microorganisms may be associated with its roots. Oechel *et al.* (1981) obtained a midsummer response surface for photosynthesis as affected by light and temperature (Fig. 5). The surface is unremarkable, with optimal photosynthetic rates of 12 mg CO_2 dm^{-2} h^{-1} near 25° C. The response is similar to that of *Colliguaya*, an evergreen species. The temperature response curve is rather flat, with < 10% declines in photosynthesis at 5 and 30° C from the optimum. At 40° C the CO_2 uptake rate is about half that at optimum temperature. Net photosynthesis is positive except at low light and high temperatures, much like the other species examined. Although the leaves have high levels of crude protein (16%) and the highest nitrogen of all the plants studied (2.7% - Shaver 1981), this did not result in higher photosynthetic rates than measured with evergreen species. The data were variable, perhaps due to onset of leaf senescence or to drought stress. Lawrence and Oechel (1981) reported a temperature response curve for this species similar to that described above (see Table 2). Dark respiration rate was low at 5 to 15° C and increased to 0.88 mg CO_2 dm^{-2} h^{-1} at 25° C.

Satureja gilliesii

This plant species has leaves which dry and shrivel under drought conditions, but that can rehydrate, regain turgidity, and probably become photosynthetically active again once there is sufficient precipitation (Montenegro *et al.* 1979b). The temperature response curve has a photosynthetic optimum of 9.0 mg CO_2 dm^{-2} h^{-1} at 25° C (Fig. 5). The decrease in rate at high temperature was the strongest found by Oechel *et al.* (1981) and similar to responses reported by Dunn (1970; see Fig. 4) for other species. At 5° C, photosynthesis decreased 58% from the optimum at 25° (at 30° C - 31%; and at 40° C - 53%). The leaves respired under conditions of low light and high temperature. At a later date, Lawrence and Oechel (1981) found almost a doubling of the photosynthetic rate at optimum temperature to 17.4 mg CO_2 dm^{-2} h^{-1} (Table 2).

Colliguaya odorifera

Colliguaya odorifera is similar to *Lithraea* in life form but more prone to drought stress. Temperature response curves were obtained by Dunn (1970) for both watered and unwatered plants (Fig. 4B). The shape of these response curves is very much like that for *Quillaja*. We may assume that the response obtained with watered plants is similar to that found in early summer, while the response obtained under the natural water regime is typical for the late summer season.

The temperature optimum decreased to lower temperature as the plants were more drought stressed.The net photosynthesis rates at the temperature optimum were 13.6 mg CO_2 dm^{-2} h^{-1} at 25° C in watered plants and 3.6 mg CO_2, dm^{-2} h^{-1} at 15° C in unwatered plants.

Similar temperature response curves of photosynthesis were found with *Colliguaya* in early summer by Oechel *et al* . (1981; Fig. 5). The CO_2 uptake rate at optimum temperature and under saturating PAR was 13.8 mg CO_2 dm^{-2} h^{-1}. This was equal to that discussed above but occurred at a slightly cooler temperature (20° C). A 15° decrease in temperature from the optimum to 5° C resulted in only an 18% decrease in photosynthetic rate, however, while increases in temperature to 30 and 40° C caused depressions of 19 and 70% respectively. Net CO2 loss was measured at PAR levels below 400 µmols m^{-2} s^{-1} and temperatures of 35° C and above. Oechel and Lawrence (1981) found a photosynthetic optimum at 25° C with maximum net photosynthesis rates of 23.8 mg CO_2 dm^{-2} h^{-1} at the same site (Table 2). Dark respiration had a Q_{10} of 2.5 to 3, with a rate of 5.5 mg CO_2 dm^{-2} h^{-1} at 25° C.

Quillaja

Only Dunn (1970) measured temperature response curves in this species. Plants were not irrigated, and gas exchange rates were very low compared to *Lithraea* even during early summer. Temperature response curves were measured between 5 or 10 to 33° C in January and again in May (Fig. 4C). The temperature optimum was near 20° C in early summer, and then shifted to 10° C or below late in the summer season. Photosynthetic rates were approximately 3.5 and 2.4 mg CO_2 dm^{-2} h^{-1} respectively at these optima. Comparing the two optima, the early season rate is 46% higher than four months later in autumn. The 20° C rate in May was 4-fold lower than in January. The plant leaves were respiring at temperatures above 32° C in January, and above 23° C in May. A seasonal analysis of the temperature response of dark respiration indicated higher rates in January than those in March, with a recovery to intermediate rates in May. A dark respiration rate of 2.5 mg CO_2 dm^{-2} h^{-1} was measured at 35° C in May.

MAXIMUM PHOTOSYNTHETIC RATES

Constructing a table of maximum photosynthetic rates (P_m) and conditions under which they are achieved is difficult for Chilean matorral species because recording of maximum rates has not always been a primary goal of the studies mentioned above. Usually, many of the key ecophysiological parameters like optimum temperature, leaf conductance, or water potential have not been recorded or indicated in publications, or maxima may only be derived from regressions of photosynthetic rate against light or temperature, e.g. data of

Oechel and Lawrence (1981). Nevertheless, an attempt is made here (Table 3). The work of Dunn (1970) and Gigon (1979) provides some detail on seasonal changes in the photosynthetic maxima which is required to gauge potentials for carbon acquisition in these shrubs. Actual observations are limited to the spring through early autumn period. The photosynthetic potential of these plants during the cool, wet seasons has not been explored. Since water relations information is not firmly tied to the gas exchange observations in Table 3, seasonal courses of xylem pressure potentials collected from equator- and pole-facing slopes at Santa Laura are presented in Fig 6 which document typical water regimes for at least four of the major species discussed.

The absolute maxima for the individual species range from a low of 8.4 mg CO_2 dm^{-2} h^{-1} in *Satureja* to a high of 19.9 mg CO_2 dm^{-2} h^{-1} in *Lepechinia*. However, caution is in order in interpreting these rates, since any species not under irrigation may have been under drought stress during the period of measurement. There is especially good agreement between authors and sites on the maximum rates in July and August for *Lithraea* (11.8-12.2 mg CO_2 dm^{-2} h^{-1}) and *Colliguaya* in the early and late summer season (14.2-16.9 and 7.8-8.4 mg CO_2 dm^{-2} h^{-1}) when neither species was under irrigation.

It is interesting that Gigon (1979) found that evergreen shrubs usually had more negative xylem pressure potentials when they were not irrigated but that water stress did not necessarily result in a decrease in P_m values. This was especially

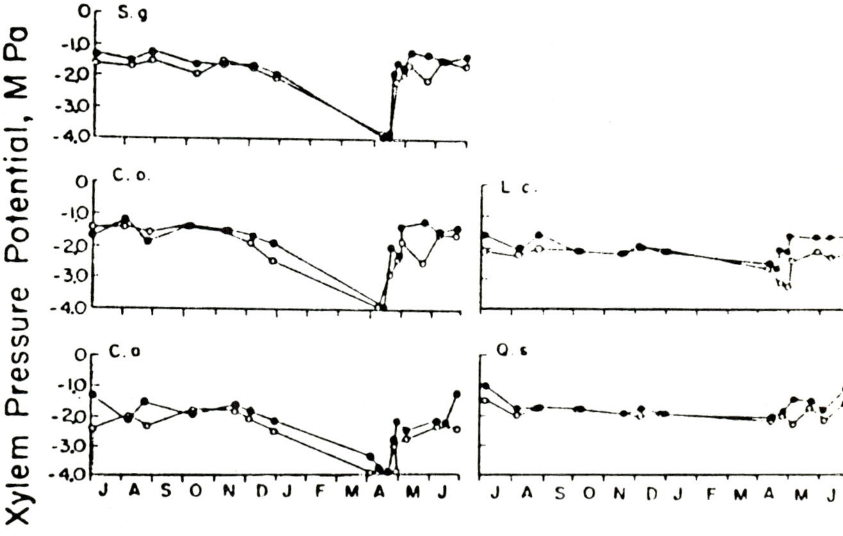

Fig. 6. Midday xylem pressure potentials from pole (-o-) and equator-facing (-•-) slopes at Fundo Santa Laura. Species are *Satureja* (S.g.), *Colliguaya* (C.o.), *Lithraea* (L.c.), *Cryptocarya* (C.a.), and *Quillaja* (Q.s.). From Giliberto and Estay (1978)

Table 3. Maximum observed photosynthetic rate (P_m), dark respiration rate (R_d), temperature during measurement (T), and xylem pressure potential (Ψ) for Chilean matorral shrubs. Water potential measurements made concurrently with photosynthetic measurements were reported only by Gigon (1979). Source refers to literature citation[1], Month is number of month[2], and Stat is whether the plants were grown under Natural or Irrigated conditions. Gas exchange units are mg CO_2 dm^{-2} h^{-1} and water potential is in MPa.

Species	Source	Month	Stat	P_m	R_d	T	$-\Psi$
		Evergreen Shrubs					
Lithraea caustica	2	5	N	7.9	4.1	25.0	
	2	5	I	7.9	5.2	25.0	
	3	6	N		2.1	30.0	
	3	6	N	8.4		20.0	
	1	7	N	12.2	3.0	24.0	
	1	7	I	13.2	2.2	19.5	
	2	8	N	11.8	2.0	25.0	1.0
	2	8	I	17.7	4.3	25.0	
	1	9	N	8.5	2.3	24.5	
	1	11	N	8.0	2.3	20.5	
	1	11	I	10.8	2.0	19.0	
Quillaja saponaria	1	7	N	4.5	1.9	19.5	
	1	9	N	3.4	1.1	8.0	
	1	11	N	3.5	1.6	21.5	
Colliguaya odorifera	2	5	N	16.9	4.1	25.0	1.6
	2	5	I	16.9	5.2	25.0	1.1
	3	6	N	14.2		18.0	
	3	6	N		4.0	30.0	
	2	7	N	15.9	1.9	25.0	1.2
	2	7	I	15.9	3.9	25.0	1.2
	1	7	N	4.6	3.7	24.0	
	1	7	I	13.1	3.7	25.5	
	2	8	N	12.0	1.8	25.0	1.3
	2	8	I	15.5	3.0	25.0	0.9
	2	9	N	13.0	2.3	25.0	1.6
	2	9	I	13.0	3.0	25.0	1.2
	1	9	N	2.7	2.2	18.0	
	2	10	N	10.7	2.2	25.0	1.1
	2	10	I	10.7	3.1	25.0	1.1
	2	11	N	7.8	3.0	25.0	1.2
	2	11	I	7.8	3.0	25.0	0.8
	1	11	N	8.4	2.6	22.0	
	1	11	I	12.8	3.6	17.0	
Kageneckia oblonga	2	5	N	13.1	4.2	25.0	1.2
	2	5	I	9.9	4.2	25.0	1.2
	2	6	N	13.4	2.2	25.0	1.1
	2	6	I	13.4	3.9	25.0	1.1
	2	7	N	12.6	2.8	25.0	1.2
	2	7	I	12.6	2.8	25.0	0.6
	1	7	N	5.1	1.7	21.0	

Species	Source	Month	Stat	P_m	R_d	T	$-\Psi$
			Evergreen Shrubs (continued)				
Kageneckia oblonga	2	8	N	10.5	1.9	25.0	2.4
	2	8	I	12.6	1.9	25.0	1.1
	2	9	N	9.9	1.6	25.0	1.9
	2	9	I	7.7	2.8	25.0	1.0
	1	9	N	0.8	1.0	17.0	
	2	10	N	8.0	2.4	25.0	1.5
	2	10	I	6.4	2.4	25.0	0.8
	2	11	N	5.1	2.4	25.0	1.5
	2	11	I	6.4	2.4	25.0	0.8
	1	11	N	2.7	1.6	7.5	
			Drought Deciduous or -Evading Shrubs				
Podanthus mitique	2	5	N	14.3	3.4	25.0	1.3
	2	5	I	17.9	4.3	25.0	0.9
	2	6	N	14.5	1.4	25.0	1.9
	2	6	I	18.2	3.8	25.0	1.2
	2	7	N	11.6	1.7	25.0	4.0
	2	7	I	15.8	4.4	25.0	1.6
	2	8	N	9.0	1.7	25.0	4.5
	2	8	I	16.3	2.8	25.0	1.6
	2	9	N	-2.8	1.7	25.0	5.0
	2	9	I	14.3	2.5	25.0	1.6
	2	10	N	--	--	25.0	4.8
	2	10	I	14.6	4.5	25.0	1.2
	2	11	N	--	--	25.0	4.6
	2	11	I	9.8	3.8	25.0	1.3
Lepechinia salviae	2	8	I	19.9	3.4	25.0	0.8
	2	9	I	15.2	2.3	25.0	0.6
	2	10	I	12.9	2.3	25.0	0.7
	2	11	I	9.6	2.3	25.0	0.7
Trevoa trinervis	3	6	N	12.0		18.0	
	3	6	N		3.4	30.0	
Satureja gilliesii	3	6	N	8.4		23.0	
	3	6	N		2.6	30.0	

[1] Dunn (1970) is cited as 1, Gigon (1979) as 2, and Oechel and Lawrence as 3.

[2] Months are numbered in order from January (1) through December (12), but data acquired in Chile has its month offset by six months to correspond with the climatically equivalent Northern Hemisphere month. This artifice has been adopted to ease the identification of seasons for northern scientists.

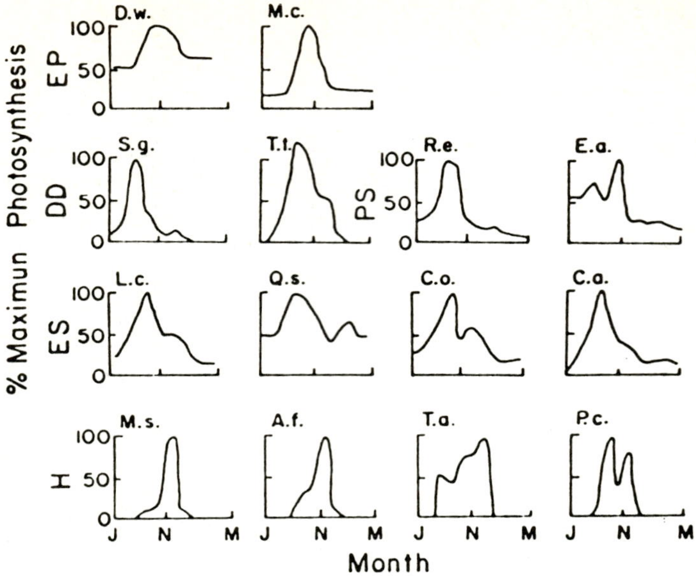

Fig. 7. Seasonal pattern of % Maximum Photosynthesis for five growth forms at Fundo Santa Laura. The growth forms are evergreen phreatophytes (EP), deciduous and drought deciduous shrubs (DD), photosynthetic stem shrubs (PS), evergreen sclerophyllous shrubs (EV), and herbs (H). Species key is as follows: D.w., *Drimys winteri*; M.c., *Myrceugenella chequen*; S.g., *Satureja gillesii*; T.t., *Trevoa trinervis*; R.e., *Retanilla ephedra*; E.a., *Ephedra andina*; L.c., *Lithraea caustica*; Q.s., *Quillaja saponaria*; C.o., *Colliguaya odorifera*; C.a., *Cryptocaria alba*; M.s., *Madia sativa*; A.f., *Avena fatua*; T.a., *Trichoclineu aurea*; and P.c., *Pasithea coerulea* . From Oechel *et al.* (1981)

apparent in *Kageneckia* where higher maxima were found in unwatered individuals, even though the water potential was twice as negative as that of watered plants.

Since sampling of gas exchange rates is slow with infrared gas analyzers and temperature controlled cuvette systems, a $^{14}CO_2$ measurement system was used to obtain seasonal patterns in maximum photosynthesis rate (Fig. 7 - Oechel, *et al.* 1981) for five growth forms from the Santa Laura site. This system also allowed a high degree of portability and access to many remote sites where IRGA systems could not have been operated. The maxima in photosynthesis observed early in the summer and the later steep declines in photosynthesis are difficult to explain, based on the response surface data in Fig. 5. In both *Podanthus* and *Lepechinia*, the drought deciduous species studied by Gigon (1979), there was a rapid seasonal decline in photosynthetic maxima, without a corresponding change in water potential in the irrigated individuals.

GAS EXCHANGE OF STEMS

Very few ecophysiological studies have considered the carbon costs or benefits of the tremendous mass of stem material found in the canopy in mediterranean region shrubs. In the matorral of Chile, the stems can comprise from 75% to over 85% of the total plant biomass (Kummerow, Montenegro, and Krause 1981). Stem area indices may be as high as 32% of the leaf area index (Lawrence 1975). The stems are clearly an important factor in the plant carbon balance, especially since the stems have actively metabolizing tissues through most of the year (Avila *et al.* 1975) and are such a significant portion of the plant biomass. To gauge the role of stems in plant carbon balance, Oechel and Lawrence (1981) and Lawrence and Oechel (1981) measured stem gas exchange with infrared gas analysis techniques under both controlled and ambient temperature conditions at the Santa Laura site. They used the same cuvette technique as for their work with foliage (Fig. 2), and measurements were made with stems of various age and size classes for several of the woody shrub species. They found that for all the shrubs studied (Table 2) the current year's stems showed positive carbon fixation under appropriate light and temperature conditions. Even though net photosynthesis was not apparent for all stems of all species, the gas exchange rates were always either positive or less negative in the light than in the dark. Among the species studied, *Trevoa* was remarkable in that it has twigs which exhibit 'leaf-like' gas exchange behavior, complete with light compensation and saturation PAR levels that are similar to those of of evergreen shrub leaves (Fig. 8A). Even the net photosythesis rates of *Trevoa* twigs approach that of some evergreen leaves.

These results suggest another mechanism utilized by plants which survive in marginal or severely stressed systems. Even considering the contribution of stem photosynthesis, the mass of stems in the matorral can constitute a strong respiratory carbon sink. Dark respiration was shown to be a decreasing function of stem diameter (Oechel and Lawrence 1981) that was similar for the stems of at least three species (Fig. 8B). A strong temperature dependence was found for current and previous year's twigs by Lawrence and Oechel (1981; Table 2), with Q_{10} values exceeding 2.0 in all cases. The results of this work still leave unsolved, the problem of the best way to present stem gas exchange data; whether outer stem area, dry weight, or even some other units are most appropriate. What is needed is a function that would embrace the living tissue volume and activity of a stem as well as relate to the photosynthetic potential of the tissue. Changes in photosynthetic competency of stems according to age and diameter class and immediate environmental conditions complicate attempts at modeling ecosystem carbon flow.

CARBON BALANCE

To accurately evaluate the carbon balance for a plant stand, information on environmental conditions, canopy architecture, and standing biomass are needed.

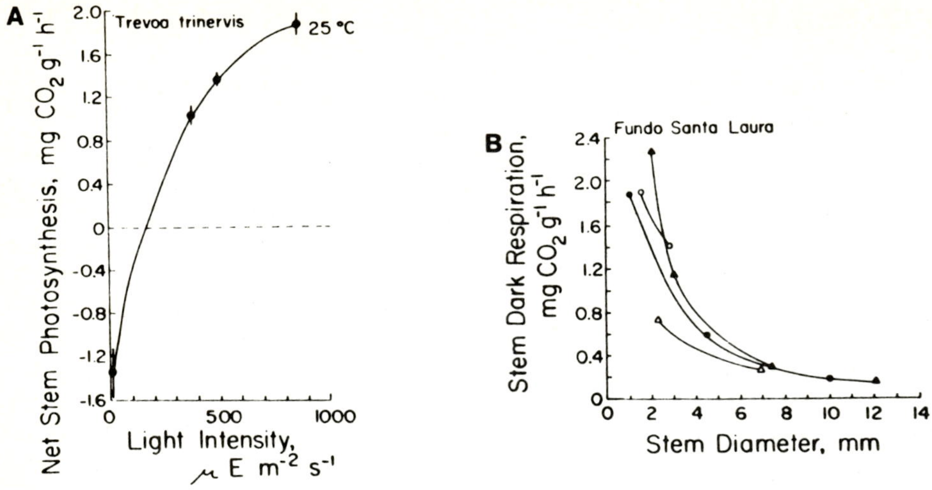

Fig. 8. A. Stem net photosynthesis for *Trevoa trinervis* from darkness to 1000 μmol quanta m^{-2} s^{-1}. Bar denotes ± one standard error. B. Relationship of stem diameter to dark respiration rate for *Lithraea* (solid triangles), *Trevoa* (solid circles), and *Satureja* (open triangles) at 25° C. Measurements for both figures made in November and December 1979. From Oechel and Lawrence (1981)

Models are required to simulate environmental conditions for each stem, leaf, and root throughout each day of all seasons. In addition, gas exchange responses for all plant parts must be adjusted for seasonal change. Then in simulation, carbon fluxes may be summed, carbon may be allocated for growth or maintenance, and a carbon balance is obtained. This process is detailed for both Chilean matorral and Californian chaparral shrubs in Miller (1981) and discussed further by Jacobson *et al.* (1981).

Carbon Acquisition Models

Dunn (1970) calculated the carbon fixation by shrubs in his study based on leaf photosynthetic and dark respiration rates measured at Quebrada La Plata. His daily, monthly and seasonal carbon gain values showed that there are great differences in carbon availability for maintenance and growth during the five month period studied. Dunn's estimates are summarized here:

	Grams CO_2 Acquired			
	January	March	May	Season
Lithraea				
per day	0.103	0.051	0.043	
per month	3.09	1.53	1.29	11.82

Kageneckia				
per day	0.021	-0.006	0.007	
per month	0.64	-0.19	0.20	1.31
Quillaja				
per day	-0.005	0.006	0.008	
per month	-0.15	0.18	0.24	0.55
Colliguaya				
per day	-0.002	-0.003	-0.007	
per month	-0.05	-0.09	-0.21	-0.68

These data were of course strongly influenced by the severe drought during the measurement period, but indicate that several of the plants at this site would be capable of a positive annual carbon balance only through strong carbon acquisition in other times of the year.

Simulation-Based Balances

In several chapters of Miller (1981) the simulation models CAPS and MEDECS are used extensively and their output compared with field observation (Jacobson *et al.* 1981). This approach produced the carbon balances for four principal matorral species shown in Fig. 9. The first noteworthy fact illustrated in this figure is the relative amount of carbon available to each of the species annually. *Colliguaya* is estimated to fix the most in a year, resulting in a gram dry weight growth equivalent of 1843 gm C m^{-2} $year^{-1}$, followed by another evergreen, *Lithraea*, at 1087 gm C m^{-2} $year^{-1}$, and two drought sensitive plants, *Satureja* and *Trevoa* at 633 and 556 gm C m^{-2} $year^{-1}$ respectively. Of this potentially allocated carbon, new biomass accounts for 36, 45, 40, and 41% respectively of the total. It is clear that respiratory functions are costly, but essential. In terms of the allocation of carbon from the growth and respiration pools to above- and belowground systems, there is close to a 50:50 split in the two evergreen species *Colliguaya* and *Lithraea*, with only slightly less carbon being allocated below- compared to aboveground. But in the drought evading species *Satureja* and *Trevoa*, 27% or less goes belowground in the former, and about 36% in the latter. The primary difference between these growth forms seems to be in the proportion of main roots. The evergreens have considerably more main and absorbing roots than the drought evaders, whose root systems may be largely sloughed at season's end.

CONCLUSIONS

We have developed an understanding of matorral shrubs based largely on theory and on only a few intensive field studies. The investigations undertaken have all

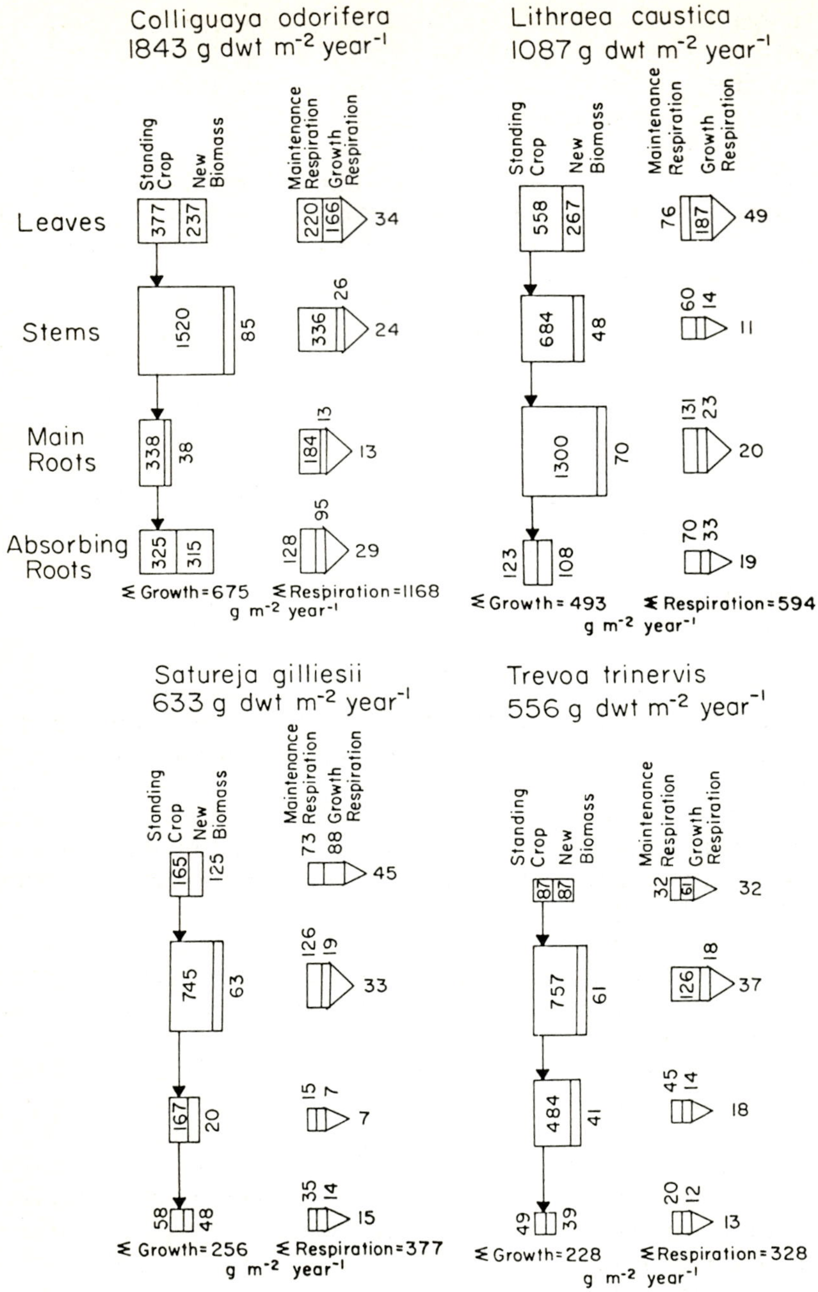

Fig. 9. Yearly carbon balance for dominant matorral shrubs. In this figure can be found the standing crop of leaves, stems, and roots, and the pattern of allocation of the total yearly pool of photosynthetically acquired carbon to the growth of new biomass and respiration (from Oechel and Lawrence 1981)

been excellent, but have been limited in scope due to lack of time, funding, and manpower. New studies are required using the techniques of the 1980's discussed in other chapters of this volume. The major attributes, findings, and faults of existing data sets are those I have listed below:

1) The most complete data sets describing gas exchange characteristics of branches of matorral shrubs are represented by the response surfaces presented by Oechel and Lawrence (1981) These, however, lack information on seasonal changes in gas exchange characteristics.

2) Drought stress changes the shape of gas exchange response surfaces for plants of all growth forms (Dunn 1970; Gigon 1979) as well as their production capabilities (Jaksic and Montenegro 1979). We need to conduct detailed laboratory studies of the physiology of these responses.

3) With nearly 50% of all carbon going to belowground growth and respiration, more detailed work is needed to determine growth, gas exchange characteristics, and turnover of roots.

4) All work needs a greater simultaneity. Ecophysiological measurements of different types should be made on the same day.

5) Much higher replication of measurements must be made so that we may begin to gauge the natural within and between species variation in matorral ecosystems.

6) Work in experimental gardens might help accomplish some of these goals, especially by offering uniformity of soil, climate, protection from gross disturbance, and security for instrumentation.

7) The data in published studies on matorral shrubs is fragmentary at best. More data exists than has been published. In the future, the power of computers may allow greater access to large data bases put together by various research groups. These well organized and indexed files could be stored on magnetic media and shared with colleagues.

8) Experimental protocols should be established to guarantee data comparability. Detailed protocols are needed for standardization of infrared gas analysis work, temperature and light measurements, water potential measurements, conductance measurements, phenological observations, and chemical and nutrient extractions and analyses.

REFERENCES

Aschmann H, Bahre C (1977) Man's impact on the wild landscape. In: Mooney, HA (ed) Convergent Evolution in Chile and California Mediterranean Climate Ecosystems. Dowden, Hutchinson and Ross, Stroudsburg, PA

Armesto J, Martínez JA (1978) Relations between vegetation structure and slope aspect in the mediterranean region of Chile. J Ecol 66:881-889

Avila G, Aljaro M, Araya S, Montenegro G (1975) The seasonal cambium activity of Chilean and Californian shrubs. Am J Bot 62:473-478

Cody ML, Mooney HA (1978) Convergence versus nonconvergence in mediterranean-climate ecosystems. Annu Rev Ecol Syst 9:265-321

Dunn E (1970) Seasonal patterns of carbon dioxide metabolism in evergreen sclerophylls in California and Chile. PhD Thesis, University of California, Los Angeles, 139 p

di Castri F, Mooney HA (1973) Mediterranean type ecosystems. Origin and structure. Springer-Verlag, Berlin-Heidelberg-New York, p 405

Gigon A (1979) CO_2-gas exchange, water relations and convergence of mediterranean shrub-types from California and Chile. Oecol Plant 14:129-150

Giliberto J, Estay H (1978) Seasonal water stress in selected shrub species in the mediterranean region of central Chile. Bot Gaz 139:236-240

Hoffmann A (1972) Morphology and histology of *Trevoa trinervis* (Rhamnaceae), a drought deciduous shrub from the Chilean matorral. Flora 161:527-538

Hoffmann A, Hoffmann AE (1976) Growth pattern and seasonal behavior of buds of *Colliguaya odorifera* a shrub from the Chilean mediterranean vegetation. Can J Bot 54:1767-1774

Hoffmann A, Kummerow J (1978) Root studies in the Chilean matorral. Oecologia (Berl) 32:57-69

Jacobson MB, Stoner WA, Richards SP (1981) Models of plant and soil processes. In: Miller PC (ed) Resource Use by Chaparral and Matorral. A Comparison of Vegetation Function in Two Mediterranean Type Ecosystems. Ecological Studies 39. Springer-Verlag, Berlin-Heidelberg-New York, pp 287-368

Jaksic FM, Montenegro G (1979) Resource allocation of Chilean herbs in response to climatic and microclimatic factors. Oecologia (Berl) 40:81-89

Kummerow J, Montenegro G, Krause D (1981) Biomass, phenology, and growth. In: Miller PC (ed) Resource Use by Chaparral and Matorral. A Comparison of Vegetation Function in Two Mediterranean Type Ecosystems. Ecological Studies 39. Springer-Verlag, Berlin-Heidelberg-New York, pp 69-96

Lawrence WT (1975) A radiation model for chaparral canopies. MSc Thesis, San Diego State University, San Diego, California, p 155

Lawrence WT, Oechel WC (1981) Photosynthetic production of perennial species in the mediterranean zone of central Chile. In: Conrad CE, Oechel WC (Tech Coord) Dynamics and Management of Mediterranean-type Ecosystems. Gen Tech Rpt PSW-58, Pacific Southwest Forest and Range Experimental Station, Berkeley, p 615

Miller PC (1981) Resource use by chaparral and mattoral. A comparison of vegetation function in two mediterranean type ecosystems. Springer-Verlag, Berlin-Heidelberg-New York, p 455

Miller PC, Hajek E (1981) Resource availability and environmental characteristics of mediterranean type ecosystems. In: Miller PC (ed) Resource Use by Chaparral and Matorral. A Comparison of Vegetation Function in Two Mediterranean Type Ecosystems. Springer-Verlag, Berlin-Heidelberg-New York, pp 17-41

Montenegro G, Rivera O, Bas F (1978) Herbaceous vegetation in the Chilean matorral. Oecologia (Berl) 36:237-244

Montenegro G, Aljaro ME, Kummerow J (1979a) Growth dynamics of Chilean matorral shrubs. Bot Gaz 140:114-119

Montenegro G, Hoffmann AJ, Aljaro ME, Hoffmann AE (1979b) *Satureja gilliesii*, a poikilohydric shrub from the Chilean mediterranean vegetation. Can J Bot 57:1206-1213

Montenegro G, Aljaro ME, Walkowiak A, Saenger R (1981) Seasonality, growth, and net productivity of herbs and shrubs of the Chilean matorral. In: Conrad CE, Oechel WC (Tech Coord) Dynamics and Management of Mediterranean-type Ecosystems. Gen Tech Rpt PSW-58, Pacific Southwest Forset and Range Experimental Station, Berkeley

Mooney HA (1972) The carbon balance of plants. Annu Rev Ecol Syst 3:315-346

Mooney HA (1977) Convergent evolution in Chile and California mediterranean climate ecosystems. Dowden, Hutchinson, Ross, Shroudsburg, PA

Mooney, HA (1977) The carbon cycle in mediterranean-climate evergreen scrub communities. In: Mooney HA, Conrad CE (eds) Proceedings of the Symposium on the Environmental Consequences of Fire and Fuel Management in Mediterranean Ecosystems. Gen Tech Rpt WO-3, USDA Forest Service, Washington.

Mooney HA, Wright RD, Strain BR (1964) The gas exchange capacity of plants in relation to vegetation zonation in the White Mountains of California. Amer Midland Natur 72:281-297

Mooney HA, Dunn EL (1970a) Convergent evolution of mediterranean climate evergreen shrubs. Evolution 24:292-303

Mooney HA, Dunn EL (1970b) Photosynthetic systems of mediterranean climate shrubs and trees of California and Chile. Am Nat 104:447-453

Mooney HA, Dunn EL, Shropshire F, Song L (1970) Vegetation comparisons between the mediterranean climate areas of California and Chile. Flora 159:480-496

Mooney HA, Dunn EL, Harrison AT, Morrow PA, Bartholomew B, Hays RL (1971) A mobile laboratory for gas exchange measurements. Photosynthetica 5:128-132

Mooney HA, Gulmon S (1978) Environmental and evolutionary constraints on the photosynthetic characteristics of higher plants. In: Plant Population Biology, Columbia University Press

Muñoz-Pizarro C (1966) Sinopsis de la flora Chilena. Second Edition, Universidad de Chile, Santiago, Chile

Mustafa J (1978) The effect of growth and species specific variability on photosynthesis along an elevational gradient in the chaparral. MSc Thesis, McGill University, Montreal, Quebec

Oechel WC, Lawrence WT (1979) Energy utilization and carbon metabolism in mediterranean scrub vegetation of Chile and California. I. Methods: A transportable cuvette field photosynthesis and data acquisition system and representative results for *Ceanothus greggii*. Oecologia (Berl) 39:321-335

Oechel WC, Mustafa J (1979) Energy utilization and carbon metabolism in mediterranean scrub vegetation of Chile and California. II. The relationship between photosynthesis and cover in chaparral evergreen shrubs. Oecologia (Berl) 41:305-315

Oechel WC, Lawrence WT (1981) Carbon allocation and utilization. In: Miller PC (ed) Resource Use by Chaparral and Matorral. A Comparison of Vegetation Function in Two Mediterranean Type Ecosystems. Ecological Studies 39. Springer-Verlag, New York

Oechel WC, Lawrence WT, Mustafa J, Martínez JA (1981) Energy and carbon acquisition. In: Miller PC (ed) Resource Use by Chaparral and Matorral. A Comparison of Vegetation Function in Two Mediterranean Type Ecosystems. Ecological Studies 39. Springer-Verlag, Berlin-Heidelberg-New York, pp 151-183

Rundel PW (1981) The matorral zone of Central Chile. In: Di Castri F, Goodall DW, Specht RL (eds) Mediterranean Type Shrubland. Elsevier, The Hague

Riveros F, Hoffmann A, Avila G, Aljaro ME, Araya S, Hoffmann AE (1976) Comparative morphological and ecophysiological aspects of two sclerophyllous Chilean shrubs. Flora 165:223-234

Sestak Z, Catsky J, Jarvis PG (eds) (1971) Plant Photosynthetic Production: Manual of Methods. Junk, The Hague, 818 p

Shaver GA (1981) Mineral nutrient and nonstructural carbon utilization. In: Miller PC (ed) Resource Use by Chaparral and Matorral. A Comparison of Vegetation Function in Two Mediterranean Type Ecosystems. Ecological Studies 39. Springer-Verlag, Berlin-Heidelberg-New York, pp 237-257

Strain BR (1969) Seasonal adaptations in photosynthesis and respiration in four desert shrubs growing *in situ*. Ecology 50:511-513

Tieszen LL, Johnson DA, Caldwell MM (1974) A portable system for the measurement of photosynthesis using 14-carbon dioxide. Photosynthetica 3:151-160

Thrower JKW, Bradbury DW (1977) Chile-California Mediterranean scrub atlas: a comparative analysis. Dowden, Hutchinson, Ross, Shroudsburg, PA, p 237

Changes during summer drought in leaf CO_2 uptake rates of macchia shrubs growing in Portugal: Limitations due to photosynthetic capacity, carboxylation efficiency, and stomatal conductance

J.D. Tenhunen, W. Beyschlag,* O.L. Lange* and P.C. Harley

Systems Ecology Research Group, San Diego State University, San Diego, California, 92182 USA

INTRODUCTION

The classically described stresses to which plants are subjected in mediterranean type climates are water stress and heat stress during the summer and cold stress during the winter. The general behavioral tendencies in gas exchange of leaves of shrubs subjected to these climate conditions were summarized nicely by Larcher (1961).

'During summer, all of these plants, depending on their own ability to reach stored water and depending on the length of the drought period, are forced to conserve water by means of stomatal closure. Such closure means that either leaves draw down internal CO_2 pressure to the compensation point and recycle respired CO_2 or that carboxylation efficiency decreases and photosynthesis is temporarily interrupted as suggested by a series of earlier investigations (Gaastra 1959; Walter 1931; Romose 1940; Montfort and Hahn 1950; Stocker 1954, 1956).'

Continuing, Larcher summarizes that 'during the winter, photosynthetic response is of a type which we normally observe with C_3 leaves of species from the temperate zone, transpiration peaking at midday with net photosynthesis and stomata remaining essentially open. Nevertheless, cold temperatures reduce photosynthetic production below levels observed in the late spring'...and we would add after recovery from drought in the fall.

In recent studies carried out in the scrub vegetation in Sobreda, Portugal near Lisbon, we analyzed photosynthetic responses that occur in leaves of Mediterranean sclerophylls under mild temperate climate conditions during the spring and fall. We then observed changes in leaf behavior in gas exchange that occurred under extreme conditions and due to the imposition of stress. In the present paper, we summarize the general changes found to occur in the potential photosynthetic capacity of sclerophyllous leaves under summer drought conditions.

*Lehrstuhl für Botanik II, Universität Würzburg, Mittlerer Dallenbergweg 64, D-8700, West Germany

NATO ASI Series, Vol. G15
Plant Response to Stress. Edited by J. D. Tenhunen et al.

FIXATION RATES OBTAINED WITH AMBIENT CO_2

Comparative studies of net photosynthesis in sclerophyll shrubs from California and Chile were undertaken within the framework of the International Biological Program (see Lawrence in this volume). Data obtained in these studies detailed changes in carbon acquisition by leaves as affected by seasonal changes in light, temperature, and water availability (see review by Oechel *et al.* 1981). Our recent studies of net photosynthesis in leaves of shrubs growing in a macchia vegetation (related to the *Quercetalia ilicis* of Braun-Blanquet, 1952) near Lisbon, Portugal, have shown the characteristics of CO_2 uptake for these plants to be similar. As illustrated in Fig. 1 for sun leaves of *Quercus coccifera* and *Arbutus unedo* in ambient air, relatively high photosynthesis rates (NP) are maintained over a wide range in temperature when plants are not water-stressed (spring in Fig. 1). Although not shown clearly in these data, the optimum for CO_2 uptake, as in the case of previous studies of shrubs of mediterranean zones, occurred between 25 and 30° C. During the experimental period in which these data were recorded, increases in temperature to 25° C associated with high light conditions increased net CO_2 uptake.

During summer, the maximum photosynthesis rate obtained during the day (NP_{max}) and the maximum leaf conductance (G_{max}) was lowered depending on the degree of water stress (see also Fig. 2). Increase in leaf temperature above 30° C led to sharp decreases in net CO_2 uptake (Fig. 1). This decrease in net photosynthesis occurs at high temperature and simultaneously at low humidity, since the dew point of the air remains relatively constant. Due to high leaf temperatures and large leaf to air vapor pressure differences at midday during summer drought, the rate of net photosynthesis is depressed and diurnal time courses for net CO_2 uptake are typically characterized by a two peaked pattern (Tenhunen et al. 1984a).

The decrease in carbon dioxide uptake rates induced by drought varied among species (Figs. 2 and 3). The type of response to decreases in predawn water potential shown in Fig. 2 is typical for arid zone plants as has been discussed by Schulze and Hall (1982). Both NP_{max} and G_{max} for *Quercus coccifera* and *Quercus suber* decreased rapidly when pre-dawn water potential initially decreased. In the case of *Arbutus unedo*, decreases in predawn water potential led to a more gradual and continuous decrease in NP_{max}. Nevertheless, the differences among these species in NP_{max} and G_{max} at predawn water potentials between -10 and -30 bar are small. Most important in influencing seasonal pattern of gas exchange in any particular species is the rapidity with which water potentials decrease, i.e. the ability or inability a species has to maintain water supply to the leaf tissues (this fits nicely with the conclusions of Kaiser reported in this volume).

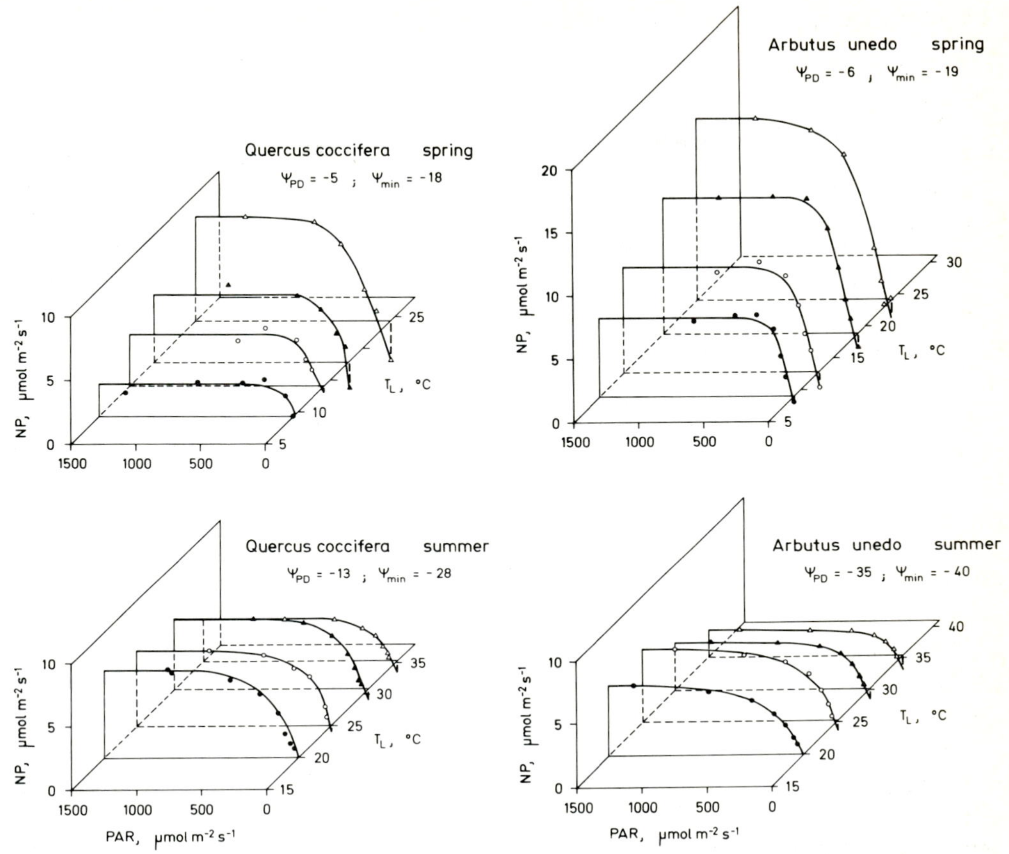

Fig. 1. The dependency of steady-state net photosynthesis rate, NP, on incident PAR at a series of leaf temperatures within the range naturally experienced by *Quercus coccifera* and *Arbutus unedo* during the spring (May 26-27, 1983) when plants experienced no water stress and after 6 weeks of drought during mid-summer (July 24-25, 1983). Leaf predawn xylem potential, Ψ_{PD}, and minimum observed xylem potential, Ψ_{MIN}, as shown. Dew point of the air was 10° C. P_a = 350 μbar. O_2 partial pressure was 210 mbar. Solid curves were freely drawn and do not indicate regression fits to the data

Seasonal changes in NP_{max} and G_{max} are shown for the three species in Fig. 3. While the two *Quercus* species appear to respond similarly to changes in predawn water potential, predawn water potential of *Quercus suber* decreased more rapidly after the onset of drought than that of *Quercus coccifera*, leading to an earlier reduction in leaf conductance and maximum photosynthesis rate. Predawn water potential of *Arbutus unedo* decreased approximately linearly throughout the drought period, resulting in a continuous decrease in conductance

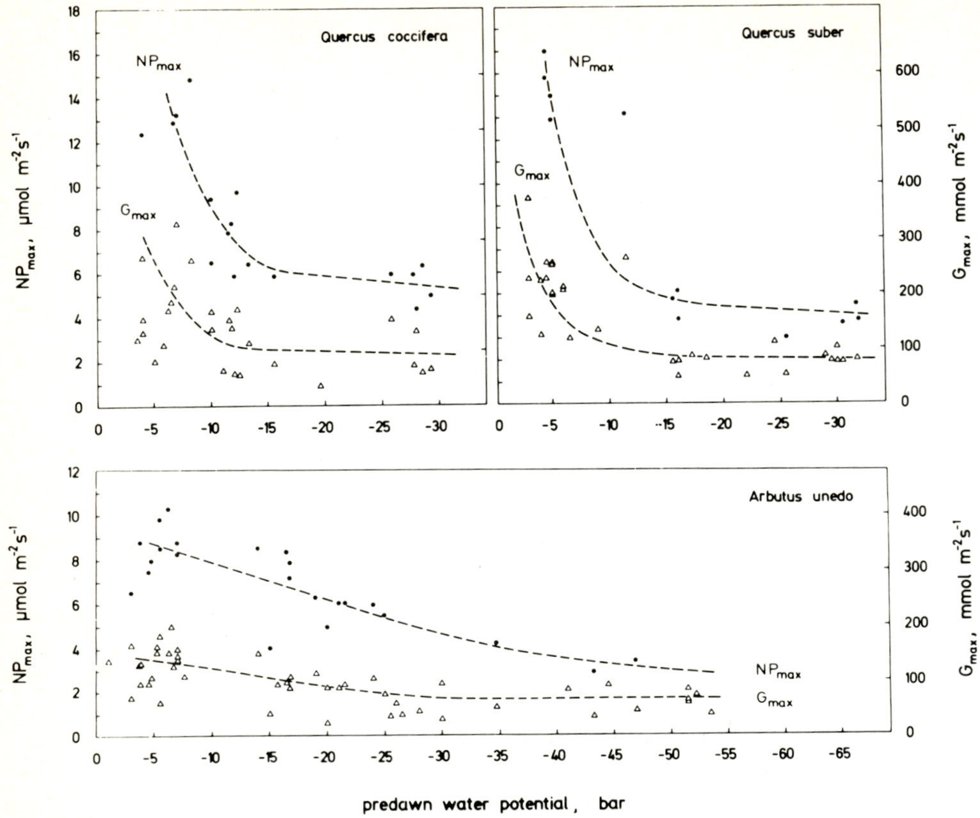

Fig. 2. The maximum net photosynthesis rate, NP_{max} - circles, and maximum stomatal conductance, G_{max} - triangles, observed during monitoring of diurnal courses of gas exchange in all seasons of the year as a function of the predawn leaf xylem water potential found on those days. Comparable data is shown for *Quercus coccifera*, *Quercus suber*, and *Arbutus unedo*. Dashed lines are freely drawn to emphasize the trends

and maximum photosynthesis rate. The NP_{max} of *Quercus coccifera* at the end of the summer drought remained higher than that of the cork oak, *Quercus suber*, or *Arbutus unedo*. Minimum predawn water potentials reached by the end of the drought in early September were approximately -30 bar for *Quercus coccifera* and *Quercus suber* and greater than -50 bar for *Arbutus unedo*.

These observations are similar to those made in the California chaparral (Oechel *et al.* 1981). The behavior of *Quercus coccifera* appears to be similar to that of the California scrub oak *Quercus dumosa*. Both appear to be deep rooted species which exploit deep soil water reserves (Hellmers *et al.* 1955; Rambal 1984; see also Rambal in this volume). The response to water stress of *Arbutus unedo* is similar to that of *Ceanothus greggii* and *Ceanothus verrucosus*. All three of these

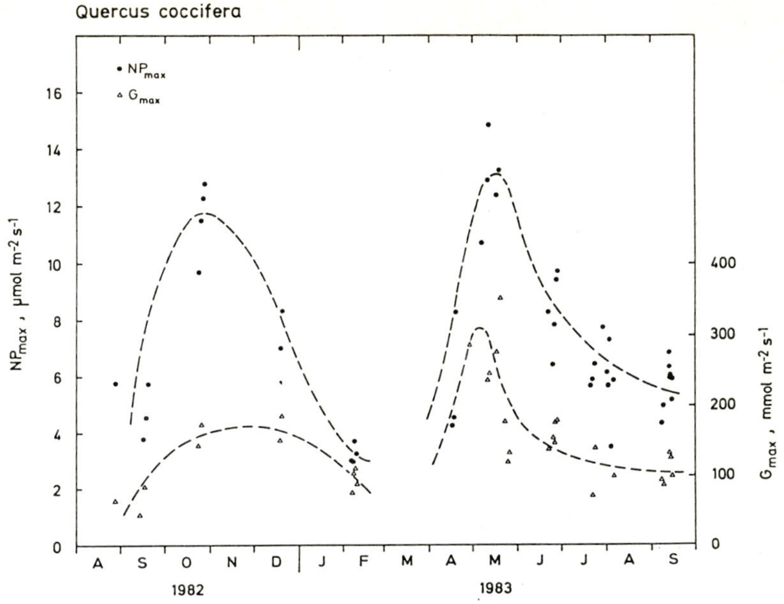

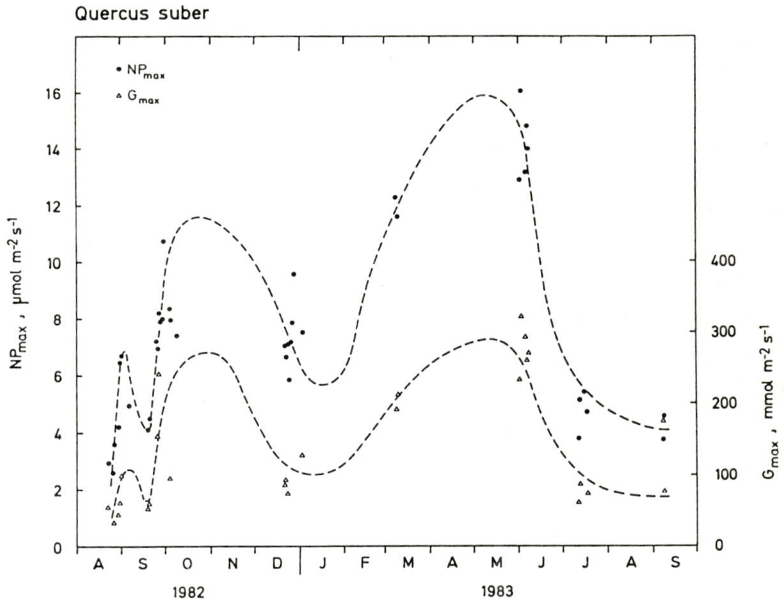

Figs. 3A and 3B. The maximum net photosynthesis rate, NP_{max} - circles, and maximum stomatal conductance, G_{max} - triangles, observed during monitoring of diurnal courses of gas exchange at different times of the year. Comparable data are shown for *Quercus coccifera* and *Quercus suber* (see also Fig. 3C - next page). Dashed lines are freely drawn to emphasize the trends

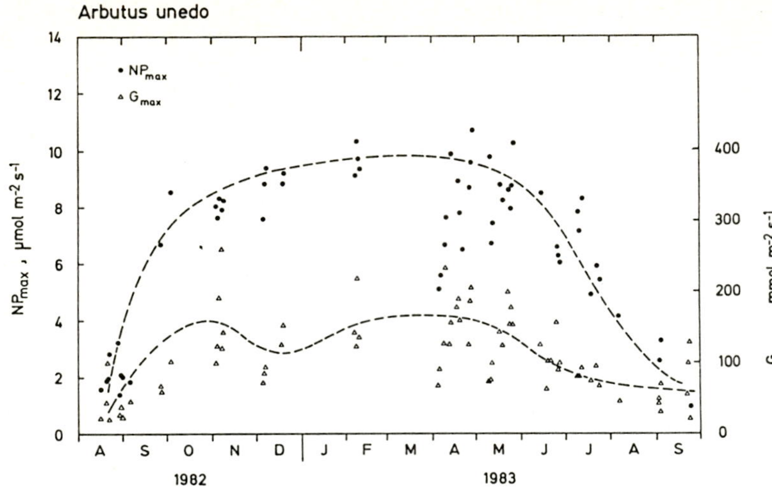

Fig. 3C. The maximum net photosynthesis rate, NP_{max} - circles, and maximum stomatal conductance, G_{max} - triangles, observed during monitoring of diurnal courses of gas exchange at different times of the year. Comparable data to those in Figs. 3A and 3B are shown for *Arbutus unedo*. Dashed lines are freely drawn to emphasize the trends

species appear to be more shallowly rooted. The response to water stress of *Quercus suber* is similar to that of the closely related species *Quercus coccifera*. On the other hand, *Quercus suber* appears less able to tap soil water reserves. Under moderate water stress conditions in July, *Quercus coccifera* leaves fixed approximately 0.2 mol CO_2 m^{-2} day $^{-1}$ as compared to 0.12 for *Quercus suber* and 0.1 for *Arbutus unedo*. Water transpired by *Quercus coccifera* was approximately 70 mol H_2O m^{-2} day^{-1}, while *Quercus suber* and *Arbutus unedo* used approximately 30 mol H_2O m^{-2} day $^{-1}$. Under extreme drought conditions in September, *Quercus coccifera* still transpired 50% more water than *Quercus suber* and more than twice as much as *Arbutus unedo*. Total daily CO_2 fixation was 1.6 times that of *Quercus suber* and three times that of *Arbutus unedo*.

The average daily rate of CO_2 fixation was determined from diurnal courses of net photosynthesis observed for these European Mediterranean shrub species (see Tenhunen *et al.* 1984a, 1985; Beyschlag 1984). The average daily rate of CO_2 fixation standardized on an hourly basis (NP_{toth}) was linearly related to the

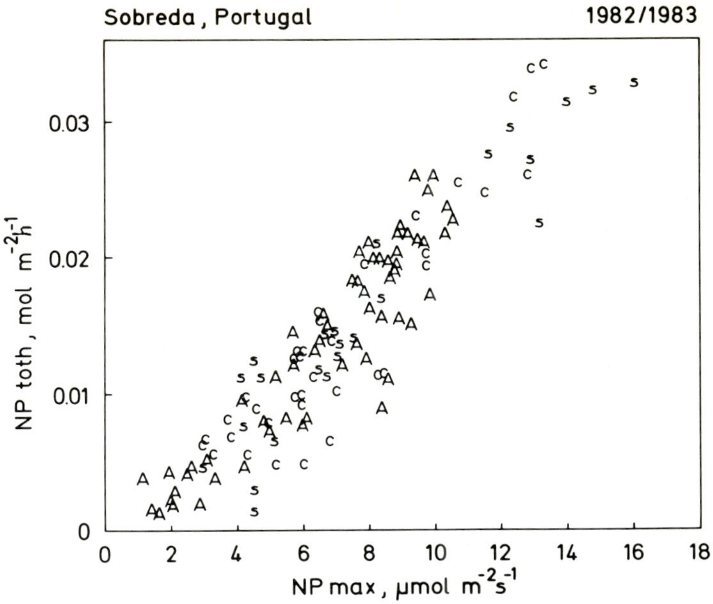

Fig. 4. Average hourly rate of carbon fixation, NP_{toth}, during the period between light compensation of photosynthesis in the morning and evening related to the maximum net photosynthesis rate, NP_{max}, observed during the day for *Quercus coccifera* (c), *Quercus suber* (s), and *Arbutus unedo* (A) as described in the text

momentary daily maximum rate NP_{max} (Fig. 4):

$$NP_{toth} = 0.0025 * NP_{max} - .0025 \qquad (1)$$

where: NP_{toth} is in mol m^{-2} $hour^{-1}$ and NP_{max} is in µmol m^{-2} s^{-1}.

No significant differences were found for the coefficients of this regression among the different species. Total daily CO_2 fixation may be obtained by multiplying NP_{toth} by the number of hours occurring between the morning and evening light compensation points. This direct relation of NP_{max} to total daily CO_2 fixation independent of species indicates the strong convergence in leaf physiological characteristics that has occurred. It further indicates that the day to day variations in weather within a particular season that might shift correlations between environmental factors controlling leaf function are less important in determining overall production than are changes in leaf mesophyll photosynthetic capacity or in stomatal behavior.

NET PHOTOSYNTHETIC POTENTIAL NP_{pot}

Faced on the one hand with similarities in response and on the other with clear species-specific differences (Fig. 2 and 3), further means of comparison are desirable which will identify special autecological characteristics of individual species and will help to explain their situation within the community. With regard to photosynthesis, experimentation under controlled environmental conditions will allow us to determine at any point in time the genetically determined photosynthetic potential that may be expressed by a plant growing in a habitat with a particular average level of available nutrients and in the absence of other stress factors (see also Tenhunen *et al.* 1984b). This potential photosynthesis of the mesophyll (NP_{pot}) may be described in terms of a mathematical model where photosynthesis rate is a function of light intensity (PAR), leaf temperature (T_L), and leaf air space CO_2 partial pressure (P_i). The photosynthetic function of any particular leaf is understood well if the carbon dioxide response or relationship of net photosynthesis rate to P_i is clearly expressed for all relevant combinations of constant light intensity and leaf temperature (e.g. Harley *et al.* 1985, Tenhunen *et al.* this volume). The photosynthetic potential is modified over the long-term due to changes in water availability, changes in nutrient availability, developmental effects, or response to stress conditions. Most common developmental effects will be the result of environmental history, canopy position, or leaf age. In our studies of gas exchange carried out with sclerophyll shrubs in Portugal, we studied photosynthetic potentials during different seasons. These studies have demonstrated the time periods in which photosynthetic potentials of various species remain relatively constant and those periods in which photosynthetic potential is altered due to the influence of environmental or endogenous events. A schematic representation of these changes in photosynthetic potential over the annual cycle for a generalized sclerophyll was discussed by Tenhunen *et al.* (1984b; see also Lange *et al.* and Harley *et al.* this volume).

EVALUATING FIXATION RATES DURING SUMMER DROUGHT IN TERMS OF NP_{pot}

Diurnal time courses of net photosynthesis rates (NP) and transpiration rates (Tr) of leaves are shown in Fig. 5 as observed with *Quercus coccifera* and *Arbutus unedo* on a series of days in 1983. The measurements were conducted with gas exchange cuvettes controlled to track ambient temperature and humidity and data were recorded every four minutes (see Tenhunen *et al.* 1984a for details on methods). Time courses of leaf temperature and leaf to air water mole-fraction difference were essentially the same in both cuvettes on any particular day. On relatively cool days during the spring and in the absence of water stress (May 10), leaf conductance of *Quercus coccifera* was higher than that of *Arbutus unedo* as were rates of water loss and CO_2 uptake. On warm days (May 5), stomata of

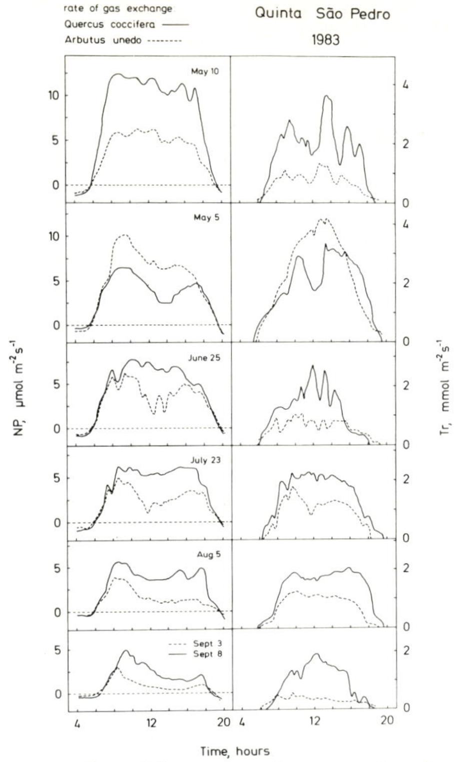

Fig. 5. Diurnal courses of ambient net photosynthesis rate, NP (left panel), and transpiration rate, Tr (right panel), of horizontal leaves of *Quercus coccifera* and *Arbutus unedo* on the dates shown during spring and summer. Environmental conditions were similar in both cuvettes

Arbutus remained open and CO_2 uptake and water loss was high. In the case of *Quercus*, stomatal conductance was lower and a marked midday closure was observed. Rates of gas exchange were restricted. With the onset of drought, the maximum conductance for leaves of *Quercus coccifera* was reduced further. With similar temperature and humidity conditions on June 25 and on May 5, the transpiration rate on June 25 was only 60% of that on May 5. Nevertheless, photosynthesis rates at midday were higher than on May 5 and no midday depression occurred. The gas exchange rates of *Arbutus* were decreased substantially in comparison with those on May 5 and a strong midday depression of net photosynthesis was observed. During the latter part of the summer, NP and Tr remained at approximately the same levels in the case of *Quercus*. The net photosynthesis rate of *Arbutus* continued to decline. The observations are compatible with data already presented in Fig. 2. While *Quercus coccifera* appears to be less sensitive to water stress, there is no indication in either type of data as to whether the mechanisms responsible for the seasonal changes represented in Fig. 2 are the same or fundamentally differ.

CO_2-Saturated Photosynthesis Rates, P_M

By defining changes in NP_{pot} which occur during the course of summer drought, much greater insight into individual species ecology is gained. In our studies, characteristics of the leaf net photosynthesis response to carbon dioxide partial pressure were identified and compared. Extensive measurements were made of CO_2-saturated carbon dioxide uptake rates (P_M) using leaf material from *Quercus suber* as well as *Quercus coccifera* and *Arbutus unedo*. This parameter is interpreted as a limitation set by the rate of regeneration of CO_2 acceptor in the leaf mesophyll (Farquhar and von Caemmerer, 1982). The effect of light intensity and leaf temperature on P_M is shown for *Quercus coccifera* and *Arbutus unedo* in Fig. 6 during the spring after full expansion of the new leaves (pre-dawn water potential, Ψ_{PD}, equal to -6 and -6.5 bar respectively) and in mid-summer at the end of July with moderate water stress (Ψ_{PD} equal to -13 and -35 bar respectively). Symbols shown indicate the individual measurements; solid lines were freely drawn to illustrate the three-dimensional response surface. The temperature range of measurements is slightly different for the two time periods, having been selected to encompass the temperatures prevailing naturally. Air dew point was approximately 10° C. Maximal values of ΔW for the response curves shown were 20 mbar bar^{-1} in the spring and 45 mbar bar^{-1} during summer. No differences were found in the case of *Quercus coccifera*, when responses at those temperatures common to both periods were compared. However, in the case of *Arbutus unedo*, the level of mesophyll photosynthetic capacity decreased to approximately one-third by mid-summer. Due to the large differences in Ψ_{PD} in mid-summer, it seems clear that the photosynthetic systems of shallow rooted *Arbutus* shrubs are more subject to damage. However, NP_{max} (under ambient conditions) of both species decreased by the end of July to the lowest level observed. The data presented here suggest that the mesophyll photosynthetic capacity of *Quercus coccifera* did not decline during this period. Thus, the decrease in NP_{max} of *Quercus coccifera* to approximately 50% of that observed in spring requires explanation and will be discussed further below. In both *Quercus* and *Arbutus*, the CO_2- and light-saturated photosynthesis rate increases with leaf temperature to approximately 30° C, above which it decreases.

Changes in the mesophyll photosynthetic capacity as seen from CO_2-saturated rates of *Quercus suber* were similar to those observed with *Arbutus unedo*. The CO_2- (CO_2 partial pressure external to the leaf, P_a, equal to 2500 µbar) and light-saturated photosynthesis rate (P_{ML}) of *Quercus suber* as affected by leaf temperature is compared to the ambient photosynthesis rate (P_a = 350 µbar) in Fig. 7 for three stages of water stress. Comparison is further made of the temperature response curves obtained during the morning and during the

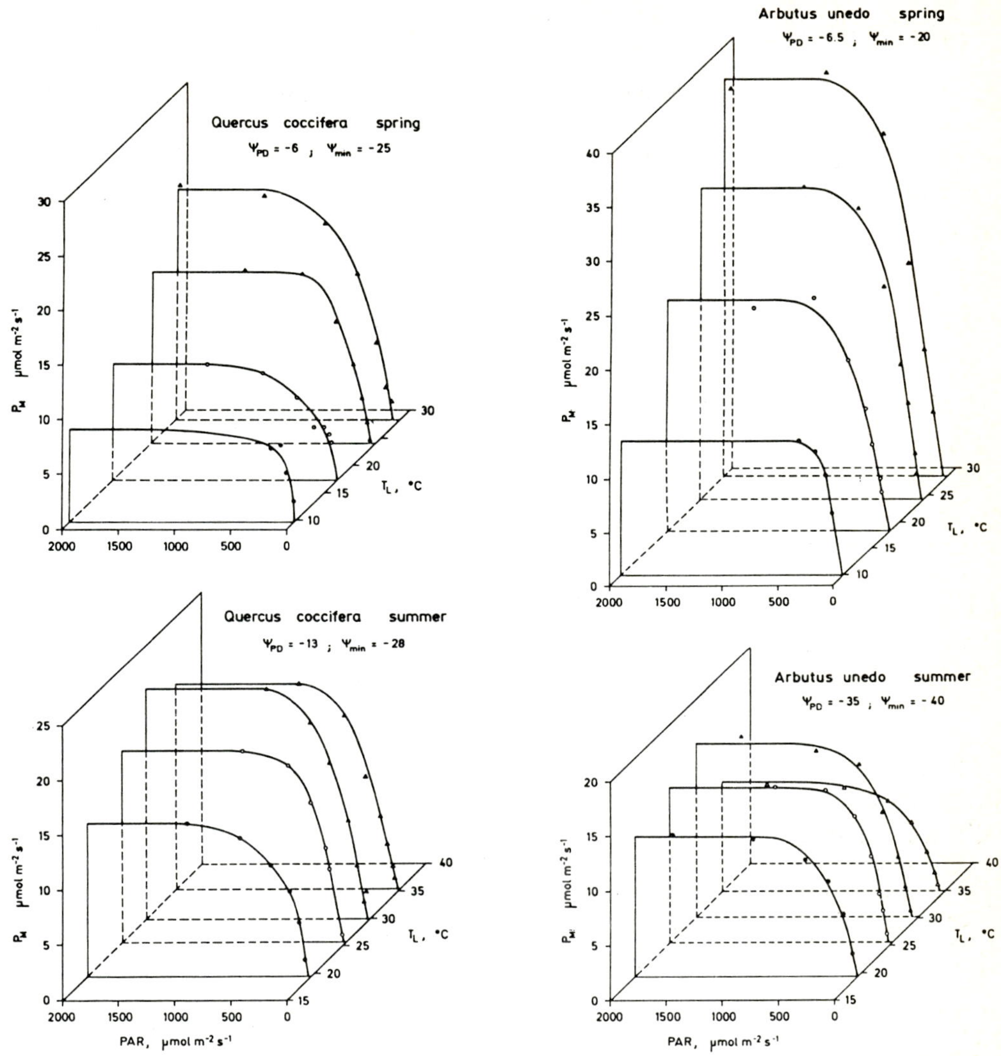

Fig. 6. The dependency of the CO_2-saturated net photosynthesis rate, P_M, on incident PAR at a series of leaf temperatures within the range naturally experienced by *Quercus coccifera* and *Arbutus unedo* during the spring (April 28-May 5, 1983) when plants experienced no water stress and after 7 weeks of drought during mid-summer (July 31-August 3, 1983). Leaf predawn xylem potential, Ψ_{PD}, and minimum observed xylem potential, Ψ_{MIN}, as shown. Dew point of the air was 10° C. P_a = 2500 µbar. O_2 partial pressure was 210 mbar. Solid curves were freely drawn and do not indicate regression fits to the data

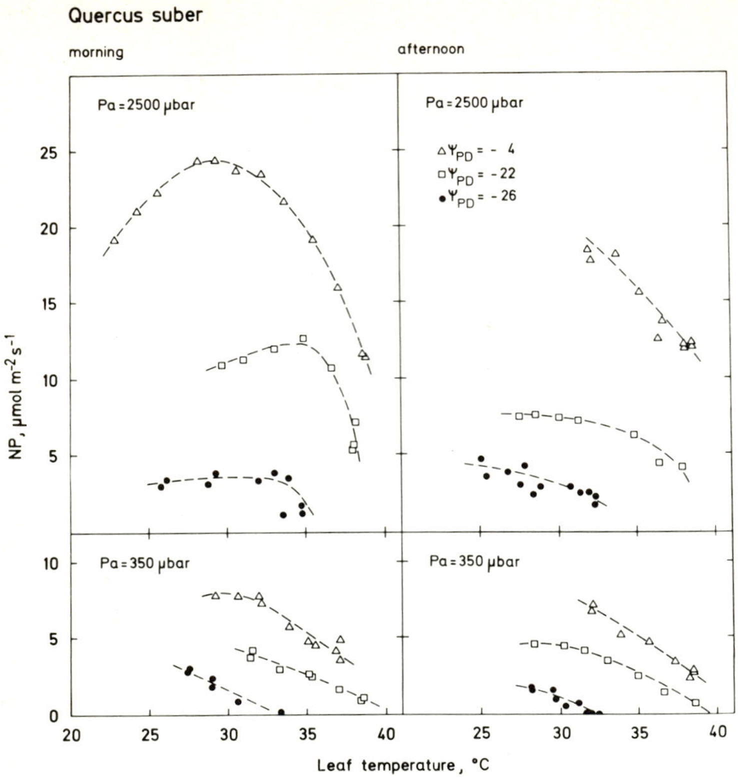

Fig. 7. Carbon dioxide saturated (P_a = 2500 µbar) and ambient (P_a = 350 µbar) net photosynthesis rates of *Quercus suber* as a function of leaf temperature. Observations were made at the three predawn water potentials indicated. Temperature was increased during the morning and decreased during the afternoon. Light intensity was saturating for photosynthesis

afternoon. With a decrease in Ψ_{PD} from -4 to -22 bar, CO_2-saturated rates decrease by 50%. Rates observed at any particular temperature were often lower in the afternoon than during the morning. The decline in CO_2-saturated photosynthesis during the afternoon did not always occur. In fact, in the case of *Arbutus unedo*, an afternoon increase in photosynthetic capacity was often observed (see Lange et al. 1985). Decrease in Ψ_{PD} to -26 bar again led to a large decrease in photosynthetic capacity. A strong influence of high temperature is seen on CO_2-saturated photosynthetic rate. With P_a equal to 350 µbar, CO_2 uptake decreased linearly with increase in temperature above approximately 30° C. The upper temperature compensation point occurred at lower temperatures as water stress increased. There were no apparent differences in rates of CO_2

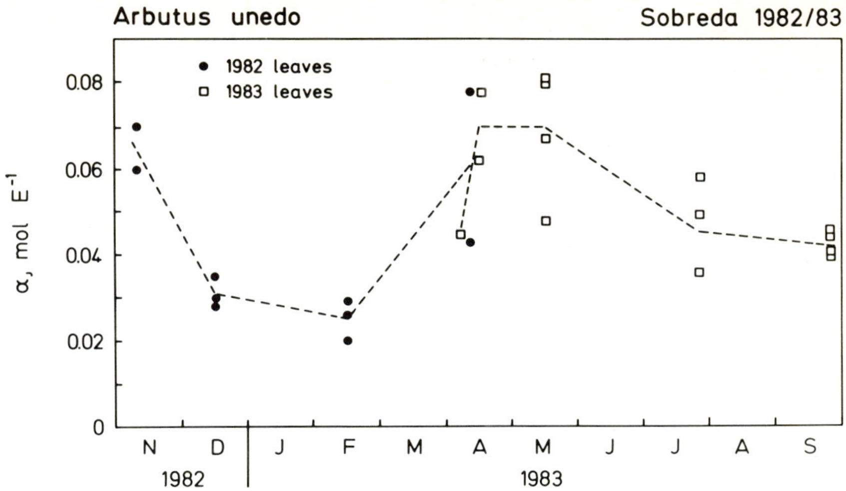

Fig. 8. Seasonal change in the parameter α (initial slope of the response of photosynthesis to incident light normal to the upper surface of the leaf measured at 2500 bar CO_2 and 21% oxygen) observed for *Arbutus unedo* in the scrub vegetation in Sobreda

uptake with ambient gas concentrations during the morning and during the afternoon periods.

We have termed the initial slope of light response curves, measured at saturating CO_2 partial pressure and on the basis of incident PAR, the light utilization efficiency denoted with the symbol α. Alpha is directly related to the quantum efficiency except that the latter is measured on the basis of absorbed PAR. The quantum efficiency has been used as an important indicator of drought stress and cold stress and demonstrates at any time the stability or state of chloroplast function in electron transport. In our studies, changes in α were best documented in *Arbutus unedo* (Fig. 8 adapted from Beyschlag 1984) where α decreased from approximately 0.07 in spring to approximately 0.04 at the end of the drought period in September 1983. A strong decrease in α due to cold stress in the winter occurred in this and other species as well but will not be discussed further here.

The Operating Point

Diurnal courses of net carbon dioxide uptake were observed while tracking ambient temperature and humidity with P_a = 350 μbar in the measurements cuvettes. On other days, the cuvettes were controlled to obtain the same time courses for temperature and humidity but P_a was held at 2500 μbar. The diurnal patterns in net photosynthesis from such experiments are shown in Fig. 9 along with diurnal courses of leaf conductance determined for ambient air with

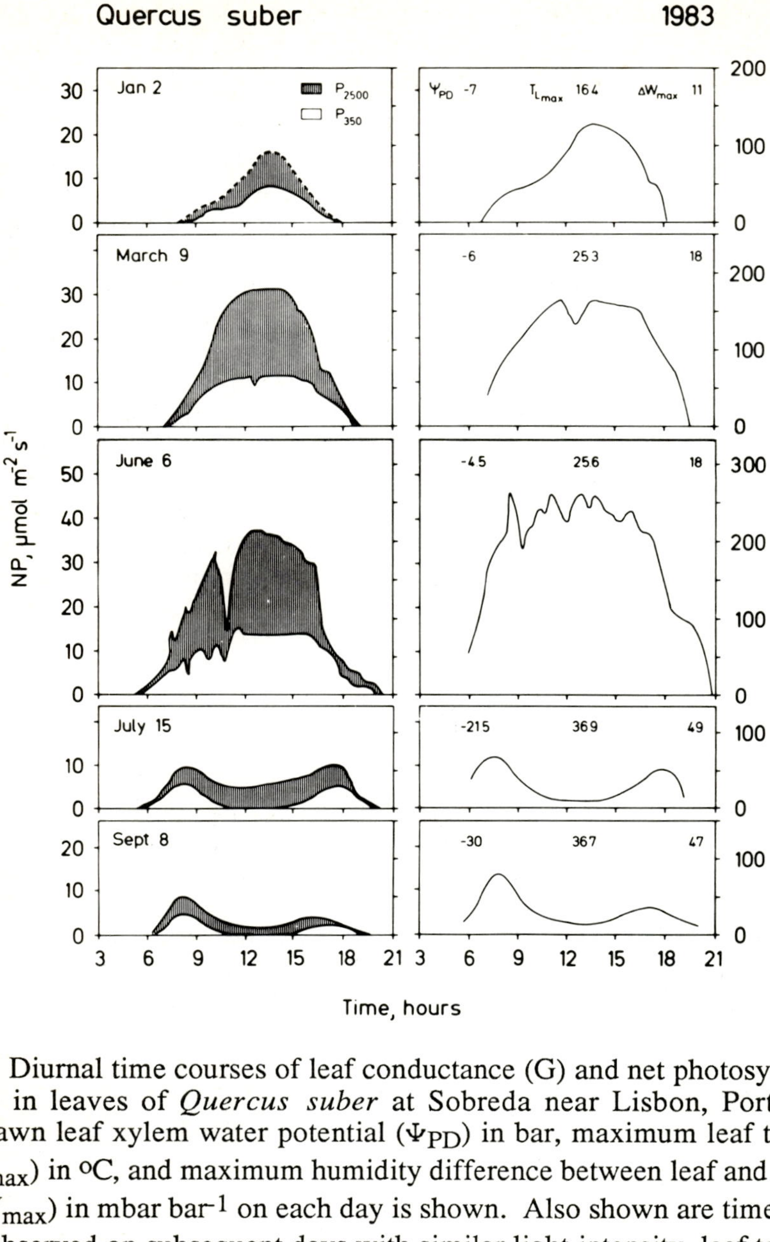

Fig. 9. Diurnal time courses of leaf conductance (G) and net photosynthesis rate (NP) in leaves of *Quercus suber* at Sobreda near Lisbon, Portugal. The predawn leaf xylem water potential (Ψ_{PD}) in bar, maximum leaf temperature (T_{Lmax}) in °C, and maximum humidity difference between leaf and external air (ΔW_{max}) in mbar bar^{-1} on each day is shown. Also shown are time courses of NP observed on subsequent days with similar light intensity, leaf temperature, and cuvette humidity, but with 2500 μbar rather than normal ambient CO_2 partial pressure maintained in the cuvette The high-CO_2 time course on January 2 (dashed line) was constructed for ambient light and temperature conditions from light and temperature response curves of NP measured at 2500 μbar.

Quercus suber. We have attempted to define diurnal changes in the operating point on the CO_2 response curve by determining the ratio of the photosynthesis rates obtained in normal air (P_{350}) and in air with high CO_2 partial pressure (P_{2500}). The data have indicated an operating point low on the CO_2 response curve. Maximal values were approximately 0.45 for *Quercus suber* in the absence of water stress and during peak photosynthesis in the early morning during the summer. These values decreased to 0.15 during midday depression of gas exchange. Similar results were obtained for *Arbutus unedo* (Beyschlag 1984) and *Quercus coccifera* (Tenhunen *et al.* 1985) in diurnal time course experiments and with steady-state CO_2 response curves measured directly.

As compared to results reported in the literature for a wide variety of plant species (average value from 33 CO_2 response curves = 0.69), the ratio P_{350}/P_{2500} is quite low in these sclerophyll species. This suggests a relatively high investment of nitrogen in RuBP regeneration components of the photosynthetic system. The reason for this investment may be that water use efficiency of the plants in question will be improved. Using a biochemically based model of leaf photosynthesis (Farquhar and von Caemmerer 1982) that had been adapted on the basis of field experimentation to reproduce photosynthetic responses and diurnal patterns of *Arbutus unedo* (Harley *et al.* 1986), the consequences of nitrogen

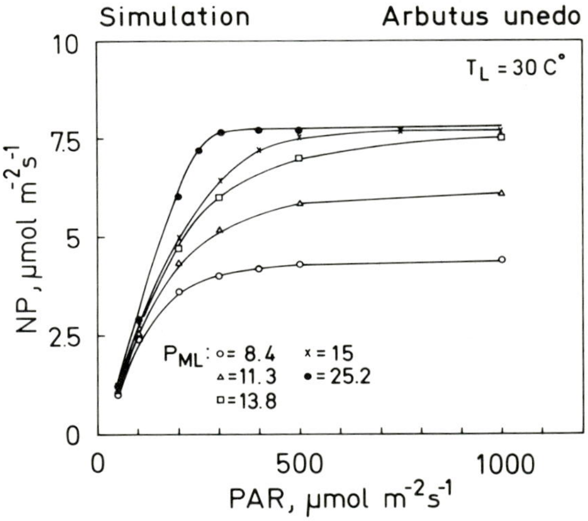

Fig. 10. The simulated effect of varying P_{ML} on the light response of net photosynthesis of the leaf mesophyll with ambient gas concentrations. Simulations were conducted with the model developed for *Arbutus unedo* by Harley et al. (1986) using the kinetic equations of Farquhar and von Caemmerer (1982). All factors other than P_{ML} were held constant. Leaf temperature was assumed to be 30° C

investments in RuBP regeneration were examined. Shown in Fig. 10 are the results of model simulations in which P_{ML} or the RuBP regeneration limitation was varied independently of other factors. The simulated light response curves shown, indicate that as P_{ML} is increased, NP at low values of PAR also increases substantially. Thus, we might hypothesize that a major share of photosynthetic production may occur under low light conditions in these species. Tuning of the photosynthetic system seems appropriate to maximize CO_2 uptake early and late in the day during summer drought when stomata open as opposed to during midday when stomatal closure often occurs. Furthermore, inner canopy leaves might also be expected to contribute importantly to carbon gain despite low incident light intensities (see Meister *et al.* in this volume).

Carboxylation Efficiency, CE, and Compensation Point

The initial slope of the carbon dioxide response curve of net photosynthesis at light saturation, carboxylation efficiency (CE), is determined primarily by kinetic characteristics of RuBP-carboxylase but also by a wall and liquid phase diffusion conductance (Raven and Glidewell 1981; Evans 1983) which has been little studied to date. For non-water-stressed plants, seasonal modification of NP_{pot} is primarily due to shifts in temperature, lower temperature decreasing P_M and CE (see Tenhunen *et al.* this volume; see also Tenhunen *et al.* 1985). At high temperatures, increasing temperature decreases P_M and CE as the optimum for photosynthesis is exceeded. Examples of temperature increases stimulating photosynthesis in winter and inhibiting photosynthesis in summer are illustrated by the CO_2 responses shown in Fig. 11 measured with leaves of *Quercus suber* for the temperature ranges normally experienced in the natural habitat during these seasons. The magnitude of P_M and CE achieved at any particular temperature decreases with increasing water stress and depending on species sensitivity to change in Ψ_{PD} (Fig. 2).

These general conclusions are supported both by measurement of steady-state CO_2 responses and when a picture of change in the CO_2 response during the course of the day is developed. The latter is accomplished by exposing leaves in the field to different gas concentrations on consecutive days during which light and leaf temperature change continuously but in a manner that is identical on each day. With an air CO_2 partial pressure of 340 μbar, a time course for net photosynthesis is obtained which reveals the natural behavior of the leaves in gas exchange. By experimentally decreasing P_a and then increasing P_a to 2500 μbar on subsequent days, CE and P_M may be derived (Tenhunen *et al.* 1984a). Results from such an experiment with *Arbutus unedo* conducted during July of 1983 are shown in Fig. 12 (Beyschlag 1984). PAR was that naturally occurring and leaf temperature was controlled as shown. The inset in the figure illustrates the diurnal course of change in CE. The parameters CE and P_M have been used to

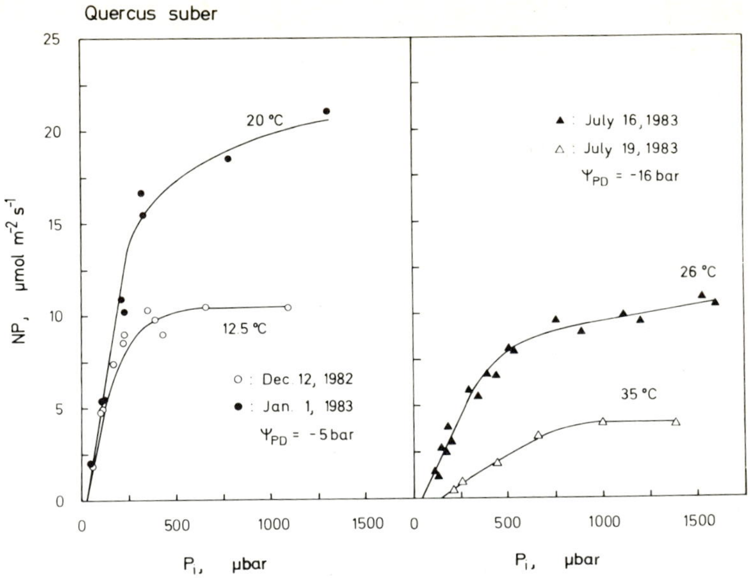

Fig. 11. Net photosynthesis rate of leaves of Quercus suber (saturating light) as a function of internal CO_2 partial pressure (P_i) observed during winter and summer at the temperatures and predawn water potentials indicated

generate an approximation of the CO_2 response at half-hour intervals according to a nonlinear equation previously shown to fit such data well (see model presented by Tenhunen *et al.* this volume).

Change in CE under similar conditions to those shown in Fig. 12 is illustrated for *Quercus coccifera* and *Quercus suber* in Fig. 13 Data are shown as a function of leaf temperature as well as leaf to air H_2O mole fraction difference. Values of CE under conditions of water stress were significantly higher for *Quercus coccifera* than for *Quercus suber*. The data in Fig. 13 suggest that CE is lower in the afternoon than in the morning under the same conditions. The lesser tendency for midday decrease in net photosynthesis rate in *Quercus coccifera* under water stress conditions compared to *Arbutus unedo* (Fig. 5) may in part be due to maintenance of a higher CE at temperatures between 30 and 35° C by the former. A relatively high carboxylation efficiency at leaf temperatures between 35 and 40° C and ΔW between 40 and 70 mbar bar^{-1} was also found for leaves of *Pistacia lentiscus*. (unpublished) and *Phillyrea angustifolia* (Harley *et al.* this volume). The differences in ability to maintain high values of CE during summer drought may be due in part to adaptations in root morphology and partitioning of biomass which allow access to available water supplies (Rambal 1984; Tenhunen *et al.* 1984b). Our observations consistently showed that predawn water potentials remain higher in *Quercus coccifera* than in *Quercus suber* or *Arbutus*

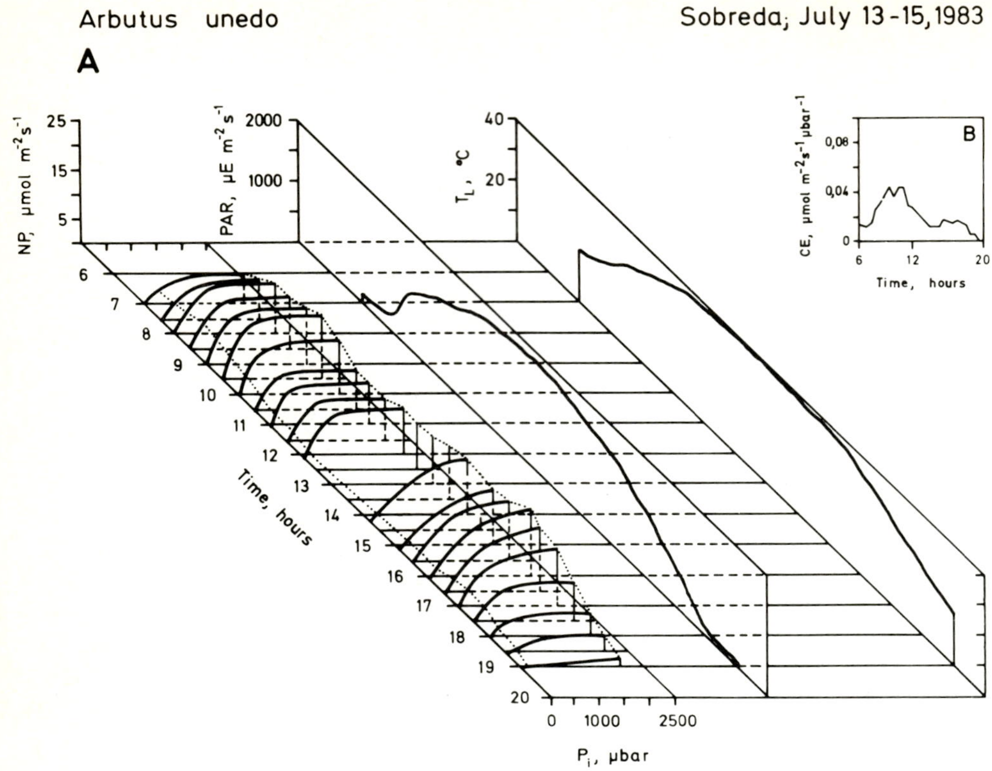

Fig. 12. A: Carbon dioxide response of net photosynthesis as a function of time of day (see text) and with changes in incident PAR and leaf temperature, T_L, as shown. Natural course of ambient photosynthesis is indicated as a dotted line. B: Diurnal course of carboxylation efficiency determined in the experiment

unedo under drought stress as discussed above. The same is probably true for *Pistacia lentiscus* although water potentials were not studied in this species due to problems with phloem exudation.

CO_2 exchange rate under ambient conditions is determined by CE, P_i (see below), and the CO_2 compensation point. The CO_2 compensation point increased strongly in these species with increase in leaf temperature at midday during summer drought. The data suggest a greater increase in compensation point in leaves of *Quercus suber* at a particular stage in the drought cycle and under similar high temperature and low humidity conditions than in *Quercus coccifera*.

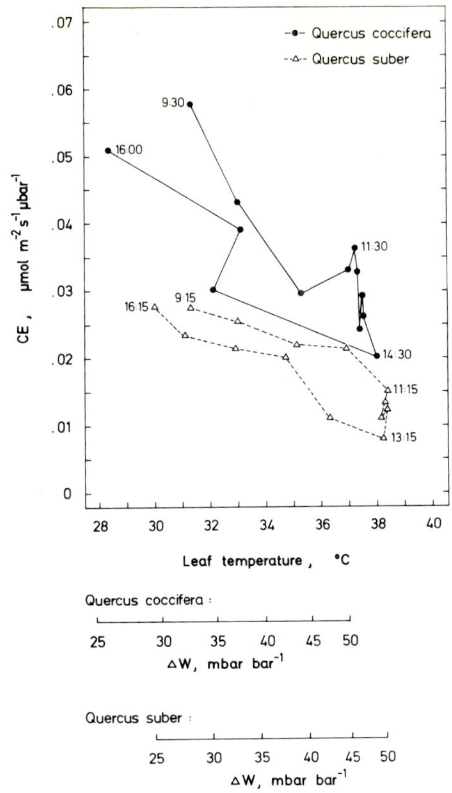

Fig. 13. Initial slope of the CO_2 response curve (carboxylation efficiency - CE) obtained at different times of day during summer drought with *Quercus coccifera* and *Quercus suber* plotted against leaf temperature and water vapor mole-fraction difference between leaf and air (ΔW)

Stomatal Conductance, G, and Leaf Internal CO_2 pressure, P_i

Regulation of stomatal conductance was similar in all species studied. Stomatal conductance is shown as affected by light intensity and leaf temperature in Fig. 14 during periods without water stress (winter and spring) and with moderate water stress at the end of July (mid-summer). With no water stress and at high light intensity, stomata remained relatively open at all temperatures. At 10° C and lower, this maximum conductance was only reached after very long periods (several hours). With the onset of dry summer conditions, maximum conductance decreased. Regulation of stomatal conductance by *Quercus coccifera* was similar to that of *Arbutus unedo* even though the degree of water stress to which *Arbutus* was subjected was much greater than that of *Quercus*. Increase in leaf temperature and in ΔW resulted in strong stomatal closure.

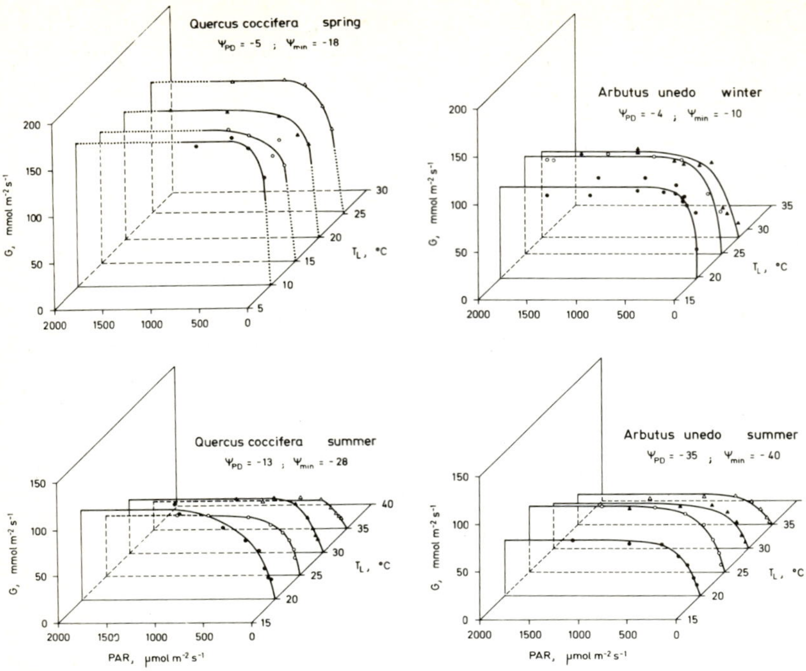

Fig. 14. The dependency of stomatal conductance, G, on incident PAR at a series of leaf temperatures within the range naturally experienced by *Quercus coccifera* and *Arbutus unedo* during the spring (April 28-May 5, 1983) when plants experienced no water stress and after 7 weeks of drought during mid-summer (July 31-August 3, 1983). Leaf predawn xylem potential, Ψ_{PD}, and minimum observed xylem potential, Ψ_{MIN}, as shown. Dew point of the air was 10° C. P_a = 350 μbar. O_2 partial pressure was 210 mbar. Solid curves were freely drawn and do not indicate regression fits to the data

Correlated change in NP and G under water stress conditions was such that leaf internal CO_2 pressure remained relatively constant in *Quercus suber* (Tenhunen *et al.* 1984a) and *Arbutus unedo* (Beyschlag 1984; Beyschlag *et al.* this volume). The apparent insensitivity of mesophyll photosynthetic capacity to moderate water stress in the case of *Quercus coccifera* along with stomatal closure (Tenhunen *et al.* 1985) resulted in decrease of P_i at times to as low as 150 μbar. If the momentary water potential is a measure of hydration of leaf material, then stomatal closure at early stages of water stress in *Quercus coccifera* along with deep rooting combine to hold leaf water content above a critical level (correlated with -30 bar Ψ_{PD}) for maintenance of mesophyll photosynthetic capacity. Under these hydration conditions, decrease of internal CO_2 apparently has no detrimental effect on mesophyll photosynthetic potential, e.g. *via* photoinhibitory processes (Björkman and Powles 1984). The susceptibility for photosynthetic damage, such as by photoinhibition, and the behavior of leaves with respect to maintenance of P_i levels may differ greatly between species that are differently suited to dry habitats.

CONCLUSION

The physiological adjustment of photosynthesis to summer drought in leaves of sclerophyll species is not fully understood. Nevertheless, the suggestions of researchers, who have postulated water stress induced changes in biochemical components and complex behavioral changes, are borne out by observations. In general, our studies in Portugal agree with other published results that show increasing water stress leading to a decrease in net photosynthesis and transpiration rates, a decrease in leaf mesophyll photosynthetic capacity, lower quantum efficiency, and lower leaf conductance. The results, however, indicate as well that species differ greatly in the degree to which these characteristic changes are expressed. Importantly, it is seen that the linkage between these behavioral components varies and that one change may take place independent of change in others.

The results outlined above identify areas which require further investigation. Most significantly, we must obtain further information on nitrogen investments by the plant in enzyme systems related to RuBP regeneration and to carboxylation and determine how these particular patterns in nitrogen investment affect water use efficiency. A second challenge is to determine the degree to which photoinhibitory processes affect production of sclerophylls and to compare the susceptibility of different species. A better understanding is required of the interrelationship between root function and aboveground behavior of leaves in gas exchange. Clearly the ability to tap underground water supplies must determine characteristics of regulation of leaf function as well as the plant's ability to avoid damage to the photosynthetic apparatus under drought stress.

REFERENCES

Braun-Blanquet J (1952) Les groupements végétaux de la France méditerranéenne. Centre National de la Recherche Scientifique, Montpellier

Beyschlag W (1984) Photosynthese und Wasserhaushalt von *Arbutus unedo* L. im Jahreslauf am Freilandstandort in Portugal. Gaswechselmessungen unter natürlichen Bedingungen und experimentelle Faktorenanalyse. PhD Thesis. University of Würzburg

Björkman O, Powles SB (1984) Inhibition of photosynthetic reactions under water stress: Interaction with light level. Planta 161:490-504

Evans JR (1983) Nitrogen and photosynthesis in the flag leaf of wheat (*Triticum aestivum* L). Plant Physiol 72:297-302

Farquhar GD, von Caemmerer S (1982) Modelling of photosynthetic response to environment. In: Lange OL, Nobel PS, Osmond CB, Ziegler H (eds), Encyclopedia of Plant Physiology, NS vol 12B: Physiological Plant Ecology II. Springer, Berlin-Heidelberg-New York-Tokyo pp 549-587

Gaastra P (1959) Photosynthesis of crop plants as influenced by light, carbon dioxide, temperature, and stomatal diffusion resistance. Medd Landbouwhogesch Wageningen 59:1-68

Harley PC, Tenhunen JD, Lange OL (1986) Modeling net photosynthesis in *Arbutus unedo* under field conditions. Oecologia (Berl) 70:393-401

Hellmers H, Horton JS, Juhren G, O'Keefe J (1955) Root systems of some chaparral plants in southern California. Ecology 36:667-678

Lange OL, Tenhunen JD, Beyschlag W (1985) Effects of humidity during diurnal courses on the CO_2- and light-saturated rate of net CO_2 uptake in the sclerophyllous leaves of *Arbutus unedo*. Oecologia (Berl) 67:301-304

Larcher W (1961) Zur Assimilationsökologie der immergrünen *Olea europaea* und *Quercus ilex* und der sommergrünen *Quercus pubescens* im nördlichen Gardaseegebiet. Planta 56:607-617

Montfort C, Hahn H (1950) Atmung und Assimilation als dynamisches Kennzeichen abgestufter Trockenresistenz bei Farnen und höheren Pflanzen. Planta (Berlin) 38:505-515

Oechel WC, Lawrence W, Mustafa J, Martinez J (1981) Energy and carbon allocation. In: Miller PC (ed) Resource use by Chaparral and Matorral. , Springer, Berlin-Heidelberg-New York, pp 151-183

Rambal S (1984) Water balance and pattern of root water uptake by a *Quercus coccifera* L. evergreen scrub. Oecologia (Berlin) 62:18-25

Raven JA, Glidewell SM (1981) Processes limiting photosynthetic conductance. In: CB Johnson (ed) Processes Limiting Plant Productivity. Butterworths, London, pp 289-301

Romose V (1940) Oekologische Untersuchungen über *Homalothecium sericeum*, seine Wachstumsperioden und seine Stoffproduktion. Dansk bot Ark 10:1-138

Stocker O (1954) Der Wasser- und Assimilationshaushalt südalgerischer Wüstenpflanzen. Ber dtsch bot Ges 67:289-299

Stocker O (1956) Wassermangel und Zellaktivität. In: W. Ruhland (ed), Springer Verlag. Handbuch der Pflanzenphysiologie, Bd. II, Berlin-Göttingen-Heidelberg, pp 639-654

Schulze E-D, Hall AE (1982) Stomatal responses, water loss and CO_2 assimilation rates of plants in contrasting environments. In: Lange OL, Nobel PS, Osmond CB, Ziegler H (eds), Encyclopedia of Plant Physiology, NS vol 12B: Physiological Plant Ecology II. Springer, Berlin-Heidelberg-New York-Tokyo pp 181-230

Tenhunen JD, Lange OL, Gebel J, Beyschlag W, Weber JA (1984a) Changes in photosynthetic capacity, carboxylation efficiency, and CO_2 compensation point associated with midday stomatal closure and midday depression of net CO_2 exchange of leaves of *Quercus suber*. Planta 162:193-203

Tenhunen JD, Meister HP, Caldwell MM, Lange OL (1984b) Environmental constraints on productivity of the Mediterranean sclerophyll shrub *Quercus coccifera*. Proceedings of INTECOL workshop - Rates of Natural Primary Productivity and Agricultural Production. Options méditerranéennes (Instituto Agronomico Mediterraneo de Zaragoza) 84/1:33-53

Tenhunen JD, Lange OL, Harley PC, Beyschlag W, Meyer A (1985) Limitations due to water stress on leaf net photosynthesis of *Quercus coccifera* in the Portuguese evergreen scrub. Oecologia (Berl) 67:23-30

Walter H (1931) Die Hydratur der Pflanze. Jena, Gustav Fischer

Seasonal and diurnal patterns in leaf gas exchange of *Phillyrea angustifolia* growing in Portugal

P.C. Harley, J.D. Tenhunen, O.L. Lange* and W. Beyschlag*

Systems Ecology Research Group, San Diego State University, San Diego, CA 92182 USA

INTRODUCTION

Weber *et al.* (1985) and Tenhunen *et al.* (this volume) have proposed that net photosynthesis (NP) may be described empirically as a function of the leaf internal partial pressure of carbon dioxide (P_i), according to the relationship,

$$NP = \frac{CE \cdot Pi \cdot Pm}{(P_m{}^2 + CE^2 \cdot P_i{}^2)^{1/2}} - R_d \qquad (1)$$

where CE, carboxylation efficiency, is the initial slope of the NP vs. P_i response curve under saturating irradiance; P_m, photosynthetic capacity, is the CO_2-saturated rate of assimilation; and R_d is dark respiration.

Tenhunen *et al.* (1984a) demonstrated that CE and P_m are strongly correlated and that changes in both may explain diurnal and seasonal variation in NP in the evergreen sclerophyll *Quercus coccifera.* They proposed a hypothetical scheme, reproduced in the lower half of Fig. 1, illustrating changes in CE over the annual cycle as the result of leaf phenology, water stress and both high and low temperature extremes.

Because CE assumes light saturation, they defined a new parameter, MF, or marginal CO_2 fixation, as the slope of the NP vs. P_i response curve between P_i values of 150 and 300 µbar, i.e. in the normal operating range of C-3 plants, at any values of irradiance and leaf temperature. Shown in the upper panel of Fig. 1 are typical diurnal time courses of MF in *Q. coccifera* during different seasons under clear weather conditions. Note that if P_i remains approximately constant through the course of the day (Wong *et al.* 1979; Beyschlag *et al.* this volume), MF is roughly proportional to NP, and the diurnal patterns of NP will be similar in shape to those of MF. We sought to determine whether the hypothetical

*Lehrstuhl für Botanik II, Universität Würzburg, Mittlerer Dallenbergweg 64, D-8700 Würzburg, West Germany

NATO ASI Series, Vol. G15
Plant Response to Stress. Edited by J.D. Tenhunen et al.

patterns presented in Fig. 1 apply to a co-occurring sclerophyllous shrub, *Phillyrea angustifolia.*

METHODS

The data reported here were collected in a natural macchia at Quinta São Pedro, Sobreda, Portugal, about 15 km south of Lisbon, between September 1982 and September 1983. Branches of *Phillyrea* were enclosed in a temperature- and humidity-controlled cuvette (Walz Mess- und Regeltechnik, Effeltrich, FRG) which was part of a mobile gas-exchange laboratory. Methods for measuring both photosynthesis and transpiration were essentially those of Schulze and Küppers (1979). To obtain the diurnal courses of gas exchange reported here, air temperature and relative humidity within the cuvette tracked ambient conditions. Leaf xylem water potentials were measured with a pressure chamber.

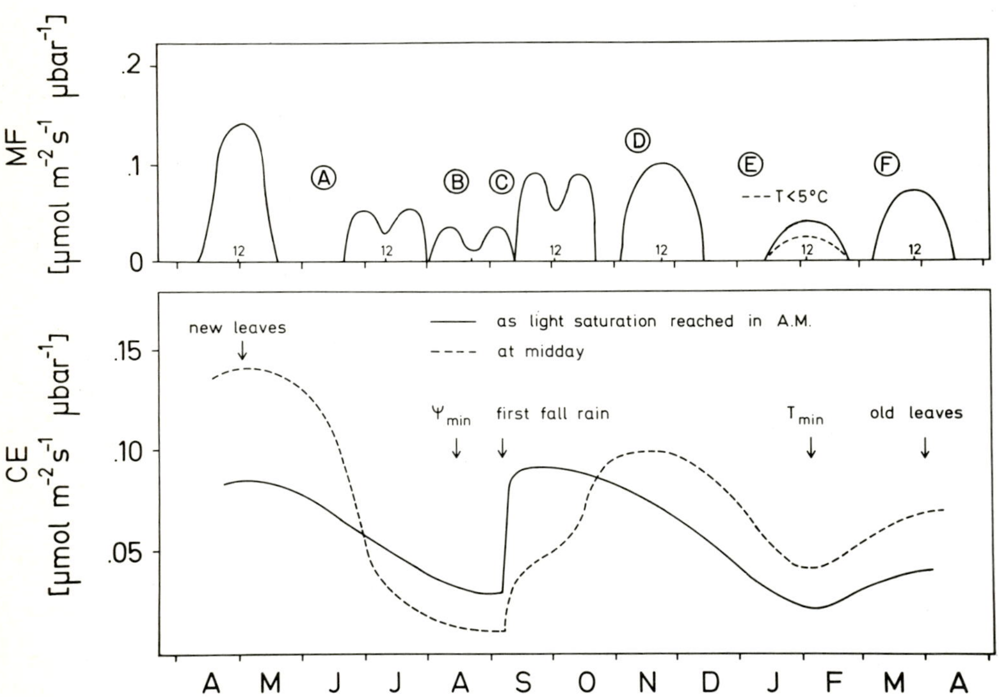

Figure 1. Hypothetical scheme illustrating changes in carboxylation efficiency (CE) over an annual cycle and in marginal CO_2 fixation (MF) during different seasons for leaves of mediterranean sclerophyll shrubs. *Solid line* for CE is the value as light saturation is reached in the morning; *dashed line* is the value of CE at midday. The *dashed line* time course in February indicates response to low temperature (from Tenhunen *et al.* 1984a). *Circled letters* (A-F) in top panel indicate positioning of time courses in Fig. 2 on annual scale

RESULTS

During 1982-83, diurnal responses of NP under naturally occurring conditions were obtained on *Phillyrea angustifolia* at different seasons (Fig. 2, A-F). Also shown in Fig. 2 are measured leaf temperature (T) and incident photosynthetically active radiation (PAR), as well as the daily maxima (Ψ_{pd}) and

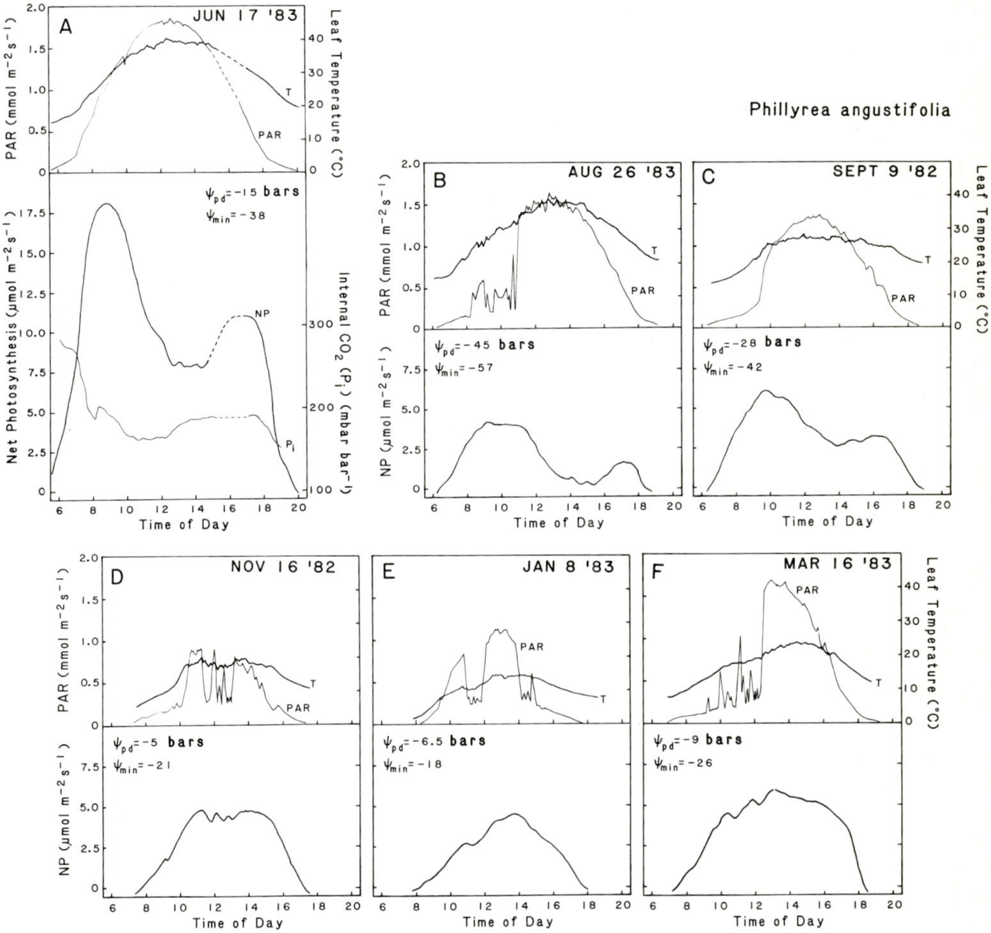

Fig. 2. Six diurnal responses of net photosynthesis (NP) and leaf temperature (T) in *Phillyrea angustifolia*. Also shown are the daily course of incident photosynthetically active radiation (PAR) and the pre-dawn (Ψ_{pd}) and midday (Ψ_{min}) leaf xylem water potentials. Data for June 17 include daily pattern of internal partial pressure of CO_2 (P_i)

minima (Ψ_{min}) of leaf xylem water potential. The *encircled letters* (A-F) in the upper panel of Fig. 1 place each of these diurnal responses in their approximate position on the annual time scale.

The annual pattern of measured diurnal responses of NP in *Phillyrea* is similar to the hypothetical pattern of MF in *Quercus*. New leaves are produced in *Phillyrea* in early May, and rates measured on young leaves in mid-June were much higher than those obtained at any other time, though midday depression of photosynthesis was already apparent. It seems likely that measurements obtained in late May or early June, before leaf temperatures exceeded 35° C, would have resulted in a single-peaked diurnal response with high rates, as suggested for *Q. coccifera* in Fig. 1.

Data are lacking for the mid-summer period, but midday depression with a small afternoon recovery is still apparent in late August 1983, with Ψ_{pd} = -45 bar and leaf temperatures approaching 40° C. Maximum rates, however, were less than 5 μmol m^{-2} s^{-1}. In late August 1982, a single rainfall event occurred, ending the summer drought and recharging soil moisture. Thus, on 9 September 1982, with Ψ_{pd}=-28 bar and leaf temperature below 30° C, maximum rates exceeded those during the extreme stress of August 1983 and midday depression of NP was less, resulting in substantially greater integrated daily carbon gain. Soil moisture content continued to increase with fall rains in 1982, resulting in Ψ_{pd} values >-10 bar through March 1983. This, coupled with leaf temperatures below 20° C led to single-peaked diurnal responses throughout the fall and winter, with maximum rates apparently limited primarily by low temperatures. No obvious effect of low nighttime temperatures on rates the following day, suggested by Tenhunen *et al.* (1984a) for *Quercus*, was apparent.

Some insight into the nature of the midday decline in NP during the summer months may be gained by comparing three consecutive days in June (Fig. 2A; Fig. 3), in which the degree of midday depression was successively reduced. As seen for all three days, stomatal closure resulting in lower values of P_i cannot explain the midday decline, as P_i decreased only slightly as NP fell sharply. Thus, CE must decline during this period, as suggested by Tenhunen *et al.* (1984b, 1985), with a simultaneous decrease in conductance. A careful comparison of June 17 and June 18 reveals that leaf temperature on June 18 fell several degrees below that on June 17, beginning about noon, just as NP began to recover. And on June 19, when only a slight midday decline in NP occurred, morning fog and a cooling trend kept leaf temperatures approximately 5° C below those of the previous two days. It is thus strongly suggested that high leaf temperatures (>30° C), or accompanying high values of leaf to air water vapor pressure deficit (Δw>35 mbar bar^{-1}) are associated with this phenomenon, as previously suggested for *Quercus suber* (Tenhunen *et al.* 1984b) and *Arbutus unedo* (Lange *et al.* 1985). Note, however, that midday depression occurred in September

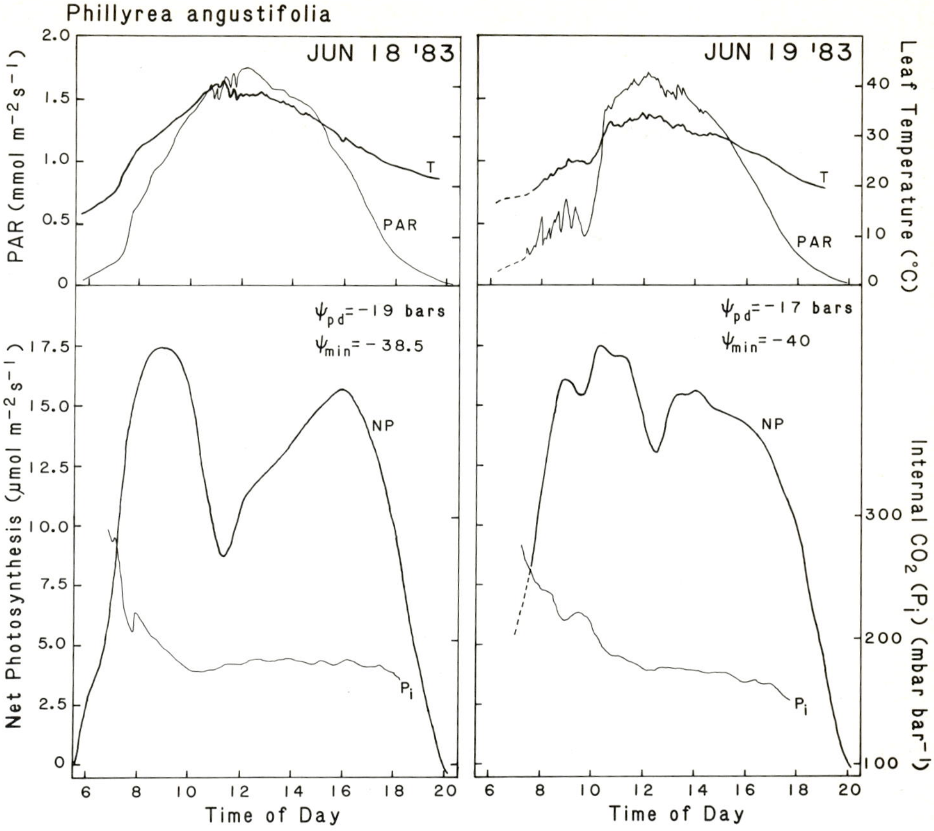

Fig. 3. Two diurnal responses of net photosynthesis (NP), leaf temperature (T) and internal partial pressure of CO_2 (P_i) in *Phillyrea angustifolia.* Also shown are the daily course of incident photosynthetically active radiation (PAR) and pre-dawn (Ψ_{pd}) and midday (Ψ_{min}) leaf xylem water potentials

despite leaf temperatures below 30° C, indicating a more complicated causal mechanism, perhaps involving an interaction between leaf temperature (or Δw) and leaf water potential or associated with a hormonally-mediated signal from a stressed root system (Davies *et al.* this volume).

Further evidence that stomatal conductance and decreasing P_i are not solely responsible for midday depression of NP is shown in Fig. 4, in which the external partial pressure of CO_2 was maintained at the saturating level of 2500 μbar. At 6:30 am, air temperature and irradiance were increased in a single step to 38° C and 0.8 mmol (photons) m^{-2} s^{-1}, respectively, and then held constant for the remainder of the day. Despite saturating P_i, NP fell very sharply after reaching a

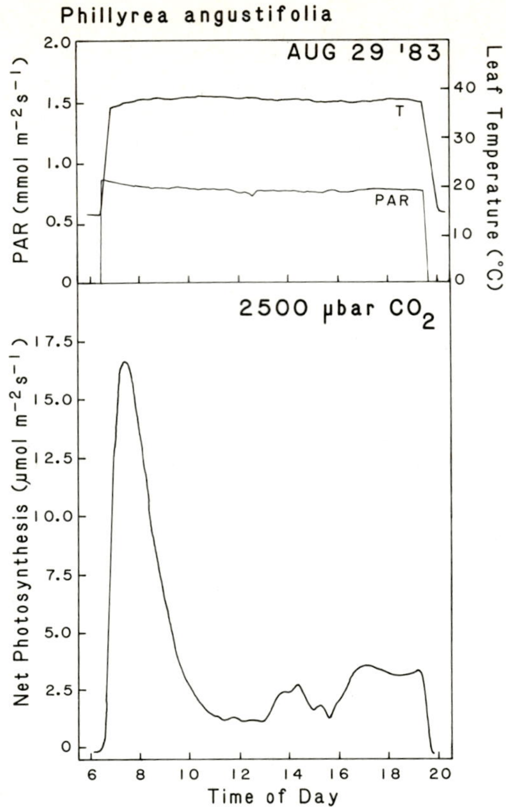

Fig. 4. Diurnal pattern of net photosynthesis (NP) in *Phillyrea angustifolia* under CO_2-saturating conditions (2500 µbars CO_2 entering cuvette). Leaf temperature (T) and photosynthetically active radiation (PAR) were held constant at approximately 38° C and 0.8 mmol m^{-2} s^{-1}, respectively.

high morning peak. The slight recovery during the afternoon, in spite of constant air temperature and Δw, provides further evidence that changes in leaf temperature and Δw alone are not sufficient to explain the observed midday depression of NP or its subsequent recovery (Lange *et al.* 1985; Tenhunen *et al.* 1985).

Finally, a comparison of the high early morning rates attained in Fig. 4 at saturating P_i (≈16.5 µmol m^{-2} s^{-1}) with those in Fig. 2B (≈4.0 µmol m^{-2} s^{-1}) demonstrates that there is a large excess of photosynthetic capacity in *Phillyrea* at this time of year, suggesting that the leaves are operating well down on the initial linear portion of the NP vs P_i response curve, as is the case for the co-occurring sclerophylls *Quercus coccifera* (Tenhunen *et al.* 1985), *Quercus suber* (Tenhunen *et al.* 1984b) and *Arbutus unedo* (Beyschlag 1984; Harley *et al.* 1986), and that stomatal limitation of NP may be quite significant.

DISCUSSION

Lange *et al.* (1982) subjected 10 species of mediterranean sclerophyll to simulated summer day conditions in the laboratory, and concluded that *Phillyrea angustifolia* was among those most sensitive to atmospheric stress, exhibiting midday stomatal closure and afternoon recovery. The data reported here confirm that midday stomatal closure, accompanied by depression of NP, occurs in *Phillyrea* under field conditions, apparently allowing the plant to regulate water loss through the summer drought by reducing transpiration during periods of maximal evaporative demand.

In general, midday depression has been shown to involve simultaneous changes in both stomatal conductance and CE, such that P_i remains relatively constant. For both *Quercus suber* (Tenhunen *et al.* 1984b) and *Q. coccifera* (Tenhunen *et al.* 1985), decreases in CE during midday and subsequent increases during afternoon recovery have been measured directly. Resemann and Raschke (1984) obtained similar results with *Arbutus*. Simultaneous changes in conductance and CE may also be inferred from large changes in NP accompanied by relatively small changes in P_i, as found for *Phillyrea* in this study, and for *Arbutus unedo* under field conditions (Beyschlag this volume). Tenhunen *et al.* (1984b) suggested that this decline in CE, by allowing P_i to remain high as conductance drops, may reduce the possibility of photoinhibitory damage to the photosynthetic apparatus (Osmond and Björkman 1972).

The cause of these reversible changes in photosynthetic characteristics, manifested as changes in CE, is unknown, but turgor related changes in cell volume, affecting intercellular solute concentrations (Kaiser 1982 this volume), and the development of localized areas of water stress within the leaf, independent of bulk leaf water potential (Sharkey 1984) have been suggested. In addition, abscisic acid has been shown to affect both stomatal conductance and photosynthetic characteristics such that P_i may remain constant (Raschke and Hedrich 1985).

Comparison of these results with similar data for *Arbutus unedo* and *Quercus coccifera* suggests that *Phillyrea* is a relatively successful drought tolerator. During the summer of 1983, values of Ψ_{pd} fell from -15 to -45 bar, values almost identical to those reported during the same period for *Arbutus* (Beyschlag 1984). *Quercus coccifera*, perhaps having access to deeper water supplies, declined from about -12 to -30 bar. Between June and September, the maximum daily rate of NP in *Arbutus* declined linearly with Ψ_{pd}, from about 6 μmol m^{-2} s^{-1} to less than 2 (Beyschlag 1984). Tenhunen *et al.* (1985) reported that total daily CO_2 assimilation in *Q. coccifera* was twice that of *Arbutus* under moderate stress conditions and 1.6 times higher under severe drought, and concluded that *Q. coccifera* was better adapted to arid conditions than *Arbutus*, the range of which is confined to more coastal and maritime regions than is that of *Quercus*, which also occupies much more arid interior habitats. *Phillyrea* co-occurs with *Q. coccifera* over much of its range, and also seems to be more drought tolerant than *Arbutus*, at least on the basis of these limited data. Under severe stress, on August 26, *Phillyrea* maintains a substantial positive carbon balance, similar to that reported for *Quercus*, despite considerably lower water potentials, and even manages a slight afternoon recovery.

In summary, these results confirm for a second mediterranean sclerophyll genus the behavior hypothesized by Tenhunen *et al.* (1984a). Periods of minimal soil water stress and relatively low atmospheric stress are characterized by single-peaked diurnal photosynthetic responses, in which photosynthesis is limited by irradiance and temperature. New growth is characterized by extremely high rates of NP. With the onset of summer drought, the daily course of NP is characterized by more and more severe midday depression, as NP and transpiration are increasingly confined to periods of lower evaporative demand, until fall rains recharge soil moisture, completing the cycle. Within this general pattern, one expects to find species-specific differences in drought tolerance, manifested in the extent to which photosynthetic capacity and daily maxima of NP are affected by decreasing soil mositure, and in the severity of midday depression. These data suggest that *Phillyrea* is relatively tolerant, maintaining a substantial positive carbon balance under conditions of severe soil and atmospheric stress. It remains unclear to what extent the very large drop in photosynthetic capacity between mid-June and late August is the result of leaf phenology effects and the development of sclerophylly and to what extent it is due to increasing levels of water stress.

REFERENCES

Beyschlag W (1984) Photosynthese und Wasserhaushalt von *Arbutus unedo* L. im Jahreslauf am Freidandstandort in Portugal. Gaswechselmessungen unter natürlichen Bedingungen und experimentelle Faktorenanalyse. Thesis, Universität Würzburg

Harley PC, Tenhunen JD, Lange OL (1986) Use of an analytical model to study limitations to net photosynthesis in *Arbutus unedo* under field conditions. Oecologia (Berl) in press

Kaiser W (1982) Correlation between changes in photosynthetic activity and changes in total protoplast volume in leaf tissue from hygro-, meso-, and xerophytes under osmotic stress. Planta 154:538-545

Lange OL, Tenhunen JD, Beyschlag W (1985) Effects of humidity during diurnal courses of the CO_2- and light-saturated rate of net CO_2 uptake in the sclerophyllous leaves of *Arbutus unedo*. Oecologia (Berl) 67:301-304

Lange OL, Tenhunen JD, Braun M (1982) Midday stomatal closure in mediterranean type sclerophylls under simulated habitat conditions in an environmental chamber. I. Comparison of the behaviour of various European mediterranean species. Flora 172:563-579

Raschke K, Hedrich R (1985) Simultaneous and independent effects of abscisic acid on stomata and the photosynthetic apparatus in whole leaves. Planta 163:105-118

Resemann A, Raschke K (1984) Midday depression in stomatal and photosynthetic activity of leaves of *Arbutus unedo* are caused by large water-vapor pressure differences between leaf and air. Plant Physiol 75 (Suppl):66

Schulze E-D, Küppers M (1979) Short-term and long-term effects of plant water deficits on stomatal response to humidity in *Corylus avellana* L. Planta 146:319-326

Sharkey T (1984) Transpiration-induced changes in the photosynthetic capacity of leaves. Planta 160:143-150

Tenhunen JD, Lange OL, Gebel J, Beyschlag W, Weber JA (1984b) Changes in photosynthetic capacity, carboxylation efficiency and CO_2 compensation point associated with midday stomatal closure and midday depression of net CO_2 exchange of leaves of *Quercus suber*. Planta 162:193-203

Tenhunen JD, Lange OL, Harley PC, Beyschlag W, Meyer A (1985) Limitations due to water stress on leaf net photosynthesis of *Quercus coccifera* in the Portuguese evergreen scrub. Oecologia (Berl) 67:23-30

Tenhunen JD, Meister HP, Caldwell MM, Lange OL (1984a) Environmental constraints on productivity of the Mediterranean sclerophyll shrub *Quercus coccifera*. Proceedings of INTECOL workshop - Rates of Natural Primary Productivity and Agricultural Production. Options méditerranéennes (Instituto Agronomico Mediterraneo de Zaragoza) 84/1:33-53

Weber JA, Jurik TW, Tenhunen JD, Gates DM (1985) Analysis of gas exchange in seedlings of *Acer saccharum*: integration of field and laboratory studies. Oecologia (Berl) 65:338-347

Wong SC, Cowan IR, Farquhar GD (1979) Stomatal conductance correlates with photosynthetic capacity. Nature 82:424-426

A model of net photosynthesis for leaves of the sclerophyll *Quercus coccifera*

J.D. Tenhunen, P.C. Harley, W. Beyschlag* and O.L. Lange*

Systems Ecology Research Group, San Diego State University, San Diego, California 92182 USA

INTRODUCTION

In the past decade, a considerable research effort has been invested in defining single factor dependencies of leaf net photosynthesis mathematically and in terms of physiologically meaningful parameters (Tenhunen *et al.* 1980a; Farquhar and von Caemmerer 1982; Harley *et al.* 1985, 1986; Weber *et al.* 1985). The single factor dependencies have been combined to define leaf CO_2 gas exchange rate as a function of incident light intensity, leaf temperature, and leaf internal CO_2 partial pressure (Tenhunen and Westrin 1979). One set of information which allows description of CO_2 uptake by leaves in the natural environment includes 1) the temperature dependency of net photosynthesis rate at light and CO_2 saturation (P_{ML}), 2) light dependencies of net photosynthesis rate at CO_2 saturation (P_M) at a series of temperatures, 3) light and temperature dependencies of leaf carboxylation efficiency CE - the initial slope of the net photosynthesis response (NP) to internal leaf air space CO_2 pressure and 4) the temperature dependency of dark respiration (R_d - assumed to continue in the light). While data describing all four of these is essential, the fundamental dependency in an appropriate photosynthesis model is the leaf response to internal CO_2 pressure. The other dependencies are utilized to describe variation in those parameters which establish an appropriate response to internal CO_2 pressure for particular light and temperature conditions prevailing at any point in time.

In several studies of net photosynthesis under fluctuating environmental conditions with leaves of *Prunus armeniaca* (Tenhunen *et al.* 1980b), *Glycine max* (Tenhunen *et al* . 1980c), *Acer saccharum* (Weber *et al.* 1985), and *Arbutus unedo* (Harley *et al.* 1986), we demonstrated clearly that descriptions of this sort could be used to effectively describe daily time courses of net photosynthesis in the natural habitat and that they allowed a more complete analysis of such data than was previously possible. In the present paper, this work is extended and a description of leaf net photosynthesis is reported for *Quercus coccifera*. The model proved to be applicable over the greater part of an annual cycle in vegetation development, with little change in leaf mesophyll photosynthetic

*Lehrstuhl für Botanik II, Universität Würzburg, Mittlerer Dallenbergweg 64, D-8700 Würzburg, West Germany

NATO ASI Series, Vol. G15
Plant Response to Stress. Edited by J. D. Tenhunen et al.

characteristics occurring except under conditions of extreme drought and cold stress. Since the net photosynthesis rate of a leaf is known when we have determined the operating point on the mesophyll CO_2 response curve (see also Lange *et al.* this volume) and the CO_2 response may be described in terms of the leaf maximum photosynthetic capacity (P_M) and carboxylation efficiency (CE), the problem involves description over time of natural variation in these parameters. A solution requires a knowledge of leaf conductance for CO_2 to the mesophyll as well. In the present paper, the photosynthesis model was applied where leaf conductance was simultaneously measured along with leaf CO_2 uptake rate.

MATERIALS AND METHODS

The experiments were conducted as described by Tenhunen *et al.* (1985) between September 1982 and September 1983 at the Quinta São Pedro Research Station near Lisbon, Portugal in natural scrub vegetation. Shrubs of *Quercus coccifera* chosen for experimental study were approximately 1 to 1.5 m in height. Branch ends with several leaves were enclosed in gas exchange cuvettes (Walz Meß- und Regeltechnik, Effeltrich, FRG) which were part of a mobile laboratory. The system was controlled to track outside air temperature or a set value was imposed. Dew point of the air supplied to the cuvettes was maintained close to that occurring naturally. Leaf material was illuminated naturally or with nine quartz halide projector lamps. Gas mixtures with partial pressures of CO_2 between 0 and 3000 μbar were supplied to the cuvettes from a mixing system. Experiments were conducted 1) to establish light, temperature, and CO_2 responses of net photosynthesis rate and stomatal conductance and 2) to observe typical diurnal time courses of leaf gas exchange. All gas exchange rates are expressed on a leaf area basis. Model parameters such as P_M and CE were determined from single factor dependencies and used to complete the mathematical description of mesophyll potential photosynthesis (NP_{pot} - see Tenhunen *et al.* 1984a; also Tenhunen *et al.* this volume) as described in detail in the appendix. Observed diurnal courses of incident photosynthetically active radiation (PAR), leaf temperature (T_L), partial pressure of CO_2 in the gas exchange chamber (P_a), and leaf conductance (G) on a four minute basis were used as input to the model to predict diurnal courses of net CO_2 uptake.

RESULTS

General Experimental Approach

Experiments were carried out during all seasons of the year. Nevertheless, a comprehensive plan of experimentation, designed to elaborate typical single factor dependencies of photosynthesis as well as characterize natural leaf behavior in gas exchange over diurnal courses, requires a period of several weeks for completion. Periods during which more rapid changes in photosynthetic characteristics of leaves occur, due to either phenological events or exposure of leaves to extreme stress, still present difficulties for such an analysis with presently available methods. The results reported in this paper, therefore, describe mesophyll photosynthesis for leaves of *Quercus coccifera* during mild fall and winter conditions which occurred in October through December 1982 and for warm summer conditions without extreme water stress which occurred between May and August of 1983. During these two periods, mesophyll photosynthetic characteristics remained constant over relatively long periods of time. The photosynthesis model derived from information obtained during these periods of stable response, however, was utilized to examine those changes in mesophyll photosynthesis that occur during shorter periods, i.e. the model provides a baseline of information on rates of mesophyll photosynthesis which may be compared to measured rates over the short-term which are higher or lower than expected.

As illustrated by Lange *et al.* (this volume) and described further by Tenhunen *et al.* (1984a; see also in this volume), carbon dioxide uptake by the leaf (NP) may be understood as depending on 1) the carbon dioxide response curve of the leaf mesophyll and 2) the actual leaf air space carbon dioxide partial pressure (P_i). The carbon dioxide response of the leaf mesophyll may be described mathematically in terms of its initial slope or carboxylation efficiency (CE) and the carbon dioxide saturated rate of carbon dioxide uptake or photosynthetic capacity (P_M).

The form of the equation describing NP as affected by P_i is:

$$NP = \frac{CE \ P_i}{(1 + CE^2 \ P_i^2 / P_M^2)^{1/2}} - R_d \qquad (1)$$

where R_d is the rate of dark respiration (see Weber *et al.* 1985 for a more detailed explanation). As described below, at less than saturating light intensity CE is replaced by MF, the marginal CO_2 fixation efficiency. A model of net photosynthesis of leaves of any species is obtained by defining variation of CE, P_M, and R_d in response to daily and seasonal changes in atmospheric environmental factors (such as light, temperature, and humidity) and to plant physiological state (as determined by phenological events or stress).

Variations in P_M

Variation in P_M during any season may be studied by observing light response curves of photosynthesis at different temperatures and saturating carbon dioxide as well as temperature response curves at saturating light and carbon dioxide. Temperature responses of P_{ML} (the special set of P_M values obtained at light saturation) obtained with *Quercus coccifera* during different times of year are illustrated in Fig. 1. Measurements conducted in the absence of water stress between October 1982 and February 1983 (curve A) indicated no seasonal shift in photosynthetic capacity during these months. The temperature optimum was above 30° C. New leaves developed in March of 1983. Measurements on mature new leaves in the late spring (May) and midsummer (August) of 1983 (*curve* B) suggest that these leaves may have had a slightly lower photosynthetic capacity than the previous year's leaves. The temperature optimum was not observed during measurements conducted in the spring, but in midsummer it was near 30° C. Below 30° C, rates obtained in spring and midsummer did not differ. The temperature dependency of P_{ML} for winter conditions between October 1982 and February 1983 is described in the appendix with an Arrhenius function (*curve* A). For summer conditions before extreme water stress occurred, this dependency (*curve* B) is described according to an equation describing activation and denaturation of an enzymatically regulated process (see Tenhunen *et al.* 1976; Johnson *et al.* 1942).

The value of P_M decreases as light intensity decreases below saturating levels. This is illustrated in the three-dimensional response surfaces presented for *Quercus coccifera* and *Arbutus unedo* (Tenhunen *et al.* this volume), where P_M is affected by light intensity and temperature . As shown in the appendix and discussed by Tenhunen *et al.* (1976), this dependency may be described according to an empirical equation of Smith (1937). The convenience of this equation is that one need only specify the value of P_{ML} for a particular temperature (as in Fig. 1) and the value of the initial slope of the light response of photosynthesis at CO_2 saturation (α - the maximum efficiency of incident light energy conversion).The parameter α was measured to be 60 in October 1982, decreased slightly to 40 in December 1982, and was determined as 45 mmol CO_2/mol photons during the summer period in 1983.

Variations in CE

As in the case of P_M, variation in the slope of the CO_2 response curve must be considered as affected by light and temperature. In the case of temperature, the dependency of CE during any season may be determined from carbon dioxide response curves of photosynthesis (based on leaf internal CO_2 pressure) obtained

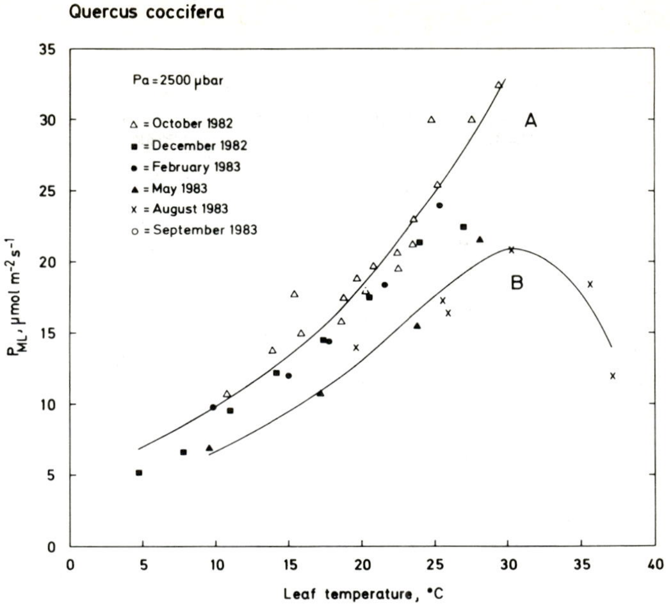

Fig. 1. Dependence of CO_2- and light-saturated net photosynthesis rate (P_{ML}) of *Quercus coccifera* on leaf temperature at different times of year (*symbols shown*). *Solid lines* indicate regressions fit to the data obtained between October 1982 and February 1983 from leaves which developed in the spring of 1982 (A) and in May and early August from leaves which developed in March 1983 (B). From Tenhunen *et al.* (1985)

at saturating light and several temperatures. Leaf temperatures varied between approximately 0 and 25° C during the winter at our study site in Sobreda near Lisbon. Increasing temperature increased CE within this temperature range and stimulated leaf carbon dioxide uptake. During the summer, increase in temperature between 0 and 25° C similarly led to an increase in CE. At midday, however, leaf temperatures were commonly in the range of 28 to 40° C. Increase in leaf temperature in this range decreased both CE (Tenhunen *et al.* 1985) and P_M in leaves of *Quercus coccifera* and decreased carbon dioxide uptake (see also Tenhunen *et al.* 1984a, b; Beyschlag 1984). The temperature dependency of CE for winter conditions between October 1982 and February 1983 is described in the photosynthesis model in the appendix with a linear function. For summer conditions before extreme water stress occurred, this dependency is described with two linear equations, allowing first and increase to an optimum CE and then a decrease. Available information was not adequate to describe change in CE near the temperature optimum in a more detailed fashion.

The slope of the carbon dioxide response of net photosynthesis decreases as light intensity decreases below saturating levels in much the same manner as does P_M (Weber *et al.* 1986). Since the term carboxylation efficiency technically refers only to the slope of the response at light saturation, we have referred to the slope of the NP versus P_i response curve estimated from determinations of NP as P_i varies between 150 and 300 μbar CO_2 (in the region relevant to leaf function in normal air) and with any constant light intensity and temperature as the marginal CO_2 fixation efficiency, MF (Tenhunen *et al.* 1984a). Changes in MF with light intensity were not determined with *Quercus coccifera*. In the present model, MF was assumed to decrease with decreasing light intensity as found for *Arbutus unedo*, a sclerophyllous species which has been investigated in detail (Weber *et al*. unpublished). Thus, with increase in PAR, MF initially increases linearly. Above approximately 500 μmol m^{-2} s^{-1} PAR it remains constant and is equal to CE. This dependency is described in the appendix according to the empirical equation of Smith (1937) using the parameters CE and β (initial increase in MF with increase in PAR).

Leaf Respiration Rate, R_d

Leaf mitochondrial respiration (dark respiration rate - R_d) was assumed to continue in the light without change (see Farquhar and von Caemmerer 1982 and Harley *et al.* 1986 for discussion). Leaf respiration rates were measured in the dark at a series of temperatures between 5° C and 40° C. Rates were similar in all seasons. The data were pooled and described with the Arrhenius equation as indicated in the appendix.

Diurnal Courses of Gas Exchange

Typical diurnal courses of change in atmospheric environmental factors, rates of leaf gas exchange, total leaf conductance for water vapor, and leaf water potential are shown for the winter and summer periods in Figs. 2 and 3. On December 18, 1982 (Fig. 2), incident photosynthetically active radiation (PAR) remained below 1000 μmol m^{-2} s^{-1} and exceeded 400 μmol m^{-2} s^{-1} only for short periods. Passing clouds decreased PAR to approximately 100 μmol m^{-2} s^{-1}, well below saturation for NP. As a result of these decreases in PAR, leaf temperatures (T_L) decreased several degrees. Based on other data, small changes in T_L would have little direct affect on NP. Stomatal response was sluggish below 15° C and leaf conductance, G, changed little in response to changes in PAR. Nevertheless, transpiration rate, Tr, fluctuated with changes in water vapor mole-fraction difference between leaf air spaces and external chamber air, ΔW. The decreases in NP, accompanied by little change in G, led to 50 μbar changes in P_i during cloudy periods (see also Beyschlag *et al.* this volume). Although predawn water potential was high (approximately -6 bar) and transpiration rate was low, leaf

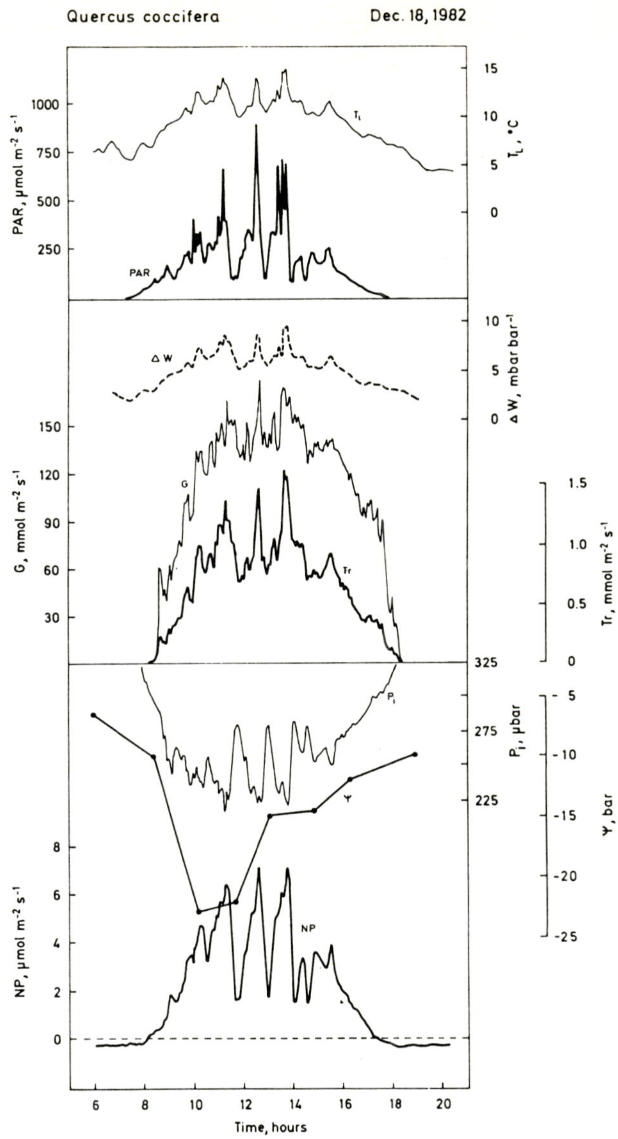

Fig. 2. Diurnal time course of gas exchange of *Quercus coccifera* leaves observed on December 18, 1982. Shown are the naturally occurring time courses for photosynthetically active radiation flux (PAR) incident on the leaves in the gas exchange cuvette, leaf temperature (T_L), water vapor mole-fraction difference between leaf air spaces and external chamber air (ΔW), total leaf conductance for water vapor (G), leaf transpiration rate (Tr), leaf net photosynthesis rate (NP), internal leaf CO_2 pressure (P_i), and leaf xylem water potential (Ψ)

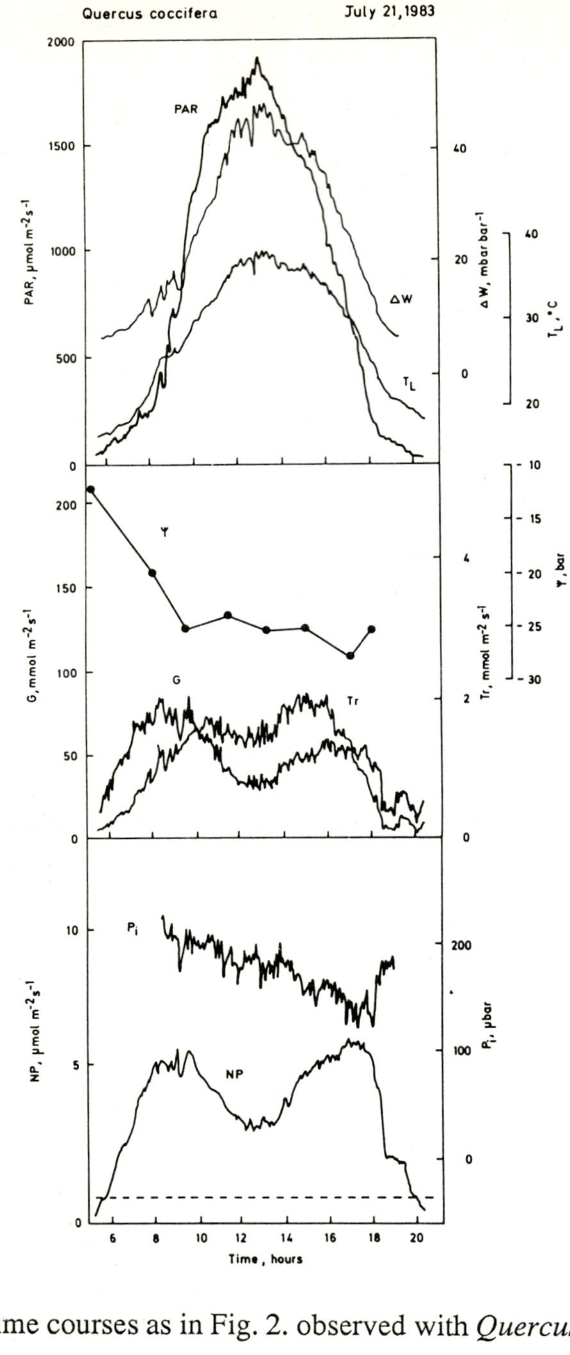

Fig. 3. Diurnal time courses as in Fig. 2. observed with *Quercus coccifera* on July 21, 1983

water potential decreased to -25 bar in the late morning, suggesting substantial resistance for resupply of water to the leaves from the cold soil.

While highest rates of CO_2 uptake were observed in the late spring (see Tenhunen *et al.* this volume), rates of NP measured on July 21, 1983 under mid-summer conditions (Fig. 3) were roughly the same as those observed on the mild winter day depicted in Fig. 2. CE is approximately the same during these periods, leaf temperature being well above optimum for CE on July 21. While ΔW was approximately four times higher on July 21 than on December 18 due to high leaf temperature and low external humidity, Tr was only twice as high due to a decrease in maximum leaf conductance that occurred immediately with the onset of dry atmospheric conditions in June 1983 and due to stomatal closure in response to increase in T_L above 30° C and ΔW above 30 mbar bar^{-1} at midday. The low leaf conductance for CO_2 coupled with a relatively high mesophyll photosynthetic capacity results in a decrease in P_i to as low as 150 μbar during the afternoon on July 21 (for further discussion see Tenhunen *et al.* 1985). Despite six weeks of drought, leaf water potential did not decrease much below values observed under mild winter conditions.

Simulation of NP Time Courses

Measured values of PAR, T_L, P_a, and G as shown in Figs. 2 and 3 were input on a four-minute basis to the model as described in the appendix. The parameter values P_{ML}, P_M, CE, α, β, and R_d varied as a function of light and temperature as has been described above. NP and P_i were determined iteratively at each point in time. To do this, the result of equation 1 assuming a known value of P_i was checked for agreement with an estimate of Pi using Fick's Law, P_a, the leaf conductance G, and the predicted value of NP (see appendix). P_i was initially set at 100 μbar and then increased in steps of 5 μbar until the value input to the model and that calculated from Fick's Law were within 5 μbar of each other. Simulated and measured courses for net CO_2 uptake are compared for December 18, 1982 and July 21, 1983 in the lower part of Figs. 4 and 5. Diurnal courses for observed and calculated net photosynthesis rates were in agreement during both winter and summer periods. The simulated diurnal courses for the light and carbon dioxide saturated photosynthesis rate P_{ML}, the carbon dioxide saturated rate at ambient light P_M, and the ambient light and carbon dioxide limited uptake rate P (equal to that shown in the lower part of the figure) are indicated in the upper part of the figures to illustrate characteristics of the model. Light saturated portions of the day occur during those periods when P_M is equal to P_{ML}.

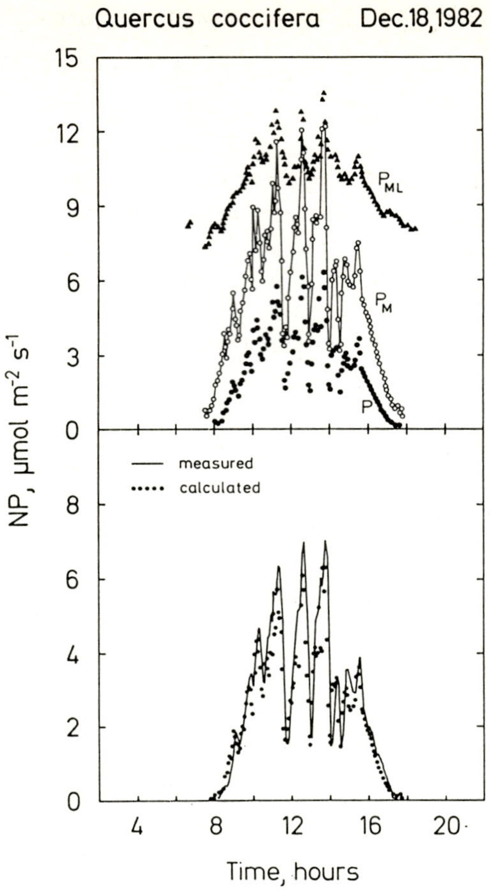

Fig. 4. Diurnal time courses of net photosynthesis rates observed in *Quercus coccifera* leaves (*solid line in lower part of figure*) compared to calculated net photosynthesis rates obtained with the model (*symbols in lower part of figure*). Also shown are diurnal courses of the ambient net photosynthesis rate (P), the CO_2-saturated photosynthetic rate (P_M), and the light and CO_2 saturated photosynthesis rate (P_{ML}) as predicted by the model. Simulation conducted with data from Fig. 2 for December 18, 1982

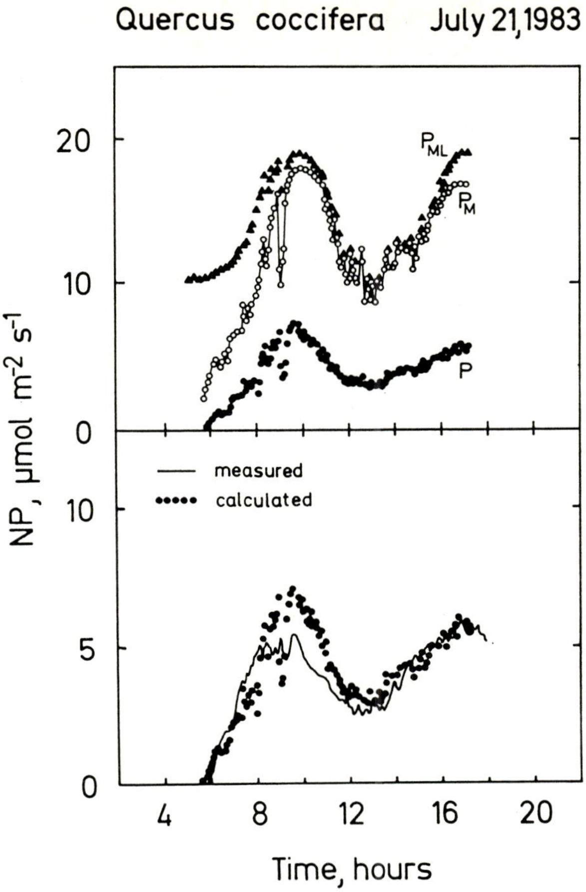

Fig. 5. Diurnal time courses as in Fig. 4. of observed and calculated net photosynthesis rates in *Quercus coccifera*. Simulation conducted with data from Fig. 3 for July 21, 1983

DISCUSSION AND CONCLUSIONS

In previous descriptions of the gas exchange response of leaves under naturally fluctuating environmental conditions (Tenhunen *et al.* 1980b, c; Harley *et al.* 1986; Weber *et al.* 1985), we have progressively worked to develop our models from steady-state single factor dependencies of net photosynthesis determined independently of measured diurnal course responses. Diurnal course measurements, therefore, provide an independent data set for model testing. In addition, an experimental methodology results which may allow very effective investment of experimental time and resources (see also Lange *et al.* this volume). Thus, with minimal effort, a maximum in understanding of leaf function in the natural habitat is obtained. The results presented in this paper demonstrate that steady-state determinations of carbon dioxide response curves may be used to structure a model for the purpose of diurnal time course simulations of net photosynthesis in sclerophyllous leaves. During a large part of the year and over a wide range of conditions, a single set of equations may be used to calculate carbon dioxide uptake rates in leaves of *Quercus coccifera.*

While the correspondence of predicted and measured photosynthesis rates was in general good during long periods in winter and summer, the model was further useful in identifying time periods during which leaf behavior deviated from that expected. For example, during periods in which leaves of *Quercus coccifera* were exposed to cold stress (leaf temperatures during the day below 5° C), observed net photosynthesis rates during the morning were lower than predicted by the model. During the afternoon after temperatures had increased to approximately 10° C, observations and model predictions agreed more closely suggesting that damage due to stress effects may have been repaired. Similarly, during periods of extreme water stress at the end of the summer when predawn water potential decreased below -30 bar, measured CO_2 uptake was lower than predicted by the model due to decreases in mesophyll photosynthetic capacity (Tenhunen *et al.* 1985).

The midday depression of photosynthesis observed during July and early August (see Fig. 3) does not appear to result from a long-term water stress induced reduction of mesophyll photosynthetic capacity (Tenhunen *et al.* 1985). In the present model, midday depression of NP is described successfully by reducing carboxylation efficiency at high temperature (see Tenhunen et al. 1984b), although decrease in leaf internal CO_2 pressure due to stomatal limitation was also important.

The steady-state response curves upon which the model was based were necessarily obtained over rather long periods of time (several hours in the case of individual CO_2 responses). As such, they tend to average in changes that occur in photosynthetic processes during the course of a day. The ability to accurately simulate diurnal courses of gas exchange will depend on just how rapidly time dependent endogenous effects influence leaf response. Nevertheless, the present results indicate that this model is useful to characterize leaf behavior, to identify stress induced responses under field conditions, and to help clarify the nature of time dependent endogenous effects which influence photosynthesis.

APPENDIX: Calculation Procedure

Net photosynthesis rate, NP in μmol m^{-2} s^{-1}, was calculated in the following manner. Input to the model are values of leaf temperature, T_L in degrees C, from which is also obtained the absolute leaf temperature, T_K in degrees K, incident PAR on the leaves, L in μmol m^{-2} s^{-1}, chamber air CO_2 pressure external to the leaf, P_a in μbar, the leaf conductance for CO_2, G_C in μmol m^{-2} s^{-1} (equal to G for water in Figs. 2 and 3 times 1000/1.56), the gas constant, R equal to 1.98646 cal K^{-1} mol^{-1}, and the light utilization efficiency, α see text section entitled*Variations in P_M*.

Analyses of temperature response curves of net photosynthesis measured at saturating light and CO_2 (Fig. 1) provide the activation energy for the overall photosynthetic process, ΔH_o = 10340 cal mol^{-1} in winter and 10380 in summer, an activation energy describing decrease in photosynthesis at high temperature in summer, ΔH_1 = 59860 cal mol^{-1}, scaling constants for the equations describing the temperature dependencies, C_1 = 20.68 and C_2 = 14.71, and a parameter describing entropy change with temperature applicable to the summer data, ΔS = 195 entropy units (see also Tenhunen et al. 1976).

The CO_2- and light-saturated photosynthesis rate for a particular temperature, P_{ML}, was then obtained from the Arrhenius equation for winter conditions:

$$P_{ML} = e^{-\Delta H_o / R\,T_K \; + \; C_1} \qquad \text{(A.1a)}$$

or from the equation of Johnson *et al.* (1942) for summer conditions:

$$P_{ML} = \frac{e^{C_2}\; T_K \; e^{-\Delta H_o / R\,T_K}}{1 + e^{-\Delta H_1 / R\,T_K}\; e^{\Delta S / R}} \qquad \text{(A.1b)}$$

Next, the CO_2-saturated photosynthesis rate is adjusted for light conditions below saturation for photosynthesis (see discussion in text). For a particular temperature and light intensity:

$$P_M = \frac{\alpha \; L}{(1 + \alpha^2 L^2 / P_{ML}^2)^{1/2}} \qquad (A.2)$$

A similar procedure is followed to establish the appropriate value for carboxylation efficiency. CE in μmol m^{-2} s^{-1} μbar^{-1} was described during winter as:

$$CE = .0032\, T_L - .014 \qquad (A.3a)$$

Since measured data were not available from the summer period below 28° C, whereas mesophyll photosynthetic capacity in general appeared to be slightly less than in winter (Fig. 1), CE at low temperature during the summer was related to P_{ML} based on winter measurements:

$$CE = .0026\; P_{ML} \qquad (A.3b)$$

The decrease in CE above 28° C during summer was described as:

$$CE = -.003\, (T_L - 28) + .05 \qquad (A.3c)$$

The decrease in efficiency below light saturation was described with MF, where MF is a fraction of CE and has the same units. The following empirical formulation is based on data obtained with *Arbutus unedo* by Weber et al. (1986). In this equation, CE is multiplied by a factor which varies between 0 and 1 according to the equation of Smith (1937) and depending on light intensity. Beta was equal to 0.004:

$$MF = \frac{\beta \; L \; CE}{(1 + \beta^2 L^2 / 1.0609)^{1/2}} \qquad (A.4)$$

Having established values for PM and MF, carbon dioxide uptake by the leaf may be described with :

$$P = \frac{MF \ P_i}{(1 + MF^2 \ P_i^2 / P_M^2)^{1/2}} - R_d \tag{A.5}$$

The activation energy for dark respiration, ΔH_r was 9790 cal mol^{-1} and the scaling constant, C_3, was 15.71. Dark respiration was described with the Arrhenius equation:

$$R_d = e^{-\Delta H_r / R T_K + C_3} \tag{A.6}$$

The value for P_i in equation A.5 must be obtained iteratively as discussed in the text (see also Tenhunen and Westrin 1979), using values for P_a and G_C and by balancing equation A.5 with Fick's Law where:

$$P_i = P_a - (10^6 NP / G_C) \tag{A.7}$$

REFERENCES

Beyschlag W (1984) Photosynthese und Wasserhaushalt von *Arbutus unedo* L. im Jahreslauf am Freilandstandort in Portugal. Gaswechselmessungen unter natürlichen Bedingungen und experimentelle Faktorenanalyse. PhD Thesis, Würzburg

Farquhar GD, von Caemmerer S (1982) Modeling of photosynthetic response to environmental conditions. In: Lange OL, Nobel PS, Osmond CB, Ziegler H (eds), Encyclopedia of Plant Physiology, NS vol 12B: Physiological Plant Ecology II. Springer, Berlin-Heidelberg-New York-Tokyo, pp 549-587

Harley PC, Tenhunen JD, Lange OL (1986) Use of an analytical model to study limitations to net photosynthesis in *Arbutus unedo* under field conditions. Oecologia (Berl) in press

Harley PC, Weber JA, Gates DM (1985) Interactive effects of light, leaf temperature, CO_2 and O_2 on photosynthesis in soybean. Planta 165: 249-263

Johnson F, Eyring H, Williams R (1942) The nature of enzyme inhibitions in bacterial luminescence: Sulfanilamide, urethane, temperature, and pressure. J Cell Comp Physiol 20: 247-268

Smith E (1937) The influence of light and carbon dioxide on photosynthesis. Gen Physiol 20: 807-830

Tenhunen JD, Hesketh JD, Gates DM (1980a) Leaf photosynthesis models. In: JD Hesketh and JW Jones (eds) Predicting Photosynthesis for Ecosystem Models, vol 1. CRC Press, Boca Raton, Florida, pp. 121-181

Tenhunen JD, Lange OL, Gebel J, Beyschlag W, Weber JA (1984b) Changes in photosynthetic capacity, carboxylation efficiency, and CO_2 compensation point associated with midday stomatal closure and midday depression of net CO_2 exchange of leaves of *Quercus suber*. Planta 162: 193-203

Tenhunen JD, Lange OL, Harley PC, Beyschlag W, Meyer A (1985) Limitations due to water stress on leaf net photosynthesis of *Quercus coccifera* in the Portuguese evergreen scrub. Oecologia (Berl) 67: 23-30

Tenhunen JD, Lange OL, Harley PC, Meyer A, Gates DM (1980c) The diurnal time course of net photosynthesis of soybean leaves: Analysis with a physiologically based steady-state photosynthesis model. Oecologia (Berl) 46: 314-321

Tenhunen JD, Meister HP, Caldwell MM, Lange OL (1984a) Environmental constraints on productivity of the Mediterranean sclerophyll shrub *Quercus coccifera*. Proceedings of INTECOL workshop - Rates of Natural Primary Productivity and Agricultural Production. Options méditerranéennes (Instituto Agronomico Mediterraneo de Zaragoza) 84/1:33-53

Tenhunen JD, Meyer A, Lange OL, Gates DM (1980b) Development of a photosynthesis model with an emphasis on ecological applications. V. Test of the applicability of a steady-state model to description of net photosynthesis of *Prunus armeniaca* under field conditions. Oecologia (Berl) 45: 147-155

Tenhunen JD, Weber J, Yocum C, Gates DM (1976) Development of a photosynthesis model with an emphasis on ecological applications. II. Analysis of a data set describing the P_M surface. Oecologia (Berl) 26:101-119

Tenhunen JD, Westrin SS (1979) Development of a photosynthesis model with an emphasis on ecological applications. IV. WHOLEPHOT - Whole leaf photosynthesis in response to four independent variables. Oecologia (Berl) 41: 145-162

Weber JA, Jurik TW, Tenhunen JD, Gates DM (1985) Analysis of gas exchange in seedlings of *Acer saccharum*: Integration of field and laboratory studies. Oecologia (Berl) 65: 338-347

Weber JA, Tenhunen JD, Gates DM, Lange OL (1986) Interaction of PPFD and CO_2 on photosynthesis response in intact leaves. Submitted to Plant Physiol

Diurnal patterns of leaf internal CO_2 partial pressure of the sclerophyll shrub *Arbutus unedo* growing in Portugal

W. Beyschlag, O.L. Lange and J.D. Tenhunen*

Lehrstuhl für Botanik II, Universität Würzburg, Mittlerer Dallenbergweg 64, 8700 Würzburg, West Germany

INTRODUCTION

Investigations of the seasonal change in diurnal patterns of leaf gas exchange revealed for a variety of plants growing in arid environments, that midday stomatal closure combined with a midday depression in CO_2 and H_2O gas exchange occurs during periods in which leaves experience high temperatures and large water vapor pressure differences between leaf air spaces and the external air (Guttenberg and Buhr 1935; Eckardt 1975; Tenhunen, Lange and Braun 1981; Lange, Tenhunen and Braun 1982). Schulze *et al.* (1974) found in experiments with desert plants, that stomata respond strongly to air humidity (see also Lange *et al.* 1971). In addition, results obtained by Lange and Meyer (1979) and Beyschlag (1984) indicated, that at least with well watered *Prunus armeniaca* and *Arbutus unedo* plants, short term changes in the internal water status of the plants seem to be of secondary importance in regulating this behavior. Thus, as a simple explanation of the midday depression of gas exchange it was suggested, that transpiration is lowered at midday by humidity controlled stomatal closure, which then restricts CO_2 supply to the mesophyll and decreases net photosynthesis rate (NP). However, the field measurements of Schulze *et al.* (1975) with *Prunus armeniaca* and those of Tenhunen *et al.* (1984) with *Quercus suber* under mediterranean summer conditions demonstrated that stomatal aperture and the photosynthetic capacity of the mesophyll of the leaves can change in a correlated manner. Thus, the internal air space CO_2 partial pressure of the leaves remained more or less constant (approximately 200 μbar) during the entire light saturation phase of the day. Wong, Cowan and Farquhar (1978, 1979) have pointed out a general correlation in stomatal conductance (G) and mesophyll photosynthetic capacity found in leaves of widely varying plant types and independent of growth conditions.

The present paper examines for leaves of *Arbutus unedo*, a common species of the coastal macchia vegetation of the European Mediterranean region (Braun-Blanquet, 1952), the degree to which NP and G varied together on days distributed throughout an annual cycle. Diurnal changes in CO_2 partial pressure

*Systems Ecology Research Group, San Diego State University, San Diego, California, USA

NATO ASI Series, Vol. G15
Plant Response to Stress. Edited by J. D. Tenhunen et al.

in the mesophyll of the leaf were scrutinized as an indicator of correlated function of photosynthetic and stomatal processes. The gas exchange measurements discussed here were carried out in Portugal between August 1982 and September 1983. Examples are selected which illustrate patterns in internal CO_2 pressure observed during different seasons and with weather conditions characteristic for each period.

METHODS

The field experiments with *Arbutus unedo* were conducted in a natural mediterranean scrub vegetation at the Research Station Quinta São Pedro, Sobreda, Portugal. Diurnal time courses of CO_2 and H_2O gas exchange in response to changes in natural environmental conditions were observed. Single leaves or branch ends with several leaves were enclosed in conditioned gas exchange cuvettes which were part of a mobile field laboratory. The methods applied for measurement of gas exchange and recording of weather parameters were described in detail by Tenhunen *et al.* (1984, 1985) and Beyschlag (1984). All equipment used was designed and built by Walz, Meß- und Regeltechnik, Effeltrich, FRG.

All measured data were recorded at intervals of four minutes throughout the day. In those cases where twigs were necessarily included in the cuvette, net photosynthesis rates of the leaves were obtained by correction for the temperature-dependent twig respiration determined after removal of the leaves. Leaf internal CO_2 partial pressure (p_i) was estimated from leaf conductance and net photosynthesis rates using the equations of von Caemmerer and Farquhar (1981) and as discussed by Sharkey *et al.* (1982). Rates are expressed on a projected leaf area basis. Parallel to all daily time course measurements, the xylem leaf water potential of comparable leaves outside of the cuvettes was measured with a pressure chamber (Scholander *et al.* 1965).

Leaf material was illuminated naturally. The measurement cuvettes were controlled either to track outside air temperature and humidity conditons or the same standardized symmetrical diurnal temperature course, based on typical measured summer microclimate conditions (for details see Tenhunen, Lange and Braun 1981; Tenhunen *et al.* 1980, 1984) was imposed on consecutive days. Since the dew point in the chamber was held constant, VPD varied simultaneously with and in a similar pattern to temperature. Naturally incident light followed a similar diurnal course on these consecutive days. In the experiments with standardized time courses different partial pressures of CO_2 (150, 340, 2500 µbar) were supplied on each day to the cuvettes and the data from the 100 and the 330 µbar days were combined to derive the carboxylation efficiency (CE) from the initial slope of the CO_2 response curve at different times of day (Tenhunen *et al.* 1984; Beyschlag 1984).

RESULTS

From 56 measured diurnal courses distributed throughout the annual cycle those were identified in which p_i remained relatively constant during light saturation of NP. This behavior occurred during approximately 60% of all measured diurnal courses. A daily time course of p_i was classified into this category, if the standard error of the mean value of all available p_i values during the phase of light saturation of NP (PAR > 500 μE m^{-2} s^{-1}) was less than +/- 25 μbar. The mean p_i values of all days of this group are shown in Fig. 1A and the respective p_i/p_e ratios in Fig. 1B (p_e = mean value of CO_2 partial pressure of the air inside the gas exchange cuvettes) as distributed throughout the year. Relatively constant p_i seems to be common during spring and summer but occurs also on sunny days in autumn and winter. The mean annual p_i value (dashed line in Fig. 1A) was 197 μbar (standard error +/- 28.5 μbar). Somewhat lower values were observed during the winter while higher values were found in spring and summer. The

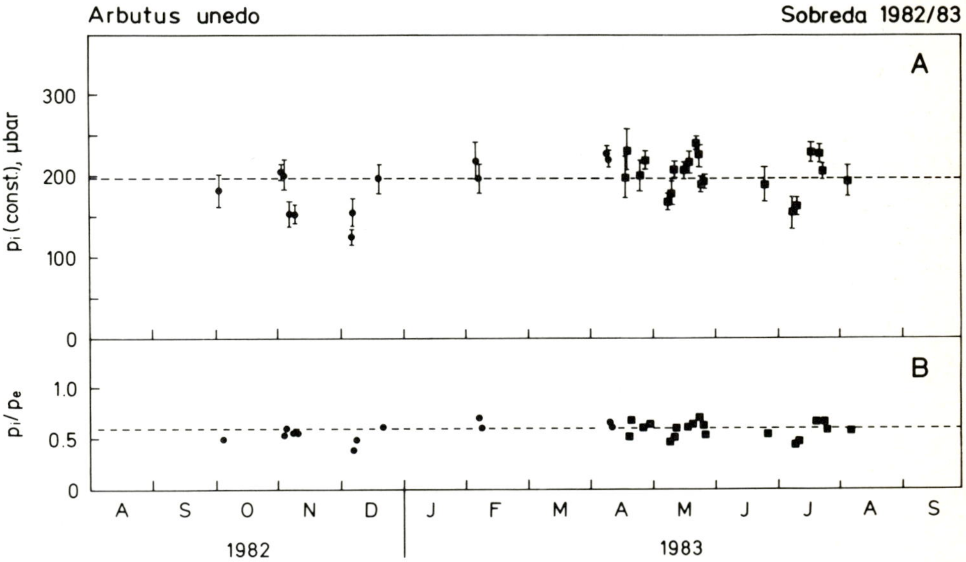

Fig. 1. A) Mean p_i values (with standard errors) estimated during the light saturation phase of all measurement days, in which p_i was within a standard error of +/- 25 μbar. B) Correspondent p_i/p_e ratios. The dotted line in each panel shows the mean value of all measurements. Solid circles are data from leaves which flushed in 1982. Solid squares are for leaves which flushed in 1983

mean p_i/p_e ratio (dashed line in Fig. 1B) was 0.59 (standard error +/- 0.08). No p_i data are available from the driest season of the year (end of August until end of September), since transpiration rates of the leaves were too low to obtain a reliable estimate.

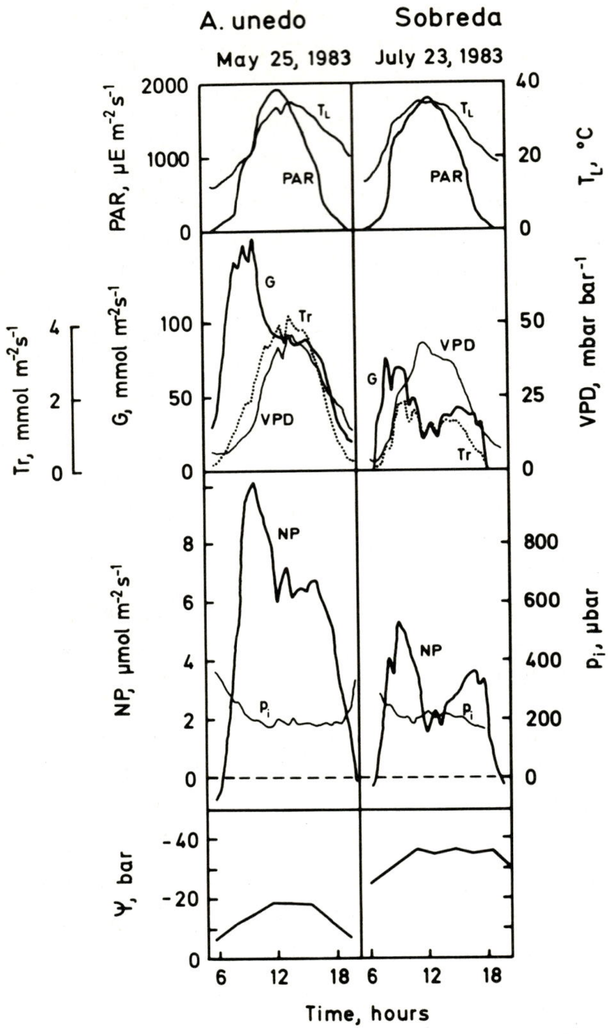

Fig. 2. Daily time courses of photosynthetic active radiation (PAR), leaf temperature (T_L), transpiration rate (Tr), water vapor mole fraction difference between leaf and air (VPD), stomatal conductance (G), net photosynthesis (NP), leaf internal CO_2 partial pressure (p_i) and xylem water potential (Ψ) of naturally grown *Arbutus unedo* leaves in the field. Group 1: Examples with relatively constant p_i (see text)

Representative diurnal courses with relatively constant p_i are presented in Fig. 2 from the late spring (May 25) and the summer dry period (July 23). The climatic conditions were very similar on these two days (daily maximum of PAR in May: 1950 $\mu E\ m^{-2}\ s^{-1}$, in July: 1730 $\mu E\ m^{-2}\ s^{-1}$; daily maximum T_L in May: 35.4° C, in July: 32° C; daily maximum of VPD in May: 45.5 mbar bar^{-1}, in July: 41 mbar bar^{-1}). In contrast the water status of the plants was much better at the end of spring (predawn water potential: - 5.5 bar; daily minimum: - 19 bar) than in midsummer (predawn water potential: - 25 bar; daily minimum - 36.5 bar). The diurnal courses of G and NP on May 25 represent a transitional stage in leaf behavior between the more or less dome shaped curves, characteristic for sunny days in the winter and the early spring (where photosynthesis is mainly limited by low temperatures) and the two peaked curves (Fig. 2 right panel) which are typical for leaves of evergreen sclerophylls under mediterranean summer conditions (see also discussion by Harley *et al.* this volume). Long term soil water stress decreased G, Tr, and NP in July under similar ambient conditions to half of the value found in May (daily maximum of G in May: 152 mmol $m^{-2}\ s^{-1}$, in July: 77 mmol $m^{-2}\ s^{-1}$; daily maximum of Tr in May: 4.2 mmol $m^{-2}\ s^{-1}$, in July: 1.9 mmol $m^{-2}\ s^{-1}$; daily maximum of NP in May: 10.2 μmol $m^{-2}\ s^{-1}$, in July: 5.3 μmol $m^{-2}\ s^{-1}$). The similar p_i levels of the two days (daily mean value on May 25: 190 μbar, on July 23: 207 μbar) demonstrate, that the observed difference in the level of NP is not caused by the difference in G, rather a decrease in the photosynthetic activity of the mesophyll occurs. Measurements of the maximum photosynthetic capacity P_{ML} (NP measured under saturating PAR and saturating CO_2 partial pressure) of the same leaves indicated a strong decrease during summer drought (Beyschlag, 1984; Tenhunen *et al.* this volume) and confirm this conclusion.

Fig. 3 shows the results of an experiment in July 1983 using the method described above with "standardized" diurnal courses of ambient conditions and different CO_2 partial pressures on consecutive days. Both, the response at saturating 2500 μbar CO_2 (P_{ML}) and the time course of CE derived from the NP data indicate a midday depression in the photosynthetic activity of the mesophyll. The response obtained at 2500 μbar CO_2 indicates further an increase in P_{ML} during the afternoon. In contrast CE shows higher values (i.e. higher photosynthetic capacity) under the same conditions in the morning than in the afternoon after the midday depression. In a similar fashion to CE, the diurnal courses of NP with p_e = 150 μbar as well as with p_e = 340 μbar are characterized by lower NP values in the afternoon than in the morning under very similar conditions of PAR, T_L and VPD. Since G was lower on both days in the afternoon as well, similar p_i values are obtained throughout the day (morning: 117 μbar, afternoon: 121 μbar with p_e= 150 μbar; morning: 200 μbar, afternoon: 197 μbar with p_e =

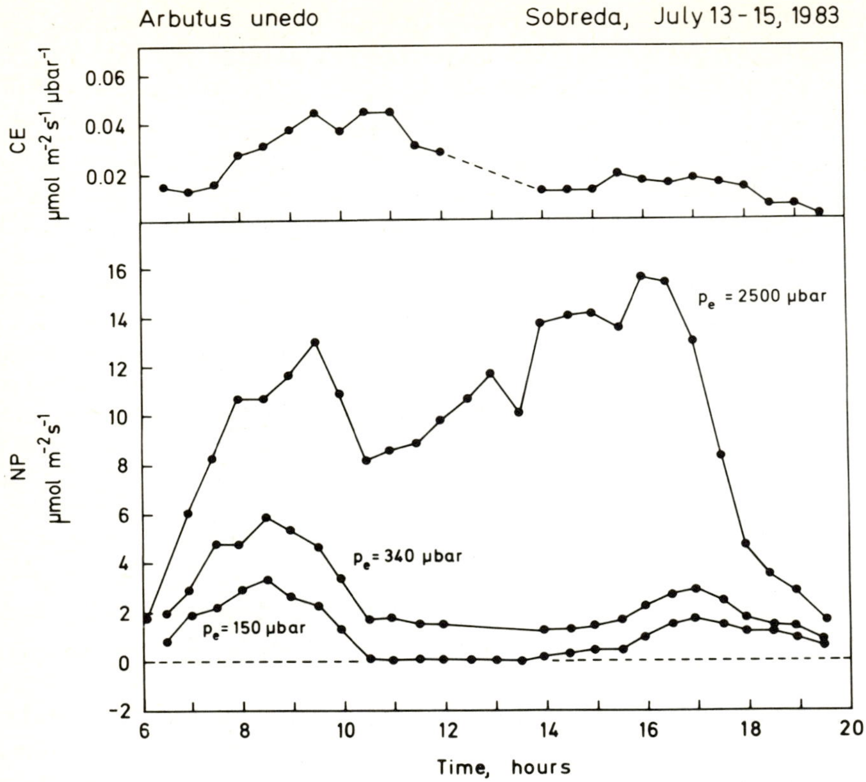

Fig. 3. Time course of rate of CO_2 uptake of *Arbutus unedo* leaves exposed to "standard" day conditions of PAR, T_L and VPD and supplied with either 2500 µbar, 340 µbar and 150 µbar CO_2 in the incoming air stream to the cuvette (B). Panel A shows the time course of the carboxylation efficiency (CE), derived from the 150 µbar and 340 µbar curves from panel B (see text)

340 µbar). Thus, the photosynthetic capacity of the leaves must be lower in the afternoon than morning. The decrease in photosynthetic capacity may possibly be explained as due to photoinhibition which occurs at 340 and 150 µbar CO_2. Powles (1984) showed that photoinhibition is dependent on the level of p_i (high p_i protects against photoinhibition) during exposure to high light. It increases under conditions of high temperature and water stress. Possibly photoinhibition (particularly because the leaves were forced into a horizontal position during the experiments) was less on the 2500 µbar day than on the other days with lower CO_2 partial pressures. This is suggested by lower afternoon values of NP and CE of the latter in the afternoon. Since such effects of photoinhibition are reversible during the night (Powles 1984), a long term influence on the photosynthetic behavior of the leaves on the following day would not be seen.

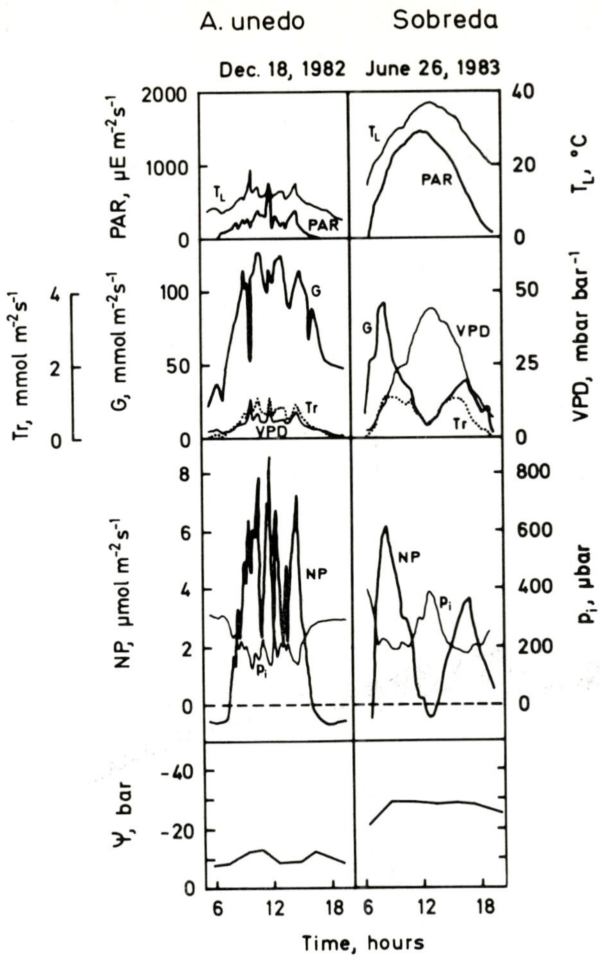

Fig. 4. See Fig. 2. Examples of group 2: p_i is not constant as described in text

Fig. 4 shows two diurnal times courses, which are examples for the second category of days, i.e. with a variation in p_i greater than +/- 25 μbar. Diurnal courses falling into this group occurred during all seasons of the year but were found with a higher frequency in winter. The left panel of Fig. 4 illustrates such a characteristic winter time course of photosynthetic response. Depending on the degree of cloudiness, PAR changed throughout the day. In response to PAR, NP fluctuated greatly. Due to light limiting conditions photosynthesis responded immediately to changes in PAR. The stomata on the other hand, required a much longer time to respond. Because of this time lag in stomatal response, the maxima in NP were accompanied by minima in p_i. As in the category of days discussed above, p_i varied around a value of approximately 200 μbar but the amplitude was much higher than +/- 25 μbar.

On June 26, 1983 (Fig. 4 right panel) NP was strongly depressed at midday and decreased below the compensation point. This resulted in a marked increase in p_i up to 380 μbar. As NP approaches zero, respiratory processes predominantly determine p_i along with stomatal conductance. In the morning and in the afternoon, p_i remained relatively constant at approximately 200 μbar despite large changes in G and NP. The fact, that on June 26 NP reaches the compensation point, while this does not occur on July 23 (Fig. 2 right panel) with very similar ambient conditions could be explained by a "hardening" in response to high temperature conditions and low plant water potential. In addition, dark respiration was found to be higher at the beginning of the summer. The leaves may respond in June more sensitively to ambient stress and show a greater decrease in CE at noontime than in July (Beyschlag 1984).

DISCUSSION

Leaf internal CO_2 partial pressure depends on CO_2 fixation and CO_2 evolution in the mesophyll on the one hand and on leaf stomatal conductance on the other. Analyses of diurnal courses of gas exchange showed that both carboxylation efficiency and leaf conductance changed considerably throughout the day when midday depression of net photosynthesis occurs. In spite of these changes, p_i remained surprisingly constant. Stomatal closure which could result in a decrease of conductance of more than 50% apparently must have been compensated by a concomitant drop in carboxylation efficiency and vice versa so that internal CO_2 was not affected. Control of this behavior is still poorly understood. We must consider the mechanisms which may be responsible for diurnal variations in photosynthetic capacity, CO_2 evolution and stomatal conductance and explain how these processes act in an apparently coordinated fashion.

Induction of midday depression of leaf conductance and CO_2 assimilation in *Arbutus unedo* is correlated with high temperatures and high VPD. In addition, it takes place under conditions of high solar radiation. Temperature and humidity are known to be factors which have a direct influence on stomatal aperture (Schulze *et al.* 1974, 1975; Lösch 1977; Raschke 1979). The observed daily time courses of G may be understood as a response to the diurnal course of VPD and temperature. If CE in C_3 plants is controlled by the properties of the enzyme ribulose-1,5-bisphosphate carboxylase as proposed by von Caemmerer and Farquhar (1981), then it should be possible to explain the observed midday depression in CE as resulting from the temperature dependency of this enzymatic reaction. Nevertheless most other published reports indicate little effect of temperature on CE over a broad range of temperatures (Badger and Collatz 1977; Ku and Edwards 1977; Peisker, Tichá and Apel 1979; von Caemmerer and Farquhar 1981; Farquhar and von Caemmerer 1982). Detailed *in vitro* studies with the isolated enzyme (Weis 1981a) showed similar results. On the other hand, Weis (1981b) described a reversible heat-inactivation of the Calvin-cycle as a possible mechanism of the temperature regulation of photosynthesis and Brooks and Farquhar (1985) showed with spinach leaves that the CO_2/O_2 specificity of the enzyme decreases markedly between 10 and 30° C. Thus, the observed diurnal CE changes may possibly be attributed to temperature effects at the enzyme level not understood at the present time. Diurnal changes in RubP-carboxylase activity as suggested above were found in a study in which enzyme was isolated at different times of day from *Glycine soja* and *Elaeis guineensis* (Vieira da Silva *et al.* 1984). The results revealed two peaked diurnal activity courses with a marked depression at noontime. The authors could further show, that diurnal changes of the water status of the leaves were not responsible for this behavior. They stated that a circadian rhythm might be responsible. This is a further possibility for consideration in the case of *Arbutus unedo*.

As another possibility, diurnal changes in the respiratory activity of the leaves must be considered. Since all respiration processes are temperature dependent, one would expect an increase in leaf CO_2 production at noontime. The results of Tenhunen *et al.* (1984, 1985) with *Quercus suber* and *Quercus coccifera*, measured under mediterranean climate conditions as well as those of Weber, Tenhunen and Lange (1985) with *Spinacia oleracea*, *Arbutus unedo* and *Quercus suber* show a marked increase of the CO_2 compensation point during midday, i.e. at high leaf temperature or with exposure to low humidity. The data from June 26 of the present paper (Fig. 4 right panel) also illustrate an increase in CO_2 production at that time of day. Thus, a part of the midday depression of photosynthesis might be due to an increase in respiration. Temperature is not the only factor, which can cause such an increase. Krampitz and Fock (1985) and Krampitz, Klug and Fock (1985) have shown with sunflower leaves, that increasing water stress suppressed carbon flux in the Calvin cycle more than in the glycolate pathway.

Photoinhibition may also play a role in regulating photosynthetic behavior during the course of a day under mediterranean summer conditions and with normal external CO_2 partial pressure. The lower CE values in the afternoon in Fig. 3, may result due to photoinhibitory effects. However, the general importance of photoinhibition in regulating gas exchange of mediterranean sclerophylls during summer is still not known.

In the last few years, it was shown, that VPD, in addition to influencing the stomata, can also directly affect the photosynthetic capacity of the mesophyll (Forseth and Ehleringer 1983; Bunce 1984; Ball and Farquhar 1984; Sharkey

1984; Lange, Tenhunen and Beyschlag 1985). Most of these authors found that the influence of VPD was restricted to P_{ML} while CE remained unchanged. However, Resemann and Raschke (1984) indicated that in *Arbutus unedo* leaves both CE and P_{ML} can decrease with increasing VPD. Under natural conditions, temperature and VPD change in parallel and it is very difficult to separate their influence on stomatal movement or mesophyll photosynthetic reactions (Weber, Tenhunen and Lange 1985). However, we can conclude, that diurnal changes of temperature and humidity can play the major role in producing such changes in G and in the photosynthetic and respiratory activity of the mesophyll as those we have observed in the field.

There are several possible means by which the observed correlated changes of G and photosynthetic capacity of the mesophyll might be effected. One possibility is, that the correlated behavior is caused by parallel, but independent, reactions of stomata and mesophyll to diurnal change in ambient conditions. A decrease in CO_2 supply of a leaf, caused by a VPD-dependent stomatal closure at noontime could, for instance, result in constant p_i if there is a parallel VPD-dependent decrease of the mesophyll photosynthetic activity and a temperature dependent increase of respiratory activity inside the leaf. In this case, p_i could remain relatively constant within a certain range of ambient conditions. On the other hand, there is some evidence that a direct coupling of function between stomata and mesophyll might exist. Wong, Cowan and Farquhar (1979) suggested that the stomata might be regulated by levels of intermediates of photosynthesis (ATP, NADPH, ribulose-bisphosphate). This would allow stomata to compensate (in terms of p_i) for changes in the mesophyll photosynthetic activity. On the other hand, the data from June 26 of the present paper show, that there are situations in which changes in G compensate only partially for changes in mesophyll CO_2 uptake. Also in the case of Quercus coccifera (Tenhunen et al. 1985), stomatal control during early summer is independent of changes in P_{ML} and NP. As a result, water appears to be conserved but p_i decreases strongly during the course of a summer day. At least in principle, this suggests that any linking of stomatal and photosynthetic functioning may be overridden depending on circumstances.

Finally, Raschke, Hanebuth and Farquhar (1979) and Raschke and Hedrich (1985) showed, that the stress hormone abscisic acid (ABA) which is well known to have a strong influence on stomatal regulation can directly influence the photosynthetic activity of the mesophyll. They discussed a negative feedback loop between photosynthesis and stomata with ABA as a signal transmitter. Burschka, Lange and Hartung (1985) showed with *Arbutus unedo* leaves in the field, that during stomatal closure, which was induced by an injection of ABA into the xylem of an intact leaf, p_i remained almost constant, when NP decreased markedly. Thus, from our present knowledge about ABA it is reasonable that a parallel behavior of mesophyll and stomata in response to changes in air humidity might be coordinated by ABA.

The further analysis of these questions should (1) separate the influence of the important environmental factors on the observed correlation between stomata and mesophyll, (2) determine whether or not photoinhibition plays a major role in photosynthesis of evergreen plants under mediterranean summer conditions and (3) investigate the significance of a possible circadian rhythm involvment in the regulation process.

ACKNOWLEDGEMENTS

We thank Dr. P. Harley, Dr. A. Meyer, Dipl. Biol. H.P. Meister and Mr. D. Jahner for experimental help in Portugal and Ms. D. Faltenbacher, Ms. W. Frosch and Ms. W. Samfaß for technical help. Dr. O. Schulz-Kampfhenkel (Institut für Internationale Zusammenarbeit in Bildung und Forschung, Hamburg) provided use of the facilities of the research station Quinta São Pedro. Calculations were made at the Rechenzentrum der Universität Würzburg. The research was supported by the Deutsche Forschungsgemeinschaft.

REFERENCES

Badger M, Collatz GJ (1977) Studies on the kinetic mechanism of ribulose-1,5-bisphosphate carboxylase and oxygenase reactions, with particular reference to the effect of temperature on kinetic parameters. Carnegie Inst Washington Yearb 76:355-361

Ball, MC, Farquhar GD (1984) Photosynthetic and stomatal responses of two mangrove species, *Aegiceras corniculatum* and *Avicennia marina*, to long term salinity and humidity conditions. Plant Physiol 74:1-6

Beyschlag W (1984) Photosynthese und Wasserhaushalt von *Arbutus unedo* L. im Jahreslauf am Freilandstandort in Portugal. Gaswechselmessungen unter natürlichen Bedingungen und experimentelle Faktorenanalyse. Thesis, Würzburg

Braun-Blanquet J (1952) Les groupements vegetaux de la France méditerranéenne. Centre National de la Recherche Scientifique, Montpellier

Brooks A, Farquhar GD (1985) Effect of temperature on the CO_2/O_2 specificity of ribulose-1,5-bisphosphate carboxylase/oxygenase and the rate of respiration in the light. Estimates from gas exchange measurements on spinach. Planta 165:397-406

Bunce JA (1984) Effects of humidity on photosynthesis. J Exp Bot 35:1245-1251

Burschka C, Lange OL, Hartung W (1985) Effects of abscisic acid on stomatal conductance and photosynthesis in leaves of intact *Arbutus unedo* plants under natural conditions. Oecologia (Berl) 67:593-595

Eckardt FE, Heim G, Methy M, Sauvezon R (1975) Interception de l'énergie rayonnante, échanges gazeux et croissance dans une forét méditerranéenne à feulliage persistant (*Quercetum ilicis*). Photosynthetica 9:145-156

Farquhar GD, von Caemmerer S (1982) Modelling of photosynthetic response to environment. In: Lange OL, Nobel PS, Osmond CB, Ziegler H (eds), Encyclopedia of Plant Physiology, NS vol 12B: Physiological Plant Ecology II. Springer, Berlin-Heidelberg-New York-Tokyo, pp 549-587

Forseth IN, Ehleringer JR (1983) Ecophysiology of two solar tracking desert winter annuals. III. Gas exchange responses to light, CO_2 and VPD in relation to long term drought. Oecologia (Berl) 57:344-351

Guttenberg H von, Buhr H (1935) Studien über die Assimilation und Atmung mediterraner Macchiapflanzen während der Regen- und Trockenzeit. Planta 24:163-265

Körner C, Schell JA, Bauer H (1979) Maximum leaf diffusive conductance in vascular plants. Photosynthetica 13:45-82

Krampitz MJ, Fock HP (1984) $^{14}CO_2$ assimilation and carbon flux in the Calvin cycle and the glycolate pathway in water-stressed sunflower and bean leaves. Photosynthetica 18:329-337

Krampitz MJ, Klug K, Fock HP (1984) The rates of photosynthetic CO_2 uptake, photorespiratory CO_2 evolution and dark respiration in water-stressed sunflower and bean leaves. Photosynthetica 18:322-328

Ku SB, Edwards G (1977) Oxygen inhibition of photosynthesis. II. Kinetic characteristics as affected by temperature. Plant Physiol 59:991-999

Lange OL, Meyer A (1979) Mittäglicher Stomataschluß bei Aprikose (*Prunus armeniaca*) und Wein (*Vitis vinifera*) im Freiland trotz guter Bodenwasserversorgung. Flora 168:511-528

Lange OL, Lösch R, Schulze ED, Kappen L (1971) Responses of stomata to changes in humidity. Planta 100:76-86

Lange OL, Tenhunen JD, Braun M (1982) Midday stomatal closure in mediterranean type sclerophylls under simulated habitat conditions in an environmental chamber. I. Comparison of the behaviour of various European mediterranean species. Flora 172:563-579

Lange OL, Tenhunen JD, Beyschlag W (1985) Effects of humidity during diurnal courses on the CO_2- and light-saturated rate of net CO_2 uptake in the sclerophyllous leaves of *Arbutus unedo*. Oecologia (Berl) 67:301-304

Peisker M, Tichá I, Apel P (1979) Variations in the effect of temperature on oxygen dependence of CO_2 gas exchange in wheat leaves. Biochem Physiol Pflanz 174:391-397

Powles SB (1984) Photoinhibition of photosynthesis induced by visible light. Ann Rev Plant Physiol 35:15-44

Raschke K (1979) Movements of stomata. In: Haupt W, Feinleib ME (eds), Encyclopedia of Plant Physiology, NS vol 7. Springer, Berlin-Heidelberg-New York, pp 383-441

Raschke K, Hedrich R (1985) Simultaneous and independent effects of abscisic acid on stomata and the photosynthetic apparatus in whole leaves. Planta 163:105-118

Raschke K, Hanebuth WF, Farquhar GD (1978) Relationship between stomatal conductance and light intensity in leaves of *Zea mays* L., derived from experiments using the mesophyll as shade. Planta 139:73-77

Resemann A, Raschke K (1984) Midday depressions in stomatal and photosynthetic activity of leaves of *Arbutus unedo* are caused by large water-vapor pressure differences between leaf and air. Plant Physiol 75(Suppl):66

Scholander P, Hammel H, Bradstreet E, Hemmingsen E (1965) Sap pressure in vascular plants. Science 148:339-345

Schulze ED, Lange OL, Evenari M, Kappen L, Buschbom U (1974) The role of air humidity and leaf temperature in controlling stomatal resistance of *Prunus armeniaca* L. under desert conditions. I. A simulation of the daily time course of stomatal resistance. Oecologia (Berl) 17:159-170

Schulze ED, Lange OL, Kappen L, Buschbom U (1975) The role of air humidity and leaf temperature in controlling stomatal resistance of *Prunus armeniaca* L. under desert conditions. II. The significance of leaf water status and internal carbon dioxide concentration. Oecologia (Berl) 18:219-233

Sharkey TD (1984) Transpiration induced changes in the photosynthetic capacity of leaves. Planta 160:143-150

Sharkey TD, Imai K, Farquhar GD, Cowan IR (1982) A direct confirmation of the standard method of estimating intercellular partial pressure of CO_2. Plant Physiol 69:657-659

Tenhunen JD, Lange OL, Braun M, Meyer A, Lösch R, Pereira JS (1980) Midday stomatal closure in *Arbutus unedo* leaves in a natural macchia and under simulated habitat conditions in an environmental chamber. Oecologia (Berl) 147:365-367

Tenhunen JD, Lange OL, Braun M (1981) Midday stomatal closure in mediterranean type sclerophylls under simulated habitat conditions in an environmental chamber. II. Effect of the complex of leaf temperature and air humidity on gas exchange of *Arbutus unedo* and *Quercus ilex*. Oecologia (Berl) 50:5-11

Tenhunen JD, Lange OL, Gebel J, Beyschlag W, Weber JA (1984) Changes in photosynthetic capacity, carboxylation efficiency and CO_2 compensation point, associated with midday stomatal closure and midday depression of net CO_2 exchange of leaves of *Quercus suber*. Planta 162:193-203

Tenhunen JD, Lange OL, Harley PC, Beyschlag W, Meyer A (1985) Limitations due to water stress on leaf net photosynthesis of *Quercus coccifera* in the Portuguese evergreen scrub. Oecologia (Berl) 67:23-30

Vieira da Silva J, Rouault O, Kpavode H, Pham Thi A (1984) Effect of drought on ribulose-bisphosphate carboxylase/oxygenase activity in soya bean and oil palm leaf tissues. In: Sybesma C (ed) Advances in Photosynthetic Research Vol. IV, Proc of the VIth Int Congr on Photosynthesis, Brussels, Belgium, Aug 1-6, 1983. Junk, The Hague

von Caemmerer S, Farquhar GD (1981) Some relationships between the biochemistry of photosynthesis and the gas exchange of leaves. Planta 153:376-387

Weber JA, Tenhunen JD, Lange OL (1985) Effects of temperature at constant air dew point on leaf carboxylation efficiency and CO_2 compensation point of different leaf types. Planta 166:81-88

Weis E (1981a) The temperature sensitivity of dark inactivation and light activation of the RubP-1,5-bisphosphate carboxylase in spinach chloroplasts. FEBS letters 129:197-200

Weis E (1981b) Reversible heat-inactivation of the Calvin cycle: A possible mechanism of the temperature regulation of photosynthesis. Planta 151:33-39

Wong SC, Cowan IR, Farquhar GD (1978) Leaf conductance in relation to assimilation in *Eucalyptus pauciflora* Sieb ex Spreng. Influence of irradiance and partial pressure of carbon dioxide. Plant Physiol 62:670-674

Wong SC, Cowan IR, Farquhar GD (1979) Stomatal conductance correlates with photosynthetic capacity. Nature 282:424-426

Effect of light on gas exchange parameters of sun and shade adapted leaves of *Ceratonia siliqua, Coffea arabica* and *Malus domestica*

M. Antonieta Nunes and M. Céu Matos*

Centro de Estudos de Produção e Tecnologia Agrícolas, Tapada da Ajuda, 1300 Lisboa, Portugal

INTRODUCTION

Wong *et al.* (1978, 1979) hypothesized that leaf photosynthesis and stomatal conductance are regulated in such a manner that Ci remains fairly constant despite changes in environmental conditions (see Beyschlag *et al.* in this volume). Nevertheless, studies in which carbon dioxide in the external air (Ca) and consequently carbon dioxide concentration in the mesophyll intercellular spaces (Ci) are varied, help determine the manner in which mesophyll photosynthetic activity, in large part determined by ribulose bisphosphate carboxylase activity, affects photosynthetic CO_2 uptake. While measurements of Ci are critical, only indirect methods are available to accomplish them, and it is controversial whether the molar concentration of carbon in solution (Ku and Edwards 1977) or the equilibrium partial pressure in the air space is the appropriate measure of CO_2 available for photosynthesis. In the present work, changes in photosynthetic activity of the leaf mesophyll were related to Ci, calculated from stomatal diffusion resistance, the rate of net photosynthetic CO_2 flux and Ca (using volume fraction as concentration units, $\mu l\ l^{-1}$).

Analysis of the curve relating net carbon assimilated during photosynthesis (NP) to internal CO_2 (Ci) reveals an initial linear portion (see Lange *et al.* in this volume). The slope of this line has been termed carboxylation efficiency (CE). The portion of the curve obtained under CO_2 saturating conditions is referred to here as photosynthetic capacity (PC). The highest values for PC are found when both saturating CO_2 and light conditions are imposed during measurement.

Carboxylation efficiency is a function of the balance in the carboxylase and oxygenase activities of ribulose bisphosphate carboxylase. Correlations in change of CE with change in environmental factors and among genotypes have been of recent interest (see Harley *et al.* in this volume). In the present work CE and PC of three crop species were evaluated for leaves that developed under saturating light conditions and with shading. The results reported here may contribute to the understanding of the physiology of sun and shade leaves of

*Estação Agronómica Nacional, 2780 Oeiras, Portugal

NATO ASI Series, Vol. G15
Plant Response to Stress. Edited by J. D. Tenhunen et al.

Ceratonia siliqua, a component of the European mediterranean ecosystem (see also Meister *et al.* in this volume). *Coffea arabica*, a shade plant from the humid tropics and *Malus domestica*, a common crop all over Europe were used to provide a contrast.

MATERIALS AND METHODS

Plants of *Ceratonia siliqua, Coffea arabica* cv. Caturra and *Malus domestica* cv. Golden Delicious EM 27 were grown in pots of large size and were watered periodically with Hoagland solution to prevent nutrient stress. The plants were well watered at all times. The experimental *Ceratonia siliqua* (carob) and *Malus domestica* (apple) trees were grown outside and were three years old, whereas the *Coffea arabica* (coffee) plants were two years old and had grown under natural light (maximum photon flux density about 600 μE $m^{-2}s^{-1}$) in a greenhouse heated during the winter period.

Approximately two months before the measurements were begun, branches of three apple trees were covered with netting and 6 carob and 4 coffee plants were placed under shade. The reduction of incident light intensity on *Ceratonia* was to approximately 1/4 of full sunlight. In the case of *Coffea*, incident light intensity was decreased to about 50 μE $m^{-2}s^{-1}$. Leaves which formed and expanded under these low light conditions were submitted to the treatments indicated in Table 1 and are referred to as "shade leaves". These are compared with leaves which formed under normal light ("sun leaves").

Table 1. Treatments applied to "sun" and "shade" leaves. At each incident measurement light intensity, measurements of gas exchange were carried out at all indicated CO_2 concentrations in the external air flushing the cuvette. In the case of *Malus domestica,* only sun leaves were studied.

	Photon flux density μE $m^{-2}s^{-1}$			CO_2 concentration μl l^{-1}				
Ceratonia siliqua	150	500	full sun	50	100	240	330	2200
Coffea arabica	150			50	100	240	330	2200
Malus domestica	150	500	full sun				330	2200

Gas Exchange Measurements

Measurements of gas exchange were made during the summer months (June to September). A small transparent "perspex" chamber with air diffusion plates inside (15 x 15 x 2.5 cm) was used to enclose the leaves and was flushed by air with varying CO_2 concentration (700 ml/min). The leaf surface boundary layer conductance was estimated at 265 mmol m^{-2} s^{-1}. Low CO_2 concentrations were obtained by dilution of normal air with a stream of CO_2 free air. The dilutions were obtained by employing a series of previously standardized capillary resistances. The CO_2 concentration to saturate photosynthesis (2200 μl l^{-1}) was obtained by dilution of 2% CO_2 supplied from a cylinder. The absolute CO_2 concentration of chamber air and the CO_2 concentration difference across the chamber were measured with a BINOS infrared gas analyser (Leybold-Heraeus, FRG).

A ventilated steady-state porometer (LiCor 1600) was used for measurements of transpiration rate, leaf resistance, photon flux density, leaf and air temperatures and the relative humidity in the external air. Between treatments, the leaf was removed from the chamber and its conductance for water vapor diffusion was immediately measured to correlate with the previously measured photosynthesis rate.

Conditions During the Measurements

Temperatures of the external air during the study were approximately 24° C for *Coffea,* 22 to 27° C for *Ceratonia* and 30° C for *Malus*. The air temperature inside the cuvette was monitored separately with resistance thermometers. Unfortunately due to ineffective cooling at the highest light intensities, it rose gradually to as high as 38° C when external air temperature was 27° C. To minimize increases in leaf temperature and air-to-leaf vapor pressure deficit over time, the period of leaf enclosure was shortened to 10 min. The relative humidity in the external air fluctuated between 30 and 60%.

Ceratonia siliqua

Light was provided as full sun intensity or decreased with neutral density filters. CO_2 concentration was kept constant throughout the day and 3 to 4 leaves were observed under medium, high and low irradiance respectively in the morning, during midday and in the afternoon. One half hour was allowed before measurements began, for plant material to adjust to the light levels imposed.

Coffea arabica

Before measurements began, the coffee plants were placed in a growth room with artificial light (Phillips HPLRG 400 W lamps); 24±2° C temperature and 50% relative humidity. To avoid water deficits due to sub-optimal relative humidity in the room, the crown of the plant (except the leaf to be studied) was sprayed with distilled water.

A photon flux density of 150 $\mu E\ m^{-2}s^{-1}$ had previously been found to be saturating for photosynthesis of coffee plant leaves (Nunes, Bierhuizen and Ploegman 1968; Rhizopoulou and Nunes 1981). Measurements were conducted between 10:00 and 13:00 hours, beginning with the leaf in normal air and then decreasing CO_2 concentration in steps. The final measurement of series was made at the highest CO_2 concentration (2200 $\mu l\ 1^{-1}$).

Malus domestica

Low light treatments were obtained by shading the leaf chamber with layers of green netting. The temperature of the chamber under full sunlight was kept at 30±1.5° C by means of circulating water.

Calculations

All rates are expressed on a projected leaf area basis. Leaf internal CO_2 concentration (Ci) was calculated as Ci = Ca - NP x R (R being stomatal + boundary layer CO_2 diffusion resistance). The compensation point (Γ) was taken as the intercept with the horizontal axis of the regression line of NP on Ci and carboxylation efficiency (CE) as the initial slope of this line.

RESULTS AND DISCUSSION

Carboxylation Efficiency

Carboxylation efficiency of *Ceratonia* was similar in sun and shade leaves under low and medium light intensities and increased with increasing light intensity (Fig. 1). CE decreased, however, in shade leaves under full light (1200 $\mu E\ m^{-2}\ s^{-1}$), while it increased further in sun leaves. Since temperature as well as leaf-to-air water vapor deficit changed during exposure to the different light treatments, this decrease in CE in shade leaves at high light intensity was most probably caused by increased leaf temperature in combination with low air humidity rather than by increased photon flux density directly.

Fig. 2 shows that leaf conductance to CO_2 (stomatal + boundary layer resistances included) measured in natural ambient air increased with increasing light

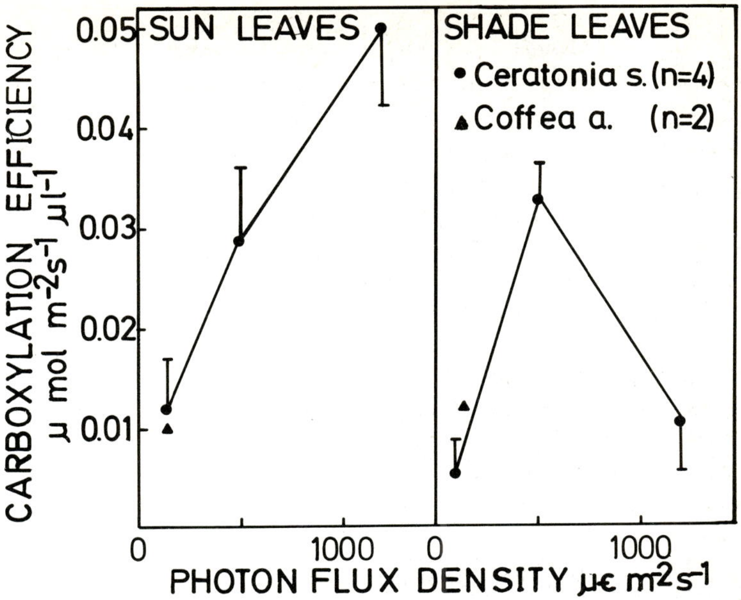

Fig. 1. Dependence of carboxylation efficiency on photon flux density incident on sun and shade leaves of *Ceratonia siliqua* and *Coffea arabica.* Average and standard errors shown for *Ceratonia*

intensities in sun but not in shade leaves from *Ceratonia* which exhibited lower conductances under full sunlight. Light conditions during growth of the leaf do not seem to have significantly affected average leaf conductance under low and medium light, although shade leaves showed a tendency to have higher conductance under low light conditions. Attention should be paid to the high conductance of *Coffea* leaves under low light and the still higher conductances of *Malus* with medium and high light intensities. With 2200 $\mu l\ l^{-1}$ CO_2 in the external air, leaf conductance was reduced in all species tested. In *Coffea,* the stomata were fully closed after an exposure of one half-hour. The values shown in Fig. 2 were obtained after an exposure of 10 min to this high CO_2 concentration. Net photosynthesis was linearly related to leaf conductance ($r \geq 0.9$).

Internal CO_2 Concentration

CO_2 compensation points showed an increase under full sunlight conditions (Table 2). As reported above, this is interpreted as an effect of heating of the chamber which may have increased temperature to 38° C (see Tenhunen *et al.*

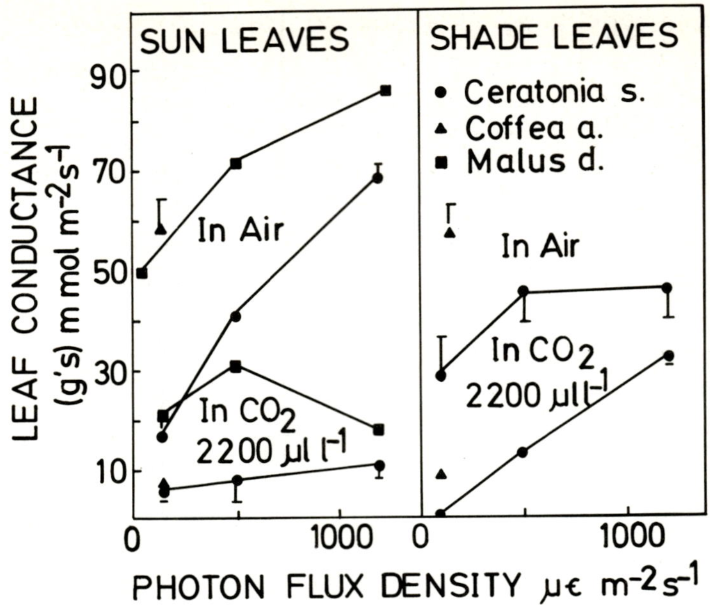

Fig. 2. Dependence of the leaf CO_2 diffusive conductance on incident photon flux density in sun and shade leaves exposed to normal air and to 2200 µl l^{-1} CO_2 concentration in air

Table 2. CO_2 compensation points (µl l^{-1}) of *Ceratonia siliqua* and *Coffea arabica* leaves measured under Low (150) Medium (500) and High (1200) photon flux densities (µE m^{-2} s^{-1}). In the case of high light, the two extreme values observed are presented.

	Sun leaves			Shade leaves		
Light	Low	Medium	High	Low	Medium	High
Ceratonia siliqua	30	36	85 200	30	32	50 172
Coffea arabica	28			46		

Table 3. Average internal CO_2 concentrations, Ci ($\mu l\ l^{-1}$), and internal to external CO_2 concentrations ratio, Ci/Ca, for *Ceratonia siliqua, Coffea arabica* and *Malus domestica* leaves measured under Low (150) Medium (500) and High (1200) photon flux densities ($\mu E\ m^{-2}\ s^{-1}$) combined with either 330 $\mu l\ l^{-1}$ and with 220 $\mu l\ l^{-1}$ CO_2 concentrations in the air flushing the cuvette.

	Sun leaves						Shade leaves					
Ca ($\mu l\ l^{-1}$)	330			220			330			220		
Light	L	M	H	L	M	H	L	M	H	L	M	H
C. siliqua												
Ci	157	163	215	140	145	195	233	179	267	176	150	198
Ci/Ca	.48	.49	.65	.62	.66	.87	.72	.54	.81	.73	.66	.88
C. arabica												
Ci	298			197			285			163		
Ci/Ca	.89			.83			.85			.83		
M. domestica												
Ci	265	237	205									
Ci/Ca	.73	.65	.56									

1984). In accordance with this interpretation, an increase in Ci under full light conditions was found to occur (Table 3). Thus, high values for the Ci/Ca ratio were found in *Ceratonia* but not in *Malus* where leaf temperature was maintained at 30° C. Perhaps high temperature associated with high light favours the oxygenase activity of ribulose bisphosphate carboxylase/oxygenase due to different temperature dependencies of its kinetic constants. Some support for this is found in the results of Badger and Collatz (1977). Alternatively, or in addition, change in the thylakoid membranes induced by such temperatures may have affected the carboxylation reaction (Weis 1981). This might result in an increase of photorespiration relative to photosynthesis. Dark respiration must also increase with temperature but the relative participation of this cannot be estimated.

Table 4. Photosynthetic capacity: maximal values of net photosynthesis rate (μmol $m^{-2}s^{-1}$) measured in leaf fed with 2200 μl^{-1} CO_2 in the cuvette, under Low (150), Medium (500) and High (1200) incident photon flux densities (μE $m^{-2}s^{-1}$). NP_{air}/PC:ratio of net photosynthesis in the air to PC measured under equivalent light conditions.

	Sun leaves			Shade leaves		
Light	Low	Medium	High	Low	Medium	High
C. siliqua						
PC	20	29	35	9	28	43
NP_{air}/PC	.10	.29	.24	.10	.23	.29
C. arabica						
PC	9			7		
NP_{air}/PC	.40			.46		
M. domestica						
PC	5	32	37			
NP_{air}/PC	.40	.35	.37			

Photosynthetic Capacity

As previously defined the photosynthetic capacity (PC) was considered as the photosynthesis rate attained when CO_2 is not limiting the carboxylation reaction. According to Farquhar and von Caemmerer (1982), the rate of carbon uptake under CO_2 saturating conditions depends on the rate of regeneration of ribulose bisphosphate, which in turn depends on the rate of electron transport. Table 4 shows that the maximal value of PC in shade leaves was obtained under high light. *Ceratonia* and *Malus* exhibited similar PC, whereas values for *Coffea* were lower. Net photosynthesis was not reduced by full sunlight as it was under normal air as reported in Fig. 3. This may be interpreted as a result of inhibition of photorespiration by high CO_2 concentrations. Examining the ratio of net photosynthesis in air to PC, sun leaves of *Ceratonia* and *Malus* operate in ambient air only at 24-37% of its maximum capacity. *Coffea* leaves benefit less from high CO_2 concentrations.

Reduction in NP during midday in summer associated with stomatal closure has been described by Tenhunen *et al.* (1984) in leaves of *Quercus suber* and by Beyschlag (1984) in *Arbutus unedo* under natural conditions in a scrub vegetation near Lisbon. Decrease in CE and PC were also detected at midday by these authors (see related papers in this volume). Although the conditions were different (e.g. period of exposure to light) we did not find evidence of closed stomata in well watered *Ceratonia* or reduced PC under high light, but CE was reduced in shade leaves. Because temperature increase in the cuvette (up to 38° C) was associated with our full sunlight treatments, high temperature accompanied by low humidity may be considered the cause of such an effect in the experiments reported here.

ACKNOWLEDGEMENT

Our appreciation to Gilberto da Conceição for his valuable help.

REFERENCES

Badger MR, Collatz GJ (1977) Studies on the kinetic mechanism of ribulose-1.5-bisphosphate carboxylase and oxygenase reactions with particular reference to the effect of temperature on kinetic parameters. Carnegie Ins Washington Yearb 76:355-361

Beyschlag W (1984) Photosynthese und Wasserhaushalt von *Arbutus unedo* im Jahreslauf am Freilandstandort in Portugal. PhD Thesis, University Würzburg

Farquhar GD, von Caemmerer S (1982) Modelling of photosynthetic response to environment. In: Lange OL, Nobel PS, Osmond CB, Ziegler H (eds), Encyclopedia of Plant Physiology, NS vol 12B: Physiological Plant Ecology II. Springer, Berlin-Heidelberg-New York-Tokyo pp 549-587

Ku SB, Edwards GE (1977) Oxygene inhibition of photosynthesis. I. Temperature dependence and relation to O_2/CO_2 solubility. Plant Physiol 59:986-990

Nunes MA, Bierhuizen JF, Ploegman C (1968) Studies on productivity of coffee. I. Effect of light, temperature and CO_2 concentration on photosynthesis of *Coffea arabica*. Acta Bot Neerl 17(2):93-102

Rhizopoulou S, Nunes MA (1981) Some adaptative photosynthetic characteristics of a sun plant, *Ceratonia siliqua* and a shade plant (*Coffea arabica*). In: Margaris NS, Mooney HA (eds) Components of Productivity of Mediterranean Climate Regions. Basic and Applied Aspects. DRW Junk Publishers, The Hague p 85

Tenhunen, JD, Lange OL, Beyschlag W, Weber JA (1984) Changes in photosynthetic capacity, carboxylation efficiency and CO_2 compensation point associated with midday stomatal closure and midday depression of net CO_2 exchange of leaves of *Quercus suber*. Planta 162:193-203

Weis E (1981) Reversible heat inactivation of the Calvin cycle: A possible mechanism of the temperature regulation of photosynthesis. Planta 151:33-39

Wong SC, Cowan IR, Farquhar GD (1978) Leaf conductance in relation to assimilation in *Eucalyptus pauciflora* Sieb. ex Spreng. Influence of irradiance and partial pressure of carbon dioxide. Plant Physiol 62:670-674

Wong SC, Cowan IR, Farquhar GD (1979) Stomatal conductance correlates with photosynthetic capacity. Nature (London) 282:424-426

Photosynthesis and water relations of grapevines growing in Portugal - response to environmental factors

M. M. Chaves and M. L. Rodrigues

Department of Botany, Instituto Superior de Agronomia, Tapada da Ajuda, 1399 Lisboa Codex, Portugal

INTRODUCTION

Grapevine (*Vitis vinifera*) is a perennial crop plant grown mostly in mediterranean type climate zones. The water requirements of the crop are provided in most situations by winter rains. Deep root systems are a feature of grapevines which allow ground water absorption during the dry period. Nevertheless, plants in many vineyards in Portugal are exposed to conditions of water and temperature stress during the summer period. We can assume that grapevine cultivars, adapted to hot dry summer conditions, have been selected for naturally and by man since early periods of cultivation, in the Douro Valley possibly as early as A.D. 55 (Olmo 1976).

Through empirical viticultural practices developed over centuries, a large volume of information has accumulated on procedures for maintaining crop quality and yield. The physiological bases for these proven cultural practices are largely unknown. The study of grapevine water relations and gas exchange responses, especially as affected by seasonal environmental change, will help improve our understanding of the strategies employed by this plant species that lead to favorable growth and production. Of particular relevance is the behavioral response of grapevine to conditions of water shortage.

Photosynthetic characteristics of grapevine under laboratory conditions have been reported by Kriedemann (1977) and Stoev and Slavtcheva (1982) and under natural conditions by Carbonneau (1978), Lange and Meyer (1979), and Chaves (1981). Studies of water relations have been conducted by Smart (1974), Liu *et al.* (1978), and Rodrigues (1984). Alleweldt *et al.* (1982) and Düring and Loveys (1982) emphasized the effects of water supply on plant growth and gas exchange.

In the present paper we summarize the information that we have gathered on leaf CO_2 exchange and plant water relationships in grapevines growing under natural conditions in Portugal. Diurnal patterns observed under different environmental conditions are analyzed and seasonal changes in parameters are summarized. Particular attention is given to changes which occur in physiological processes during exposure to drought and to the ability of the plant to recover after stress is relieved.

NATO ASI Series, Vol. G15
Plant Response to Stress. Edited by J. D. Tenhunen et al.

DIURNAL PATTERNS IN LEAF PHYSIOLOGICAL PARAMETERS

Typical diurnal patterns of leaf gas exchange obtained in early summer with well-watered, potted grapevine plants in Portugal are illustrated in Fig. 1 for three types of days. Measurements were made with a CO_2/H_2O porometer constructed by H. Walz (Effeltrich, W. Germany - for further experimental details see Chaves *et al.* 1986). On a cloudy day (July 3), with maximum air temperature (T_a) of 22°C and maximum incident photosynthetically active radiation (PAR) of about 800 μmol m^{-2} s^{-1}, the diurnal pattern of net photosynthesis rate (P_N) was "dome shaped". P_N was correlated with PAR and T_a and maximal rates occurred at approximately midday (14.00 hours). With higher air temperatures (up to 31°C) but with passing clouds (June 24), an assymetry became apparent in the diurnal response ofP_N with highest rates occurring during the morning period. On a bright sunny day (July 1) with maximum air temperature of 32°C and a PAR of 1950 μmol $m^{-2}s^{-1}$, this assymetry increased with peak photosynthetic rates taking place at about 10.00 hours, three hours before the maxima in air temperature and PAR, and with a continual decrease in P_N thereafter.

Changes in stomatal conductance to water vapor (g_s) were very similar to those occurring in P_N which might imply a strong diffusion limitation of photosynthesis rate. However, photosynthetic activity is limited not only by g_s but also endogenously, since leaf internal CO_2 pressure (p_i) remained relatively constant at light saturation (Fig. 2). These types of results were also obtained in studies of diurnal time courses of photosynthesis of *Quercus suber* under mediterranean climate conditions (Tenhunen *et al.* 1984, 1987). In the latter study, P_N was measured independently of CO_2 diffusion limitation by utilizing an ambient partial pressure of CO_2 of 2500 μbar. The CO_2 saturated photosynthesis rate and g_s responded in parallel to changes in temperature, leaf to air water vapour gradient (Δw), and light. This strong correlation between P_N and g_s (correlation coefficient of 0.96 in our grapevine data) has been interpreted by some authors as an indicator of a direct response of g_s to changes in P_N (Wong *et al.* 1979).

Transpiration rates (T_r in Fig. 1) were equally high in the morning and later in the day, despite the decreases in conductance during the afternoon. This is due to increases in leaf to air water vapour mole fraction gradient (Δw), which attains its highest value in the early afternoon (around 30 Pa kPa^{-1} for the sunny day and 20 Pa kPa^{-1} for the cloudy days). Water use efficiency (P_N/T_r in Fig. 2) was similarly quite constant except at the beginning and at the end of the day. Thus, the stomatal behavior discussed above effectively regulates water use.

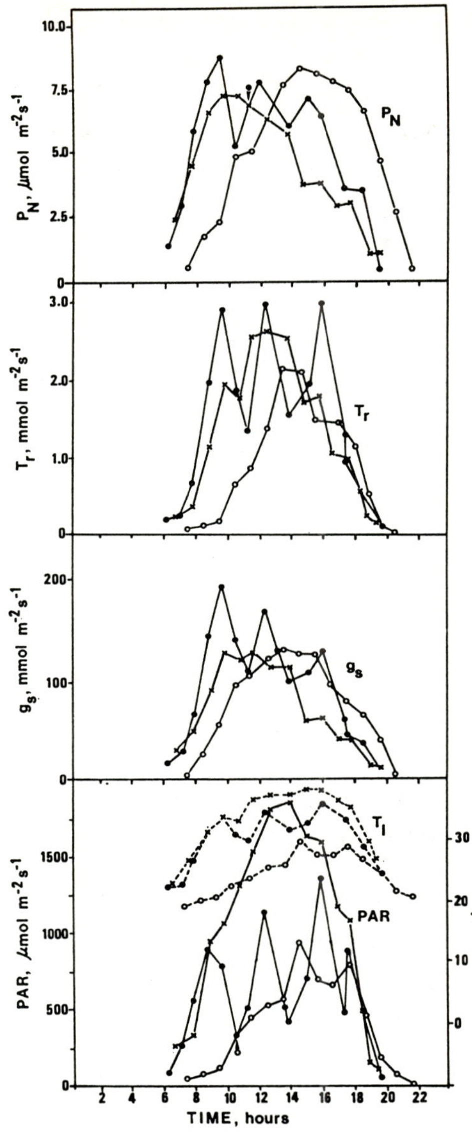

Fig. 1. Diurnal time courses of net photosynthesis rate (P_N), transpiration rate (T_r), stomatal conductance for water vapor (g_s), leaf temperature (T_1) and incident photosynthetically active radiation (PAR) on a cloudy day, July 3, 1983 (open circles); on a temporarily cloudy day, June 24, 1983 (closed circles) and on a bright sunny day, July 1, 1983 (X). Symbols indicate average values for 2 or 3 mature leaves of grapevines from cv. T. Amarela (Adapted from Chaves *et al.* 1986)

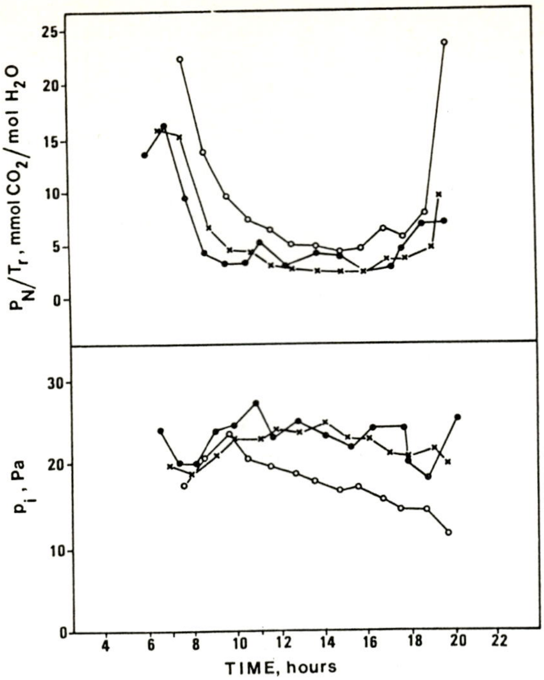

Fig. 2. Diurnal time courses of water use efficiency, expressed as P_N/T_r ratio and intercellular CO_2 partial pressure (p_i) on a cloudy day, July 3, 1983 (open circles); on a temporarily cloudy day, June 24, 1983 (closed circles) and on a bright sunny day, July 1, 1983 (X)

Diurnal patterns of leaf water potential (Ψ) and leaf osmotic potential (Ψ_s) obtained with field grown grapevines on a relatively cool, cloudy early summer day (June 27) and a warm, clear mid-summer day (August 13) are illustrated in Fig. 3. The osmotic potential of leaf sap (Ψ_s) was measured with a Knauer Semi-Micro Osmometer and leaf water potentials with a pressure chamber. On both days minimal ψ values were obtained around 13.00-15.00 hr, with a recovery thereafter. A rapid decrease in Ψ values observed during the morning of August 13 compared to June 27 occurs in response to changes in Δw, in spite of lower stomatal conductance. Diurnal patterns in Ψ_s were similar to those found for ψ, with maxima and minima of -0.60 MPa and -1.15 MPa on June 27 and of -1.05 MPa and -1.60 MPa on August 13. On August 13, the values of Ψ_s and Ψ approach each other at midday, suggesting that low turgor pressures occurred.

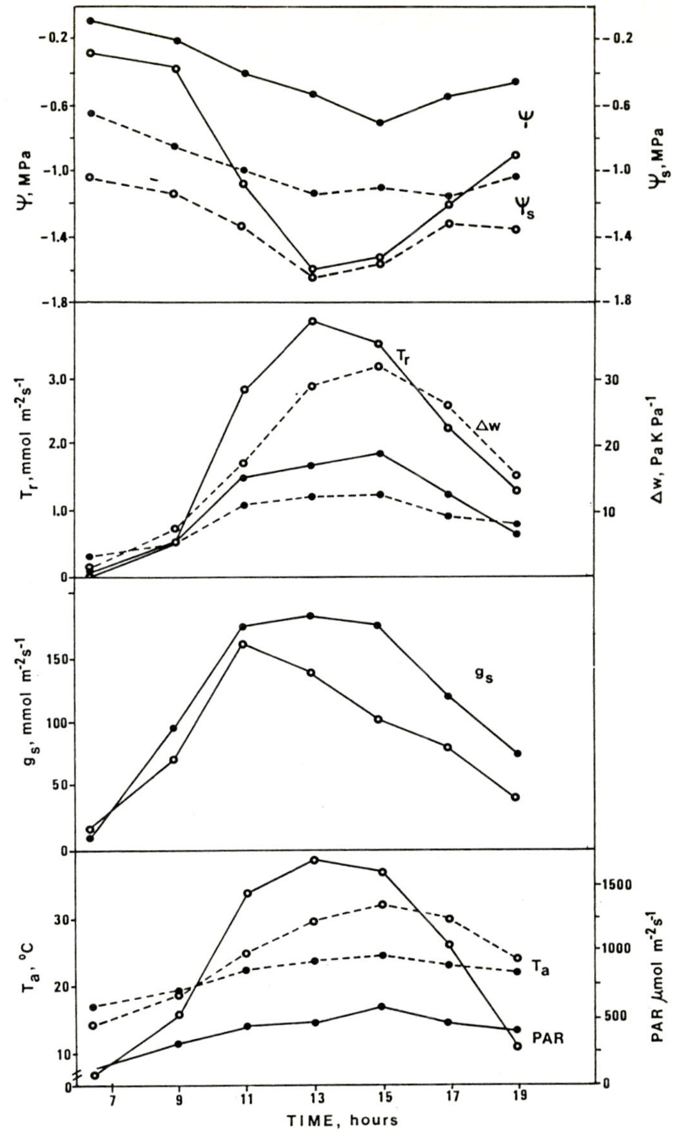

Fig. 3. Diurnal time courses of leaf water potential (Ψ), osmotic potential (Ψ_S), transpiration rate (T_r), leaf to air water vapor mole fraction gradient (Δw), stomatal conductance for water vapor (g_S), air temperature (T_a) and incident photosynthetically active radiation (PAR) measured for mature leaves of field grown grapevines (cv. Dona Maria) on June 27, 1984 (solid circles) and August 13, 1984 (open circles). Symbols indicate the mean values for 4 leaves

SEASONAL PATTERNS IN GAS EXCHANGE AND WATER RELATIONS

Leaf predawn water potential of field grown plants decreased throughout the growing season as soil water availability decreased (Fig. 4). The minimum water potential reached in early afternoon (Ψ_{min}) also decreased as the season progressed and the lowest values observed were -1.9 MPa. Nevertheless, moderate stomatal conductances were often found under conditions of water stress (Rodrigues unpublished). Two peaked diurnal response patterns for g_s with a midday stomatal closure, as frequently observed during summer for many mediterranean species (see Lange *et al.* and Tenhunen *et al* . in this volume), were not detected in our experiments with grapevines. However, the assymetry in g_s and P_N diurnal patterns, illustrated in Fig. 1 for potted plants, was also observed under field conditions (Fig. 3) when moderate water stress had developed and/or as the crop matured.

Maximal leaf photosynthesis rates measured in field grown grapevines increased as the season progressed until mid-August, despite increasing water stress (Fig. 5), then decreased in September. Young leaves and older basal leaves initially exhibited lower values of gross photosynthesis (P_G) than fully enlarged leaves in mid-position on the stocks. Lower rates of basal leaves can partly be accounted for due to natural shading. In August, the youngest terminal leaves had the highest photosynthesis rates. In September, all leaves responded more or less similarly with low positive uptake of CO_2.

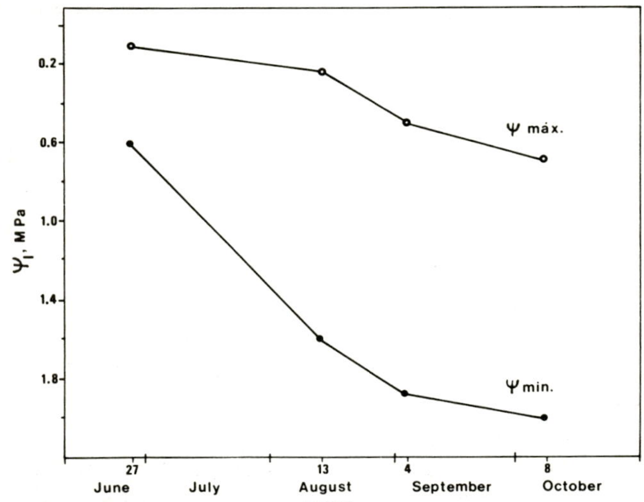

Fig. 4. Seasonal courses in predawn leaf water potential (open circles) and the minimum values observed in early afternoon (14.00-15.00 hours) with field grown grapevines (cv. Dona Maria). Symbols indicate mean values for 6 leaves

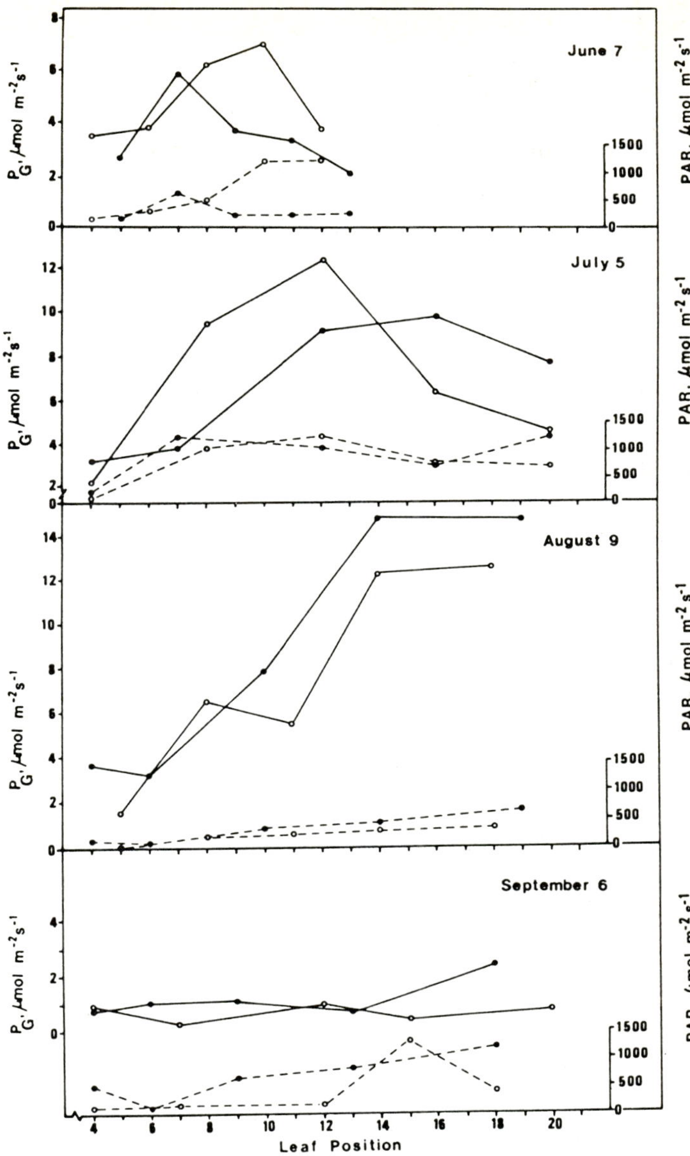

Fig. 5. Influence of leaf position and time during growing season (dates in panels of figure) on gross photosynthesis rate (P_G - solid line) of leaves in two shoots of field grown grapevine (cv. Periquita). Photosynthetically active radiation (PAR - dashed line) intercepted by the leaves in their natural position are indicated for each shoot. Numbering indicated for leaves is from the bottom of the shoot. Symbols indicate average values for 2 samples. Measurements were made with a $^{14}CO_2$ porometer

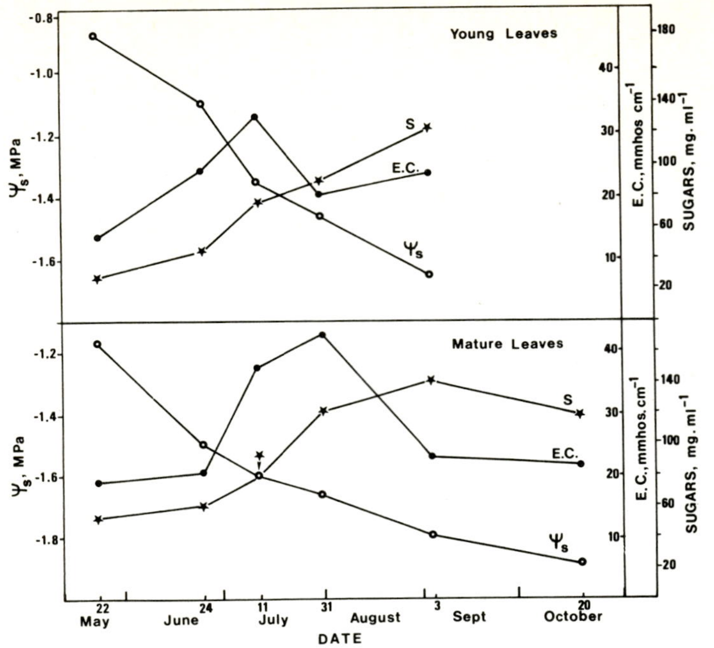

Fig. 6. Seasonal changes of leaf sap osmotic potential (Ψ_s - open circles), sap electrical conductivity (E.C. - solid circles), and total soluble sugars of sap (S - solid stars) measured in young and mature leaves of field grown grapevines (cv. Rosaki)

Osmotic potential of the sap, measured in the morning in both young and mature grapevine leaves (9.00 hours), became increasingly more negative as the season progressed (Fig. 6). Thus, osmotic adjustment took place as observed by Düring and Loveys (1982). The concentration of soluble sugars in the leaves increased and these metabolites probably play an important role in determining osmotic potentials. Sap electrical conductivity also increased in early summer and decreased thereafter.

RESPONSE TO WATER STRESS IN SEMI-CONTROLLED EXPERIMENTS

Semi-controlled experiments were set up at the Institute of Agronomy in Lisbon to more closely examine the response of grapevine plants to water stress and to guide future experimentation. Soil water shortage was slowly imposed on three grapevine cultivars planted in large containers. Observed changes in photosynthetic rate (P_N), respiration rate (R_D), stomatal conductance for water vapor (g_s) and leaf water potential (Ψ_1) during the drying cycle and after rewatering are illustrated along with control values in Fig. 7. Non-watered plants of all cultivars decreased stomatal conductance sharply two weeks after

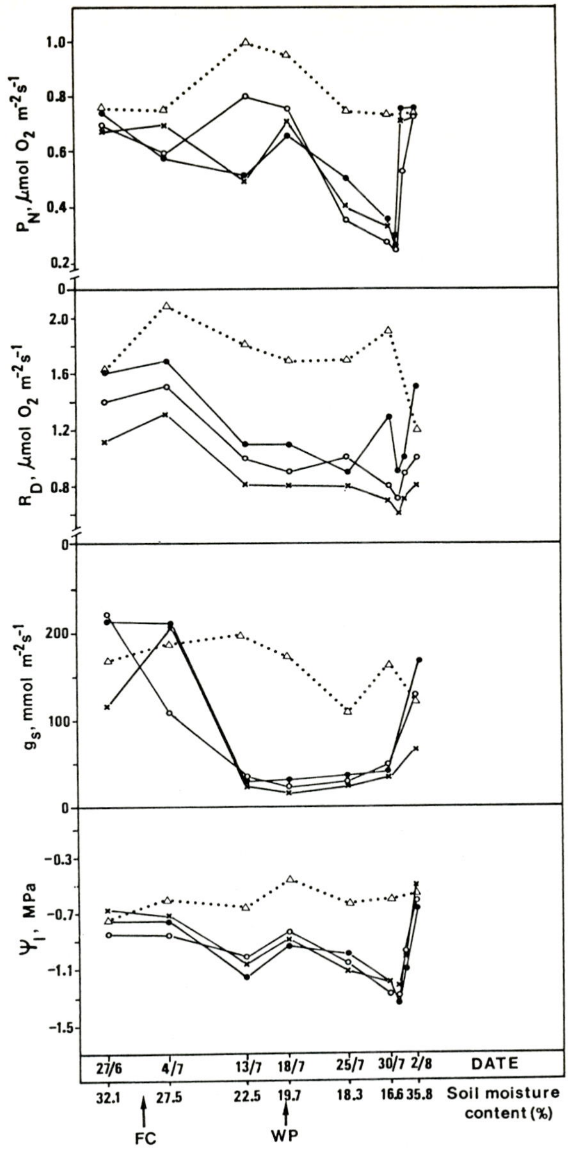

Fig. 7. Changes in photosynthetic rate (P_N), respiration rate (R_D), stomatal conductance for water vapor (g_s) and leaf water potential (Ψ_1) of three grapevines cultivars {Rosaki (closed circles), Sultana (X), Alphonse Lavallée (open circles)} during a water stress and recovery cycle. Plants were grown in containers and under field conditions. Triangle symbols indicate the mean values of the respective variables for control plants of the three cultivars. The arrows indicate field capacity (F.C.) and wilting point (W. P.)

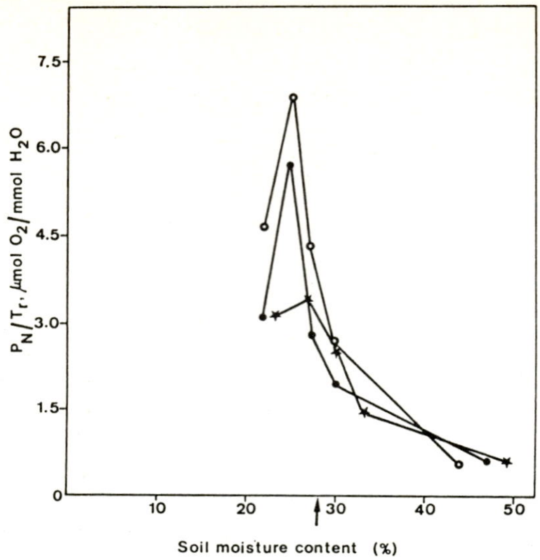

Fig. 8. Change in water use efficiency of different grapevine cultivars (Rosaki - open circles; Alicante Branco - solid circles; Dona Maria - solid stars) observed over a water stress cycle on potted plants grown in a greenhouse. The arrow indicates the soil moisture corresponding to field capacity

irrigation ceased and when early morning leaf water potential was approximately -1.1 MPa (measurements taken at 9.00 hours). At this early stage of the treatment, photosynthetic activity was not affected. Similar results were reported by Smart (1974). Tenhunen *et al.* (1985) found an early sensing of summer drought conditions as evidenced by decreases in maximal conductance in leaves of *Quercus coccifera* accompanying initial changes in predawn water potential and before leaf mesophyll photosynthetic characteristics were affected. Potential signals from the roots with partial soil drying that might be responsible for regulation of activity in the shoots are discussed by Davies *et al.* (this volume). Dark respiration (R_D) similarly decreased by more than 50% during this early stage of treatment. Since photosynthesis remained high, the decrease in respiration was not apparently related to maintenance of mesophyll photosynthetic capacity.

When Ψ reached -1.3 MPa, photosynthesis was depressed to approximately 30% of that of control plants. Thus, the plants studied in these experiments are much more sensitive to water potential decreases than plants established in vineyards (Figs. 4 and 5). One day after rewatering, a rapid recovery occurred and all variables studied reverted to levels approximately the same as control plants. Irreversible damage to cellular components apparently had not occurred.

The effects of drought on water use efficiency (expressed as P_N/T_r) are indicated in Fig.8. The highest water use efficiency was observed with mild water deficiency and resulted from the abrupt stomatal closure and decrease in T_r at this stage. In fact, the primary effect of drought was in stomata (see Fig.7) which affected more pronouncely transpiration than photosynthetic rates. However, as stress increased, P_N/T_r fell.

CONCLUSIONS

Grapevines growing in Portugal exhibit physiological adaptations to high temperatures and water shortages characteristic of mediterranean summer conditions. Stomatal conductance is maximal in the middle of the morning. This is more frequently the case as the season progresses and/or water stress develops. Stomatal conductance changes do not seem to be closely related to changes in leaf water potential. The seasonal decrease in osmotic potential (osmotic adjustment) allows grapevines to maintain a positive turgor as water potentials become more negative.

Future studies must include observations of seasonal changes in dark respiration and water use efficiency in established vineyards to clarify the importance of these during development of the crop and in response to naturally imposed summer drought. Controlled experimentation must examine the links between osmotic adjustment, seasonal increase in photosynthetic capacity and drought induced changes in physiological behavior.

REFERENCES

Alleweldt G, Eibach R, Rühl E (1982) Investigations on gas exchange in grapevine. I. Influence of temperature, leaf age and daytime on net photosynthesis and transpiration. Vitis 21:93-100

Carbonneau A (1978) Essai de détermination en biologie de la plante entière, de relations essentielles entre le bioclimat naturel, la physiologie de la vigne et la composition du raisin. Ann. Amélior. Plantes 28:195-221

Chaves MM (1981) Photosynthetic characteristics and environmental control in some grapevines cultivars. In: Margaris NG and Mooney HA (eds) Components of Productivity of Mediterranean Climate Regions - Basic and Applied Aspects. Dr. W. Junk, Publ.,The Hague, pp 105-113

Chaves MM, Harley P, Tenhunen JD, Lange OL (1986) Gas exchange studies in two Portuguese grapevine cultivars. In preparation

Düring H, Loveys BR (1982) Diurnal changes in water relations and abcisic acid in field grown *Vitis vinifera* cvs. I. Leaf water potential components and leaf conductance under humid temperate and semiarid conditions. Vitis 21:223-232

Kriedemann RE (1977) Vineleaf photosynthesis. In: International Symposium on the Quality of the Vintage. Capetown, South Africa, pp 67-87

Lange OL, Meyer A (1979) Midday stomatal closure observed in apricot (*Prunus armeniaca*) and grapevine (*Vitis vinifera*) under field conditions in spite of non-limiting soil-water supply. Flora 168:511-528

Liu WT, Pool R, Wenkert W, Kriedemann PE (1978) Changes in photosynthesis, stomatal resistance and abcisic acid of *Vitis labruscana* through drought and irrigation cycles. Ann J Enol Vitic 29:239-246

Olmo HP (1976) Grapes. In: Simmonds NW (ed) Evolution of Crop Plants. Longman, New York, pp 294-298

Rodrigues ML (1984) Water relations of grapevines grown under different irrigation treatments. Proc. 4th Congress Federation European Societies Plant Physiology, Strasbourg, France, pp 305-306

Smart RE (1974) Aspects of water relations of the grapevine (*Vitis vinifera*). Amer J Enol Viticult 25:84-91

Stoev K, Slavtcheva T (1982) La photosynthèse nette chez la vigne (*Vitis vinifera* L.) et les facteurs écologiques. Connaissance de la Vigne et du Vin 16:171-185

Tenhunen JD, Lange OL, Gebel J, Beyschlag W, Weber JA (1984) Changes in photosynthetic capacity, carboxylation efficiency, and CO_2 compensation point associated with midday stomatal closure and midday depression of net CO_2 exchange of leaves of *Quercus suber*. Planta 162:193-203

Tenhunen JD, Lange OL, Harley PC, Beyschlag W, Meyer A (1985) Limitations due to water stress on leaf net photosynthesis of *Quercus coccifera* in the Portuguese evergreen scrub. Oecologia (Berl) 67: 23-30

Tenhunen JD, Pearcy RW, Lange OL (1987) Diurnal variations in leaf conductance and gas exchange in natural environments. In: E Zeiger, G. Farquhar and I. Cowan (eds) Stomatal function. Stanford University Press, Stanford, California, in press

Wong SC, Cowan IR, Farquhar GD (1979) Stomatal conductance correlates with photosynthetic capacity. Nature 282:424-426

Chlorophyll fluorescence as an indicator of heat induced limitation of photosynthesis in *Arbutus unedo* L.

W. Bilger, U. Schreiber and O.L. Lange

Institut für Botanik und Pharmazeutische Biologie der Universität Würzburg, Mittlerer Dallenbergweg 64, D-8700 Würzburg, Federal Republic of Germany

INTRODUCTION

Chlorophyll fluorescence analysis may provide us with a sensitive indicator of stress induced limitations of photosynthesis (for recent reviews see Smillie and Hetherington (1983) and Schreiber and Bilger, this volume). High temperature treatment causes a variety of fluorescence changes: (1) At elevated temperatures the dark fluorescence level, F_o, is increased several-fold (Schreiber and Berry 1977), the critical temperature for the F_o-increase being closely correlated with the critical temperature for irreversible damage to the leaf tissue as determined from leaf necrosis (Bilger, Schreiber and Lange 1984). (2) Following heat treatment, there is a decrease in variable fluorescence, F_{var}, as measured upon illumination at room temperature (Krause and Santarius 1975; Santarius and Müller 1979). In isolated chloroplasts this decrease in variable fluorescence has been correlated with damage to the watersplitting system (Yamashita and Butler 1968). (3) Mild heat treatment, which does not cause irreversible tissue damage, induces changes in the low temperature fluorescence emission spectra indicative of a heat induced state I - state II shift, i.e. of a change in energy distribution in favor of photosystem I (Weis 1984, 1985).

Here we wish to report on heat induced changes in fluorescence response of *Arbutus unedo* leaves observed with a new modulation fluorometer (Schreiber, Schliwa and Bilger 1986) and by utilizing the saturation pulse method (Bradbury and Baker 1981; Quick and Horton 1984). By differentiating between photochemical and nonphotochemical fluorescence quenching (see Schreiber and Bilger, this volume), a sensitive means of determining heat induced limitations of photosynthesis is established. It will be shown that heating affects photochemical and nonphotochemical quenching before structural damage is indicated by F_o and F_{var} changes. The changes in fluorescence quenching are closely correlated with a heat induced decrease of CO_2-fixation rate.

MATERIALS AND METHODS

The experiments were carried out with fully expanded leaves of *Arbutus unedo*, grown as potted plants at the University of Würzburg Botanical Garden. During

NATO ASI Series, Vol. G 15
Plant Response to Stress. Edited by J. D. Tenhunen et al.

summer when this study was conducted, the plants were standing outside. Detached leaves were dark adapted for 2 to 4 hours, before being heat treated in darkness by immersion in a stirred waterbath for 5 min at temperatures between 38 and 54° C. Immediately after heat treatment, fluorescence induction curves were recorded using a modulation fluorometer (PAM 101 Chlorophyll Fluorometer, H. Walz, Effeltrich, Germany). White actinic light was supplied by a halogen lamp fiber-illuminator (Schott KL 1500). Saturation pulses of strong white light (10,000 μmol m^{-2} s^{-1} photosynthetically active radiation, PAR, for 0.7 sec) were obtained from a halogen lamp pulse source (FL 103, Walz). Saturation pulse light passed through a short-pass filter ($\lambda < 700$ nm; Balzers DT Cyan) before entering the fiberoptics. The quenching coefficients were calculated as outlined by Schreiber *et al.* (1986; see also Schreiber and Bilger, this volume). Oxygen evolution was measured polarographically with a leaf disc electrode (Type LD2, Hansatech, Ltd., Norfolk, U.K.) at 5% CO_2 and 20° C, the leaf discs being cut out from a freshly heated leaf. Light intensity was measured with a quantum-radiometer (LI 185B, LI-COR, Lincoln, Nebraska). Chlorophyll content was determined according to the method of Arnon (1949).

RESULTS

Original traces of fluorescence induction curves of *Arbutus unedo* control and heat treated leaves as measured with the modulation fluorometer are shown in Fig. 1. The intensity of the modulated measuring beam is sufficiently low to prevent a fluorescence increase beyond the dark level, F_o. Upon application of 0.7 second light pulses of saturating intensity, photochemical quenching is transiently eliminated with complete reduction of PS II acceptors and the modulated fluorescence signal rises to a saturation level. When saturation pulses are repeatedly applied and superimposed on otherwise continuous actinic light, the resulting fluorescence induction curves provide information on the state of the photosynthetic apparatus. Heat pretreatment strongly affects the pattern of such induction curves, particularly with respect to the fluorescence yield during saturation pulses.

The changes observed in fluorescence yield provide information on the effects of temperature on photochemical quenching, q_Q, and energy-dependent quenching, q_E (q-analysis, see Schreiber and Bilger, this volume). In Fig. 2, time courses of q_Q and q_E are compared with oxygen exchange rates for control and heat treated leaves. It is apparent that at approximately 2 minutes after onset of illumination, the rise in O_2-evolution rate in a control leaf is associated with an increase in q_Q and a decrease in q_E. The oscillations are typical for experiments conducted at high CO_2 partial pressure (Walker *et al.* 1983) used here to avoid stomatal limitation of gas exchange. In the heat treated leaf, an initial burst in O_2 is followed by a significant decline in O_2-evolution, then a slow rise sets in

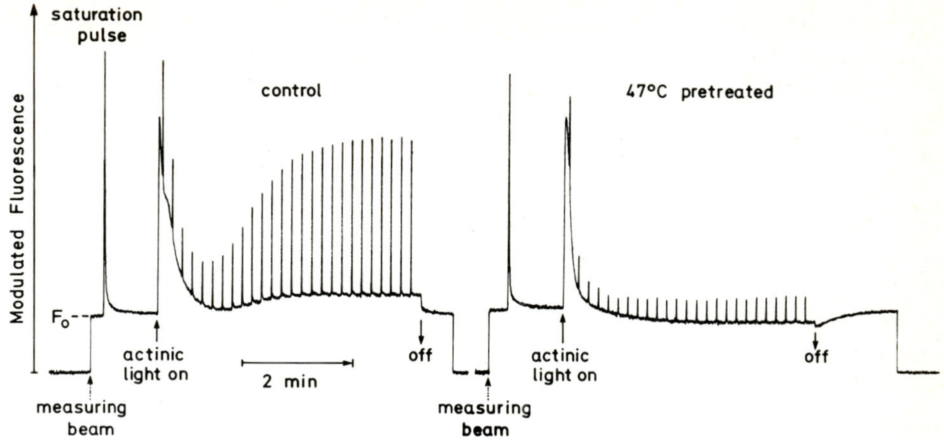

Fig. 1. Original recordings of fluorescence induction curves of predarkened *Arbutus unedo* leaves measured with the PAM modulation fluorometer. One minute before and every 10 seconds during actinic illumination (125 µmol m^{-2} s^{-1}) saturation pulses (0.7 second duration, 10,000 µmol m^{-2} s^{-1}) were given. The samples were enclosed in a leaf chamber with 5% CO_2 in air. Heat pretreatment was carried out as described in MATERIALS AND METHODS

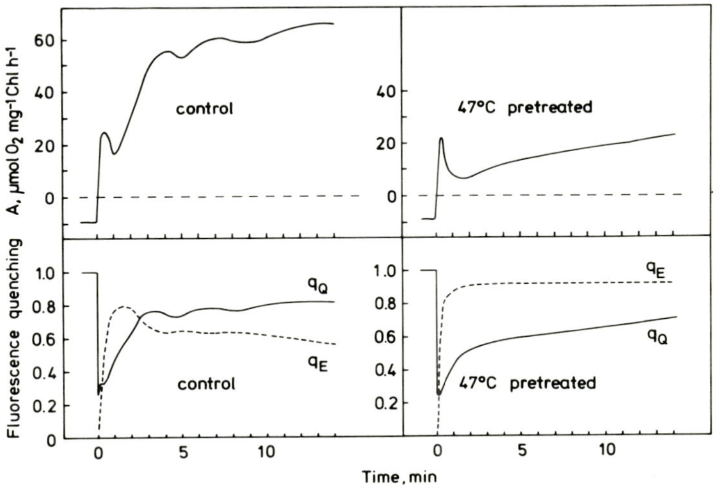

Fig. 2. Comparison between the time dependent changes of assimilation rate and quenching coefficients in a control and a heated leaf disc. At time zero, actinic light (125 µmol m^{-2} s^{-1}) was switched on. Fluorescence and oxygen evolution were measured simultaneously in a leaf disc electrode at 20° C and 5% CO_2 in air. Every 10 seconds a saturation pulse was given

accompanied by a slow increase in q_Q. At the same time, close to maximal q_E is observed. Compared to the control leaf, the heat treated leaf is severely limited in the use of NADPH and ATP in the Calvin cycle, leading to accumulation of electrons and of membrane energy. This is already apparent from the original induction curves depicted in Fig. 1.

The dependence of steady state Q-quenching (panel A) and E-quenching (panel B) on the temperature of a preceding heat treatment is illustrated in Fig. 3. In the same figure, the corresponding temperature dependencies of initial fluorescence, F_o (panel C), and of variable fluorescence, F_{var} (panel D), are depicted. For comparison, the dependence of CO_2 assimilation rate on pretreatment temperature is shown in Fig. 4. The decline in overall rate is most closely matched by the decrease in Q-quenching (Fig. 3A) in the temperature range between 44 and 48° C. E-quenching starts increasing at considerably lower temperatures (Fig. 3B), while the changes in F_o (Fig. 3C) and F_{var} (Fig. 3D) occur at higher temperatures (critical temperature about 47° C) than changes in q_Q, q_E and assimilation rate.

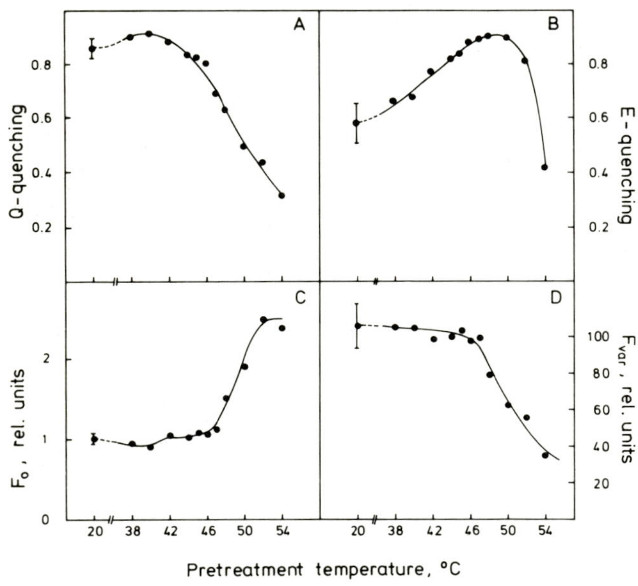

Fig. 3. Dependencies of different fluorescence parameters on pretreatment temperature. The error bars represent standard deviations (n=4), points without error bars represent mean values from two separate measurements. Q- and E-quenching are determined in the steady state after 8 min. F_{var} is the difference between F_o and the maximal fluorescence level obtained in the first saturation pulse

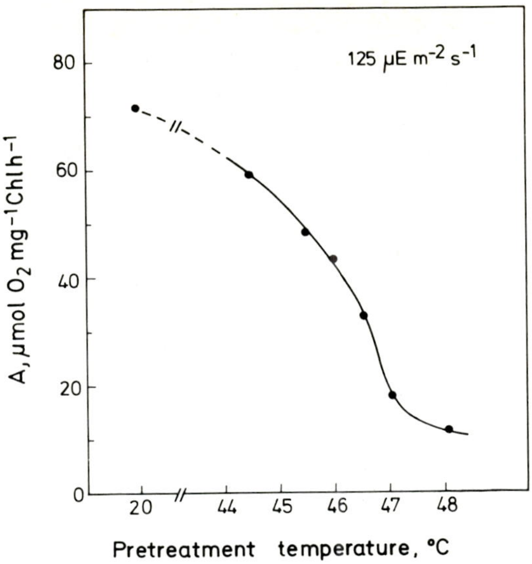

Fig. 4. Dependence of assimilation rate at steady state, measured as oxygen evolution rate in a leaf disc electrode at 5% CO_2 in air, 20° C and 125 μmol m^{-2} s^{-1}, on pretreatment temperature. Steady state was reached after 15 min

The close correlation between Q-quenching and assimilation rate is further demonstrated in Fig. 5. Assimilation rate and q_Q are linearly related, when photosynthetic capacity is limited to a variable extent by heat pretreatment at different temperatures.

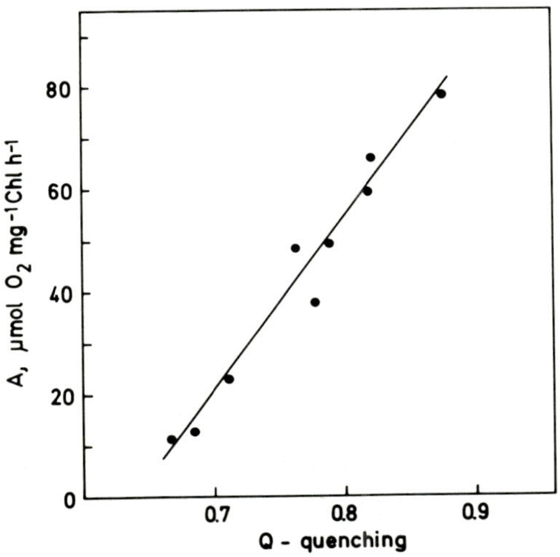

Fig. 5. Relationship between assimilation rate and Q-quenching determined simultaneously after different heat pretreatments. Measuring conditions as in Fig. 4

Returning to the original recordings of fluorescence induction curves in Fig. 1, fluorescence yield drops below the F_o-level during illumination in the heat treated sample. When the actinic light is turned off, there is a small, rapid decrease of fluorescence followed by a slower increase to the original F_o-level. Energy quenching does not only affect variable fluorescence, but F_o as well, as already described for strong energization at high light intensities and in the absence of CO_2 (Bilger and Schreiber 1987). The energy-quenched dark fluorescence level F_6, which is observed briefly after darkening, has been interpreted as indicating a shift of energy distribution in favor of PS I. In Fig. 6, F_6 is plotted as a function of pretreatment temperature and compared with F_o measured in pretreated leaves but before exposure to actinic light. With increasing temperature, actinic light causes a decrease in F_6, which reaches a minimum value at 47° C. This minimum is identical to the value observed in a control leaf exposed to strong illumination in the absence of CO_2 (marked by an asterisk in Fig. 6).

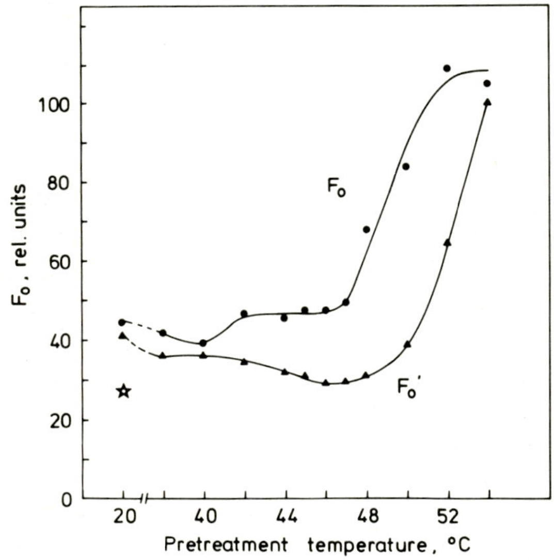

Fig. 6. Dependence of F_o and the minimal level of fluorescence, F_6, after turning-off the actinic light on pretreatment temperature. The asterisk (☆) denotes the F_o-value of a control leaf observed following illumination with 1,500 μmol m^{-2} s^{-1} PAR in the absence of CO_2. F_o-values were determined after 8 minutes illumination at 125 μmol m^{-2} s^{-1} PAR in the presence of air level CO_2

DISCUSSION

Detailed information on heat induced limitation of photosynthesis may be obtained by measurements of modulated chlorophyll fluorescence and application of the saturation pulse method. By differentiating between photochemical and non-photochemical quenching, clearcut insights are gained into limitations on electron transport and membrane energy utilization. With the formalism of q-analysis, at least semiquantitative data are obtained as indicated by the linear dependence between q_Q and assimilation rate (Fig. 5).

In previous work (Schreiber and Berry 1977; Smillie 1979; Bilger *et al.* 1984; Berry, Fork and Garrison 1975; Santarius and Müller 1979), heat damage has been evaluated by study of the increase in F_o and the decrease in F_{var}. While these parameters may well describe the extent of irreversible damage to the overall leaf tissue, the present data suggest that assimilation rate is already affected at considerably lower temperatures (Figs. 3C, 3D and 4). The steady state value of q_Q (Figs. 3A and 5) is a reliable indicator of this type of heat induced limitation. This means that NADPH consumption in the Calvin cycle is inhibited at lower temperatures than electron transport capacity. Since q_Q may be readily determined by the saturation pulse method with plants in their natural environment, a rapid means for assessing heat induced limitation of photosynthesis *in situ* is now available.

Even at pretreatment temperatures lower than those affecting assimilation rate and q_Q, a significant increase in non-photochemical quenching occurs (q_E, Fig. 3B) which is reflected in the decrease of the dark fluorescence level following illumination (F_6, Fig. 6). This increase in quenching appears to involve two types of heat induced changes. First, mild heating seems to cause a shift of energy distribution in favor of PS I. Such a shift has also been shown to occur in the dark when longer periods of heating are applied (Weis 1984, 1985). The present results suggest that pigment state shifts are enhanced upon illumination. The consequences are a decrease in F_o to F_6 (Fig. 6), a stimulation of PS II reoxidation (see small increase of q_Q in Fig. 3A) and a stimulation of cyclic flow around PS I which causes extra-energization. Second, mild heating appears to cause deactivation of Calvin cycle enzymes (Weis 1981), curbing ATP consumption and leading to accumulation of stored energy in membrane gradients and increase in q_E.

Both quenching coefficients, q_Q and q_E, are more sensitive to heat induced changes of the photosynthetic apparatus than the fluorescence parameters used in the past, F_o and F_{var}. The former seem to be more indicative of functional change which is readily reversible (see Weis 1981), whereas the latter parameters reflect structural damage, repair of which requires at least several days (Bilger, unpublished data). Through use of the newly developed modulation fluorometer in combination with the saturation pulse method (Schreiber *et al.* 1986), a reliable means is obtained of determining physiologically relevant limitations of photosynthesis. Since this measuring system is not affected by external ambient light, it is well suited for field research. Hence, it should be possible to determine the effect of heat in combination with other environmental stress factors on photosynthesis of attached leaves *in situ.* Such work would complement corresponding research done so far by gas exchange measurements (see papers of Lange, Tenhunen, Beyschlag and Harley, this volume).

ACKNOWLEDGEMENT

We wish to thank the Deutsche Forschungsgemeinschaft for financial support.

REFERENCES

Arnon DJ (1949) Copper enzymes in isolated chloroplasts. Polyphenoloxidase in *Beta vulgaris.* Plant Physiol 24:1-15

Berry JA, Fork DC, Garrison S (1975) Mechanistic studies of thermal damage to leaves. Carnegie Inst Year Book 74:751-759

Bilger W, Schreiber U (1987) Energy dependent quenching of dark-level chlorophyll fluorescence in intact leaves. In: Govindjee et al. (eds), Excitation Energy and Electron Transfer in Photosynthesis, Photosynthesis Research, suppl vol. Martinus Nijhoff/Dr. W. Junk, Dordrecht, in press

Bilger W, Schreiber U, Lange OL (1984) Determination of leaf heat resistance: Comparative investigation of chlorophyll fluorescence changes and tissue necrosis methods. Oecologia (Berl) 63:256-262

Bradbury M, Baker N (1981) Analysis of the slow phases of the *in vivo* chlorophyll fluorescence induction curve. Changes in the redox state of photosystem II electron acceptors and fluorescence emission from photosystem I and II. Biochim Biophys Acta 635:542-551

Krause GH, Santarius KA (1975) Relative thermostability of the chloroplast envelope. Planta 127:285-299

Quick WP, Horton P (1984) Studies on the induction of chlorophyll fluorescence quenching by redox state and transthylakoid pH gradient. Proc R Soc Lond B 220:371-382

Santarius KA, Müller M (1979) Investigations on heat resistance of spinach leaves. Planta 146:529-538

Smillie RM (1979) Coloured components of chloroplast membranes as intrinsic membrane probes for monitoring the development of heat injury in intact tissues. Aust J Plant Physiol 6:121-133

Smillie RM, Hetherington SE (1983) Stress tolerance and stress-induced injury in crop plants measured by chlorophyll fluorescence *in vivo.* Plant Physiol 72:1043-1050

Schreiber U, Berry JA (1977) Heat-induced changes of chlorophyll fluorescence in intact leaves correlated with damage of the photosynthetic apparatus. Planta 136:233-238

Schreiber U, Schliwa U, Bilger W (1986) Continuous recording of photochemical and non-photochemical chlorophyll fluorescence quenching with a new type of modulation fluorometer. Photosynth Res in press

Walker DA, Sivak MN, Prinsley RT, Cheesbrough JK (1983) Simultaneous measurement of oscillations in oxygen evolution and chlorophyll a fluorescence in leaf pieces. Plant Physiol 73:542-549

Weis E (1981) Reversible heat-inactivation of the Calvin-cycle: A possible mechanism of the temperature regulation of photosynthesis. Planta 151:33-39

Weis E (1984) Short term acclimation of spinach to high temperatures. Effect on chlorophyll fluorescence at 293 and 77K in intact leaves. Plant Physiol 74:402-407

Weis E (1985) Light- and temperature-induced changes in the distribution of excitation energy between PS I and PS II in spinach leaves. Biochim Biophys Acta 807:118-126

Yamashita T, Butler WL (1968) Inhibition of chloroplasts by UV-irradiation and heat-treatment. Plant Physiol 43:2037-2040

Ecological implications of sun/shade-leaf differentiation in sclerophyllous canopies: Assessment by canopy modeling

H.P. Meister, M.M. Caldwell* , J.D. Tenhunen** and O.L. Lange

Lehrstuhl für Botanik II der Universität Würzburg, Mittlerer Dallenbergweg, 8700 Würzburg, West Germany

INTRODUCTION

In mediterranean sclerophyllous shrubs with high leaf area indices, light can be a limiting factor for canopy photosynthesis in the spring when water availability exerts little limitation (Miller 1983, Tenhunen *et al.* 1984). Light penetration within stands is determined by structural characteristics of the plant, primarily the distribution of leaf area index (LAI), stem area index (SAI), and leaf and stem inclination angles at various levels in the canopies. The resulting light extinction patterns can result in the development of leaves of different morphological and physiological characteristics known as "sun/shade leaf" adaptations. Very different degrees of adaptation have been reported in the literature (e.g. Björkman 1981, Björkman *et al.* 1972, Schulze 1970, Syvertsen 1984) but data on hard-leaf vegetation have been only rarely available. Also, quantitative estimates of the significance of these shade adaptations for canopy photosynthesis and water use efficiency are seldom attempted.

We determined the degree of sun/shade leaf differentiation for *Quercus coccifera* a shrub commonly found in mediterranean areas of southwest Europe. With respect to light extinction, four different canopy types were chosen in order to assess the significance of sun/shade differentiation for a range of different canopy structures. We used a canopy model that was developed for *Quercus coccifera* following the principles outlined by Duncan *et al.* (1967), Miller (1972) and Norman (1980) and incorporates the complex interrelations between plant structure, microclimate profiles, individual leaf processes and canopy gas exchange.

*Department of Range Science, Utah State University, Logan, Utah 84322, USA

**Systems Ecology Research Group, San Diego State University, San Diego, California 92182, USA

NATO ASI Series, Vol. G 15
Plant Response to Stress. Edited by J. D. Tenhunen et al.

METHODS

Measurements

Dark respiration, light compensation points, and daily photosynthesis and conductance profiles were measured with a CO_2-Porometer (Schulze *et al.* 1982, Lange *et al.* 1985) in *Quercus coccifera* canopies at the Research Station Quinta São Pedro in Portugal. For each of four shrubs which represent different types of canopies, structural analyses were performed in 5-cm intervals, including LAI, SAI, leaf inclination, stem inclination and leaf azimuth angle. These parameters were assessed for canopy segments covering 300 to 900 cm^2 ground area. Simultaneously, microclimatic data necessary for both model input and model verification were collected. For further verification purposes a "traversing light sensor method" (Roberts and Miller 1977, Baldocci *et al.* 1984 Seeber 1984) was employed using a horizontal moving sensor for photosynthetically active radiation (PAR). Continuous recording during each pass in different canopy layers provided profiles of actual sunlit and shaded foliage areas and extinction data for direct beam and diffuse photosynthetic active radiation. (PAR).

The Model

The model calculates aboveground microclimatic and CO_2 and H_2O gas exchange profiles of a horizontal *Quercus coccifera* canopy, which is divided into 3 to 10 layers. Input and output parameters of the modeled system are shown in Fig. 1. Shrub structure is characterized by values of leaf area index (LAI), stem area index (SAI), leaf inclination and stem inclination for each layer. A gap-probability calculation approach following Monsi and Saeki (1953), Warren Wilson (1960) and Duncan *et al.* (1967) is used to compute direct shortwave radiation (SWR) extinction by leaves and twigs, treating PAR and total SWR separately. Uniform sky brightness and the sun as a point source of direct beam are assumed. Diffuse SWR penetration from the sky is computed analogously by assuming radial sky segments to be the radiation source, whereas leaf diffuse SWR is handled by application of a cone-geometrical scattering approach (Duncan *et al.* 1967). A leaf conductance submodel is driven by light, water vapor pressure deficit and predawn water potential values according to common assumptions of stomatal behaviour (Hall *et al.* 1976, Schulze and Hall 1982). Wind attenuation follows a LAI-dependent logarithmic decline with depth without further consideration of turbulent processes (Campbell 1977).

Penetration and scattering of longwave radiation (LWR) are treated in the same manner as diffuse SWR, longwave radiation emittance by leaves is a function of previously calculated leaf temperatures. Since several factors are interdependent, such as LWR, conductance, transpiration and leaf temperature, these calculations must be conducted iteratively. This proceeds until individual leaf energy budgets (Gates 1980) in all canopy layers are balanced. Sunlit and

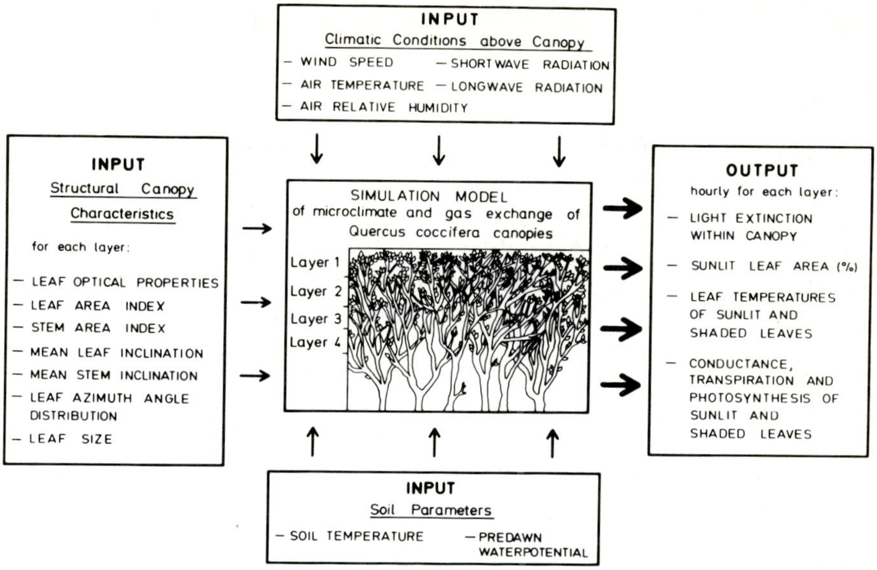

Fig. 1. Input and output parameters of the canopy microclimate and gas exchange model

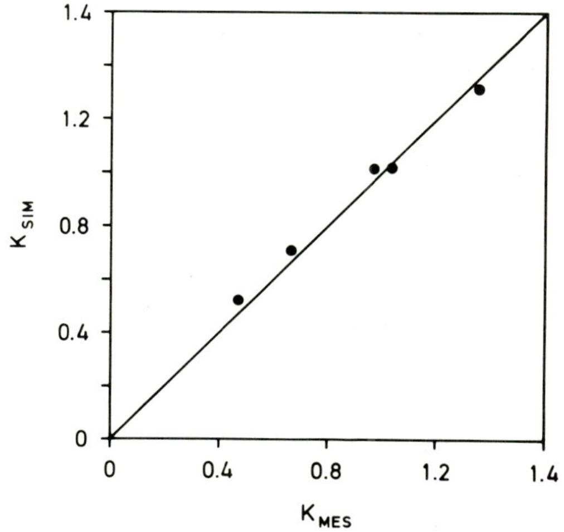

Fig. 2. Comparison of simulated versus measured PAR extinction coefficients for several canopies

shaded leaves at each depth are treated separately. Leaf temperature, PAR and conductance values finally enter a photosynthesis submodel that contains a Johnson equation for temperature dependency and two Smith equations for light and net CO_2 responsivity similar to the approach of Weber *et al.* (1985, see also Tenhunen *et al.* 1976 and immediately preceeding in this volume) without a separate calculation of photorespiration. Dark respiration follows an Arrhenius equation. Model parameters were derived from gas exchange measurements with large cuvettes under controlled conditions in the field (see Tenhunen *et al.* 1985, also Tenhunen, Beyschlag *et al.* this volume). Gas exchange for each canopy layer is summed for the entire canopy for each hour.

Verifications of the major simulated microclimatic and gas exchange values were obtained in the field by continuous measurements within the canopy. An example of measured and simulated PAR extinction coefficients for different shrubs is shown in Fig. 2. A regression with a slope of 1.01 and r^2 of 0.99 indicates the degree of correspondence obtained in these simulations. Verifications of gas exchange profiles tend to show less precision. This is due to the variability in photosynthetic capacity among individual leaves and changes in the basic photosynthetic capacity during the course of the day which are not understood well enough at present to be included in the model.

Simulations

Four different sets of shrub structural characteristics were used with different LAI values: 8.2 (shrub 1), 5.4 (shrub 2), 3.3 (shrub 3) and 2.1 (shrub 4). For each shrub, simulations were performed for three different cases using the same microclimatic input data for a warm, clear spring day in March:

- Case 1: sun/shade leaf differentiation corresponding to actually measured values were used.
- Case 2: 50% of the above differentiation was assumed.
- Case 3: no differences between leaves were assumed.

Characteristics of shade adaptation considered were dark respiration, light compensation points, leaf conductance and photosynthetic capacity.

RESULTS AND DISCUSSION

Measurements within a 2-m tall canopy with LAI of 8.2 and average leaf inclination decreasing from 60 degrees at the top to 40 degrees in lower layers, (shrub 1) showed pronounced differences between sun and shade leaves. Less differentiation was found in shrub 4 (Table 1). This canopy had significantly less LAI (2.1), leaf inclinations decreasing from 72 to 56 degrees from upper to lower layers, respectively, and consequently less light extinction in the canopy. The most obvious distinction between the two shrubs was in light compensation

Table 1. Characteristics of the lowermost shade leaves as a percent of the average sun leaf values in the spring.

	Shrub 1	Shrub 4
Light compensation at 18° C	15%	69%
Dark respiration at 20° C	33	58
Specific leaf weight	71	79
Leaf length	77	70
Photosynthetic capacity Coefficient*	60	45
Maximal conductance Coefficent*	10	78

*Scaling coefficients reflecting maximum photosynthetic capacity and maximum leaf conductance used in the photosynthesis and conductance submodels. These were derived for the submodels from measured gas exchange behaviour in the spring under ambient conditions in the field.

Table 2. Simulated daily gas exchange balances of *Quercus coccifera* canopies.

		P (μmol m^{-2} s^{-1})	P/T (μmol/mol)
Shrub 1	Case 1	364	2.8
	Case 2	328	1.9
	Case 3	226	1.1
Shrub 2	Case 1	366	2.5
	Case 2	348	2.0
	Case 3	297	1.5
Shrub 3	Case 1	320	2.8
	Case 2	316	2.5
	Case 3	298	2.1
Shrub 4	Case 1	220	3.5
	Case 2	237	3.0
	Case 3	247	2.7

point of shade leaves which was, in comparison to that of sun leaves, much lower in shrub 1 than in shrub 2. Dark respiration rates and leaf conductance capacities showed similar patterns. Morphological characteristics (leaf specific weight and leaf size) and photosynthetic capacity appeared to differ less between the shrubs.

When the leaf characteristics measured in the field were used in the simulations, with a gradation with depth in the canopy (case 1), the highest net CO_2-uptake during a 24-hour period was achieved compared to case 2 and 3 (as shown in Table 2 for shrubs 1, 2 and 3). For shrub 1, shade acclimation appeared to provide the greatest advantages: daily net CO_2-gain is 61% and water use efficiency is 155% greater for case 1 than for case 2. These differences are mainly due to changes in the most light-limited layers (Fig. 3), in which sun leaves cannot effectively use low light intensities. With a decreasing degree of differentiation, the contribution of lower layer leaves to total canopy photosynthesis declines, and in case 3, a positive carbon balance cannot be achieved in the lower layers.

Obviously, with decreasing LAI, shade acclimation becomes less important for CO_2-uptake balance as a consequence of greater light penetration to lower layers. The trend mentioned above is, in fact, reversed for shrub 4 where case 3 leads to higher CO_2-uptake than case 1 (Table 2). This is due to the lower photosynthetic capacity in shade leaves that does not enable them to exploit the available light energy completely. In shrub 4, even in the lowest layer, leaves are exposed to light intensities above light saturation during most of the day, whereas in the other shrubs, shade leaves frequently operate close to light compensation points. From these results, it appears that at approximately an LAI of 2.3, sun/shade differentiation becomes advantageous for carbon gain under the conditions of these simulations.

The ratio of photosynthetic CO_2 uptake to transpirational water loss, on the other hand, benefits from sun/shade differentiation in all four shrubs. Daily water use efficiency, P/T, is always greatest for case 1. The largest differences occur for shrub 1 where in case 2 a decrease of 32%, and in case 3, a decrease of 46% compared to case 1 is found. This is mainly due to increased transpiration in lower layers as a consequence of higher conductance values. Total transpiration in shrub 1 is 33% higher in case 2, and 58% higher in case 3, than in case 1. Conductance of both sunlit and shaded leaves in lower layers are affected. At a given light intensity, shade-adapted leaves have lower stomatal conductance. This leads to water conservation for shaded leaves in the lower canopy layers. However, this also results in lower conductance when these lower leaves are in sunflecks. In this case, conductance may limit the capacity of the leaf to utilize the light energy. In the springtime, when water is less limiting, somewhat higher conductance of lower leaves may be beneficial allowing exploitation of sunflecks, whereas in the dry summer period, this would lead to excessive water loss. Our measurements in different times of the year indeed indicate a gradual decline in

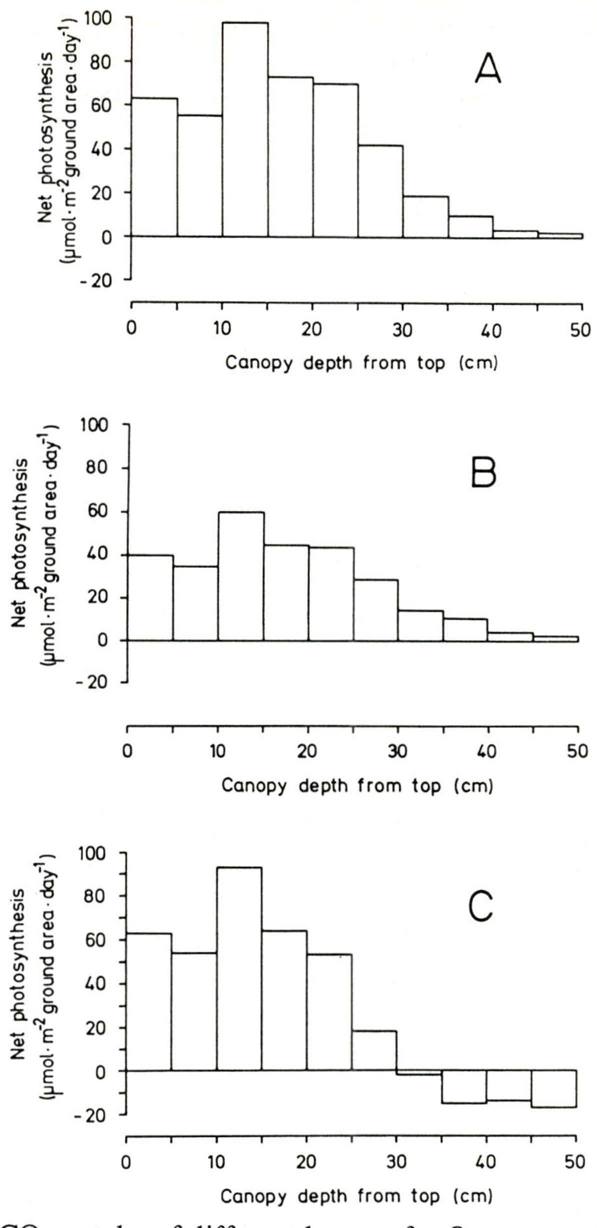

Fig. 3. Daily net CO_2 uptake of different layers of a *Quercus coccifera* canopy for normal (case 1), reduced (case 2) and no (case 3) differentiation between sun and shade leaves

shade leaf conductances and shade leaf light responsivity with increasing water stress.

The ecological significance of our simulations on a larger scale depends on the occurrence of different sized shrubs. Our observations indicated that LAI values of 1- to 2-m tall *Quercus coccifera* fall in the range of 3 and 8 which is consistent with observations for other species with similar appearance in southern France (Eckhardt *et al.* 1977) and in California (Kummerow *et al.* 1981, Jacobson *et al.* 1981). Mediterranean shrubs in areas of high annual precipitation (650 mm per year in Lisbon) have LAI values between 3 and 4 (Cody and Mooney 1978). In these cases, considerable light extinction within shrubs is expected and sun/shade leaf adaptations could provide a mechanism to increase not only canopy CO_2-uptake but also daily water use efficiency. *Quercus coccifera* shrubs with lower LAI between 1 and 2 are also commonly found in Portugal (Meister unpubl.) and southern France (Leterme 1983). In those cases, total canopy height is usually much less than 1 m. Nevertheless, light extinction could be comparable to shrubs with higher LAI due to other structural characteristics such as leaf inclination, but no quantitative data are available.

There are also naturally occurring fluctuations in canopy structure which occur each year. Neither total LAI nor leaf inclinations are static (see also Ehleringer and Comstock in this volume), the highest LAI occurs after flushing of new leaves in spring, and the LAI declines during summer. Secondary flushing of leaves in the fall can also occur. These are typical patterns for mediterranean evergreen shrub phenology (Kummerow *et al.* 1981, Hollinger 1984). This suggests changes in light penetration to lower layers during the course of a year which in turn should affect physiological adaptations of shade leaves. Not only will leaves acclimate to different light conditions depending on their time of flushing, but seasonal changes in solar intensity and weather patterns will also influence the light climate. Flushing of leaves in the spring will itself cause a large change in light penetration to lower leaf layers. Mature leaves are known to change their properties within several weeks in response to sudden changes in the light environment (Björkman 1981, Syvertsen 1984). The results presented here lead to the hypothesis that acclimation must take place rapidly in *Quercus coccifera* shrubs after spring flushing, otherwise negative daily CO_2-balances would occur in lower shaded leaves.

Our simulations have shown that sun/shade leaf differentiation is important for canopy gas exchange budgets. Therefore, physiological differences between sun and shade leaves need to be included in canopy models if those budgets are of interest. This is especially true for analyses of "optimal LAI" where it is a common procedure to optimize total canopy LAI in relation to other variables (e.g. Monsi and Saeki 1953, Duncan *et al.* 1967, Hollinger 1984). Failure to account for sun/shade differentiation can lead to underestimation of "optimal LAI" values.

ACKNOWLEDGEMENTS

We are grateful to Dr. Schulz-Kampfhenkel (Institut für internationale Zusammenarbeit in Bildung und Forschung, Hamburg) for providing use of the facilities at the research station Quinta São Pedro. Computations were performed at the Rechenzentrum der Universität Würzburg. Financial support from the Deutsche Forschungsgemeinschaft and the Alexander von Humboldt-Stiftung (to MMC) is also gratefully acknowledged.

REFERENCES

Baldocci DD, Hutchison BA, Matt DR, McMillen RT (1984) Seasonal variations in the radiation regime within an oak-hickory forest. Agr Forest Meteorol 33(2-3):177-192

Björkman O (1981) Responses to different quantum flux densities. In: Lange OL, Nobel PS, Osmond CB, Ziegler H (eds), Encyclopedia of Plant Physiology, NS vol 12A: Physiological Plant Ecology I. Springer, Berlin-Heidelberg-New York-Tokyo, pp 57-107

Björkman O, Ludlow MM, Morrow PA (1972) Photosynthetic performance of two rainforest species in their native habitat and analysis of their gas exchange. Carnegie Inst. Washington Yearb 71:94-102

Campbell GS (1977) An introduction to environmental biophysics. Springer, Berlin-Heidelberg-New York

Cody ML, Mooney HA (1978) Convergence versus nonconvergence: mediterranean-climate ecosystems. Ann Rev Ecol Syst 9:265-321

Duncan WG, Loomis RS, Williams WA, Hanau R (1967) A model for simulating photosynthesis in plant communities. Hilgardia 38(4):181-205

Eckardt FE, Berger A, Methy M, Heim G, Sauvezon R (1977) Interception de l'énergie rayonnante, échanges de CO_2, régime hydrique et production chez différents types de végétation sous climat méditerranéen. In: Moyse A (ed) Les processus de la production végétale primaire. Gauthier-Villars, Paris, pp 1-75

Gates DM (1980) Biophysical ecology. Springer, Berlin-Heidelberg-New York

Hall AE, Schulze ED, Lange OL (1976) Current perspectives of steady-state stomatal responses to environment. In: Lange OL, Kappen L, Schulze E-D (eds) Water and Plant Life, Ecological Studies 19. Springer, Berlin-Heidelberg-New York, pp 169-188

Hollinger DY (1983) Photosynthesis, water relations, and herbivory in co-occuring deciduous and evergreen California oaks. Ph.D. Dissertation, Stanford University

Jacobson MB, Stoner WA, Richards SP (1981) Models of plant and soil processes. In: Miller PC (ed) Resource Use by Chaparral and Matorral, Ecological Studies 39. Springer, Berlin-Heidelberg-New York, pp 287-433

Kummerow J, Montenegro G, Krause D (1981) Biomass, phenology and growth. In: Miller PC (ed) Resource Use by Chaparral and Matorral, Ecological Studies 39. Springer, Berlin-Heidelberg-New York, pp 49-96

Lange OL, Tenhunen JD, Harley PC, Walz H (1985) Method for field measurements of CO_2-exchange. The diurnal changes in net photosynthesis and photosynthetic capacity of lichens under mediterranean climatic conditions. In: Brown DH (ed) Lichen Physiology and Cell Biology. Plenum, New York, pp 23-39

Leterme J (1983) Structure et comportement hydrique de *Quercus coccifera* L. dans un gradient climatique. Diplome d'Université des Sciences et Techniques du Languedoc, Montpellier

Miller PC (1972) Bioclimate, leaf temperature and primary production in red mangrove canopies in South Florida. Ecology 53(1): 22-45

Miller PC (1983) Canopy structure of mediterranean-type shrubs in relation to heat and moisture. In: Kruger FJ, Mitchel DT, Jarvis JUM (eds) Mediterranean-type Ecosystems. The Role of Nutrients. Ecological Studies 43. Springer, Berlin-Heidelberg-New York, pp 133-166

Miller PC, Hajek E, Poole DK, Roberts SW (1981) Microclimate and energy exchange. In: Miller PC (ed) Resource Use by Chaparral and Matorral, Ecological Studies 39, Springer, Berlin, pp 49-96

Monsi M, Saeki T (1953) Über den Lichtfaktor in den Pflanzengesellschaften und seine Bedeutung für die Stoffproduktion. Jap J Bot 14:22-52

Norman J (1980) Interfacing leaf and canopy light interception models. In: JD Hesketh and JW Jones (eds) Predicting Photosynthesis for Ecosystem Models, vol 1. CRC Press, Boca Raton, Florida, pp 49-67

Roberts SW, Miller PC (1977) Interception of solar radiation as affected by canopy organization in two mediterranean shrubs. Oecol Plant 12(3):273-290

Schulze, ED (1970) Der CO_2-Gaswechsel der Buche (*Fagus sylvatica L.*) in Abhaengigkeit von den Klimafaktoren im Freiland. Flora 159:177-232

Schulze ED, Hall AE, Lange OL, Walz H (1982) A portable steady-state porometer for measuring the carbon dioxide and water vapor exchanges of leaves under natural conditions. Oecologia (Berl) 53:141-145

Seeber, MC (1984) Bestandestruktur, Mikroklima und Energiehaushalt von Grasbestaenden zwischen 1500 und 2500m MH. Dissertation, Leopold-Franzens-Universität Innsbruck

Syvertsen JP (1984) Light acclimation in citrus leaves. II. CO_2 assimilation and light, water, and nitrogen use efficiency. J Amer Hort Sci 109(6):812-817

Tenhunen JD, Lange OL, Harley PC, Beyschlag W, Meyer A (1985) Limitations due to water stress on leaf net photosynthesis of *Quercus coccifera* in the Portuguese evergreen scrub. Oecologia (Berl) 67, 23-30

Tenhunen JD, Meister HP, Caldwell MM, Lange OL (1984) Environmental constraints on productivity of the Mediterranean sclerophyll shrub *Quercus coccifera*. Proceedings of INTECOL workshop - Rates of Natural Primary Productivity and Agricultural Production. Options méditerranéennes (Instituto Agronomico Mediterraneo de Zaragoza) 84/1:33-53

Tenhunen JD, Weber JA, Yocum CS, Gates DM (1976) Development of a photosynthesis model with an emphasis on ecological applications. II Analysis of a data set describing the P_M surface. Oecologia (Berl) 26:101-119

Warren Wilson J (1960) Inclined point quadrats. New Phytol 59:1-8

Weber JA, Jurick TW, Tenhunen JD, Gates DM (1985) Analysis of gas exchange in seedling of *Acer saccharum*: Integration of field and laboratory studies. Oecologia (Berl.) 65: 338-34

4. Carbon Cycling and Plant Response to Limited Nutrient Supply

Responses of plants to nutrient supply in mediterranean-type ecosystems

F.J. Kruger

South African Forestry Research Institute, P.O. Box 727, Pretoria, South Africa 0001

INTRODUCTION

Habitats in mediterranean-type ecosystems vary from those which are relatively fertile to those markedly deficient in mineral nutrients (Specht and Moll 1983). The broad patterns of spatial and temporal variation in soil nutrients and their correlates with plant form and nutrient status in mediterranean-type ecosystems have been well documented (Specht and Moll 1983). Responses to availability of mineral nutrients and how these vary among plant species have also received attention and have been related to the successional status, life-history characters, and other ecological traits of the species concerned. The ecophysiology of mineral nutrition has received relatively little attention, especially regarding the relationships between nutrient uptake, carbon assimilation, and growth (e.g. Barrow 1977).

Chapin (1980) concluded from a study of the literature on mineral nutrition of wild plants that natural selection in fertile environments leads to a plant form with rapid resource capture and growth, whereas infertile environments give rise to plant species with both slow growth and slow resource capture. Hypothetically, the two extremes can be described as in Table 1.

In this paper I examine the validity of certain aspects of this model with respect to sclerophylls of mediterranean type ecosystems. In doing so, I give attention specifically to the literature as it relates to native species with regard to (a) interspecific variation in uptake (b) patterns of variation within the species and (c) photosynthesis response to nutrient status. Some concluding comments are made regarding necessary experimental programmes.

VARIATIONS AMONG SPECIES IN UPTAKE PATTERNS

The variations in nutrient status among plant species of mediterranean environments as they relate to spatial variation in soil and successional changes in nutrient availability have been well documented (e.g. Specht and Moll 1983).

NATO ASI Series, Vol. G15
Plant Response to Stress. Edited by J.D. Tenhunen et al.

Table 1. Contrasting sets of plant traits, hypothetically adaptive in habitats of opposite fertility status (from Chapin 1980). Plants characteristic of habitats of high and moderate fertility may either be competitive or ruderal species in Grime's (1979) terms, while those of infertile habitats would be stress-tolerant species.

Traits	
Habitats of high and moderate fertility	Habitats of low fertility
- High relative growth-rates.	- Low relative growth rates.
- High absorptive capacity in roots responsive to external concentrations.	- Low root absorptivity and photosynthesis rates suffice to maintain growth rates.
- Root absorptivity sensitive to and requires high photosynthesis rate, because of high respiration rates and required carbohydrate reserves.	- High root-shoot ratios and presence of mycorrhizae are the consequence of low root turnover and serve to maximise uptake of minerals.
- High photosynthesis rate, responsive to leaf N content, sometimes to P.	- Capacity to acquire and maintain nutrient reserves.
- Photosynthesis rates and respiration rates respectively of leaves and roots decline with age of the organ, requiring a high rate of tissue turnover, with consequently high rates of nutrient loss.	- Stress induces slight loss in tissue nutrient concentrations.
	- Normal metabolism maintained at low nutrient concentrations in tissues.
	- At low fertility, higher production is maintained through longer growth periods and higher survivorship.

Species of the same habitat also differ in their response to seasonal availability of minerals. Gray (1983), for example, has compared the nitrogen use by evergreen and semi-deciduous shrubs of California chaparral, the two forms occupying adjacent patches of a mosaic within the same habitat. Foliar nutrient concentrations in the semi-deciduous species were higher but more variable than in the evergreen. Uptake by the semi-deciduous species responded strongly to seasonal nutrient flushes. The evergreen shrub took up nutrients in excess of growth requirements when levels of availability permitted, and this was interpreted as storage for growth during periods when nitrogen could less readily be taken up.

Fires caused significant non-seasonal temporal changes in nutrient availability (Gray and Schlesinger 1981; Rundel 1983). Pate and Dell (1984), for example, have estimated the effect of fire on nutrient pools in soils grossly deficient in mineral nutrients in the sand-plain kwongan of Western Australia. The effects of a simulated fire were to increase the pools in the upper 400 mm of soil by 56.9% in the case of potassium, 15.9% in calcium, 13.4% in magnesium, 4.9% in phosphorus, and 2.5% in nitrogen. The amounts involved were about sufficient for the growth demand during the first 12 to 24 months after fire.

Variations in nutrient uptake among species in similar habitats were explored here by analysis of data for nutrient concentrations of shoots in species from nutrient-poor soils of the kwongan in Western Australia and fynbos in the southwestern Cape. Data for Western Australia were from Pate and Dell (1984; see also Pate *et al.* 1985) and those for the Cape from Kruger (unpublished). In both cases the species investigated included short-lived fire-ephemeral herbs and shrubs, as well as longer-lived herbs and shrubs of the later stages in pyric succession (Table 2). The patterns of correlation between nutrition, plant form, and successional status were examined by means of principal components analysis. Data were tissue concentrations of nitrogen, phosphorus, potassium, calcium and magnesium for 33 species, nine from the Cape and 24 from Australia.

There was considerable variation among species in the concentrations of mineral nutrients (Table 3). Eighty-nine per cent of the variation among species could be accounted for by the first three principal components. The first component was more or less evenly loaded for each of the nutrients, and summarised overall variation in foliage concentrations of nutrients. The second was positively loaded for calcium and magnesium and negatively for the others, representing variation in the contrast between concentrations of calcium and magnesium and the other elements. The third was positively loaded for nitrogen, moderately for calcium, and negatively for the other elements (Table 4).

The patterns of variation among species are illustrated in the ordination diagram (Fig. 1) showing also the classes of species determined by cluster analysis of the data using Ward's minimum variance algorithm for agglomeration of clusters (SAS Inc. 1985). The analysis distinguished a category of 13 species with high overall foliage nutrient concentrations, all but four of which being fire-ephemeral shrubs and herbs (groups 1 and 2 in Table 2). In the first group of three shrubs and one herb (group 1 in Table 1; stars in Fig. 1), all but one were classified as fire ephemerals. Members of the cluster were distinguished by moderate overall

Table 2. List of species used in analysis of variations in concentrations of mineral nutrients in shoots. All analyses were for plants 18-24 months old, except annual fire-ephemeral herbs from Western Australia, which were sampled at around six months of age. Data are from Pate and Dell (1984) and Kruger (unpublished).

Species	Family	Origin	Cluster in which classified
Monocarpic fire ephemerals			
Calandrinia corrigioloides	Portulacaceae	W.A.	2
Trachymene pilosa	Apiaceae	W.A.	2
Stipa elegantissima	Poaceae	W.A.	2
Polycarpic fire ephemerals			
Gyrostemon ramulosus	Gyrostemonaceae	W.A.	2
Tersonia brevipes	Gyrostemonaceae	W.A.	1
Scaevola phlebopetala	Goodeniaceae	W.A.	4
Aspalathus globosa	Fabaceae	Cape	2
Ehrharta ramosa	Poaceae	Cape	3
Hypocalyptus oxalidifolius	Fabaceae	Cape	2
Othonna quinquedentata	Asteraceae	Cape	1
Ursinia crithmoides	Asteraceae	Cape	2
Long-lived, obligate-seeding shrubs			
Adenanthos cygnorum	Proteaceae	W.A.	4
Beaufortia elegans	Myrtaceae	W.A.	4
Comesperma calymega	Polygalaceae	W.A.	3
Daviesia preissii	Fabaceae	W.A.	3
Hakea obliqua	Proteaceae	W.A.	3
Spyridium oligocephalum	Rhamnaceae	W.A.	4
Stackhousia brunonis	Stackhousiacea	W.A.	2
Erica sessiliflora	Ericaceae	Cape	3
Leucadendron xanthoconus	Proteaceae	Cape	3
Long-lived, sprouting herbs			
Isotropis cuneifolia	Fabaceae	W.A.	1
Scaevola canescens	Goodeniaceae	W.A.	1
Elegia juncea	Restionaceae	Cape	3
Tetraria bromoides	Cyperaceae	Cape	3
Long-lived, sprouting shrubs			
Banksia attenuata	Proteaceae	W.A.	3
Conosperum triplinervium	Proteaceae	W.A.	4
Daviesia nudiflora	Fabaceae	W.A.	3
Daviesia quadrilatera	Fabaceae	W.A.	4
Dryandra nivea	Proteaceae	W.A.	3
Jacksonia floribunda	Fabaceae	W.A.	4
Pityrodia bartlingii	Chloanthaceae	W.A.	2
Stirlingia latifolia	Proteaceae	W.A.	4
Verreauxia reinwardtii	Goodeniaceae	W.A.	4

Table 3. Statistics for concentrations of mineral nutrients in shoots of 33 species from the Kwongan of Western Australia and the Fynbos of the Cape Province. Units are $mg.g^{-1}$. See text for further details.

Element	Mean	Standard deviation	Minimum	Maximum
Nitrogen	12.6	5.30	5.90	29.0
Phosphorus	0.578	0.413	0.060	1.87
Potassium	11.47	8.49	2.10	36.0
Calcium	14.0	9.92	0.970	42.5
Magnesium	3.30	2.54	0.500	14.4

Table 4. Eigenvectors of the principal components analysis of mineral nutrient concentrations in shoots of nine plant species of the Cape Fynbos and 22 from the Kwongan in Western Australia. See text for further details.

Nutrient	Principal component			
	First	Second	Third	Fourth
Nitrogen	0.435	-0.232	0.827	-0.054
Phosphorus	0.545	-0.369	-0.169	-0.148
Potassium	0.546	-0.040	-0.384	0.630
Calcium	0.152	0.765	0.293	0.359
Magnesium	0.438	0.472	-0.233	-0.670

concentrations of nutrients. The cluster was distinguished from the low-concentration group by relatively high levels of potassium (14.4 - 17.5 $mg.g^{-1}$) and magnesium (3.6 - 4.5 $mg.g^{-1}$). Members were distinguished from other members in the high-nutrient group (group 2 in Table 1; circles in Fig. 1) by lower phosphorus and calcium concentrations.

The second group included five Australian and three Cape species, of which all but two were fire ephemerals. A sub-group within this included two leguminous shrubs from the Cape which were richer in nitrogen than their counterparts. A second sub-group in this category included the herb, *Isotropis cuneifolia*, and two shrubs, *Othonna quinquedentata* and *Tersonia brevipes*, and was distinguished by moderate to high potassium (15.9 - 23.0 $mg.g^{-1}$) and calcium (16.0 - 34.5 $mg.g^{-1}$) levels. *Othonna* had an exceptionally high concentration of magnesium (14.4

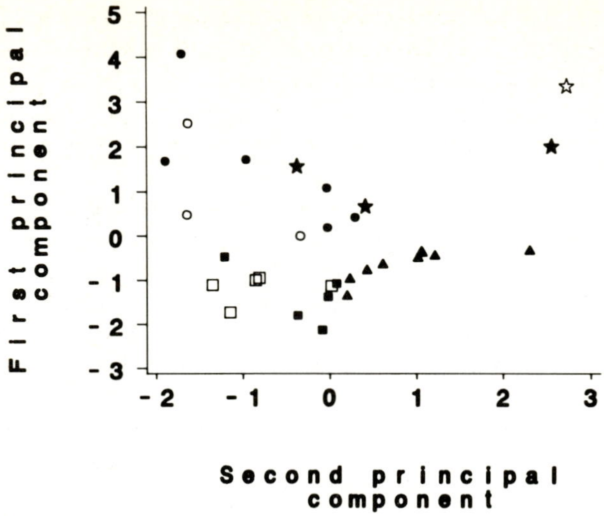

Fig. 1. Groups of species defined in terms of foliar nutrient concentration by Ward's agglomerative clustering algorithm (see Table 2), and plotted on the first and second principal components. Stars represent the first group, circles the second, squares the third, and triangles, the fourth group. Solid symbols represent species from Australia and open symbols those from South Africa

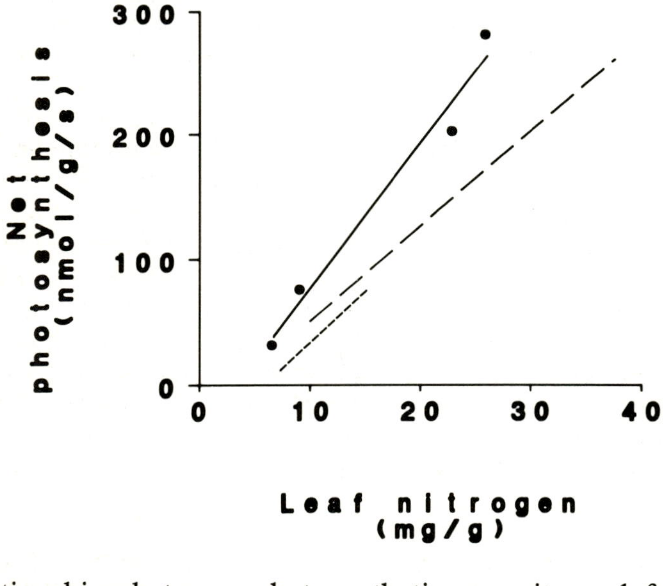

Fig. 2. Relationships between photosynthetic capacity and foliar nitrogen concentration in *Eucalyptus maculata* (solid circles: Cromer 1984), several species of *Protea* (dotted line: Mooney *et al.* 1983) and *Betula pendula* (dashed line: Linder, MacDonald and Lohammer 1981)

$mg.g^{-1}$). Two Australian fire-ephemeral herbs, *Stipa elegantissima* and *Calandrinia corrigioloides*, were classified in a sub-group distinguished by high phosphorus (1.3 - 1.9 $mg.g^{-1}$) and potassium (26.6 - 36.0 $mg.g^{-1}$) and low calcium (2.5 - 2.7 $mg.g^{-1}$). *Scaevola canescens* was uniquely distinguished by high concentrations of potassium and calcium (28.6 and 42.5 $mg.g^{-1}$ respectively).

The second category of 20 species, tended to have low nutrient concentrations overall, especially of nitrogen, i.e. less than or equal to 16.0 $mg.g^{-1}$, and phosphorus, which did not exceed 0.8 $mg.g^{-1}$. These were all longer-lived species which persist until late in the pyric succession, except the grass, *Ehrharta ramosa* from the Cape, and the short-lived shrub, *Scaevola phlebopetala*, from Western Australia. Two principal sub-groupings were detected.

Group number 3 (squares in Fig. 1) with six Australian and five Cape species, three of which were herbs, tended to have lowest concentrations of nitrogen, calcium, and magnesium. Shrubs were mainly in the Proteaceae but two legumes were included. This contrasted with group number 4 (triangles in Fig. 1) consisting of eight Australian shrubs, all but one of which were classified as long-lived. Tissue concentrations of nitrogen in this group were moderate (8.0 - 16.0 $mg.g^{-1}$), of phosphorus were low (0.16 - 0.76 $mg.g^{-1}$), and of calcium were high (14.6 - 32.1 $mg.g^{-1}$). The group was intermediate in terms of nutrient concentration, but clearly distinct from the longer-lived species with low overall concentrations. It included species which were apparently classified as intermediate in the pyric succession, such as *Stirlingia latifolia*, and others such as *Adenathos cygnorum* which are frequent on roadsides and in other disturbed habitats (personal observation).

Plants in their native habitats vary less in their tissue nutrient concentrations with variation in soil nutrient availability than do cultivated plants (Chapin 1980). Nevertheless, this analysis indicates that species exposed to the same availabilities of nutrients have markedly different uptake capacities and suggests great diversity of nutrition among the species within these infertile habitats. Similar life-forms in the Cape and Australia tend to resemble each other in their foliar nutrients (cf Specht and Moll 1983 for species with low concentrations). Uptake capacities are correlated with life-history strategy. But species within a class discriminated in terms of overall mineral nutrient contents of ten differ substantially with respect to concentrations of one or more nutrients, and species within growth-form categories within the same habitat apparently may differ substantially in their uptake capacities.

Uptake capacities in short-lived species were related to morphological features such as shallow, fibrous root systems. These plants were plastic, tolerating low levels of nutrients in the absence of competition but responding markedly as resource availability was increased (Pate and Dell 1984; Pate *et al.* 1985). Relative growth-rates of perennial fire-ephemerals were an order of magnitude greater than those of long-lived shrubs.

The species which grow and reproduce under conditions where the post-fire nutrient flush has been depleted have root-morphological and physiological adaptations to enhance uptake from dilute soil solutions (Lamont 1982). These species have a limited response to resource availability, nevertheless. For example, Lewis and Stock (1978) and Stock and Lewis (1984) have provided experimental evidence for fynbos Proteaceae showing that xylem transport mechanisms are limited in the rate at which nitrogen may be transported from root to shoot, and the plants generally have low rates of nitrogen metabolism. Measured rates of nitrogen uptake were as little as one-fiftieth of those observed in crop plants. This was associated with slow growth, and seen to be adaptive in low-nutrient environments. Evidence of this kind from mediterranean-type ecosystems tends to support the model proposed by Chapin (1980). A proper test requires information on the mechanisms involved in uptake, allocation, and assimilation.

VARIATION IN MINERAL NUTRIENT UPTAKE WITHIN THE SPECIES AND ITS RELATION TO GROWTH

Species adapted thus to different levels of fertility tend to vary in their capacity of plants for uptake at optimal and sub-optimal levels of mineral nutrient supply and those of fertile habitats apparently tend to be plastic in this respect (Chapin 1980; Table 1). Ingestad (1982) has analysed the relationships between nutrient availability and plant growth during the phase of exponential growth under conditions when only minerals are the limiting factors.

Responses of plants to nutrient availability are related to the rate at which ions are released or added to the external nutrient medium provided by the soil solution or the experimental solution, because growth depletes the supply in the rhizosphere. Responses cannot be analysed in terms of instantaneous external concentrations in the rooting medium. Rather, the rate at which nutrient mass is added to the medium must be determined in relation to the rate at which it is taken up. The nutrient additions required to maintain exponential growth of plant tissues at constant nutrient concentrations must also be increased exponentially. With experiments of appropriate design, Ingestad (1982, and earlier papers) found that relative growth rates, relative rates of mineral uptake, and internal nutrient concentrations in seedlings of species in *Betula*, *Alnus*, and several conifers were directly proportional to relative rates of supply. The response curve relating plant growth to nutrient supply was not continuous but appeared to consist of a phase in the sub-optimum range where relative growth-rates were directly proportional to nutrient supply, a second phase of saturation where relative growth-rate was constant, and a third phase of toxicity caused by increasing external nutrient concentration as such. The threshold between the first and the second phases, at about 3-10 mmol $N.m^{-3}$, was abrupt rather than gradual.

For sclerophyllous trees, Cromer, Wheeler and Barr (1984) have examined these relationships in three species of *Eucalyptus*. After an initial lag phase of 14 to 42 days or so during which cotyledonary reserves were diluted, internal nutrient concentrations adjusted and stabilized at a level directly proportional to the rate of nutrient addition to the rooting medium. The exception was the large-seeded species, which would have required a longer period of growth before growth-rate adjusted to external rates of supply (Barrow 1977). At this stage of steady-state growth, relative growth-rates also became linearly related to the relative rates of nutrient addition.

How general are these relationships among sclerophylls? Gray and Schlesinger (1983) have conducted a similar experiment with shrubs of the California chaparral, one, *Ceanothus megacarpus*, being an evergreen sclerophyll and the other, *Salvia leucophylla*, being semi-deciduous. Their experiment involved nitrogen concentrations near or beyond optimum (21 to 168 $mg.l^{-1}$ or 1.5 to 12 $mmol.m^{-3}$), and their data do not permit analysis of rates of growth and uptake in relation to mineral supply rates. Their results for growth and for relative root absorption rates indicate saturation (supra-optimality) in *C. megacarpus*, but a continued capacity to respond to nitrogen in *S. leucophylla*. These results tend to support Chapin's model - *S. leucophylla* is a species of successionally fertile habitats, *C. megacarpus* is not. However, the plot of the relationship between tissue nitrogen concentration and relative growth-rate is nonlinear, indicating saturation. An experiment which included sub-optimal rates of nitrogen supply is required to resolve the issue. A rather similar experiment was conducted by Grundon (1972) with shrubs of Australian heathland but the data necessary for growth analysis were not reported.

There have been many studies of variations in uptake of phosphorus in relation to supply at sub-optimal levels. Experiments by Grundon (1972), Barrow (1977) and others have indicated the low capacity for absorption of phosphorus among sclerophylls but the patterns of uptake can not be related to growth (Cromer *et al.* 1984). This is not only because the treatments did not allow for nutrient depletion (except in Grundon's experiment) but also because of the confounding effect of seed reserves. Large-seeded sclerophylls, especially those rich in phosphorus (cf Kuo, Hocking and Pate 1982), contain reserves sufficient for 100-500 days of seedling growth without the necessity to take up minerals (e.g. Grundon 1972; Barrow 1977). In such cases there is no net uptake by the plant for the duration of many experiments and the maximum rate of uptake by such species was not determined. Though experimental work has shown clearly the small response of such sclerophylls to and the toxicity of added phosphorus (Ozanne and Specht 1981) the effects are obtained at levels not experienced in the field, despite ashbeds after fire.

VARIATIONS IN SPECIFIC RESPONSES OF PHOTOSYNTHESIS TO INTERNAL NUTRIENT LEVELS

The response of net photosynthesis to foliar nitrogen content is usually linear, in deciduous as well as evergreen sclerophyllous species (Mooney 1983; Cromer 1984), at least in the optimal to sub-optimal range of nitrogen concentrations. Certain sclerophylls have an extraordinary capacity for adjustment to nitrogen content. *Eucalyptus* species maintain higher levels of net photosynthesis at low foliar nitrogen concentrations than do species of *Protea* and yet are able to increase photosynthetic capacity linearly with an increase in nitrogen at least three-fold and perhaps to 40 $mg.g^{-1}$ (Fig. 2; Mooney, Ferrar and Slatyer 1978; Cromer 1984; cf Mooney *et al.* 1983). *Eucalyptus* species such as *E. cladocalyx* from infertile habitats have evidently retained these capacities for uptake (Groves 1967) and, presumably, a similar photosynthesis response. Species of Proteaceae are unlikely to increase foliar nitrogen to such levels (Stock and Lewis 1984). There is preliminary evidence that the responses in photosynthesis to tissue nitrogen, though linear, are not uniform (Fig. 2). The line for *Protea* spp. apparently has the same slope as for *Betula* spp. and others, but a lower intercept. From Cromer's (1984) work, it appears that at least in some *Eucalyptus* spp. the response is stronger, and the rates at low nitrogen levels higher than in other species.

The evidence is therefore of substantial variation within sclerophylls in their photosynthetic response to foliar nitrogen. Though *Eucalyptus* is taken to be typical of a more fertile habitat than *Protea* and its analogues (Specht and Moll 1983), the genus is able to invade fynbos and heath habitats, indicating perhaps that the members of the genus have more efficient nutrient uptake mechanisms or nutrient-use efficiencies than the sclerophylls native to these habitats. A search of the recent literature indicated that relationships between foliar phosphorus and photosynthetic capacity in mediterranean sclerophylls have not received explicit attention.

CONCLUSION

There is a great diversity of syndromes of plant mineral nutrition in mediterranean-type ecosystems, perhaps moreso in the highly leached ecosystems of Australia and South Africa. Species vary in their capacities for response to intermittently increased supplies of mineral nutrients. Nutrient stress has evidently selected ultimately for forms with a low capacity for response to enhanced supply (Table 1; Specht *et al.* 1983) or alternatively to forms with life-histories and physiologies permitting exploitation of transient supplies. This tends to support general aspects of the model proposed by Chapin (1980) but nutrition of *Eucalyptus* species and several other woody forms evidently differs markedly from that of other, slower-growing sclerophylls despite the ability to occupy similar habitats. Comparative experiments, designed to allow critical comparison of responses in photosynthesis, assimilation and growth to varying rates of nutrient supply, are needed to appreciate properly the mechanisms involved. An aspect requiring attention is the interaction between soil moisture and nutrition. Although mass flow arising from the transpiration stream cannot be critical in uptake of minerals (Chapin 1980) it is likely that lateral mass flow through the soil is the cause of major patterns of productivity in the landscapes of especially the Cape fynbos (e.g. Mooney 1983). Soil moisture availability has complex effects on nutrient availability and uptake (Lamont 1982) and these as well as competition between plants for these resources are strongly implicated in productivity of forest ecosystems in mediterranean zones (e.g. Bowen and Nambiar 1984).

Waring and coworkers (e.g. Waring 1983) have demonstrated simple relationships between leaf-area index and productivity of forests and shrublands, in turn dependent on water balance, but noted that the relationships evidently do not hold for mediterranean ecosystems. Specht (e.g. Specht and Moll 1983) has developed a rather similar though more complex explicit construct for mediterranean vegetation, which attempts to take account of the effects of both water balance and nutrients (see Specht in this volume). Understanding the causal relations involved in such constructs will help to find the general models. For this we will require critical experiments to determine how assimilation and hence growth and productivity per unit leaf area is determined by the interaction between water and nutrients.

Such understanding is of practical as well as scientific value. In forestry as well as other fields it is necessary to manage multispecies ecosystems. Theories about plant form and nutrition will and often do underly management strategies. For example, Specht (1975; also Groves and Keraitis 1976) has proposed a model invoking seven 'forms' of plants distinct in their life-history and nutritional characteristics as a basis for decision in rehabilitation of open-cast mining. Management of such systems can improve along with understanding of the underlying functional relationships.

REFERENCES

Bowen GD, Nambiar EKS (1984) Future directions in plantation nutrition research. In: Bowen GD, Nambiar EKS (eds) Nutrition of Plantation Forests. London, Academic Press, pp 489-505

Barrow NJ (1977) Phosphorus uptake and utilization by tree seedlings. Australian Journal of Botany 25:571-584

Chapin FS (1980) The mineral nutrition of wild plants. Annual Review of Ecology and Systematics 11:233-260

Cromer RN (1984) The influence of nutrition on growth and photosynthesis in *Eucalyptus*. In: Grey DC, Schonau APG, Schutz CJ, Van Laar A (eds) Symposium on Site and Productivity in Fast Growing Plantations. IUFRO, Pretoria and Pietermaritzburg, pp 669-678

Cromer RN, Wheeler AM, Barr NJ (1984) Mineral nutrition and growth of *Eucalyptus*. New Zealand Journal of Forest Science 14:229-239

Gray JT (1983) Nutrient use by evergreen and deciduous shrubs in southern California. I. Community nutrient cycling and nutrient-use efficiency. Journal of Ecology 71:21-41

Gray JT, Schlesinger WH (1981) Nutrient cycling in mediterranean type ecosystems. In: Miller PC (ed) Resource Use by Chaparral and Matorral. Springer-Verlag, Berling-Heidelberg-New York, pp 259-285

Gray JT, Schlesinger WH (1983) Nutrient use by evergreen and deciduous shrubs in southern California. II. Experimental investigations of the relationships between growth, nitrogen uptake and nitrogen availability. Journal of Ecology 71:43-56

Grime JP (1979) Plant strategies and vegetation processes. New York, Wiley, p 222

Groves RH (1967) Within-species variation in the growth and nutrition of *Eucalyptus cladocalyx*. Australian Journal of Botany 15:161-173

Groves RH, Keraitis K (1976) Survival and growth of seedlings of three sclerophyll species at high levels of phosphorus and nitrogen. Australian Journal of Botany 24:681-690

Grundon NJ (1972) Mineral nutrition of some Queensland heath plants. Journal of Ecology 60:171-181

Ingestad T (1982) Relative addition rate and external concentration; driving variables in plant nutrition research. Plant Cell and Environment 5:443-453

Kuo J, Hocking PJ, Pate JS (1982) Nutrient reserves in seeds of selected Proteaceous species from southwestern Australia. Australian Journal of Botany 30:231-249

Lamont B (1982) Mechanisms for enhancing nutrient uptake in plants, with particular reference to mediterranean South Africa and Western Australia. The Botanical Review 48:597-689

Lewis OAM, Stock WD (1978) A preliminary study of the nitrogen nutritional status of members of the South African Proteaceae. Journal of South African Botany 44:143-151

Linder S, MacDonald J, Lohammer T (1981) Effect of nitrogen status and irradiance during cultivation on photosynthesis and respiration in birth seedlings. Energy Forestry Project, Swedish University for Agricultural Science, Uppsala, Sweden, Technical Report 16, 29 p

Mooney HA (1983) Carbon-gaining capacity and allocation patterns of mediterranean-climate plants. In: Kruger FJ, Mitchell DT, Jarvis JUM (eds) Mediterranean-Type Ecosystems: The Role of Nutrients. Springer-Verlag, Berlin-Heidelberg-New York, pp 103-119

Mooney HA, Ferrar PJ, Slatyer R (1978) Photosynthetic capacity and carbon allocation patterns in diverse growth forms of *Eucalyptus*. Oecologia (Berl) 36:103-111

Mooney HA, Field C, Gulmon SL, Rundel P, Kruger FJ (1983) Photosynthetic characteristics of South African sclerophylls. Oecologia (Berl) 58:398-401

Ozanne PG, Specht RL (1981) Mineral nutrition of heathlands: phosphorus toxicity. In: Specht RL (ed) Ecosystems of the World, Vol. 9B, Heathlands and Related Shrublands: Analytic Studies. Elsevier, Amsterdam, pp 209-213

Pate JS, Dell B (1984) Economy of mineral nutrients in sandplain species. In: Pate JS, Beard JS (eds) Kwongan: Plant Life in the Sandplain. University of Western Australia Press, Nedlands, pp 227-252

Pate JS, Casson WE, Rullo J, Kuo J (1985) Biology of fire ephemerals of the sand plains of the Kwongan of southwestern Australia. Australian Journal of Plant Physiology 12:641-655

Rundel PW (1983) Impact of fire on nutrient cycles in mediterranean-type ecosystems with reference to chaparral. In: Kruger FJ, Mitchell DT, Jarvis JUM (eds) Mediterranean-Type Ecosystems: The Role of Nutrients. Springer-Verlag, Berlin-Heidelberg-New York, pp 192-207

SAS Institute Inc. (1985) SAS Users' Guide: Statistics, Version 5 Edition. Cary, NC: SAS Institute Inc., 956 p

Specht RL (1975) The effect of fertilizers on sclerophyll (heath) vegetation - the problems of revegetation after sand-mining of high dunes. Search 6:459-461

Specht RL, Moll EJ (1983) Mediterranean-type heathlands and sclerophyllous shrublands of the world: an overview. In: Kruger FJ, Mitchell DT, Jarvis JUM (eds) Mediterranean-Type Ecosystems: The Role of Nutrients. Springer-Verlag, Berlin-Heidelberg-New York, pp 41-65

Stock WD, Lewis OAM (1984) Uptake and assimilation of nitrate and ammonium by an evergreen fynbos shrub species *Protea repens* L. (Proteaceae). New Phytologist 97:261-268

Specht RL, Moll EJ, Pressinger F, Sommerville J (1983) Moisture regime and nutrient control of seasonal growth in mediterranean ecosystems. In: Kruger FJ, Mitchell DT, Jarvis JUM (eds) Mediterranean-Type Ecosystems: The Role of Nutrients. Springer-Verlag, Berlin-Heidelberg-New York, pp 120-132

Waring RH (1983) Estimating forest growth and efficiency in relation to canopy leaf area. Advances in Ecological Research 13:327-354

Nutrient use strategy by evergreen-oak (*Quercus ilex* ssp. *ilex*) in NE Spain

A. Escarré, Ll. Ferrés*, R. Lopez*, J. Martin, F. Rodá* and J. Terrades*

Departamento de Biología, Universidad de Alicante, Alicante, Spain

INTRODUCTION

Quercus ilex is adapted to the low winter temperatures characteristic of high elevation sites (see Comin *et al.* in this volume) and low soil moisture caused by the lack of summer precipitation in the mediterranean basin. Extremes in these conditions in mediterranean climate zones are more unpredictable than in other climate regions and adaptative strategies that minimize the effects of extreme cold or drought stress are probably important for survival. This paper is a first attempt at describing possible nutrient use strategies of the evergreen-oak that may help overcome mineral resource fluctuations due to large discontinuities in the hydrological fluxes (see Comin *et al.* in this volume). The biogeochemistry at La Castanya experimental station is described and mineral nutrient fluxes between stems and leaves of *Q. ilex.* are estimated.

BIOMASS, PRODUCTION, AND BIOGEOCHEMISTRY OF THE FOREST AT LA CASTANYA

Comin *et al.* (this volume) described the topography, orientation, local site characteristics, and climate of forest sites at La Castanya and L'Avic. The evergreen-oak forest at La Castanya has a total biomass of 230 t ha^{-1}, 165 of which is aboveground and the remaining 65 t ha^{-1} estimated for roots. A more detailed description of the biomass distribution in various plant organs is given in Table 1. Net production is estimated at 11.2 t $ha^{-1}y^{-1}$ which may be separated into 3.1 t $ha^{-1}y^{-1}$ of leaves, 2.36 t $ha^{-1}y^{-1}$ of branches, 3.4 t $ha^{-1}y^{-1}$ of trunks and 0.58 t $ha^{-1}y^{-1}$ of fruits. These values are high compared with previous observations in *Q. ilex* forests (Lossaint & Rapp 1971, Susmel *et al.* 1976).

Nutrient pools for five elements in various organs and biomass compartments are also given in Table 1. Although leaves make up only 2.9% of the biomass, they have 14.7% of the total N. More than 50% of the total Ca is stored in the aboveground bark which amounts to only 13% of the biomass. Twigs and stems smaller than 1 cm in diameter, including bark, make up 7.4% of the total biomass but accumulate 18.8% of P and between 10-12% of K, Ca and Mg.

---

*Departamento de Ecología, Universidad Autónoma Barcelona, Bellaterra, Barcelona, Spain

NATO ASI Series, Vol. G15
Plant Response to Stress. Edited by J. D. Tenhunen et al.

Table 1. Biomass and nutrient pools of five elements in various plant organs in the *Quercus ilex* forest at La Castanya.

Fraction	Biomass t ha-1	N	P	Ca kg ha-1	K	Mg
Trees						
Leaves	6.1	82.6	7.6	41.0	29.4	8.5
New twigs and leaves	0.8	5.9	1.0	6.2	5.2	1.4
Wood	127.0	195.6	65.4	362.2	260.5	58.8
Bark	27.0	126.6	9.1	786.5	89.2	38.2
Root Wood (Ø>1 cm)	37.5	103.0	23.1	150.2	136.1	37.1
Root Bark (Ø>1 cm)	11.2	47.4	2.1	120.2	46.5	11.3
Roots (Ø≤1 cm)	16.3	---	---	---	---	---
Shrubs	4.6	17.0	3.5	39.5	12.9	3.2
Herbs	0.14	1.5	0.2	1.2	1.9	0.3
Above-ground total	165	429	87	1237	399	110
Underground total	48.7	150	25	270	183	48
Grand total	213.7	579	112	1507	582	158

In Table 2, nutrient fluxes, nutrient pools other than those described in Table 1, and parameters of nutrient cycling are summarized for five key elements of the evergreen-oak forest at La Castanya.

Table 2. Biogeochemical fluxes, nutrient pools and nutrient cycling parameters in the evergreen-oak forest at La Castanya.

	N	P	Ca	K	Mg
Aerial nutrient pool*	429	87	1237	399	110
Subterranean nutrient pool*	150	25	270	183	48
Aerial pool increment**	9.2	2.5	23.2	14.7	3.3
Subterranean pool increment**	4.7	0.8	8.4	5.7	1.5
Global deposition**	2.26	0.05	10.5	1.2	1.3
Net throughfall** (1)	0.41	0.65	4.1	17.8	1.8
Litterfall** (2)	35	4.26	47.80	21.9	5.86
Return: (1) + (2)**	35.41	4.91	51.9	39.7	7.66
Retention (Aerial+subterranean) (3)**	22.27	4.28	51.90	23.26	6.31
Absorption: (1) + (2) + (3)**	57.68	9.19	103.80	62.96	13.97
Requirements**	68.67	8.78	68.5	41.76	10.81
Cycling rate+	0.06	0.04	0.03	0.07	0.05
Nutrient cycling rate/ organic matter cycling rate	2.47	1.77	1.39	2.75	1.95
Efficiency++	317	2283	216	282	1463

Units: *Kg ha^{-1} **Kg ha^{-1} y^{-1} +y^{-1}

++Kg ha^{-1} y^{-1} OM(NPP)/Kg ha^{-1} y^{-1} nutrient return

COMPARISON WITH OTHER FORESTS

The nutrient content of evergreen-oak leaves is similar to that of conifer needles and less rich than in deciduous leaves. As for the nutrient contents in trunks and branches, *Quercus ilex* is again more similar to resinous trees than deciduous trees. However, the branches and truck typically have a high P content as well as a high Ca and K content (primarily in the bark). The mass of mineral nutrients in the evergreen-oak forest is high relative to the total biomass present because of the rich stem fractions. The relatively low nutrient content of leaves is interpreted as an adaptation to soils with scarce, fluctuating mineral resource. The abundance of some mineral elements in stems may indicate a storage strategy for coping with this type of environment (Ferrés 1984). The nutrient cycling rate in *Quercus ilex* forests is low (small amounts of nutrient per unit of organic matter cycled) and more similar to that of the coniferous forest. Statistical differences cannot be demonstrated, when comparing the cycling efficiencies at La Castanya with values for coniferous and deciduous forests [calculated according to the definition proposed by Gray (1983)]. Efficiency for cycling N lies close the upper limit for conifers.

Data on nutrient cycling in forest ecosystems compiled by Cole & Rapp (1981) were used together with data from the experimental stations at La Castanya and L'Avic to obtain with principal components analysis a multivariate ordination of forest types. The similarities and differences of 25 forests characterized by 30 variables (biomass, production, and nutrient contents of various fractions) are shown on the first two principal axes of the ordination in Fig. 1. The first principal component is related to the nutrient content of leaves and discriminates the deciduous (high concentration) from the coniferous (low concentration) forests. The evergreen-oak forests occupy an intermediate position but are more closely related to the coniferous forests. The second principal component refers to the chemical characteristics of twigs and thin branches. The highest values on this axis are found for evergreen-oak forest due to the high concentration of nutrients in stem materials as discussed above.

RETRANSLOCATION

Cole & Rapp (1981) defined retranslocation as the difference between requirements and absorption. If this definition is applied in our current study of the evergreen-oak forest, retranslocation on an annual basis occurs only in the case of N. Similar indications were obtained at La Castanya by Verdú (1984) who used the difference between nutrient concentration in living and dead leaves to estimate retranslocation. As for other elements (P, K and Mg), the results obtained at La Castanya are less clearcut. Rodá (1983) demonstrated increase of these elements in net throughfall during leaf senescence, suggesting a dominance of leaching rather than retranslocation processes.

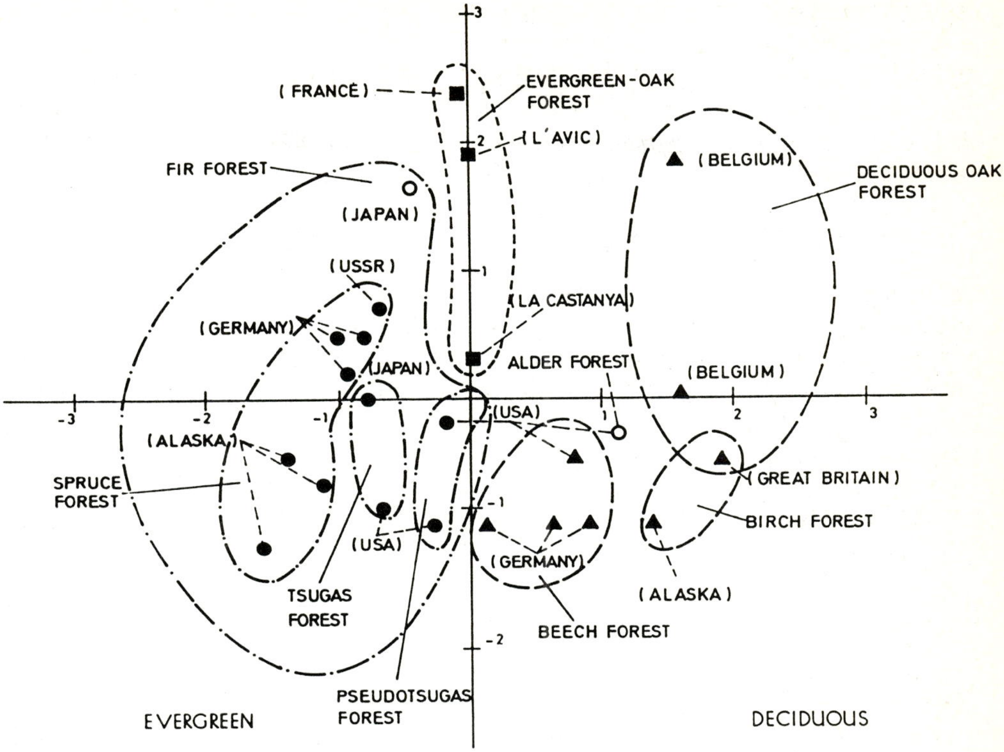

Figure 1. Ordination of 25 forest according to the first two principal components. The variables used are biomass, production and nutrient content of different fractions. The ordination axes are related to nutrient content of leaves (horizontal) and nutrient content of branches (vertical - see text)

To obtain a more detailed description of nutrient distribution between leaves and stems during the period of bud formation, branches from four trees were sampled on April 21, May 22, June 12, July 1 and August 21. The samples were separated into young leaves and twigs and leaves and branches of one, two and three years of age. Each fraction was dried and weighed. Nutrient contents for each tree, date, and fraction were determined and the total mass of N, P, K and Ca were estimated.

Nutrient dynamics were analyzed in the following manner. If M_i represents the nutrient mass for a fraction at time i, we can write $M_i = M_{i-1} \pm \Delta B \cdot C_{i-1} \pm X_i$, where ΔB stands for the change in biomass and C_{i-1} for the concentration at time i-1. Positive values of X_i indicate that nutrient molecules are imported or concentrated within a fraction and negative values show that nutrients are exported or diluted. Results of the analysis showed similar patterns for N, P and K of young leaves. Nutrients appeared to be mobilized from one year old leaves, to reside in one year old branches just before bud development and then to pass to the young leaves. Changes in carbon metabolism during this spring and summer period are described by Mauffette and Oechel (this volume) for old and new leaves of a Californian oak species similar to *Quercus ilex* .

Consistent patterns were not found for the various nutrients in other fractions. Trends suggested a general export of N, little change in P, and accumulation of K. Ca was found to accumulate except in one and two year old branches. If a nutrient budget is constructed using the first and last samples as endpoints, 48% of the N used in new structures is retranslocated from other fractions: 18% from one year old leaves, 16.7% from dead leaves and the remaining 12% from branches. With regard to P, retranslocation accounts for 31% of the P used in new structures: 17.2% from the dead leaves and 13.9% from one year old leaves None of the P is obtained from branches. Retransloction of K and Ca could not be estimated by this method. At least at the time of bud formation, fluxes of P and K from the hypothetical reserve in young stems to the bud were not apparent.

CONCLUSIONS

Information on the biogeochemistry of the evergreen-oak forest suggests characteristics intermediate between coniferous and deciduous forests, with a closer affinity to the first type, especially with respect to the leaf longevity. The fact that the *Quercus ilex* has a high nutrient content in some fine branches may be interpreted as an adaptation to fluctuating nutrient availability, provided that these reserves can be easily mobilized. Monitoring the nutrient distribution between leaves and stems during the period of bud formation revealed certain retranslocation patterns, but no evidence was found for mobility or transport to the leaves of the elements more concentrated in branches (P, Ca and K).

REFERENCES

Cole DW, Rapp M (1981) Elemental cycling in forest ecosystems. In: Reichle DE (ed), Dynamic Properties of Forest Ecosystems. Cambridge University Press, Cambridge, England, p 341

Ferrés L (1984) Biomasa, producción y mineralomasas del encinar montano de La Castanya (Montseny). Doctoral thesis, Departamento de Ecología, Universidad Autónoma de Barcelona, Bellaterra

Gray JT (1983) Nutrient use by evergreen and deciduous shrubs in southern California. I. Community nutrient cycling and nutrient-use efficiency. J Ecol 71:21-41

Lossaint P, Rapp M (1971) Répartition de la matière organique, productivité et cycles des élèments minéraux dans des écosystèmes du climat méditerranéen.

In: Duvigneaud P (ed) Productivity of Forest Ecosystems. Unesco, Paris, p 597

Rodá F (1983) Biogeoquímica de les aigües de plutja i de drenatje d'algúns ecosistemes forestals del Montseny. Doctoral thesis, Departament d'Ecología, Universidad Autónoma de Barcelona, Bellaterra

Susmel L, Viola F, Bassato F (1976) Ecología della lecceta del Supramonte di Orgosolo (Sardegna centro-orientale). Annali del Centro di Economia Montana delle Venezie 10: 1-261

Verdú AMC (1984) Circulació de nutrients en tres ecosistemes forestals del Montseny: caiguda de virosta i descomposició de la fullaraca. Doctoral thesis, Departamento de Ecología, Universidad Autónoma de Barcelona, Bellaterra

Seasonal photosynthate allocation of the Californian coast live oak *Quercus agrifolia*

Y. Mauffette and W.C. Oechel*

Département des Sciences Biologiques, Université du Quebec à Montréal, C.P. 8888, Succ. "A", Montréal, Québec H3C 3P8, Canada

INTRODUCTION

The growth and competitive success of a plant species involves the integrated use of a number of resources which are in limited supply, including carbon, water, and nutrients. While many studies have reported rates of carbon uptake in photosynthesis, only rarely have they simultaneously described subsequent carbon allocation processes. The timing of carbon allocation to and the forms and amount of carbon allocated to different functions such as growth, storage, maintenance and protection are important to fitness (Mooney 1972). Studies of two evergreen shrub species, *Heteromeles arbutifolia* (Mooney and Chu 1974) and *Larrea divaricata* (Oechel *et al.* 1972), have provided basic descriptions of the seasonal changes in partitioning of photosynthate in shrubs from semiarid and arid regions. Because assimilate partitioning is strongly influenced by both plant size and sink distribution, allocation patterns of trees may differ considerably from shrubs. Since trees have higher maintenance demands than shrubs, this too may lead to important difference in their allocation patterns for photosynthate (McLaughlin *et al.* 1979).

This paper describes research undertaken to examine seasonal patterns in photosynthate allocation to leaf and branch materials in the coast live oak (*Quercus agrifolia*). This evergreen tree species is most abundant in valleys and on lower slopes as part of foothill woodlands and mixed evergreen forests (Griffin 1977). It is distributed along the coast of California from Mendocino County to the Sierra San Pedro Martir of Baja California Norte over an area 1120 km long and 96 km wide (Griffin and Critchfield 1972). Seasonal variations in photosynthate incorporation into different leaf metabolic fractions are described and these variations observed for the coast live oak are compared to those found in evergreen shrubs growing in arid and semiarid regions of the southwestern United States.

*Systems Ecology Research Group, San Diego State University, San Diego, California 92182 USA

NATO ASI Series, Vol. G15
Plant Response to Stress. Edited by J.D. Tenhunen et al.

MATERIALS AND METHODS

In April of 1982, a total of 15 individuals with a diameter at breast height ranging from 5.4 to 113.5 cm, were selected at the Sky Oaks Biological Field Station located 132 km NE of San Diego at an elevation of 1500 m. At the study site, bud break, with the emergence of new shoots and leaves, began in early June. Leaves reached full expansion two to four weeks later. The growth of new shoots at the apex of the previous year's growth, permitted a separation of the leaves into two different age classes: current year's (new) and previous year's (old). Leaves of *Q. agrifolia* normally persist for 12 to 18 months, although in some cases up to 24 months pass before leaf abscission (Opler 1974). Two major leaf falls were recorded during early summer and late fall.

To determine patterns of carbon allocation of current assimilate during different times of the year, four 20 cm branches of each of the 15 trees were labeled with ^{14}C at monthly intervals from July 1982 to July 1983. For each tree, one branch at each of four cardinal points, was enclosed in a polyethylene bag and incubated with $^{14}CO_2$ in air during the early morning. Mooney and Chu (1974) reported that during a period of 4 weeks after labeling, carbon equilibrated with the various chemical fractions in leaves. Thus, the branches were harvested 4 weeks after labeling and the four branches pooled for each tree. The labeled tissues were divided into their respective parts (new and old leaves and new and old stems) and then frozen. The samples were stored at -10° C until processed.

The harvested tissues were oven-dried and ground (to 40 mesh). Several classes of chemical components were isolated from leaf tissue samples using a modified sequential extraction scheme based primarily on solubility difference (Dickson 1979). The dried leaf material was homogenized and extracted with a methanol:chloroform:water mix (MCW) to remove the soluble constituents. The soluble constituents were then separated into a chloroform fraction that contained pigments, oils, and other $CHCl_3$-soluble compounds (described as lipid fraction). An aliquot of the $CHCl_3$ fraction was assayed for ^{14}C activity using liquid scintillation counting techniques (LSC). Phenolic compounds were precipitated from the water-methanol fraction using polyvinylpyrrolidine (PVP) (Loomis and Battaille 1966). The remaining extract was then separated into organic-amino acids and sugars using ion exchange resins. The ^{14}C activity present in phenolics and organic-amino acids fraction was determined by comparing the LSC activity in the extract before and after adding the PVP or passing through anion-cation exchange resins.

The pellet from the MCW extract was washed with ethyl acetate to remove non-water soluble tannins (Swain 1979). The ^{14}C activity present in the ethyl acetate was determined by LSC, and this value added to the activity present for phenolics. The pellet remaining after the ethyl acetate wash was then treated with enzymes to remove proteins and reserve polysaccharides (starch). After protein and starch extractions, the residue fraction which contains primarily hemicellulose, cellulose, and lignin was dried and combusted using a Packard Tri-Carb Model 306. The $^{14}CO_2$ released in the combustion process was trapped in carbo-sorb II and counted a using scintillation spectrometer. To determine total ^{14}C activity present in tissues, leaf and stem sub-samples were also combusted prior to the sequential extraction. All counts were corrected for background and quenching.

RESULTS AND DISCUSSION

Seasonal changes in percent distribution between sample classes of total ^{14}C label retained by branches is given in Table 1. The data from the different labeling events are grouped into the six periods shown, based on time of harvest. The time interval prior to and immediately after bud break is a period when large percentages of labeled carbon are used in the construction of new stems and leaves. The greatest percent retention of total ^{14}C label in new stems (55.9%) is observed in June and in the new leaves during July (59.8%). It is important to note that results obtained with branches harvested in June reflect the utilization of photosynthate produced in May, prior to bud break. Thus, the high levels of activity of ^{14}C observed in the new stems represents carbon incorporated by old leaves and then translocated to the growing points. In July, carbon is utilized for leaf growth and development and relatively little is exported out of the new leaves.

Table 1. Seasonal changes in percent distribution between sample classes of total ^{14}C label retained by branches. The months indicate the time of harvest (August 1982 to July 1983). Emergence of new stems and flushing of leaves occurs in early June. The n refers to the number of oak trees sampled; mean $\pm$ 1 standard error are presented.

Month	n	New Leaves	Old Leaves	New Stems	Old Stems
August	14	33.2 ± 3.0	6.5 ± 0.5	42.9 ± 2.0	17.4 ± 1.8
November	15	21.6 ± 1.3	16.9 ± 1.1	37.4 ± 2.0	24.1 ± 1.6
February	12	37.4 ± 6.5	25.0 ± 3.2	20.4 ± 3.0	17.1 ± 2.3
May	15	35.9 ± 1.5	28.5 ± 1.5	19.9 ± 1.7	15.7 ± 1.5
June	13	18.0 ± 1.3	16.7 ± 1.9	55.9 ± 3.6	9.4 ± 1.6
July	15	59.8 ± 3.6	5.4 ± 0.7	29.3 ± 2.9	5.5 ± 0.8

Significant differences are found in percentage of label in the lipid fraction between the new and old leaves during the summer months, while during the remainder of the year a similar percentage of label (16-21%) is found in lipids

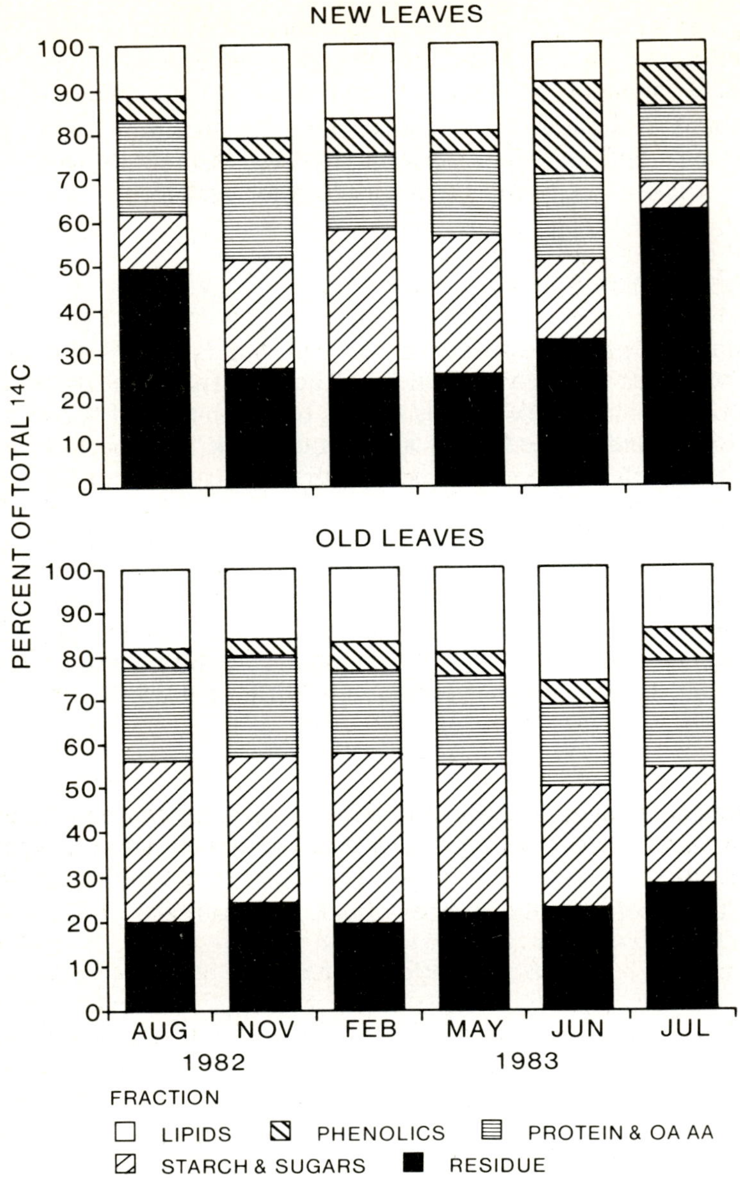

Fig. 1. The relative seasonal distribution of photosynthetically incorporated ^{14}C into five different chemical fractions in new and old leaves of *Quercus agrifolia*. The values presented are percentages of the total label in all fractions. The months refer to the time of harvest which was one month after labeling. Young leaf tissue analyzed in June 1983 was from leaves which emerged after labeling.

for both ages of leaves (Fig. 1). The increase in ^{14}C allocation to lipids observed in June for old leaves, corresponds to a decrease in the proportion of label in the sugar fraction, which may be the result of a sugar to lipid conversion as suggested by Oechel *et al.* (1972) for *L. divaricata*. A similar maximum in the lipid fraction was observed for *L. divaricata* in the summer, while a constant allocation to lipids was reported for *H. arbutifolia* (Mooney and Chu 1974).

The pattern of retention of label in the phenolic fraction indicates a proportionally greater percentage in the new leaves during early leaf growth (Fig. 1). These results are contrary to those of Mooney and Chu (1974) for *H. arbutifolia* who observed greater allocation to phenolics prior to new leaf growth. They suggested that during the growth period, a large fraction of the available carbon went to metabolic compounds and cell wall materials required for new growth, rather than being used to construct protective substances. However, the timing and the differential costs associated with synthesizing and storing these natural products (phenolics) may be governed by the plants' physiological constraints (McKey 1979; Mooney *et al.* 1983; Prudhomme 1983). For example, plants require considerable amounts of carbon for lignin formation. The aromatic alcohols required for lignification are derived from the shikimic acid pathway (Swain 1979), and these phenolic acids may earlier serve as defensive chemicals prior to lignification. Thus, the maxima in percent of label in phenolics observed in young oak leaves may indicate that they initially play a defensive role and later a structural one.

In this study, starch and sugar fractions were combined into one component. Although greater changes are observed in new leaves, the seasonal pattern in label retention in sugars and starch for new and old leaves show a similar trend. Allocation to starch and sugars is high during the winter months and decreases in the summer. The ^{14}C incorporated into the starch and sugar fraction range from 6% to 34% for new leaves and 27% to 39% in old leaves. Comparable results have been observed with *H. arbutifolia* where the combined starch and sugar fraction was higher during non-growing periods. These patterns indicate the importance of the fall and winter months when significant amounts of reserve materials are built up by these evergreen species. In contrast, *L. divaricata* allocated a constant proportion of photosynthate to sugar and starch throughout the year.

When protein, amino acid and organic acid fractions are grouped, similar percentages of label are found in this component (17-24%) throughout the year in both leaf classes. These data suggest a constant allocation for metabolic processes such as maintenance. However, the diverse products present in the amino and organic acids fraction could possibly be involved in other metabolic processes. In comparing the results obtained with oaks with those from *L. divaricata*, slight differences are observed between species. *L. divaricata* exhibited relatively constant proportions of label in protein and organic acids through time, but amino acid levels were highest in the fall and lowest in the spring. In

H. arbutifolia, greater seasonal fluctuations of these fractions were found, and maximum allocation to these fractions are observed during the growth period.

The residue fraction is a combination of lignin, hemicellulose and cellulose. These chemicals are primarily constituents of cell walls and can be used as an indicator of growth (Oechel *et al.* 1972). The highest levels of ^{14}C labeled photosynthate in the residue fraction (62%) are in new leaves during the July period. This correlates with the high retention of ^{14}C in new leaves as shown in Table 1. Similar results have been observed for *L. divaricata* and *H. arbutifolia* where peaks in allocation to these fractions were correlated with their growth period.

CONCLUSIONS

The carbon allocation patterns of the coast live oak described here are different from those reported for the desert evergreen shrub *Larrea divaricata* and for the chaparral evergreen shrub *Heteromeles arbutifolia*. *L. divaricata* exhibits a fairly uniform partitioning of carbon throughout the year. The constant metabolic activity and the lack of dormancy displayed by *L. divaricata* in its arid habitat may explain the pattern of carbon allocation observed in this species (Oechel *et al.* 1972). *H. arbutifolia*, on the other hand, exhibits strong seasonal changes in carbon allocation (Mooney and Chu 1974), such that allocation to phenolics was higher prior to the growing season, and the peak photosynthate allocation to metabolic components occurred during periods of active growth. *H. arbutifolia* also showed an increase in the allocation to storage (starch and sugar fractions) during the non-growth period. The patterns observed for the coast live oak are somewhat different from either of the above shrub species.

Several biotic or abiotic factors may play a role in determining the allocation patterns observed for this evergreen tree species. The substantial amounts of carbon allocated to phenolics and structural compounds in the young leaves of this oak may serve to protect them against numerous spring herbivores (Opler 1974; Brown 1980). Losses of newly produced tissues to herbivores in mediterranean shrub species have been shown to be great (Montenegro, this volume). Large amounts of carbon were required to effect a high degree of sclerophylly. The highly sclerophyllous leaves of *Q. agrifolia* may enable it to be more drought resistant in late season (Rundel 1980).

The partitioning of carbon is also intimately related to the growth form of the plant. Shrubs have relatively more leaf biomass than do trees (Rundel 1980). Therefore, trees may potentially invest more carbon in defensive compounds on a per leaf basis than shrubs which divert more photosynthate to a proportionally larger canopy. Shrubs generally lack a central trunk and have a smaller requirement for woody tissues than trees (Rundel 1980). The high levels of ^{14}C present in stems of *Q. agrifolia* may be indicative of the important requirements of carbon for the formation of woody tissues. Storage in leaves and stems may be more important in trees compared to shrubs which may rely on their roots for storage. The oaks channeled carbon to sugars and starch, suggesting that during the non-growth period a large proportion of carbohydrates were stored in the leaves and these were used to supplement current photosynthate during early season growth.

ACKNOWLEDGEMENTS

We thank P. Harley, D. Hilbert, R. Karban, S. Sparks, and P. Zedler for critically reviewing an earlier version of this manuscript. This study was supported by the National Science Foundation, San Diego State University, and Sigma Xi.

REFERENCES

Brown LR (1980) Insects feeding on California oak trees. In: Plumb TR. Proceedings of the Symposium on the Ecology Management, and Utilization of California Oaks, June 26-28, Claremont, CA. Gen Tech Rep PSW-44. Berkeley, CA: Pacific Southwest Forest and Range Experiment Station, Forest Service, US Department of Agriculture. pp 184-194.

Dickson RE (1979) Analytical procedures for the sequential extraction of ^{14}C-labeled constituents from leaves, bark and wood of cottonwood plants. Physiol Plant 45:480-488

Griffin JR (1977) Oak woodland. In: Barbour MG, Major J (eds) Terrestrial Vegetation of California. Wiley-Interscience, New York. pp 383-415.

Griffin JR, Critchfield WB (1972) The distribution of forest trees in California. Pacific Southwest Forest and Range Experiment Station 82/1972, Forest Service, US Department of Agriculture, Berkeley, California, USA

Loomis WD and Battaille J (1966) Plant phenolics compounds and the isolation of plant enzymes. Phytochemistry 5:41-58

McKey D (1979) The distribution of secondary compounds within plants. In: Rosenthal GA, Janzen DH (eds) Herbivores: Their Interaction with Secondary Plant Metabolites. Academic Press, New York. pp 55-133.

McLaughlin SB, McConathy RK, Beste B (1979) Seasonal changes in within-canopy allocation of ^{14}C-photosynthate by white oak. Forest Sci 25:361-370

Mooney HA (1972) The carbon balance of plants. Ann Rev Ecol Syst 3:315-346

Mooney HA, Chu C (1974) Seasonal carbon allocation in *Heteromeles arbutifolia*, a California evergreen shrub. Oecologia (Berl) 14:295-306

Mooney HA, Gulmon SL, Johnson ND (1983) Physiological constraints on plant chemical defenses. In: Resistance to Insects. American Chemical Society. pp 21-35.

Oechel WC, Strain BR, Odening WR (1972) Tissue water potential, photosynthesis, ^{14}C-labeled photosynthate utilization, and growth in the desert shrub *Larrea divaricata*. Ecol Mono 42:127-141

Opler PA (1974) Biology, ecology, and host specificity of microlepidoptera associated with *Quercus agrifolia* (Fagaceae). University of California Press, Berkeley, California, USA

Prudhomme TI (1983) Carbon allocation to antiherbivore compounds in a deciduous and an evergreen subarctic shrub species. Oikos 40:344-356

Rundel PW (1980) Adaptations of Mediterranean-climate oaks to environmental stress. In: Plumb TR, Proceedings of the Symposium on the Ecology, Management, and Utilization of California Oaks, June 26-28, 1979, Claremont, CA. Gen Tech Rep PSW-44. Berkeley, CA: Pacific Southwest Forest and Range Experiment Station, Forest Service, US Department of Agriculture. pp 43-54.

Swain T (1979) Tannins and lignins. In: Rosenthal GA, Janzen DH (eds) Herbivores: Their Interaction with Secondary Plant Metabolites. Academic Press, New York, pp 657-682

Biomass, nitrogen, and phosphorus accumulation over a southern California fire cycle chronosequence.

C.H. Black

Systems Ecology Research Group, San Diego State University, San Diego, CA 92182 USA

INTRODUCTION

The structural and functional similarities of mediterranean ecosystems of the world impressed early plant ecologists (e.g. Grisebach 1872, Schimper 1898) and have provided the subject for recent comparative studies (di Castri and Mooney 1973, Miller 1981, Specht 1969a,b). These systems are typically fire dominated: at regular intervals intense fires consume most above-ground organic material and initiate succession. The vegetation on these sites, stimulated by ash-deposited nutrients and released to some extent from soil moisture limitations of growth shows high initial rates of biomass accretion and nutrient uptake. Net productivity and nutrient uptake rates decline after the first decade of recovery as nutrients become immobilized in standing biomass, and in litter and soil organic matter, and as transpiration of mature canopies reduces the amount of water available.

The time scale of development of mediterranean vegetation, short in comparison to many other systems, and the accessible nature of this vegetaion make mediterranean systems amenable to close scrutiny. In particular these systems offer excellent opportunities for testing process oriented models of stand growth and productivity (Jacobson, Stoner, and Richards 1981, Lawrence 1975, Miller and Mooney 1976, Miller and Stoner 1979). Frequently these systems are dominated by only one or two species so that the seasonal course of only a few species' plant gas and water exchange characteristics can be integrated with a knowledge of plant architecture and stand structure to produce models which describe stand net photosynthesis and biomass accumulation. Comparison of these models with empirical studies of biomass accumulation such as the one described here allows identification of strengths and weaknesses in our understanding of structure and function in mediterranean ecosystems.

The present study was part of an overall research program to further examine the structure and function of California chaparral stands as they develop with time. We were particularly concerned with carbon, nitrogen, and phosphorus pools, and how these were distributed in ecosystems as a function of time after fire. We present data on the variation we encountered between closely situated stands of different ages and between stands on north and south slopes which had recovered from the same fire event.

NATO ASI Series, Vol. G15
Plant Response to Stress. Edited by J. D. Tenhunen et al.

MATERIALS AND METHODS

Methods applied in this study will be described in greater detail elsewhere (Marion and Black in prep.). Briefly, seven stands of ages 4, 11, 11, 40, 54, 80, and 85 years with matching north and south aspects were located on the western slopes of the mountains bordering the western edge of the Anza-Borrego Desert in northern San Diego County, at the Sky Oaks Biological Station and in the Cleveland National Forest. The stands were situated between 900 and 1500 m above sea level, and the farthest distance between stands was approximately 15 km. All stands were on the Sheephead soil series (a loamy, mixed, mesic, shallow Ultic Haploxeroll) except for the 54-year-old north aspect stand, which was on the Tollhouse series (a loamy, mixed, mesic, shallow Entic Haploxeroll). On each aspect of each stand duplicate 5 x 5 m plots were established. All aboveground standing vegetation within each plot was harvested and weighed at the site. Representative stems of each species found in a plot were separated into leaf and stem material, dried to a constant weight at 65° and weighed. Litter was collected on duplicate 1 m^2 areas within each plot. Organic material > 2 mm in size was separated from these collections by flotation and sieving and dried to a constant weight. Nitrogen and phosphorus contents of plant tissues and litter were determined by acid digestion and analysis with a Technicon Autoanalyzer. Total soil pools of carbon, N, and P were determined by combustion in a LECO induction furnace, Kjeldahl digestion, and perchloric digestion, respectively, using four samples each of soil from the A and B horizons collected in each plot. Plant potentially available N and P levels were determined with both chemical extractions and a biological assay (Marion and Black in prep.). Soil carbon and mineral nutrient levels were adjusted for soil moisture, gravel content, bulk density, and soil horizon depth to estimate pools on a per area basis.

STANDING DEAD BIOMASS

Immediately after a fire these ecosystems may have highly variable amounts of organic carbon and nutrients in standing dead material (Fig. 1). One of the north slope 4-year-old plots had 12600 kg ha^{-1} of partially burned large dead stems, while the other had only 1390 kg ha^{-1} standing dead biomass. Quantities of organic carbon and minerals left in the standing dead biomass component at the time of burning is dependent on 1) the nature (e.g. live or dead) and size class distribution of plants in the vegetation existing at the time of the fire (see Rundel *et al.* in this volume), and 2) the heat of combustion and duration of the fire. Rundel and Parsons (1979), for example, found that chamise stand structure shifted with age, with a majority of stem biomass in the first decade of growth in size classes that are usually consumed by fire, to larger size classes in later decades which are not generally consumed. Biomass of standing dead material

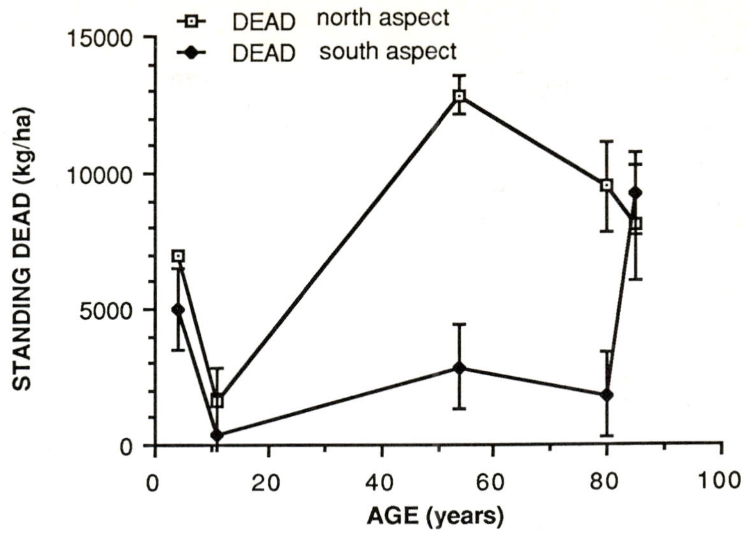

Fig. 1. Mean standing dead biomass pools for north and south aspects of stands. Bars indicate ± one standard error

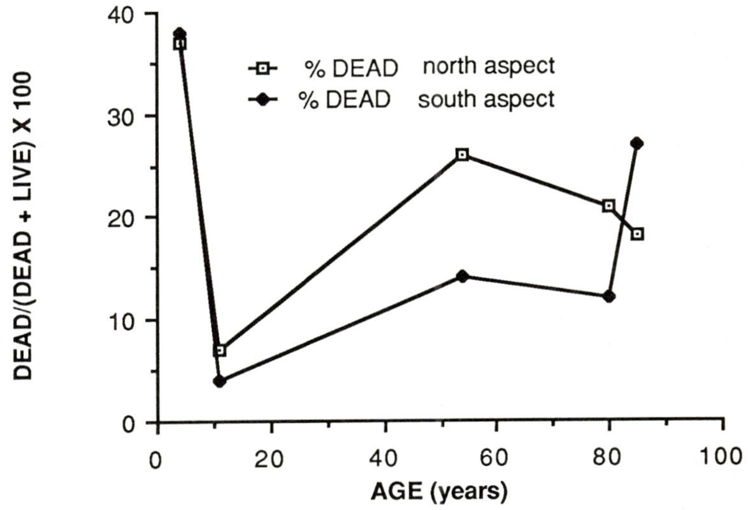

Fig. 2. Standing dead biomass as a percentage of total standing biomass for stands

remained highly variable over the age sequence: coefficients of variation averaged 33% for north aspect slopes and 30% for south aspect plots.

For our stands, the lowest amounts of standing dead biomass occurred in the 11-year-old stands. Although one oak on the north facing slope had a large amount of partially burned dead woody stems, large stems from other species that may have survived the previous fire appear to have decomposed. Decomposition rates of chaparral litter have been investigated (Schlesinger 1985, Schlesinger and Hasey 1981), but decay rates of large woody debris in these ecosystems have not been studied.

By stand age of 40 years, most standing dead biomass consisted of dead limbs and individuals of plants which had initiated growth with the previous fire. The steep increase in biomass of standing dead in north aspect plots at age 40 corresponds to the death of *Ceanothus* individuals, which begin to have a high mortality rate near this age (Hanes 1971, personal observation). Different mortality patterns of different species on the opposing slopes may contribute to the difference in standing dead biomass accumulation patterns observed: for north aspects a logarithmic increase reaching an asymptote after age 40, and for south aspects a gradual linear increase with time. The high variability in standing dead biomass makes it impossible to show a statistically significant difference in these patterns, however.

Accumulated standing dead biomass on north aspect slopes is three to four times that on south aspects. A single exception, in the 85-year-old south aspect plots, was due to the presence of a large quantity of dead redshank from a single plant. This is not typical of most areas of the 85-year-old stand. North slopes had a larger percentage of total biomass in standing dead material also (Fig. 2). The range of percent biomass in standing dead found in this study encompasses the values found in other studies of biomass pools in mediterranean ecosystems (see Table 10.4 in Gray and Schlesinger 1981). There are no significant increases in standing dead biomass accumulation after stand ages of 40 years, in part because after this time recruitment to the standing dead pool has ceased, and in part because decomposition and transformation to other pools (i.e. large woody debris) is balancing recruitment to this pool.

STANDING LIVE BIOMASS

Standing live biomass is shown as a function of stand age in Fig. 3. Accumulation rates are maximal in the first decade, and decline steeply thereafter. The different accumulation rates for the north and south aspects during the first decade encompass all California chaparral values (see below) except those of the highly productive *Ceanothus megacarpus* stands studied by Schlesinger and Gill (1980), and demonstrate the importance of aspect in affecting net productivity in these areas, a fact previously recognized by Hanes (1971). In contrast

Schlesinger *et al.* (1982) found no apparent effect of aspect on biomass accumulation rates for highly productive stands of *Ceanothus megacarpus* in the Santa Ynez Mountains. They present evidence that these stands are not as nutrient or water limited as many other mediterranean systems.

Standing live biomass distribution was variable in all stands, and variability showed no pronounced trend with age. Coefficients of variation averaged 5% for north aspects and 11% for south aspects. These are similar to average values of 5% and 6% for 8-year-old and younger stands of chamise and mixed chaparral, and 12% and 11% for old stands of chamise and mixed chaparral studied by Sampson (1944). They are smaller than the 14% and 21% values found by Schlesinger *et al.* (1982), who stated that for their sites, variation increased with stand age due to differences in site quality. They did not test the significance of this increase. The close proximity of these stands and their locations on a similar soil type would minimize differences in biomass due to site differences.

Net live biomass accumulation continued until after 40 years of age. After 50 years the vegetation appeared to reach a steady state, with net accumulation rates near zero. The variability in standing biomass on plots in this study indicates that at least ten plots on each aspect would be needed to show a significant change of 10% in biomass amounts for older stands. Leaf weight as a percentage of total aboveground biomass (Fig. 4) was initially high, and declined with age to a relatively steady state after 40 years of age.

Old stands of mediterranean vegetation are uncommon due to frequent fires in these regions, and therefore few other studies have examined biomass accumulation in stands older than two or three decades. Rundel and Parsons (1979) found that net biomass accumulation was only about six percent less on stands greater than 60 years old than on 37-year-old stands for montane chamise chaparral. Net accumulation in heath scrub studied by Specht and Rayson (1958) had apparently stopped by 50 years of age. Sampson described "old, unburned" stands of chamise and mixed chaparral which contained approximately 30,000 kg ha^{-1}, a value similar to that found in the 40-year-old north aspect stand in this study. Matorral vegetation studied by Merino and Vicente (1981) showed biomass peaks near 20 years of age, followed by apparent but not statistically significant declines. The above mentioned studies of biomass accumulation with age, as well as the present results on accumulation and percent leaf tissue, do not demonstrate the pronounced decline in standing live biomass, or the accumulation of standing dead material either on an absolute or a percentage basis that have traditionally been thought to occur as these mediterranean ecosystems pass from maturity into senescense (Hanes 1971).

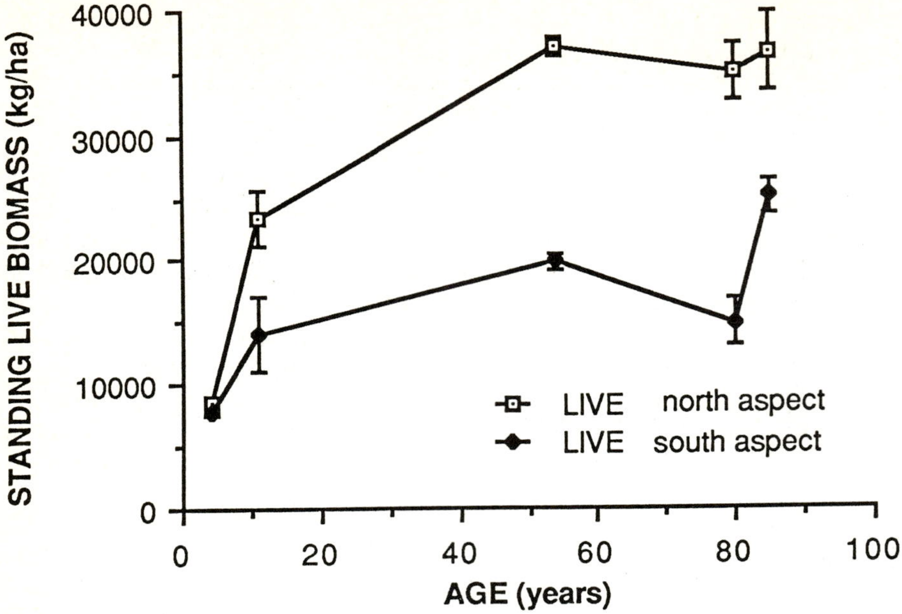

Fig. 3. Standing live biomass pools for north and south aspects of stands

LITTER ACCUMULATION

Litter accumulation in these stands and litterfall for a small subset of the stands is described by Marion, Black, and Rhese (1984). Litter accumulation (Fig. 5) was highly variable, with coefficients of variation averaging 19.7% and 29.6% on north and south aspects, respectively. Large variation in young stands is due in part to incomplete canopy closure, and irregular distribution of litterfall. In old stands it is partially due to very localized input of large woody debris. In all areas the irregular surface topography resulted in litter collecting in depressions, which further contributed to variability. Unlike standing live and standing dead biomass, which are best described by semilogarithmic curves, net litter accumulation appeared to increase linearly with age.

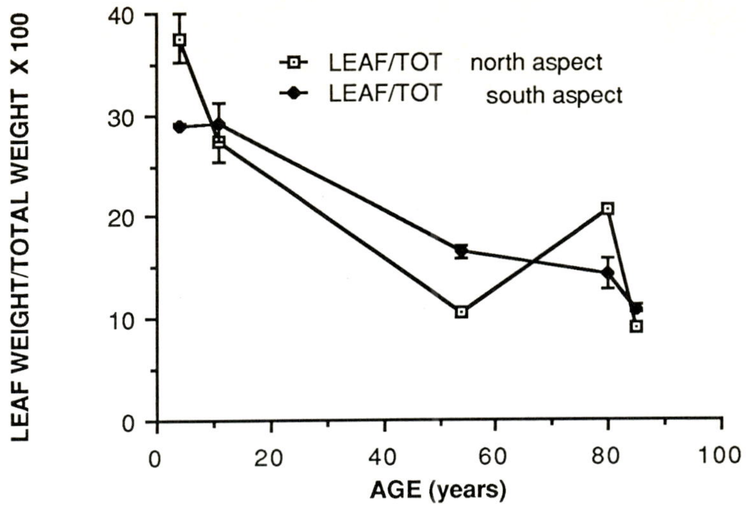

Fig. 4. Leaf weight as a proportion of total standing live biomass for the diffferent stands

MINERAL NUTRIENT ACCUMULATION

Aboveground mineral nutrient accumulation was the product of nutrient concentrations and amounts of aboveground biomass in the different categories. N and P concentrations did not vary in a systematic matter in standing litter (Marion *et al.* 1984) or in standing dead wood (Marion and Black in prep.). N concentrations in live tissue increased slightly to approximately 55 years, then declined; P concentrations showed a marked logarithmic decline with age (Marion and Black in prep.). Net effects of these concentration changes are shown in Figs. 6 and 7, which show total aboveground N and P pools as a function of age.

Total ecosystem N except for N in large roots is shown as a function of age in Fig. 8. This data indicates that nitrogen is accumulating in these ecosystems at a rate of approximately 16 kg ha^{-1} yr^{-1} on north aspect slopes and 8 kg ha^{-1} yr^{-1} on south aspect slopes. These values fall between the low levels of N input from deposition (Ellis, Verfaillie, and Kummerow 1983) plus symbiotic fixation (Kummerow *et al.* 1978) found in these mountainous areas of San Diego County, California and the higher levels in the mountains west of Los Angeles (as high as 23 kg N ha^{-1} yr^{-1}, Riggan personal communication). Estimates of N loss from California chaparral due to volatilization by fire and subsequent erosional losses range from 156 kg ha^{-1} (DeBano and Conrad 1978) to 572 kg ha^{-1} (Riggan, personal communication). These N input and output data from chaparral systems give a range of between 10 and 70 years required for a system to gain N lost due to a fire event, a value considerably less than the 150 years predicted by

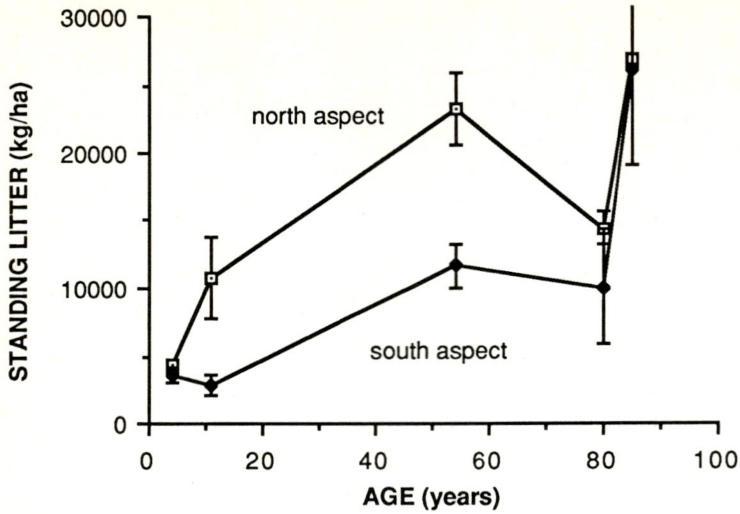

Fig. 5. Standing litter for the different stands. Data from Marion *et al*. 1984

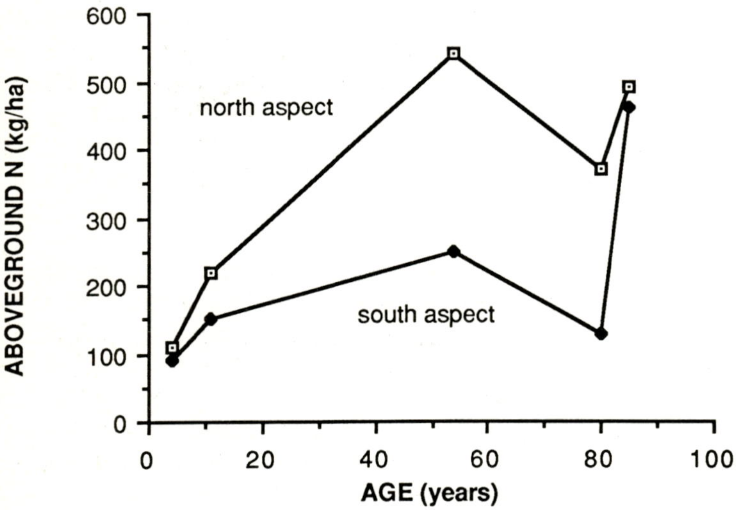

Fig 6. Total aboveground nitrogen (in standing dead + standing live + litter) for the different stands

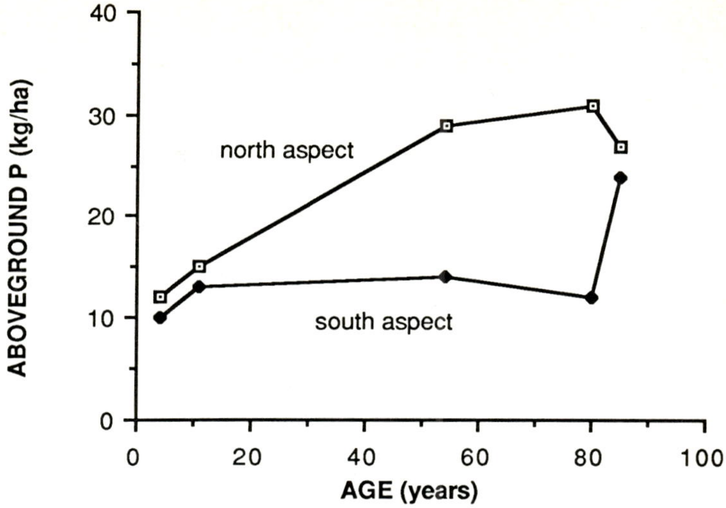

Fig. 7. Total aboveground phosphorus (in standing dead + standing live + litter) for the stands

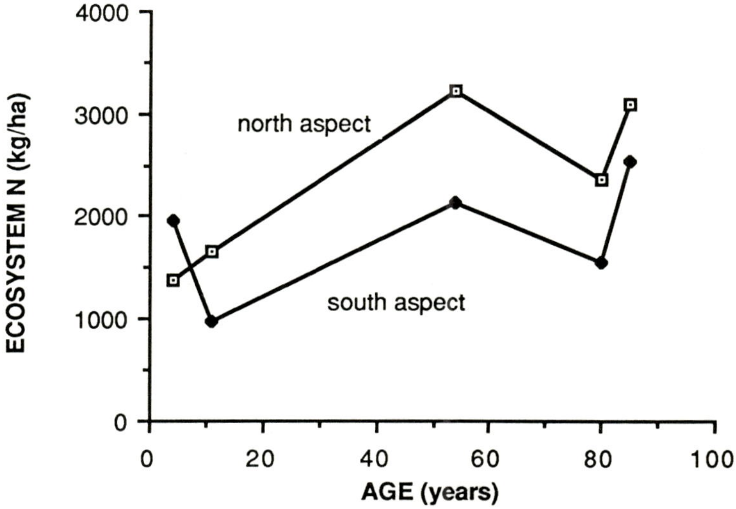

Fig. 8. Total ecosystem nitrogen (except for N in large roots) for the stands

Table 1. Semilogarithmic equations describing standing live biomass as a function of stand age (in years) for several different mediterranean ecosystems.

Ecosystem	Biomass Equation	r^2	Reference
garrigue	B = 199 + 16665 ln (age)	.906	Specht 1969
chaparral	B = -1510 + 6775 ln (age)	.881	Specht 1969
Ceanothus megacarpus chaparral	B = -15726 + 19814 ln (age)	.480	Schlesinger and Gill (1980)
chaparral	B = -13620 + 8696 ln (age)	.979	Rundel and Parsons (1979)
chaparral	B = -2664 + 8288 ln (age)	.930	Sampson (1944)
chaparral, north	B = 1161 + 8988 ln (age)	.901	this study
chaparral, south	B = 2997 + 4057 ln (age)	.682	this study

Schlesinger and Gill (1980) and encompassing the 'average' California fire frequency of every 30 years (Hanes 1971). Total ecosystem phosphorus levels did not vary in a systematic manner with stand age, as might be expected with a nutrient less subject to volatilization losses or movement in soil profiles.

COMPARISON WITH OTHER SYSTEMS

Rates of biomass accumulation in mediterranean ecosystems can generally be described with semilogarithmic equations (Table 1). These regression models were used to illustrate changes in average biomass accumulation rate over time for several decades of vegetation development (Fig. 9). The different systems included in this comparison vary widely in initial productivity, but all show the general pattern of rapidly declining biomass accumulation in the second or third decade of growth, as nutrient and water limitations ensue. The two systems with unusually high early productivity and large accumulation of standing biomass are the French garrigue studied by Lossaint and Rapp (1971) and Lossaint (1973) and the *Ceanothus megacarpus* chaparral studied by Schlesinger and Gill (1980). The garrigue is characterized by levels of atmospheric N deposition as great as 14.6 kg ha^{-1} yr^{-1}. The *Ceanothus megacarpus* chaparral may be characterized by a high N fixing capability, and older stands appear to be less water limited than other areas of California chaparral (Schlesinger *et al.* 1982).

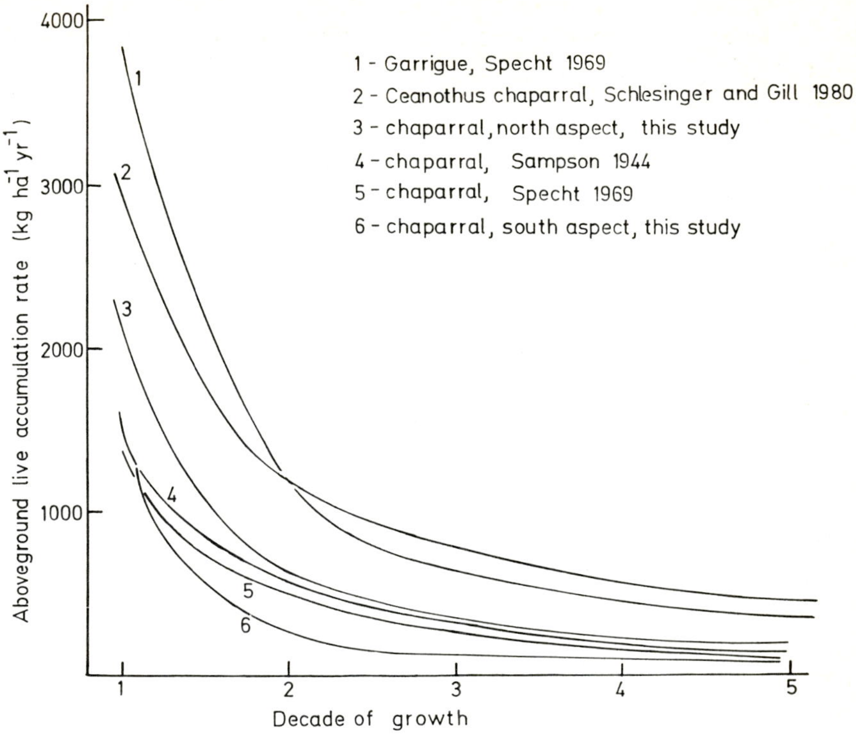

Fig. 9. Rate of accumulation of biomass for different mediterranean ecosystems as a function of decade of growth. Rates were averaged over each decade using beginning and ending pools of the specific decade calculated from the data present in the various studies. Equations are given in Table 2

Table 2 presents data comparing biomass, N, and P accumulation in stands considered in this study with other California chaparral stands studied by a variety of authors. Values of biomass and nutrient pools at 5 years (representing young stands) and the oldest age reported from each respective study were determined (either from actual values, with the logarithmic models of Table 1, or from a logarithmic model for N and P accumulation rates and calculated from data on standing pools presented in the respective studies). Biomass values in "mature stands", which often appear to be in a steady state with regard to biomass and nutrient levels, range from 1.7 to 3.1 times the values occurring at 5 years of age. In systems composed to a large extent of nitrogen fixing *Ceanothus* species (*Ceanothus megacarpus* stands studied by Schlesinger and Gill [1980], and the north aspect stands of this study), the increase in N pools from young to old stands exceeds or is equal to the biomass increase, and the P increase is considerably less. In the pure chamise chaparral stands studied by Rundel and Parsons (1979, 1980),

Table 2. Pools of biomass, nitrogen (N) and phosphorus (P) at age of 5 years and at older stages ("Old"), and -fold increase between these two times. [5] indicates a value calculated for 5-year-old stands by a semilogarithmic fit to data as described in Table 1.

Ecosystem	Pool	(Kg ha^{-1})	X Increase	Reference
Chamise chaparral	5-Year	"Old"		
Biomass	9,990	31,438	3.1	Sampson 1944
N	N.M.	N.M.		
P	4	14	3.5	
Mixed chaparral	5-Year	"Old"		
Biomass	10,102	30,979	3.1	Sampson 1944
N	N.M.	N.M.		
P	5	16	3.2	
Chamise chaparral	[5]	≥ 60		
Biomass	11,949	20,553	1.7	Rundel and
N	52	90	1.7	Parsons, 1979a,b
P	8.8	19.0	2.2	
Ceanothus megacarpus chaparral	[5]	21		
Biomass	20,000	48,000	2.4	Schlesinger and
N	14	42	3.0	and Gill, 1980
P	25	29	1.2	
Mixed chaparral	[5]	37		
Biomass	12,016	27,255	2.3	Specht 1969a
N	32	142	4.4	
P	8	10	1.3	
Mixed chaparral N aspect	[5]	85		
Biomass	15,627	40,547	2.6	This study
N	183	484	2.6	
P	34	42	1.2	
Chamise chapparal S aspect	[5]	85		
Biomass	9,526	20,775	2.2	This study
N	224	291	1.3	
P	20	24	1.2	

N.M. = Not measured

Sampson (1944), and the south aspect stands of this study, percent P increases are similar to or greater than N and biomass increases. These differences suggest that even though one study (Kummerow *et al.* 1978) found low amounts of nitrogen fixed by *Ceanothus* at specific sites, this fixation may be an important determinant of ecosystem biomass and nutrient accumulation rates. Furthermore, the general restriction of *Ceanothus* to north aspect slopes could lead to differences in the roles of N and P in limiting productivity and in nutrient accumulation patterns on different aspects of stands initiated by the same fire event.

REFERENCES

DeBano LF, Conrad CE (1978) The effect of fire on nutrients in a chaparral ecosystem. Ecology 59:489-497

di Castri E, Mooney HA (eds) (1973) Mediterranean Type Ecosystems: Origin and Structure. Springer, Berlin-Heidelberg-New York, 405 p

Ellis BA, Verfaillie JR, Kummerow J (1983) Nutrient gain from wet and dry atmospheric deposition and rainfall acidity in southern California chaparral. Oecologia (Berl) 60:118-121

Gray JT, Schlesinger WH (1981) Nutrient cycling in mediterranean type ecosystems. In: Miller PC (ed) Resource Use by Chaparral and Matorral. Springer, Berlin-Heidelberg-New York, pp 259-285

Grisebach AHR (1872) Die Vegetation der Erde nach ihre klimatischen Anordnung. W. Engelmann, Leipzig

Hanes TL (1971) Succession after fire in the chaparral of southern California. Ecol Monogr 35:213-235

Jacobson MB, Stoner WA, Richards SP (1981) Models of plant and soil processes. In: Miller PC (ed) Resource Use by Chaparral and Matorral. Springer, Berlin-Heidelberg-New York, pp 287-368

Kummerow J, Alexander JV, Neel JW, Fishbeck K (1978) Symbiotic nitrogen fixation in *Ceanothus* roots. Amer J Bot 65:63-69

Lawrence WT (1975) A radiation model for chaparral canopies. MS thesis, San Diego State University, San Diego, California

Loissant P (1973) Soil-vegetation relationships in mediterranean ecosystems of southern France. In: di Castri E, Mooney HA (eds) Mediterranean Type Ecosystems: Origin and Structure. Springer, Berlin-Heidelberg-New York, pp 199-210

Loissant P, Rapp M (1971) Repartition de la matiere organique, productivite et cycles des elements mineraux dan des ecosystemes de climat mediterraneen. In: Duvigneaud P (ed) Productivity of Forest Ecosystems. UNESCO, Paris, pp 597-617

Marion GM, Black CH, Rhese M (1984) Litter accumulation and litterfall along a chaparral fire cycle chronosequence. In: Dell B (ed) Medecos IV: Proceedings of the 4th International Conference on Mediterranean Ecosystems, University of Western Australia, Nelands, Western Australia, pp 107-108

Merino J, Martin Vicente A (1981) Biomass, productivity and succession in the scrub of the Doñana Biological Reserve in Southwest Spain. In: Margaris NS, Mooney HA (eds) Components of Productivity of Mediterranean-climate Regions: Basic and Applied Aspects. Dr. W. Junk Publishers, The Hague, pp 197-203

Miller PC (ed) (1981) Resource Use by Chaparral and Matorral. Springer, Berlin-Heidelberg-New York, 455 p

Miller PC, Mooney HA (1976) The origin and structure of American arid-zone ecosystems. In: de Wit CT, Arnold GW (eds) Critical Evaluation of Systems Analysis in Ecosystems Research and Management. Center for Agricultural Publishing and Documentation, Wageningen pp 38-56

Miller PC, Stoner WA (1979) Canopy structure and environmental interactions. In: Solbrig O, Jain S, Johnson GB, Raven PH (eds) Plant Population Biology. Columbia University Press, New York, pp 163-173

Rundel PW, Parsons DJ (1979) Structural changes in chamise (*Adenostoma fasciculatum*) along a fire induced age gradient. J Range Manage 32:462-466

Rundel PW, Parsons DJ (1980) Nutrient changes in two chaparral shrubs along a fire induced age gradient. Am J Bot 67:51-58

Sampson AW (1944) Plant Succession on Burned Chaparral Lands in Northern California. Calif Agric Exp Stn Bull 685, Berkeley

Schimper AFW (1898) Pflanzen-Geographie auf physiologischer Grundlage. G Fisher, Jena

Schlesinger WH (1985) Decomposition of chaparral shrub foliage. Ecol 66:1353-1359

Schlesinger WH, Gray JT, Gill DS, Mahall BE (1982) *Ceanothus megacarpus* chaparral: a synthesis of ecosystem processes during development and annual growth. Bot Rev 48:71-117

Schlesinger, WH, Gill DS (1980) Biomass, production, and changes in the availability of light, water, and nutrients during development of pure stands of the chaparral shrub, *Ceanothus megacarpus*, in the Santa Ynez Mountains, California. Ecol 59:1256-1263

Schlesinger WH, Hasey MM (1981) Decomposition of chaparral shrub foiage: Losses of organic and inorganic constituents from deciduous and evergreen leaves. Ecol 62:762-774

Specht RL (1969a) A comparison of the sclerophyllous vegetation characteristic of mediterranean type climates in France, California, and southern Australia. I Structure, morphology, and succession. Aust J Bot 17:277-292

Specht RL (1969b) A comparison of the sclerophyllous vegetation characteristic of mediterranean type climates in France, California, and southern Australia. II Dry matter, energy, and nutrient accumulation. Aust J Bot 17:293-308

Specht RL, Rayson P (1958) Dark Island Heath (Ninety-Mile Plain, South Australia). VI Pyric succession: Changes in composition, coverage, dry weight, and mineral nutrient status. Aust J Bot 6:59-88

Effect of copper on water relations and growth of *Triticum*

A. Casimiro

Departamento de Biologia Vegetal, Faculdade de Ciências Universidade de Lisboa, 1294 Lisboa Codex, Portugal

INTRODUCTION

Plants are subjected to a variety of stresses including abnormal levels of nutrients. The essentiality of copper in the growth of higher plants was established by Sommer (1931). Copper deficiencies have been reported for many plant species growing in a variety of habitats around the world. Cereals, particularly *Triticum aestivum* L., are highly sensitive to copper deficiency. The symptoms of copper deficiency have been described in numerous publications, Piper (1942), Graham (1975), Gartrell *et al.* (1979). The loss of turgor - "wilting" - is the most common symptom (Rahimin and Bussler 1973; King 1974; Graham 1976).

From the works of Lipman and Mackinney (1931) it is known that copper deficiency depresses vegetative growth, lowers root/shoot ratio and delays maturity. This delay under mediterranean climate conditions, extends the growing season into the summer and consequently may result in a significant loss of grain yield (Graham *et al.* 1981). In order to better understand the wilting phenomenon and its effect on growth and yield of *Triticum* under mediterranean climate conditions, the effect of copper deficiency on water relations and lignification were studied.

MATERIALS AND METHODS

Wheat plants (*Triticum aestivum* L., var. *nazareno Stramppeli*) were grown in a growth cabinet at 22 $\pm$ 2° C in complete and copper-deficient nutrient solutions, under a photosynthetically active radiation of 150 μE m^{-2} s^{-1}, and with a photoperiod of 12 hours. Each pyrex pot culture contained 2 liters of nitrate type nutrient solution (Hewitt 1966). The water used for nutrient solutions was treated with Chelex 100 resin. Culture pots were covered with black polyethylene material and each contained 10 wheat plants. Culture solutions were renewed weekly. The volume of nutrient solution uptake in culture pots was measured weekly, when nutrient solutions were changed. Roots and shoots were dried at 80° C for 48 h for dry weight determination.

Water potential was determined on the youngest fully expanded leaves (Y.F.E.L.) with a pressure bomb (Scholander *et al.* 1965). Leaf conductance was measured with a ventilated steady-state porometer (LiCor 1600). Upper surfaces of the

NATO ASI Series, Vol. G15
Plant Response to Stress. Edited by J. D. Tenhunen et al.

youngest fully expanded leaves were used. Experiments were repeated every two weeks for control and copper-deficient plants. The results are means of 5 to 10 replicates in 5 parallel assays.

Anatomical observations were made on diaphanized Y.F.E.L. (13 weeks old). The xylem lignification in control and copper-deficient plants was compared. Copper concentration in roots and shoots was determined by atomic absorption spectometry (Gladstones *et al.* 1975).

RESULTS

Plants grown in copper-deficient solutions morphologically exhibit symptoms of deficiency. Wilting is extremely prevalent after 7 weeks (Fig. 1).

Fig. 1. Effects of copper deficiency on wilting of plants (age 13 weeks)

Fig. 2 shows that in both, control plants and copper-deficient plants, vegetative growth follows a sigmoidal pattern. However, dry matter production and root/shoot ratio are significantly depressed in copper-deficient plants.

The effect of copper deficiency on the uptake of nutrient solution is shown in Fig. 3. Copper-deficient plants used less nutrient solution than controls. Nevertheless, in either situation the pattern of solution uptake followed that of growth (Fig. 2). Stomatal conductance and water potential were markedly reduced by copper deficiency (Fig. 4).

Optical microscopy of diaphanized leaves revealed that lignification of xylem vessels was reduced in copper-deficient plants (Fig. 5). The measured concentration of copper in roots and shoots, control plants and deficient plants throughout the experiment are shown in Fig. 6.

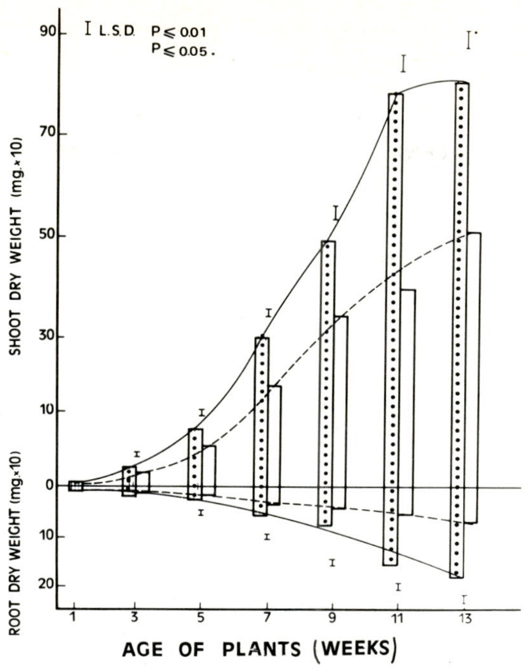

Fig. 2. Root and shoot dry weight: solid squares - control plants (+ Cu); open squares - deficient plants (- Cu) throughout the course of the experiment

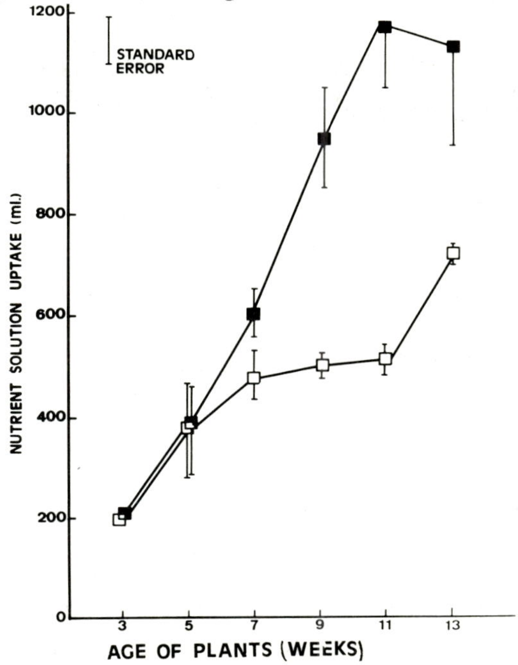

Fig. 3. Nutrient solution uptake: solid squares - control plants (+ Cu); open squares - deficient plants (- Cu) throughout the course of the experiment

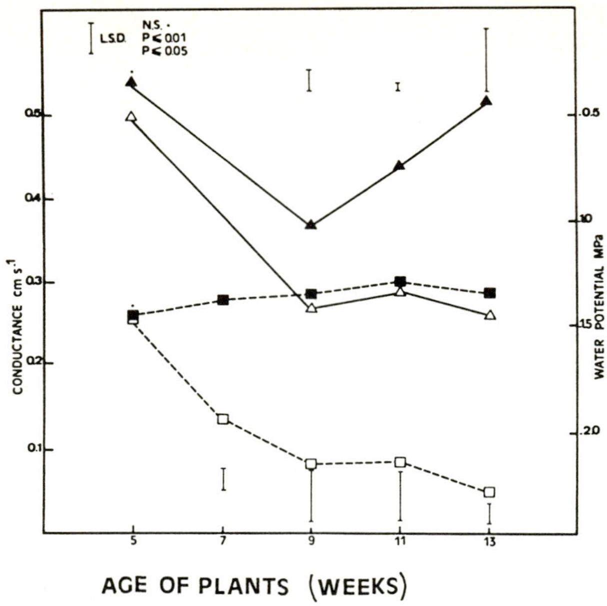

Fig. 4. Water potential (dashed lines): solid squares - control plants (+ Cu); open squares - deficient plants (- Cu); and stomatal conductance: solid triangles - control plants (+ Cu); open triangles - deficient plants (- Cu) throughout the course of the experiment

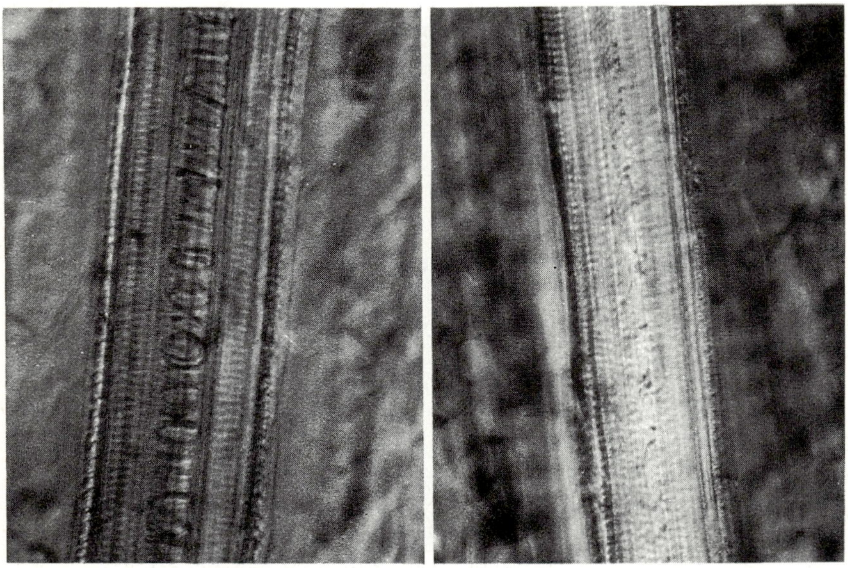

Fig. 5. Stem cross sections of control plants (left) and deficient plants (right) showing xylem vessels

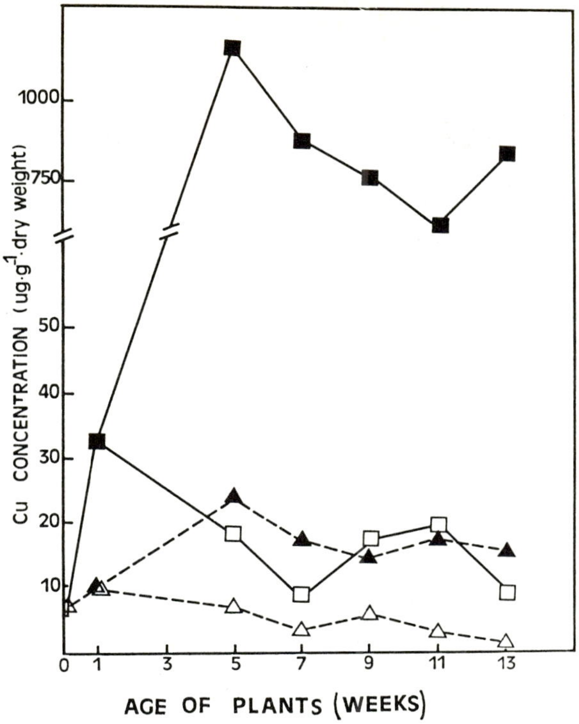

Fig. 6. Copper concentration in roots (solid lines): solid squares - control plants (+ Cu); open squares - deficient plants (- Cu); and shoots (dashed line): solid triangles - control plants (+ Cu); open triangles - deficient plants (- Cu) throughout the course of the experiment.

DISCUSSION

Copper deficiency markedly affects the vegetative growth of wheat plants. The results show that copper deficiency reduces root development more than shoot and hence decreases the root/shoot ratio in agreement with the findings of Lipman and Mackinney (1931). Some authors (Piper 1942; Chaudry and Loneragan 1970; King 1974) have suggested that a decrease in root/shoot ratio may restrict soil solution uptake and adversely affect water relations in plants. We would then expect to find lower leaf water potentials and lower leaf conductances in copper-deficient wheat plants. The results presented here agree. However, Graham (1976) found high leaf water potentials together with low stomatal conductances. The authors suggested that there were no significant effects of copper on plant water relations. It was further suggested that stomatal closure in deficient plants is due to a specific effect of copper on guard cells, in which inadequate levels of ATP are produced to effect full opening.

Graham (1976) also considered that the "wilting" phenomenon might be explained as a structural weakness due to reduced lignification of cell walls and xylem vessels. The results reported here confirm that copper deficiency affects lignification of xylem vessels. This supports the hypothesis of Pissarek (1974) that water supply to the shoots of copper-deficient plants may be restricted due to the collapse of weakly lignified xylem vessels. From our results, it seems clear that both decreased xylem lignification and adverse effects on plant water relations are involved in the "wilting" phenomenon.

Delays in maturity, due to mild copper deficiencies, may be of agronomic importance in mediterranean climates. The extension of the growing season into the water stressed summer period may lead to loss of grain yield. According to Graham (1976), wilting symptoms observed under these conditions may be due to additive effects of "true" and "apparent" wilting.

ACKNOWLEDGEMENT

I thank Prof. Dr. F. Catarino for the kind invitation to participate in this meeting.

REFERENCES

Bussler W (1981) Physiological functions and utilization of copper. In: Loneragan JF, Robson AD and Graham RD (eds) Copper in Soils and Plants. Academic Press, Sydney, pp 213-234

Chaudry FM, Loneragan JF (1970) Effects of nitrogen, copper, zinc fertilizers on the copper and zinc nutrition of wheat plants. Aust J Agric Res 21:865-879

Gartrell JW, Brennan RF, Robson AD (1979) Symptoms and treatment of copper deficiency in wheat. J Agric West Aust 20:18-20

Gladstones JS, Loneragan JF, Simmons WJ (1975) Mineral elements in temperature crop and pasture plants. III. Copper. Aust J Agric Res 26:113-126

Graham RD (1975) Male sterility in wheat plants deficient in copper. Nature 254:514-516

Graham RD (1976) Anomalous water relations in copper-deficient wheat plants. Aus J Plant Physiol 3:229-236

Graham RD, Nambiar EKS (1981) Advances in research on copper deficiency in cereals. Aust J Agric Res 32:1009-1037

Hewitt EJ (1966) Sand and water culture. Methods used in the study of plant nutrition. Revised 2nd edn Comm Bureau of Horticulture and Plantation Crops, East Malling Tech Comm, pp 431-Table 40

King PM (1974) Copper deficiency symptoms in wheat. J Agric South Aust 77:96-99

Lipman CB and Mackinney G (1931) Proof of the essential nature of copper for higher green plants. Plant Physiol 6:593-599

Piper CS (1942) Investigations on copper deficiency in plants. J Agric Sci 32:143-183
Pissarek (1974) Investigation of the anatomical changes in oats and sunflower caused by copper deficiency. Z Planzenernaehr 137:224-234
Rahimi A, Bussler W (1973) The effect of copper deficiency on the tissue structure of higher plants. Z Planzenernaehr 135:185-195
Scholander PF, Hammel HT, Bradstreet D, Hemmingsen EA (1965) Sap pressure in vascular plants. Science 148:339-346
Sommer AL (1931) Copper as an essential element for plant growth. Plant Physiol 6:339-345

5. Mediterranean Plant Growth, Structure, and Development

Quantification of mediterranean plant phenology and growth

G. Montenegro

Laboratorio de Botánica Facultad de Ciencias Biológicas, Pontificia Universidad Católica de Chile, Casilla 114-D Santiago, Chile

INTRODUCTION

Cool, moist winters and hot, dry summers characterize the climate of mediterranean-type ecosystems (Aschmann 1973; Thrower and Bradbury 1977). Mediterranean shrublands are typical in these regions, i.e. scrub formations of woody shrubs with mainly evergreen assimilating organs and semi-sclerophyll to sclerophyll leaf consistency (e.g. chaparral, matorral, maquis). An overstory of small trees and an understory of annuals and herbaceous perennials may also be present (Specht 1969; Montenegro 1978; Di Castri 1981).

A number of previous studies have compared the vegetation of several of these regions from an ecological point of view (Specht 1969; Mooney *et al.* 1970; Quezel 1979) but attempts to cover all five mediterranean-climate regions of the world are rare (Di Castri 1981). One factor that has made vegetation comparisons in these ecosystems difficult has been the use of different methodologies. A close collaboration in comparative studies of two mediterranean-climate regions was first undertaken within the framework of the International Biological Program (I.B.P.), beginning in 1970, to analyze structural, functional, and evolutionary features of mediterranean ecosystems of Chile and California. Methodologies were standardized in order to maximize the degree of comparability of the results. The principal results of the I.B.P. mediterranean ecosystems project were summarized by Mooney *et al.* in 1977. A comparison of vegetation function in the chaparral of southern California and the matorral of central Chile was undertaken by Miller and co-workers in which the main objective was to analyze and compare resource use by primary producers (Miller 1981).

The objective of the present paper is to review methods applied in the quantification of plant phenology and growth in mediterranean-type systems, with special emphasis on comparative studies in the matorral of central Chile and the chaparral of southern California. Quantification of the rates of increase in leaf number, leaf area and stem length helps clarify the seasonal dynamics of vegetative growth. The rates of vegetative growth depend both on genetic factors and on the resource environment under which the plant is growing (Borchert 1975). The measured growth data, combined with information on resource availability, provide a basis for determining species and community resource use, and the efficiency of resource use (Miller 1981). Fig. 1 is a simplified scheme

NATO ASI Series, Vol. G15
Plant Response to Stress. Edited by J. D. Tenhunen et al.

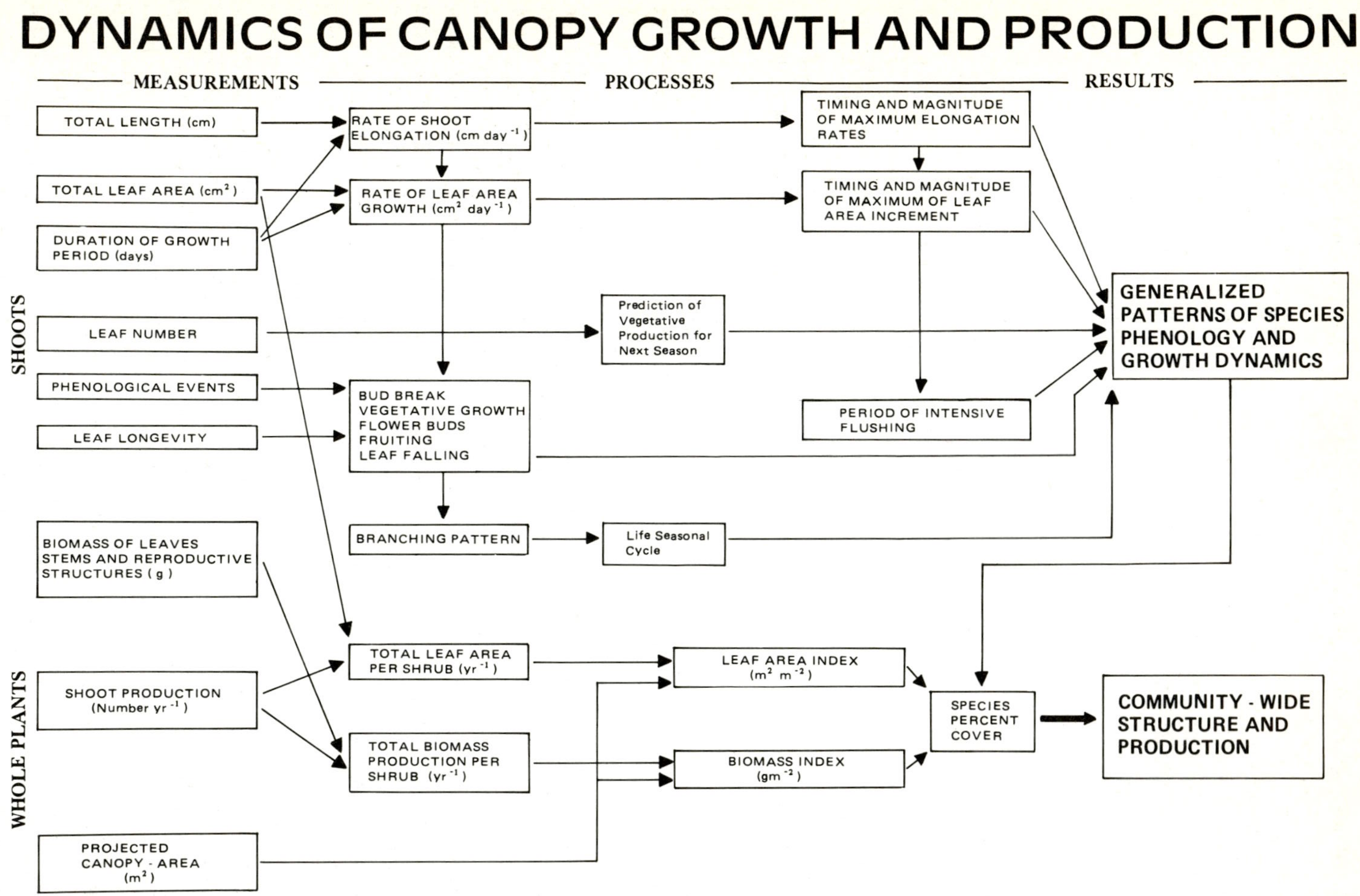

Fig. 1. Scheme indicating linkages between quantitative aspects of plant growth and phenology and community structure and function

indicating linkages between quantitative aspects of plant growth and phenology and community structure and productivity. The organization is generally toward more integrated levels of function from left to right, and toward increasing levels of system structure from top to bottom. Although crude, Fig. 1 may be useful as a basic framework within which the phenology and growth studies discussed here may be related to other aspects of community structure and function. Such studies of growth analysis contribute to the construction of precise models of production within communities or ecosystems.

QUANTITATIVE PHENOLOGY

The timing of biological events at the producer level depends on more or less regular seasonal or annual changes in habitat conditions. Seasonal environmental trends may limit plant developmental events to certain periods (Mooney *et al.* 1974). Phenological events such as stem and leaf initiation and elongation, flower bud formation, flowering, fruiting, fruit dispersal and leaf death or abscission are processes closely tied to species specific carbon allocation strategies (Mooney 1972, 1974; Mooney *et al.* 1977). Comparative studies during I.B.P. on the origin and structure of ecosystems in the mediterranean-climate regions of California and Chile included a detailed quantitative assessment of phenological events as well as of climate, in order to determine whether phenological sequences and plant development occur in these different regions in response to similar environmental trends and as a result of similar adaptations (Mooney *et al.* 1977).

The amount of carbon fixed by plants is determined by the availability of resources including light, water, temperature, and nutrients (Mooney 1981). In mediterranean-climate regions nutrients are often limiting throughout much of the year (Specht 1979; Mooney 1981). Deficiencies of nitrogen and phosphorous are common in mediterranean shrub regions (Miller *et al.* 1977). Fertilization experiments with N and N + P in the Chilean matorral resulted in a significant leaf biomass increase and higher number of sylleptic shoots developing from axillary buds that normally remain dormant during the growing season (Kummerow *et al.* 1982). A nutrient limitation on primary production was demonstrated by fertilization experiments in California (Hellmers *et al.* 1955) and in Australia (Specht 1963; Specht and Groves 1966; see also discussion by Kruger in this volume). However, most analyses of plant development and primary production in mediterranean-type climate regions, especially in Chile and California, have emphasized the interaction between temperature and water availability as limiting factors (Mooney *et al.* 1977; Miller 1981).

In the I.B.P. phenology study, a number of dominant species from the Chilean matorral and California chaparral were investigated. The net energy captured by the shrubs was estimated indirectly by measuring the amount of biomass that accumulated during the growing season. The method consisted of harvesting standard sample branches of approximately 30 cm length (this usually included all the leaves on the branch) at biweekly intervals. Sample branches were randomly taken from different sides of the shrubs, from random positions in the shrub canopy, and from at least 25 different individuals of each species. The branches were separated into new leaves, old leaves, new stems, old stems, flowers and fruits and all components were then oven dried and weighed. The number of sample branches per projected square meter of shrub canopy was utilized together with information on species relative cover in order to estimate the biomass of a particular plant component within the entire community at a given time (Mooney *et al.* 1974). Litter production was measured monthly using plastic bowls (30 cm in diameter) randomly distributed under the canopy of the same 25 shrubs used for branch sampling (Mooney and Kummerow 1977).

Seasonal canopy growth and litter production of the dominant shrub species of mattoral and chaparral are shown in Figs. 2 and 3, respectively (Mooney and Kummerow 1977). A pronounced seasonality occurred in the California species as compared with the Chilean matorral. Canopy growth in California in the spring and summer of 1973 (Fig. 3) appeared strongly limited by low water availability. Data from Miller *et al.* (1977) for the previous winter 1971-1972 showed that precipitation was about 250 mm for the year, the driest winter recorded during the four years that quantitative phenology was measured. Evergreen shrubs from the Chilean matorral (Fig. 2) such as *Colliguaya odorifera, Cryptocarya alba, Kageneckia oblonga* and *Lithraea caustica* showed an increment in canopy growth during the spring and summer 1972-1973, probably in response to the unusually wet winter of 1972. Fundo Santa Laura, the Chilean primary study site, received 2260 mm of precipitation (Miller *et al.* 1977). Litter fall for many species in both Chile and California was maximal soon after the initiation of new leaves. Certain species, however, such as *Quercus* species in California and *Satureja gilliesii* in Chile, exhibit a fairly uniform litter production year-round (Mooney *et al.* 1977). Total vegetative production was very similar on the two continents (Mooney *et al.* 1977) when expressed as the ratio of above-ground biomass to annual above-ground production. However total above-ground biomass and net-productivity in Chile is only one-third that of the California chaparral due to the lower shrub cover in Chile (Table 1).

Plant interactions appear to play a role in addition to mineral nutrient limitation as factors controlling shrub productivity in these two mediterranean regions. Mineral nutrient availability, phosphorous in particular, is lower at the California primary study site, Echo Valley, than at Fundo Santa Laura in Chile (Miller *et al.* 1977). While the yearly production at Fundo Santa Laura was significantly lower than at Echo Valley, herb cover and productivity of herbaceous plants at Fundo Santa Laura is significantly higher (Montenegro *et al.* 1978; Steward and Webber 1981; Kummerow *et al.* 1981). Thus, the higher

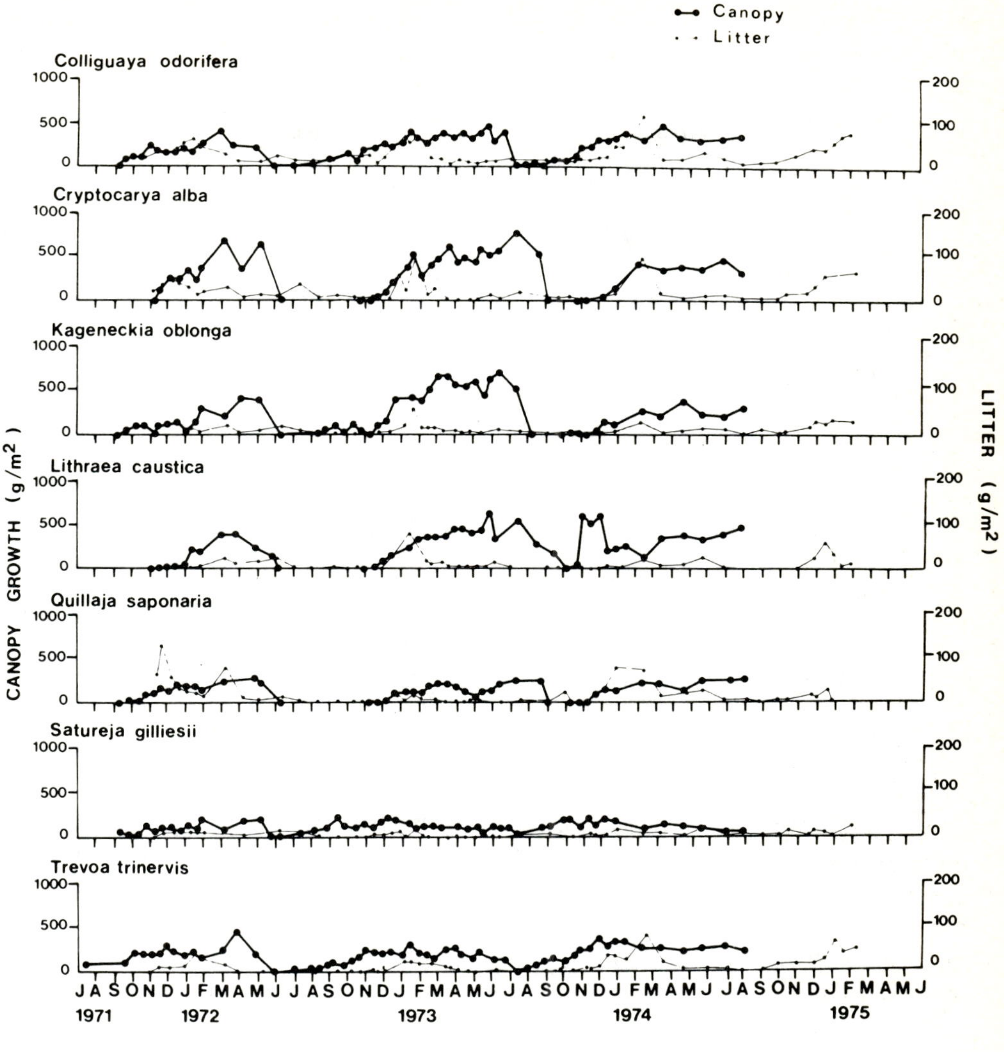

Fig. 2. Seasonal canopy and litter biomass of the dominant plants of the Chilean matorral. The values for litter are the amounts collected on the dates indicated. Canopy figures are the average dry weight of current twig growth on the plant at the indicated time (Mooney and Kummerow 1977)

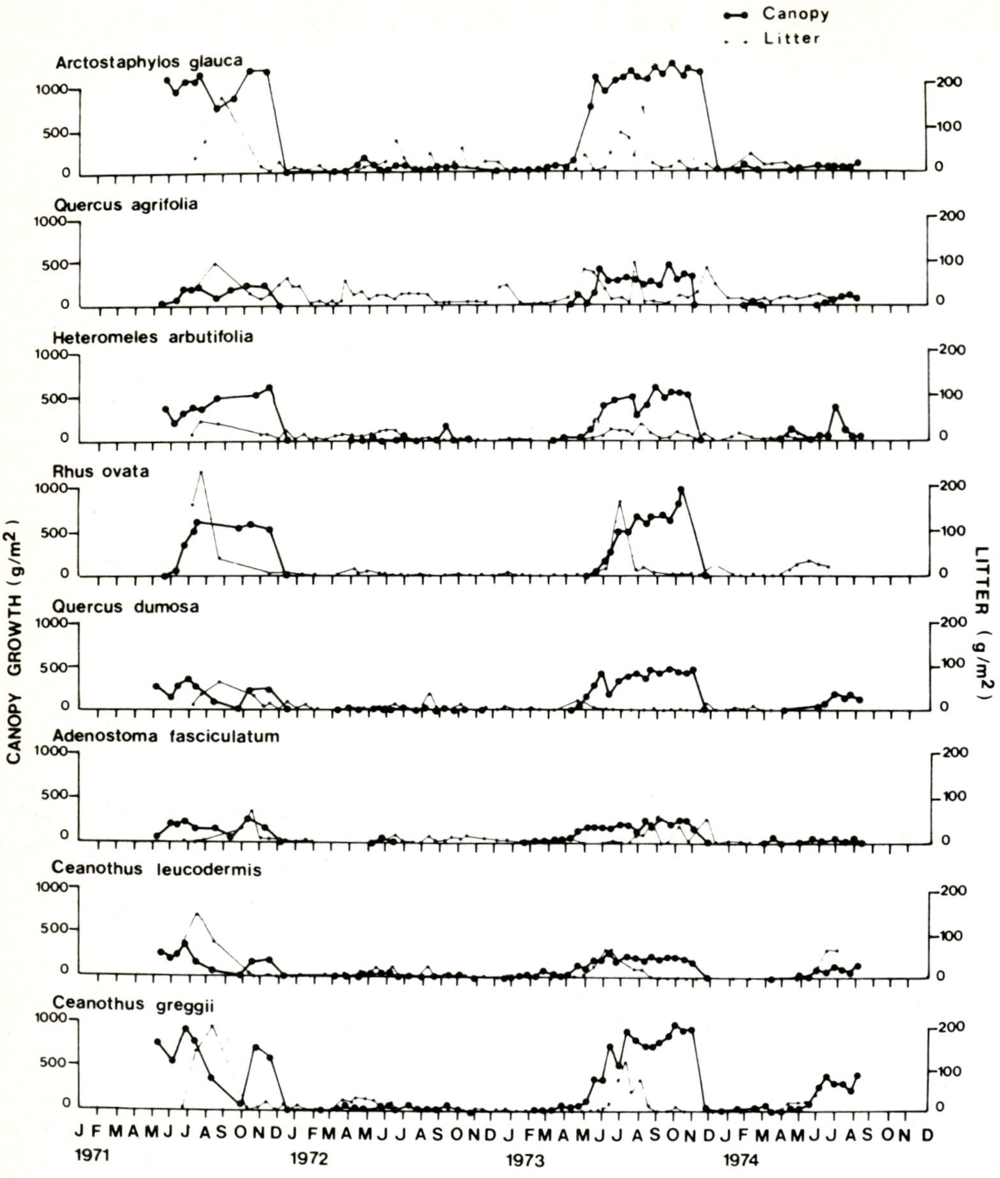

Fig. 3. Seasonal canopy and litter biomass of the dominant plants of the Californian chaparral. The values for litter are the amounts collected on the dates indicated. Canopy figures are the average dry weight of current twig growth on the plant at the indicated time (Mooney and Kummerow 1977)

Table 1. Standing biomass and net productivity for chaparral (C - Echo Valley) and matorral (M - Santa Laura) communities.

	Percent cover in community	Above-ground biomass within community (g m^{-2})	Above-ground twig production within community (g m^{-2} yr^{-1})	Biomass production ratio
C	91.5	2,307.9	670.9	3.4
M	59.3	738.4	252.5	2.9

Data is from Mooney *et al.* (1977). Estimated values for dominant species in both communities.

herbaceous cover at Fundo Santa Laura may reduce nutrient availability to Fundo Santa Laura shrubs to a lower effective level compared to Echo Valley shrubs.

DYNAMICS OF SHOOT GROWTH

A detailed quantification of leaf growth and stem elongation was approached by monitoring the seasonal course of development of representative branches of dominant shrub species in the research areas of Chile and California. The dynamics of shoot elongation and leaf expansion were studied between February 1976 and August 1977 at Echo Valley (Kummerow *et al.* 1981) and between October 1975 and April 1977 at Fundo Santa Laura (Montenegro *et al.* 1979). At both research sites identical methodologies were used. Ten vigorous branches from ten different individuals of each species were tagged at the level of the last leaf formed during the previous growing period. Shrubs were sampled on both pole-facing and equator-facing slopes. After growth began, the length of the whole shoot and that of each leaf appearing on the newly-formed shoot were measured at weekly intervals.

Sylleptic axillary shoots arising from axillary buds of the current years' growth were included in the measurements in sequence as they appeared. In the middle of the growing season, 150 leaves from several shrubs of each species were harvested. The area and length of each leaf was measured to establish a correlation between leaf length and leaf area. This relation was then used to estimate areas of each leaf growing on the tagged branches, allowing estimates of total leaf area per shoot at weekly intervals. During the observation period, all leaf senescence, death, or abscission was recorded, for both current seasons' leaves and older leaves.

Data describing the growth dynamics of Chilean shrubs were gathered for the dominant species growing in matorral communities at the Chilean primary site, located on the slopes of the Coastal Range (Montenegro *et al.* 1979, 1982) as well as for species of foothill and montane communities of the western slopes of the Andean Cordillera (Montenegro *et al.* 1981; Aljaro and Montenegro 1981). The timing of growth and specific rates of growth for each species are summarized in Tables 2 and 3. *Cryptocarya alba* and *Lithraea caustica*, the last species to initiate growth, pass through the period of flushing in about 5 weeks, and exhibit a significantly greater increment in leaf area per shoot during this period. In contrast, the deciduous or semideciduous species that initiate growth early in the season maintain low rates of leaf area increase throughout their entire extended growing period.

Table 2. Period of bud break (a), period of most intensive flushing (b), days duration of leaf growth (c), and highest weekly leaf area increment (d) in cm^2 $week^{-1}$ of matorral shrubs.

		(a)	(b)	(c)	(d)
S. gilliesii	*Ls*	May 22 - Jun 5	Jul 3 - Oct 19	108	1.8 ± 1.4
	Ss	Aug 8 - Aug 20	Oct 19 - Nov 27	39	.6 ± .4
T. trinervis	*Ls*	Jun 28 - Jul 3	Sep 21 - Oct 19	16	2.5 ± .6
	Ss	Jul 4 - Jul 10	Jul 21 - Oct 7	78	.4 ± .02
C. odorifera		Jul 7 - Jul 21	Aug 18 - Nov 21	95	9.1 ± 8.0
Q. saponaria		Oct 6 - Nov 2	Oct 25 - Dec 31	67	9.9 ± 5.7
C. alba		Nov 4 - Dec 2	Nov 27 - Dec 31	34	22.3 ± 8.1
L. caustica		Nov. 7 - Dec 6	Nov 21 - Dec 23	32	28.9 ±27.0

Note - On shoots of *S. gilliesii* and *T. trinervis*, leaf area is included for both long shoots and sylleptic short shoots (Montenegro *et al.* 1979)

Figures 4 and 5 show the time course of leaf expansion observed for Chilean matorral shrubs during the 1976 growing season (Montenegro *et al.* 1979). The curves represent the cumulative leaf area of a representative shoot, chosen from among those shoots that most closely matched the final calculated per shoot mean leaf area values (Table 3).

Table 3. Total leaf area accumulated per shoot during the growing season, rate of leaf area increase on a single shoot, and growing periods for leaves of matorral shrubs on pole-facing slopes (PFS) and equator-facing slopes (EFS).

		Accumulated leaf area/shoot (cm^2)		Leaf area increase/shoot (cm^2/day)		Leaf growth period (days)	
		PFS	EFS	PFS	EFS	PFS	EFS
Colliguaya odorifera		43 ± 22 (5)	23 ± 17 (5)	$.26 \pm .13$	$.17 \pm .09$	167 ± 3	134 ± 28
Quillaja saponaria		33 ± 34 (5)	40 ± 18 (5)	$.43 \pm .28$	$.42 \pm .17$	76 ± 28	96 ± 12
Cryptocarya alba		138 ± 56 (5)	95 ± 31 (5)	$1.33 \pm .49$	$1.03 \pm .34$	104 ± 6	93 ± 14
Lithraea caustica		113 ± 134 (4)	79 ± 80 (6)	1.18 ± 1.13	$.77 \pm .70$	96 ± 19	102 ± 16
Satureja gilliesii	Ss	8(2)	2(2)	.05	.02	142	142
	Ls	18(2)	12(2)	.10	.06	176	179
Trevoa trinervis [a]	Ss	4(8)		.02		151	
	Ls	18(2)		.12		177	

Note: Ss = short shoot; Ls = long shoot. Numbers in parentheses are numbers of shoots.

[a]This species occurred only on PFS in the research area (Montenegro *et al.* 1979)

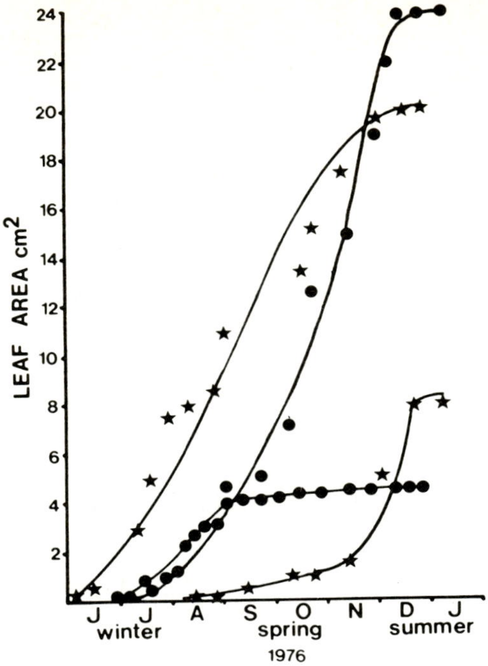

Fig. 4. Time course of leaf area increase observed for shoots of deciduous shrubs from the Chilean matorral during the 1976 growing season; *Trevoa trinervis* (circles) and *Satureja gilliesii* (stars). The upper two curves show the cumulative leaf area of one representative dolichoblast of each species, the lower curves show corresponding values for the brachyblasts (Montenegro *et al.* 1979)

A comparison of the biomass production per shoot (Table 4) shows that average shoots of evergreen sclerophyll shrubs produced substantially more leaf biomass per unit stem weight than did deciduous shrubs. The sprouting behavior of every species must be considered in order to calculate community biomass production (see Fig. 1). Kummerow *et al.* (1981) compared growth pattern dynamics of individual shoots of chaparral and matorral shrubs in an attempt to determine whether growth processes were similarly regulated in both countries. The results of this comparison are summarized in Table 5. The study indicates that the Chilean shrub species showed in general a longer vegetative growth period, produced longer shoots and had a higher leaf area per shoot. However the growth rates observed in California and Chile were not significantly different.

Comparing the results with those obtained by Mooney *et al.* (1977) leads to the conclusion that the low net productivity of Chilean shrubs compared to shrubs of the California chaparral may be due to differences in growth caused by

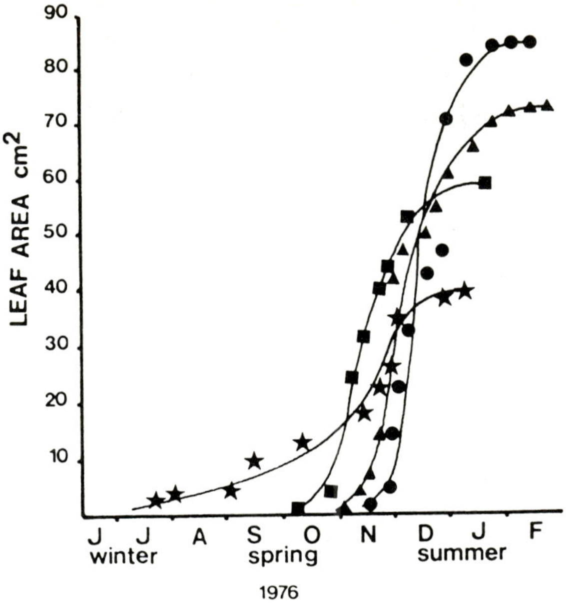

Fig. 5. Time course of leaf area increase observed for shoots of matorral evergreen shrubs during the 1976 growing season; *Cryptocarya alba* (circles), *Lithraea caustica* (triangles, *Quillaja saponaria* (squares) and *Colliguaya odorifera* (stars). Each curve shows the cumulative leaf area of one representative shoot (Montenegro *et al.* 1979).

Table 4. Gram dry weight of leaves, stems, and reproductive structures of average shoots of matorral shrubs.

	Leaves	Stems	Reproductive structures
Satureja gilliesii [b]	.28 ± .12	.35 ± .29	.30 ± .06
Trevoa trinervis	.13 ± .16	.17 ± .27	.03 ± .02
Colliguaya odorifera	.41 ± .18	.09 ± .03	.57 ± .48
Quillaja saponaria	.46 ± .33	.06 ± .05	.41 ± .18
Cryptocarya alba	1.07 ± .61	.18 ± .12	.11 ± .0
Lithraea caustica	1.25 ± 1.23	.32 ± .39	.42 ±·.43

Note: Reproductive structures = flowers, fruits, seeds, and inflorescences.

[b]Dry weight of stems for *S. gilliesii* includes the short shoots (Montenegro *et al.* 1979)

Table 5. Total leaf area accumulated per shoot during the growing season in cm^2 (a); rate of leaf area increase in cm^2 day^{-1} (b); shoot lengths at end of growing season in cm (c); shoot elongation rates in cm day^{-1} (d); and the length of growth period in days (e); of four chaparral shrub species at Echo Valley and six matorral shrub species at Fundo Santa Laura ($\bar{X} \pm SE$).

	(a)	(b)	(c)	(d)	(e)
Southern California					
A. fasciculatum (7)	2.5 ± 0.45	0.03 ± .01	5.2 ± 1.70	.09 ± .03	69 ± 26.08
A. glauca (17)	33.0 ± 6.50	0.82 ± .13	6.9 ± 1.24	.13 ± .02	48 ± 3.64
C. greggii (20)	7.9 ± 1.52	0.12 ± .02	4.2 ± 0.83	.06 ± .01	68 ± 3.58
R. ovata (6)	43.5 ± 24.05	1.19 ± .43	2.9 ± 1.18	.08 ± .06	31 ± 5.31
Central Chile					
S. gilliesii (4)	14.9 ± 2.92	0.08 ± .02	25.9 ± 3.25	.14 ± .12	178 ± 3.85
T. trinervis (2)	18 ± 5.33	0.095 ± .04	21.6 ± 7.40	.12 ± .04	169 ± 16.97
C. odorifera (10)	33 ± 6.74	0.22 ± .04	8.4 ± 1.49	.05 ± .01	151 ± 8.10
Q. saponaria (10)	37 ± 8.19	0.43 ± .21	5.3 ± 1.14	.05 ± .01	86 ± 7.24
C. alba (9)	117 ± 16.10	1.18 ± .14	12.2 ± 1.77	.12 ± .16	99 ± 3.9
L. caustica (7)	93 ± 37.49	0.93 ± .33	13.8 ± 5.75	.12 ± .05	100 ± 6.24

Note: Number of observations is in parentheses. The rates are calculated on the basis of the total growth period (adapted from Kummerow *et al.* 1981) factors other than resource availability. For example, Aljaro and Montenegro (1981) observed that growth of *Escallonia myrtoidea*, an evergreen sclerophyll shrub of the Chilean montane scrub community, depends on the position of the branches. Branches that initiate growth are found in the upper part of the canopy. Similar phenomena have been reported for *Pinus taeda* and *P. caribea* where upper shoots flushed three times while other shoots failed to elongate at all (Eggler 1961). *Lithraea caustica* is also a shrub species with very erratic sprouting behavior (Montenegro *et al.* 1979). Of the buds tagged for the study of growth dynamics, a large number remained dormant for a full year.

Man's impact in Chile has been and currently still is more pervasive than in California (Di Castri 1981). This includes wood-cutting and charcoal production in addition to grazing. Recent studies (Montenegro *et al.* unpublished) have shown that at the Chilean Santa Laura primary site, the following shrub species resprout from basal underground buds located in lignotubers: *Lithraea caustica, Podanthus mitique, Muehlenbeckia hastulata, Colliguaya odorifera, Quillaja saponaria, Cryptocarya alba, Satureja gilliesii, Trevoa trinervis, Kageneckia oblonga, Colletia spinosissima, Retanilla ephedra, Talguenea quinquenervia, Escallonia pulverulenta, Maytenus boaria* and *Acacia caven.* The list includes species that were selected for quantitative phenology studies and for analysis of the dynamics of growth. Chilean shrubs may produce longer shoots and higher leaf area per shoot as compared with Californian shrubs because of the individual plant history in relation to man's utilization.

Thus, the matorral of Chile may be more heterogeneous and may be composed of individuals at different stages of recovery from relatively frequent disturbance (see also discussion by Trabaud in this volume). Montenegro *et al.* (1983) compared leaf area increase of shoots arising from lignotubers and shoots which develop from axillary buds. They found significantly higher values for regenerating shoots. However the shoot biomass on plants developing from above-ground buds and those developing from below-ground buds were not significantly different. Primary production studies after fire in mediterranean sclerophyllous shrublands in southern Australia, California and France (review by Specht 1981), have shown that the annual growth increment decreases with increasing stand age. The growth dynamics of stands discussed by Specht are similar, perhaps because the stands tend to be rather homogeneous due to primary disturbance by fire. The heterogeneity of the matorral may explain in part the large variability described in Table 5.

Biomass production of individual shoots are not available for Californian shrubs, to determine whether the biomass per shoot is higher than in Chile, despite the lower values in shoot leaf area and shoot length. However, the average sclerophyll index of current year's leaves is lower in matorral shrub species than in chaparral species (Oechel and Lawrence 1981), which may indicate lower biomass in current shoots. These authors also suggested a larger variability in

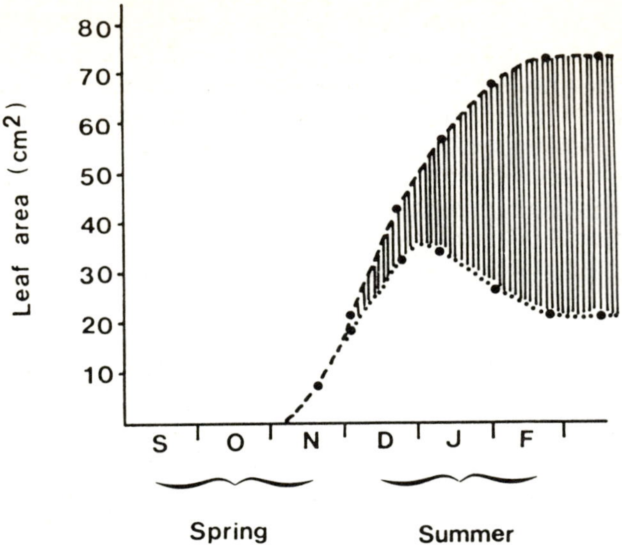

Fig. 6. Cumulative leaf area increase of one representative shoot of *L. caustica,* during the 1978 growing season. The shaded area represents leaf area consumed or damaged by insects (Montenegro *et al.* 1980)

sclerophyll index among matorral species. *Lithraea caustica* showed the highest sclerophyll index of any species (Oechel and Lawrence 1981). Significantly, as discussed below, *L. caustica* is also the species that showed the greatest loss of leaf area due to herbivorous-insects. As shown in Table 1, the values of productivity are rather similar on the two continents when they are expressed as the ratio of standing biomass to net productivity (Mooney *et al.* 1977). One should note, that the matorral as a system is more efficient than the chaparral, since less standing biomass (2.9 g) is required to produce one gram of new biomass.

By observing the dynamics of shoot growth, the seasonality and amount of defoliation of Chilean shrubs by phytophagous insects was studied (Montenegro *et al.* 1981; Fuentes *et al.* 1981). Observations were made from April 1978 through March 1979 at Fundo Santa Laura. These showed up to 70% loss of new leaf area due to insect attack for *Lithraea caustica* shoots (Fig. 6). In all shrubs studied, approximately 60% of the newly formed branches exhibited predation effects (Montenegro *et al.* 1981). Considering that *Lithraea caustica* has the highest productivity per shoot found for matorral shrubs (Montenegro *et al.* 1979; Table 4), the total net annual production may be shifted significantly due to the effects of herbivores. In fact, it has been shown that insect grazers in the Chilean matorral cause 2 to 18% leaf area loss in some of the dominant shrub species (Fuentes *et al.* 1981; Montenegro *et al.* 1981; Walkowiak *et al.* 1984).

The approach outlined above for developing a picture of the dynamics of shoot growth appears to be a potentially important tool for examining the role of consumer organisms in mediterranean ecosystems. Since all harvest methods ignore consumption by various herbivores, these methods provide an added dimension in our understanding of production processes.

CAMBIAL ACTIVITY

Comparative studies of cambial activity throughout the growing season were conducted with pairs of shrub species from sites closely matched with respect to climate and topography in the Chilean matorral and Californian chaparral (Avila *et al.* 1975; Mooney *et al.* 1977). The main criteria for the pairings of Chilean and Californian shrub species were morphological similarities based on leaf size, texture, branching pattern, and anatomical structure of the leaves and stems (Fishbeck and Kummerow 1977). Cambial activity was analyzed using free-hand cross sections of branches in the second year of growth. From an anatomical point of view the commencement of cambial activity consists usually of two stages; radial expansion of cambial cells and continuous division of those cells. The resulting cells then undergo gradual differentiation (Fahn 1982).

The study of Avila *et al.* (1975) was restricted to histological observations and measurements, utilizing the number of cell layers in the cambial zone and the amount of unlignified xylem adjacent to the cambial cylinder as an index of cambial activity. Fahn *et al.* (1968), using simultaneous histological and radioactive labelling methods to measure cambial activity, found a positive correlation between histologically observed cambial activity and the incorporation of radioactive label. Measurements of the width of newly-formed xylem, easily distinguished with safranin-fast green stain, were made in radial planes at 15 randomly chosen positions along the circumference of cambium rings in at least five different branches of each species. The monthly increases in new xylem were expressed as percentages of the total seasonal increase in radial xylem growth.

The seasonal pattern of cambial activity is shown in Fig. 7 for two shrub pairs: *Lithraea caustica* (Chile) and *Rhus ovata* (California); and *Quillaja saponaria* (Chile) and *Quercus dumosa* (California). *Lithraea caustica* and *R. ovata* maintain cambial activity throughout the year, even during the cool winter and during the drought period from late summer through fall. Both species are in the *Anacardiaceae*, and the similarities in cambial activity may in part be phylogenetically based. However, the deep root system of *L. caustica* (Giliberto and Estay 1978) and the presence of a lignotuber (Montenegro *et al.* 1983) may significantly affect function and allow this species to absorb and conserve enough water to maintain cambial activity throughout the year. *Rhus ovata* is also a deep-rooting species with apparent access to water unavailable to other chaparral shrubs (Kummerow *et al.* 1981). *Rhus ovata* maintains xylem pressure

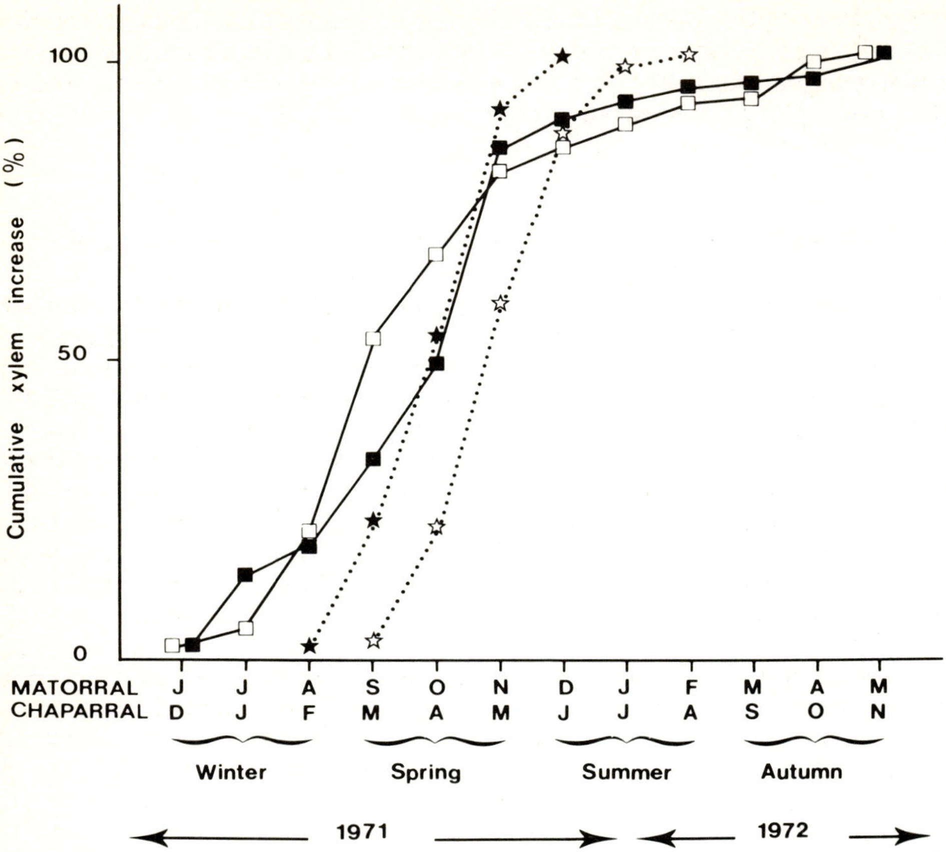

Fig. 7. The cumulative increase in xylem diameter in monthly percentages of the total seasonal increase for two matorral shrubs *Lithraea caustica* (solid squares) and *Quillaja saponaria* (solid stars) and two chaparral shrubs *Rhus ovata* (open squares) and *Quercus dumosa* (open stars)

potentials of > -2.0 MPa during the driest period of the year (Poole and Miller 1975), similar to potentials reported for *L. caustica* (Poole *et al.* 1981).

Quercus dumosa and *Q. saponaria*, two species which are phylogenetically unrelated but which are morphologically similar, also exhibit very similar seasonal patterns of cambial activity (Fig. 7). Both species have a characteristic midsummer and autumn dormancy, resuming growth at the end of winter with a brief but intense period of xylem formation. The similar cambial activity patterns displayed by these members of two geographically separated communities may be interpreted as indicating evolutionary convergence. Initiation of cambial activity is triggered and at least partly regulated by environmental factors (Waisel and Fahn 1965; Fahn *et al.* 1968; Waisel *et al.*

1970). The regular experience of a prolonged drought period imposes constraints on plants that may result in complex physiological and morphological adaptations which include cell and tissue adaptations to endure very negative water potentials (Poole and Miller 1975; Roberts *et al.* 1981).

A criticism of the methodology applied in these comparative analyses is that measurements of the increment in cambium activity involved only radial growth. Future investigations must consider the area of the cambial surface occupied by ray initials. Margaris and Papadogianni (1977) analyzed the size and distribution of ray initial units in species dominating maquis and phryganic formations, using camara lucida diagrams of cambial tangential strips. According to their observations the area occupied by ray initials is reduced in species of the more xeric phrygana communities compared to the area occupied by ray initials in the maquis species. The adaptive significance of the low ratio of ray and fusiform initial cells has been interpreted as a feature to reduce water diffusion through intercellular spaces, with a resulting reduction in loss of xylem water to stem surfaces (Hook and Brown 1972). However, a larger number of fusiform initials will result in a larger number of vessels, which should ultimately increase the vertical water transport capacity of the plant. During the periods of water availability in winter and in spring, a greater water conducting capacity may confer a competitive advantage in terms of a capacity to rapidly take up water and nutrients during these periods. Since water is often limiting in mediterranean systems (Miller 1981), the evolution of xylem anatomy to maximize water flow may represent a significant competitive advantage during those times of year when water is available.

ACKNOWLEDGEMENTS

This work is dedicated to the late Professor P.C. Miller. Appreciation is due Dr. Stephen Roberts for reading and commenting on this manuscript and for improving its English presentation. I would like to thank Mr. Miguel Gómez who drew all the figures for this paper. The work was supported by the Catholic University of Chile Funds DIUC 79/85 and UNESCO-MAB 6 Program 155/84 to G. Montenegro.

REFERENCES

Aljaro ME, Avila G, Hoffmann A, Kummerow J (1972) The annual rhythm of cambial activity in two woody species of the Chilean matorral. Amer J Bot 59:879-885

Aljaro ME, Montenegro G (1981) Growth of dominant Chilean shrubs in the Andean Cordillera. Mount Research Develop 1:287-291

Aschmann H (1973) Distribution and peculiarity of mediterranean ecosystems. In: Di Castri F, Mooney HA (eds) Mediterranean-Type Ecosystems: Origin and Structure. Springer, Berlin-Heidelberg-New York, p 11

Avila G, Aljaro ME, Araya S, Montenegro G, Kummerow J (1975) The seasonal cambium activity of Chilean and Californian shrubs. Amer J Bot 62:473-478

Borchert R (1975) Endogenous shoot growth rhythms and indeterminate shoot growth in Oak. Physiol Plant 35:152-157

Daget P (1977) Le bioclimat méditerranéen: Caractères généraux, modes de caractérisation. Vegetatio 34:1-20

Di Castri F (1981) Mediterranean-type shrublands of the world. In: Di Castri F, Goodall DW, Specht RL (eds) Mediterranean-Type Shrublands. Elsevier, Amsterdam, p 1

Eggler WA (1961) Stem elongation and time of cone initiation in southern pines. Forest Sci 7:149-158

Fahn A, Waisel Y, Benjamin L (1968) Cambial activity in *Acacia raddiana Sari*. Ann Bot 32:677-686

Fahn A (1982) Plant Anatomy, 3rd edn, Pergamon, Oxford, New York, Toronto, Sydney, Paris, Frankfurt

Fishbeck K, Kummerow J (1977) Comparative wood and leaf anatomy. In: Thrower NHW, Bradbury DE (eds) Chile-California Mediterranean Scrub Atlas. A Comparative Analysis. Dowden, Hutchinson and Ross, Stroudsburg, Pennsylvania, p 148

Fuentes ER, Etchégaray J, Aljaro ME, Montenegro G (1981) Shrub defoliation by matorral insects. In: Di Castri F, Goodall DW, Specht RL (eds) Mediterranean-Type Shrublands. Elsevier, Amsterdam, p 345

Giliberto J, Estay H (1978) Seasonal water stress in selected shrub species in the mediterranean region of Central Chile. Bot Gaz 139:236-240

Hellmers H, Bonner JF, Kelleher JM (1955) Soil fertility: a watershed management problem in the San Gabriel Mountains of southern California. Soil Sci 80:189-197

Hook DD, Brown CL (1972) Permeability of the cambium to air in trees adapted to wet habitats. Bot Gaz 133:304-310

Kummerow J, Montenegro G, Krause D (1981) Biomass, phenology and growth. In: Miller PC (ed) Resource Use by Chaparral and Matorral. A Comparison of Vegetation Function in Two Mediterranean-Type Ecosystems. Springer, Berlin-Heidelberg-New York, p 69

Kummerow J, Avila G, Aljaro ME, Araya S, Montenegro G (1982) Effect of fertilizer on fine root density and shoot growth in Chilean matorral. Bot Gaz 143:498-504

Margaris NS, Papadogianni P (1977) The ratio of ray and fusiform initials in some plants dominating mediterranean formations in Greece. Flora 166:219-222

Miller PC, Bradbury DE, Hajek E, Lamarche V, Thrower NJW (1977) Past and present environment. In: Mooney HA (ed) Convergent Evolution in Chile and California. Dowden, Hutchinson and Ross, Stroudsburg, Pennsylvania, p 27

Miller PC (1981) Editor: Resource Use by Chaparral and Matorral. A Comparison of Vegetation Function in Two Mediterranean-Type Ecosystems. Springer, Berlin-Heidelberg-New York, p 455

Montenegro G, Rivera O, Bas F (1978) Herbaceous vegetation in the Chilean matorral. Oecologia (Berl) 36:237-244

Montenegro G, Aljaro ME, Kummerow J (1979) Growth dynamics of Chilean matorral shrubs. Bot Gaz 140:114-119

Montenegro G, Jordán M, Aljaro ME (1980) Interactions between Chilean matorral shrubs and phytophagous insects. Oecologia (Berl) 45:346-349

Montenegro G, Aljaro ME, Walkowiak A, Saenger R (1981) Seasonality, growth and net-productivity of herbs and shrubs of the Chilean matorral. Proc Symp Dynamics and Management of Mediterranean-Type Ecosystems. Gen Tech Rep PSW, Berkeley, p 135

Montenegro G, Araya S, Aljaro ME, Avila G (1982) Seasonal fluctuations of vegetative growth in roots and shoots of Central Chilean shrubs. Oecologia (Berl) 53:235-237

Montenegro G, Avila G, Schatte P (1983) Presence and development of lignotubers in shrubs of the Chilean matorral. Can J Bot 61:1804-1808

Mooney HA, Dunn EL, Shropshire F, Song L (1970) Vegetation comparisons between the mediterranean climatic areas of California and Chile. Flora 159:480-496

Mooney HA (1972) The carbon balance of plants. Ann Rev Ecol Syst 3:315-346

Mooney HA, Parsons DJ, Kummerow J (1974) Plant development in mediterranean climate. In: Lieth H (ed) Phenology and Seasonality Modeling. Springer, Berlin-Heidelberg-New York, p 254

Mooney HA (1974) Seasonal variation in the production of tannins and cyanogenic glucosides in the chaparral shrub, *Heteromeles arbustifolia*. Oecologia (Berl) 15:65-76

Mooney HA, Kummerow J (1977) Quantitative phenology. In: Thrower NJW, Bradbury DE (eds) Chile-California Mediterranean Scrub Atlas. A Comparative Analysis. Dowden, Hutchinson and Ross, Stroudsburg, Pennsylvania, p 121

Mooney HA, Johnson A, Parsons D, Keeley S, Hoffman A, Hays R, Giliberto J, Chu C (1977) The producers--their resources and adaptive response. In: Mooney HA (ed) Convergent Evolution in Chile and California Mediterranean Climate Ecosystems. Dowden, Hutchinson and Ross, Stroudsburg, Pennsylvania, p 224

Mooney HA (1981) Primary production in mediterranean climate regions. In: Di Castri F, Goodall DW, Specht RL (eds) Mediterranean-Type Shrublands. Elsevier, Amsterdam, p 249

Oechel WC, Lawrence W (1981) Carbon allocation and utilization. In: Miller PC (ed) Resource Use by Chaparral and Matorral. A Comparison of Vegetation Function in Two Mediterranean-Type Ecosystems. Springer, Berlin-Heidelberg-New York, p 186

Poole DK, Miller PC (1975) Water relations of selected species of chaparral and coastal sage communities. Ecology 56:1118-1128

Poole DK, Roberts SW, Miller PC (1981) Water utilization. In: Miller PC (ed) Resource Use by Chaparral and Matorral. A Comparison of Vegetation Function in Two Mediterranean-Type Ecosystems. Springer, Berlin-Heidelberg-New York, p 123

Quezel P (1979) "Matorrals" mediterranees et "Chaparrals" Californiens. Quelques aspects comparatif de leur dynamicque, de leurs structures et de leur signification ecologique. Ann Sci Fores 36:1-12

Roberts SW, Miller PC, Valamanesh A (1981) Comparative field water relations of four co-occurring chaparral shrub species. Oecologia (Berl) 48:360-363

Specht RL (1963) Dark Island heath (Ninety-mile Plain, South Australia). VII. The effect of fertilizers on composition and growth, 1950-60. Aust J Bot 11:67-94

Specht RL (1969) A comparison of the sclerophyllous vegetation characteristic of mediterranean-type climates in France, California and Southern Australia. I. Structure, morphology and succession. Aust J Bot 17:277-292

Specht RL, Grooves GH (1966) A comparison of the phosphorus nutrition of Australian heath plants and introduced economic plants. Aust J Bot 14:201-221

Specht RL (1973) Structure and functional response of ecosystems in the mediterranean climate of Australia. In: Di Castri F, Mooney HA (eds) Mediterranean-Type Ecosystems: Origin and Structure. Springer, Berlin-Heidelberg-New York, p 113

Specht RL (1979) (ed) Heathlands and Related Shrublands. Elsevier, Amsterdam, p 497

Specht RL (1981) Primary production in mediterranean-climate ecosystems regenerating after fire. In: Di Castri F, Goodall DW, Specht RL (eds) Mediterranean-Type Shrublands. Elsevier, Amsterdam, p 257

Steward D, Webber PJ (1981) The plant communities and their environments. In: Miller PC (ed) Resource Use by Chaparral and Matorral. A Comparison of Vegetation Function in Two Mediterranean-Type Ecosystems. Springer, Berlin-Heidelberg-New York, p 44

Thrower NJW, Bradbury DE (1977) Chile-California scrub. A comparative analysis atlas. Dowden, Hutchinson and Ross, Stroudsburg, Pennsylvania, p 237

Waisel Y, Fahn A (1965) The effects of environment on wood formation and cambial activity in *Robinia pseudoacacia* L. New Phytol 64:436-442

Waisel Y, Liphschitz N, Fahn A (1970) Cambial activity in *Zygophyllum dumosun* Boiss. Ann Bot 34:409-414

Walkowiak AM, Simonetti JA, Serey I, Jordán M, Arranz R, Montenegro G (1984) Defensive patterns in shrubs of Central Chile; a common strategy? Ecol Plant 15:191-199

Phenology of mediterranean plants in relation to fire season: with special reference to the Cape Province South Africa

E. J. Moll

Department of Botany, University of Cape Town, Private Bag, Rondebosch, 7700, Republic of South Africa

INTRODUCTION

A number of different shrubland formations occur in the mediterranean climate zone of South Africa (see Figs. 1 and 2). According to the classification of Acocks (1953) the "Veld Types" represented include West Coast Strandveld, Coastal and Mountain Renosterveld, Coastal Macchia and Macchia. More recently the floristic affinities and the understanding of the phytogeographic relationships of these "Veld Types," as well as their nomenclature has altered somewhat (Moll and Bossi 1983, Moll *et al.* 1984). Today we more correctly recognize two basic shrubland formations that occur in the mediterranean climate zone of the southwestern Cape Province. These are true heathlands, or Cape Fynbos communities (Specht 1979, Moll and Jarman 1984a and 1984b), which belong to the Cape Floristic Kingdom; the two sub-types that occur are Mountain Fynbos (=Macchia) and Lowland Fynbos (=Coastal Macchia). The second formation is a non-heath, truly mediterranean shrubland formation, which consists of Cape Transitional Small Leafed Shrublands with essentially Cape/Karoo-Namib phtyogeographic affinities known locally as Renosterveld, and Cape Transitional Large Leafed Shrublands with essentially Tongaland-Pondoland/Cape/Karoo-Namib phytogeographic affinities known locally as Strandveld (Moll *et al.* 1984). These two non-heath shrublands have both Cape and Palaeotropic Florisitic Kingdom affinities (see Fig. 1).

The importance of the distinction between heathlands and non-heathlands, which in the Cape coincidentally occupy the mediterranean climate zone, is that they both respond differently to fire and other environmental factors. These differences have not been adequately researched and are, therefore, not well known. Some of these differences are listed in Table 1.

It is now generally conceded that heathlands are edaphically controlled and occur independently of climate (Specht 1979), whilst the physiognomically similar mediterranean shrublands (non-heathlands) are climatically determined (di Castri, Goodall and Specht 1981). So heathlands should not be compared to mediterranean shrublands in terms of the theory of convergence. However, the South African situation is still confused both locally and internationally because the Cape shrublands have, until recently, never been properly understood (Moll 1986). Central to the problem in the Cape is that both shrubland formations are

NATO ASI Series, Vol. G15
Plant Response to Stress. Edited by J. D. Tenhunen et al.

Table 1. Similarities and differences between the heathland and non-heath shrublands in the mediterranean climate zone of the Southwestern Cape.

HEATHLAND	NON-HEATH SHRUBLAND
Soils oligotrophic or dystrophic[a]	Soils mesotrophic or dystrophic[a]
Plants essentially evergreen with few annuals; except at the semi-arid fringe[a]	Plants mostly evergreen, a deciduous component is conspicuous and occasionally prominent. Many annuals occur[a]
Leaves generally iso-bilateral[b]	Leaves dorsiventral[b]
Mainly C_3 grasses [b c]	C_3 and C_4 grassses are co-dominant[b]
Fairly stable composition[a d]	Major vegetation changes in last ±2000 years (man induced) [a d]
Geophytes restricted	Geophytes widespread, common and even locally abundant
Restioid, ericoid and proteoid growth forms diagnostic[a]	Growth forms shrubs, succulents, annuals, etc. Restioid, ericoid and proteoid forms generally absent[a]
Most woody plants obligative re-seeders after fire	Most woody plants re-sprout after fire
Mostly intact today[d]	Mostly cleared for agriculture[d]
All endemic Cape families occur here[e]	Many endemic genera and species, but no families[e]
All species belong to the Cape Floristic Kingdom[e]	Species belong to the Cape and Palaeotropic Floristic Kingdoms [e f]

[a] Moll *et al.* 1984, and Moll and Jarman 1984a and 1984b
[b] Cowling 1983
[c] Vogel, Fuls and Ellis 1978
[d] Moll and Bossi 1984
[e] Moll, Hilton-Taylor and Jarman 1984
[f] Bayer 1984

superficially similar and often occur in close proximity to each other (and even intergrade). The region also coincides with the Cape Floristic Kingdom (Takhtajan 1969, Moll 1986) which is world renowned for its floristic uniqueness. It is this floristic uniqueness which has probably contributed to the great lack of ecological understanding of Cape vegetation, by capturing our imagination florally and blinding us to the ecological complexity of the region! Thus, any discussion of how phenology relates to fire seasons needs to be carefully researched with respect to the two different shrubland types. Our current state of knowledge is poor, but some good data are becoming available as part of the Fynbos Biome co-operative research program (Jarman 1982). The main emphasis in this paper is on the comparison of the similarity and differences of the Cape's shrublands, with only brief reference as to how these shrublands relate to the other mediterranean shrublands in the world.

CLIMATE AND FIRE HAZARD

Aschmann's (1984) restrictive definition of mediterranean climate states that at least 65% of annual precipitation must fall in the winter half year (May through October in the southern hemisphere), and that the mean annual rainfall is between 250 mm (or up to 350 mm in the hot, more arid interior) to 800 mm (or up to 1,200 mm depending on the mean annual temperature range of 15 to 24°C respectively). By this definition only a very small region of South Africa has a mediterranean climate (Fig. 1). Even within this restricted area most high ground (above some 500 m) has too much rain, and in addition is often mist covered (particularly in the summer half year, when it has been estimated that considerable moisture, not recorded in rain gauges, is precipitated (Marloth 1903 and 1905, Nagel 1956 and 1961)). The vegetation of even the drier mountains is now known not to demonstrate the characteristic low summer plant water potentials characteristic of true mediterranean plants (Miller 1985). This further demonstrates that these heathlands (=mountain fynbos) are not responding as predicted by the theory of convergent evolution, despite their occurrence in a mediterranean climate zone. The obvious explanation is that one should be comparing the Cape's mediterranean "non-heath" shrublands with other mediterranean shrublands of the world, and not the Cape's "heathlands" with other mediterranean shrublands (which is what has been attempted in the past (see also Cowling and Campbell 1980).

The vegetation of the mediterranean climate zone *sensu stricto* of the Cape is not recognized locally as being unique. It is rather seen to be part of the various shrubland formations, known collectively as the "Fynbos Biome," which extend from the mediterranean climate zone in the west, through the all year rainfall zone, to the summer rainfall zone in the east (Day *et al.* 1979); an area that more-or-less corresponds to the Cape Floristic Kingdom (Bond and Goldblatt 1984). The distribution of the major shrublands occupying the Fynbos Biome are shown in Fig. 1.

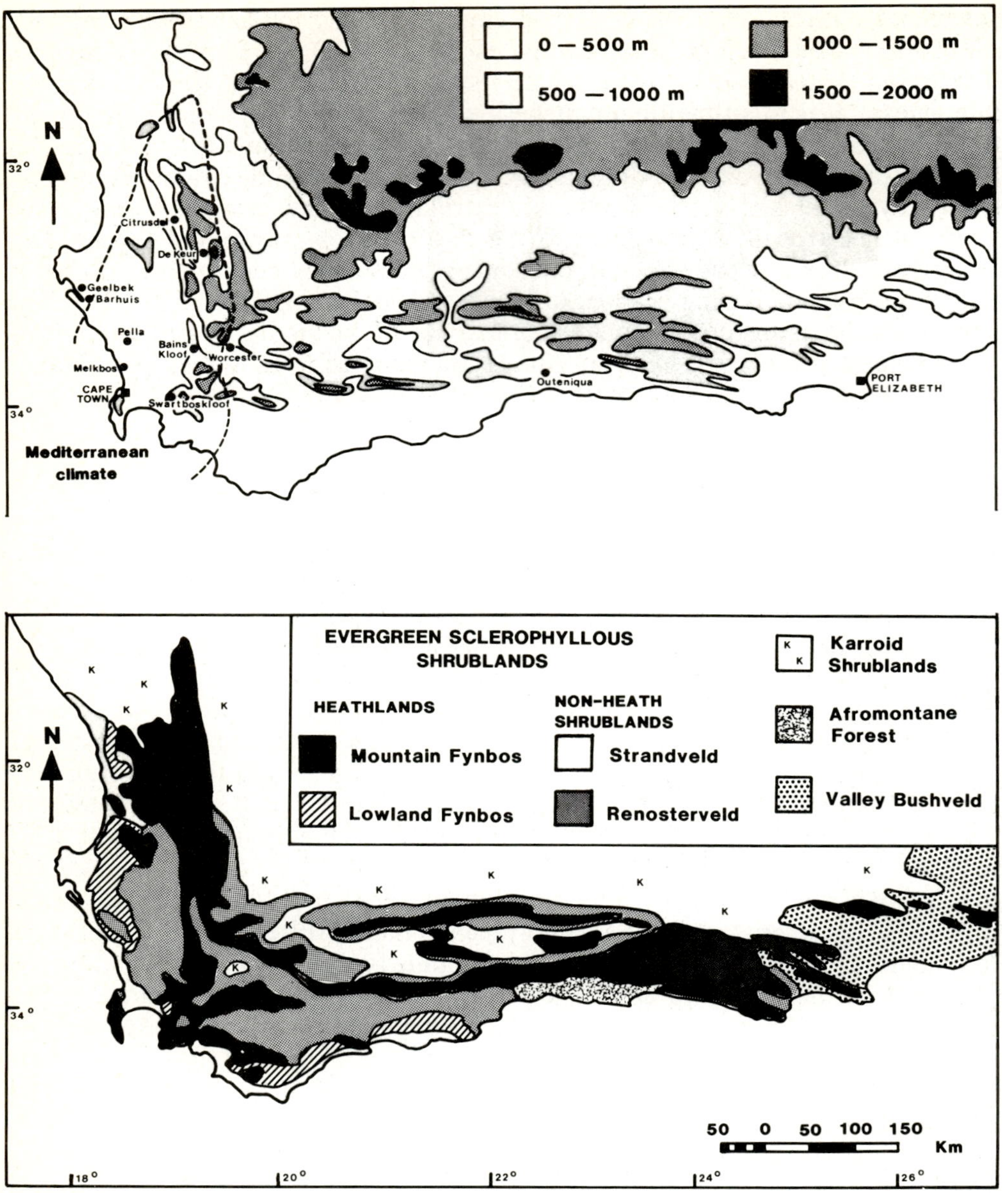

Fig 1. The top map shows the major topographic features, place names and approximate extent of the mediterranean climate zone in the southwestern and southern Cape (after Aschmann 1984). The bottom map gives the distribution of the major plant communities that occur in the fynbos biome. Note that the areas of Karroid Shrublands, Afromontane Forest and Valley Bushveld shown are not considered as components of the Fynbos Biome (Day *et al.* 1979, Moll *et al.* 1984)

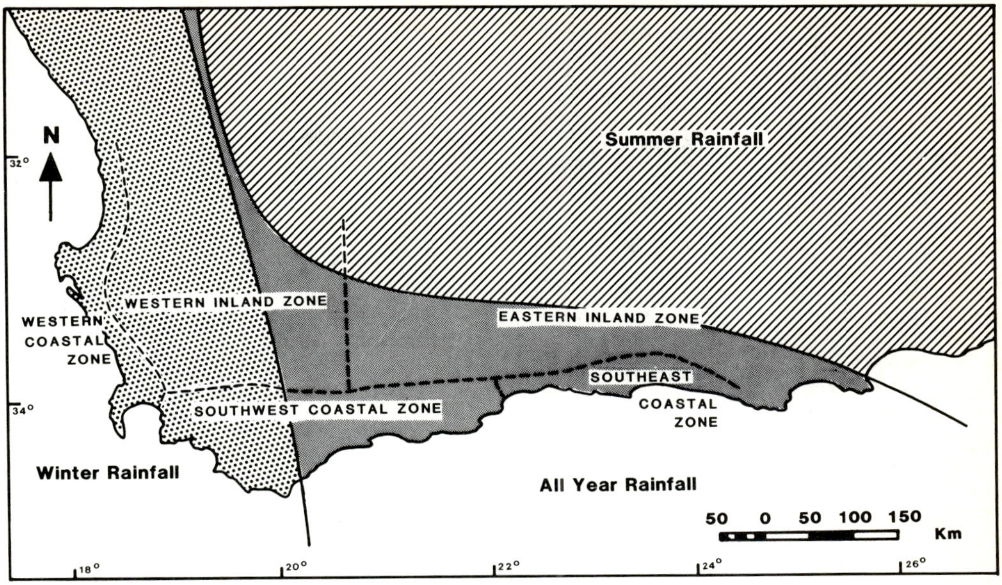

Fig. 2. Seasonal distribution of rainfall in geographical area of the Fynbos Biome. The dotted lines deliniate van Wilgen's (1984) fire climate zones

Based on fire potential during the fire season, van Wilgen (1984) has recognized five major fire climate zones (Fig. 2) in the Fynbos Biome. None of these zones coincide directly with the mediterranean climate area, but rather to coastal and inland weather conditions over the whole region. By comparing van Wilgen's fire climate zones with the vegetation types, there is also no correlation. This is perhaps not really surprising since van Wilgen was concerned essentially with Mountain Fynbos fire management, so all his work relates to heathlands. Nevertheless van Wilgen's data are useful to compare the monthly progression of fire danger in the mediterranean area (Citrusdal), to that of the all year rainfall area (Outeniqua - Fig. 3). This comparison indicates a very high summer fire hazard in the Citrusdal.

At this stage, some observations with respect to fire hazard are necessary to demonstrate that there is a marked difference in flammability between heathlands, and non-heath mediterranean shrublands. Cape heathlands, occurring over a range of climatic zones, will burn at any season as long as the fire is preceded by a few days of dry weather (van Wilgen 1985). On the other hand I have observed that non-heath mediterranean shrublands will not burn in

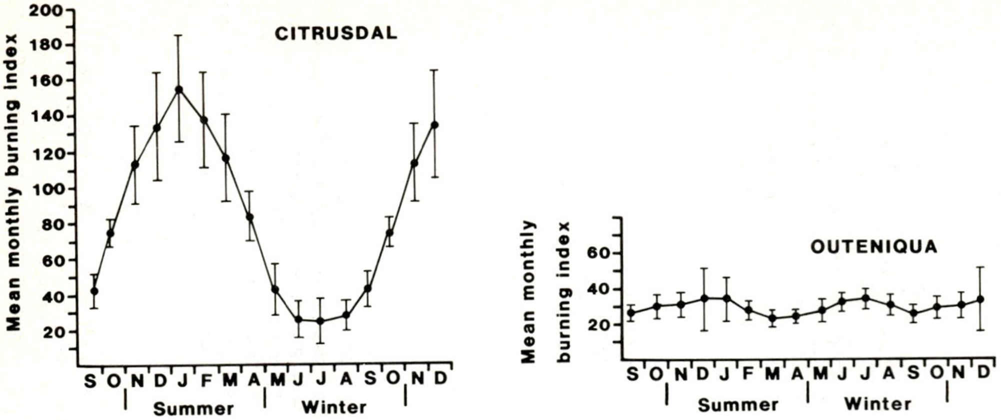

Fig. 3. Seasonal march of the mean monthly burning index for Citrusdal (in the mediterranean climate zone) and Outeniqua (in the all year rainfall zone). Data from van Wilgen (1985)

early autumn, winter or early spring as the fuel bed is too wet to carry a fire (in South Africa as in California). Thus in an average year, the only season when mediterranean non-heath shrublands will burn is summer.

PHENOLOGY

In relating phenology to fire season it must be remembered that plants survive fire best if burnt when dormant. However, if plants survive the fire as seeds, then the season of burn must be made when it is most favourable for post-fire germination and establishment to occur. Thus any assessment of phenology in relation to fire season in a mediterranean climate zone must take cognisance of the post-fire regeneration strategies of the plants.

In the non-heath shrublands of the mediterranean climate zone of South Africa most perennial herbaceous and woody species resprout after fire. The major exception is one species, *Elytropappus rhinocerotis* (the renosterbos), which covers extensive areas today and is a re-seeeder after fire. It is, however, generally conceded that the dominance of this species is entirely due to encroachment caused by poor veld management over perhaps the last 2,000 years (Boucher and Moll 1981), so the present extent is an artifact. Thus in the non-heath shrublands (mainly resprouters) one would expect that the main growth flush would occur outside the greatest fire hazard period (late summer). On the other hand the woody heathland plants (mainly reseeders) are killed by fire so season of growth need not be fire related. However, the herbaceous heathland

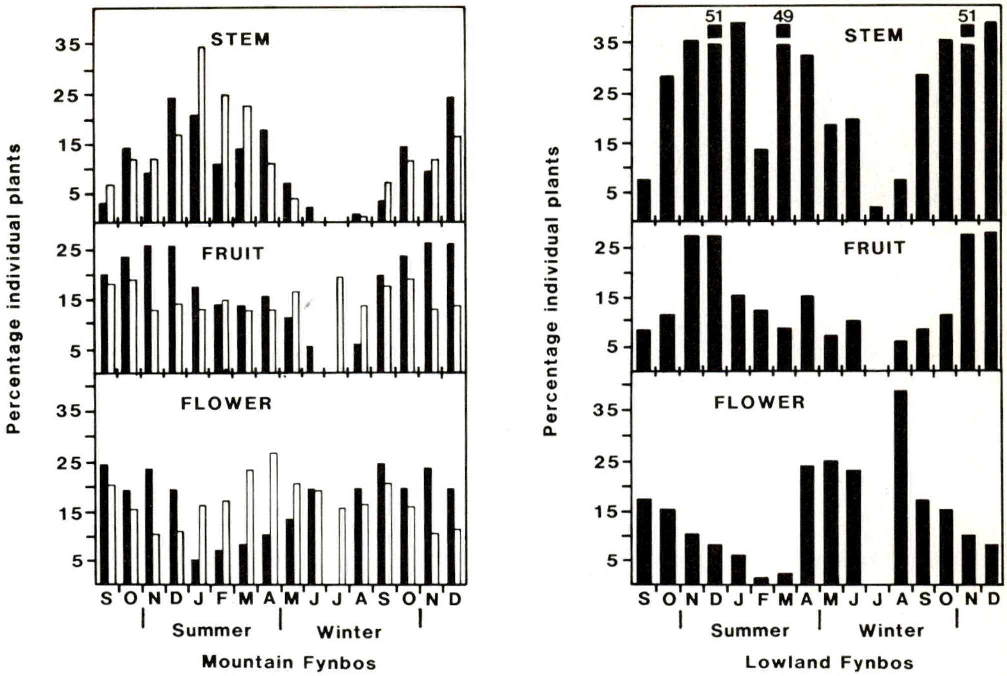

Fig. 4. Vegetative growth, fruiting and flowering phenology from Mountain Fynbos (shaded columns from Swartboschkloof where 404 individuals of 50 species were sampled, and unshaded columns from Sir Lowry's Pass 15 km to the south where 543 individuals of 61 species were sampled), and from Lowland Fynbos (from Barhuis)

plants (mainly resprouters) should grow outside the greatest fire hazard period. Thus, a partitioning of growth seasons can be expected in the heathlands.

A final factor that must be considered is fire frequency. The Cape shrublands cannot be burned more often than once every four or five years, since it takes that long for sufficient fuel to accumulate. In fact, it is thought that the "natural" fire régime in fynbos was perhaps every 15 to 40 years (Moll *et al.* 1980, van Wilgen 1985). As far as the non-heath shrublands are concerned there is no information on what a "natural" fire régime may have been. The present communities, however, probably could not be burned more often than once every four or five years, except in exceptional areas where the grass component remains high and fires would carry annually.

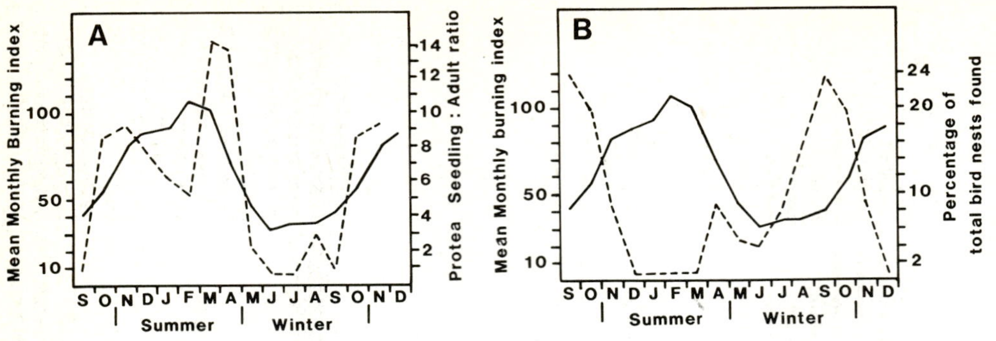

Fig. 5 Monthly fire hazard (solid line) in relation to A: fire survival of *Protea* spp at de Keur in the Cedarberg and B: total bird nests found (from van Wilgen 1985)

Figures 4 and 6 summarize the only comparable phenological data from all four shrubland communities in the mediterranean climate zone of the Cape Province. These data, collected by Sommerville in 1978-1979 (Siegfried 1981) were made on monthly visits to each site when growth, flowering and fruiting observations were made on a number of individuals of as many perennial species as possible. Thus the data set represents the community response and must be interpreted with great care since the individual species' phenophases are not given. Pierce (1984), in a synthesis of phenology in the Fynbos Biome, summarized all available phenological information (published and unpublished). This information can be used to modify the conclusions drawn from Sommerville's data.

The data for Mountain Fynbos came from two sites of different post fire age; Sir Lowry's Pass less than five years old, and Swartboschkloof about 20 years old. At Sir Lowry's Pass stem growth peaks in mid-summer while at Swartboschkloof it is earlier. My own data from a 35 year old community at Bains Kloof show that the resprouters, which form the understorey, grow in late spring and early summer, while the tall re-seeders in the overstorey grow later. Data on flowering from Sommerville (Fig. 4), Rycroft (1950) and van der Merwe (1966) indicate that in older Mountain Fynbos, most flowering takes place in late winter and spring, though some species flower throughout the year. In young vegetation maximum flowering is in late summer, but it must be remembered that the possibility of fire is small because of insufficient fuel. In both young and old communities most seeds are ripe in late spring and early summer. These data all support the hypothesis that the resprouters tend to grow in the season when fire hazard is low, while the re-seeders tend to grow when the fire hazard is high.

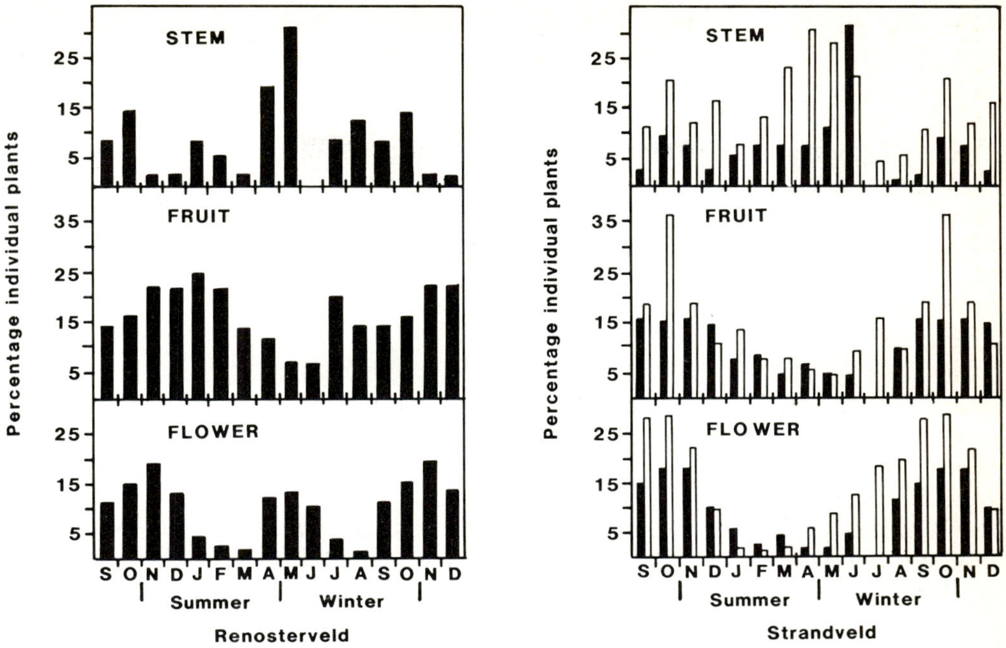

Fig. 6 Vegetative growth, fruiting and flowering phenology from Renosterveld (data from de Grendel where 275 individuals of 32 species were sampled from a 15 year old community) and from Strandveld (shaded columns from Geelbek where 410 individuals of 44 species were sampled from a 15 year old community at the arid fringe, and unshaded columns from Melkbos where 733 individuals of 52 species were sampled from a ± 75 year old community

This hypothesis requires further verification, particularly needed is a comparison between the same species that occur in both the mediterranean and summer rainfall zones.

The data for Lowland Fynbos, from a more arid site (Barhuis), show that most vegetative growth occurs in summer (Fig. 4), though much activity is apparently concentrated in early summer. Additional detailed data from eleven species that were measured on a monthly basis over 15 months (Sommerville 1983) at a wetter lowland site (Pella), show that most growth occurs in late spring and early summer. In both of Sommerville's studies the maturation of seed occurred in late spring or early summer.

All these phenological observations and measurements lend support to the hypothesis that fire in mid- to late-summer is least harmful for heathlands (=fynbos). These are further supported by van Wilgen and Viviers' (1985) demonstration that fires in late summer lead to maximum germination and establishment of obligative, serotinous seeded, Proteaceae. Further support is also given by van Wilgen (1985) who showed that birds tend to nest in late winter and early spring when the fire hazard is low (Fig. 5).

Figure 6 shows data from Renosterveld and Strandveld, both non-heath shrublands in the mediterranean climate zone. The data for Renosterveld show that shoot growth is more-or-less bimodal, occurring in spring and autumn when the fire hazard is low. Flower production in Renosterveld is also bimodal, following vegetative growth. Fruit production is high in winter and early summer, tending to be lower in late summer, which is the season of maximum fire hazard.

Data in Pierce (1984) for *Themeda triandra*, which was probably the dominant grass understorey in Renosterveld prior to encroachment by *Elytropappus rhinocerotis* (Boucher and Moll 1981), indicate that in the mediterranean climate zone *Themeda* grows through winter to spring; so would suffer least from fire in late summer and early autumn. In the all year rainfall zone *Themeda* has a bimodal growth season, winter and summer, which from the Outeniqua fire danger data (see Fig. 3) coincides with the most hazardous fire period. However, the fire hazard in every month of the year in this region is relatively high, so *Themeda's* growth is more likely to be regulated by rainfall which peaks in winter and summer. Personal observations from the summer rainfall area of Natal, where the fire season is restricted to late autumn through to early spring. confirm that *Themeda* grows in spring and early summer. These data for *Themeda* suggest that growth is moisture controlled and that it is perhaps coincidental that the wet season in the mediterranean and summer climate zones, when *Themeda* is most susceptible to damage by fire is the lowest fire hazard period.

In Strandveld vegetative growth is also bimodal; more strongly so at Melkbosch, which is a wetter site, than Geelbek. Flower and fruit production at both sites peaks strongly at the end of winter to early summer, thus avoiding the real fire risk period in mid- to late summer.

The data from both the non-heath shrubland formations in South African mediterranean climate zone, which in their undisturbed state are dominated by resprouters, indicate that there is least phenological activity in the period of greatest fire hazard. Thus supporting the contention that resprouting plants tend to avoid fire when they would be most susceptible to damage.

COMPARISON WITH OTHER MEDITERRANEAN SYSTEMS

Specht (1979) showed that Australian heathlands in the mediterranean climate zone show similar phenophases to their South African counterparts (=fynbos). Descriptions from Mallee communities in Australia (Specht 1981a) indicate that their phenology is bimodal with the understorey growing in spring and early summer, and the overstorey growing in late summer and autumn. These Mallee communities have no physiognomic counterparts in mediterranean climate areas so it is difficult to compare them. Perhaps they are most similar to Renosterveld?

Data from California (Mooney and Kummerow 1981) indicate that Californian chaparral tends to grow in spring, and Chilean matorral is described as having a much more protracted growth phase than chaparral, perhaps due to the greater diversityof growth forms in matorral. Thus matorral is apparently more similar to Strandveld, which is also characterized by a diversity of growth forms and which, like matorral, does not burn easily in any season. There are apparently no analogous vegetation types in Australia.

CONCLUSIONS

There is little doubt that the phenology of the shrubland communities of the mediterranean climate zone of South Africa and in California can be related to fire season. Mooney and Kummerow (1981) concluded that the phenological patterns between Chile, California and the Mediterranean are basically similar, and the the plants of Australia and South Africa differ substantially in their phenological patterns "because their shrubs grow in summer as well as in spring." An explanation of this misunderstanding is that Mooney and Kummerow (1981) did not apparently realize that they were comparing heathlands in Australia and South Africa, with non-heath shrublands in Chile, California and the Mediterranean. Additional phenological data are now available from South Africa (see Pierce 1984), particularly new data from non-heath shrublands in the mediterranean climate zone. If the Cape heathland phenological data are interpretated in terms of seasonal growth of the understorey (essentially resprouters) and overstorey (essentially re-seeders), then the apparent anomalous community phenology is readily understood.

What remains difficult to explain is why the resprouting overstorey heathland plants in Australia grow in summer? My interpretation of this is that the *Eucalyptus* spp. are so tall (and dense?), and the heathland understorey is so sparse, that the fire intensities in the understorey are low. In addition, heathland fires are more frequent and this also lowers the intensity. Thus, the Australian heathlands, dominated by emergent *Eucalyptus* spp., burn like savannas; where fire in late winter or early spring, in the intial stages of the growing season, is not too damaging.

It may be argued that some South African Renosterveld shrubs grow in late summer and autumn. *Elytropappus rhinocerotis* is a good example of a species that apparently grows out-of-phase with the climate; which has been related to its tropical origin (Levyns 1956). This late summer growth phase is when the fire hazard is great. But since *E. rhinocerotis* is a re-seeder, and its dominance is of recent origin, this is not biologically important in terms of fire seasons and phenology in mediterranean shrublands.

Finally much has been written about the theory of convergent evolution in mediterranean ecosystems, and a number of references have pointed to the strong similarity between Chile, California and the Mediterranean, and that South Africa and Australia are often different (Cody and Mooney 1978). The basis for this dissimilarity has been related to the different origins of the biota (Specht 1973), or to a dissimilar pattern of resource availability caused by the very limited nutrients available in the soils of South Africa and Australia (Specht 1979 and 1981b). However, this study indicates that the proponents of the convergence theory have not been aware that they were comparing non-heath shrublands in Chile, California and the Mediterranean, with heathlands in South Africa and Australia. If one compares the non-heath shrublands of the mediterranean climate zone of South Africa with their analogues in Chile, California and the Mediterranean, there are strong similarities which lend better support to the theory of convergence than was previously thought.

ACKNOWLEDGEMENTS

The Foundation for Research Development of the CSIR and the University of Cape Town are acknowledged for financial support, Ms.E.Mantle for typing the manuscript and Ms.P.Chesslet for preparing the figures.

REFERENCES

Acocks JPH (1953) Veld Types of South Africa. Mem Bot Surv S Afr 28

Aschmann H (1984) A restrictive definition of Mediterranean climates. Bull Soc bot Fr 131, Actual bot 2/3/4:21-30

Bayer MB (1984) The Cape Flora and the Karoo - a winter rainfall biome versus a fynbos biome? Veld & Flora 70(1):17-19

Bond P and Goldblatt P (1984) Plants of the Cape Flora: a descriptive catalogue. Supp to J S Afr Bot 13

Boucher C and Moll EJ (1981) South African mediterranean shrublands. In: di Castri F, Goodall DW and Specht RL (eds) Mediterranean-type Shrublands. Ecosystems of the World Vol 11 Elsevier, Amsterdam

Cody ML and Mooney HA (1978) Convergence versus non-convergence in mediterranean-climate ecosystems. Ann Rev Ecol Syst 9:265-321

Cowling RM (1983) The occurrence of C_3 and C_4 grasses in fynbos and allied shrublands in the southwestern Cape, South Africa. Oecologia (Berl) 58:121-127

Cowling RM and Campbell BM (1980) Convergence in vegetation structure in the mediterranean communities of California, Chile and South Africa. Vegetatio:191-197

Day J, Siegfried WF, Louw GN and Jarman ML (eds) (1979) Fynbos ecology: a preliminary synthesis. S Afr Nat Sci Prog Report No 40

di Castri F, Goodall DW and Specht RL (eds) (1981) Mediterranean-type Shrublands. Ecosystems of the World Vol 11 Elsevier, Amsterdam

Fuggle RF (1981) Macro-climatic patterns within the fynbos biome. National Programme for Environmental Science, Fynbos Biome Project

Jarman ML (1982) A look at the littlest floral kingdom. Scientiae 23(3):9-19

Levyns MR (1956) Notes on the biology and distribution of the rhenoster bush. S Afr J Sci 52:141-143

Marloth R (1903) Results of experiments on Table Mountain for ascertaining the amount of moisture deposited from the south east clouds. Trans S A Phil Soc 14:403-408

Marloth R (1905) Results of further experiments on Table Mountain for ascertaining the amount of moisture deposited from the south east clouds. Trans S A Phil Soc 16:97-105.

Moll EJ (1986) A review of some new concepts in "fynbos" ecology. In: J Parkington *et al.* (ed) People and Places: Papers in the Prehistory of the Western Cape. BAR International Series

Moll EJ and Bossi L (1983) Vegetation of the Fynbos Biome: a 1:1 000 000 vegetation map. Govt Printer Pretoria

Moll EJ and Bossi L (1984) Assessment of the extent of the natural vegetation of the fynbos biome of South Africa. S Afr J Sci 80:355-358

Moll EJ, Campbell BM, Cowling RM, Bossi L, Karman ML and Boucher C (1984) A description of major vegetation categories in and adjacent to the Fynbos Biome. S Afr Nat Sci Prog Report No 83

Moll EJ, Hilton-Taylor C and Jarman ML (1984) Where are the Cape's Mediterranean Shrublands? pp 115-116 In: Dell B (ed) Proceedings of the Fourth International Conference on Mediterranean Ecosystems. Nedlands, Western Australia

Moll EJ and Jarman ML (1984a) Clarification of term Fynbos. S Afr J Sci 80: 351-352

Moll EJ and Jarman ML (1984b) Is Fynbos a Heathland? S Afr J Sci 80:352-355

Moll EJ, McKenzie B and McLachlan D (1980) A possible explanation for the lack of trees in the Fynbos, Cape Province, South Africa. Biol Conserv 17:221-228

Mooney HA and Kummerow J (1981) Phenological development of plants in mediterranean-climate regions. In: di Castri F, Goodall DW and Specht RL (eds) Mediterranean-type Shrublands. Ecosystems of the World Vol 11 Elsevier, Amsterdam

Nagel JF (1956) Fog precipitation on Table Mountain. Quart J Roy Met Soc 82:452-460

Nagel JF (1961) Rainfall in the Table Mountain area with special reference to orographic effects. Notos, Weather Bureau Pretoria, 10:59-73

Pierce SH (1984) A synthesis of plant phenology in the Fynbos Biome. S Afr Nat Sci Prog Report 88

Rycroft HB (1950) A quantitative ecological study of the vegetation of Biesievlei, Jonkershoek. Unpublished Ph.D. Thesis, University of Cape Town

Siegfried WR (1981) Trophic structure of some communities of fynbos birds. In: Moll EJ (ed) Proceedings of a Symposium on Coastal Lowlands of the Western Cape. University of Western Cape, Bellville

Sommerville JEM (1983) Aspects of coastal fynbos phenology. Unpublished M.Sc. Thesis, University of Cape Town

Specht RL (1973) Structure and functional response of ecosystems in the mediterranean climate of Australia. In: di Castri F and Mooney HA (eds) Mediterranean Type Ecosystems. Springer-Verlag Berlin

Specht RL (ed) (1979) Heathlands and related shrublands. Ecosystems of the World Vol 9A Elsevier, Amsterdam

Specht RL (1981a) Mallee ecosystems in southern Australia. In: di Castri F, Goodall DW and Specht RL (eds) Mediterranean-type Shrublands. Ecosystems of the World Vol 11 Elsevier, Amsterdam

Specht RL (1981b) Heathlands and related shrublands. Ecosystems of the World Vol 9B Elsevier, Amsterdam

Takhtajan A (1969) Flowering plants: origin and dispersal. Oliver and Boyd, Edinburgh

van der Merwe P (1966) Die flora van Swartboschkloof, Stellenbosch en die herstel van die soorte na 'n brand. Annale van die Universiteit van Stellenbosch 41, Serie A (14):691-736

van Wilgen BW (1984) Fire climates in the southern and western Cape Province and their potential use in fire control and management. S Afr J Sci 80:358-362

van Wilgen BW (1985) The derivation of fire hazard indices and burning prescriptions from climatic and ecological features of the Fynbos Biome. Unpublished Ph.D. Thesis, University of Cape Town

van Wilgen BW and Viviers M (1985) The effect of season of fire on serotinous Proteaceae in the western Cape and the implications for fynbos management. S Afr Forestry J 133:49-53

Vogel JC, Fuls A and Ellis RP (1978) The geographical distribution of kranz grasses in South Africa. S Afr J Sci 74:209-215

Comparative phenology of four mediterranean shrub species growing in Portugal

J.S. Pereira, G. Beyschlag* , O.L. Lange*, W. Beyschlag* and J.D. Tenhunen**

Instituto Superior de Agronomia, Tapada da Ajuda, P-1399 Lisboa codex, Portugal

INTRODUCTION

Knowledge of the phenology of woody plants is essential in order to understand their above- and belowground function, resource allocation and performance under natural conditions. Vegetation in areas with mediterranean-type climates is usually rich in species and plant life forms (see Trabaud in this volume) and although a considerable amount of work has been devoted to the study of phenology of mediterranean-type plant communities (see the contributions in this volume by Montenegro; Arianoutsou and Mardiris, and Moll), more information is needed in special cases to fully characterize plant function within these ecosystems. The objective of the present work was to examine the phenology of reproduction and aboveground growth of several characteristic scrub species growing under a mediterranean-type of climate in Portugal as part of a larger scope study of gas exchange and water relations (see the chapters by Beyschlag *et al.*, Correia *et al.*, Harley *et al.*, Lange *et al.*, Meister *et al.*, and Tenhunen *et al.*).

MATERIALS AND METHODS

Site Characteristics

The study was conducted in 1982 and 1983 at the Research Station Quinta de São Pedro, at Sobreda da Caparica, near Lisbon. The climate is of the mediterranean-type tempered by oceanic influence. The average annual temperature and precipitation (from 1931 to 1960) in the nearby meteorological station of Tapada da Ajuda in Lisbon are 16.9° C and 693.5 mm respectively. Drought usually extends from May through September. Frost is possible from December through January. Figure 1 illustrates the course of rainfall and air temperatures during 1982 and 1983. The study years were drier than normal: total rainfall was 558 and 674 mm respectively. Average temperature in both years was 16.5° C.

---

*Lehrstuhl für Botanik II der Universität Würzburg, Mittlerer Dallenbergweg 64, D-8700 Würzburg, West Germany

**Systems Ecology Research Group, San Diego State University, San Diego, CA 92182 USA

NATO ASI Series, Vol. G15
Plant Response to Stress. Edited by J. D. Tenhunen et al.

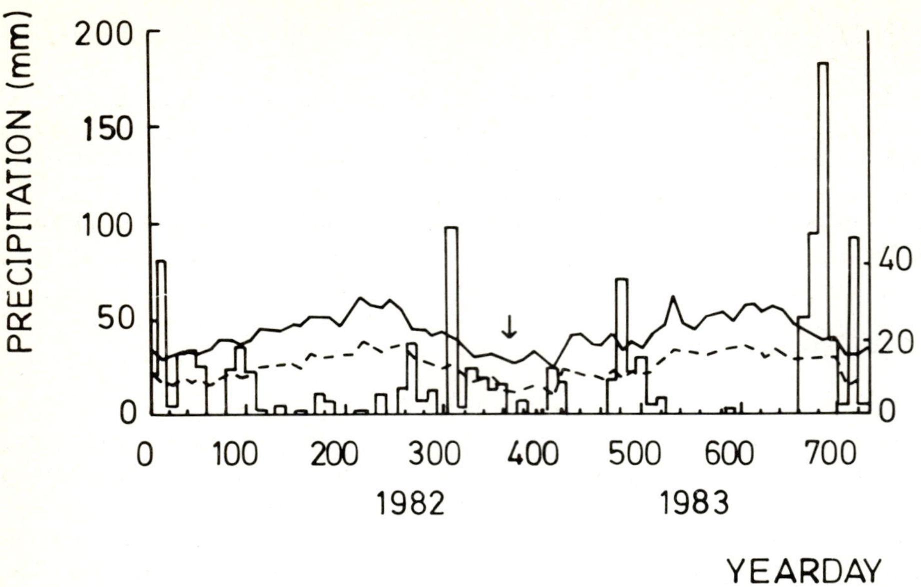

Fig. 1. The course during 1982 and 1983 of 10-day period sums of precipitation (bar diagram) and 10-day period averages of maximum and minimum air temperatures (T_{max}; T_{min} respectively) at Tapada da Ajuda, Lisbon. Days are numbered sequentially from January 1, 1982 through the two years (yearday)

The soil at the experimental site is of the type Cambisols (FAO) with mineral soil horizons low in organic matter (0.8 to 1.125% on a dry weight basis) and in nitrogen content (0.05 to 0.09%). The soil pH in water is approximately 5.1 to 5.3. The upper mineral horizon (7 to 15 cm in thickness) had a higher clay content (18.5 to 21.5%) than the underlying less altered material which had approximately 4% clay. The difference was due to the larger sand fraction in the latter since the loam fraction was constant (ca. 7.5%) throughout the profile. The organic horizon composed of debris and decomposing organic matter had an average thickness of 6 cm.

The vegetation on the experimental plot is a regrowth of an evergreen macchia scrub, and was 2 to 4 m in height. It is dominated by sclerophyllous species such as *Quercus coccifera, Pistacia lentiscus, Phillyrea angustifolia, Arbutus unedo* and *Myrtus communis* with scattered trees of *Quercus suber* and *Pinus pinea*. The atlantic character of this mediterranean scrub as well as the low soil pH of the habitat is reflected by the presence of numerous *Ulex parviflorus* shrubs.

Bi-weekly Measurements

The species examined were *Quercus coccifera*, *Q. suber, Arbutus unedo* and *Cistus salvifolius*. Three plants of each species, typical in appearance and located in different parts of the experimental plot were chosen for study. Phenological observations at the whole plant level on flowering, fruit development, leaf fall and presence or absence of growing shoots, were made every second week from January 28, of 1982 until December 13, of 1983. During the same period, shoot growth was measured on selected individually marked twigs (3 per tree; 9 per species). The following parameters were measured: number of leaves, leaf dimensions (length and width), twig length and twig diameter (1 cm above insertion).

The area and the biomass of individual leaves were estimated from relationships between (leaf length times leaf width) and leaf area and between leaf area and leaf mass. The relationships were derived by destructive sampling. Each harvested sample was divided into groups of approximately 100 leaves similar in shape but of different size, and average leaf length, width, and area were determined. The changes in twig biomass as a function of time were estimated using regression equations relating biomass to the product of twig length and the squared diameter 1 cm above the base of the twig. Marked twigs were harvested and their biomass determined at the end of the experiment.

RESULTS

The species studied were all evergreen. Figure 2 shows the general pattern (based on both years of observation) found for the occurrence of flowering, fruit growth and maturation, shoot growth (the combination of leaf expansion and internode extension) and leaf fall in each species.

Quercus coccifera

Flowering occurs very early in the spring and female flowers were visible in late February and early March. The fruits were noticeable by mid-May and grew until the autumn when maturation occurred. As shown in Fig. 3, extension of the internodes of twigs started early in the spring, immediately after the appearance of flowers and was rapid. It was followed by leaf expansion which occurred for at least two weeks after internode expansion had stopped. Although most leaves were shed approximately 13 months after emergence in the spring of the second year when new shoots were growing (Fig. 4A), some current-year leaves fall at any time of the year and others are retained longer than 17 months. In addition to the major flush of shoot growth in the spring, there was a secondary flush of growth in the autumn. This occurred in 1982 but not in 1983. Autumn growth was much less important for the plant than the spring growth. Only ca. 5% of the branches at the top of the canopy of the shrubs studied had the

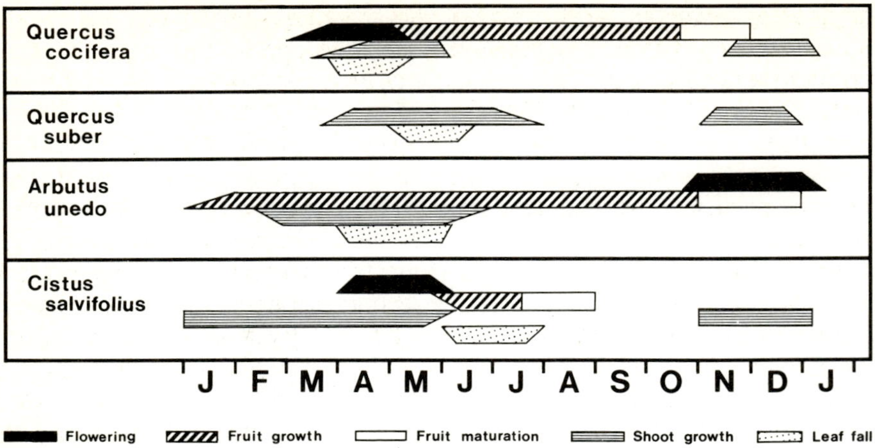

Fig. 2. The general pattern in phenological events observed for *Quercus coccifera*, *Quercus suber*, *Arbutus unedo* and *Cistus salvifolius* at Quinta de São Pedro during an annual course

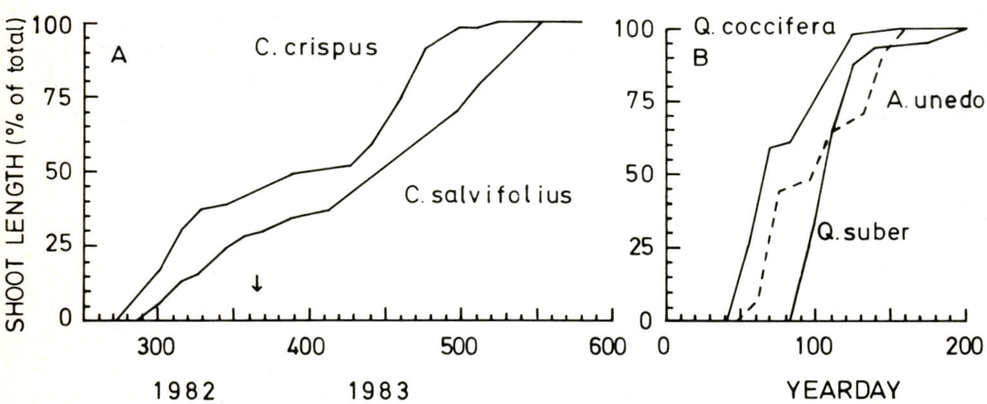

Fig. 3. Accumulated internode extension as a percent of total for selected individual twigs of *Cistus salvifolius* and *Cistus crispus* (left panel) and *Quercus coccifera*, *Quercus suber*, and *Arbutus unedo* (right panel). Days on time axis are the same as in Fig. 1

autumn growth flush. Not only was the shoot biomass produced less but also the number of leaves produced per twig and the average leaf area were less in the autumn than in spring (Table 1).

Quercus suber

The *Quercus suber* trees at the experimental site had not yet reached the reproductive stage. The pattern in vegetative growth was very similar to that found for *Quercus coccifera*, even though shoot growth started later and lasted longer (Figs. 2 and 3). In the case of *Q. suber*, the average longevity of leaves was 252 days (Fig. 4B). Since we did not detect any leaves living longer than 17 months, the cohort of leaves observed was a sample of the whole population of leaves produced by the plants studied. Therefore, 252 days may be taken as the absolute average longevity of *Q. suber* leaves. This resulted from a significant "juvenile mortality" and the "life expectation" of leaves 1 month old was 339 days. 70% of these leaves lived longer than 1 year. As in the case of *Q. coccifera*, the flush of growth in autumn was less important than the flush in spring (Table 1). It only occurred in 1982 and in not more than ca. 10% of the branches.

Arbutus unedo

In *Arbutus unedo*, vegetative growth began earlier than in the two *Quercus* species (Figs. 2 and 3) and leaf expansion continued two weeks longer than did the extension of internodes. Growth of the flower stalk took place throughout the summer and flowering occurred in the winter. Fruit began to mature during November and December but fruit growth and development progressed from the time of fruit set until late in the next autumn or early winter. As seen in Fig. 4C, 1-year-old leaves were retained by the plants along with newly produced leaves until near the middle of the growing season. About 35% of the leaves persisted for longer than 17 months.

Cistus salvifolius

This species differed greatly from the others by exhibiting a long period during which shoots elongated. Shoot elongation occurred through most of the wet season (Figs. 2 and 3). Flowering in spring resulted in termination of extension in flower-bearing branches. Extension continued until early summer in those shoots that did not bear flowers. Fruit development occurred simultaneously with leaf fall. As seen in Fig. 5, a large percentage of the leaves which emerged in winter were lost during the following summer. Although 24% of the leaves produced during one growing season survived until the next growing period, most leaves were much shorter lived. The leaves produced in the beginning of the growing season had an average longevity of only 131 days. In general, the leaves that survived the summer were those produced latest during the growth period. This species also differed from the other species studied by having a much higher "specific leaf area" (area per unit of leaf biomass).

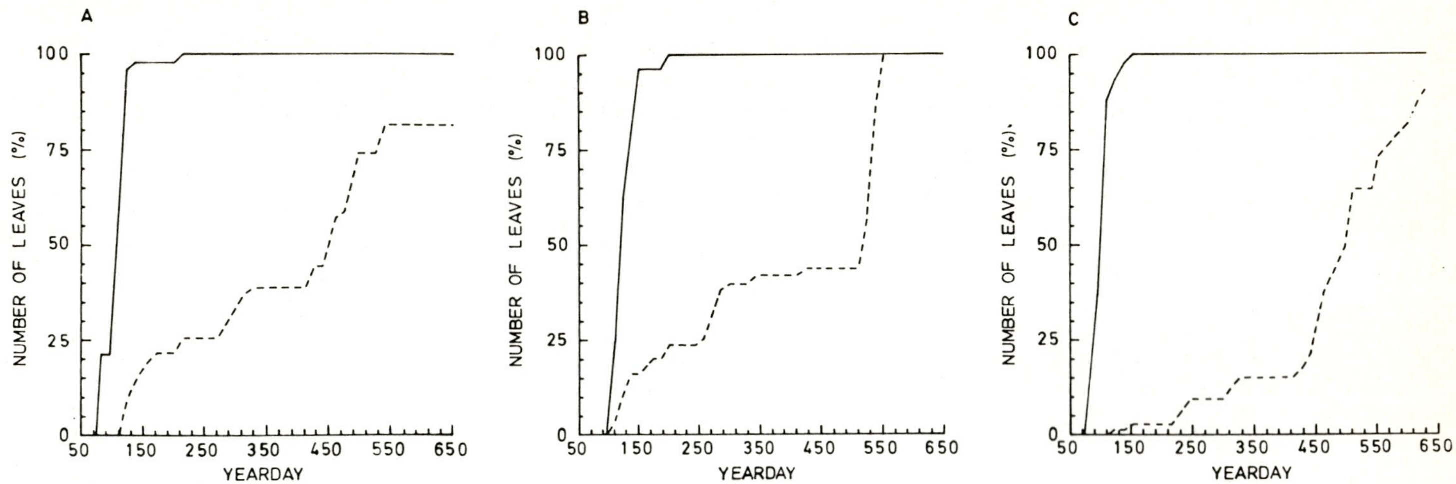

Fig. 4. Accumulated number of leaves emerged (solid line) and shed (broken line) as a percent of total for selected twigs of (A) *Quercus coccifera*, (B) *Quercus suber*, and (C) *Arbutus unedo*. The percentage of leaves attached to the shoot at any moment may be obtained as the difference between the 2 values shown. Days on time axis are the same as in Fig. 1

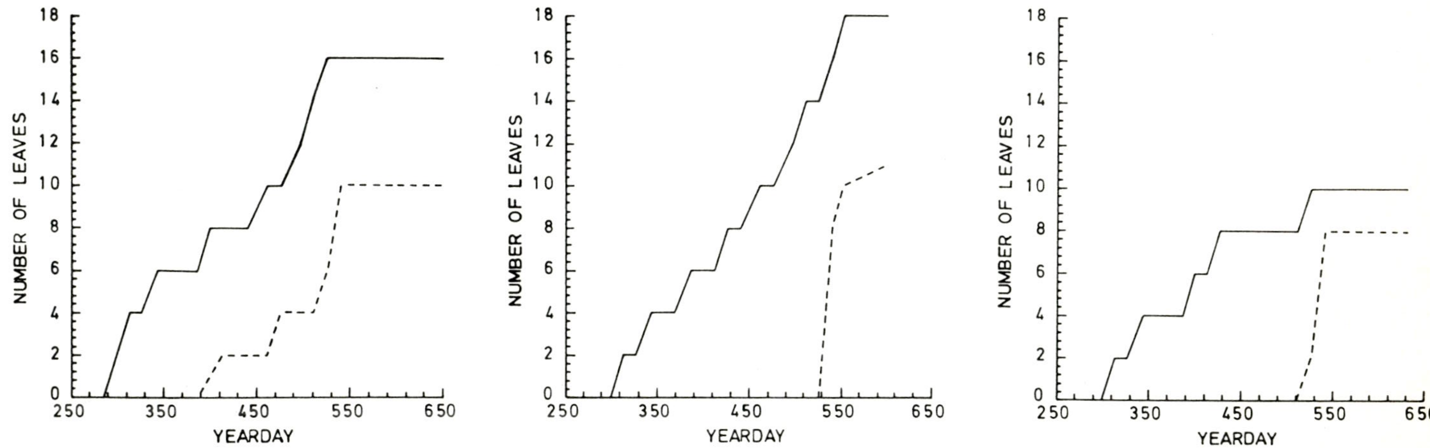

Fig. 5. Accumulated number of leaves emerged (solid line) and shed (broken line) in 3 selected twigs of *Cistus salvifolius*. The percentage of leaves attached to the shoot at any moment may be obtained as the difference between the 2 values shown. Days on time axis are the same as in Fig. 1

Table 1. Average total leaf area per twig, total biomass of individual shoots at the end of the growing period, average area of mature individual leaves, and "specific leaf area" (SLA) of *Quercus coccifera* and *Quercus suber* (growth flushes of spring and autumn), *Arbutus unedo,* and *Cistus salvifolius.* Averages are shown for 9 twigs per species ± standard errors.

	Leaf area per twig cm^2	Shoot biomass (leaves + twig) g	Areas of individual leaves; cm^2	SLA $cm^2\ g^{-1}$
Quercus coccifera				
Spring	25.7 ± 9.6	0.53 ±0.22	2.48 ±0.54	62.7
Autumn*	13.5 ± 6.9	0.53 ±0.23	1.05	54.3
Quercus suber				
Spring	61.1 ±34.9	1.04 ±0.60	5.95 ±2.88	79.6
Autumn*	6.7 ± 2.6	0.24 ±0.11	1.42	
Arbutus unedo	117.0 ±66.7	1.97 ±1.12	7.67 ±3.57	65.8
Cistus salvifolius	13.6 ± 9.2	.23 ±0.09	1.23 ±0.38	71.6

*For these measurements twigs which had actual autumn growth were newly selected.

DISCUSSION

Flowers were present during all times of the year except the driest months of the year in late summer and early autumn and the coldest months during winter (as mentioned above, the *Q. suber* plants under observation had not yet reached the reproductive stage). Thus, flowering of the different species was not synchronized among these species growing near Lisbon as has been described for other areas with mediterranean-type vegetation (Kummerow *et al.* 1981). With the exception of *C. salvifolius*, fruits had large quantities of food reserves and fruit growth continued over long periods (at least more than 4 months). In general, fruit maturation occurred in the autumn. The time between fruit set and maturation was much shorter for the dry dehiscent fruits of *C. salvifolius*.

Three different patterns of shoot growth were found. The two *Quercus* species exhibited a major growth flush for a short period in spring (ca. 6 weeks in *Q. coccifera* and ca. 10 weeks in *Q. suber*). In addition, a low percentage of buds which would sprout in the next spring grew already in the autumn. Nevertheless, this was dependent on the specific climatic characteristics of the year. The growth of *A. unedo* occurred in late winter and spring and extended over a longer period (ca. 12 weeks). Shoots of *C. salvifolius* grew during the rainy season from autumn throughout winter and spring for a total duration of approximately 26 weeks. In all species studied, shoot growth occurred during the water-stress-free period of the year, assuming that predawn leaf water potential lower than -1 MPa represents the onset of severe plant water stress. As shown in Tenhunen *et al.* (this volume), such predawn leaf water potentials were reached by late June irrespective of the species considered.

The pattern of growth in the shrubs that accumulate biomass in woody stems and may reach the size of trees in favorable locations (the two *Quercus* species and *Arbutus unedo*) differed from that of the smaller sub-shrub *Cistus*. Shoot growth occurred at high rates for a fairly short period in spring in the former whereas it continued during the entire rainy season in *Cistus*. The amount of biomass accumulated in branches of the two *Quercus* species and *A. unedo* was greater than that accumulated by *Cistus*. It is likely that the intense growth of new shoots in spring in the former group of species depends strongly on reserves accumulated from autumn and winter photosynthesis whereas in *Cistus* shoot growth may depend more on current photosynthesis. It is likely in *Cistus* that large reductions in whole-plant photosynthesis (loss of leaves and low assimilation rates, cf. Harley *et al.* 1986) due to water stress lead to the depletion of food reserves during the dry summer season.

Most of the 1-year-old leaves of the evergreen shrubs were shed as abscision was stimulated at the time when cambial growth of 1-year-old twigs started (Kozlowski 1971). The timing of leaf shedding of other evergreen oaks similarly has been shown to occur during the period of new shoot growth (Lossaint and Rapp 1978; Ebel, Kümmel and Walther 1980). Nevertheless, drought may induce further leaf fall and possibly some leaves were shed in response to it in both *Q. coccifera* and *A. unedo*. In *Cistus*, a large proportion of the current season leaves were shed during summer. This drastically reduces the evaporative surface of these winter growing plants as also reported by Orshan (1963). *Cistus crispus* (unpublished) showed a parallel behavior in all respects.

The lifespan of individual leaves of mediterranean sclerophyll evergreens is quite variable (Orshan 1982). As we found in *Quercus coccifera* and in *Arbutus unedo*, about one-third of the leaves produced in one growing season remained attached to the 1-year-old branches through the next growing season. However, the average longevity of individual leaves was less than 12 months and in *Quercus suber* the absolute longevity of the longest living leaves was not more than 17 months as has been found for other mediterranean species (Orshan 1982).

The impact of leaf longevity on carbon budgets of plants is of great importance (Mooney and Gulmon 1982). Evergreen mediterranean sclerophyll shrubs behave with respect to foliage production in a manner which is intermediate between coniferous plants with long-lived foliage and deciduous or semi-deciduous species (e.g. *Quercus faginea*). They photosynthesize year-round (Tenhunen *et al.*, this volume) but must continue to replace a significant proportion of leaves (about 70% in *Quercus suber*) while very slowly incrementing plant leaf area. In *Cistus salvifolius* the amount of new leaves required to maintain the same leaf area per plant during the growing season is comparatively greater than in the other species. However, given the much smaller amount of biomass per unit of leaf area in this species, the absolute requirements in terms of fixed carbon to produce these leaves may be smaller than at first expected. On the other hand, water consumption during the dry season is drastically reduced as a consequence of the loss of the leaves produced early in the growing period and stomatal closure (Harley *et al.* 1986).

Seasonal phenological patterns of internode extension and leaf formation and expansion are expressions of genetically controlled and environmentally mediated changes in plant physiology. They reveal differences in adaptive strategies of the individual plant species with respect to the mediterranean-type climate which exhibit a moist, mild winter and a hot, dry summer period. The phenological events are integrated with species-characteristic seasonal patterns of water use and primary production and may provide clues that will help us understand both the seasonal regulation of these processes and whole plant response in the natural habitat. As pointed out by Mooney (this volume), a large challenge remains with regard to understanding the ecology of mediterranean and arid zone plants and this challenge can only be met by applying a diversity of approaches. With regard to evaluating the competitive potential of particular plant species, phenological information such as that presented here may be used together with detailed studies of gas exchange (see the chapters by Beyschlag *et al.*, Correia *et al.*, Harley *et al.*, Lange *et al.*, and Tenhunen *et al.*) and information on canopy structure and function (Meister *et al.*) to estimate important seasonal trends in whole plant water use and carbon fixation.

ACKNOWLEDGEMENTS

We would like to acknowledge the help of Mr. Páscoa de Matos with field data collection and Mr. G. Radermacher with computer programming. Dr. O. Schulz-Kampfhenkel (Institut für Internationale Zusammenarbeit in Bildung und Forschung-IZBF) provided the use of the facilities at the research station Quinta de São Pedro. Calculations were made at the Rechenzentrum der Universität Würzburg. The research was supported by the Instituto Nacional de Investigação Cientifica of Lisbon and the Deutsche Forschungsgemeinschaft. The support of the Alexander von Humboldt-Stiftung of Bonn to one of us (JSP) is also acknowledged.

REFERENCES

Ebel F, Kümmel F, Walther H. (1980) Beobachtungen über die Jahresrhythmik des Laubfalls und die Lebensdauer der Blätter bei einigen immergünen Gehölzen. In: 100 Jahre Arboretum (1879-1979) Berlin, 101-115. Mittleitungen aus dem Botanischen Garten der Sektion Biowissenschaften Halle, Nr. 41

Harley PC, Tenhunen JD, Pereira JS, Beyschlag W, Lange OL (1986) Seasonal changes in net photosynthesis rates and photosynthetic capacity of leaves of *Cistus salvifolius*, a European semi-evergreen shrub. Oecologia (Berl) in press

Kozlowski TT (1971) Growth and Development of Trees. Vol. I. Academic Press, New York

Kummerow J, Montenegro G, Krause D (1981) Biomass phenology and growth. In: Miller PC (ed) Resource Use by Chaparral and Matorral. Springer-Verlag, Berlin-Heidelberg-New York, pp 69-96

Lossaint P, Rapp M (1978) La forêt méditerranéenne de chênes verts. In: Lamotte M, Bourlière F (eds) Problemes d'Ecologie Structure et Functionnement des Ecosystémes Terrestres. Masson, Paris, pp. 129-185

Mooney HA, Gulmon SL (1982) Constraints to leaf structure and function in reference to herbivory. Bioscience 32:198-206

Orshan G (1963) Seasonal dimorphism of desert and mediterranean chamaephytes and its significance as a factor in their water economy. In: Ruther AJ, Whitehead FH (eds) The Water Relations of Plants. Blackwell, Oxford, UK, pp. 206-222

Orshan G (1982) Monocharacter growth from types as a tool in a analytical-synthetic study of growth forms in mediterranean type ecosystems. A proposal for an inter-regional program. Ecologia Mediterranea 8:159-171

Observations on the phenology of two dominant plants of the Greek maquis

M. Arianoutsou and Th. A. Mardiris

Division of Ecology, Department of Biology, School of Sciences, University of Thessaloniki, 540 06 Thessaloniki, Greece

INTRODUCTION

A great part of the land surface of Greece (approximately 40%) is covered by natural mediterranean type ecosystems. In this area warm and dry summers alternate with mildly cold and wet winters (Aschmann 1973). Rainfall occurs primarily during the late autumn and early winter months. The period of greatest water availability coincides with the period of lowest temperatures. During the dry summer season, temperature are highest and relative humidity is lowest which results in large summer water deficits. Evidence for adaptation to mediterranean climate conditions has been sought through study of both the morphology and the phenology of the mediterranean vegetation. Several phenological studies have been undertaken in mediterranean-type ecosystems of other parts of the world, e.g. California and Chile (Mooney *et al.* 1974; Hoffmann and Hoffmann 1976; Mooney and Kummerow 1977; Montenegro *et al*. 1979; Mooney and Kummerow 1981; Kummerow *et al.* 1981). In all these studies the phenological changes of the species (evergreen sclerophylls or drought-deciduous) have been found to change characteristically in response to seasonal environmental events in the native habitats.

Several studies on the structure and function of the mediterranean-type ecosystems of Greece were started during the last decade. Most of the information collected to date has dealt with phryganic ecosystems, in which seasonally dimorphic plants dominate. The ecology of the Greek maquis ecosystems, where the dominant plant species are evergreen sclerophylls, is less well known. Phenological study is one of the first steps in understanding the function of ecosystems. This is because phenological events reflect the way in which the species (and/or the system) exploit the so-called favourable periods in order to gain carbon and to allocate photosynthetic products for growth and reproduction. This paper reports results of phenological studies in Greek maquis ecosystems obtained during a two-year period in northern Greece.

SITE DESCRIPTION

The site used for this investigation is located in northern Greece, in Stavros-Halkidiki, close to the peninsula of Mount Athos. The site was chosen primarily because of the high density and cover of the species *Quercus coccifera* L. and

NATO ASI Series, Vol. G 15
Plant Response to Stress. Edited by J. D. Tenhunen et al.

Arbutus unedo L., which are considered typical representatives of the evergreen sclerophyll vegetation in Greece. The plants occurred on a north-east facing slope on metamorphic rocks of biotitic gneiss and amphiboles. The slope was less than 10° at 100 m above sea level. Vegetation at the site has not burned for at least 20 to 25 years. No grazing or other human activities occur there at present. The dominant plant species based on biomass are *Quercus coccifera* L. (32%) and *Arbutus unedo* L. (15%). Other woody species occurring in the area are *Phillyrea media* L., *Erica arborea* L., *Quercus ilex* L. and *Cistus* sp.

FIELD PHENOLOGY METHODS

Detailed phenological observations were made between April 1981 and April 1983 at monthly intervals on randomly chosen *Quercus coccifera* and *Arbutus unedo* individuals. On each plant, one representative branch was labeled with a permanent nylon ribbon for the phenology census. Each branch was monitored throughout the year for the following.

1) Total number of leaves on the branch
2) Total length of the main branch and the lateral branches
3) Time of leaf emergence and leaf fall
4) Presence of flower buds, flowers, or fruits

RESULTS AND DISCUSSION

The characteristic seasonal changes in daily air temperature and rainfall at the site are shown for the two-year study period in Fig. 1. Temperature increased during the growing season (spring to early summer) along with photoperiod. Rainfall in general decreased by late spring. Occasional summer thunderstorms occurred during both 1981 and 1982. As far as seasonal fluctuations in factors which drive phenological events, such as light, water, and heat energy, are concerned, water availability is the most unpredictable. Incident radiation and, to a somewhat lesser extent, mean air temperature fluctuated with a fairly predictable seasonal pattern.

Plant phenological responses to the physical environment were not synchronous for the two species studied. This is assumed to occur because the species utilize resources differently in this multispecies community. *Quercus coccifera* flowers in spring and *Arbutus unedo* flowers from October to early spring (April or March). In *Arbutus unedo* flower bud differentiation occurred almost 12 months before flowering (Fig. 2, lower part). These buds remained on the branches from October 1981 until the flowers formed in October 1982. In *Q. coccifera*, the flower buds differentiated between October and March. Development continued on through anthesis during a 4 to 5 month period in spring and early summer. The timing of flowering in *A. unedo* in the fall may depend on time of

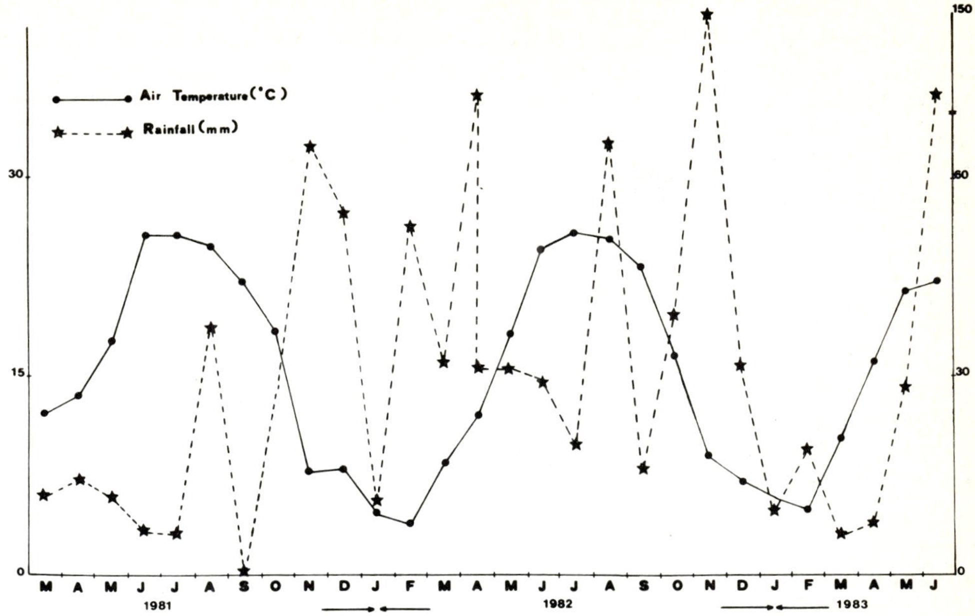

Fig. 1. Climate diagram of the area where the experimental site is located

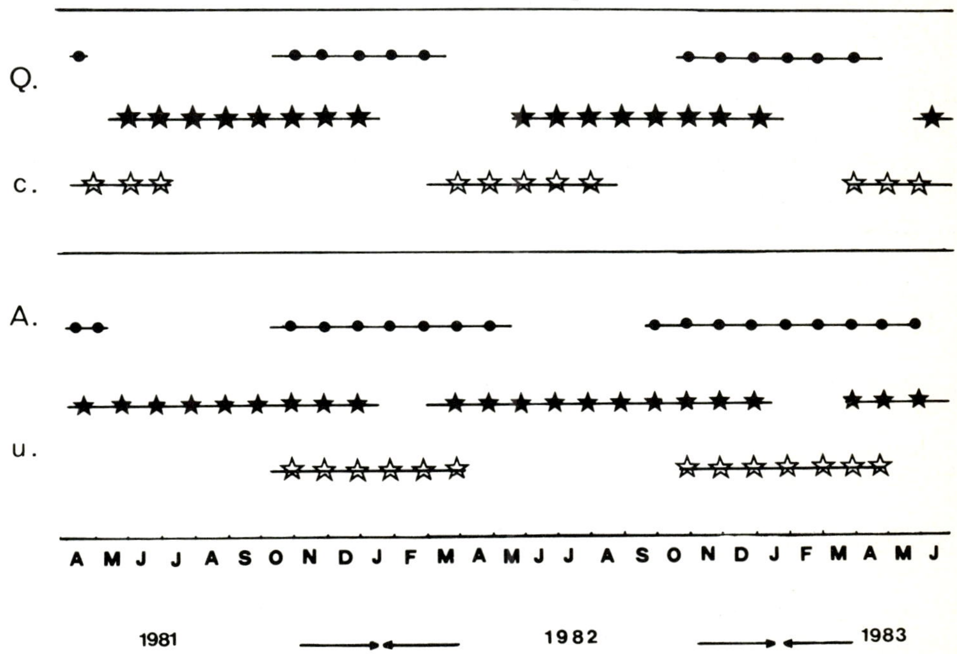

Fig. 2. Periods during which bud formation (•), fruit formation (blackened stars), and flowering (open stars) occurred in *Quercus coccifera* and *Arbutus unedo*

occurrence of autumn rains and subsequent increase in water availability. A somewhat similar but even more complicated situation was reported by Kummerow *et al.* (1981) for two of the dominant species of the California chaparral. Flower development in *Adenostoma fasciculatum* depended on moisture availability just preceding flowering in April and May. Flowering in *Arctostaphylos glauca* was affected by water availability during springtime one year before.

Formation of fruits in *Quercus coccifera* started late in the spring. Fruit ripening occurred over a period of 3-4 months. Fruits began dropping in late summer and early autumn but many continued expanding until spring. In 1981, 88% of the fruits were shed between November and February, while in 1982, 63% of the total were shed between September and October. The difference may have been related to changes in water availability, since during the summer months of 1981 rainfall was lower than in 1982 (Fig. 1) and rainfall may induce the stimulus for fruit abscission. In *Arbutus unedo*, fruits are formed in early spring. The maturation of these fruits occurs over more than one year. Fruits start to fall in late autumn. The majority, 97% in 1981 and 90% in 1982, were shed between December and January.

Seasonal changes in shoot elongation of *Q. coccifera* and *A. unedo* were studied between April 1981 and June 1983 at the Stavros-Halkidiki site. Growth was initiated in the spring in both species. 40-45% in *A. unedo* and 45-50% in *Q. coccifera* of the total yearly shoot elongation occurred in spring. A very small amount of growth (<10% of annual elongation) was observed during the autumn months. Mean leaf sizes of *Q. coccifera* and *A. unedo* are shown in Table 1.

Table 1. Leaf Dimensions of *Q. coccifera* and *A. unedo*.

Species	Length cm	Width cm	Surface cm^2	Size Class
	5.00	2.50	12.50	maximum
Q. cociffera	3.00	2.00	6.00	mean
	0.50	0.30	0.15	minimum
	10.00	4.80	48.00	maximum
A. unedo	6.80	3.40	23.12	mean
	2.00	1.00	2.00	minimum

Cambial activity in stems of *A. unedo* and *Q. coccifera* does not occur at the same time as stem elongation (Arianoutsou-Faraggitaki *et al.* 1984). In the former species, cambial activity occurs between October and December, while in the latter species, it is initiated in June and completed by the end of August. Cambial activity of *Q. coccifera* appears less affected by moisture availability. Avila *et al.* (1975) state that in both Californian and Chilean evergreen species, cambial activity of shrub species occurs throughout the year, with a maximum observed either in spring or in early summer.

REFERENCES

Arianoutsou-Faraggitaki M, Psaras G, Christodoulakis N (1984) The annual rhythm of cambial activity in two woody species of the Greek "maquis". Flora 175:221-229

Aschmann H (1973) Distribution and peculiarity of mediterranean ecosystems. In: Di Castri F, Mooney HA (eds) Mediterranean-Type Ecosystems. Springer-Verlag, Berlin, p 11

Avila G, Aljaro ME, Araya S, Montenegro G, Kummerow J (1975) The seasonal cambium activity of Chilean and Californian shrubs. Amer J Bot 62:473-476

Hoffmann AJ, Hoffmann EA (1976) Growth pattern and seasonal behavior of buds of *Colliguaya odorifera*, a shrub from the Chilean mediterranean vegetation. Can J Bot 54:1767-1774

Kummerow J, Montenegro G, Krause D (1981) Biomass, phenology and growth. In: Miller PC (ed) Resource Use by Chaparral and Matorral. Springer-Verlag, New York, Heidelberg, Berlin, p 69

Montenegro G, Aljaro ME, Kummerow J (1979) Growth dynamics of Chilean matorral shrubs. Bot Gaz 114-119

Mooney HA, Kummerow J (1977) Quantitative phenology. In: Thrower JW, Bradburgy D (eds) Chile-California Mediterranean Scrub Atlas. Dowden, Hutchinson, and Ross, Stroudsburg, Pennsylvania, p 121

Mooney HA, Kummerow J (1981) Phenological development of plants in mediterranean-climate regions. In: Di Castri F, Goodall DW, Specht RL (eds) Mediterranean-Type Shrublands. Elsevier, Amsterdam, Oxford, New York, p 303.

Mooney HA, Parsons DJ, Kummerow J (1974) Plant development in mediterranean climates. In: Lieth H (ed) Phenology and Seasonality Modeling. Springer-Verlag, Amsterdam, Oxford, New York, p 255

Crown structure of *Eucalyptus globulus* Labill. in a coppiced plantation

J.S. Pereira, C.C. Araújo and N. Borralho

Dept. E. Florestal, Instituto Superior de Agronomia, Tapada da Ajuda, P-1399 Lisboa codex, Portugal

INTRODUCTION

Forest biomass production is a function of photosynthetically active radiation intercepted by the foliage (Jarvis and Leverenz 1983). It has been shown that total aboveground biomass production by *Eucalyptus globulus* stands is linearly dependent on the solar radiation intercepted by the canopy (Linder 1985). Canopy characteristics such as leaf area index, vertical distribution of foliage, branch and leaf inclination angles, distribution of leaf azimuthal angles and grouping of foliage affect the interception of radiation and therefore forest primary productivity (Jarvis and Leverenz 1983). These canopy characteristics result from the interaction between two types of processes: 1) the genetically determined "growth plan" that may be reiterated in response to accidental damage incurred in the crown (Tomlinson 1983) and 2) environmental influences on the growth and development of branches and leaves. The objective of this work was to elaborate the crown structure of trees in a *Eucalyptus globulus* coppiced plantation as part of a larger study of production in these plantations.

MATERIALS AND METHODS

The study was conducted at Quinta de Santo Antonio near Cadaval, Portugal (39° 14' lat. N and 9° 2' long. W). The soil originated from quartz sandstone. Climate was of the mediterranean type tempered by maritime influence (average annual air temperature 15° C; average yearly precipitation 745 mm and average sunshine duration 55% of the possible maximum). The experimental stand was planted with a spacing of 2 x 2 m and was in the second coppice rotation (shoots 8 years old). The number of stumps per ha was approximately 1668 with an average number of stems per stump of 1.9 $\pm$ 0.2 and average d.b.h. of 10.1= $\pm$ 0.6 cm.

Trees 1 to 4 (Table 1) were felled and sectioned at 1.3 m aboveground and subsequently in 2 m long logs up to a stem diameter of 2.5 cm (over bark). The part of the tree with a stem diameter less than 2.5 cm was divided into four equal segments (50 to 80 cm per section depending on specific trees). Considering the order of insertion, branches were numbered centrifugally by assigning order 1 to the main stem.

NATO ASI Series, Vol. G15
Plant Response to Stress. Edited by J.D. Tenhunen et al.

Fresh weights of trunk, branches of each order, and leaves associated with each log were measured in the field. Subsamples of each of these biomass components were taken to the laboratory for determination of oven dry weights and leaf areas. Fresh to dry weight and leaf fresh weight to leaf projected area ratios, were used to calculate the total biomass of every aboveground component and the total leaf area. The angle of insertion (as the positive angle formed between its axis and the axis of the next lower order branch) and the angle of the branch with respect to the horizontal of every second branch were measured throughout the crown.

Table 1. Dendrometric characteristics, biomass distribution, and leaf area of the five trees studied.

	Tree number				
	1	2	3	4	5
d. b. h. (cm)	7.3	5.9	5.8	7.9	7.0
Height (m)	12.40	9.40	9.14	12.04	10.10
Height of first living branch (m)	6.90	5.77	4.30	5.38	6.65
Total biomass (kg):					
in trunk	15.27	6.58	7.67	14.65	--
2nd order branches	0.4	0.36	0.57	0.6	--
3rd order branches	0.09	0.11	0.25	0.31	--
4th order branches	0.01	0.01	0.02	0.05	--
Total leaf area (m^2)	5.2	6.2	6.3	8.2	--

All the plants in the stand had only the adult type of leaves. Leaf orientation with respect to the points of the compass (leaf azimuthal angle in degrees from the north), leaf inclination (as the angle of the most vertical aspect of the leaf lamina with respect to the horizontal) and leaf petiole insertion angle (measured as a positive angle formed between its axis and the axis of the branch) were measured with a protractor. These measurements were not made in the field because of difficulties arising from movements caused by the wind. Instead one extra tree was felled (number 5 in Table 1; a sprout of the same stump as, and thus genetically indentical to tree number 4). The cardinal points of the compass were previously marked at the base of the crown and the tree taken to the laboratory with the cut section in water. The cut section of the stem was maintained in water and leaf water potential was determined periodically with a pressure bomb. All measurements were made after the crown was repositioned in its natural orientation and before any significant water stress was detected.

RESULTS

Most of the branch biomass (approximately 70% of total) was found in branches of order 2 (Table 1). Branches of orders 3 and 4 represented ca. 27% and 3% of total branch biomass respectively. The highest foliage "concentration" (leaf area per unit of trunk height) was near the middle of the crown. Fig. 1 shows that cumulative leaf area reached at least 90% of total at approximately 50% of tree height.

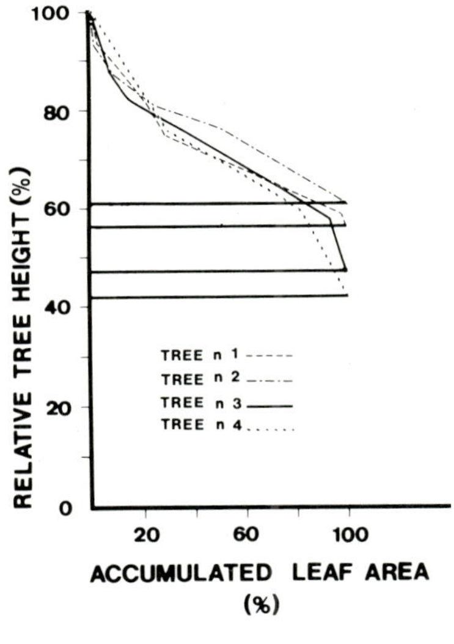

Fig. 1. Accumulated leaf area of *E. globulus* (as percent of total) summed moving from the top of the crown toward the bottom. Crown position is expressed relative to the top of individual trees which is set at relative tree height of 100%

The area specific weight of individual mature leaves expressed per unit of projected area did not vary significantly with position within the crown or among trees. The mean value was 19.1 ± 0.6 mg cm^{-2}. On the average, each biomass unit of branches of order 2 supported approximately one third (range 0.24 to 0.30) of the leaf area (A_L) supported by one biomass unit of branches of order 3

(Table 2). Variability was high as shown by the confidence interval ($p = 0.95$) in Table 2. The leaf area per unit of biomass of branches of order 4 was ca. 5.26 times higher than that supported by branches of order 2.

Table 2. Branch characteristics of *E. globulus* trees (means and confidence intervals at $p=0.95$ are indicated).

Characteristic	Branch order	Mean		Confidence interval
Leaf area supported per unit dry weight of branch ($cm^2\ g^{-1}$)				
	Branches of 2nd order	54	±	15.4
	Branches of 3rd order	188	±	40.7
	Branches of 4th order	298	±	61.6
Branch insertion angle (degrees)	Branches of 2nd order	40.7	±	2.7
	Branches of 3rd order	40.3	±	1.6
	Branches of 4th order	38.7	±	3.2
Branch angle with respect to the horizontal (degrees)				
	Branches of 2nd order	49.3	+	2.7
	Branches of 3rd order	-3.4	+	12.6
	Branches of 4th order	24.7	+	18.8

The xylem cross-sectional area (A_{SW}) at the base of the crown per unit of A_L was on the average $0.02 \pm 0.002\ mm^2\ cm^{-2}$. It was assumed that the entire xylem cross-sectional area at the base of the crown consisted of conducting xylem. Within the crown, A_{SW} at a given internode was linearly related to the cumulative A_L above that internode:

$$A_{SW} = 38.603 + 0.019\ A_L \qquad (1)$$

where A_{SW} is in mm^2 and A_L in cm^2 ($r^2 = 0.903$, $n=17$).

The ratio of A_{SW} to A_L varied little throughout the crown except near the tip of the main stem where the axis itself supported most leaves. Here A_{SW}/A_L was rather high.

The angles of insertion of the branches were rather constant (Table 2) and were not significantly different for branches of different orders or from various levels of the crown. Although the angle with the vertical of branches of order 2 equaled the angle of insertion at the proximal end (inclination angle of 49.3°), these branches were more horizontal in their distal portion (angle with the vertical at the distal end of the branch 14.8 ± 8.1). Branches of order 3 tended to be either horizontal (originating from the lower side of the main branch), vertical (originating from the upper side of the main branch) or with an angle similar to that of the main branch (lateral in origin). Nevertheless, the frequency of upright third order branches was negligible, and therefore, the average branch was nearly horizontal (Table 2).

Petiole insertion angles were quite constant and there were no significant changes with branch order or height of branch origin. The average petiole insertion angle was 43.7° (Table 3). This, however, does not adequately describe the distribution of the inclination angle of the leaf laminas necessary for consideration of solar radiation interception.

Table 3. Leaf inclination, angle of petiole insertion and leaf azimuth of *E. globulus* (means and confidence interval at p=0.95 are indicated).

Average leaf petiole insertion angle (degrees)	43.7	±	1.7
Average leaf inclination angle (degrees)	80.3	±	1.8
Leaf azimuth	No preference		

The distribution of leaf inclination angles is shown in Fig. 2 for the lower, middle and upper parts of the crown. There were no significant differences among crown levels and leaf inclinations tended toward the vertical (75% inclined at angles equal or greater than 80°). Part of the variation in leaf inclination angle can be explained by phylotaxis. As *E. globulus* leaves grow, the petioles become twisted so that most leaves are pendent. It was found that if leaves were grouped according to origin on the branch, those originating from the upper side had angles further from the vertical than the leaves originating laterally or from the lower side of the branch (Table 4). The leaves borne on the upper side of the branch were fewer (Fig. 3) and were smaller than leaves originating laterally or from the lower side of the branch. The former had only 30% of the biomass per leaf found for other groups. As shown in Fig. 4, leaves were not oriented preferentially with respect to azimuth angles.

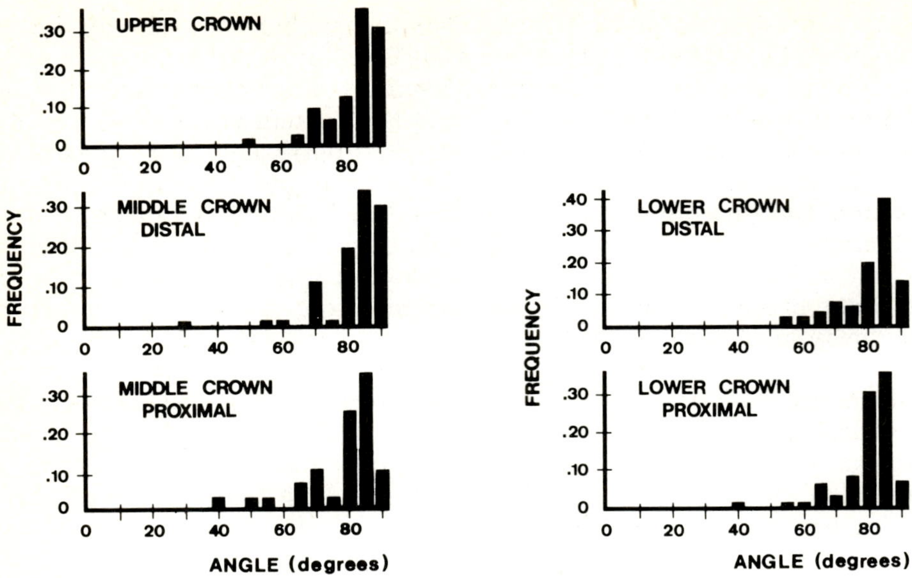

Fig. 2. Fraction of leaves found in different classes of leaf inclination angles for the lower, middle and upper parts of the crown of *E globulus*

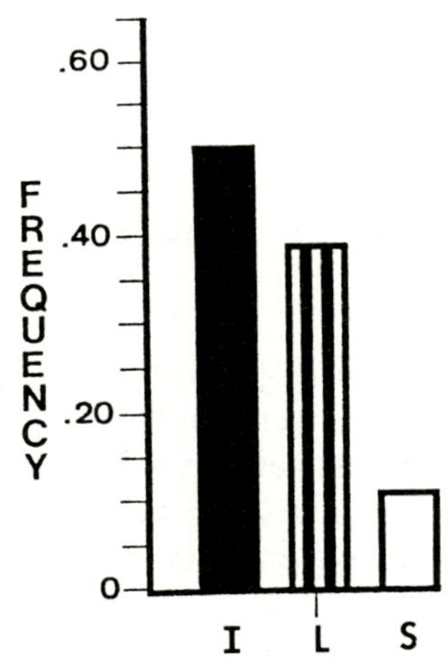

Fig. 3. Fraction of leaves borne on different sides of branches of *E. globulus*: I, inferior; L, lateral; and S, superior

Table 4. Leaf lamina inclination angle according to phylotaxy. The means ± confidence intervals (P=0.95) are indicated.

Leaves originating from the lower side of the branch (degrees)	85.2	±	0.8
Leaves originating laterally on the branch (degrees)	82.3	±	1.0
Leaves originating from the upper side of the branch (degrees)	69.5	±	3.2

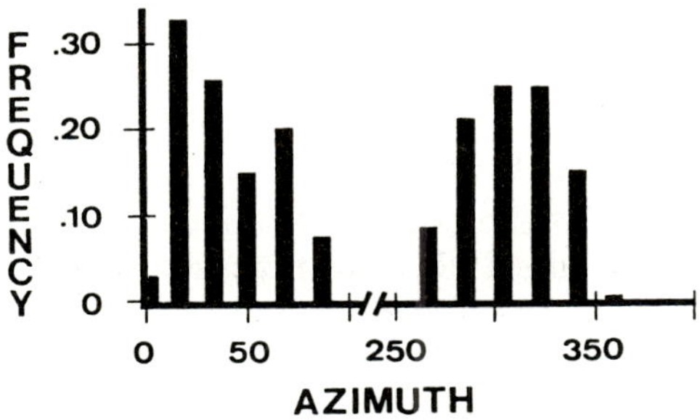

Fig. 4. Fraction of leaves found in different classes of leaf azimuth angle (degrees from north in the first and fourth quadrants) in *E. globulus*

DISCUSSION

Leaf area distribution (A_L) with height of the trees had a maximum near the mid-point of the live crown as found for many coniferous species (Jarvis and Leverenz 1984). The total A_L per tree was relatively small. Assuming a stand density of 3 169 stems ha^{-1}, the leaf area index (L) based on this limited sample was 2.1. In more productive stands of *E. globulus* in Portugal, L may reach 3.5 or even higher (Fabião personal communication). These values lie within the range reported by Linder (1985) for mature eucalypt forests in Australia.

Since A_{SW} and A_L are linearly related (Grier and Waring 1974; Waring *et al.* 1977; Kaufmann and Troendle 1981), it has been suggested that the leaf area of woody plants may be estimated accurately by the ratio of A_{SW} to the A_L above that stem section. The A_{SW}/A_L ratio in *E. globulus*, calculated at the base of the live crown assuming that all the stem xylem was functional (Zimmermann 1983), was similar to that found for other woody angiosperms with diffuse porous wood, such as *Acer macrophyllum* and *A. circinatum* (Waring *et al.* 1977). It was much lower than that found for *Populus tremuloides* (Kaufmann and Troendle 1981). In conifers, the xylem cross-sectional area per unit of leaf biomass ranges from ca. 0.5 to 2.1 $mm^2\ g^{-1}$ (Long *et al.* 1981; Kaufmann and Troendle 1981; Zimmermann 1983) which is slightly higher than was found in this study for *E. globulus* ($0.92 \pm 0.04\ mm^2\ g^{-1}$). This may be the result of lower hydraulic conductivity of coniferous wood in comparison to most broad-leaved angiosperms (Zimmermann 1983). Whitehead and Jarvis (1981) demonstrated that the slope of the relationship betwen A_L and A_{SW} is proportional to hydraulic conductivity of the wood and inversely proportional to the average transpiration rate at the growth site.

The slope of dA_L/dA_{SW} within the crown of *E. globulus* was very similar to dA_L/dA_{SW} (A_{SW} measured at 1.3 m) obtained for 17 trees of *Acer macrophylum* by Waring *et al.* (1977). The A_{SW} that supplied one unit of A_L remained nearly constant throughout most of the crown as found for other species (Long *et al.* 1981; Kaufmann and Troendle 1981). It increased markedly, however, near the tip of the main stem, where the branch system was not yet fully developed and most leaves were attached to the stem axis. This may be of significance for apical dominance since leaves on the leader branch are better supplied by the xylem than those on lateral branches (Zimmermann 1983; Tomlinson 1983).

The number and biomass of branches, as well as the bifurcation ratio, are influenced by environment (Borchert and Slade 1981) but branching angles are under genetic control (Jankiewicz and Stecki 1976; Kramer and Kozlowski 1979). At least for some tree species, branching patterns are such that shading within the canopy is minimized (Honda and Fisher 1978). In *E. globulus*, despite a constant insertion angle, the frequency of vertical 3rd order branches, which would "crowd" the inner canopy is quite low. On the contrary, the occurrence of horizontal 3rd order branches which fill gaps in the outer canopy is high. This type of branching likely has a significant influence on canopy carbon fixation of this shade intolerant species.

In *E. globulus,* leaf insertion and inclination angles did not vary with position in the crown. This suggests, along with the constancy of leaf specific area throughout the crown, that *E. globulus* has a low capacity to produce shade leaves. In comparison, temperate zone deciduous angiosperms, as well as some evergreen mediterranean shrubs have shade leaves which are significantly more horizontal than sun leaves in the outer canopy (Ford and Newbould 1971; McMillen and McClendon 1979; Tenhunen *et al.* 1984, see also Meister *et al.* in this volume). At latitudes of 30 to 40° N the solar elevation angle is high during the growing season. Therefore, nearly vertical leaves tend to maximize the penetration of light through the canopy. In *E. globulus* growing at these latitudes, it may be of adaptative value to have mostly vertical leaves. This would allow higher L and a higher primary productivity than with more horizontal leaves of the same type.

ACKNOWLEDGEMENTS

This research was supported by the Instituto Nacional de Investigação Cientifica of Lisbon, Portugal. We wish to thank Prof. P.G. Jarvis and Dr. M.M. Chaves for their comments on the manuscript. The help of H. Miguel, C. Barrento and C. Alegria with the collection of field data and of I. Leitão who typed the manuscript is also gratefully acknowledged.

REFERENCES

Borchert R, Slad NA (1981) Bifurcation ratios and the adaptative geometry of trees. Bot Gaz 142:394-401

Ford ED, Newbould PJ (1971) The leaf canopy of coppiced deciduous woodland. I. Development and structure. J Ecol 59:843-862

Grier CC, Waring RH (1974) Conifer foliage mass related to sapwood area. Forest Sci 20:205-206

Honda H, Fisher JB (1978) Tree branch angle: maximizing effective leaf area. Science 199:888-889

Jankiewicz LS, Stecki ZJ (1976) Some mechanisms responsible for differences in tree form. In: Cannell MGR and Last FT (eds) Tree Physiology and Yield Improvement. Academic Press, New York, pp 157-172

Jarvis PG, Leverenz JW (1983) Productivity of temperate, deciduous and evergreen forests. In: Lange OL, Nobel PS, Osmond CB, Zieger H (eds) Encyclopedia of Plant Physiology, NS vol 12D: Physiological Plant Ecology IV. Springer, Berlin-Heidelberg-New York-Tokyo, pp 233-280

Kaufmann MR, Troendle CA (1981) The relationship of leaf area and foliage biomass to sapwood conducting area in four subalpine forest tree species. Forest Sci 27:477-482

Kramer PJ, Kozlowski TT (1979) Physiology of Woody Plants. Academic Press, New York

Linder S (1985) Potential and actual production in Australian forest stands. In: Landsberg JJ, Parsons W (eds) Research for Forest Management. CSIRO, Melbourne, pp 11-55

Long, JN, Smith FW, Scott DRM (1981) The role of Douglas-fir stem sapwood and heartwood in the mechanical and physiological support of crowns and development of stem form. Can J Forest Res 11:459-464

McMillen GG, McClendon JH (1979) Leaf angle: an adaptative feature of sun and shade leaves. Bot Gaz 140:437-442

Tenhunen JD, Meister HP, Caldwell MM, Lange OL (1984) Environmental constraints on productivity of the Mediterranean sclerophyll shrub *Quercus coccifera*. Proceedings of INTECOL workshop - Rates of Natural Primary Productivity and Agricultural Production. Options méditerranéennes (Instituto Agronomico Mediterraneo de Zaragoza) 84/1:33-53

Tomlinson (1983) Tree architecture. Amer Sci 71:141-149

Waring RH, Gholz HL, Grier CC, Plummer ML (1977) Evaluating stem conducting tissue as an estimator of leaf area in four woody angiosperms. Can J Bot 55:1474-1477

Whitehead D, Jarvis PJ (1981) Coniferous forests and plantations. In: Kozlowski TT (ed) Water Deficits and Plant Growth. VI. Academic Press, New York, pp 49-152

Zimmermann MH (1983) Xylem structure and the ascent of sap. Springer-Verlag, Berlin-Heidelberg-New York

Vertical structure of *Halimium halimifolium* shrubs in Doñana National Park (SW Spain)

M.C. Diaz Barradas and F. Garcia Novo

Departamento de Ecología, Facultad de Biología, Universidad de Sevilla, Apartado 1.095, 41080 Sevilla, Spain

INTRODUCTION

In the areas of the world with a mediterranean climate, similar vegetation types occur that are characterized by sclerophyllous evergreen shrubs with a variable understorey of annual herbs (Naveh 1967; Specht 1969; Mooney and Parsons 1973; Parsons 1975; Orshan 1983; Miller 1981). In the Iberian Peninsula this mediterranean vegetation is widespread on degraded forest areas, abandoned fields and pasturelands, and on mountain ridges. In Doñana National Park in southwestern Spain the clearance of cork oak (*Quercus suber*) forests, growing on sandy soils, led to the development of dense understorey vegetation. Overgrazing and wild fires resulted in soil degradation which favoured the expansion of oligotrophic and pyrophitic shrubs. These now appear as major components of the Park vegetation (Garcia Novo 1977, 1979).

The structure of plant canopies determines the way in which many environmental factors influence canopy function. The architecture of the plants is the result of evolutionary adaptation to environmental conditions during the geological past (Monsi, Uchijima and Oikawa 1973; Smith 1984). The interception of radiation, convective exchange of heat, transport and uptake of carbon dioxide, loss of water vapour and other processes are affected by canopy size and the geometrical distribution of plant organs (Monsi *et al.* 1973; Miller *et al.* 1981). In this paper the vertical distribution of organs in *Halimium halimifolium* is described.

METHODS

The study site is located at the "Corral Largo" in Doñana National Park. This is the largest slack in the mobile dune system. Slack vegetation is dominated by umbrella pine forest (*Pinus pinea*) with an understorey formed by *Ulex parviflorus* and *Stauracanthus genistoides*. Bare sand surfaces have primarily been colonized by a perennial vegetation consisting of *Scrophularia frutescens, Artemisia campestris* ssp. *maritima* and *Halimium halimifolium* (Ramirez and Torres 1977; Rivas-Martinez *et al.* 1980). This latter shrub of the family Cistaceae is widespread in the secondary scrub of the Park.

NATO ASI Series, Vol. G15
Plant Response to Stress. Edited by J.D. Tenhunen et al.

The vertical distribution of organs (leaves, branches, stems, flowers and fruits), of leaf area, and of chlorophyll content were determined in 55 *Halimium halimifolium* plants by harvest and frequency methods. In the field, each plant was harvested in 5" (approximately 12.5 cm) vertical sections. In the laboratory, each layer was separated into leaves and stems of the present year, leaves and stems of the previous year, other stems according to diameter class (less than 3, 3-7, 7-15 and over 15 mm in diameter), dead leaves and stems, and inflorescences. The samples were dried at 80° C for 48 hours, to determine the dry weight.

The surface area to the nearest square mm and specific leaf weight of 10 leaves randomly chosen from each 5 inch level were measured and used to estimate the total leaf area for the level based on total dry weight of leaves. Three samples of leaves were taken from each level and fresh weight was determined. One sample was dried (80° C, 48 hours) to determine the dry weight/fresh weight ratio. The other two samples were ground and extracted for 48 hours with cold acetone (80%) in darkness. Leaf pigments in the extract were estimated by spectrophotometry, using the formulas developed by Arnon (1949), Holden (1965), and Strain and Svec (1966).

For each plant, frequency distributions of organs were determined using a vertical prism of 20 x 50 cm base and with a vertical grid in one of its faces and a vertical squared panel in the opposite face. The occurrence in 22 columns of every organ of the plant within the prism are recorded with the aid of a small parallelepiped of 12.5 x 12.5 x 200 mm. The parallelepiped is placed horizontally and is perpendicular to both grid and panel surfaces. The presence or absence of plant organs is recorded for every parallelepiped and frequencies are computed over columns and rows (layers in the vegetation 2.5 cm thick as indicated in figures). The number of rows depends on plant height; the number of columns is kept to a maximum of 22 that always correspond to the center of the plant. This method allows a more refined analysis of structure than is possible with traditional harvesting methods. Also, since the method is non-destructive, it allows repeated sampling of the same plant over time. In this case, both the grid and panel are placed in a fixed position using a permanent base.

RESULTS AND DISCUSSION

The studied area at "Corral Largo" is a pioneer zone of primary colonization, where stresses are severe and the plants grow as isolated individuals. Depending on their size, it is possible to distinguish two plant types in *Halimium halimifolium.* Small individuals (< 70 cm height) exhibit a unimodal distribution of photosynthetic biomass (leaves), and the photosynthetic/live wood biomass ratio lies between 0.5 and 1.5. Large individuals (> 100 cm height) exhibit a bimodal distribution of leaf biomass. These are old plants that were strongly affected by severe drought in the years 1979 through 1983, and they have a very low photosynthetic/live wood ratio, with values below 0.1.

In Figs. 1 and 2, vertical distribution profiles for photosynthetic and non-photosynthetic biomass are depicted for "small" and "large" type plants. Stems show a well defined pattern of decreasing biomass from base to apex for all plants. Leaves show a unimodal distribution for "small" plants and a bimodel distribution for "large" plants, with maxima at 20-30 cm from the soil surface and at 15-25 cm below the top of the plant. The vertical profiles of accumulated biomass are considerably different between "small" and "large" plants. In the former, the photosynthetic biomass predominates, especially in the upper levels. For larger plants photosynthetic biomass is a much smaller component and only predominates in the upper levels.

Tenhunen *et al.* (1984; see also Meister et al. this volume) studied the vertical distribution of aboveground biomass of *Quercus coccifera* shrub near Lisbon (Portugal), about 500 km NW of Doñana. The authors found a well defined photosynthetic layer in the top of the canopy. This pattern is described as the "monolayer" model by Horn (1971). In *Halimium halimifolium*, at the Corral Largo, the distribution of leaf biomass is more similar to that of Horn's multilayer model. The general shrub architecture of *Halimium* is similar to species studied in Echo Valley of California and Fundo Santa Laura of Chile (Giliberto, Mooney, Kummerow 1977; Kummerow, Montenegro and Krause 1981).

The average values of biomass and photosynthetic/non-photosynthetic biomass ratio for 5 plants of each type are shown in Table 1. The biomass ratio of photosynthetic/non-photosynthetic material has been computed averaging the ratio for each plant. Dead organs have not been included. Leaf and new stem biomass is very similar in both plant types of *H. halimifolium* and independent of the amount of live wood dry weight and plant size. The average value of the photosynthetic/non-photosynthetic ratio, thus varies from 0.96 in small plants to 0.06 in large plants. The values for this ratio found in *H. halimifolium* is within the range quoted for other mediterranean areas of the world. Basanta (1982) at "Coto Nacional La Pata del Caballo" Sierra Morena (SW Spain) found values between 0.07-0.28 depending on the topography. Gray (1982) reported values of 0.09 in the *Ceanothus* chaparral of California to 0.18 in areas dominated by *Salvia leucophylla*. The value of this ratio was 2.0 at the top of ridges in the stabilized sands at Doñana during the first stages (1-3 years) of secondary succession after fire. It decreased to a constant value of 0.13 in subsequent years (Merino and Martin 1981; Martin 1982).

The average leaf area index (LAI) in *H. halimifolium* is 1.73, with no differences between small and large plants (average LAI for small plants 1.77; large plants 1.65). The vertical distribution of LAI is similar in small and large individuals as shown in Fig. 3. Kummerow *et al.* (1981) found similar values 1.0-4.4 in the matorral (Chile) and 1.0-3.6 in the chaparral (California). Gray (1982) found values of 1.3 in the coastal sage scrub (California). Studying the succession after fire in *Ceanothus megacarpus* chaparral, Schlesinger and Gill

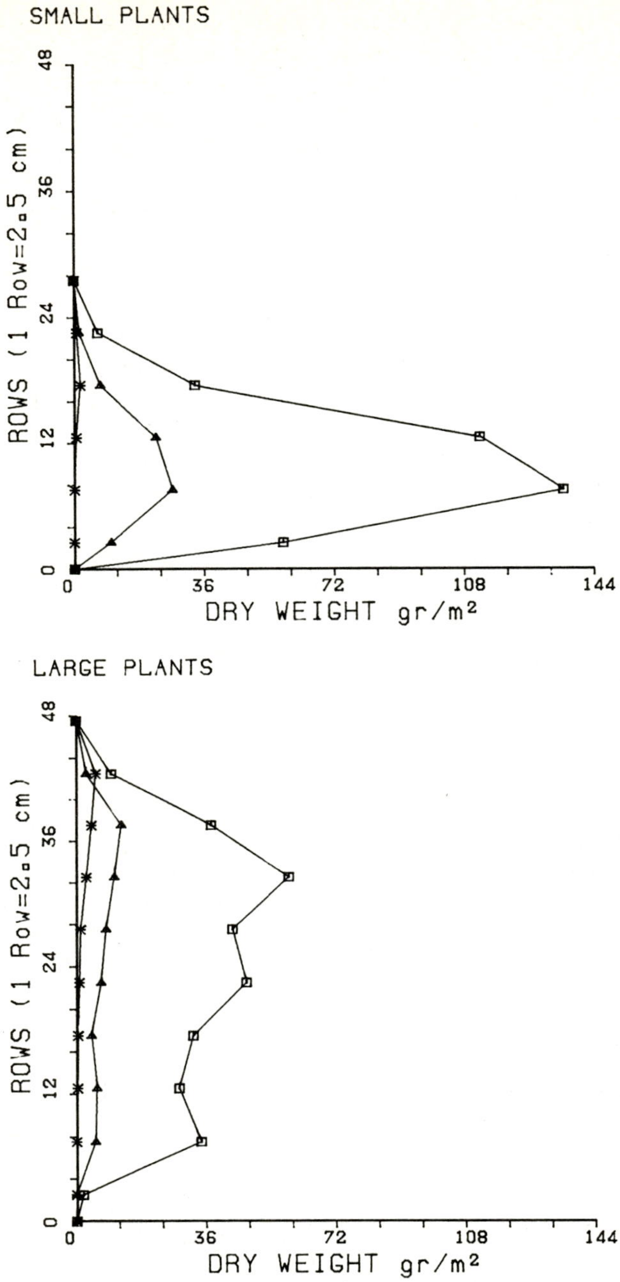

Fig. 1. Vertical distribution of photosynthetic organs (leaves - squares, new stems - triangles, and flowers - asterisks) in shrubs of *Halimium halimifolium* as described in the text. Upper figure is for "small" shrubs and lower figure for "large" shrubs

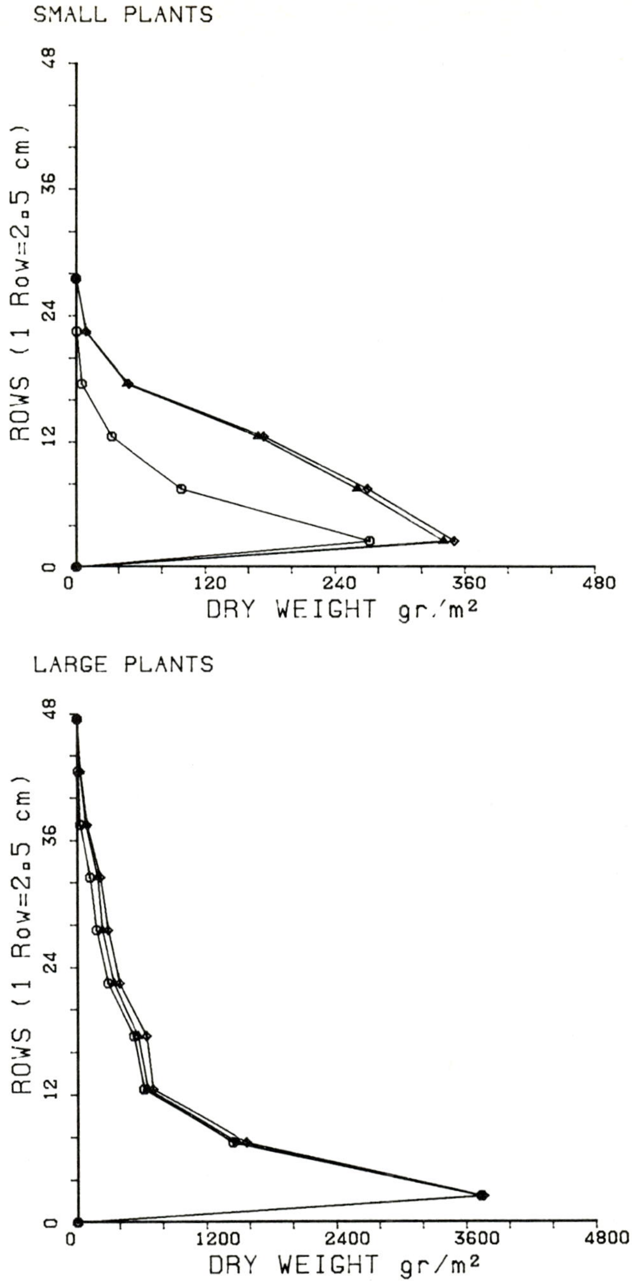

Fig. 2. Vertical distribution of woody biomass (circles), woody + photosynthetic biomass (triangles), and total biomass (including standing dead - diamonds) in shrubs of *Halimium halimifolium* as described in the text. Upper figure is for "small" shrubs and lower figure for "large" shrubs

Table 1. Average values of biomass (x - dry weight per unit projected area; g/m^2; s - standard deviation) and average ratio of photosynthetic to non-photosynthetic biomass in small and large plants of *Halimium halimifolium.*

	Small plants		Large plants	
	x	s	x	s
Leaves	313	(98)	293	(65)
New stems	68	(20)	56	(31)
Live wood	428	(180)	6976	(2648)
Dead organs	22	(18)	335	(144)
Photosynthetic / Non-photosynth.	0.96	(0.36)	0.06	(0.03)

(1980) found that this index increases from 0.8 (after 5 years) to 1.6 (after 21 years). The LAI of the vegetation in mediterranean regions commonly is in the 1-2 range, increasing to 3-4 as precipitation increases or the site water balance becomes more favourable (Loissaint 1973; Miller 1981, 1983; Cody and Mooney 1978 - see also Rambal in this volume).

The leaf specific weight (mg d.w./cm^2 leaf area) was computed for each level in all plants. The general values of leaf specific weight in *H. halimifolium* are similar to Californian and Chilean species (Kummerow *et al.* 1981; Gray 1982). In small plants, the value of this characteristic (15-20 mg/cm^2) did not change at different positions along the vertical profile. In larger plants, it decreased from higher to lower levels, and it was possible to distinguish two types of leaves: sun leaves with a leaf specific weight between 18 and 20 mg/cm^2 and shade leaves between 10 and 12 mg/cm^2. These differences have been repeatedly mentioned by other authors (Ford and Newbould 1971; Horn 1971; Dale 1983). Tenhunen *et al.* (1984) show that the specific leaf weight in *Quercus coccifera* decreases with depth in the canopy from approximately 17 mg/cm^2 to 12 mg/cm^2 at 30 cm depth where the innermost leaves occur. This decrease has in some cases been paralleled by a proportional decrease in leaf photosynthetic capacity measured at optimal temperature, light and CO_2 saturation (Tenhunen *et al.* 1984).

The distribution of chlorophyll within plant canopies of mediterranean forests has been described (Tuba and Nylas 1980; Tuba 1981; Ida 1981; Ford and Newbould 1971; Gracia 1981, 1983; Escarre *et al.* 1984), but little information has been summarized for shrub canopies. The leaf concentration of chlorophyll in *H. halimifolium* during the period of our study was on the average 1.76 mg/g dry weight for small plants and 1.62 mg/g dry weight for large ones. Values

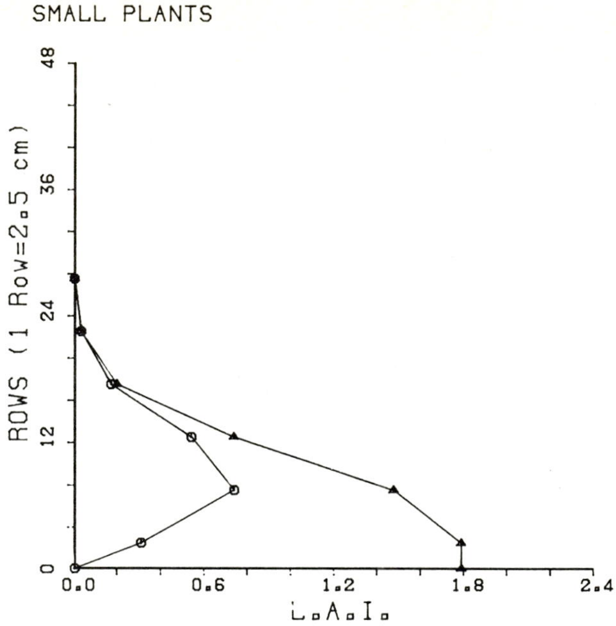

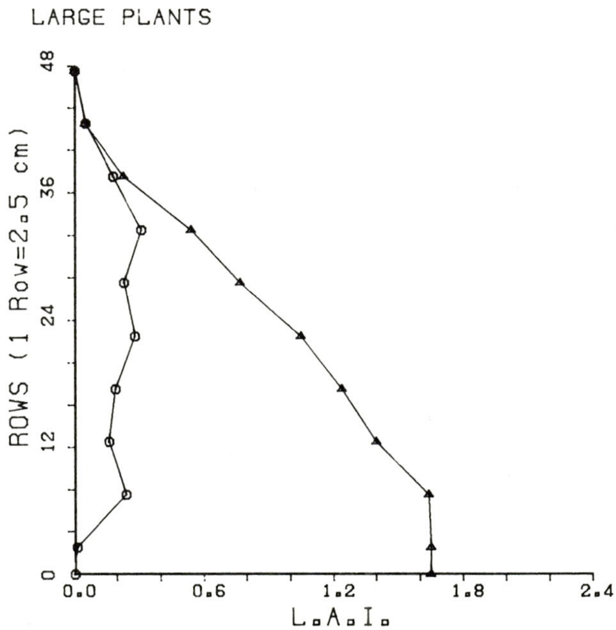

Fig. 3. Vertical distribution of leaf area index (LAI) in shrubs of *Halimium halimifolium* as described in the text. Upper figure is for "small" shrubs and lower figure for "large" shrubs. LAI of each level (circles) as well as the accumulated LAI from the top to bottom of shrubs (triangles) is shown

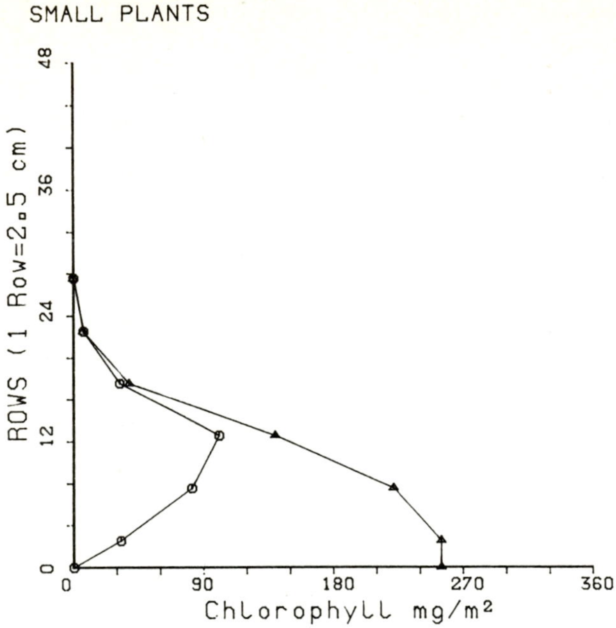

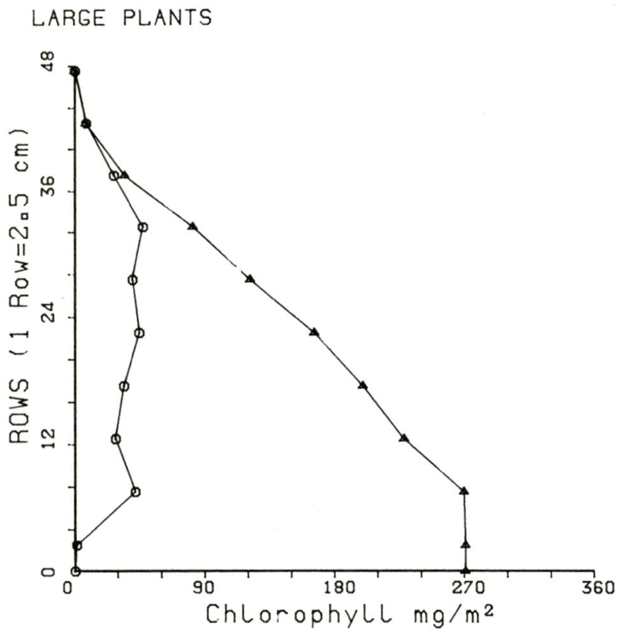

Fig. 4. Vertical distribution of chlorophyll in shrubs of *Halimium halimifolium* as described in the text. Upper figure is for "small" shrubs and lower figure for "large" shrubs. Chlorophyll content of each level (circles) as well as the accumulated chlorophyll from the top to bottom of shrubs (triangles) is shown

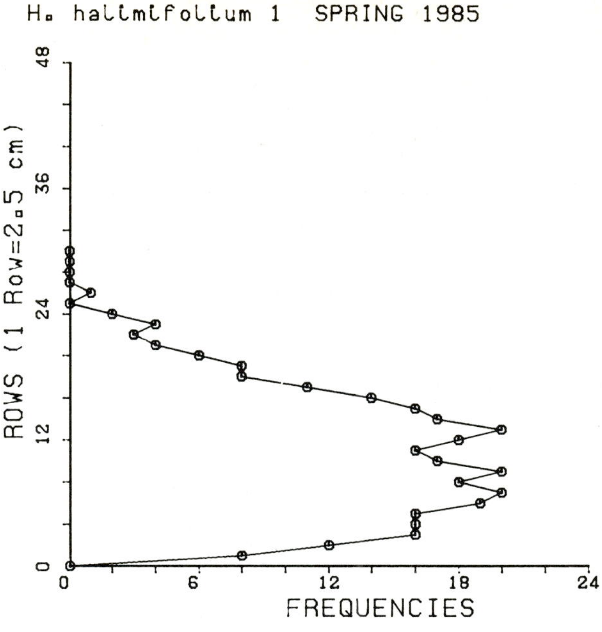

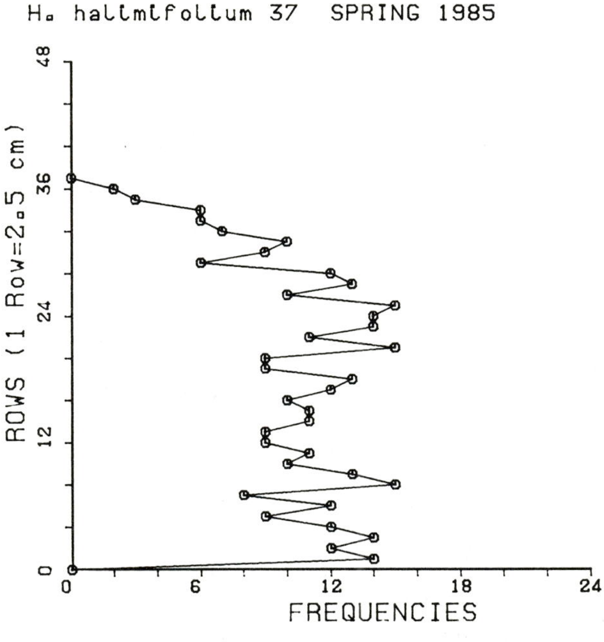

Fig. 5. Vertical profile of frequency of occurrence of leaves in foliage layers of shrubs of *Halimium halimifolium* as described in the text. Upper figure is for "small" shrubs and lower figure for "large" shrubs

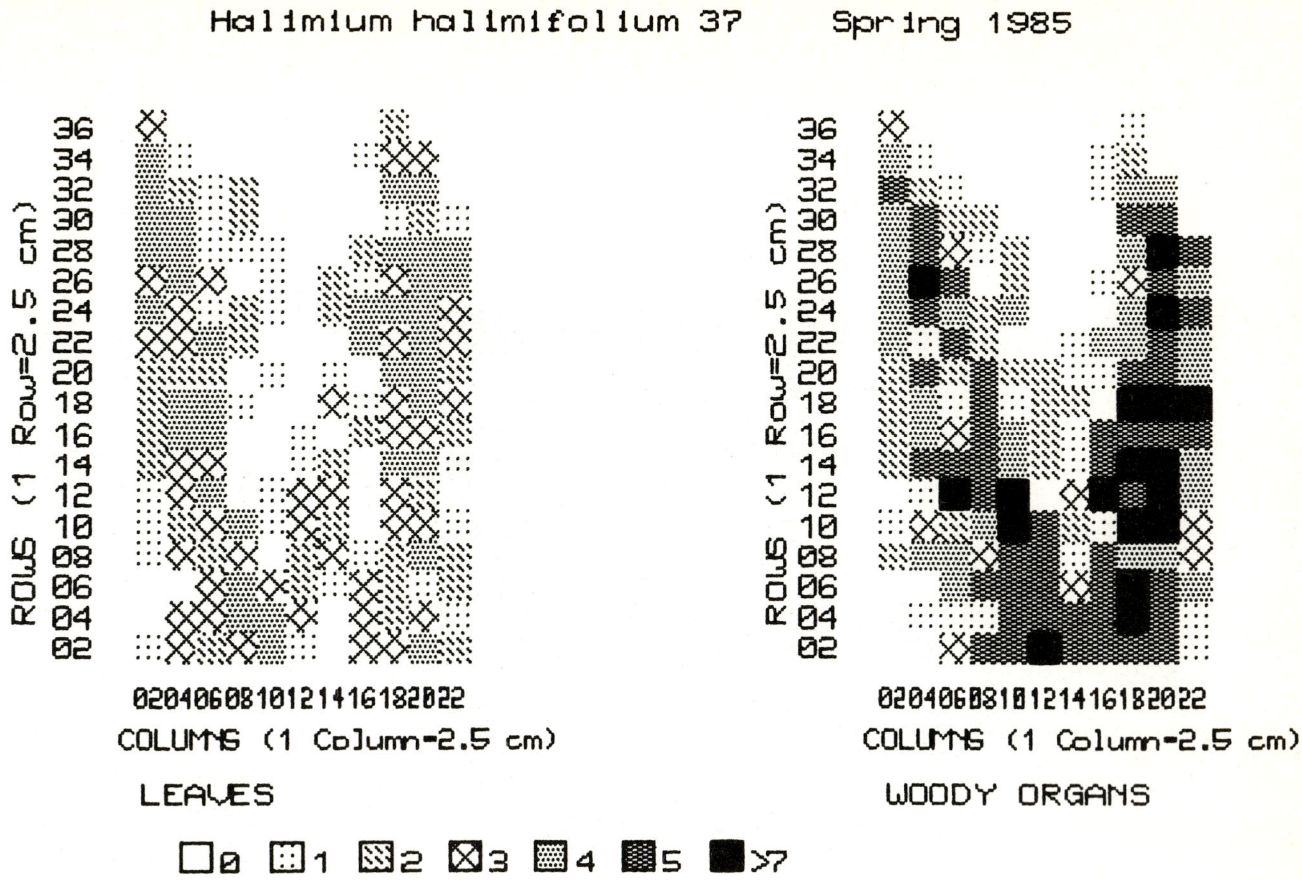

Fig. 6. Frequency distribution of leaves and woody twigs within the canopy of *Halimium halimifolium* determined according to the methods outlined in the text during the spring of 1985

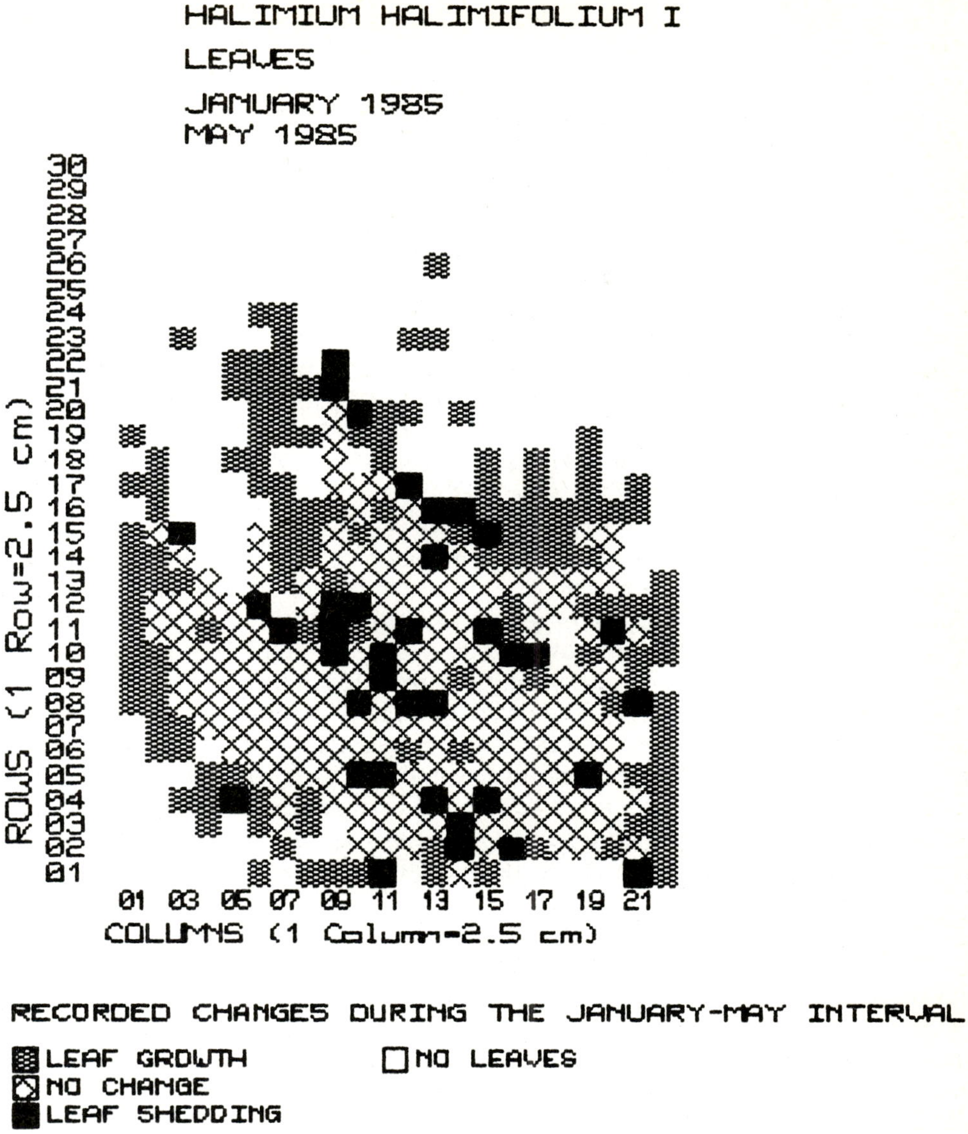

Fig. 7. Distribution of leaf turnover events identified with the methods described in the text within the canopy of *Halimium halimifolium* during the interval January to May 1985

ranged between 0.97 and 2.3 mg/g dry weight. Estimates of the total chlorophyll weight found in different canopy levels are presented in Fig. 4. The vertical distribution of chlorophyll is similar to that of leaf biomass and LAI. The average values of total pigment density are 292 mg/m^2 in small plants and 270 mg/m^2 in larger ones.

In recent years, a considerable effort has been made by investigators to estimate the vertical structure of plant canopies by non-destructive methods. The first who described a point-quadrat sampling method was Warren Wilson (1958, 1959, 1965). Other attempts to apply this approach have been described by Ford and Newbould (1971), MacArthur and Horn (1969) and Aber (1979 a,b). In the mediterranean area, several such studies have been undertaken by Debussche (1978); Debussche, Escarre and Lepart (1980); Houssard, Escarre and Roman (1980); Houssard and Escarre (1981). The work of Meister *et al.* reported in this volume, similarly utilizes the Warren Wilson approach for estimating certain aspects of canopy structure.

The application of our frequency method in canopies of *H. halimifolium* appears to have some advantages over other non-destructive methods that have been tried, because a better definition of organ distribution within the plant canopy is achieved. Detailed profiles are obtained by averaging organ frequencies by rows (Fig. 5). These frequencies are strongly correlated with the biomass of plant organs found in various levels. The layers of a plant canopy are not homogenous. Roberts and Miller (1977) studied the three dimensional heterogeneity of the LAI and light interception in isolated individuals of two mediterranean species (*Lithraea caustica* and *Colliguaya odorifera*) using a destructive method. With the frequency method reported here, we were able to analyse the two dimensional heterogeneity of canopy structure of the plant non-destructively (Fig. 6).

This analysis may be performed repeatedly over time, thus allowing one to follow dynamic changes. This may prove to be quite interesting in mediterranean regions due to the strong seasonal shift in environmental stresses (Aschmann 1973; Miller and Hajek 1981; Orshan 1983). Preliminary results with *H. halimifolium* show that the addition of new leaves during winter and spring takes place in the periphery of the canopy (Fig. 7) in both large and small plants. This increase in peripheral leaves is coupled to an inner thinning process so that old leaves within the canopy are shed but not replaced. Thin photosynthetic twigs grow with the new leaves in the periphery, but in the inner part of the canopy they die off after losing their leaves. This repeating process results in the two-fold increase in leaf biomass, leaf surface and pigment weight of isolated *H. halimifolium* shrubs as they increase in size as compared to a ten-fold increase in plant biomass. The effects of light, and water availability on dynamic changes in canopy structure of *H. halimifolium* plants are currently being investigated.

REFERENCES

Aber JD (1979a) A method for estimating foliage-height profiles in broad leaved forests. J Ecol 67:35-40

Aber JD (1979b) Foliage height profiles and succession in northern hardwood forests. Ecology 60(1):18-23

Arnon DI (1949) Copper enzymes in isolated chloroplasts. Polyphenoloxidase in *Beta vulgaris*. Plant Physiol 24:1-15

Aschmann H (1973) Distribution and peculiarity of mediterranean ecosystems. In: di Castri F, Mooney HA (eds) Mediterranean Type Ecosystem: Origin and Structure. Springer-Verlag, Berlin-Heidelberg-New York, pp 11-19

Basanta A (1982) Vegetación seral de Sierra Moreno. Tésis Doctoral, Universidad de Sevilla

Cody ML, Mooney HA (1978) Convergence and dissimilarities of mediterranean-climate ecosystems. Ann Rev Ecol Syst 9:263-321

Dale JE (1983) El crecimiento de las hojas. Omega, Barcelona

Debussche M (1978) Etude de la dynamique de la végétation sur le versant nord-ouest du Mont Aigoual. Thèse Doct (Académie de Montpellier, Université des Sciences et Thechniques du Languedoc), Montpellier

Debussche M, Escarre J, Lepart J (1980) Changes in mediterranean shrub communities with *Cytisus purgans* and *Genista scorpius*. Vegetatio 43:73-82

Escarre A, Gracia C, Roda F, Terradas J (1984) Ecología del bosque esclerófilo mediterráneo. Investigación y Ciencia 95:68-78

Ford ED, Newbould PJ (1971) The leaf canopy of a deciduous woodland. J Ecol 59:843-862

Garcia Novo F (1977) The effects of fire on the vegetation of Doñana National Park. In: Mooney HA, Conrad CE (eds) Proceedings of the Symposium on the Environmental Consequences of Fire and Fuel Management in Mediterranean Ecosystems. USDA Forest Service. General Technical Report WO-3:318-325

Garcia Novo F (1979) The ecology of dune vegetation of Doñana National Park. In: Jeffries RL, Davy AJ (eds) Ecological Processes in Coastal Environments. Blackwell's, Oxford, pp 571-592

Giliberto J, Mooney HA, Kummerow J (1977) Shrub structure analysis. In: Thrower NJW, Bradbury DE (eds) Chile-California Mediterranean Scrub Atlas. Dowden, Hutchinson and Ross, Stroudsburg, Pennsylvania, pp 144-147

Gracia CA (1981) El modelo de lago como técnica de interpretación forestal. Actas I. Congreso Español de Limnología. Barcelona:125-129

Gracia CA (1983) La clorofila en los encinares del Montseny. Tésis Doctoral, Universidad de Barcelona

Gray JT (1982) Community structure and productivity in *Ceànothus* chaparral and coastal sage scrub of southern California. Ecological Monographs 52(4):415-435

Holden M (1965) Chlorophylls. In: Goodwin TW (ed) Chemistry and Biochemistry of Plant Pigments. Academic Press, London, New York, pp 461-488

Horn H (1971) The adaptive geometry of trees. Princeton Univ Press, Princeton, New Jersey

Houssard C, Escarre J, Romane F (1980) Development of species diversity in some mediterranean plant communities. Vetatio 43(1-2):59-72

Houssard C, Escarre J (1981) Mesures de la quantité relative de feuillage strates dans deux taillis de chêne pubescent. Ann Sci Forest 38(4):449-468

Kummerow J, Montenegro G, Krause D (1981) Biomass, phenology and growth. In: Miller PC (ed) Resource use by Chaparral and Matorral. Springer-Verlag, Berlin-Heidelberg-New York, pp 69-96

Lossaint P (1973) Soil-vegetation relationships in mediterranean ecosystems of southern France. In: di Castri F, Mooney HA (eds) Mediterranean Type Ecosystems: Origin and Structure. Springer-Verlag, Berlin-Heidelberg-New York 199-210

MacArthur RH, Horn HS (1969) Foliage profile by vertical measurements. Ecology 50:802-804

Martin Vicente A (1982) Sucesión tras el fuego del matorral de las arenas estabilizadas de la Reserva Biológica de Doñana. Tésis Doctoral, Universidad de Sevilla

Merino J, Martin Vicente A (1981) Biomass, productivity and succession in the scrub of the Doñana Biological Reserve in south-west Spain. In: Mooney HA and Margaris NS (eds) Components of Productivity of Mediterranean-climate Regions. Basic and Applied Aspects. Dr. W Junk, The Hague, pp 197-203

Miller PC (1981) Similarities and limitations of resource utilization in mediterranean type ecosystems. In: Miller PC (ed) Resource use by Chaparral and Matorral. Springer-Verlag, Berling-Heidelberg-New York, pp 369-407

Miller PC (1983) Canopy structure of mediterranean-type shrubs in relation to heat and moisture. In: Kruger FJ, Mitchell DT, Jarvis JUM (eds) Mediterranean Type Ecosystems- The Role of Nutrients. Springer-Verlag, Berlin-Heidelberg-New York, pp 134-166

Miller PC and Hajek E (1981) Resource availability and environmental characteristic of mediterranean type ecosystems. In: Miller PC (ed) Resource use by Chaparral and Mattoral. Springer-Verlag, Berlin-Heidelberg-New York, pp 17-41

Miller PC, Hajek E, Poole DK, Roberts SW (1981) Microclimate and energy exchange. In: Miller PC (ed) Resource use by Chaparral and Matorral. Springer-Verlag, Berlin-Heidelberg-New York, pp 97-121

Monsi M, Uchijima Z, Oikawa T (1973) Structure of foliage canopies and photosynthesis. Ann Rev of Ecol System 4:301-327

Mooney HA, Parsons DJ (1973) Structure and function in the California chaparral- an example from San Dimas. In: di Castri F, Mooney HA (eds) Mediterranean Type Ecosystems, Origin and Structure. Springer-Verlag, Berlin-Heidelberg-New York, pp 83-112

Naveh Z (1967) Mediterranean ecosystems and vegetation types in California and Israel. Ecology 48:445-459

Orshan G (1983) Approaches to the definition of mediterranean growth forms. In: Kruger FJ, Mitchell DT, Jarvis JUM (eds) Mediterranean Type

Ecosystems- The Role of Nutrients. Springer-Verlag, Berlin-Heidelberg-New York, pp 86-102

Parsons DJ (1976) Vegetation structure in the mediterranean scrub communities of California and Chile. J Ecol 64:435-447

Ramirez L, Torres A (1977) Tipología y dinámica de los complejos ambientales del sistema de dunas móviles en la Reserva Biológica de Doñana (huelva). Boletín de la Estación Central de Ecología (VI) No 11:3-11

Rivas-Martinez S, Costa M, Castroviejo S, Valdes E (1980) Vegetación de Doñana (Huelva, España). Lazaroa, Vol 2:189 p

Roberts SW, Miller PC (1977) Interception of solar radiation as affected by canopy organization of two mediterranean shrubs. Oecol Plant 12:273-290

Smith BH (1984) The optimal design of a herbaceous body. The American Naturalist 123, No 2:197-211

Specht RL (1969) A comparison of the sclerophyllous vegetation characteristic of mediterranean type climates in France, California and southern Australia. I. Structure, morphology and Succession. Aust J Bot 17:277-292

Strain HH, Svec WA (1966) Extraction, separation, estimation and isolation of the chlorophylls. Vernon LP, Seely GR (eds) The Chlorophylls. Academic Press, London-New York, pp 21-66

Tenhunen JD, Meister HP, Caldwell MM, Lange OL (1984) Environmental constraints on productivity of the Mediterranean sclerophyll shrub *Quercus coccifera*. Proceedings of INTECOL workshop - Rates of Natural Primary Productivity and Agricultural Production. Options méditerranéennes (Instituto Agronomico Mediterraneo de Zaragoza) 84/1:33-53

Tuba Z (1981) The changes of the photosynthetic pigment systems of two paprika (red pepper) varieties from the fully developed vegetative stage to the ripening of the fruit. Bot Közlem 68. Kötet 1-2:123-131

Tuba Z, Nyilas J (1980) Correlation between the vertical light and photosynthetical pigment structure of a multilayer community (turkey-oak forest). Acta Botanica Academiae Scientiarum Hungaricae 26(3-4):447-454

Warren Wilson J (1958) Analysis of the spatial distribution of foliage by two dimensional point quadrats. New Phytologist 59:92-101

Warren Wilson J (1959) Analysis of the distribution of foliage in grassland. In: Ivins JD (ed) The Measurement of Grassland Productivity. Academic Press, London-New York, pp 51-61

Warren Wilson J (1965) Stand structure and light penetration. I. Analysis by point-quadrats. Journal of Applied Ecology 2:383-390

Structural analysis of sclerophylly in eleven evergreen phanerophytes in Greece

N.S. Christodoulakis and K.A. Mitrakos

Institute of General Botany, University of Athens, Athens 157 01, Greece

INTRODUCTION

In the past (Grisebach 1872; Schimper 1898; Turrill 1929) as well as recently (Johnson 1973; Mooney *et al.* 1974), an evolutionary and ecological importance has been attributed to the leaf structure of the mediterranean evergreen sclerophyllous phanerophytes. Nevertheless, anatomical data on these leaves are sporadic and rather superficial (Kummerow 1973).

During the last five years we studied the leaf structure in 11 of approximately 15 evergreen sclerophyllous shrub species growing in Greece, on a macro-, micro- and submicroscopical level (Christodoulakis 1984). Some of our data are presented in this paper.

MATERIALS AND METHODS

Leaves were collected from naturally established plants in Malakasa, a site located 37 km north of Athens at an elevation of 400 meters and covered with a typical east mediterranean maquis vegetation. (See Diamantoglou and Mitrakos 1981; Diamantoglou and Kull 1984). Eight evergreen sclerophyllous phanerophytes are present in Malakasa (*Arbutus andrachne* L., *Arbutus unedo* L., *Myrtus communis* L., *Olea europaea* L. ssp. *oleaster* (Hoffm. and Link), *Phillyrea media* L., *Pistacia lentiscus* L., *Quercus coccifera* L. and *Quercus ilex* L.). Three further sclerophyllous species were studied (*Ceratonia siliqua* L., *Laurus nobilis* L. and *Nerium oleander* L.) but their leaves were collected from plants growing in Kesariani, 5 km east of Athens at an altitude of 400 m. The first of these had been planted and the other two were growing wild.

Three-hundred mature leaves from three plants were used for measuring dimensions and obtaining leaf discs for measuring dry weight. Thirty leaves from the same individuals were immediately fixed in the field in 3% phosphate buffered glutaraldehyde at 0° C for 2 hours (Sabatini *et al.* 1963). These were then placed in phosphate buffered 1% osmium tetroxide at 4° C for 2 hours (Ledbetter and Porter 1963). Dehydration and embedding in plastic (Durcupan ACM, Fluka) followed. Semi-thin sections were obtained on an LKB Ultrotome. A Zeiss light microscope was used for the observations.

NATO ASI Series, Vol. G 15
Plant Response to Stress. Edited by J.D. Tenhunen et al.

RESULTS AND DISCUSSION

Among the species studied, all but two (*C. siliqua* and *P. lentiscus*) bear simple leaves. Mean values for the length, width, surface and thickness, are given in Table 1, together with the corresponding scale drawings of a "mean leaf." The outline of a "mean leaf" was established as the average outline obtained with ten leaves having length and width identical to the calculated mean values.

Table 1. "Mean leaves" and their dimensions for the eleven species.

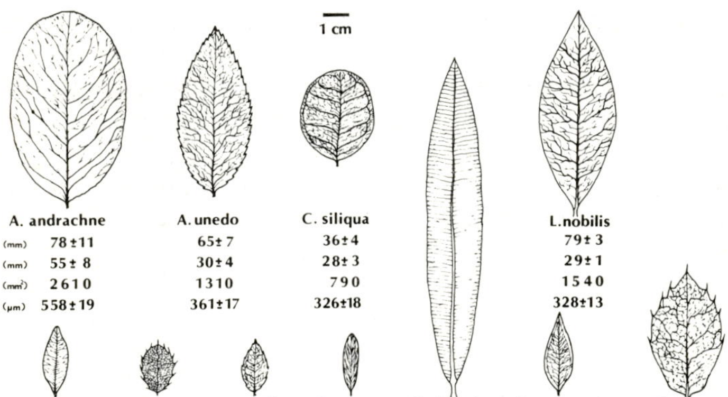

	A. andrachne	A. unedo	C. siliqua	L. nobilis
LENGTH (mm)	78±11	65±7	36±4	79±3
WIDTH (mm)	55±8	30±4	28±3	29±1
SURFACE (mm^2)	2610	1310	790	1540
THICKNESS (μm)	558±19	361±17	326±18	328±13

	P. lentiscus	Q. coccifera	P. media	O. europaea	N. oleander	M. communis	Q. ilex
LENGTH (mm)	28±3	21±3	21±1	24±5	131±5	32±3	51±7
WIDTH (mm)	11±1	14±1	11±1	6±05	22±1	12±2	28±4
SURFACE (mm^2)	200	190	150	110	2075	250	994
THICKNESS (μm)	430±32	304±12	302±11	470±13	465±31	312±14	240±27

A measure of convergence or divergence for the four leaf size parameters among the eleven species is given by the ratios of the highest to the lowest values measured, being for length 1:6.2, width 1:9.1, surface 1:23.7 and thickness 1:2.3.

On the basis of leaf surface (Raunkiaer 1916; Orshan 1982) the species studied are grouped as follows:

nanophylls (26-225 mm^2): *P. lentiscus, Q. coccifera, P. media, O. europaea*
nano-microphylls (226-1225 mm^2): *C. siliqua, M. communis, Q. ilex*
microphylls (1226-2025 mm^2): *A. unedo, L. nobilis*
micro-mesophylls (2026-5625 mm^2): *A. andrachne, N. oleander*

Obviously the "small leaf strategy" (Mooney and Dunn 1970) varies considerably within mediterranean evergreen sclerophylls. The length to width ratio varies from 1.3 for the rather rounded leaves of *C. siliqua*, to 5.9 for the elongated leaves of *N. oleander*.

The microscopical analysis of the leaf histological components revealed interesting variations as well (Table 2). Concerning epidermal tissue (Table 2) we can see that the thickness of the upper epidermis varies form 19 to 66 µm and that of the lower from 10 to 46 µm. The ratios of upper to lower epidermis vary from 1.1 (*Q. coccifera*) to 2.0 (*Q. ilex*). The thickness sum of both upper and lower epidermis is from 30 µm (*Q. coccifera*) to 112 µm (*N. oleander*). The percentage epidermis is of leaf thickness varies from 7.4% (*O. europaea*) to 24.0% (*N. oleander*). The ratio epidermal cell thickness to epidermal outer cell-wall thickness is negatively correlated (x=0.75) in the species studied to the ratio leaf thickness to epidermal cell thickness. This means that whenever epidermal tissue is relatively thin, the outer cell wall is relatively thick. Exceptionally thick cuticle covers both epidermes in the leaves of *N. oleander*.

The thickness of the photosynthetic tissue (mesophyll) in most of the species is approximately 160 µm. In *N. oleander* this value is highest (224 µm) while in *Q. coccifera* it is lowest (89 µm). The ratio of palisade parenchyma thickness to the mesophyll thickness reveals that all species except *N. oleander* are xeromorphic to very xeromorphic (Pyykko 1966).

Table 2. Quantitative leaf histological data for the eleven species.

	epidermis						palisade				
	cell thickness (µm) upper	cell thickness (µm) lower	cell wall – cuticle thickness (µm) upper: exter. wall	upper: cuticle	lower: exter. wall	lower: cuticle	cell length (µm)	no of layers	% of leaf thickness	stomatal frequency	stomatal index
Arbutus andrachne	31±4	20±4	11±3	3	5±2	2	74±19	3	45	155±39	10
Arbutus unedo	37±6	32±6	13±5	4	11±4	3	75±12	2	42	190±40	10
Ceratonia siliqua	40±7	21±6	3±2	1	2±1	1	54±26	2	34	224±32	7
Laurus nobilis	24±3	20±3	4±1	2	4±1	4	62±11	3	44	305±38	5
Myrtus communis	19±2	14±3	7±2	3	5±1	2	54±19	2	35	456±75	8
Nerium oleander	66±8	46±6	5±2	9	3±1	5	66±16	2	27	stomata in crypts	
Olea europea	20±3	15±3	9±2	1	4±1	1	57±25	3	36	386±40	9
Phillyrea media	19±3	13±2	12±2	1	7±1	1	40±9	3	40	474±51	8
Pistacia lentiscus	28±5	22±6	10±2	2	8±2	2	69±30	2	39	234±29	12
Quercus coccifera	19±2	17±3	7±2	2	4±2	2	44±14	3	49	228±31	16
Quercus ilex	20±4	10±2	6±2	3	2±1	2	71±16	2	52	289±51	–

$x \pm t\sigma_{n-1}$

Stomatal frequency is generally high with the lower values in the two *Arbutus* species, a fact which is in accordance with the generalization that small leaves are characterized by high stomatal frequency.

Sclerophylly Index (Gigon 1979) was also determined by calculating the leaf dry weight per surface area (g/cm^2). Since leaf thickness varies with a ratio of 1:2.3 among the eleven species, we introduced (Christodoulakis 1984) the Leaf Density Index (LDI) in order to express leaf density or compactness. LDI, given in grams of dry weight cm^{-3} of leaf tissue is presented, together with the corresponding sclerophylly index, in Table 2.

Table 3. Sclerophylly index (Gigon 1979) and leaf density index.

	Quercus coccifera	Ceratonia siliqua	Arbutus unedo	Phillyrea media	Quercus ilex	Pistacia lentiscus	Olea europaea	Myrtus communis	Nerium oleander	Arbutus andrachne	Laurus nobilis	Citrus limone	Pistacia terebinthus	Platanus orientalis	Populus deltoides	Pyrus communis	Morus alba
Sclerophylly index $\times 10^{-4}$ gr dr w/cm^2	216	203	222	171	125	221	229	138	204	220	123	170	158	132	165	105	140
Leaf density gr dr w/cm^3	0.71	0.62	0.61	0.56	0.52	0.51	0.48	0.44	0.44	0.39	0.37	0.42	0.37	0.39	0.38	0.39	0.33
												mesophyte leaves					

REFERENCES

Christodoulakis NS (1984) Structural analysis of sclerophylly in evergreen species of "maquis" vegetation. PhD Thesis. University of Athens

Diamantoglou S, Mitrakos K (1981) Leaf longevity in mediterranean evergreen sclerophylls. In: Margaris NS and Mooney HA (eds) Components of Productivity of Mediterranean Climate Regions. Basic and Applied Aspects. Junk, The Hague, pp 17-19

Diamantoglou S, Kull U (1984) Carbohydrate content and osmotic conditions of leaves and barks of *Arbutus unedo* L. and *Arbutus andrachne* L. throughout the year. Ber Deutsch Bot Ges Bd 97:433-441

Gigon A (1979) CO_2-gas exchange, water relations and convergence of mediterranean shrub-types from California and Chile. Oecol Plant 14(2):129-150

Grisebach A (1872) Die Vegetation der Erde nach ihrer klimatischen Anordnung. Engelmann, Leipzig

Johnson A (1973) Historical view of the concept of ecosystem convergence. In: diCastri F and Mooney HA (eds) Mediterranean Type Ecosystems. Origin and Structure. Springer Verlag, Berlin-Heidelberg-New York, pp 3-7

Kummerow J (1973) Comparative anatomy of sclerophylls of mediterranean climatic areas. In: diCastri F and Mooney HA (eds) Mediterranean Type Ecosystems. Origin and Structure. Springer Verlag, Berlin-Heidelberg-New York, pp157-167

Ledbetter MC, Porter KR (1963) A "microtubule" in plant cell fine structure. J Cell Biol 19:239-250

Mooney HA, Parsons DJ, Kummerow J (1974) Plant development in mediterranean climates. In: Lieth H (ed) Phenology and Seasonality Modeling. Springer Verlag, Berlin-Heidelberg-New York, pp 255-267

Orshan G (1982) Monocaracter growth for m types as a tool in an analytic-synthetic study of growth forms in mediterranean type ecosystems. A proposal for an interegional program. Ecologia mediterranea. Definition and localization of terrestrian mediterranean biota. Tome VIII - Fascicule 1/2: 113-119

Pyykko M (1966) The leaf anatomy of East Patagonian xeromorphic plants. Ann Bot Fennici 3:453-622

Raunkiaer C (1916) The use of leaf size in biological plant geography. In: Raunkiaer C (ed) Life Form and Terrestrial Plant Geography. Oxford Clarendon Press, Oxford, pp 220-232

Sabatini DD, Bensch K, Barrnett BJ (1963) Cytochemistry and Electron microscopy. The preservation of cellular ultrastructure and enzymatic activity by aldehyde fixation. J Cell Biol 17:19-58

Schimper AFW (1898) Pflanzen Geographie auf physiologischer Grundlage. Gustav Fisher, Jena

Turrill WB (1929) The plant-life of the Balkan Peninsula. A phytogeographical study. Clarendon Press, Oxford

The costs of growing and maintaining leaves of mediterranean plants

J. Merino

Department of Ecology, Faculty of Biology, University of Sevilla, 41080 Sevilla, Spain

INTRODUCTION

The costs in terms of respired photosynthate involved in growing and maintaining leaves is an important consideration for understanding the overall carbon balance of plants. Estimates of these costs have been made for a number of crop species (Penning de Vries 1975; Hunt and Loomis 1979). In the case of plants growing in natural ecosystems, only a few estimates have been made for chaparral and tundra species (Miller and Stoner 1979; Merino *et al.* 1982).

Detailed analyses of leaf biochemical composition in previous studies of crop species (McDermitt and Loomis 1981) and chaparral shrubs (Merino *et al.* 1982, 1984) have suggested that leaf caloric content, ash content after combustion, and total leaf nitrogen may be used to develop simple indices related to growth and maintenance costs. This approach is applied here in a broad survey, to estimate the cost of growing and maintaining leaves of a number of common plant species of Central and Southern Spain, and to determine whether any recognizable patterns of variation exist between species, life forms, leaf types, or leaves of plants developing on substrates of differing fertility.

METHODS

Sampling

Materials were collected from woody species growing in areas of Spain with a mediterranean climate but differing in fertility (water and nutrient availability). For each species (a full list is given in the Appendix), leaves of different ages on the plant were harvested during early spring (evergreen) and early summer (winter deciduous). After harvesting, the samples were stored on ice and brought to the laboratory. The petioles were separated from the samples and discarded. The stems were oven dried at 80 ^{o}C, ground and stored for subsequent analyses.

Analytical Procedures

Ash content was determined gravimetrically after combustion of the samples for 3 hours at 500° C. Total nitrogen was determined with an elemental-analyzer (Carlo Herba, Milano, Italy). Heat content was determined with an adiabatic

NATO ASI Series, Vol. G15
Plant Response to Stress. Edited by J.D. Tenhunen et al.

bomb calorimeter (Gallenkamp, London, England) following the methods outlined by Lieth (1968). Two measurements were made per sample and when the variation was larger than 5%, a third one was made. In most cases, the variation was less than 2%.

Estimation of Cost for Leaf Growth

Various approaches have been used to calculate the cost involved in growing a leaf. Estimates have been based on the chemical composition (Penning de Vries *et al.* 1974), on respiration rates and growth analysis (Kimura *et al.* 1978), and on elemental analysis (McDermitt and Loomis 1981). In the present study, the growth cost is estimated as:

$$T = [(0.06967\ \Delta H_c - 0.064)/e]\ (1-A) + 2.6795\ N$$

where T is the growth cost of the tissue in g of glucose g dw^{-1}, ΔH_c is the ash free heat of combustion of the tissue in KJ, A is the ash content in g g dw^{-1} and N is the organic nitrogen content in g g dw^{-1}. The parameter e is a measure of growth efficiency; for plant leaves this value has been estimated at 0.88 from leaves of known composition (unpublished data).

This index of growth cost was developed by K. Williams and colleagues (unpublished) and is an extension of the approach taken by McDermitt-Loomis (1981). It is assumed that all the nitrogen in the tissue comes from nitrate which is most commonly the source of nitrogen in mediterranean ecosystems (Ellenberg 1971).

Estimation of Cost for Leaf Maintenance

The requirements for leaf maintenance have been estimated by the methods of Penning de Vries (1975). Accordingly, the average cost of protein turnover was assumed to be 0.0405 g glucose per g of protein per day, for maintenance of cellular ion concentrations is 0.008 g glucose per g dry weight per day, and for turnover of membrane lipids is 0.0017 g glucose per g dry weight per day. The maintenance cost of leaves in each sample was calculated by multiplying the coefficient for protein maintenance by the protein concentration in each sample (protein = nitrogen concentration in the sample x 6.25) and adding to it an estimated cost of maintaining both ionic concentration and membrane lipids.

The samples were grouped and compared according to life forms, leaf size, evergreen or deciduousness, individuals growing in arid or more mesic habitats, and individuals growing in very unfertile or more fertile habitats.

RESULTS

Growth Cost

The values calculated for growth cost of leaves for each species, the number of areas sampled and the standard deviation for each species are listed in the Appendix. The average value for all species was 1.56 g glucose g dw^{-1}, ranging from 1.23 for *Ficus carica* to 1.91 for *Erica scoparia.* The difference between the two extremes is large. The variability within a species growing under different conditions (see for example *Quercus rotundifolia* or *Phillyrea angustifolia*) was, however, found to be small. The results of the statistical comparisons between species groups are shown in Tables 1 and 2.

When comparing different sets of species there was not a significant difference in growth cost between life forms (Table 1a), between individuals growing on very dry substrates and individuals growing on more mesic ones (Table 2a), or between individuals growing on very oligotrophic soils (dune sand and sandstone) and those growing on less oligotrophic soils (Table 2b, c). When comparing species with different leaf sizes (Table 1c) the leptophyllous leaves (1.68 g glucose g dw^{-1}) were more costly to grow than nanophyllous (1.54), microphyllous (1.54), and mesophyllous (1.52). The difference in growth cost of leaves between evergreen (1.59) and deciduous species (1.48) was also significant (Table 1d); however, the difference between deciduous and evergreen leaves, similar in size (microphyllous), was not significant (Table 1e).

Maintenance Cost

The values calculated for maintenance cost of leaves for each species, the number of areas sampled and the standard deviation for each species are listed in the Appendix. The average value for all species was 0.0136 g glucose g dw^{-1} day^{-1} ranging from 0.0113 for *Calluna vulgaris* to 0.0161 for *Phlomis purpurea.* The results of the statistical comparisons between species groups are shown in Tables 3 and 4.

Related to life form, shrub leaves (0.0134 g glucose g dw^{-1} day^{-1}) are less expensive to maintain than vine (0.0140) and tree (0.0138) leaves (Table 3 a). Deciduous tree leaves (0.0146) are more expensive to maintain than evergreen leaves (0.0135) or gymnospermous leaves (0.0117); and gymnospermous leaves are less expensive than evergreen leaves (Table 3 b). Considering all the samples, deciduous leaves are more expensive to maintain (0.0146) than evergreen leaves (0.0133). The result is the same when comparing leaves of similar size (Table 3 d, e). Mesophyllous (0.0148) and microphyllous (0.0141) leaves are more expensive to maintain than the nanophyllous (0.0132) and leptophyllous (0.0129) ones (Table 3 c). There were no significant differences found when comparing native plants from substrates with different degrees of stress or fertility (water and nutrients) (Table 3 a, b, c).

Table 1. Significance of differences in mean values (*t* test) determined for growth cost of leaves in different life forms (a), different tree types (b), plants with different leaf sizes (c), plants with different leaf types considering all the samples (d), and plants with different microphyllous leaf types (e). Numbers indicate the mean growth cost in g glucose g dw^{-1}. The number in parentheses indicates the number of samples compared (significance levels: *** = $p<0.01$, ** = $p<0.05$, * = $p<0.09$, ns = not significant).

	Shrubs (26)	Trees (19)	Vines (4)	
Shrubs 1.59		ns	ns	(a)
Trees 1.54			ns	
Vines 1.53				

	Gymnosperms (3)	Evergreen (5)	Deciduous (10)	
Gymnosperms 1.65		ns	**	(b)
Evergreen 1.56			ns	
Deciduous 1.48				

	L (9)	N (7)	Mi. (23)	Me. (4)	
Leptophyllous 1.68		**	***	*	
Nanophyllous 1.54			ns	ns	(c)
Microphyllous 1.54				ns	
Mesophyllous 1.52					

Evergreen 1.59 (29)	Deciduous 1.48 (11)	***	(d)
Evergreen 1.57 (15)	Deciduous 1.50 (8)	ns	(e)

Table 2. Significance of differences in mean values (*t* test) determined for growth cost of leaves in individuals under different degree of stress: water stress (a), nutrient stress considering only the species common in the areas compared (b), nutrient stress considering all the species present in the area (c). Numbers indicate the mean growth cost in g glucose g dw^{-1}. The number in parentheses indicates the number of samples compared (significance levels: *** = $p<0.01$, ** = $p<0.05$, * = $p<0.09$, ns = not significant).

Xeric	1.62 (5)	More mesic	1.62 (5)	ns	(a)
Unfertile	1.55 (17)	More fertile	1.57 (17)	ns	(b)
Unfertile	1.56 (35)	More fertile	1.55 (30)	ns	(c)

Table 3. Significance of differences in mean values (t test) determined for maintenance cost of leaves in different life forms (a), different tree types (b), plants with different leaf sizes (c), plants with different leaf types considering all the samples (d), and plants with different microphyllous leaf types (e). Numbers indicate the mean growth cost in g glucose g dw^{-1}. The number in parentheses indicates the number of samples compared (significance levels: *** = $p<0.01$, ** = $p<0.05$, * = $p<0.09$, ns = not significant).

		Vines (4)	Trees (19)	Shrubs (26)	
Vines	0.0140		ns	**	
Trees	0.0138			***	(a)
Shrubs	0.0134				

		Deciduous (11)	Evergreen (5)	Gymnosperms (3)	
Deciduous	0.0146		**	***	
Evergreen	0.0135			***	(b)
Gymnosperms	0.0117				

		Me. (4)	Mi. (27)	N. (7)	L. (7)	
Mesophyllous	0.0148		ns	**	***	
Microphyllous	0.0141			**	***	(c)
Nanophyllous	0.0132				ns	
Leptophyllous	0.0129					

Evergreen	0.0133 (28)	Deciduous	0.0146 (14)	***	(d)
Evergreen	0.0137 (17)	Deciduous	0.0146 (10)	**	(e)

Table 4. Significance of differences in mean values (t test) determined for maintenance cost of leaves in individuals under different degree of stress: water stress (a), nutrient stress considering only the species common in the areas compared (b), nutrient stress considering all the species present in the area (c). Numbers indicate the mean growth cost in g glucose g dw^{-1}. The number in parentheses indicates the number of samples compared (significance levels: *** = $p<0.01$, ** = $p<0.05$, * = $p<0.09$, ns = not significant).

Xeric	0.0130	(5)	More mesic	0.0133	(5)	ns	(a)
Unfertile	0.0137	(16)	More fertile	0.0137	(16)	ns	(b)
Unfertile	0.0139	(33)	More fertile	0.0140	(30)	ns	(c)

DISCUSSION

Growth Cost

The growth costs of leaves estimated in this study of species from the mediterranean flora of Spain were similar but lower than those calculated for chaparral species of California using somewhat different methods (Merino *et al.* 1982, 1984). In the case of some deciduous species the growth cost of their leaves was estimated to be similar to that of evergreen plants. However, as a general rule, the leaves of evergreen plants are more expensive to grow than those of deciduous species. This agrees with hypotheses put forward by Orians and Solbrig (1977). The occurrence of no significant differences in growth cost when comparing leaves of similar size or when comparing deciduous or evergreen trees is difficult to explain.

Because several authors have reported changes in leaf characteristics as a consequence of stress (Elias 1979; Karlsson 1981), it is difficult to explain the absence of significant differences in the leaf growth cost between populations growing on different substrates of differing fertility (water and nutrients). In addition, growth cost is very similar when comparing plants of the same species but growing at different sites. The similarity in leaf growth cost is not just a consequence of taxonomical similarity of the various sites, since it is also apparent when comparing all the species present. It follows that the cost of growing leaves in the mediterranean climate ecosystems of Southern Spain is independent of the conditions under which the plants are grown. At the species level, this similarity in growth cost suggests a genetic control. In the broader context of the mediterranean flora, the high similarity in growth cost suggests the existence of limits on the amount of energy which can be spent in constructing leaves.

Leaf size emerges as an important feature determining leaf cost. In effect, since smaller leaves are associated with the evergreen character and the growth cost seems related to leaf size, the higher cost of the evergreen leaves could be more a consequence of their smaller size than a general feature of the evergreen character. In mediterranean ecosystems, leaves are usually rich in very expensive secondary compounds laid on their surfaces. Since the surface/weight ratio increases inversely to size, the content of expensive products also increases (weight basis) as does the growth cost per gram of leaf tissue in smaller leaves.

Maintenance Cost

The values for maintenance costs of leaves of deciduous species reported here are considerably lower than those reported for chaparral species obtained by other methods (Merino *et al.* 1982, 1984). They are also lower than those reported for crops (Penning de Vries 1975). The values found here for the evergreen species are more similar but also lower than previously reported (Chung and Barnes

1977; Merino *et al.* 1982, 1984). The values are approximately 85% of what might be expected of leaf maintenance cost were determined by gas exchange (Merino *et al.* 1982). Although the average maintenance cost of leaves of different life forms are very similar, leaves of shrubs are significantly less costly to maintain than those of trees and vines. These differences do not appear related to the life form characteristics but to particular features of the species compared. When considering leaves of the different kinds of trees, leaves of deciduous trees are more expensive to maintain than leaves of vines, the leaves of evergreen trees are as expensive to maintain as those of shrubs (almost all the shrubs considered in this study are evergreen) and gymnospermous leaves are even less expensive to maintain than leaves of shrubs.

Since the differences in maintenance cost between evergreen and deciduous leaves are statistically significant whether considering all the species or when comparing a more homogeneous set of leaves such as the microphyllous ones, the evergreen or deciduous character appears to be a strong determining feature of leaf cost maintenance. These results support the hypothesis that low maintenance costs associated with evergreen leaves may be important in the long-term success of mediterranean species (Mooney and Dunn 1970; Miller 1983). The importance of size in determining maintenance cost is related to the ratio of functional to structural components of the leaf. This ratio is lower in smaller leaves (weight basis) and they are less costly to maintain.

Although some evidence suggesting that water stress decreases the maintenance cost and nutrient stress increases it (Penning de Vries 1975), none of the comparisons indicated a significant difference in leaf maintenance cost between plants growing under more stressed or less stressed conditions. This lack of correlation of estimated maintenance cost with environmental stress is probably a direct consequence of the way in which the maintenance costs reported here were calculated. Maintenance costs are directly dependent on the leaf nitrogen concentration which was found to be similar in all species. Considering the reputed difference in fertility of the substrates compared, and that usually average leaf nitrogen concentration is correlated with substrate fertility (Gulmon and Chu 1983), it is difficult to understand this similarity in leaf nitrogen concentration. Compensatory effects in the regulation of nitrogen investments by the plant may underlie this phenomenon (Chapin 1980).

Total nitrogen concentration may not be a good index of maintenance respiration for a comparison of the diverse species studied here. Total nitrogen may be distributed in several fractions, each of which requires a different maintenance cost (Merino *et al.* 1984). The proportion of cellular constituents in these fractions may differ among species.

ACKNOWLEDGEMENTS

I would like to thank A. González and V. Ruiz for their support, and to Nini Márquez for typing the text. This study was supported by the Spanish CAICYT, Grant No. 2896.

APPENDIX

Estimated leaf growth costs: s - standard deviation, n - number of areas sampled.

Species	Growth cost g glucose g dw^{-1}	s	n
CISTACEAE			
Cistus albidus	1.46	0.071	5
laurifolius	1.63	0.046	4
ladanifer	1.57	0.049	3
libanotis	1.59	0.007	3
monspeliensis	1.56	0.019	2
salvifolius	1.45	0.036	4
populifolius	1.50	0.064	3
Halimium commutatum	1.47	0.007	3
halimifolium	1.53	0.021	3
umbellatum	1.56	0.021	3
LABIATAE			
Rosmarinus officinalis	1.75	0.000	4
Phlomis purpurea	1.49	0.015	2
Lavandula stoechas	1.74	0.007	2
OLEACEAE			
Phillyrea angustifolia	1.74	0.034	5
Olea europaea	1.66	0.013	2
Fraxinus ornus	1.58	0.016	2
angustifolia	1.51	0.000	3
ERICACEAE			
Erica scoparia	1.91	0.021	3
Calluna vulgaris	1.75	0.019	2
Arbutus unedo	1.56	0.060	4
Arctostaphylos uva-ursi	1.70	0.008	2
SALICACEAE			
Populus alba	1.37	0.043	2
Salix spp.	1.59	0.059	2
FAGACEAE			
Quercus coccifera	1.56	0.028	3
rotundifolia	1.51	0.027	7
fruticosa	1.55	0.013	2
suber	1.54	0.007	3
faginea	1.48	0.052	5

Species	Growth cost g glucose g dw^{-1}	s	n
Quercus *pyrenaica*	1.48	0.003	5
canariensis	1.48	0.008	2
ROSACEAE			
Pyrus bourgeana	1.56	0.042	3
Rubus ulmifolius	1.61	0.007	3
Crataegus monogyna	1.50	0.028	3
Rosa spp.	1.44	0.025	2
CUPRESSACEAE			
Juniperus communis	1.58	0.099	3
oophora	1.71	0.022	2
MISCELLANEOUS			
Pinus pinaster	1.68	0.032	2
Ceratonia siliqua	1.50	0.049	3
Myrtus communis	1.47	0.014	3
Nerium oleander	1.69	0.019	2
Pistacia lentiscus	1.58	0.020	5
Ficus carica	1.23	0.049	2
Rhamnus frangula	1.52	0.014	2
Aristolochia baetica	1.50	0.016	2
Smilax aspera	1.62	0.012	4
Vitis vinifera	1.50	0.007	2

Estimated leaf maintenance cost: s - standard deviation, n - number of areas sampled.

Species	Maintenance cost g glucose g dw^{-1} day^{-1}	s (x10^{6})	n
CISTACEAE			
Cistus albidus	0.0130	1410	5
laurifolius	0.0134	100	4
ladanifer	0.0135	714	3
libanotis	0.0134	233	3
monspeliensis	0.0130	187	2
salvifolius	0.0143	538	4
populifolius	0.0134	445	3
Halimium commutatum	0.0122	311	3
halimifolium	0.0132	141	3
umbellatum	0.0129	000	3
LABIATAE			
Rosmarinus officinalis	0.0131	415	4
Phlomis purpurea	0.0161	320	2
Lavandula stoechas	0.0139	339	3
OLEACEAE			
Phillyrea angustifolia	0.0137	221	5
Olea europaea	0.0136	201	2

Species	Maintenance cost g glucose g dw^{-1} day^{-1}	s ($x10^6$)	n
Fraxinus ornus	0.0153	350	2
angustifolia	0.0155	523	3
ERICACEAE			
Erica scoparia	0.0129	198	3
Calluna vulgaris	0.0113	209	2
Arbutus unedo	0.0125	337	4
Arctostaphylos uva-ursi	0.0115	138	2
SALICACEAE			
Populus alba	0.0139	176	2
Salix spp.	0.0152	189	2
FAGACEAE			
Quercus coccifera	0.0132	219	3
rotundifolia	0.0130	744	7
fruticosa	0.0133	356	2
suber	0.0144	198	3
faginea	0.0137	273	5
pyrenaica	0.0146	127	5
canariensis	0.0146	177	3
ROSACEAE			
Pyrus bourgeana	0.0143	177	3
Rubus ulmifolius	0.0146	410	3
Crataegus monogyna	0.0131	361	3
Rosa spp.	0.0135	269	2
CUPRESSACEAE			
Juniperus communis	0.0116	148	3
oophora	0.0119	106	2
MISCELLANEOUS			
Pinus pinaster	0.0115	286	2
Ceratonia siliqua	0.0135	750	3
Myrtus communis	0.0131	502	3
Nerium oleander	0.0132	139	2
Pistacia lentiscus	0.0142	1424	5
Ficus carica	0.0145	196	2
Rhamnus frangula	0.0115	105	2
Aristolochia baetica	0.0157	369	2
Smilax aspera	0.0140	894	4
Vitis vinifera	0.0158	268	2

REFERENCES

Chapin FS III (1980) The mineral nutrition of wild plants. Ann Rev Ecol Syst 11:233-260

Chung HH, Barnes RL (1977) Photosynthate allocation in *Pinus taeda*. I. Substrate requirements for synthesis of shoot biomass. Can J For Res 7:106-111

Elias P (1979) Some ecophysiological features in leaves of plants in an Oak-Hornbeam forest. Folia Geobot Phytotax 14:29-42

Ellenberg H (1971) Nitrogen content, mineralization and cycling. In: Duvigneaud P (ed) Productivity of Forest Ecosystems. UNESCO, Paris, p 509

Gulmon SL, Chu C (1983) The effects of light and nitrogen on photosynthesis, leaf characteristics, and dry matter allocation in the chaparral shrub, *Diplacus aurantiacus*. Oecologia (Berl) 49:207-212

Hunt WF, Loomis RS (1979) Respiration modeling and hypothesis testing with a dynamic model of sugar beet growth. Ann Bot 44:5-17

Karlsson S (1981) The function of leaves in relation to the environment. Medd Växtekol Inst, Lund, Sweden

Kimura M, Yokoi Y, Hogetsu K (1978) Quantitative relationships between growth and respiration. II. Evolution of constructive and maintenance respiration in growing *Helianthus tuberosus* leaves. Bot Mag 91:43-56

Lieth H (1974) The measurement of calorific values of biological material and the determination of ecological efficiency. In: Eckardt F (ed) Functioning of Terrestrial Ecosystems at the Primary Production Level. UNESCO, Paris, p 233

McDermitt DK, Loomis RS (1981) Elemental composition of biomass and its relation to energy content, growth efficiency and growth yield. Ann Bot 48:275-290

Merino J, Field C and Mooney HA (1982) Construction and maintenance costs of mediterranean-climate evergreen and deciduous leaves. I. Growth and CO_2 exchange analysis. Oecologia (Berl) 53:208-213

Merino J, Field C and Mooney HA (1984) Construction and maintenance costs of mediterranean-climate evergreen and deciduous leaves. II. Biochemical pathway analysis. Oecol Plant 19:211-229

Miller PC (1983) Canopy structure of mediterranean-type shrubs in relation to heat and moisture. In: Kruger FJ, Mitchell DT, and Jarvis JUM (eds) Mediterranean-type Ecosystems, The Role of Nutrients. Springer, Berlin-Heidelberg-New York, pp 133-166

Miller PC and Stoner WA (1979) Canopy structure and environmental interactions. In: Solbrig OT, Jain S, Johson GB, Raven PH (eds) Topics in Plant Population Biology. Columbia University Press, New York, p 428

Mooney HA, Dunn EL (1970) Photosynthetic systems of mediterranean-climate shrubs and trees of California and Chile. Am Nat 104:447-453

Orians GH, Solbrig OT (1977) A cost-income model of leaves and roots with special reference to arid and semiarid areas. Am Nat 111:670-690

Penning de Vries FWT (1975) The cost of maintenance processes in plant cells. Ann Bot 39:72-92

Penning de Vries FWT, Brusting AHM, van Laar HH (1974) Products, requirements and efficiency of biosynthesis: a quantitative approach. J Theor Biol 45:339-377

Features of seed germination in *Arbutus unedo* L.

C.P.P. Ricardo and M.M. Veloso

Departamento de Botânica, Instituto Superior de Agronomia, 1399 Lisboa Codex and Estação Agronómica Nacional, 2780 Oeiras, Portugal

INTRODUCTION

Two species of the genus *Arbutus*, *A. unedo* L. and *A. andrachne* L. are found in Europe (Polunin and Eveard 1976). While *A. andrachne* is confined to Eastern Mediterranean countries, *A. unedo* (the strawberry tree) occurs throughout the entire Mediterranean region, the southwest of Europe and as far north as southwest Ireland. In the Iberian Penlnsula *A. unedo* is a widespread shrub (or a small tree to 12 m) most common in upland or mountainous regions and less frequent in the Central Plateau (Lazaro 1921).

In Portugal *A. unedo* is widespread (Fig. 1). Several ecotypes may have evolved. In the north the plant grows under Atlantic climate conditions with snowy winters and moderately dry summers, while in the most southern province, the Algarve, the winters are mild but drought stress during the Mediterranean summers is substantial (Fig. 2). The significance of *A. unedo* for the Algarve results from its being both an important component of the scrub vegetation formations that cover large sections of the interior mountains and due to the economic importance of its ripe fruits used by local farmers to produce madroño, a typical brandy.

Farmers have contributed to propagation of the species by burying the fruit remains from distillery activities throughout nearby hills. Propagation of *A. unedo* from seed has also been achieved in nurseries of the National Forestry Services. There was originally a belief that germination was very slow and erratic, and that burying a mixture of mashed fruits with manure (or cow dung) was the best way to achieve high germination percentages. It was thought that seeds, when washed away from the fruits, would rapidly loose germinaton potential. The present paper intends to clarify these points and contribute to the understanding of propagation of *A. unedo* from seed under natural conditions.

MATERIALS AND METHODS

Fruits were harvested in late December, which corresponded to the end of the fruiting season (September to January), in three areas of the Algarve hill country: 1) Montinho de Monchique, close to Caldas de Monchique; 2) Herdade da Parra, near Silves; 3) Mata Nacional, close to Conceição de Tavira (Fig. 1). To obtain the seed, ripe fruits were mashed in water and debris was removed by decanting

NATO ASI Series, Vol. G15
Plant Response to Stress. Edited by J. D. Tenhunen et al.

Fig. 1. Distribution of *Arbutus unedo* in Portugal (darkened areas - according to J.A. Franco, unpublished survey). Numbers 1 to 3 on the map refer to plant communities from which seed was harvested (see Materials and Methods)

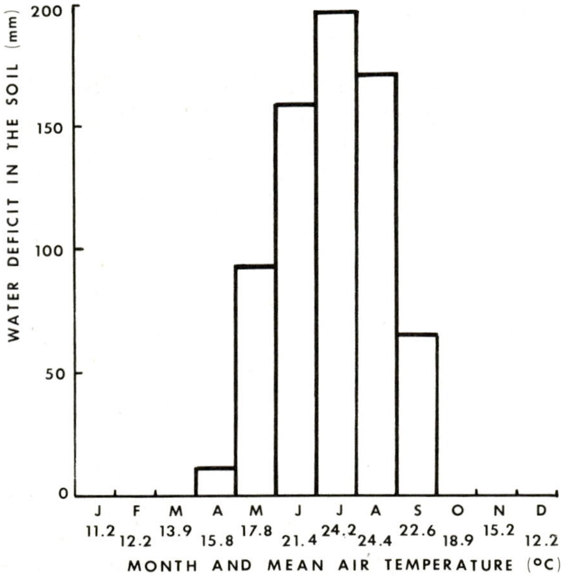

Fig. 2. Mean monthly air temperature and mean monthly water deficit in the soil at collection site number 1 in Fig. 1. Data averaged from 1935 through 1960 (from Ferreira 1970 and Bettencourt 1974)

and sieving. After thoroughly rinsing the seeds they were gently pressed against filter paper and air dried at room temperature. Until used for the germination assays they were stored in small boxes in the laboratory.

Samples of 100 to 300 seeds were tested for germination potential either on artificial substrate in Petri dishes or in soil in clay pots. The Petri dishes (9 cm in diameter) contained 6 sheets of Whatman filter paper number 4 soaked with 12 ml of the solutions indicated in the figures and were incubated in the dark at constant temperature. Seeds that had initiated germination were removed daily. In clay pots (16 cm in diameter) filled with sand or with a compound soil (two parts sand, two parts peat and one part clay soil), seeds were buried at 1 cm depth and watered every other day with tap water. The pots were placed under dim light (5 μEinstein m^{-2} s^{-1} PAR from Philips HPL bulbs) with a 12 hr photoperiod and a day/night temperature of 20/15° C. Germinated seeds were removed daily. For the germination tests with fruits, a number of them, containing approximately 300 viable seeds were gently mashed and mixed with an equal volume of soil or cow dung. The paste thus obtained, was spread on the top of the soil and covered with 1 cm of soil. Cold treatments were applied immediately after sowing and lasted 30 days at 4° C in the dark. To test hormonal stimulation,150 ml of a 500 ppm solution of GA_3 was applied to the soil at sowing.

RESULTS AND DISCUSSION

Complete germination of *A. unedo* seeds sown in the dark immediately after harvest occurred at temperatures below 15° C (Fig. 3A). No germination occurred at 30 or 25° C. Although at 20° C, the rate of germination was rapid in the first days after imbibition, it soon declined and higher final percent germination values were observed at 12 and 15° C. Prolonged exposure of seeds to temperatures of 20° C or higher has been reported to unfavorably affect germination in other species (Heydecker 1977). This could be due to unfavorable effects of high temperature on membrane stability (Hendricks and Taylorson 1979). The exogenous application of GA_3 to *A. unedo* seeds counteracts to some extent the inhibitory effect of high temperature (Fig. 3B).

In order to determine whether high temperature deleteriously and irreversibly affected ability to germinate, we imbibed seeds in the dark at 30° C and stored them in this condition for a month. Thereafter, we decreased the temperature to 15° C (Fig. 4A - open triangles). The treated seeds (open squares) did not lose the capacity to germinate, although the rate of germination was much lower than found for seeds initially imbibed at 15° C. The final percent germination value was the same in both cases (not shown). Applied GA_3 accelerated germination such that it occurred as rapidly as in the controls. The experiments shown in Fig. 4A were performed with one-year-old seeds. The germination capacity of two-

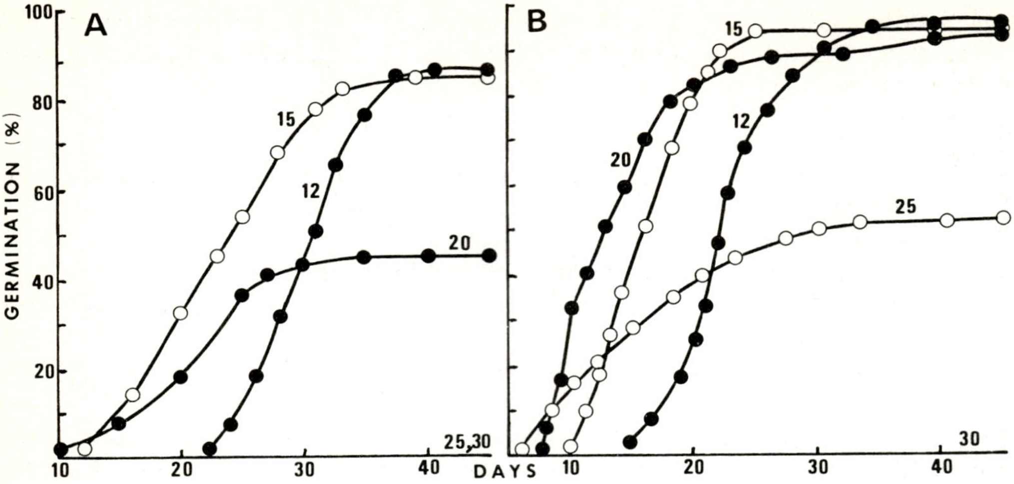

Fig. 3. Percent germination of *Arbutus unedo* seeds on filter paper as a function of time after sowing and at different temperatures (indicated next to individual curves) in the dark. A - with water added; B - with 500 ppm GA_3 added. Seeds used were from Monchique immediately after harvest in 1982

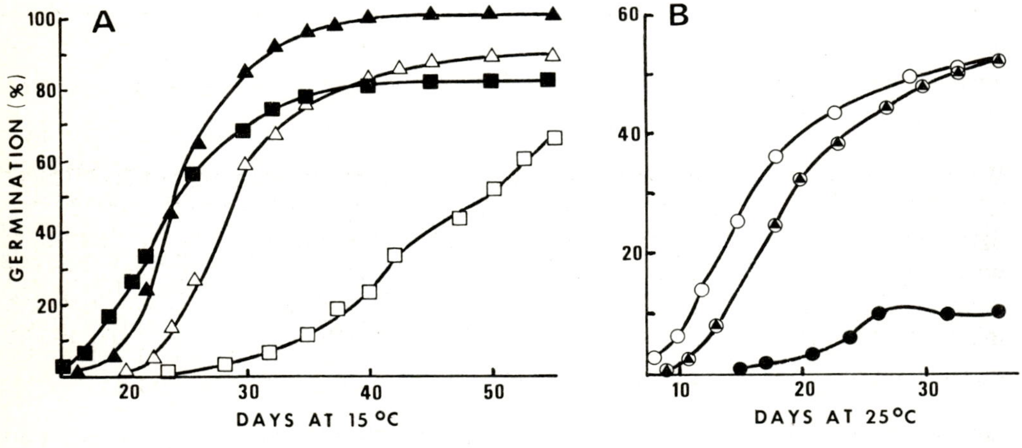

Fig. 4. Percent germination of *Arbutus unedo* seeds on filter paper as a function of time after sowing in the dark. Treatment effects of high temperature stress and seed age on germination are shown. A - (open triangles - control) one year old seeds imbibed and germinated at 15° C with water added; (closed triangles) one year old seeds imbibed and germinated at 15° C with GA_3 added; (open squares) pretreatment with water at 30° C for a month and then assayed at 15° C with water added or (closed squares) with GA_3 added. B - (open circles) seeds immediately after harvest imbibed and germinated at 25° C with GA_3 added; (circles with triangle inset) the same after one year; and (closed circles) the same after two years. Note the difference in the ordinate scales between A and B

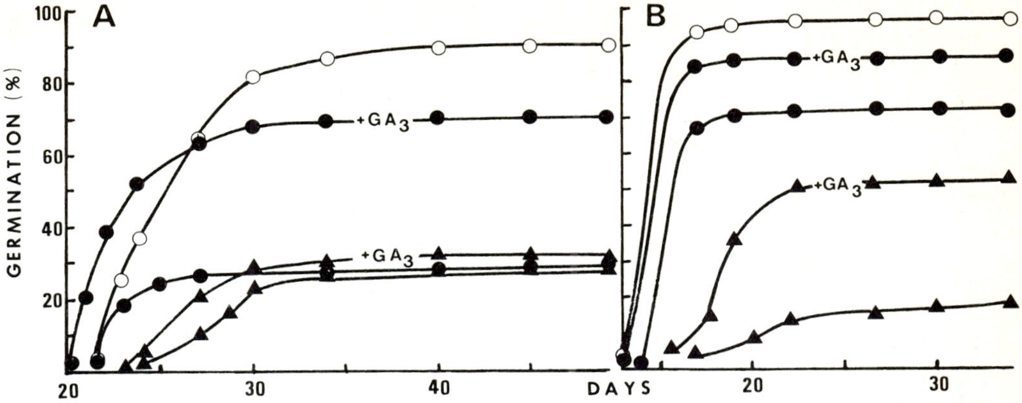

Fig. 5. Percent germination of *Arbutus unedo* seeds as a function of time after sowing in soil with day/night temperatures of 20°/15° C and 12 hour low light photoperiod. A - without cold treatment; B - after one month at 4° C; (open circles) in sand; (closed circles) compound soil; (solid triangles) compound soil plus cow dung; +GA_3 indicates additional treatment with GA_3 at 500 ppm. Seeds used were from Monchique immediately after harvest in 1982

year-old seeds was reduced, although germination did occur even at 25° C if GA_3 was applied. These results shown in Figs. 3 and 4 may explain in part the erratic germination observed in nurseries, particularly when sowing occurs in late spring of warmer years.

We observed good germination in soil with day/night temperatures of 20°/15° C (Fig. 5A) but even better results were obtained after a cold treatment at 4° C (Fig. 5B). Regarding the belief that manure is needed to achieve good germination, it was surprising to find that germination occurred best on sand and that peat and cow dung inhibited germination. We do not know why the presence of organic matter was inhibitory. However, since GA_3 counteracted that effect it appears that some metabolic imbalance developed.

When we studied germination from mashed fruits we found that only a small number of seeds had germinated after 160 days, even if a cold treatment was applied (Figs. 6A and B). The highest germination was observed in the cold treated samples on rewetting after a 10-month period in dry soil. It seems that inhibitors in the fruit were removed by the combined action of cold and drying. GA_3 was able to overcome part of the effect of these inhibitors during the initial

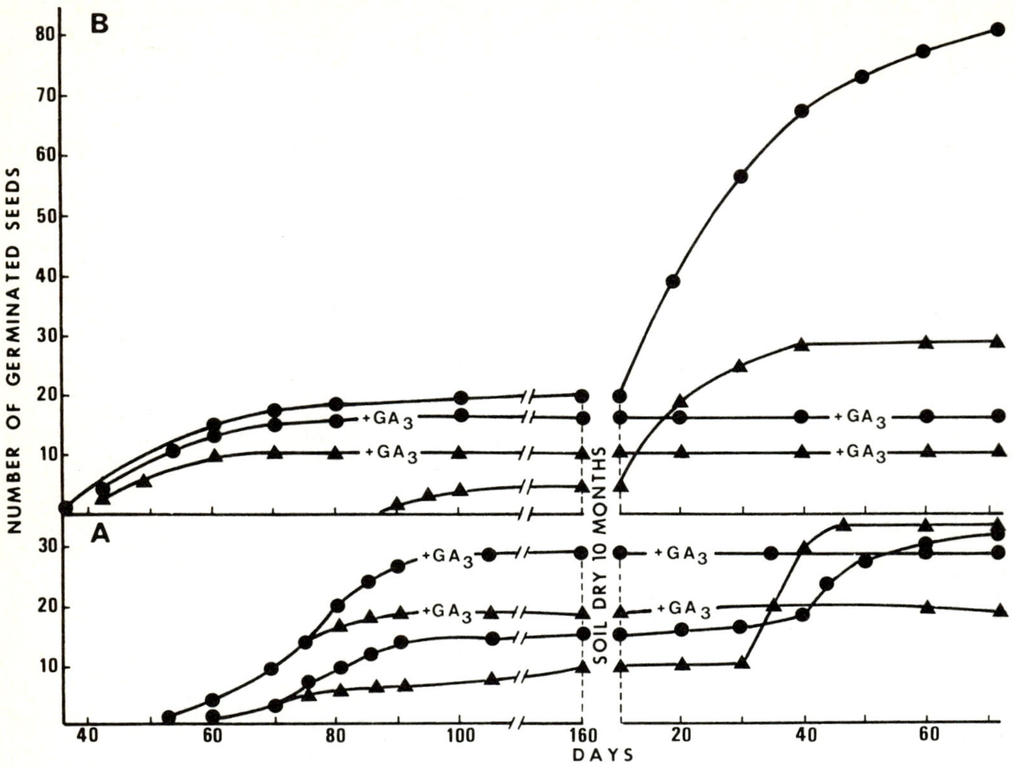

Fig. 6. Germination of *Arbutus unedo* seeds from mashed fruits as a function of time after sowing in soil with day/night temperatures of 20°/15° C and 12 hour low light photoperiod. A - without cold treatment; B - after one month at 4° C; (closed circles) compound soil; (solid triangles) compound soil plus cow dung; +GA_3 indicates additional treatment with GA_3 at 500 ppm. After 160 days, the pots were kept dry for ten months, after which germination was again monitored with moist conditions. Seeds used were from Monchique immediately after harvest in 1982

period of 160 days, particularly in the samples rich in organic matter that were not cold-treated. However, the hormone once applied prevented further germination after drying and rewetting, indicating that its action on the seeds led to their death under these circumstances.

CONCLUSIONS

Seeds of *A. unedo* germinate easily in the laboratory at a constant temperature of 10 to 20° C. In nature, germination may be decreased by temperature above

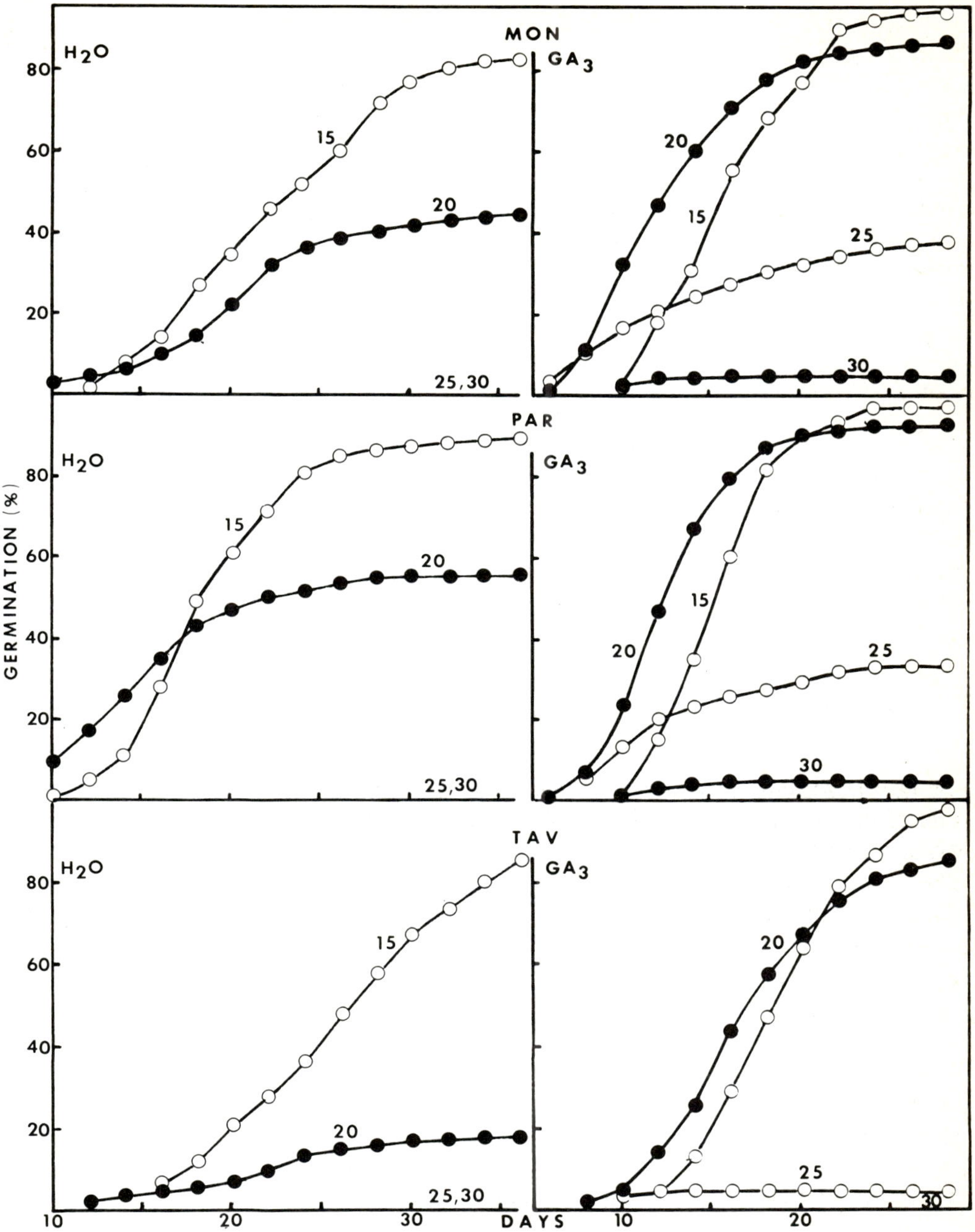

Fig. 7. Percent germination of *Arbutus unedo* seeds from three plant communities (MON - Monchique; PAR - Parra; and TAV - Tavira) as a function of time after sowing on filter paper and at different temperatures (indicated next to individual curves) in the dark. H_2O - with water added; GA_3 - with 500 ppm GA_3 added. Seeds were used immediately after harvest in 1982

20° C, the presence of inhibitors (including those of the fruits) and disadvantageous soil conditions. The sensitivity to high temperature varied with seed origin and year of production. Seeds from the Tavira plant community were more sensitive to high temperature than those from the two other communities (Fig. 7).

Considering the prevailing climatic conditions of the Algarve (Fig. 2), October and November are the months with mean air temperatures cool enough for optimal germination and the months that allow maximal seedling development before the April to September drought period. Due to the presence of inhibitors in the falling fruits, seeds will probably remain dormant for one year and germinate the next October if no other inhibitors are present in the soil. Of course, seeds eaten by birds could germinate immediately if dropped on favourable soil.

ACKNOWLEDGEMENTS

We would like to thank Prof. J. A. Franco for kindly providing the information on distribution of *Arbutus unedo* in Portugal included in Fig. 1.

REFERENCES

Bettencourt M (1974) Valores médios de evapotranspiração potencial determinados segundo os métodos de Thornthwaite, Quijano, Penmann, Turc e Blaney-Cridle. Instituto Nacional de Meteorologia e Geofísica, Lisboa

Ferreira HA (1970) O clima de Portugal. Fascículo XIII. Serviço Meteorológico Nacional, 2ª ed., Lisboa

Hendricks SH, Taylorson RB (1979) Dependence of thermal response of seeds on membrane transitions. Proc Natl Acad Sci 76:778-781

Heydecker W (1977) Stress and seed germination: an agronomic view. In: Khan AA (ed) The Physiology and Biochemistry of Seed Dormancy and Germination. North-Holland Publ. Co., Amsterdam, pp 237-282

Lazaro B (1921) Compendio de la flora Española, volume III, 3ª ed. Imprensa Classica Española, Madrid

Polunin O, Everard B (1976) Trees and bushes of Europe. Oxford Univ Press, London

6. Fire Effects on Mediterranean Plant Populations

Postfire demography of resprouting and seedling establishment by *Adenostoma fasciculatum* in the California chaparral

P. W. Rundel, G. A. Baker, D. J. Parsons*, T. J. Stohlgren*

Laboratory of Biomedical and Environmental Sciences and Department of Biology, UCLA, Los Angeles, CA 90024, USA

INTRODUCTION

Adenostoma fasciculatum (chamise) forms the dominant element of chaparral ecosystems in California. This evergreen, ericoid-leaved shrub occurs as a codominant in mixed chaparral or an overwhelming dominant in chamise chaparral, being present in over 70% of the chaparral stands in the state (Hanes 1971). No other chaparral shrub approaches *A. fasciculatum* in community importance. Unlike the majority of chaparral shrubs which respond to fire by either resprouting or reseeding, *A. fasciculatum,* utilizes both reproductive strategies (Wells 1969; Keeley and Zedler 1978; see also Hilbert in this volume). Although there have been numerous studies of succession in chamise chaparral (Horton and Kraebel 1955; Hanes 1971), no research to date has focused on the important questions of the demography of resprouting and seedling establishment by *A. fasciculatum* in the first few years following chaparral fires. Is there significant mortality of chamise burls during fires? How do fire seasonality and intensity affect resprout survival and growth? How do chamise seedlings compete with resprouts for establishment in postfire stands? What factors limit chamise seedling survival and growth? These are all important questions. In this paper we present results of a three-year study of the demographics of resprout mortality and growth and of seedling establishment for *A. fasciculatum* following burn and clip treatments at two seasons of the year in mature stands of chamise chaparral in the southern Sierra Nevada, California. These data quantitatively describe both the population dynamics and seasonal production of resprouts of *A. fasciculatum* and the patterns of postfire seedling establishment and growth. We give particular attention to a consideration of the physiological controls on rates of growth in resprouts.

Field studies were carried out in a mature community of chamise chaparral at 650 m on a south-facing slope near Shepherd Saddle, Sequoia National Park. This old-growth stand was almost a pure monoculture of *Adenostoma fasciculatum,* with a mean canopy height of approximately 2.5 m. Reliable records at Sequoia National Park indicate that no fires have occurred in the study area for at least 60

*National Park Service, Sequoia and Kings Canyon National Park, Three Rivers, CA 93271, USA

NATO ASI Series, Vol. G15
Plant Response to Stress. Edited by J. D. Tenhunen et al.

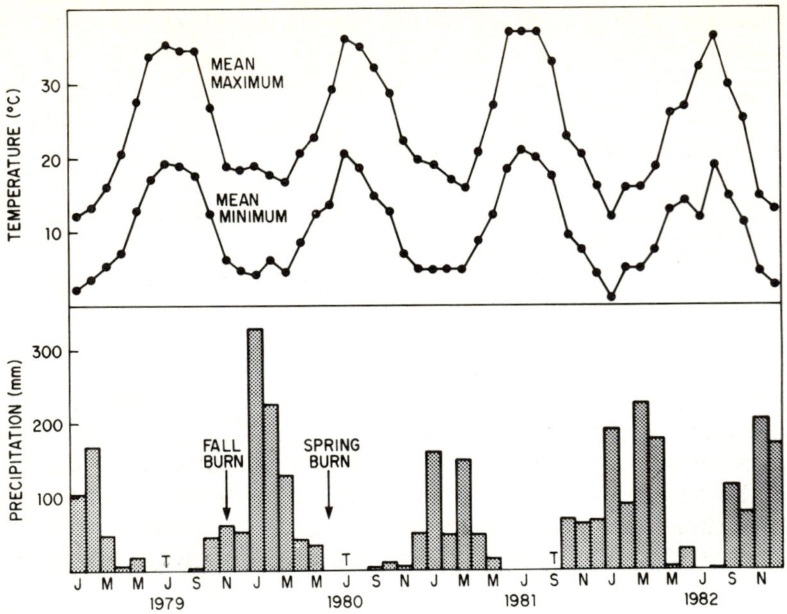

Fig. 1. Mean monthly maximum and minimum temperatures and monthly precipitation at Ash Mountain (510 m elevation), Sequoia National Park, California. Trace amounts of precipitation are indicated by a "T"

years (Parsons 1981). Additional studies were also carried out following the September 1978 and August 1981 Potwisha wildfires in old-growth chamise chaparral at 810 m. The community structure of old-growth chamise chaparral in Sequoia National Park has been described by Parsons (1976). Stand densities of *A. fasciculatum* in old-growth communities at Sequoia National Park have a mean density of 5458 shrubs ha^{-1}, with a mean stand aboveground biomass of 30,657 kg ha^{-1} (Stohlgren *et al.* 1984). The number of standing dead shrubs ranges from 8 to 22% of the total, with these primarily in the smaller size classes. Mean shrub dry weight (aboveground live and dead tissue) is 5600 g (Stohlgren *et al.* 1984).

Mean annual precipitation at Ash Mountain in Sequoia National Park, near our study site at 510 m, is 665 mm. Precipitation was 949 mm for the 1979-80 hydrologic year and 495 mm for 1980-81. Mean maximum temperatures at Ash Mountain ranged from a high of 42° C in July and August 1980, to a low of 14° C in January 1979 during our study (Fig. 1). Mean minimum temperatures ranged from a high of 25° C in July 1981 to a low of 2° C in January and February in both years.

METHODS

Experimental manipulations were initiated in November 1979 with an experimental burn of approximately 5 ha. At this time of the year, before the onset of the winter rainy season, no new growth had occurred in control shrubs of *A. fasciculatum.* A 10 x 10 m area within the boundary of the fall burn was selected for intensive studies and was designated the fall burn plot. At the same time as the fire, an unburned area of chamise chaparral within 100 m of the fall burn plot was cleared by hand and the aboveground biomass removed. This 10 x 10 m area was designated the fall clip plot.

A second set of experimental manipulations were carried out in June 1980. The spring fire burned an area of 5 ha near to the fall burn. At this time of the year, chamise shrubs were actively growing with vegetative growth just finishing and reproductive growth just beginning. A 10 x 10 m quadrat within the burn was designated as the spring burn plot, while an adjacent unburned area was clipped and cleared (8 x 8 m) as a spring clip plot. An unburned buffer zone between the fall and spring burns was used as an unburned control plot.

Regrowth of resprouts of *A. fasciculatum* in each study plot were followed at roughly 4-week intervals following burn or clip treatments. Biomass of each marked resprout was estimated from a regression using canopy volume as the independent variable ($y = \ln 942 + 0.772 \ln x$, where y = biomass in g and x = resprout volume in m^3). Canopy volume was calculated as a cylinder from measurements of mean maximum height and mean diameter. The preburn biomass of each individual was estimated from a regression based on the summed cross-sectional area of all stems at 30 cm height above the ground. Except in extremely intense burns, fires do not consume stems at this height and thus they can be readily calipered after the fire. The regression was $y = 0.18x + 0.06$ where y = shrub biomass in kg and x = stem basal area in cm^2, $r = 0.94$, $n = 39$ (Stohlgren *et al.* 1982).

Seedling establishment of *A. fasciculatum* in the experimental plots was monitored at 4-week intervals in ten 1 x 1 m quadrats in each plot. Records were kept of seedling number, mortality and herbivory levels. Seedling biomass for the spring plots was measured at the end of May 1981, 12 months after the burn and clip treatments.

Water potentials of resprouts, seedlings and control shrubs were measured with a Scholander-type pressure bomb following standard procedures. Data points are means of 3 to 5 samples from two or more individuals. Standard deviations of these values were generally less than 0.2 MPa in resprouts and control shrubs.

RESULTS

Resprout Growth

The fall burn plot included 45 shrubs of *A. fasciculatum* in 100 m^2 for a density of 4500 shrubs ha^{-1}, while the spring burn plot included 55 shrubs for a density of 5500 shrubs ha^{-1}. In both stands there was a typical skewed distribution where the smaller size classes contained the largest number of individuals (Fig. 2). Similar stand structures were present in the clip plots.

Postfire mortality of *A. fasciculatum* burls varied considerably between treatments. Mortality in the fall burn treatment was 13%, i.e. 13% of the burls failed to resprout following the experimental burn. There was zero mortality observed in the fall clip plot. Mortality in the spring burn plot was high at 75%, but by comparison only 38% in the adjacent spring clip plot. Since our postfire sampling did not distinguish between fire-induced and preburn shrub mortality in the burn plots, levels of treatment-induced mortality are likely somewhat lower than these values.

Levels of mortality in the two burn plots were not evenly distributed across preburn size-classes. Small shrubs experienced higher levels of mortality than did large shrubs. In the fall burn 80% of the small shrubs with an aboveground biomass less than 3.7 kg survived, but there was no mortality in large shrubs with 7.3 kg or more biomass (Fig. 2). The predominant mortality in small shrubs was accentuated dramatically in the spring burn plot. Only 11% of the shrubs in the smallest size class survived, while there was no mortality in the largest size class (Fig. 2).

Resprout regrowth following fire was strongly influenced by the preburn biomass of the shrub. There was a significant linear relationship between the biomass of resprouts one year after the fall burn and the estimated preburn biomass of the shrubs (Fig. 3). This relationship held through the second postfire growing season as well. The seasonal dynamics of growth in resprouts of four preburn size classes (number of size classes reduced compared to data in Fig. 2) of shrubs from the fall burn indicated similar patterns of growth during 1981 and 1982 (Fig. 4). The slopes of the four lines drawn from peak levels of biomass in June 1981 to July 1982 were not statistically different indicating rates of growth in the resprouts to be equal during this period. It would seem, therefore, that a growth spurt soon after fire leads to larger resprouts from large preburn shrubs. However, expressed on the basis of preburn biomass, the relative production of large shrubs is less than that of small shrubs over longer periods. Since our estimates of resprout biomass are based on a volumetric regression, the decline in mean shrub biomass during the late summer and fall of 1981 was largely attributable to vertebrate herbivory.

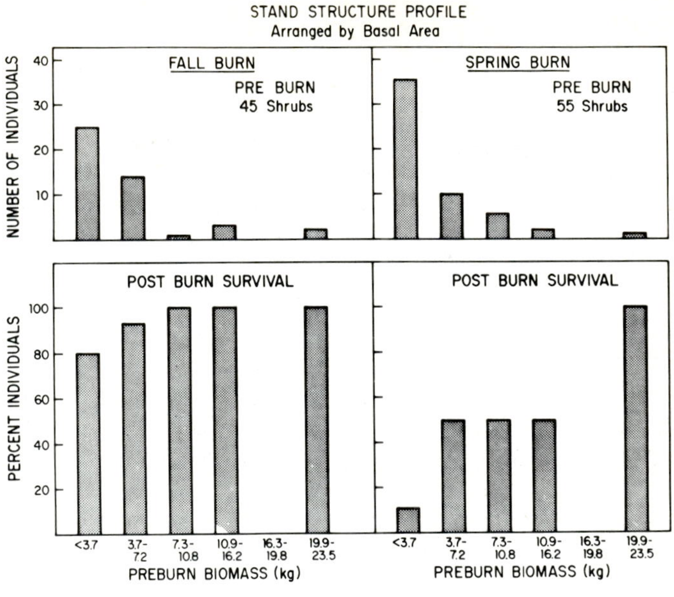

Fig. 2. Structural profile for fall and spring burn stands of *Adenostoma fasciculatum*, and survival of root burls in relation to aboveground shrub biomass before burning

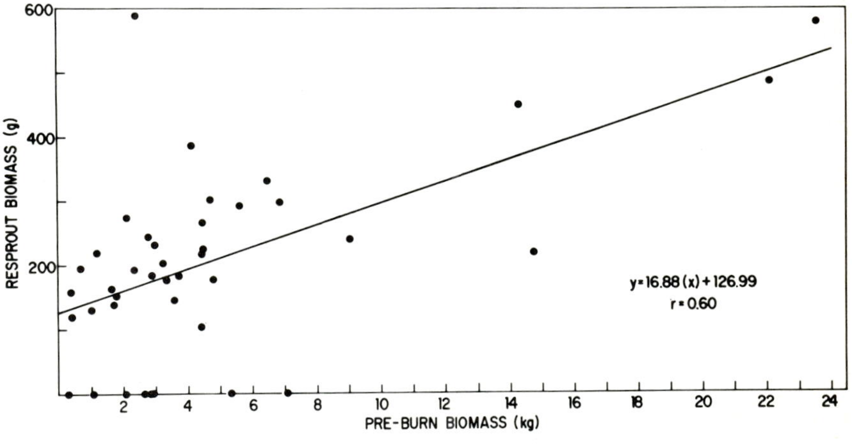

Fig. 3. Regression of resprout biomass of *Adenostoma fasciculatum* one year after a fall burn on preburn biomass

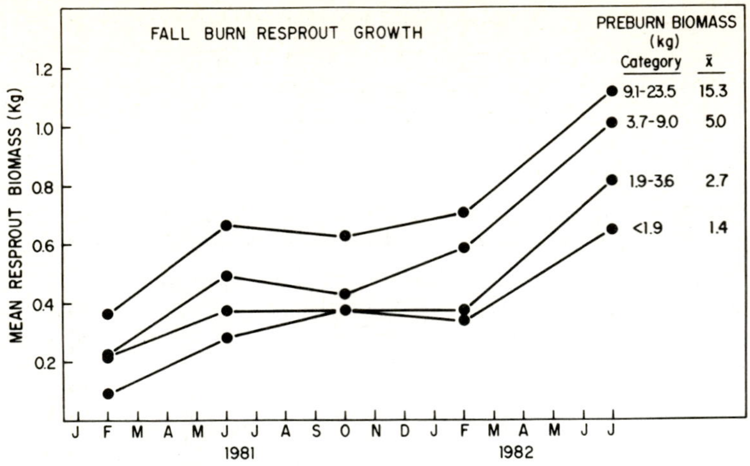

Fig. 4. Growth by resprouts in the fall-burn plot in relation to preburn biomass of *Adenostoma fasciculatum*. The experimental burn took place in November 1979

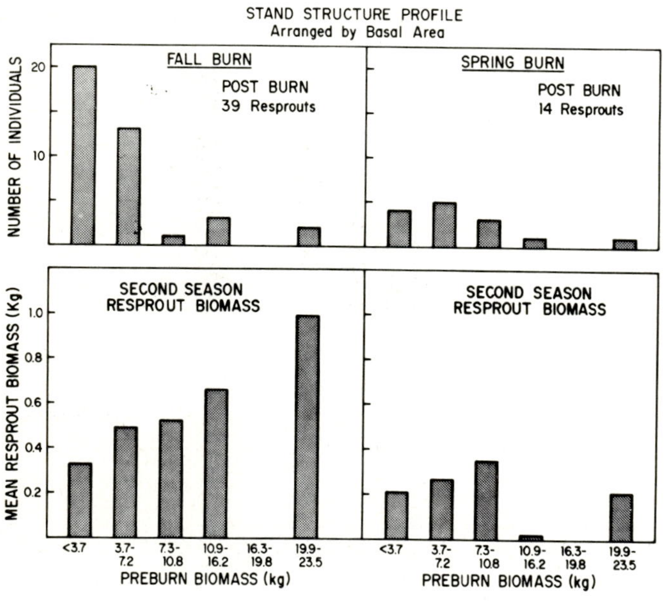

Fig. 5. Structural profile for fall and spring burn stands of preburn shrub and postburn resprout biomass of surviving *Adenostoma fasciculatum*

The pattern of regrowth by resprouts in the spring burn plot was not quite as clear as in the fall burn plot, but this may be due to the small postfire sample size of large shrubs. Mean resprout biomass increased with increasing shrub size in the three smallest size classes but was lower in the few larger shrubs sampled (Fig. 5).

Mortality in the spring clip plot showed the same pattern of relationship to preburn size of shrub as seen in the burn plots. Only 28% of the shrubs in the smallest size-class survived our clipping treatment. While all shrubs in the two largest size-classes survived (Fig. 6). Mean resprout biomass following the first two seasons of post-treatment regrowth was linearly related to preburn shrub size (Fig. 6).

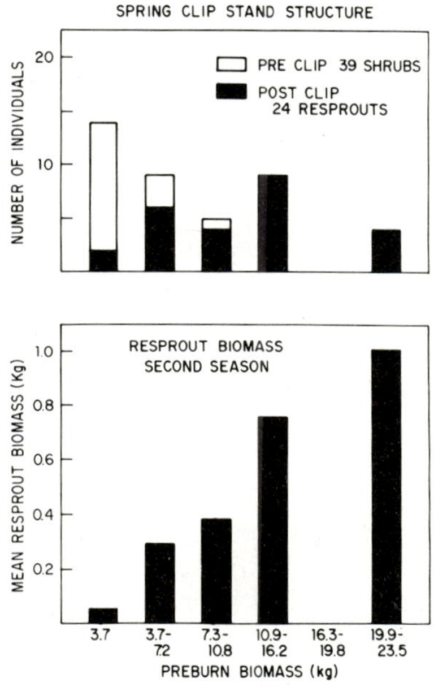

Fig. 6. Structural profile for spring clip plot of preburn shrub and postburn resprout biomass of surviving *Adenostoma fasciculatum*

Although the spring burn and clip treatments were carried out at the same time, the patterns of regrowth by resprouts were quite different. Clipped shrubs exhibited a low level of resprouting in July 1980, only one month after treatment (Fig. 7). This small stubble of growth was visible through the summer months and a rapid period of stem elongation occurred in these plants between October and December. No resprout growth was visible in the spring burn plot until January of 1981. Rapid growth and stem elongation began in April 1981, roughly at the same time as in the clip plot and control shrubs. The rate of

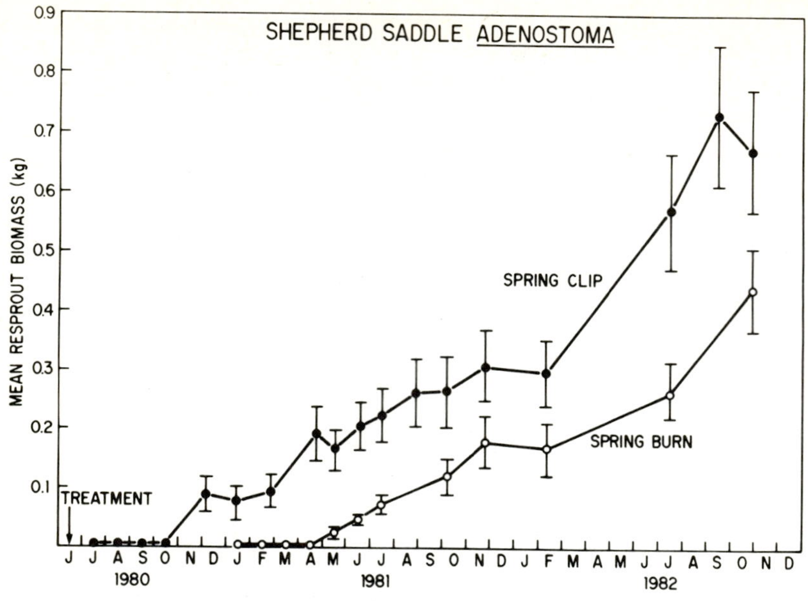

Fig. 7. Growth by resprouts of *Adenostoma fasciculatum* in spring burn and spring clip plots. Brackets indicate ± 1 standard error

resprout growth during 1981 was similar in clip and burn plots. Clip plots, however showed a more rapid rate of growth during the 1982 growing season (Fig. 7).

At the end of two seasons of post-treatment growth (1980 and 1981), the mean biomass of chamise resprouts in the fall burn plot was 497 g shrub^{-1} and 810 g shrub^{-1} in the fall clip plot. The fall burn plot at this stage had 2237 kg ha^{-1} of chamise resprouts. More rapid resprout growth in clip plots over burn plots was also observed in the spring treatment. Following two seasons of growth (1981 and 1982) the spring clip plot had 569 g shrub^{-1} versus 264 g shrub^{-1} in the spring burn plot. On a stand basis, these differences are magnified further due to the differential mortality with treatment. The spring clip plot had 1366 kg ha^{-1} of chamise resprouts while the spring burn plot had only 369 kg ha^{-1}. Although some of the differences in growth rates between fall and spring treatments can be attributed to seasonal effects, climatic differences between the three years make direct comparisons impossible.

Comparative measurements of midday water potentials in treatment and adjacent control plots indicated that resprouts in burn treatments experienced significantly higher water potentials during their first season of growth than control shrubs. During the spring of 1980, fall burn resprouts had water potentials typically 0.5 MPa higher than controls shrubs (Fig. 8A). From July to September this

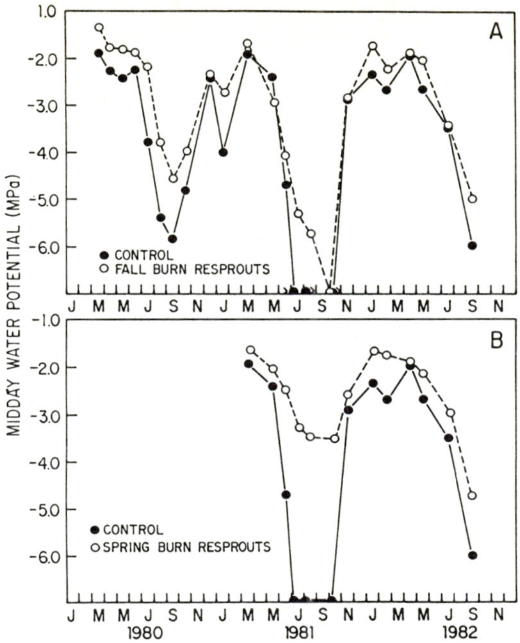

Fig. 8. Seasonal patterns of midday water potentials in *Adenostoma fasciculatum* for treatment and control plots: A. Fall burn plot; B. Spring burn plot

difference averaged nearly 1.5 MPa. The minimum water potential reached -5.8 MPa in control shrubs in September 1980, but only -4.5 MPa in fall burn resprouts. No differences were found between water potentials of shrubs in the fall burn and control plots in spring 1981 and 1982 but minimum summer water potentials were significantly lower in control shrubs in both years. Resprouts in the fall clip plot exhibited an identical pattern of seasonal water potentials to that of resprouts from the fall burn plot.

In the spring burn plot, water potentials were minimally higher (about 0.3 MPa) than control shrubs during the spring following treatment (1981) but these differences increased substantially during the summer months (Fig. 8B). The minimum water potentials in the spring burn plot reached only -3.5 MPa during this period while control shrubs experienced -7.0 MPa. The same pattern was repeated in 1982 but with smaller differences. Water potentials were minimally higher in the burn plot than in the control shrubs (a small but significant difference) during the spring and early summer months, but this difference reached -1.3 MPa in September. As in the fall treatments, the resprouts in the spring clip treatments showed seasonal patterns of water potential very similar to those of the spring burn treatments.

Seedling Establishment

Large numbers of seedlings of *Adenostoma fasciculatum* germinated in all of the treatment plots during the first spring following fire or clipping. In March of 1980, there was a mean density of 17 chamise seedlings m^{-2} (170,000 ha^{-1}) in the fall burn plot. There was a strong tendency for clumping of seedlings. Thus, the actual number per m^{-2} was quite variable, ranging from 5 to 32 m^{-2} in our samples. By early April, heavy seedling mortality was already evident due to drought. In July 1980, the mean number of seedlings had dropped to 9 m^{-2}. Seedling density continued to decline later that summer and through the following summer (Fig. 9A). In March 1981, there were 4 seedlings m^{-2} and in April 1982, only 0.5 seedling m^{-2}. Chamise seedlings growing in the shade of resprouting *A. fasciculatum* or other shrubs were often etiolated and smaller than more exposed seedlings, but did not die during the first few years from overshading. Herbivory by deer was a major factor limiting seedling growth, although grazing does not invariably cause death of the seedlings.

In the case of those seedlings which survived the first summer, growth was rapid during the second year. The mean chamise seedling dry weight (above- and belowground) was 36 mg in May of 1980 in the fall burn plot. In June of 1981 their mean dry weight was 1440 mg. By January 1983, maximum seedling height reached 80 cm, although mean height at this time was about 30 to 40 cm.

Seedling establishment following the spring burn was very high. Chamise seedlings first appeared in February 1981 and by March there were 67 chamise seedlings m^{-2} in the spring burn plot. This number declined rapidly through the spring and summer of that year to less than 5 seedlings m^{-2} in November (Fig. 9B). There was little mortality during the following winter and spring months and over 4 seedlings m^{-2} were still present in July 1982.

Initial seedling densities in the spring clip plot were significantly higher than in the burn plot, despite the lack of any heat treatment. At the end of the spring growing season in June 1981, when there were a mean of 21 chamise seedlings m^{-2} in the spring burn plot, there were 273 seedlings m^{-2} in the spring clip plot. These chamise seedlings, however, apparently experienced higher levels of intraspecific and interspecific competition than did the burn plot seedlings. The mean dry weight of burn plot chamise seedlings in late May 1981 (one year after the fire) was 31 mg, while the clip plot seedlings were only 20 mg each. Local differences in fire intensity during the spring experimental burn did not have significant effects on either densities or biomass of chamise seedlings. In May 1981, the mean densities of seedlings in the spring burn plot and in adjacent areas which had experienced somewhat higher or lower fire intensity during the original burn were 33 m^{-2}, 50 m^{-2} and 41 m^{-2} respectively. Mean dry weights of individual seedlings were 35, 38 and 37 mg, respectively.

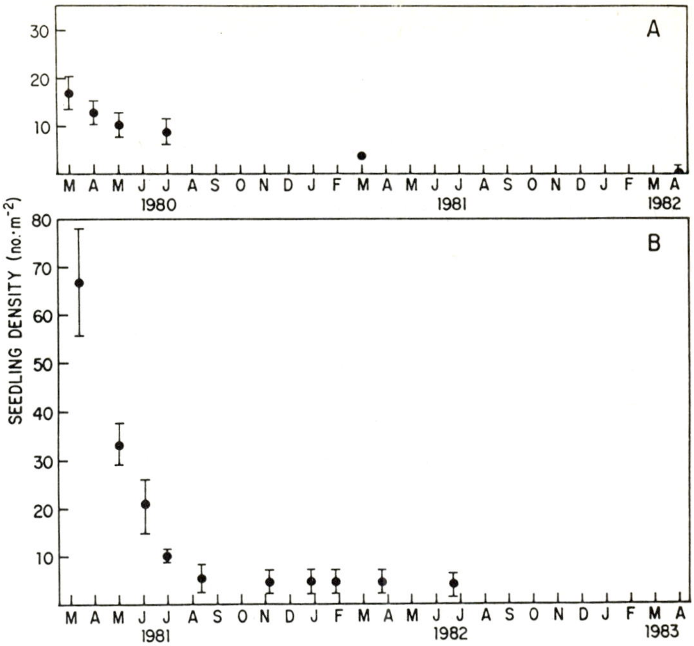

Fig. 9. Postfire densities of *Adenostoma fasciculatum* seedlings in the burn plots: A. Fall burn plot; B. Spring burn plot. Brackets indicate ± 1 standard error

Densities of seedlings on plots within the nearby area burned by the Potwisha wildfire in August 1981 were similar to those in our experimental burn plots. In May 1982, sampling in two areas of the Potwisha burn indicated densities of 52 and 32 chamise seedlings m^{-2}. These seedlings were quite large; the mean dry weights were 114 and 179 mg, respectively.

Minimum water potentials in well-established chamise seedlings were generally higher than in control shrubs and frequently lower than in resprouts. Minimum midday water potentials in vigorous fall burn seedlings were -4.6 MPa in September 1980, compared with -5.8 and -4.6 MPa for control and burn resprouts respectively. Minimum water potentials at the end of the following summer (1981) were -5.3 MPa in the seedlings and > -7.0 and -6.9 MPa in the control and burn plots respectively. Vigorous chamise seedlings in the spring burn plot at this same time (one year after fire) experienced only -3.5 MPa water potential identical to that of spring burn resprouts. In September 1982 there was no significant difference in minimum water potentials between fall burn seedlings, spring burn seedlings, fall burn resprouts and spring burn resprouts (-4.7, -5.1, -5.0 and -4.7 MPa respectively). Water potentials of control shrubs were significantly lower, however, at -6.0 MPa.

DISCUSSION

Resprout Growth

Postfire recovery in root crowns of *Adenostoma fasciculatum* at Sequoia National Park as observed, clearly demonstrates that both fire seasonality and intensity significantly affect levels of mortality. Mortality data from our two experimental burns and from the 1978 Potwisha fire are summarized in Table 1. Since temperature measurements were not made during these fires, intensities are relative values estimated from the mean minimum diameter classes of chamise branches remaining after burning. In each of the three burns, chamise mortality decreased with decreasing fire intensity. Maximum rates of mortality in local areas of intense fire in the June 1980 (spring) burn were 80%, while the clip plot (zero intensity burn) had 36% mortality. The highest level of mortality we have observed was 92% in a high intensity burn area of the 1978 Potwisha fire. This was a very local phenomenon, however, and not at all typical of the burn overall. Only 35% mortality occurred in lower intensity areas of the same fire. The pattern of mortality produced by late fall burns is very different. Our moderate intensity fall burn (November 1979) resulted in a low 14% mortality and the clip plot (zero intensity burn) had no mortality. Again, since our analyses were based on preburn stand structures estimated after the fires and a small number of standing dead shrubs were commonly present, these mortality values are maxima as previously discussed.

Table 1. Mortality of root crown of *Adenostoma fasciculatum* following fire and clip treatments on three dates in Sequoia National Park.

Date	Treatment	Root Crown Mortality(%)
Spring burn June 1980	High intensity fire	79
	Moderate intensity fire	64
	Low intensity fire	46
	Clip	36
Potwisha fire August 1978	High intensity fire	92
	Low intensity fire	35
Fall burn November 1979	Moderate intensity fire	14
	Clip	0

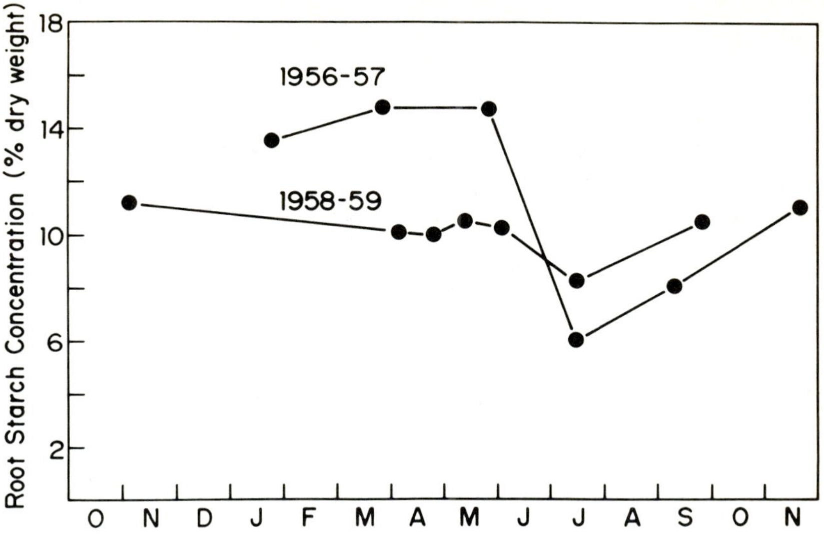

Fig. 10. Seasonal changes in concentration of starch in root tissue of *Adenostoma fasciculatum* during 1956-57 (typical year) and 1958-59 (dry year). Redrawn from Laude *et al.* (1961)

The seasonal effect of fire on burl mortality can be readily explained by physiological data collected more than 20 years ago at the Hopland Field Station of the University of California in the northern part of the state (Jones and Laude 1960; Laude *et al.* 1961). During a "typical" rainfall year such as 1956-57, high levels of starch present in root and burl tissues in the early spring are metabolized and translocated out as shoot growth occurs (Fig. 10). In late summer these tissues change from a source to a sink for non-structural carbohydrates as photosynthate is translocated back from leaf and stem tissues after new growth ceases.

Chaparral fires which burn in late spring or early summer consume aboveground tissues of chamise at a time when there are low levels of carbohydrate reserves in root tissue and thus resprouting ability is limited. Fire or clipping treatments at these times of the year induce relatively high levels of mortality of shrubs, particularly in smaller shrubs with less buffering capacity in their root crowns. By mid-fall when root carbohydrate stores have been replenished, there is little or no mortality caused by removal of aboveground tissues through burning or clipping. Seasonal patterns of carbohydrate allocation between roots and shoots may be highly variable depending on environmental conditions during growth. In dry years such as 1958-59 when there is little new growth, root starch reserves

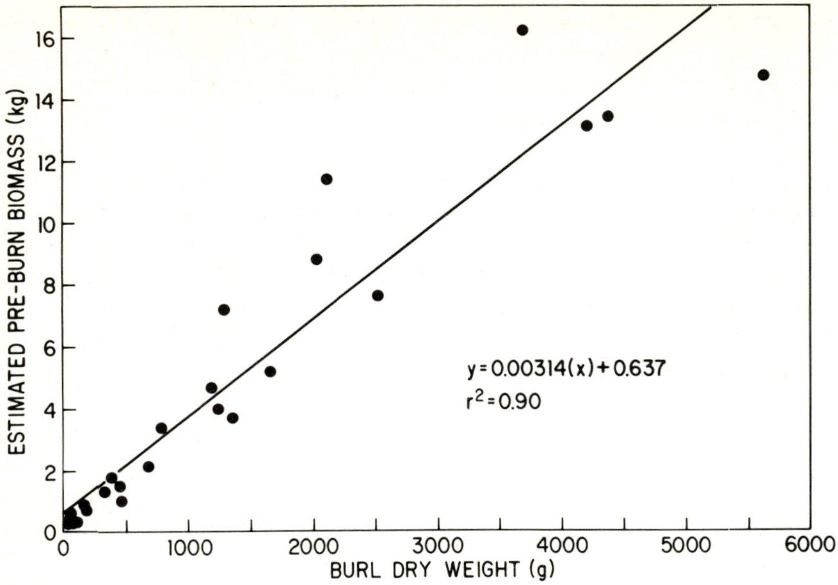

Fig. 11. Relationship between aboveground biomass of mature *Adenostoma fasciculatum* and burl dry weight

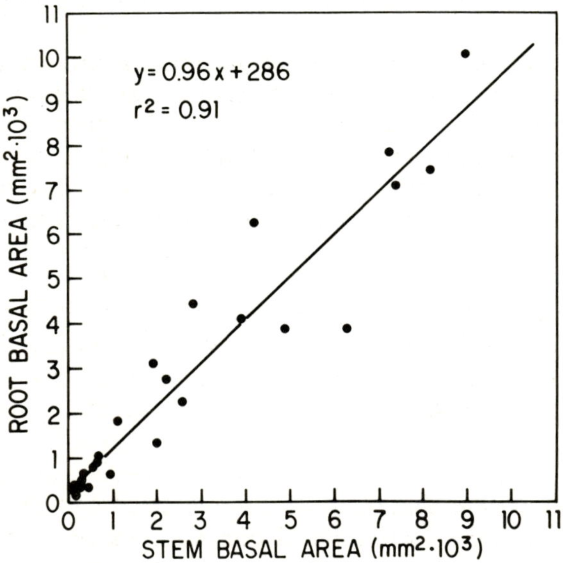

Fig. 12. Relationship between basal stem area of mature *Adenostoma fasciculatum* and basal area of major roots at their burl junction

are maintained at relatively low levels and no major translocation takes place (Fig. 10). Under these conditions when stand flammability is likely to be high, spring or summer fires would be expected to produce lower levels of mortality in *A. fasciculatum* than in wetter years.

Differential patterns of mortality within a single burn are largely attributable to fire intensity. Intense fires burn away major parts of the aboveground tissues and produce a heating front in the soil which is sufficient to kill meristematic tissues on the upper surface of the root burl. Small shrubs with shallow root crowns and less burl biomass are most sensitive to heat damage. Large burls, however, may also be sensitive if they have been exposed by erosion or have built up large areas of dead tissues which may ignite.

The magnitude of postfire resprouting in surviving shrubs of *A. fasciculatum* was clearly related to preburn shrub biomass (Fig. 3), but the factors causing this are potentially of a complex nature. Field excavations in our study areas have established that the aboveground biomass of mature chamise shrubs is linearly related to the burl dry weight (Fig. 11). Large shrubs have larger burls than do smaller shrubs. Thus more storage tissue is present in these larger individuals. Secondly, there is a significant linear correlation between shrub aboveground biomass and the basal area of major roots where they join the burl (Fig. 12). Since this basal area of roots would be expected to be linearly correlated with total root biomass, larger shrubs should have higher absolute root biomass and thus after fires have greater water absorption capacity than smaller shrubs.

Thus, three hypotheses may be invoked to explain differential initial rates of resprout growth and higher levels of net primary production by resprouts from larger preburn shrubs. First, larger shrubs may have more functional absorptive surface of root than small shrubs and thus maintain higher water potentials following fire. A second hypothesis could be that greater burl and resprout biomass in larger shrubs provides improved nitrogen availability for resprouts through either reallocation of stored nitrogen or increased uptake rates following fire (related to greater root extent or surface area). The third hypothesis is that stored carbohydrates provide the major source of carbon for resprout growth during at least the first year, and shrubs with greater preburn aboveground biomass would be able to supply larger volumes of root burl with storage carbohydrates. None of these three hypotheses are mutually exclusive.

If *A. fasciculatum* maintains its root biomass after fire, larger shrubs would be expected to have greater absorptive areas and to maintain less negative water potentials for longer periods of time than would smaller shrubs with less root biomass. Less negative water potentials in resprouts compared to control plants have been reported for *A. fasciculatum* in the coast ranges of Southern California (Hastings and Oechel 1982; Oechel 1982). Our data indicate that water potentials of resprouts were significantly higher than those of control plants throughout the first year of growth and to a lesser degree during the second year (Fig. 8). These data suggest that resprout shrubs may have slowly adjusted their root absorptive surface following fires by shedding absorptive roots. Stand leaf area index in the second year following fire is quite low in comparison to control plots and yet water potentials are very similar for much of the year. Leaf conductances were not measured, however, and thus the differential water relations cannot be fully evaluated. Parsons *et al.* (1981) found a comparable pattern of apparent root/shoot adjustments following drought stress in *Arctostaphylos viscida* in the same area. Such adjustment was not evident in the first year following fire in *A. viscida*.

We were unable to demonstrate significant differences in tissue water potentials of resprouts from the different size-classes of preburn shrubs shown in Fig. 4 at any time through the first postfire growing season. We therefore conclude that water relations are not a significant factor in promoting more rapid resprout growth in larger preburn shrubs over smaller shrubs.

Oechel (1982) and Hastings and Oechel (1982) hypothesized that higher rates of photosynthesis in postfire resprouts of *A. fasciculatum* in comparison with control plants was due to nitrogen enhancement resulting from increased availability of nutrients in ash following fire. These studies found that rates of photosynthesis in postfire resprouts were 3-6 times greater (on a g dry weight of leaf basis) than those of control plants. Resprouts from clipping treatments where there was no nutrient enhancement from ash fall had very similar photosynthetic rates to those of control plants. These data, however, are inconsistent with our observed pattern of productivity and leaf nutrient concentrations. In our studies of *A. fasciculatum*, shrubs in clip plots had consistently higher levels of productivity than did those in paired burn plots. Leaf nutrient analyses at the end of the first postfire growing season in our fall burn plot found no significant difference between nitrogen contents of resprouts in clip and burn plots, although both were much higher than nitrogen contents in control shrubs and slightly lower than those in seedlings (Table 2). Leaf phosphorus contents were actually slightly higher in clip plot resprouts than burn plot resprouts, although this difference was probably not significant. Although we have not yet looked at leaf nutrient contents in relation to shrub size-class, there are no data to suggest that differential nutrient availability is a key factor in producing differential growth responses.

All of our data strongly suggest that differential carbohydrate storage is the key factor in limiting postfire resprout growth. Larger preburn shrubs with larger burls have more available carbohydrate reserves to promote resprouting than do smaller shrubs. Our resprout growth data for the fall burn (Figure 4) suggest that much of the differential pattern of growth in relation to size class was established in the first year following fire. The slopes of biomass change during the 1981 and 1982 seasons (2 and 3 years after the fire) are very close for each size class. This pattern suggests, therefore, that productivity rates are very similar. It is in the first year following fire that the slopes for productivity are

Table 2. Leaf nitrogen and phosphorus contents of *Adenostoma fasciculatum* for first year, postfire seedlings and resprouts from burn and clip plots, and for one and two-year old foliage of adjacent mature shrubs. Data are values for bulked subsamples.

	Nitrogen (mg/g dry wt.)	Phosphorus (mg/g dry wt.)
Seedling	32	2.8
Resprout - burn	28	3.1
Resprout - clip	27	3.3
Control - 1 year old	14	1.8
- 2 year old	12	1.3

very different for respective size-classes. These first year rates are strongly influenced by available carbohydrate reserves.

Seedlings

Although the mortality of seedlings of *A. fasciculatum* was very high in the first year following fire and moderately high in the second year, seedlings which survived to a third growing season appeared to be well-established with a good potential for surviving despite water stress in subsequent years. The fall burn density of 0.5 chamise seedlings m^{-2} in April 1982 provided 5000 seedlings ha^{-1} to potentially become established in the stand. The mean density of *A. fasciculatum* in old-growth stands of chamise at Sequoia National Park is 5450 shrubs ha^{-1} (Stohlgren *et al.* 1984). Following our fall burn, the density of shrubs surviving by resprouting was 3900 shrubs ha^{-1}. Successful establishment of only 19% of the seedlings which survived to the third postfire year would be necessary to reestablish the initial stand density of 4500 shrubs ha^{-1}. To reach the mean regional density of 5450 shrubs ha^{-1} in old-growth chamise at Sequoia National Park (Stohlgren *et al.* 1984), 31% of the 5000 chamise seedlings would have to survive to maturity. In the spring burn plot, where shrub mortality was very high, the postfire density of resprouts was only 1400 shrubs ha^{-1}. The seedling population in July 1982, two years after the fire, was about 40,000 seedlings ha^{-1}. To restore mean stand densities, 3050 chamise seedlings would have to survive to maturity.

Although chamise seedlings would be expected to have significantly shallower roots than resprouts or control shrubs, our water potential data indicates that well-established seedlings, even in the first year of growth, may have minimum midday water potential equal to or higher than those of resprouts and control shrubs. Less vigorous seedlings, however, did show strong indications of drought stress and thus we view such stress as a major cause of first and second-year seedling mortality. While seedling mortality during the first three postfire years of growth are largely attributable to drought stress, shading clearly becomes a significant cause of mortality in later years. Two years following our fall burn the shrub coverage was already 58%, with nearly two-thirds of this total due to coverage by short-lived shrubs such as *Lotus scoparius*, *Malacothamnus fremontii* and *Eriodictyon californicum*. These short-lived shrubs decline rapidly in total coverage, however, and chamise resprouts become the dominant element 5 to 10 years after a fire. Shrub cover reaches 100% in stands about 20 years after fire, at which time the mean chamise canopy height is about 2 m. It is difficult for chamise seedlings to compete well for canopy space in direct competition with resprouts because of the slower seedling growth rates. Seedlings which become successfully established in large openings between resprouts, nevertheless, should compete well for canopy space. Our observations suggest that some of the small chamise shrubs we find under the canopy in mature stands are seedlings which have survived in favorable understory microenvironments.

Scattered data are available describing postfire densities of seedlings of *A. fasciculatum* in other areas. Horton and Kraebel (1955) studied survival of resprouts and seedlings in a series of stands of mixed-chamise chaparral in the San Gabriel and San Bernardino Mountains of southern California. In one of their stands 10 years following fire, there were 5000 chamise resprouts ha^{-1} and 3250 chamise seeedlings ha^{-1}. Two years after this fire there were 5250 seedlings ha^{-1} and three years after the fire 4750 seedlings ha^{-1}. At a stage 25 years following the fire, densities had declined to 4750 resprouts ha^{-1} and 1750 seedlings ha^{-1}, apparently due to herbivory and shade-induced mortality. At an age of 25 years, seedlings had reached a mean height of 81 cm, close to the resprout mean height of 103 cm. In other stands which they studied, there was a mean density of 8250 chamise seedlings ha^{-1} one year after burning. This density dropped rapidly to 4500 seedlings ha^{-1} after two years and 1750 ha^{-1} after three years. At a stage 10 years after fire there were 1250 seedlings ha^{-1}, compared with 6750 resprouts ha^{-1} in the same stands. Mature shrubs of *A. fasciculatum* in their stands were much smaller than ours at Sequoia, and as a result shrub densities were somewhat higher.

Studies of *Adenostoma fasciculatum* by Sweeney (1956) in stands of mixed chaparral in Lake County in northern California reported very high densities of chamise seedlings following fire. In late February following a fire the previous August he found an initial germination of 1.43×10^6 chamise seedlings ha^{-1}. By the end of June this density had declined sharply to 52,400 seedlings ha^{-1}. By June of the second year there was a further drop to 23,500 seedlings ha^{-1} and then to 18,200 seedlings ha^{-1} in the third year. Sweeney suggested that there was no apparent relationship in his study sites between cover of herbaceous species and mortality of chamise seedlings.

In other studies, Christensen and Muller (1975) report spring densities of up to 20,000 chamise seedlings ha^{-1} in plots of chamise chaparral near Santa Barbara in the Coast Range of Central California. Densities of up to 80,000 ha^{-1} were induced in artificially heated and fertilized soils beneath mature stands of chamise. Vogl and Schorr (1972) counted chamise seedling levels of up to 30,830 ha^{-1} one year after a chaparral fire in the San Jacinto Mountains. Howe and Carothers (1980) have described the importance of chamise seedling establishment in local areas in southern California where postfire shrub mortality was high. Keeley and Keeley (1981), however, found a remarkably low rate of chamise seedling establishment in chaparral near San Diego. In many stands they surveyed one year after fires there were fewer surviving seedlings than resprouts of *A. fasciculatum.*

Stone and Juhren (1953) suggested *Adenostoma* possesses two types of seeds with differing mechanism of germination. One group requires fire for germination while the other requires no heat pretreatment to germinate. The large numbers of chamise seedlings which we have observed in our spring clip plot (10 times the level in the adjacent burn plot) supports the conclusion that fire is not a requirement for germination. Only very rarely, however, have we observed any chamise seedling germination under the canopy of our control plot or other mature stands of chamise chaparral. The conventional wisdom has been that germination and/or seedling establishment beneath chamise canopies is limited by allelopathic compounds (Muller *et al.* 1968; McPherson and Muller 1969; Christiansen and Muller 1975b). Our observations are not completely consistent with this theory, however. Chamise seedlings and seedlings of other herbaceous species were abundant beneath chamise canopies along the sharp margins of our spring clip plot, suggesting light was possibly more a limiting factor than allelopathy.

Although we attribute the majority of seedling deaths in the first two years following germination to drought stress, there is no question that vertebrate herbivory is also an important factor. It is very difficult to accurately quantify effects of herbivory, however, because of the great variability from stand-to stand and year-to-year. Mule deer are the major herbivores in our stands at Sequoia. While many first-year chamise seedlings may be consumed completely, herbivory on second-year and older seedlings may remove a large amount of the leaf area without causing mortality. Christiansen and Muller (1975a) suggested that herbivory was extremely important and that virtually any unprotected seedling in their chamise study plots would eventually succumb to grazing. In our stands, however, herbivory appears to be secondary to drought stress in the first two years following fire. One indication of this is the lack of seedling mortality during the winter of 1981-82 in our spring burn plot (Fig. 9B). This is the season of maximum deer activity in the area.

Both resprouting and successful establishment of seedlings are necessary for the maintenance of dominance by *Adenostoma fasciculatum* in stands of chamise chaparral in the southern Sierra Nevada. Although burl mortality following fall fire is relatively low, large numbers of chamise seedlings must survive and grow to maturity if preburn densities of chamise are to be restored. Late summer and fall are the most common seasons for natural fires in this area (Parsons 1981).

Demographic patterns in these stands of chamise chaparral are very different from those in pure stands of *Ceanothus* in the Santa Inez Mountains. In pure *Ceanothus* stands with no resprouting, high densities of seedling *C. megacarpus* became established and gradually thin through shading. Schlesinger and Gill (1978) found a change from clumping to random spacing to regular spacing with decreasing stand densities of individuals. This is what would be expected if intraspecific competition for light was the dominant limiting factor. In chamise chaparral in our area, there is a very different change from a tendency toward regular spacing of *A. fasciculatum* in stands with high density through random spacing to clumped or contagious spacing at low densities in old stands (Stohlgren *et al.* 1984). This pattern reflects the importance of resprouting for stand recovery in contrast to reseeding in stands of *C. megacarpus*. Unusually intense fires or burns in spring or early summer produce high levels of burl mortality in *A. fasciculatum*, and interspecific competition among seedlings may be an important factor in influencing subsequent stand composition. Reseeding species of *Ceanothus* and *Arctostaphylos* may outcompete chamise seedlings under such circumstances and establish themselves as important canopy dominants.

ACKNOWLEDGEMENTS

This study was funded by a Research Contract CK8550-9-004 with the National Park Service. We thank M. Bivon and S. D. Bennedetti for field assistance and the staff of Sequoia and Kings Canyon National Parks for logistic support.

REFERENCES

Christensen NL, Muller CH (1975a) Relative importance of factors controlling germination and seedling survival in *Adenostoma* chaparral. Amer Midland Natur 93:71-78

Christensen NL, Muller CH (1975b) Effects of fire on factors controlling plant growth in *Adenostoma* chaparral. Ecol Monogr 45:29-55

Hanes TL (1971) Succession after fire in the chaparral of southern California. Ecol Monogr 41:27-52

Hastings SJ, Oechel WC (1982) Photosynthesis and water relations of mature and resprout chaparral vegetation. In: Dynamics and Management of Mediterranean-Type Ecosystems. USDA Forest Service Gen Tech Rep PSW-58, p. 602

Horton JS, Kraebel KJ (1955) Development of vegetation after fire in the chamise chaparral of southern California. Ecology 36:244-262

Howe GF, Carothers LE (1980) Postfire seedling reproduction of *Adenostoma fasciculatum* H & A. Bull So Calif Acad Sci 79:5-13

Jones M, Laude HM (1960) Relationships between sprouting in chamise and the physiological condition of the plant. J Range Manag 13:210-214

Keeley JE, Zedler PH (1978) Reproduction of chaparral shrubs after fire: a comparison of the sprouting and seeding strategies. Amer Midland Natur 99:142-161

Keeley JE, Keeley SC (1981) Postfire regeneration of Southern California chaparral. Amer J Bot 68:524-530

Laude HM, Jones MB, Moon WE (1961) Annual variability in indicators of sprouting potential in chamise. J Range Manag 14:23-26

McPherson JK, Muller CH (1969) Allelopathic effects of *Adenostoma fasciculatum*, "chamise," in the California chaparral. Ecol Monogr 39:177-198

Muller CH, Hanawalt RB, McPherson JK (1968) Allelopathic control of herb growth in the fire cycle of California chaparral. Bull Torrey Bot Club 95:225-231

Oechel WC (1982) Carbon balance studies in chaparral shrubs: implications for biomass production. In: Dynamics and Management of Mediterranean-Type Ecosystems. USDA Forest Service Gen Tech Rep PSW-58, p. 634

Parsons DJ (1976) The role of fire in natural communities: an example from the southern Sierra Nevada, California. Envir Conserv 3:91-99

Parsons DJ (1981) The historical role of fire in the foothill communites of Sequoia National Park. Madroño 28:111-120

Parsons DJ, Rundel PW, Hedlund RP, Baker GA (1981) Survival of severe drought by a non-sprouting chaparral shrub. Amer J Botany 68:973-979

Schlesinger WH, Gill DS (1978) Demographic studies of the chaparral shrub, *Ceanothus megacarpus*, in the Santa Ynez Mountains, California. Ecology 59:1256-1263

Stohlgren TJ, Parsons DJ, Rundel PW (1984) Population structure of *Adenostoma fasciculatum* in mature chamise chaparral in the southern Sierra Nevada, California. Oecologia (Berl) 64:87-91

Stohlgren TJ, Stephenson NL, Parsons DJ, Rundel PW (1982) Using stem basal area to determine biomass and stand structure in chamise chaparral. In: Dynamics and Management of Mediterranean-Type Ecosystems. USDA Forest Service Gen Tech Rep PSW-58, p. 634

Stone EC, Juhren G (1953) Fire-stimulated germination: effect of burning on germination of brush seed investigated. Calif Agric 7:13-14

Sweeney JR (1956) Response of vegetation to fire. A study of the herbaceous vegetation following chaparral fires. Univ Calif Publ Bot 28:143-250

Vogl RJ, Schorr PK (1972) Fire and manzanita chaparral in the San Jacinto Mountains, California. Ecology 53:1179-1188
Wells PV (1969) The relation between mode of reproduction and extent of speciation in woody genera of the California chaparral. Evolution 23:264-267

A model of life history strategies of chaparral shrubs in relation to fire frequency

David W. Hilbert

Systems Ecology Research Group, San Diego State University, San Diego, CA 92182

INTRODUCTION

A model was developed which represents the dynamics of fire adapted plants which have strong seed dormancy. This class of plants is unusual in that their population density continually decreases between fires since recruitment only occurs following a disturbance; normally fire.

This model was used to analyze the effect of fire frequency on the fitness of populations which display different life history characteristics and to determine the relative selection pressure on these characteristics as a function of the frequency of fire.

The results suggest that, for shrubs in the California chaparral, characteristics associated with seed production and seed survival are most strongly selected for at intermediate fire intervals, about 5 to 40 years. Attributes related to resprouting and high adult survivorship are selected for when the fire interval is very short, less than 5 years, or when the interval is long, greated than 60 years. When fires occur in the range of every 40 to 60 years, selection is equally strong for both seed related and resprouting related characteristics.

Plants which do not survive fires (obligate seeders) can have fitnesses equal to or greater than fire resprouters at intermediate fire frequencies if their seed production and seed survival rates are sufficiently high.

LIFE HISTORIES OF CHAPARRAL SHRUBS

Fire is a common disturbance in many ecosystems and most mediterranean type ecosystems, including California chaparral, are especially influenced by frequent fire. Consequently the plant species found in chaparral have been suggested to display many adaptations to fire. One of the most widespread and potentially adaptive traits of these plants is seed dormancy. Seed storage until fire, either on the plant or in the soil, is observed in many families of plants from mediterranean climates: Cupressaceae, Casuarinaceae, Proteaceae, and Myrtaceae in Australia (Gardner 1957); Proteaceae in South Africa (Williams 1972); Pinaceae in the Mediterranean (Naveh 1975); and Rhamnaceae, Ericaceae (Wells 1969), as well as Rosaceae and other families in California. While seeds may be produced

NATO ASI Series, Vol. G15
Plant Response to Stress. Edited by J. D. Tenhunen et al.

every year for this class of plants, it is exclusively following a fire that seedlings are produced. The populations of these plants are constantly decreasing, except for bursts of recruitment occurring in the year following fire, and the life history and population dynamics of these plants are consequently quite different than are represented by the common models (e.g. Stearns 1977, Caswell 1982, and Cole 1954) used in the analysis of life histories. These models assume continuous or yearly recruitment and are inappropriate for plants with strong seed dormancy.

Two general life history strategies can be recognized within this class of fire adapted plants with seed dormancy. These are obligate seeders which do not survive fire but generally produce many seedlings, and the resprouters which can survive fire and resprout from basal burls as well as produce seedlings (Keeley and Zedler 1982, St. John 1976, Carpenter and Recher 1979). This paper primarily deals with relatively long lived shrubs but there are also a number of short lived perennials and annuals with seed dormancy which produce seedlings immediately following fire. A recent paper (Malanson 1985) uses modeling to assess competition between coastal sage species which differ in their ability to resprout following fire. However, for these species, fire is not necessary for germination and the model assumes yearly recruitment into the population--a very different situation than is considered in this paper.

A model is described here which represents the population dynamics of plants which display seed dormancy. This model allows one to determine fitness as a function of the number of years between fires and specific life history traits. By determining the sensitivity of fitness to changes in parameters describing life history, it is also possible to assess the relative selection pressure on each trait as a function of fire frequency. The objective of this analysis is to predict the strategies one would expect to evolve under various fire regimes. A comparison of obligate seeders and resprouters is of particular interest since many genera of important chaparral shrubs (e.g. *Ceanothus* [Rhamnaceae] and *Arctostaphylos* [Ericaceae]) contain species displaying one or the other of these strategies (Wells 1969, Keeley 1977a). Resprouters should be at an advantage under most fire regimes and it is difficult to understand the success of obligate seeders (Keeley 1977b). Nonetheless, many obligate seeders in chaparral apparently evolved from sprouting ancestors (Wells 1969, Stebbins 1974). The distribution of species may also be very dependent on fire frequency and the complex environmental gradients determining fire and the analysis is useful in understanding these controls on distribution.

MODEL DERIVATION

The model developed here gives the change in population size or density occurring from the year immediately after one fire to just after another fire (F) years later. The dynamics occurring in the intervening years are not of direct interest except that the patterns of mortality and seed production determine the recruitment of new individuals after a fire. Only one species is considered and there is no competition between species with different life history traits. Rather, the model is intended to give a relative measure of success for any life history type as a function of the number of years between fires.

Two classes of individuals are defined in the model based on whether the plant began the fire free interval as a resprout (N_r) or began this interval as a seedling (N_s). Separation of these two classes is necessary since there are several differences in the characteristics of these two types of plants. Individuals starting the fire free interval, for example, may reach sexual maturity earlier than a seedling of the same species. The dynamics of the population comprising these two classes can be represented by the system of equations:

$$N_s(F) = S_r(F)N_r(0) + S_s(F)N_s(0)$$

$$N_r(F) = R_r(F)N_r(0) + R_s(F)N_s(0) \tag{1}$$

where F is the number of years between fires, $S_r(F)$ is the per capita production of seedlings by resprouts as a function of F, $S_s(F)$ is the per capita production of seedlings by plants starting the interval as seedlings, $R_r(F)$ is the proportion of initial resprouts which are alive and will resprout in year F, and $R_s(F)$ is the proportion of plants starting the interval as seedlings which will resprout after a fire in year F. $N_r(0)$ and $N_s(0)$ are the number of resprouts and seedlings at time zero (i.e. the first year following a fire).

This system of equations can be written in matrix form as

$$\bar{N}(F) = \bar{A}\bar{N}(0) \tag{2}$$

where $\bar{A}$ is the coefficient matrix

$$\begin{bmatrix} S_r(F) & S_s(F) \\ R_r(F) & R_s(F) \end{bmatrix} \tag{3}$$

This matrix is analogous to the population projection matrix in age-classified demographic models. For discreet models of this type, fitness is given by the dominant eigenvalue of A (Caswell 1982) found from the quadratic equation

$$\text{Fitness} = \{(S_r + R_s) + ((S_r - R_s)^2 - 4S_sR_r)^{1/2}\}/2 \tag{4}$$

Note that the coefficients R_r, R_s, S_r, S_s, are all functions of the length of time between fires. These functions can be found as analytic solutions of difference equations representing the yearly dynamics of seedlings, resprouts, and seeds stored in the soil.

Assuming constant mortality rates μ_r for resprouts and μ_s for plants starting as seedlings, the numbers of plants in each year are given by:

$$N_r(t+1) = \mu_r N_r(t)$$
$$N_s(t+1) = \mu_s N_s(t) \qquad (5)$$

These equations have the solutions

$$N_r(t) = N_r(0)\mu_r^t$$
$$N_s(t) = N_s(0)\mu_s^t \qquad (6)$$

The difference equation representing the accumulation of seeds (produced by resprouts) is:

$$S'_r(t+1) = \phi S'_r(t) + \sigma N_r(t+1) \qquad (7)$$

where ϕ is the yearly probability of survival for seeds and σ is the per capita seed production rate. Note that S' is used for viable seeds and S is used for seedlings. For years less than the age of first reproduction (A), the number of seeds is zero. Solving Eq. 7 for the number of viable seeds present in year F and using Eq. 6 for $N_r(t)$ gives:

$$S'_r(F) = \sigma N_r(0)(\mu_r/\phi)^A(-\phi/(\mu_r-\phi))\phi^F + \sigma N_r(0)\mu_r^{(F+1)}/(\mu_r-\phi) \qquad (8)$$

where $S'_r(F)$ is zero for all F less than the age of first reproduction. A similar equation is found for $S'_s(F)$ by substituting the appropriate parameters (μ_s and A_s).

Eqs. 6 and 8 are used to specify the coefficient matrix (Eq. 3) and determine fitness for any given fire interval and set of life history parameters from Eq. 4. Specifically:

$$R_r(F) = \lambda \mu_r^F$$
$$R_s(F) = \lambda \mu_s^F \qquad (9)$$

where λ is the probability that a shrub alive in year F will survive the fire and resprout. And also:

$$S_r(F) = \gamma S'_r(F)/N_r(0)$$
$$S_s(F) = \gamma S'_s(F)/N_s(0) \qquad (10)$$

where γ is the probability that a viable seed present in year F will survive the fire and germinate.

METHOD OF ANALYSIS

From Eq. 4 with the coefficient matrix specified by Eqs. 9 and 10 it is possible to plot fitness as a function of fire interval for any particular set of life history characteristics. An example is presented in Fig. 1. While this provides some insight into particular cases, it is difficult to draw any general conclusions from such plots. The difficulty is that the specific shape of the fitness curve is dependent on all the life history parameters. This would be much more useful if we knew these parameters for specific shrub species. Complete data of this kind are not available, however, and the aim of this paper is to draw very general conclusions. A very useful and general analysis would be to determine the sensitivity of fitness to each parameter. This sensitivity is proportional to selection pressure on each trait (Caswell 1982) and is defined as the partial derivative of fitness (from Eq. 4) with respect to the parameter of interest. These sensitivities can be obtained analytically but the results are very complex algebraic expressions involving most of the parameters. In other words, the selection pressure on each trait is a complex function of the entire suite of life history traits and fire interval.

To overcome these problems, a computer algorithm was developed which calculates the fitness of 19,694 different combinations of parameters in a complete factorial design. Low, medium, and high values were chosen for each of the nine life history parameters in Eq. 4 using reasonable estimates from the literature. Fitness for all possible combinations (3^9) of parameters were calculated and correlated with the parameter values. This analysis was performed for ten different values of F, the fire-free interval. The parameter values are presented in Table 1. Note that the range of parameters includes plants which are obligate seeders ($\lambda = 0$) with high seed production ($\sigma = 10{,}000$) and strong resprouters with low seed production ($\lambda = .80$ and $\sigma = 100$). The correlation coefficient calculated in this way for each of the parameters is an average measure of sensitivity or relative selection pressure. While not identical to standard mathematical sensitivity (Tomovick 1960), correlation has been used as a measure of parameter sensitivity (Rose 1983). Here it has the advantage of being a general statement of relative selection pressure on each trait as a function

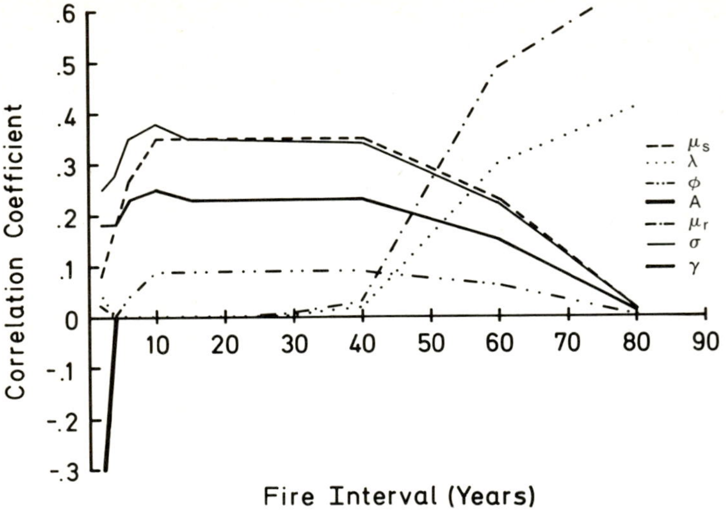

Fig. 1. Fitness as a function of length of the fire free interval for two hypothetical shrubs, one of which is a resprouter (----) and the other is an obligate seeder (- - - -). Parameters for the resprouter and seeder are:

	λ_r	λ_s	μ_r	μ_s	A	A_s	ϕ	σ	γ
resprouter	.8	.4	.995	.65	2	6	.25	1000	.6
seeder	0	0	0	.85	0	4	.60	80	.6

Notice that despite a low seed production rate and the inability to resprout, the obligate seeder is more fit at intermediate fire frequencies due to its much higher seed survivorship

of fire frequency for plants falling within the range of life histories specified in Table 1. In effect, the correlation coefficient is the average sensitivity of all 3^9 plants to changes in each parameter. The sensitivity (selection pressure) for a trait in any specific plant still depends on the value of all the other traits. By using a correlation coefficient in this way one can average out all the variability due to these specific choices of parameters and look at the average effect of each parameter individually.

The number of cases for which fitness must be calculated (nearly 20,000) makes the analytic solutions essential. If it were not for these solutions, the yearly dynamics of shrubs and seeds would have to be simulated for every case in order to determine fitness.

Table 1. Parameter values used in the factorial sensitivity analysis. The range of values is intended to include the expected life history characteristics of chaparral shrubs in general.

Parameter	Low	Medium	High
γ	.01	.30	.60
σ	100	1000	10000
μ_r	.75	.85	.99
μ_s	.45	.65	.85
A_r	1	3	5
A_s	4	6	8
λ_r	0.0	.40	.80
λ_s	0.0	.20	.40
ϕ	.10	.30	.50

Parameter definitions:

- γ the probability that a viable seed will survive a fire and germinate.
- σ the yearly per capita seed production rate.
- $\mu_{r\ \mathrm{or}\ s}$ yearly probability of survival of plants beginning the fire free period as resprouts or seedlings.
- $A_{r\ \mathrm{or}\ s}$ the age of first reproduction.
- $\lambda_{r\ \mathrm{or}\ s}$ the probability of surviving a fire and resprouting.
- ϕ the yearly probability for seed survival.

RESULTS

Results from the parameters sensitivity analysis are presented in Fig. 2. Two parameters, the probability of resprouting by an individual beginning the fire free interval as a seedling (λ_s) and the age of first reproduction of a seedling (A_s), are not included because they were of little significance at all fire intervals.

In general, life history characteristics associated with the seeder strategy (σ, μ_s, γ, and ϕ) are most selected for when the fire-free interval is short or intermediate. For this range of fire intervals, the per capita seed production rate and survivorship of seedlings are most closely correlated with fitness. This is followed by the probability of a seed surviving a fire and germinating and by the survivorship of stored seeds. For very frequent fires (fire intervals less than the

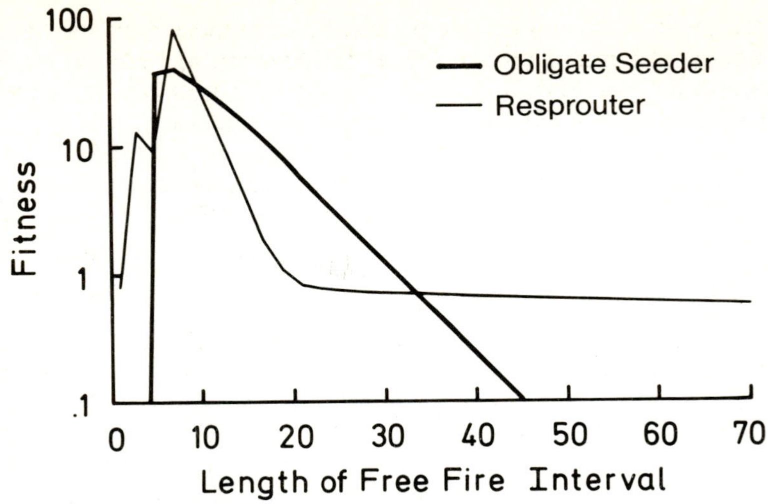

Fig. 2. Correlation of each life history parameter and fitness for a range of fire free intervals of 2 to 80 years

average age of reproduction) the age of first reproduction is highly negatively correlated with fitness. Resprouting is then of great importance.

At fire intervals greater than 40 years, the selection pressure on sprouting characteristics should increase and then surpass selection for seeder characteristics at intervals longer than 55 years. In the model, these characteristics are the survivorship of plants which start the interval as resprouts and the ability to resprout.

CONCLUSIONS

A number of conclusions can be drawn from this analysis. The model suggests that obligate seeders can have a higher fitness at intermediate fire frequencies than plants which both resprout and seed. This is possible despite lower average plant survivorship and seed production rates if the seed survivorship rate is sufficiently high. The sensitivity analysis further implies that obligate seeding characteristics should be selected when fire free intervals are in the approximate range of five to 40 years. In the years immediately following attainment of sexual maturity the density of shrubs is relatively high compared to later in the fire cycle. Consequently, total seed production by the population is large and this, coupled with the comparatively short time that seeds must survive, leads to a large stored seed pool which can produce many seedlings following a fire. Resprouting is not highly selected for at this time since the post-fire population will be dominated by seedlings. Similarly, high survivorship of resprouts is not at all significant when compared to that of seedlings.

When fires occur at intervals of 40 to 60 years, currently the typical fire frequency in the chaparral of southern California, a balance of sprouting and seed production should be selected for. Interestingly, the dominant shrub here (*Adenostoma fasciculatum*) displays this type of mixed strategy.

Resprouting and high shrub survivorship should be selected for when fires are infrequent (> 60 years apart) and when fires occasionally occur very frequently (less than the age of first reproduction). High shrub survivorship is of obvious importance when the fire interval is long. What is less obvious is that the seed pool at this time is expected to be small due to fewer individuals producing seeds and the mortality or inviability of seeds produced earlier in the fire interval when the shrub population was higher. Consequently, the post-fire population is more dominated by resprouts, and species with high resprouting ability will be relatively more successful at these long fire intervals. In the case of very short fire intervals, the ability to resprout is obviously essential if most of the plants have not yet attained sexual maturity.

Conclusions from this analysis are a direct outcome of the assumptions employed in constructing the model and are accurate only insofar as the model represents typical fire adapted plants. In this regard, there are several modifications which could be made to strengthen the model. In particular the assumption of Type 1 survivorship for plants and seeds could be changed to a more general function such as the Weibull function. Unfortunately, this would make analytic solution impossible but simulation could then be employed. Some form of competition between resprouts and seedlings of resprouting species could also be added. One may, for example, expect that individual seedlings will be more successful at long fire intervals since the density of resprouts would be low. However, density dependent controls of this type would be difficult to include in the present formulation. On the other hand, resprouting and survival of seeds in a fire could easily be made functions of the fire interval without losing the ability to find analytic solutions.

While the model is relatively simple, the use of analytic solutions and numerical techniques provide a powerful method for analysis of the life histories of fire adapted plants. Such models may also help to explain the patterns of distribution of species in fire dominated ecosystems and assist in empirical studies by providing a formal theoretical background upon which data collection can be based.

ACKNOWLEDGEMENTS

Discussions with Drs. P. Zedler, F. Kruger, E. Moll, and Mr. S. Sparks significantly aided the model development. This research was supported by a grant from the National Science Foundation DEB 80-25977.

REFERENCES

Carpenter FL, Recher HF 1979 Pollination, reproduction, and fire. Am Nat 113:871-879

Caswell H 1982 Life history theory and the equilibrium status of populations. Am Nat 120:317-337

Cole LC 1954 The population consequences of life history phenomena. Quart Rev Biol 29:103-137

Gardner CA 1957 The fire factor in relation to the vegetation of Western Australia. West Aust Nat 5:166-173

Keeley JE 1977a Fire-dependent reproductive strategies in *Arctostaphylos* and *Ceanothus*. In: Proceedings of the symposium on the environmental consequences of fire and fuel management in Mediterranean ecosystems. United States Forest Service, Washington, DC. pp 391-396

Keeley JE 1977b Seed production, seed populations in soil, and seedling production after fire for two congeneric pairs of sprouting and nonsprouting chaparral shrubs. Ecology 58:820-829

Keeley JE, Zedler PH 1978 Reproduction of chaparral shrubs after fire: A comparison of sprouting and seeding strategies. Am Midl Nat 99:142-161

Malanson, GP 1985 Simulation of competition between alternative shrub life history strategies through recurrent fires. Ecological Modelling 27:271-283

Naveh Z 1975 The evolutionary significance of fire in the Mediterranean region. Vegetatio 29:199-208

Stearns SC 1977 The evolution of life history tactics: A critique of the theory and a review of the data. Annu Rev Ecol Syst 8:145-171

Stebbins GL 1974 Flowering Plant Evolution Above the Species Level. Harvard Univ Press, Cambridge, Massachusetts

St. John TV 1976 The dependence of certain conifers on fire as a mineralizing agent. PhD Thesis, Univ California, Irvine

Wells PW 1969 The relationship between mode of reproduction and extent of speciation in woody genera of the California chaparral. Evolution 23:264-267

Williams IJM 1972 A revision of the genus *Leucadendron* (Proteaceae). Contr Bol Herb 3:1-425

Natural and prescribed fire: survival strategies of plants and equilibrium in mediterranean ecosystems

L. Trabaud

Section Etudes des Systèmes Ecologiques, Centre Emberger, CNRS, BP 5051, 34033 Montpellier Cedex, France

INTRODUCTION

Most species growing in ecosystems where fire is common, possess adaptive traits that allow them to survive fire or easily regenerate and reproduce after fire. It is often difficult to prove that known traits are really adaptive for fire, because many traits appear advantageous to organisms in their response to various disturbances and constraints of the environment as well. Since fire may occur at any stage in plant development, fire adaptations may similarly be associated with any particular phenological state and with function of all organs throughout the plant life cycle.

Nevertheless, in any particular species, certain traits may be more pronounced and may appear to be of special significance as concerns the passage of fire. Thus, various trends in modification of sexual or vegetative reproductive behavior have been associated with the frequency or the intensity. These may be grouped into the four following categories (Gill 1977, 1981):

1 - Vegetative reproduction of perennials through sprouts which may appear at any time in the life cycle depending on the location of dormant buds, their protection by soil or bark, and physiological response to the immediate environment.

2 - Sexual reproduction promoted by increased stimulation of flowering after fire.

3 - Increased seed liberation and dispersal due to fire.

4 - Stimulation of germination by fire.

Sprouting of perennials after fire depends on the survival of buds having vascular connections with the belowground organs. Subterranean organs are protected from fire by the soil which is a good insulator and conducts little heat produced by the burning vegetation (Packham 1971; Aston and Gill 1976; Trabaud 1979). Mooney and Dunn (1972) found that nearly 50% of small woody shrubs in California and Chile sprouted after fire. Kruger (1977) suggested that approximately 65% of south African fynbos species behaved in the same manner.

NATO ASI Series, Vol. G15
Plant Response to Stress. Edited by J. D. Tenhunen et al.

Naveh (1974) in Israel and Trabaud (1980a) in Southern France found that nearly all the woody species of these regions sprout after fire. Many herbaceous perennials behave in the same way (Daubenmire 1968; Vogl 1974; Trabaud 1980a). Most trees of the California chaparral (Hanes 1971) and of the Mediterranean Basin garrigues and maquis (Trabaud and Lepart 1980) produce sprouts. With the exception of perhaps 12 to 15 species, all eucalypts possess lignotubers (McArthur 1968). As the plant grows, the lignotubers also increase in size and progressively become buried in the soil. Mallee eucalypts of small size which are frequently burned may possess lignotubers that grow continuously. Many living buds located on these lignotubers survive fire due to their subterranean position and sprout after fire.

According to Wells (1969) sprouting is a very ancient adaptation and sprouting occurs only when foliage is destroyed by an external agent. As long as leaves are alive, inhibiting factors such as those active in apical dominance prevent the buds from becoming active. When foliage is removed, so is the inhibition and then dormant buds produce sprouts (Gill 1977). Bark, like soil, is an effective insulator (Hanes 1961; Martin 1963). Bark thickness is the most important factor determining fire resistance of trees. Numerous eucalypt species have fire-resistant bark (Vines 1968; Gill and Ashton 1968) as does *Quercus suber*. Large numbers of shoots are produced from the trunks and branches of these trees when fires are not extremely severe but result in defoliation.

Fire can stimulate fire resistant plants to produce inflorescences. This has expecially been noted in families of monocotyledons. Increased flowering after fire has been observed among the terrestrial Orchidaceae (Naveh 1974 in Israel; Hall 1959 in South Africa; Gill 1980 in Australia). In South Africa, flowering of geophytes in the Amaryllidaceae, Iridaceae and Liliaceae after fire is a characteristic of fynbos vegetation formations (Kruger 1977). In Australia, the stimulatory effect of fire on flowering has been studied in detail with *Xanthorrhoea australis* (Gill 1977, 1981; Gill and Ingwersen 1976). Many mechanisms seem to be involved in the process of flower induction as a result of fire. Prolific flowering is followed by increased seedling production.

Seed retention on the plant and fire stimulated dehiscence also appears to be an important adaptation to the occurrence of fire. If seedlings cannot find "safe sites" and become established, annual liberation of seeds by perennials will not ensure reproduction. If however, seed dispersal coincides with the appearance of highly favorable germination conditions, reproduction will be ensured. Thus, many species have dehiscent woody fruits that release their seeds shortly after fire. Several species of pines possess serotinous cones that dehisce only after experiencing the extreme heat of fire. Another example of massive release of seeds after fire followed by establishment of seedlings is provided by *Banksia ornata* (Proteaceae), a shrub often dominant in the mediterranean heath of southern Australia (Specht *et al.* 1958). *Eucalyptus* species release a few seeds each year. In the absence of fire, seeds fall two or four years after maturity.

Liberation of seeds can be greatly increased after fire. Seed release is most marked after the most intense fires (Cremer 1962; Mount 1969).

The high level of seed germination in recently burned plant communities can generally be attributed to the release of seeds stored by plants or to stimulated germination of seeds buried in the soil for many years. In some cases, seeds may be transported to the burned area from sources located in neighbouring unburned areas. Germination after fire of seeds stored in the soil has been attributed to two main causes: alteration of the structure of the seed coat or change in allelopathic relations (Muller *et al.* 1968; Christensen and Muller 1975). The response of seeds to high temperatures depends on the duration of exposure as well as heat intensity. In nature, the response of the buried seeds depends on soil temperature and length of exposure (Christensen and Kimber 1975) and on the position of seeds in the soil (Trabaud 1970, 1980a, b). Long periods of exposure to lower temperatures will produce the same result as short exposure to higher temperatures, whether the response is germination or death of the seed. This has been observed for forest species of eastern Australia (Floyd 1966), and the garrigues of southern France (Vuillemin and Bulard 1981). After fire, many annual or biennial species germinate and flower abundantly, then disappear in two or three years. This has been observed in the Californian chaparral (Muller *et al.* 1968), Australia (Specht *et al.* 1958), South Africa (Kruger 1977), Israel (Naveh 1974) and France (Trabaud and Lepart 1980).

The studies reported here were undertaken to determine the relative importance of these various adaptive traits for plants of the Mediterranean Basin. Two types of information are considered in this evaluation: (1) the observed development and reestablishment of the vegetation during a period of 10 years in plots burned by wildfires and (2) the appearance of newly germinated seedlings in soil samples taken from the field and studied under controlled laboratory conditions.

STUDY AREAS AND METHODS

Plots Burned by Wildfires

Vegetation development was followed on 36 recently burned plots located near Montpellier (Trabaud 1980, 1983; Trabaud and Lepart 1980) in the following plant communities representative for this area[1] :

- dense coppice-woodlands of *Quercus ilex* (7 plots);
- dense garrigues of *Quercus coccifera* (10 plots);
- pine forests of *Pinus halepensis* (7 plots);
- garrigues of *Rosmarinus officinalis* (6 plots);

[1]Nomenclature follows Tutin *et al.* (1964-1980), *Flora Europaea.*

- swards of *Brachypodium retusum* (3 plots);
- swards of *Brachypodium phoenicoides* (3 plots).

At each site, a permanent transect line was established 20 m long with reference stakes cemented in the soil. At each 10 cm along the transect, a sighting needle was used to view the presence of species along a straight line below the observation point. An estimate was obtained that described the layering of the vegetation and abundance of each species. In addition, a floristic list was made for all species present in a plot of 100 m^2 which encompassed the transect line. All observations were carried out in the spring during the first, second, fifth and tenth year after fire.

Comparative Field and Laboratory Experiments

Two areas were chosen to compare the appearance of species in the field with those appearing in soil samples from the selected sites in the laboratory. The first area included both a *Pinus halepensis* forest and a *Quercus coccifera* garrigue which burned on the same day due to a wildfire. This area was located 20 km northeast of Montpellier (Trabaud 1970). The second area chosen for study was part of an experimental *Quercus coccifera* garrigue area 10 km north of Montpellier where prescribed burn plots have been established (Trabaud 1980a, b; Boulet 1985). Floristic observations were carried out directly in the pine forest and the garrigue on areas of 100 m^2 each, 14 days, 3 months and 10 months after the wildfire. In the experimental area, floristic lists were constructed just before the prescribed burns and 12 months after on two series of three plots of 50 m^2 sorted at random.

Soil samples were taken from each of the two areas and transferred to a greenhouse. For the wildfire plots, two soil samples 50 x 50 cm and 5 cm deep, were gathered four days after the fire in each of the two communities. For the prescribed burn plots, one sample was taken from each of the six random plots in the same manner as described above immediately before and after the burns. In each case, the samples were separated into strata: litter, 0-2 cm, 2-5 cm. In the context of this paper, we discuss each of these layers as replicates. All the samples were put in large plastic pots, were set upon a sterilized sand layer, and were protected from external deposits by a gauze. The pots were held at relatively constant ambient temperature (about 20° C), were exposed to daylight, and received regular waterings. The number of seedlings germinating from the propagules present in the samples from the burned as well as from unburned plots were recorded at intervals. During the observation periods soil samples were turned over to allow new propagules to germinate.

RESULTS

Life Forms and Survival Strategies

All of the species within the communities studied, survive a more or less frequent passage of fire (Trabaud 1970, 1980; Trabaud and Lepart 1980, 1981). Their persistence depends on the survival strategies they employ. One approach to examining the reestablishment of plants after wildfires is to compare the distribution of life forms (Raunkiaer 1905) which characterize the taxa of adult size in each community. The proportion of the different life forms remain practically the same 1 year or 10 years after a wildfire in *Quercus* coppice woodlands, *Quercus* garrigue, and in the sward communities (Fig. 1). In the *Pinus halepensis* forests and the *Rosmarinus* garrigue, there is a progressive reestablishment of phanerophytes offset by a decrease in hemicryptophytes as the communities get older.

Therophytes, plants most often characterizing young communities, are practically non-existent after fire except in the communities dominated by grasses (*Brachypodium* swards). The therophytes are a natural floristic component of the swards (Bharucha 1932; Soroceanu 1936), but even here they are present in low frequency and decrease as the communities age. Therophytes have almost disappeared from the other communities five years after wildfire. Apparently therophytes can establish during the first years after fire in the absence of strong competition from species of the original communities which have not yet achieved continuous cover (Trabaud and Lepart 1980).

Table 1 shows the percent of the time (including all samplings over a 10 year period) that resprouting species were found to make up the entire sample and the estimated contribution of resprouting species to the vegetation. Only in the Aleppo pine forests and the rosemary garrigues do non-sprouters play a significant role. *Pinus halepensis* and *Rosmarinus officinalis*, the dominant species of these communities, reproduce only by seed after fire.

Germination Capacities in the Field and under Laboratory Conditions

Area 1 - pine forest and the garrigue burned by wildfire: Fourteen days after the wildfire, no seedlings or sprouts were apparent at the soil surface. Three months after fire, only a few perennial species, which later became most abundant at the site, had resprouted. Ten months after the fire, 50 species were observed in the pine forest. At the same point in time, only 30 were observed in the garrigue (Table 2). In both habitats, more than 50% of the species which had appeared reproduced both by seeds and/or sprouts. Very few were annuals or biennials. Species reproducing exclusively by seeds (either perennials, annuals or biennials) were more numerous in the pine forest than in the garrigue.

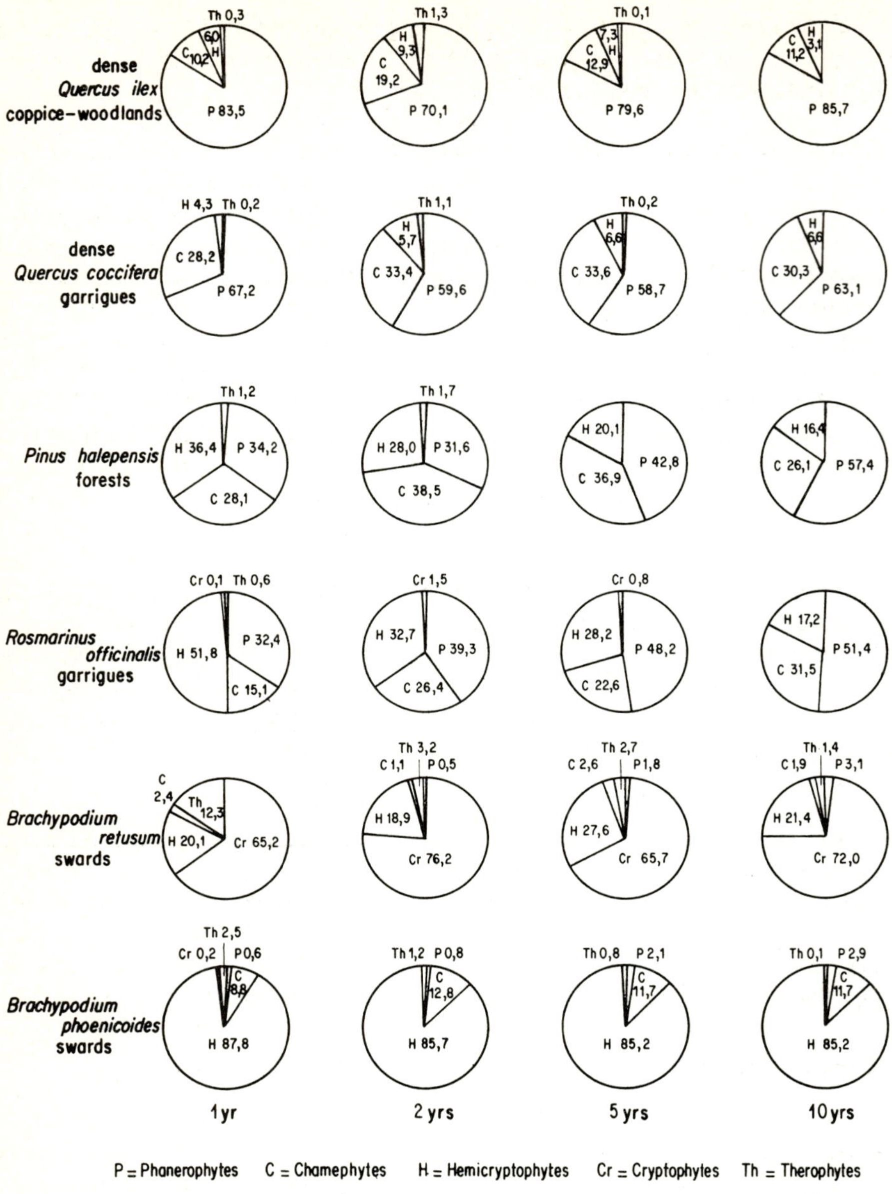

Fig. 1. Post-fire development of life forms in the various communities studied

Table 1. Percent of the time (including all samplings over a 10 year period) that resprouting species were found to make up the entire sample and estimated contribution of resprouting species to the vegetation

	% of time	Percentage of vegetation made up by sprout regenerating species
Dense coppice woodlands of *Quercus ilex*	79.5	80 - 100 %
Dense garrigues of *Quercus coccifera*	77.9	80 - 100 %
Forests of *Pinus halepensis*	52.2	40 - 60 %
Garrigues of *Rosmarinus officinalis*	64.7	20 - 50 %
Swards of *Brachypodium retusum*	99.4	70 - 90 %
Swards of *Brachypodium phoenicoides*	76.5	70 - 80 %

Table 2. Percentages of species present in the pine forest and the garrigue burned by a wildfire possessing various survival strategies.

	Pine forest		Garrigue	
Number of months after the wildfire	3	10	3	10
	%		%	
Perennials preferentially regenerating by sprouts	50.0	28.0	55.5	36.6
Perennials regenerating both by sprouts and seeds	45.4	36.0	44.5	50.0
Perennials reproducing only by seeds	4.6	18.0	0	6.7
Annuals and biennials	0	18.0	0	6.7
Total number of species	22	50	9	30

Table 3. Percentages of species and seedlings in the soil samples from the pine forest and the garrigue burned by a wildfire separated according to their most common survival strategies.

	Pine forest		Garrigue	
	Species	Seedlings	Species	Seedlings
	%		%	
Perennials preferentially regenerating by sprouts	0	0	0	0
Perennials regenerating both by sprouts and seeds	25.0	18.3	33.4	6.4
Perennials reproducing only by seeds	16.7	3.4	33.3	17.8
Annuals and biennials	58.3	78.3	33.3	75.8
Total number	12	323	15	281

In the laboratory (Table 3), 15 species germinated from the garrigue soil samples compared to 12 from the pine forest. In samples from the pine forest, 58.3% of the species which appeared were annuals or biennials; but they occurred in greater numbers so that 78.3% were annual seedlings. In samples from the garrigue, annuals and biennials comprised only 33.3% of the germinated species, but 75.4% of the seedlings, a percentage comparable to the pine forest. Most of the germinated seedlings common to both communities belong to annual or biennial species. It should be noted that there was a total absence of perennial species which regenerate preferentially by sprouts.

Area 2 - the prescribed-burn plots of the *Quercus coccifera* garrigue: This area exhibited a great floristic stability. Only a few new species appeared after burning (Table 4). Before burning, 90% of the species present in the plots were perennials regenerating by seeds and sprouts or by sprouts only. This pròportion was 80% after burning. Annuals and biennials which were practically non-existent before the prescribed burns appeared in the plots after the burns. In soil samples removed from the plots before and after burning, there was an absence of seedlings of plants which regenerate preferentially from sprouts (Table 5). The number of annuals and biennials germinating increased dramatically after burning (they represent at least 23% for the species number and 25% for the

Table 4. Percentages of the species observed in the experimental plots of the *Quercus coccifera* garrigue before and after the prescribed burns exhibiting various survival strategies.

	Plots before burns						Plots after burns					
	A	B	C	D	E	F	A	B	C	D	E	F
	%						%					
Perennials preferentially regenerating by sprouts	40.9	40.0	57.1	42.3	46.4	43.4	37.5	36.4	48.5	42.3	44.8	45.4
Perennials regenerating both by sprouts and seeds	50.0	50.0	37.7	46.1	50.0	47.8	50.0	45.4	30.3	46.1	48.3	40.9
Perennials reproducing only by seeds	9.1	6.7	7.1	11.5	3.6	8.7	12.5	9.1	6.1	3.8	0	0
Annuals or biennials	0	3.3	0	0	0	0	0	9.1	15.1	7.7	6.9	13.6
Total number of species	22	30	28	26	28	23	24	33	33	26	29	22

Table 5. Percentages of species and seedlings in the soil samples from the *Quercus coccifera* garrigue separated according to their most common survival strategies.

	Samples before burnings						Samples after burnings					
Number of Replications	6		9		6		6		9		36	
	Species	Seedlings	Species	Seedlings	Species	Seedlings	Species	Seedlings	Species	Seedlings	Species	Seedlings
	%						%					
Perennials preferentially regenerating by sprouts	0	0	3.7	0.3	6.2	0.5	0	0	4.5	0.5	6.9	0.9
Perennials regenerating both by sprouts and seeds	36.3	48.5	48.1	54.1	50.0	64.5	70.6	59.8	50.0	53.1	47.7	24.9
Perennials reproducing only by seeds	18.2	12.9	14.8	12.3	15.6	9.8	5.9	1.0	18.2	6.1	11.3	8.4
Annuals and biannuals	45.4	38.6	33.3	33.2	28.1	25.1	23.5	39.1	27.2	40.3	34.1	65.7
Total number	11	132	27	301	32	474	17	284	22	211	44	1779

seedlings). The tendency was the same for perennials reproducing only by seeds. The percentages remained identical for species regenerating by sprouts and seeds.

DISCUSSION AND CONCLUSION

Most of the perennial plants of the communities studied are able to produce sprouts; several regenerate both by seeds and sprouts. Among the woody perennials found to occur in the study plots, only *Argyrolobium zanonni, Cistus monspeliensis, Pinus halepensis* and *Rosmarinus officinalis* reproduce solely by seeds. Most of these appear only during the spring following a fire. Those species which regenerate vegetatively, rapidly reoccupy the burned areas. Due to their highly competitive trapping of resources, they prevent alien pioneer species from invading the areas. This dominance by established species is typical of most of the communities studied, since in these communities, except pine forests and *Rosmarinus* garrigue, 80% or more of the recorded species immediately regenerate by sprouts. Based on these data, it is, therefore, easy to understand why the plant communities of the mediterranean region remain after burning similar to those which existed before fire. In the field, perennial species regenerating by seeds and sprouts or sprouts only, occur in the highest percentages (always more than 60%); perennial species reproducing solely by seeds may make up as much as 18%; and annuals and biennials are practically non-existent.

The data from the soil samples indicate that perennials reproducing solely by seeds and annuals and biennials invest their effort at establishment after a burn. In the plant communities of the French Mediterranean area, there are lots of seeds buried in the soil which are able to develop when conditions are advantageous. Nevertheless, fire does not notably modify the floristic composition of the communities studied. Those species which appear somewhat successful at colonizing new areas are plants producing many wind-borne seeds. All are annuals or biennials. They only occupy a small part of the ground surface and are represented by only a few individuals. Due to rapid revegetation of the bare ground by plants of the communities existing before fire (Trabaud 1980; Trabaud and Lepart 1980), plants available as buried seeds do not appear to germinate or germinate only in very small numbers. Short-lived species are only numerous during the second and third year after fire. They do not constitute an important aboveground biomass component except in *Cistus* stands, Aleppo pine forests and *Rosmarinus* garrigues (Trabaud 1980; Trabaud and Lepart 1980).

This trend toward recovery of original community types is found in other mediterranean areas. In California, many chaparral shrub species sprout from stumps and others from rhizomes (see Rundel *et al.* this volume). Most chaparral species sprout quickly, and grow and spread rapidly. They have an advantage over species that can reproduce only by means of seeds, because shoots soon become larger and more vigourous than seedlings (Sampson 1944; Sweeney

1956; Hanes 1971). In a few years, these species regain their prefire density and coverage. Chaparral seeds remain dormant yet viable in the ground litter for a long time. Most species (*Adenostoma, Ceanothus* and *Arctostaphylos*) require the intense heat produced by fire in order to germinate (Quick 1935; Stone and Juhren 1951), but others not requiring the stimulus of heat germinate promptly. Those species that produce seeds are adapted to infrequent fires whereas those that sprout can resist repeated fire (Keeley 1981; see also Hilbert this volume). Dominant chaparral species as well as dominant garrigue species (Trabaud 1980a) are often those that adopt both strategies and can regenerate by means of either seeds or sprouts.

Lignotubers and burls are found in shrubs of other Mediterranean countries including Australian mallee, where they may be highly developed (Gill 1975, 1981) and in South Africa, where they are quite common (Kruger 1977). In Australian heath shortly after fire, a number of plants sprout from underground root systems (Specht *et al.* 1958). There are few annuals, especially grasses, and these appear only during the first year after fire. They are never abundant later as in the chaparral. Purdie (1977) found exactly the same trend in eucalypt forests of southeastern Australia. Eucalypts are well adapted to regenerate after fire. If the fire is intense, seedling regeneration can be excellent. However, if the fire burns the foliage without killing the trees, epicormic sprouts appear in large numbers (Mount 1969; Ashton 1981; Christensen *et al.* 1981). Most plants of the South African fynbos resprout after fire and many species exhibit pronounced positive reproductive response to fire through increased flowering.

The species that were studied can survive fire by:

1) having an enormous underground system which allows them to produce numerous and vigourous sprouts. These plants are chamaephytes or phanerophytes when they reach adult size; they regenerate vegetatively from burls or suckers; and immediately after fire they behave as geophytes: *Q. coccifera, Q. ilex, B. retusum,* etc.;

2) producing bulbs on rhizomes (geophytes): *Allium* spp., *Iris* spp., *Asphodelus ramosus, Narcissus requienii*, etc. These can disseminate by seeds, but do not sucker at all;

3) producing numerous seeds, as *Pinus* spp., *Cistus* spp., *Rosmarinus officinalis*, which can colonize warm dry areas after fire. Their competitive ability is generally weaker than that of the species of the first group. Since they die after fire, these plants behave as annuals;

4) completing the life cyle as an annual.

We can conclude based on the present studies that these species have a variety of strategies to cope with any type of disturbance likely to occur during their life cycles. According to survival traits expressed under a particular fire regime, plants of the French Mediterranean area can be classified into two broad life form categories:

- geophytes: subdivided into true geophytes - without burl sprouts, and false geophytes - with burl sprouts,

- therophytes: subdivided into true therophytes - annual or biennials, and false therophytes - perennials reproducing solely by seed after fire.

Rather than life form categories, however, two other systems may provide better information; that of Gill (1981) which considers reproductive phases and that of Noble and Slatyer (1977) taking vital attributes for recovery into account. These must be considered further in future studies.

REFERENCES

Ashton DH (1981) Fire in tall open forests (wet sclerophyll forests). In: Gill AM, Groves RH, Noble IR (eds) Fire and Australian Biota. Aust Acad Sci, Canberra, 339-366

Aston AR, Gill AM (1976) Coupled soil moisture, heat and water vapour transfers under simulated fire conditions. Aust J Soil Res 14:55-66

Bharucha FR (1931) Etude écologique et phytosociologique de l'association à *Brachypodium ramosum* et *Phlomis lychnitis*. Beihefte Bot Centralb 2:247-379

Boulet C (1985) Bilan floristique d'une garrigue de chene kermes soumise à deux types de perturbations controlées. These 3ème Cycle Ecol Veget, Fac Sci Tech Marseille, p 215

Christensen NL, Muller CH (1975) Effect of fire on factors controlling plant growth in Adenostoma chaparral. Ecol Monogr 45:29-55

Christensen PE, Kimber PC (1975) Effect of prescribed burning on the flora and fauna of southwest Australian forests. Proc Ecol Soc Aust 9:85-106

Christensen PE, Recher H, Hoare J (1981) Responses of open forest to fire regimes. In: Gill AM, Groves RH, Noble IR (eds) Fire and Australian Biota. Aust Acad Sci, Canberra, 367-394

Cremer KW (1962) Effects of fire on seedshed from *Eucalyptus regnans*. Aust For 29:252-262

Daubenmire R (1968) Ecology of fire in grasslands. Advances in Ecological Research, Academic Press, London, vol. 5:209-266

Dobzhansky T (1956) What is an adaptive trait? Amer Nat 90:337-347

Floyd AG (1966) Effect of fire on weed seeds in wet sclerophyll forests of northern New South Wales. Aust J Bot 14:243-256

Gill AM (1975) Fire and the Australian flora: A review. Aust For 38:4-25

Gill AM (1977) Plant traits adaptive to fires in Mediterranean land ecosystems. Proc Symp Environm Consequences Fire and Fuel Manage Medit Ecosyst. USDA For Serv Gen Tech Rep WO-3:17-26

Gill AM (1981) Adaptive responses of Australian plant species to fires. Gill AM, Groves RH, Noble IR (eds) Fire and Australian Biota. Aust Acad Sci, Canberra, 243-272

Gill AM, Ashton DH (1968) The role of bark type in relative tolerance to fire in the central Victorian eucalypts. Aust J Bot 16:491-498

Gill AM, Ingwersen F (1976) Growth of *Xanthorrhoea australis* R. Br. in relation to fire. J Appl Ecol 13:195-203

Hall AV (1959) Observations on the distribution and ecology of Orchidaceae in the Muizenberg Mountains, Cape Peninsula. J South Africa Bot 25:265-278

Hanes TL (1971) Succession after fire in the chaparral of southern California. Ecology 48:259-264

Hare RC (1961) Heat effects on living plants. USDA For Serv Occ Pap 183, Southern Forest Exp Stn, p 32

Keeley JE (1981) Reproductive cycles and fire regimes. Proc Conf Fire Regimes and Ecosyst Properties. USDA For Serv Gen Tech Rep WO-26:231-277

Kruger FJ (1977) Ecology of Cape fynbos in relation to fire. Proc Symp Environm Consequences Fire and Fuel Manage Medit Ecosyst, USDA For Serv Gen Tech Rep WO-3:230-244

McArthur AG (1968) Fire resistance of eucalypts. Proc Ecol Soc Aust 3:83-90

Martin RE (1963) Thermal properties of bark. For Prod J 13:418-426

Mooney HA, Dunn EL (1972) Land use history of California and Chile as related to the structure of the sclerophyll scrub vegetations. Madroño 21:305-319

Mount AB (1969) Eucalypt ecology as related to fire. Proc Annu Tall Timbers Fire Ecol Conf 9:75-108

Muller CJ, Hanawalt RB, McPherson JK (1968) Allelopathic control of herb growth in the fire cycle of California chaparral. Bull Torr Bot Club 95:225-231

Naveh Z (1974) The ecology of fire in Israel. Proc Annu Tall Timbers Fire Ecol Conf 13:131-170

Noble IR, Slatyer RO (1977) Postfire succession of plants in Mediterranean ecosystems. Proc Symp Environm Consequences Fire and Fuel Manage Medit Ecosyst. USDA For Serv Gen Tech Rep WO-3:27-36

Packham DR (1971) Heat transfer above a small ground fire. Aust For Res 5:19-24

Purdie RW (1977) Early stages of regeneration after burning in dry sclerophyll vegetation. 2. Regeneration by seed germination. Aust J Bot 25:35-44

Raunkiaer C (1905) Types biologiques pour la géographie botanique. Kgl Danske Videnskannes Selskab Forhandl 5:347-437

Sampson AW (1944) Plant succession on burned chaparral lands in northern California. Bull 685, Univ Calif Agric Exp Stn, Berkeley, p 104

Soroceanu E (1936) Recherches phytosociologiques sur les pelouses mésoxérophiles de la plaine languedocienne (*Brachypodietum phoenicoidis*). Comm SIGMA 40, p 250

Specht RL, Rayson P, Jackman ME (1958) Dark Island Heath (Ninety Miles Plain, South Australia). 6. Pyric Succession: Changes in composition, coverage, dry weight and mineral nutrient status. Aust J Bot 6:59-88

Stone EC, Juhren G (1953) Fire stimulated germination: Effect of burning on germination of brush seed investigated. Calif Agric 7:13-14

Sweeney JR (1956) Responses of vegetation to fire. A study of the herbaceous vegetation following chaparral fires. Bot Publ Univ Calif 28:143-250

Trabaud L (1970) Quelques valeurs et observations sur la phytodynamique des surfaces incendiées dans le Bas-Languedoc. Natur Monspel 21:231-242

Trabaud L (1979) Etude du comportement du feu dans la garrigue de chene kermes à partir des températures et des vitesses de propagation. Ann Sci Forest 36:13-38

Trabaud L (1980a) Impact biologique et écologique des feux de végétation sur l'organisation, la structure et l'évolution de la végétation des zones de garrigues du Bas-Languedoc. Thèse Doctorat Etat, Univ Sci Tech Languedoc, Montpellier, p 288

Trabaud L (1980b) Influence du feu sur les semences enfouies dans les couches superficielles du sol d'une garrigue de chene kermes. Natur Monspel 39:1-12

Trabaud (1983) Evolution après incendie de la structure de quelques phytocénoses méditerranéennes du Bas-Languedoc (Sud de la France). Ann Sci Forest 40:177-195

Trabaud L, Lepart J (1980) Diversity and stability in garrigue ecosystems after fire. Vegetatio 43:49-57

Trabaud L, Lepart J (1981) Changes in floristic composition of a *Quercus coccifera* garrigue in relation to different fire regimes. Vegetatio 46:105-116

Tutin TG, Heywood VH, Burges NA, Moore DM, Valentine DH, Walters SM, Webb DA (1964-1980) Flora Europaea, Cambridge Univ Press, 5 vol

Vines RG (1968) Heat transfer through bark and the resistance of trees to fire. Aust J Bot 16:499-514

Vogl RJ (1974) Effects of fire on grasslands. In: Kozlowski TT, Ahlgren CE (eds) Fire and Ecosystems. Academic Press, New York, 139-194

Vuillemin J, Bulard C (1981) Ecophysiologie de la germination de *Cistus albidus* et *C. monspeliensis*. Natur Monspel 46:1-11

Wells PV (1969) The relation between mode of reproduction and extent of speciation in woody genera of the Californian chaparral. Evolution 23:264-267

7. The Response to Stress at the Landscape Scale

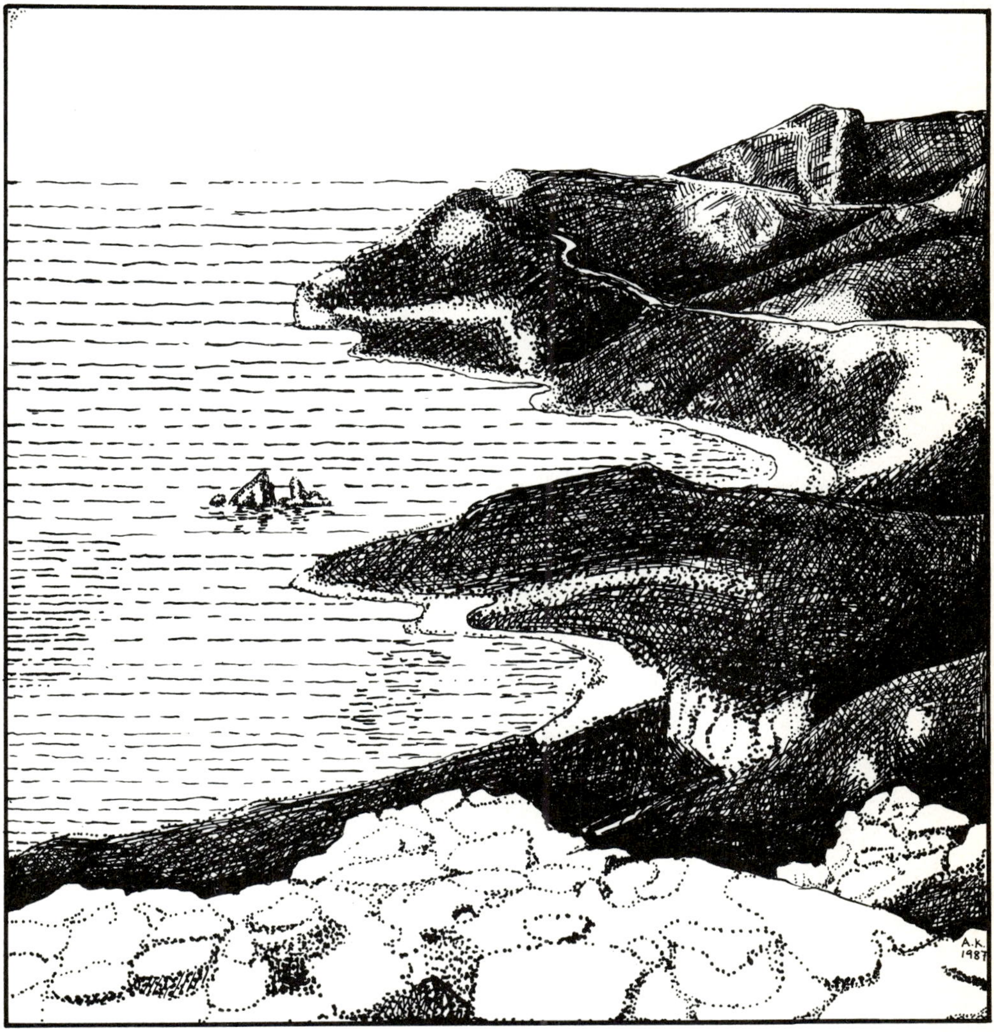

The effect of summer drought on vegetation structure in the mediterranean climate region of Australia

R.L. Specht

Botany Department, University of Queensland, St. Lucia, Queensland, Australia 4067

INTRODUCTION

The most typical mediterranean climate region of Australia, wherein a dry summer alternates with a humid winter, may be delimited by the humid to subhumid Köppen climatic-types Csa and Csb (Fig. 1a, after Dick 1975). Eastward from the South East of South Australia towards Melbourne, Victoria, summer rainfall becomes increasingly probable. Inland, winter rain alternating with a dry summer (BShs and BSks climatic-types) becomes less significant as the climate grades into the arid zone (BSh and BSk climatic-types). Throughout its extent from the perhumid to the semi-arid zone, the mediterranean climate is characterized by a period of summer drought. The spatial distribution and ecomorphological properties of the foliage in the plant community must equilibrate with this seasonal climatic cycle.

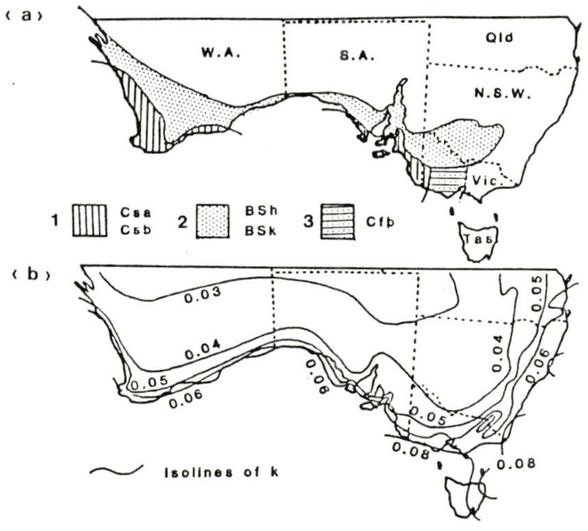

Fig. 1. (a) Distribution of mediterranean climate regions in Australia: (1) Typical humid-subhumid region, (2) Marginal semi-arid region, (3) Marginal humid-subhumid region. (b) A map showing isolines of the evaporative coefficient (k) - see text and Fig. 2

NATO ASI Series, Vol. G15
Plant Response to Stress. Edited by J. D. Tenhunen et al.

VEGETATION OF THE MEDITERRANEAN CLIMATE REGION

In the mediterranean climate region of Australia, the vegetation develops three or four strata (Specht 1972b; Beard 1981). The overstorey composed of an upper stratum, and a sparse mid-stratum, invariably possesses evergreen, sclerophyllous leaves. The foliage of the understorey of dwarf shrubs and ground plants varies markedly from evergreen sclerophyllous on soils low in plant nutrients (especially P and N) to seasonal malacophyllous (soft-leaved), savanna-type species on more fertile soils (Table 1).

Table 1. Vegetation characteristics in the mediteranean climate zone of Australia (Specht 1972b; Beard 1981).

	Soil nutrient level		
Climate	Low	Medium	High
All climates			
Overstorey	Evergreen sclerophyll	Evergreen sclerophyll	Evergreen sclerophyll
Understorey	Evergreen sclerophyll	Seasonal savanna	Seasonal savanna
Perhumid	Tall open-forest	--	--
Humid	Open-forest[a]	Open-forest[b]	--
Subhumid	Woodland[a]	Woodland[b]	Woodland Tussock grassland
Semi-arid	Tall shrubland[a] (mallee)	Tall shrubland (mallee)	Tussock grassland

[a]On soils very low in nutrients, trees may become stunted, multi-stemmed, or absent; the understorey alone persists as a heathland vegetation.

[b]Vegetation on calcareous, high pH soils forms a tall shrubland (mallee) with an understorey containing many sclerophyllous graminoids, and a few sclerophyllous sub-shrubs.

Even in the stunted communities (mediterranean heathlands and sclerophyllous shrublands) of the subhumid to semi-arid zone, three strata can be distinguished (Specht and Rayson 1957; Specht 1966; Specht and Moll 1983):

Upper stratum: Shrubs/small trees with loosely-packed, evergreen leaves.
Mid stratum: Mid-height shrubs with closely-packed, evergreen leaves.
Ground stratum: (a) Low shrubs with closely-packed, evergreen leaves. (b) Sclerophyllous graminoids with varying stem density per clump

ANNUAL WATER BALANCE OF EVERGREEN PLANT COMMUNITIES

For an evergreen plant community to survive in the mediterranean climate, it must have access to water stored at depth in the soil during the wet season to be used throughout the summer drought period, albeit at a low level. This conclusion is based on long-term water balance studies on several native plant communities in southern Australia, heathlands (Specht 1957a and b; Specht and Jones 1971), mallee-broombush (Specht 1966), and eucalypt open-forest/woodland (Martin and Specht 1962).

Monthly estimates of evapotranspiration (Ea) can be expressed relative to either potential evapotranspiration (Ep) or pan evaporation (Eo). The Moisture Index (Ea/Eo) may be plotted against the monthly estimates of available water (W) to produce a curve asymptoting to unity (Thornthwaite 1948; Thornthwaite and Mather 1954; Prescott *et al.* 1952; Butler and Prescott 1955; Specht 1957b, 1972a). It was found that, as the mediterranean climate communities studied by the author rarely reached maximal water supply even during the wettest winter, only the first, essentially linear, part of the asymptoting curve could be determined. For evergreen plant communities, the relative monthly Moisture Index may be expressed by the following equation (Specht 1972a):

$$\text{Moisture Index} = Ea/Eo = \underline{k}\,(P - R - D + Sext) \qquad (1)$$

where Ea = actual evapotranspiration,
Eo = pan evaporation (class A pan),
P = precipitation,
R = runoff,
D = drainage,
Sext = extractable soil water (beginning of month),
W = available water = (P - R - D + Sext),
$\underline{k}$ = evaporative coefficient

The evaporative coefficient $\underline{k}$ remains constant throughout the year for evergreen plant communities.

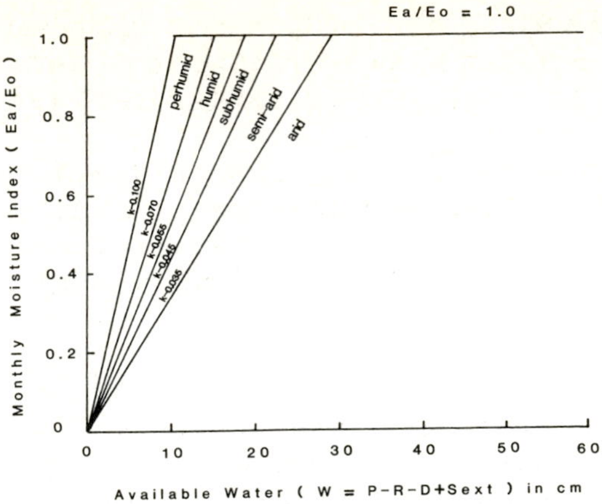

Fig. 2. The relationship between monthly Moisture Index (Evapotranspiration [Ea]/Pan evaporation [Eo]) and available water (W = P - R - D + Sext) for perhumid to arid mediterranean climates (Specht 1972a)

The following spectrum of linear regressions can be related to the following climatic zones (see also Fig. 2):

Perhumid communities	$\underline{k} > 0.070$
Humid communities	$\underline{k} > 0.055$ but < 0.070
Subhumid communities	$\underline{k} > 0.045$ but < 0.055
Semi-arid communities	$\underline{k} > 0.035$ but < 0.045
Arid communities	$\underline{k} < 0.035$

At any locality in this spectrum of mediterranean climates, the monthly values of the Moisture Index fluctuate from a high value in winter-spring to a low value during the mediterranean summer (see observed values for Dark Island heathland, plotted on Fig. 9). In all field experiments, a small amount of water was found to be available to deep-rooted overstorey species, even in the driest summer. The Moisture Index shows a trend to zero, but never equals zero. This observation can be used to estimate the optimal value of the evaporative coefficient ($\underline{k}$), for a particular locality (see Fig. 1b), using iterative computing of Equation (1) (Specht 1972a, 1981a).

It is only during the wettest period of the year in the most humid section of the mediterranean climatic region that available water (W) no longer limits evapotranspiration (Ea), and Ea/Eo approaches 1.0.

Reduction in available water (W) by increased Run-off (R), increased Drainage (D) or reduced Soil Storage (Smax), will lead to a reduced monthly Moisture

Index. Conversely, run-on and seepage from surrounding areas, increased Soil Storage (Smax) or impeded drainage (D) due to hard pan formation will lead to an increased monthly Moisture Index. In the most humid sites in the mediterranean climate, the monthly Moisture Index for every month of the year equals unity; thus the mean annual Moisture Index also equals unity (Fig. 3a). Summer drought is no longer obvious, except in extreme drought years.

FOLIAGE STRUCTURE OF EVERGREEN PLANT COMMUNITIES - OVERSTOREY

When the mean annual value of the Moisture Index equals unity in the perhumid mediterranean zone, the evergreen foliage canopy becomes "closed" with a Foliage Projective Cover of over 70 to 80 percent. In less humid mediterranean climates, a similar canopy structure can be induced in plant communities which receive increased water input from surrounding areas. This "closed" community is then sharply differentiated from the "open" communities of the surrounding dryland landscape (Fig. 3 b).

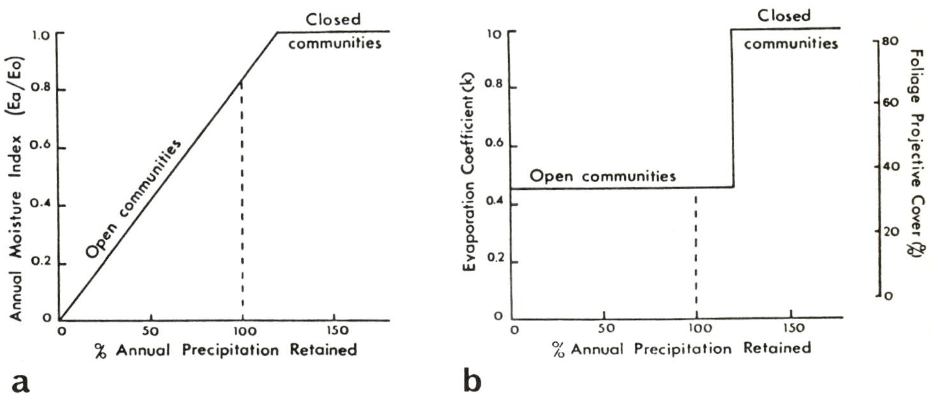

Fig. 3. (a) Mean annual Moisture Index and (b) the evaporative coefficient (k) and Foliage Projective Cover of the overstorey, plotted against the percentage of annual precipitation retained in the soil within the rooting zone (Specht 1981b)

The percentage of the land covered by evergreen overstorey foliage (the Foliage Projective Cover) decreases linearly as the evaporative coefficient (k = Relative evapotranspiration [Ea/Eo]/available water [W]) of the mediterranean climate decreases (Fig. 4). As the mediterranean climate becomes more arid, the overstorey foliage cover decreases, thus reducing the rate of evaporation of water from the ecosystem.

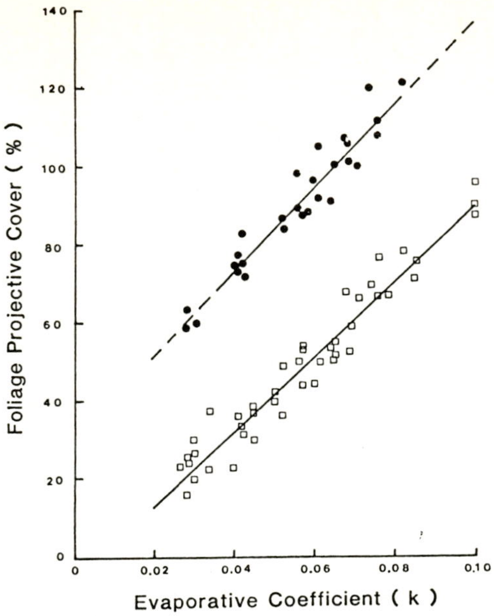

Fig. 4. Foliage Projective Covers (%) recorded for the overstorey (open squares) and overstorey + understorey (•) strata of evergreen plant communities plotted against the evaporative coefficient (k) - after Specht (1983)

Not only does the overstorey foliage cover decrease with increasing aridity, but ecomorphological characters of the leaves show similar gradients (Fig. 5, from Specht and Specht in preparation). As the climate becomes more arid, both mean leaf area (cm^2) and leaf specific area (cm^2 g^{-1}) decrease, providing further resistances to the evaporation of water from the ecosystem.

The plant community appears to develop a leaf canopy in equilibrium with the annual evaporative power of the atmosphere and water supply. This can be viewed in terms of modification of the stable, evergreen Foliage Projective Cover (of overstorey plus understorey) and degree of sclerophylly (Leaf Specific Area), both of which affect canopy resistance to water loss. A deep root system to extract water stored at depth in the soil will ensure that the evergreen foliage will continue to transpire (at a low rate) throughout summer.

More mesic mediterranean climates enable a greater horizontal cover (F.P.C.) of vegetation to develop with a larger evaporative coefficient (k) in Equation (1). The larger evaporative coefficient is associated with very rapid loss of water from the ecosystem at the onset of summer drought. Martin and Specht (1962) posed the question, "Are mesic communities less drought resistant?" While much more

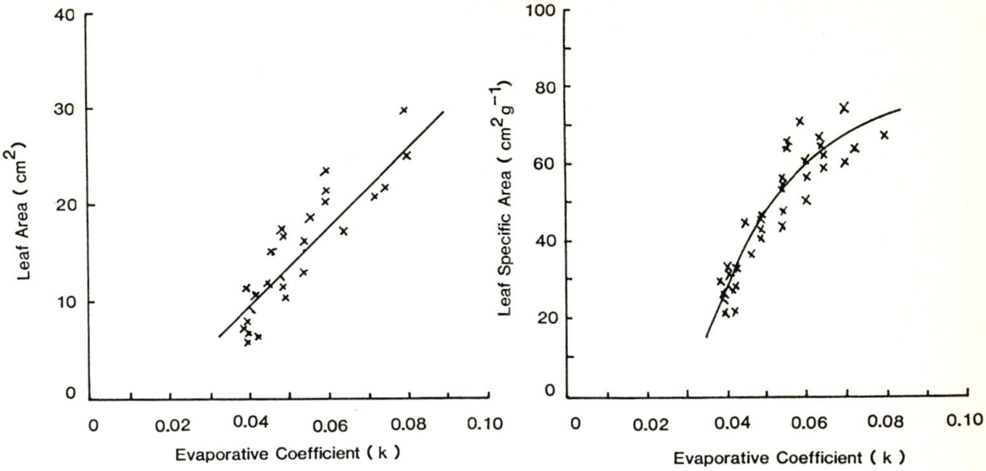

Fig. 5. Mean leaf area (cm^2) and leaf specific area (cm^2 g^{-1} dry weight) of leaves of overstorey species from evergreen plant communities plotted against the evaporative coefficient (k) - after Specht and Specht (in prep.)

information must be gathered, a few studies have been undertaken that will help lay the basis for answering this question. Myers and Neales (1984) have produced information on stomatal control of water loss along the xeric to mesic gradient from mallee to woodland eucalypt communities near Melbourne, Victoria. Sinclair (1980) has investigated the responses of water potential and stomatal conductance to summer drought in the Mount Lofty Ranges, South Australia. The balance between FPC and seasonal changes in leaf water relations in the maintenance of evergreen foliage within the mediterranean climate is a subtle interaction of aerodynamics, community structure and ecophysiology.

Mediterranean climate regions experience considerable variability in rainfall from year to year. Field studies on the water balance of a range of plant communities (Specht 1957b; Martin and Specht 1962; Specht and Jones 1971) indicate that although the monthly Moisture Index (the intensity and length of summer drought) may vary considerably from year to year, the evaporative coefficient (k) of these evergreen communities remains constant. The Foliage Projective Cover and leaf resistance characters remain relatively constant from year to year. This variability in monthly Moisture Index from year to year has a marked effect on the seasonal phenology of the community. Attributes such as seedling establishment of overstorey species (Specht 1981c) and the production of flowers and fruit (A. Specht 1985) are affected (Fig. 6).

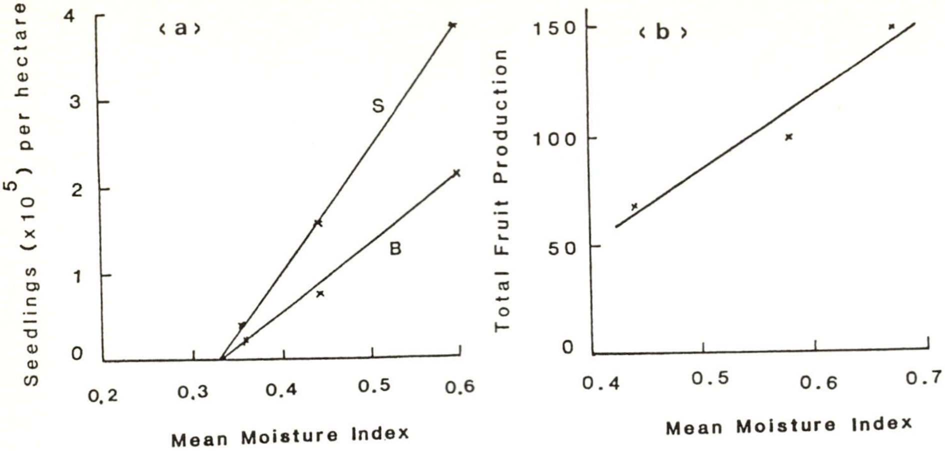

Fig. 6. (a) Seedling survival of *Banksia ornata* (B) and other shrubs (S) with increasing mean monthly Moisture Index during the growing season (Specht 1981c, after Specht *et al.* 1958). (b) Fruit production on tagged shoots of *Eucalyptus intermedia* in an open-forest, plotted against the mean monthly value of the Moisture Index for the season of shoot growth (A. Specht 1985)

FOLIAGE STRUCTURE OF EVERGREEN PLANT COMMUNITIES - UNDERSTOREY

In evergreen plant communities, regenerating after a perturbation such as fire, Total Foliage Projective Cover (the sum of overstorey and understorey F.P.C.) rapidly regenerates to reach a constant value related to the evaporative coefficient (k) of Equation (1). (Fig. 7 from Specht and Morgan 1981). At first, understorey F.P.C. develops rapidly but, with time, the slowly regenerating overstorey F.P.C. replaces the understorey F.P.C. Eventually at maturity, overstorey and understorey F.P.C.s reach constant values related to the evaporative coefficient (k) for the locality. Values of understorey F.P.C., recorded in mature communities, decrease only slightly (47 to 40%) from humid to semi-arid climates (Fig. 4). Overstorey F.P.C. of mature communities decreases from 90 to 20% over the same climatic range. The understorey differs markedly from the overstorey in foliage and rooting attributes, as summarised in Table 2.

Foliage in the understorey is more closely-packed than in the overstorey, resulting in increased temperatures in the boundary layers around the axillary and terminal buds (Fig. 8). This increased temperature leads to earlier shoot and floral growth during the spring season when available water and temperature are optimal (Yates 1981; Specht and Yates in preparation), preceding the summer drought.

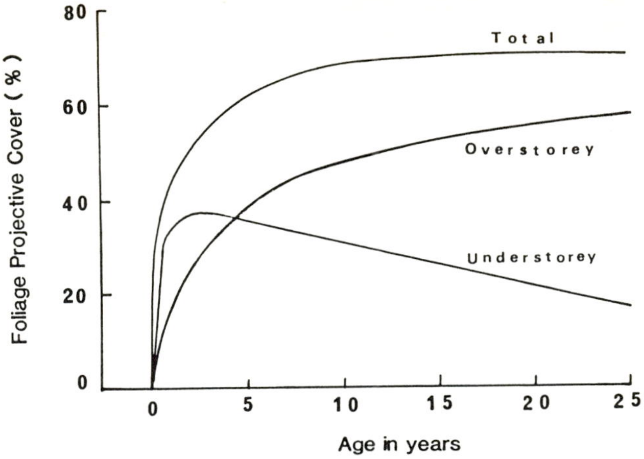

Fig. 7. Changes in Foliage Projective Covers of overstorey, understorey, and (overstorey + understorey) strata in post-fire succession in Dark Island heathland (Specht and Morgan 1981)

Table 2. Shoot and root characteristics of major strata of evergreen plant communities in the mediterranean climate zone of Australia (Specht and Rayson 1957; Specht *et al.* 1958, 1981, 1983).

Stratum	Foliage density	Shoot growth	Rooting depth
Upper	Loosely-packed	Late spring/summer	Deep
Mid	Closely-packed	Spring	Medium
Ground			
Dicot.	Closely-packed	Late winter/spring	Shallow
Monocot.	Graminoid clump	Late winter/spring	Shallow

As rooting depth of the understorey strata is shallow, with reduced available water (W) in Equation (1), summer drought must impose considerable stress on these evergreen species (see Huber transpiration estimates in Specht 1957b). Shallow-rooted ground stratum species must show some degree of desiccation-tolerance (Doley and Trivett 1974; Gaff *et al.* 1976; Mittelheuser 1977) to survive the extreme summer drought as evergreens.

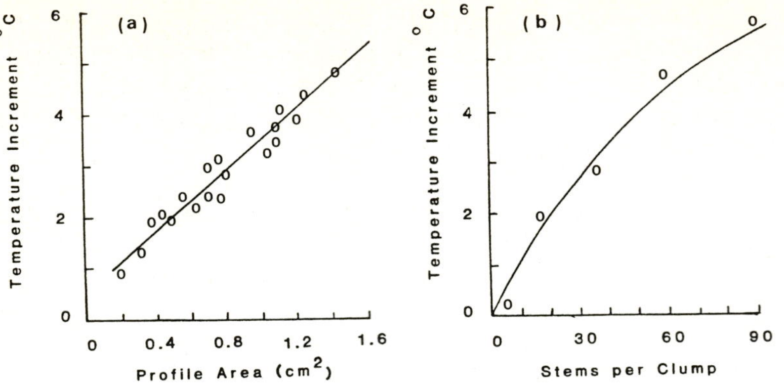

Fig. 8. (a) Increase in axillary bud temperature of dicotyledon shoots with increasing profile area (2 cm lengths of stem, 1 cm radius) intercepting solar radiation in still air. (b) Increase in temperature in the region of floral buds (near the top of stems) in sclerophyllous graminoids with increasing number of rush-like stems per clump, when exposed to solar radiation in still air.

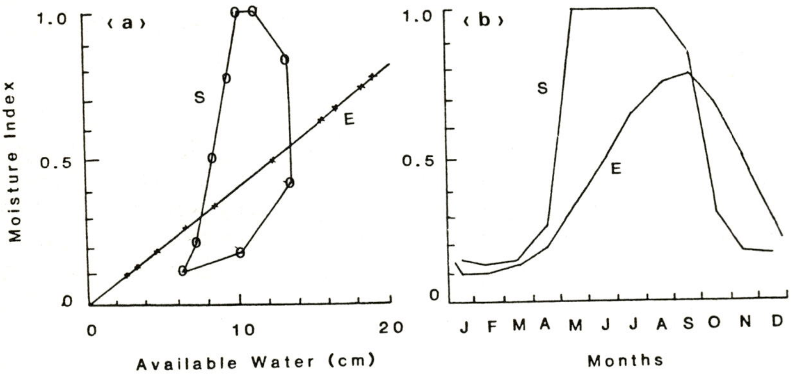

Fig. 9. Monthly values of the Moisture Index recorded (Specht 1957a and b) for an evergreen heathland (E) and seasonal pasture (S) near Keith, South Australia, (a) Plotted against available water (W = P - R - D + Sext),(b) Plotted throughout the year.

FOLIAGE STRUCTURE OF SEASONAL PLANT COMMUNITIES

A seasonal savannah understorey is characteristic of the mediterranean plant communities found on soils of medium to high nutrient status (Table 1). Overstorey trees become sparse or absent on the cracking clays of the Basalt Plains of western Victoria and in the frosted highland 150 km north of Adelaide, South Australia. Over extensive areas, wooded landscapes have been cleared to develop seasonal pastures and crops. The potential growing season of this seasonal vegetation varies from about 5 months in the semi-arid to 9-10 months in the humid section of the mediterranean-climate.

Seasonal communities rapidly utilize all wet-season water which is available within their relatively shallow rooting zone (Specht 1957b). F.P.C. and hence the value of the evaporative coefficient (k) of Equation (1) and Fig. 2, vary throughout the growing season; Moisture Index is higher throughout the wet season (especially spring) enabling more growth during that favourable period (Fig. 9). The onset of summer drought induces death in therophytes, and dormancy in hemicryptophytes and geophytes.

SUMMER DROUGHT: ITS EFFECT ON PHENOLOGY

The distinctive feature of the mediterranean climate is the extremely dry period experienced annually during summer. Annual plant species and seasonal perennials avoid this drought period; their life cycles (from germination/sprouting to flower/seed production and subsequent death) are attuned to the humid growing season.

However, much of the upper stratum of mediterranean vegetation has evergreen, sclerophyllous foliage. On soils of low nutrient status, the understorey species are also evergreen and sclerophyllous. These species must possess various characteristics to enable them to survive the dry summer season. As the mediterranean climate becomes more severe (from the perhumid to humid to subhumid to semi-arid), the length of the summer drought increases from almost zero to over 6 months in duration.

On soils of low-nutrient status, overstorey species show a shoot-growth response at a high temperature (Table 2). Foliage shoot elongation is not initiated until late spring-summer during the period of greatest water deficit. Soil water conserved deep in the soil from the wet winter/spring season enables canopy photosynthesis to continue, albeit at a low level. Reserves of photosynthates stored during the winter-spring period (Specht 1975, 1981b) provide much of the energy for summer shoot growth. Floral apices are initiated during the summer shoot-growth phenophase, but development is delayed until autumn when the break of the wet season occurs enabling canopy photosynthesis to increase.

The ecological advantage of this periodicity in shoot growth of overstorey species during the dry mediterranean summer (Specht and Rayson 1957; Specht 1969) is because: 1) efficient trapping by rootlets of mineral nutrients released from decomposing litter (Specht 1981d) during the winter-spring period is essential to prevent leakage of nutrients from nutrient-poor ecosystems; 2) the increased surface area of specialised rootlets (proteoid, cyperoid, restioid, etc. - Lamont 1981), and the development of biochemical mechanisms such as polyphosphate synthesis (Coleman and Specht 1981), ensure increased efficiency in the conservation of scarce nutrients; and 3) accumulation of nutrients during spring occurs at the same time as storage of photosynthates. Thus when shoot growth is initiated during the increasingly dry period of late spring-summer, a photosynthate and nutrient store is ready to be mobilised (Specht *et al.* 1983).

CONCLUSIONS

The spatial distribution and ecomorphological attributes of the foliage of evergreen plant communities in the mediterranean climate have evolved to achieve an evaporative balance with the environment, expressed by the evaporative coefficient (k = relative evapotranspiration [Ea/Eo]/available water [W]). Excess water is stored in the soil during the humid season, and used, albeit at a low rate, during the dry summer season. In contrast, plant communities with seasonal foliage (grasslands, etc.) tend to utilize all water available during the humid season leaving little for the dry summer season.

Mediterranean plant communities on soils of high-nutrient status are composed of evergreen overstorey and seasonal ground strata. The water relations of these ecosystems lie between those with fully-evergreen and fully-seasonal foliage. Water is rapidly utilized during spring when the high nutrient and water supplies promote growth of the whole plant; shoot growth of both overstorey and ground strata leads to increased Leaf Area Indices, while Foliage Projective Cover remains constant. Available water in the surface soil is rapidly depleted towards the end of spring and the ground stratum dies. As water deep in the soil is depleted during summer, the evergreen overstorey loses older leaves thus reducing the Leaf Area Index of this stratum (Specht *et al.* 1983).

Evergreen plant communities on soils of low nutrient status have evolved both nutrient and water conservation strategies. The overstorey species show sequential seasonal rhythms - starch and nutrient accumulation during winter-spring; shoot growth (plus flower initiation) and almost simultaneous leaf-fall during the dry summer season; flower/fruit development and stem/root growth during autumn (Specht 1975, 1981b; Specht *et al.* 1983). The seasonal rhythms of the relatively short-lived, relatively nutrient-rich, species of mid and ground strata (Specht and Moll 1983) are similar to the evergreen overstorey of species in nutrient-rich communities.

REFERENCES

Beard JS (1981) Vegetation survey of Western Australia, 1:1,000,000 series. Explanatory notes to Sheet 7, "Swan". University of Western Australia Press, Nedlands, p 222 and map

Butler PF, Prescott JA (1955) Transpiration from wheat and pasture in relation to available moisture. Aust J Agric Res 6:52-61

Coleman RG, Specht RL (1981) Mineral nutrition of heathlands: The possible role of polyphosphate in the phosphorus economy of heathland species. In: Specht RL (ed) Ecosystems of the World. Vol. 9B. Heathlands and Related Shrublands. Analytical Studies. Elsevier, Amsterdam, pp 197-207

Dick RS (1975) A map of the climates of Australia: according to Köppen's principles of definition. Qld Geogr J Ser 3,3:33-69 and map

Doley D, Trivett NBA (1974) Effects of low water potentials on transpiration and photosynthesis in Mitchell grass (*Astrebla lappacea*). Aust J Plant Physiol 1:539-550

Gaff DF, Zee SY, O'Brien TP (1976) The fine structure of dehydrated and reviving leaves of *Borya nitida* Labill. - a desiccation-tolerant plant. Aust J Bot 24:225-236

Lamont BB (1981) Specialised roots of non-symbiotic origin in heathlands. In: Specht RL (ed) Ecosystems of the World. Vol. 9B. Heathlands and Related Shrublands. Analytical Studies. Elsevier, Amsterdam, pp 183-195

Martin HA, Specht RL (1962) Are mesic communities less drought resistant? A study on moisture relationships in dry sclerophyll forest at Inglewood, South Australia. Aust J Bot 10:106-118

Mittelheuser CJ (1977) Rapid ultrastructural recovery of water stressed leaf tissue. Z Pflanzenphysiol 82:458-461

Myers BA, Neales TF (1984) Seasonal changes in the water relations of *Eucalyptus behriana* F. Muell. and *E. microcarpa* (Maiden) Maiden in the field. Aust J Bot 32:495-510

Prescott JA, Collins JA, Shirpurkar GR (1952) The comparative climatology of Australia and Argentina. Geogr Rev 42:118-133

Sinclair R (1980) Water potential and stomatal conductance of three *Eucalyptus* species in the Mount Lofty Ranges, South Australia: responses to summer drought. Aust J Bot 28:499-510

Specht A (1985) Temperature effects on eucalypt shoot growth in the Brisbane region. PhD Thesis, University of Queensland

Specht RL (1957a) Dark Island heath (Ninety-Mile Plain, South Australia). 4. Soil moisture patterns produced by rainfall interception and stem-flow. Aust J Bot 5:137-150

Specht RL (1957b) Dark Island heath (Ninety-Mile Plain, South Australia). 5. The water relationships in heath vegetation and pastures on the Makin sand. Aust J Bot 5:151-172

Specht RL (1966) The growth and distribution of mallee-broombush (*Eucalyptus incrassata-Melaleuca uncinata* association) and heath vegetation near Dark Island Soak, Ninety-Mile Plain, South Australia. Aust J Bot 14:361-371

Specht RL (1969) A comparison of the sclerophyllous vegetation characteristic of mediterranean-type climates in France, California and southern Australia. I. Structure, morphology and succession. II. Dry matter, energy and nutrient accumulation. Aust J Bot 17:277-292, 293-308

Specht RL (1972a) Water use by perennial evergreen plant communities in Australia and Papua New Guinea. Aust J Bot 20:273-299

Specht RL (1972b) The vegetation of South Australia, 2nd ed Government Printer, Adelaide, p 328

Specht RL (1975) A heritage inverted: Our flora endangered. Search 6:472-477

Specht RL (1981a) Growth indices - their role in understanding the growth, structure and distribution of Australian vegetation. Oecologia (Berl) 50:347-356

Specht RL (1981b) Ecophysiological principles determining the biogeography of major vegetation formations in Australia. In: Keast A (ed) Ecological Biogeography of Australia. Junk, The Hague, pp 301-333

Specht RL (1981c) Responses to fires in heathlands and related shrublands. In: Gill AM, Groves RH, Noble IR (eds) Fire and the Australian Biota. Australian Academy of Science, Canberra, pp 395-415

Specht RL (1981d) Nutrient release from decomposing leaf litter of *Banksia ornata*, Dark Island heathland, South Australia. Aust J Ecol 6:59-63

Specht RL (1983) Foliage projective covers of overstorey and understorey strata of mature vegetation in Australia. Aust J Ecol 8:433-439

Specht RL, Jones R (1971) A comparison of the water use by heath vegetation at Frankston, Victoria, and Dark Island Soak, South Australia. Aust J Bot 19:311-326

Specht RL, Moll EJ (1983) Mediterranean-type heathlands and sclerophyllous shrublands of the world: An overview. In: Kruger FJ, Mitchell DT, Jarvis JUM (eds) Mediterranean-Type Ecosystems. The Role of Nutrients. Springer-Verlag, Berlin-Heidelberg-New York, pp 41-65

Specht RL, Morgan DG (1981) The balance between the foliage projective covers of overstorey and understorey strata in Australian vegetation. Aust J Ecol 6:193-202

Specht RL, Rayson P (1957) Dark Island heath (Ninety-Mile Plain, South Australia). 1. Definition of the ecosystem. Aust J Bot 5:52-85

Specht RL, Specht A (in prep) Canopy structure in *Eucalyptus*-dominated communities in Australia: Shoot growth, foliage projective cover, leaf area index, and leaf specific area.

Specht RL, Yates DJ (in prep) The structure and location of foliage shoots in Australian heathlands and their effect on the temperature experienced at growth apices

Specht RL, Rayson P, Jackman ME (1958) Dark Island heath (Ninety-Mile Plain, South Australia). 6. Pyric succession: Changes in composition, coverage, dry weight, and mineral nutrient status. Aust J Bot 6:59-88

Specht RL, Rogers RW, Hopkins AJM (1981) Seasonal growth and flowering rhythms: Australian heathlands. In: Specht RL (ed) Ecosystems of the World.

Vol. 9B. Heathlands and Related Shrublands. Analytical Studies. Elsevier, Amsterdam, pp 5-13

Specht RL, Moll EJ, Pressinger F, Sommerville J (1983) Moisture regime and nutrient control of seasonal growth in mediterranean ecosystems. In: Kruger FJ, Mitchell DT, Jarvis JUM (eds) Mediterranean-Type Ecosystems. the Role of Nutrients. Springer-Verlag, Berlin, pp 120-132

Thornthwaite CW (1948) An approach toward a rational classification of climate. Geogr Rev 38:55-94

Thornthwaite CW, Mather JR (1954) Climate in relation to crops. Met Monogr 2(8):1-10

Yates DJ (1981) Optical properties of heathlands. In: Specht RL (ed) Ecosystems of the World. Vol. 9B. Heathlands and Related Shrublands. Analytical Studies. Elsevier, Amsterdam, pp 91-97

Landscape ecology, management and conservation of European and Levant Mediterranean uplands

Z. Naveh

Faculty of Agricultural Engineering, Technion, Israel Institute of Technology, Haifa 32000 Israel

INTRODUCTION

Mountainous and hilly uplands, too steep and rocky, or with soil too poor for profitable cultivation, make up great parts of the total land area in all European and Levant Mediterranean countries. They are covered, in general, by more or less severely depleted communities of sclerophyll trees and shrubs, or dwarf shrubs and grasslands which are of low economic value. Even the most mesic forests and woodlands in these countries are progressively deteriorating, and their biological diversity and scenic beauty are threatened by the combined impacts of traditional and modern agro-pastoral and neo-technological landscape despoilation (Naveh and Dan 1973; Le Houreou 1981; Thirgood 1981).

There is an urgent need for conservation and rehabilitation of these landscapes. The object of this paper is to show how this can be achieved through a holistic landscape-oriented ecological approach supporting integrated and dynamic conservation and multi-purpose management. To this end, I will first outline briefly some of the major paradigms of holistic landscape ecology, and then show how these can be applied in the attempt to save these deteriorating uplands.

LANDSCAPE ECOLOGY AS THE SCIENTIFIC BASIS FOR INTEGRATED LAND MANAGEMENT AND CONSERVATION

Landscape ecology has developed in the last 30 years in Central and Eastern Europe as an interdisciplinary branch of contemporary ecology and geography. It is considered presently in most of these countries to provide the major scientific basis for land use planning, management, conservation and reclamation. Its emergence as a formal discipline can be regarded as a significant attempt by academic and professional ecologists, geographers, planners, agronomists, foresters, conservationists, landscape architects and other land managers to replace the narrow and chiefly reductionistic tendencies in their professions with innovative, integrated and holistic approaches to landscape research, appraisal, planning and management. As is witnessed by the wide range of publications on landscape ecology studies and projects in recent years, some of these presented for the first time in the English language at a seminar of the recently founded International Association for Landscape Ecology (IALE) (Brandt and Agger

NATO ASI Series, Vol. G 15
Plant Response to Stress. Edited by J. D. Tenhunen et al.

1984), landscape ecology has become now a truly international and interdisciplinary science. By viewing the landscape as the total spatial and functional entity of the natural and human living space, and by integrating the geosphere with the biosphere and with man-made artifacts, landscape ecology has transcended the purely natural realm of classical bio-ecological and physical sciences. It has entered the realm of human-centered knowledge concerned with land use and is thereby becoming a human ecosystem science and an efficient tool in understanding and bridging the gap between natural, agricultural and urban systems.

Man, as a biological creature, is dependent for his existence on the viability of natural and agricultural ecosystems. His technoecosystems cannot survive without the biosphere, and their exponential growth may simultaneously endanger the technosphere itself (Vester 1976; Ehrlich and Ehrlich 1983). As explained in detail by Naveh and Lieberman (1984, see Fig. 1), landscape ecology could contribute to this urgently needed integration of the biosphere and technosphere at a higher level of ecosphere organization by providing scientific and educational information for holistic land appraisal and study, planning, management and conservation. Thereby, landscape ecologists would further the urgently needed reconciliation between man and nature at the critical interface of land use. This is a goal which must be achieved as part of our leap from an industrial to post-industrial society (Toffler 1980).

THE EVOLUTION AND DEGRADATION OF MEDITERRANEAN UPLAND LANDSCAPES

Full comprehension of the nature of our present Mediterranean landscapes may be achieved only by understanding their evolution and dynamics as an integral part of the Mediterranean Total Human Ecosystems. This evolution proceeded along with the major phases of cultural transition of primitive man from hunting and food-gathering (*homo erectus*) to intensive food collecting (*homo sapiens*) and finally to food producing (*homo sapiens sapiens*) from the Middle Pleistocene onwards. As a closely coupled dynamic process of physical, geomorphological, biological and cultural transition, it lead first to the creation of semi-natural landscape ecotopes and then, with accelerating speed to semi-agricultural and agro-pastoral ones. This process can therefore, rightly, be called after Di Castri (1981) "the co-evolution between Mediterranean man and landscape."

The "Co-Evolution" of Mediterranean Man and Mediterranean Landscapes

Mediterrean landscapes gained their present high, sharply folded and faulted geomorphological form, with mountains and hills often rising close to the coast, by violent uplifting in the late Tertiary and early Quarternary Periods. Being highly dissected, and complex, with many steep slopes and shallow rocky soils, they are very vulnerable to sheet and gully erosion if their protective natural

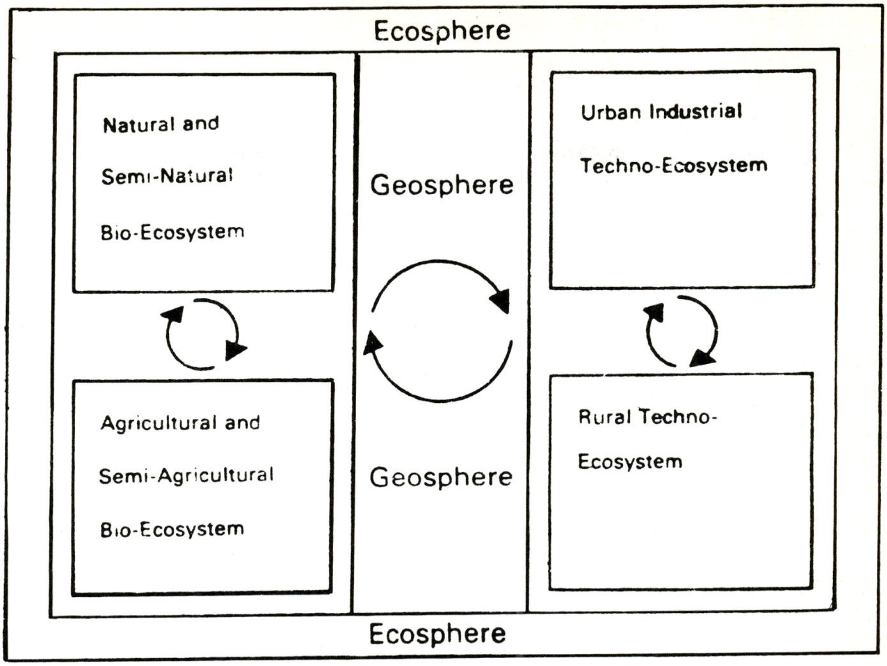

Fig. 1. Conceptual model of the ecological hierarchy of concrete ecosystems with the ecosphere as the largest system and global landscape of the total human ecosystem, integrating ecotopes of the biosphere and technosphere through the geosphere (see also Naveh and Lieberman 1984)

canopy is denuded and their shallow soil mantle is exposed to dessication in the dry summer and to torrential rains in the winter. The Mediterranean geoflora has evolved since Cretaceous times as a vegetation band between temperate and tropical floras and present species were derived from both of these. The sclerophyll woody species exhibited traits well-suited to the new climatic patterns of increasing drought and decreasing winter temperatures that developed during the Pleistocene. Most herbaceous elements, now constituting up to 50% of all species (Pignatti 1978), increased in importance during this period. The final shaping of these landscapes took place during the Quarternary in situations of climatic fluctuations, tectonic movement, and volcanic activity. Raging fires, grazing pressures of heavy ungulates and other herbivores, and human modifications increasingly diversified local site conditions.

Natural fires have continually been caused by lightning and during a period 500,000 years ago by intensive volcanic activity. As reviewed by Perles (1977), archeological findings of charred bones, charcoal, reddened hearths and fire-traced stones in Greece, South France, Spain, Syria, and Israel have shown that fire was also a tool widely used by Acheulian hunter-gatherers in the Mediterranean. It served as the first harnessed energy source and as a major driving force for biological and cultural evolution. Fire was mastered about 100,000 years ago by Mousterian Neanderthaloids who produced lamps to light their caves and torches to carry fire. With fire, they could clear dense forest and brush thickets to facilitate hunting and food collecting. They increased edible food for man and beast by encouraging the lush regenerating trees, invading grasses, bulbs and tuber plants.

From these early periods, there is palynological evidence of the abundance of *Quercus* and *Pistacia* in Israel (Horowitz 1979) and of charcoal from *Quercus* and *Phillyrea* in Southern France (Vernet 1977). We can assume that these and other woody and herbaceous genotypes developed efficient strategies (modification of reproduction and vegetative regeneration mechanisms - see also Hilbert this volume) to cope with natural and man-induced fire stress and that these responses were coupled with effective means of avoiding or tolerating other environmental stress factors (Naveh 1975).

The pre-agricultural intensive utilization of natural plant and animal resources accompanied by landscape modification, reached its peak in the last late Pluvial 10,000 to 15,000 years ago. Of primary influence were hunting, fishing and food collecting upper paleolithic and epipaleolithic cultures, such as the Natufian of Mt. Carmel. Paleontological findings made by Tchernov (1984) have been interpreted as evidence for this fire- and man-maintained equilibrium between natural and semi-natural ecotopes (Naveh 1984). This type of equilibrium was similarly achieved by fishing, hunting, and food collecting tribes of Coastal California Indians (Lewis 1973) and Aborigines in West Australia (Hallam 1979) who utilized fire under comparable environmental conditions.

The Evolution of the Agro-pastoral Mediterranean Landscapes

The final stages of this "co-evolution" of Mediterranean man and Mediterranean landscapes started several thousand years later with the domestication of crops, livestock, and (during the Bronze Age) fruit trees. The man-modified semi-natural landscapes were converted into man-dominated, terraced and intensively used agro-pastoral ones. According to recent palynological findings by Bottema and Van Zeist (1981), the landscape dessication in the early Holocene was caused chiefly by increased human agro-pastoral perturbations. The use of fire for slash and burn cultivation of the lowland forests and for the improvement of upland pastures, combined with grazing, cutting, and coppicing, continued and intensified the pressures resulting due to defoliation and encouraged further the herbaceous and woody colonizers best adapted to these combined natural and cultural perturbations.

Recent palynological findings in Israel (Baruch 1983), and in the Western Mediterranean (Pons 1981), have indicated periods of intensive agricultural land-use by dense populations (especially from the Hellenistic to the Byzantine period), during which the natural woody vegetation receded and areas under cultivation expanded. Especially the importance of olives increased. During periods of instability, warfare, agricultural and population decay, such as the last period of the Ottoman Empire and until World War I, the natural woody vegetation appeared to increase in abundance

In the coastal region of Isteria and after the first neolithic land clearing and cultivation, Beug (1967) found that evergreen sclerophylls extended inland into thinned stands of the more ancient sub-Mediterranean *Quercus pubescens* forest. Pons (1981) concluded in his review of the history of Mediterranean shrublands that these sclerophyll species expanded from small, isolated formations after the Upper Pleistocene due to human activities. He also suggests that *Quercus ilex* forests, generally considered as the Mediterranean climax vegetation, became established as demographic pressures increased, as did *Quercus coccifera*: According to these views, the formations may be considered as resulting from human intervention. The "natural development of climax communities without human interference" is only a myth. We must, however, also be aware of the dangers of generalizing. For example, the effect of fire, goat grazing and woodcutting on Mediterranean uplands has allegedly been to turn all mountain slopes that were once forest covered into rock and scrub deserts. In many countries, many of these shallow and rocky slopes were never cultivated nor terraced and the vegetation has probably not changed since prehistoric times.

Such cyclical changes in ecosystem structure and function, corresponding to different historical phases of land use have been described in detail for Israel by Naveh and Dan (1973) and referred to as multivariate "anthropogenic biofunctions." In these, man as their chief controlling factor has by his actions changed all ecosystem state factors for the worse. A possible exception is found in the case of terracing. Nevertheless, in Israel, the fertile and fine-structured brown rendzina and terra rossa soils have suffered much less from erosion than has been generally assumed. This is true if the vegetation mantle has not been disturbed and the soil altered by cultivation or other means. Due to the great resilience acquired during the above described evolutionary history, this hardy vegetation provides very efficient soil protection. The protection is effective as long as the woody plants can regenerate vegetatively from their extensive rootstock, the herbaceous perennials from their underground bulbs, rhizomes and other regenerative tissues, and the annuals can draw from sufficient seed reserves in and above the soil.

On the other hand, catastrophic soil erosion, flooding and siltation which lead to badlands and swamp formation, occurred due to the neglect of terrace walls in periods of political upheaval, warfare and population decline. This was the case over periods of many centuries during the decline of the Roman Empire and after the downfall of the Byzantine Empire. In the Levant, the general agricultural decline during the last 1,300 years led to the replacement of settled agriculture by pastoral nomadism. According to Taylor (1945), the depth of silt and gravel over Roman and Byzantine bridges indicates that between two and four billion cubic meters of soil were washed off the western side of the central Palestinian hill country, which was mostly terraced. This would have been sufficient to provide 4,000 to 8,000 square kilometers of farming land. Vita Finci (1969) has provided proof of similar erosive changes occurring in other Mediterranean countries since Roman times.

The Dynamic Agro-Pastoral Flow Equilibrium and its Present Distortion

Throughout the long history of agro-pastoral utilization, the semi-natural and pastoral ecotopes of open forests, shrublands, woodlands, and grasslands, and the agricultural ecotopes of terrace patches and hand-cultivated "rock polycultures" became closely interwoven to create a rich culturally-related landscape mosaic. Traditional grazing, burning and coppicing led to the establishment of a dynamic flow equilibrium between herbaceous and woody components and established their prevalance in upland ecotopes which were not exposed to extreme agro-pastoral pressures for prolonged periods. This has contributed much to the biological diversity and attractiveness of open Mediterranean landscapes and has become a most important asset for encouraging recreation and tourism (Naveh and Dan 1973; Di Castri 1981).

This dynamic flow equilibrium is now increasingly disrupted by the combined impacts of intensified agro-pastoral pressures, neo-technological landscape degradation, despoilation and pollution, urban-industrial sprawl and uncontrolled mass recreation and tourism (Ambio 1977). Not only the intensification of these human perturbations, but also their complete cessation in some areas, either by abandonment as a result of depopulation or by intentional protection in nature reserves, has disrupted the flow equilibrium by reducing ecological diversity, vitality, stability and attractiveness. In undisturbed and protected maquis, the vigorous vegetative regeneration of sclerophylls from stunted shoots and rootstocks, is followed by the gradual closure of the shrub canopy and almost total suppression of the herbaceous understory. The dense brush thicket seems gradually to become stagnant and possibly even senescent and is very prone to fire because of the accumulation of dry and dead branches and undecomposed litter. In Israel, such undisturbed, single-layered, tall shrub communities are dominated almost exclusively by *Quercus calliprinos* accompanied by a few co-dominant shrubs and subordinate dwarfshrubs. Very few, shade-tolerant perennial grasses and geophytes occur near open rock outcrops and shrub borders. Compared to disturbed, semi-open and multi-layered shrub communities, the floral diversity of undisturbed stands is 4 to 6 times lower. The same is true for species richness and abundance of birds, rodents, reptiles and insects (Naveh and Whittaker 1979).

Similar situations have also been described in the Western Mediterranean. Adriatic *Quercus ilex* forests are losing their open appearance and their floristic and faunistic richness due to cessation of traditional silvo-pastoral utilization (Horvat *et al.* 1974). According to Ilijanic and Hecimovic (1981), the prevention of all human disturbance since 1948 on the island of Lokum, near Dubrovnic, is causing the complete loss of the richest and most attractive lower "degradation" stages in vegetation development which were maintained under traditional agro-pastoral activities. Similar tendencies toward decreased structural and floristic diversity and increased fire vulnerability, resulting from the abandonment of *Q. ilex* and *Q. coccifera* shrublands and mixed *Pinus halepensis* forests are found in Southern France.

Where the traditional agro-pastoral utilization of forests and shrublands is still continued, the situation is very different. According to Susmel *et al.*, the 2,000-year-old *Quercus ilex* forest covering about 25,000 ha between Orgosolo and Oliena in Sardinia, is maintaining its energetic and metabolic balance. It is reproducing itself despite traditional pastoral utilization by pigs, goats, sheep and cattle. A balance has been established between the archaic pastoralism (probably with little change since Roman times) and the forest and its wildlife. This balance is preferable, both from ecological and sociological points of view, to strict conservation as might be inspired by the climax dogma which is still promoted by many prominent Mediterranean ecologists (Tomaselli 1977). Protectionist policies will not only reduce ecological and cultural diversity but increase fire hazards and endanger the existence of this unique Total Human Ecosystem.

With the decrease in population of rural areas and the abandonment of fields and pastures, followed by re-encroachment of the woody vegetation and accumulation of dry and dead wood, large amounts of fuel are present and there are few barriers to the spread of fires. In addition, recreation and tourism have increased spectacularly and resulted in a manyfold increase in the number of wild fires. The destruction of the highly flammable natural and planted pine forests has been devastating. Traditionally frequent fires have been replaced in the Mediterranean by ineffective attempts at complete fire suppression. Without introducing more rational and integrated policies of active fire and fuel management, including prescribed burning as practiced in California and Australia, these wildfires will become more and more frequent and catastrophic. Because of accelerated biological and scenic impoverishment, the impact of present-day vegetation management schemes in the Mediterranean is more and more substantial (Naveh 1982a). If these trends continue unhampered, only a few open, unspoiled landscapes will be retained in small, protected upland areas in the future. Surrounded by ugly man-made deserts of dereliction and despoilation, these will be very soon turn into overcrowded "outdoor recreation slums." This is already the case in many areas.

DIRECT AND INDIRECT BENEFITS FROM MEDITERRANEAN UPLANDS AND THEIR UTILIZATION

Direct Economic Benefits from Forestry and Pasture

The equilibrium in vegetated regions of Mediterranean uplands is further influenced by forest and pasture management policies. Problems of forest and pasture management and production and their impacts on Mediterranean uplands have been reviewed extensively by Le Houerou (1981), Thirgood (1981), and Naveh and Lieberman (1984). These will be summarized only briefly here. The most important direct benefits from these uplands - as well as the major cause for their degradation - is the conversion of their primary productivity into animal products and into firewood by uncontrolled grazing, browsing, and fuel-wood cutting. The common low productivity of these uplands has not only resulted from poor animal husbandry, poor grazing management, and poor forestry practices but stems from severe inherent edaphic and climatic limitations on phytomass production. The maximum annual wood increment per hectare from pine forests on these uplands is often less than 1.0 cubic meter and on the rockier and drier slopes even less than 0.5 cubic meter. Even these benefits are reduced by the expenses of prevention and fighting of wildfires. On the other hand, in Greece, Liacos (1977) has shown that with rational silvo-pastoral management, upland pine forest productivity can be greatly improved.

Coniferous trees and especially *Pinus halepensis* and *Pinus brutia* play an important role in afforestation projects because of their hardiness, rapid establishment and relatively rapid growth. In many Mediterranean countries such upland pine afforestations in severely impacted areas have saved thousands of hectares from further depletion, provided labor for rural populations and dramatically improved these denuded and barren mountain landscapes. In many cases they opened the way for the vegetative regeneration of the sclerophyll understory. This aspect could be utilized with proper forest management practices to achieve a semi-natural, mixed and multi-layered multiple purpose forest (see below). The early rapid development of such hardy pioneer trees has created the mistaken impression that foresters can fool nature in the Mediterranean and establish dense, stable, and productive pine forests like those growing under much more favorable conditions in Central Europe. On the contrary, even present low productivity levels are severely threatened due to air pollutants and pest infestations. Recent field and laboratory studies (Naveh *et al.* 1981) revealed that widespread chlorotic mottle is being caused in Israel (as in California) by atmospheric ozone concentrations above 0.05 ppm in all pine species. It is followed in *P. halepensis* by severe infestations by scales and mites and on ozone stressed *P. canariensis* trees in the coastal region by bark beetles. Similar damage can be expected in all European Mediterranean countries, wherever the sea breeze carries photochemical oxidants from urbanized coastal regions to the forested slopes. This will increase, without doubt, each year.

Morandini (1977) pointed out the importance of sclerophyll forests and shrublands for local firewood supply and for commerical fiber production by means of mechanized woodchopping, which could pay for the thinning and opening of dense shrubland thickets into recreational forests and woodlands. This has been carried out successfully in the "Goren Recreation Forest" in Israel. Here, by converting a dense, inpenetrable and highly flammable maquis shrub thicket into an open oak park-forest, both the carrying capacity for controlled goat grazing and the species richness of valuable pasture plants and of flowering geophytes was trebled and a much higher recreation amenity was achieved (Josephi and Naveh unpublished data). Morandini (1977) stated rightly, that much more profitable wood and fiber production could be achieved in the Mediterranean zone if intensive planting of high yielding timber and fiber trees, like *Populus* and *Eucalyptus* species, occurred in lowland areas with fertile and deeper soils or on gentle slopes of abandoned but arable agricultural land rather than on the low-potential and marginal uplands. Unfortunately, this is not yet recognized by many foresters in the Mediterranean, who are encouraged by the fact that their forestry departments are oriented towards crop production and not towards nature conservation, environmental protection and social welfare.

Since goats are well adapted to the rugged terrain and can digest of thorny leaves and stems of sclerophylls, production of goat's milk, meat, and hides have become a major source of income in many Mediterranean countries, especially in Greece, Turkey and the Levant. These range goats, which are able to survive and produce under conditions no longer suitable for cattle and sheep, have been blamed as the main cause of land ruin in the Mediterranean. In fact, they are only the last phase of the vicious cycle of indiscriminate burning, cutting, grazing, slope denudation and cultivation (French 1970). There is little hope of preventing further ecosystem degradation without strictly controlling grazing, and this can only be achieved effectively in fenced paddocks. Public and communal ownership of lands and allowing of unrestricted grazing are major constraints. If any increase in livestock and pasture production is to be expected, it must be coupled with an increase in the socio-economic status of the pastoralist and elimination of ignorance and indifference, through a coordinated effort for public education at all levels. This must be an integral part of multiple benefit landscape conservation and rehabilitation master-plans.

Through maintenance of open woodlands and grasslands, a manifold increase in pasture and livestock production could be obtained. Nevertheless, this will not decrease the dependence of pasture production on seasonal and annual climatic fluctuations but probably will even increase it (Naveh 1982b). The major climatic bottleneck in Mediterranean livestock production can only be overcome by replacement of low value woody plant species with more palatable and productive, summergreen drought tolerant shrubs and trees which provide nutritious fodder in the fall and winter and in drought years. These should be as efficient as the indigeneous shrubs and trees or moreso in tapping of water and nutrients from the shallow soil and underlaying rock layers. But they should have much higher levels of efficiency in channelling solar energy into economically useful products. Among the Mediterranean sclerophylls, the Carob tree, *Ceratonia siliqua*, is of highest fodder value, but several exotic trees and shrubs studied in long-term introduction trials in Northern Israel show great promise. Their establishment on depleted scrubland by the same techniques used for pine afforestation could lead to a manyfold increase in productivity, especially if combined with the improvement of livestock and if carried out as part of a master-plan for upland reclamation and afforestation (Naveh and Lieberman 1984).

Indirect Benefits and "Soft Values"

The greatest indirect socio-ecological and economic benefit derived from these uplands stems from their scenic value and their extraordinary wealth of plant and animal life, which are enjoyed not only by the local populations but by millions of tourists. The open woodlands, shrublands and grasslands are of special importance as a refuge for a wide variety of plant and animal species. Their high floristic diversity include not only highly attractive, ornamental geophytes and valuable pasture plants, but also progenitors of many domesticated cereals and a wealth of hardy and drought resistant herbaceous and woody plants which could serve as genetic stock for future food, fodder, industrial, pharmaceutical, cosmetic, and ornamental products. Their loss, not only by overuse of resources but also by complete protection or by planting of dense pine or *Eucalyptus* forests, would be final and irrevocable. In addition, they are "living sponges" which protect the watersheds and aid in the control of floods, erosion and environmental pollution, especially near densely populated coastal regions and intensively cultivated areas. In California, Westman (1977) has shown that these aspects of ecosystem function are worth millions of dollars.

It is obvious from above discussion, that most of these plant resources can be maintained and preserved by imposing controlled defoliation pressures in the form of grazing and browsing by livestock and/or wildlife, by periodic coppicing and in certain cases also by controlled burning. In this way, economic benefits may be derived while ensuring over the long term the integrity of this vegetation mosaic and accompanying intangible assets (Naveh and Lieberman 1984).

HOLISTIC MULTIPLE BENEFIT LAND USE

From our discussion, it is obvious that neither approaches which consider only livestock production and forestry production, nor those which espouse nature conservation and recreation amenity, can provide a satisfactory means for sustaining and improving the productivity and attractivness of the open Mediterranean landscape and at the same time provide for the needs of the people and national economies. To accomplish these purposes, an entirely new holistic approach is required in which the benefits derived from the landscape are optimized and understood in terms of comprehensive management master-plans. In these, both economic and non-economic benefits should be expressed as variables to which relative values are attached and which are functionally affected by decisions made. The practical goals of such master-plans should be the implementation and dynamic maintenance of an equilibrium in multiple benefit land use. In choosing a master-plan solution, the following relevant variables should be considered:

1) The bio-ecological domain related to those physical, chemical and biological processes which ensure ecosystem function and structure and thereby highest attainable productivity, diversity and stability.

2) The socio-ecological domain related to environmental quality, socio-hygienic, psychological and cultural requirements which ensure the greatest possible overall benefits for the largest possible human populations from a national and regional point of view.

3) The socio-economic domain related to the direct economic benefits to be derived from these uplands as pastures, forests and recreation parks under rational management.

In nature reserves the main aim should be the conservation of biological productivity, diversity and stability. In nature parks and recreation areas it should be the optimal maintenance of landscape, wildlife and recreation amenities to enable maximum enjoyment with minimum damage to natural and cultural resources. In the remaining open uplands used primarily for socio-economic benefits, two major "ecotechniques" should be applied:

1) Vegetation and ecosystem management of the existing plant cover and controlled manipulation of the soil-plant-animal complex by controlled grazing, thinning, coppicing and pruning of trees, and if feasible, controlled fire and pastoral agro-technical improvements, such as reseeding, fertilizing, selective weed control, according to site potentials and socio-economic and other requirements and conditions.

2) Multipurpose afforestation and reforestation. This is essential in degraded batha dwarfshrub and shrublands and wherever denudation has reached an advanced stage, as well as in man-created habitats on roadsides, in camping grounds and at recreation sites. It should include also the establishment of fire protection strips and buffer zones for urban and industrial complexes. In contrast to the single-layered, monotonous and vulnerable conifer plantations, the aim is the creation of a new type of semi-natural, multi-layered and stable park-forest with multiple ecological, economical and social benefits. These new forests should resemble in structure, diversity and stability the richer multi-layered park-like and semi-open Mediterranean forests and woodlands with a scattered tree-overstory, like the above-mentioned Goren Recreation

Forest. But at the same time, they should be of greater economical and ornamental value and of greater fire resistance due to the replacement of inferior local plant species with more appropriate introduced shrubs and trees.

In order to ensure the greatest possible overall benefit from what are often conflicting land use demands, flexible multi-purpose strategies should be applied according to site potentials and local and national requirements. This is realized in practice by the creation of multiple land use patterns. The newly proposed Multiple-Use Management Areas or Managed Resource Areas (IUCN, 1978) are especially designed for implementing lasting and far-reaching changes in Mediterranean upland use. They will be especially effective if combined with demonstration, research and training projects. Important guidelines for the systems approach and methodology to be applied have been outlined in the MAB projects for the transformation and development of mountain regions in Switzerland (Messerli and Brugger 1984) and in Austria (Holling 1978).

For decision making and consideration of options and management strategies, multidimensional, computerized cybernetic models can be used in which the mutual influences of all relevant production, protection, and regulation variables as well as their compatibility can be assessed and quantified. For our studies in the Galilee in Israel, such models were developed (Naveh 1978; Naveh and Lieberman 1984), in which two major block variables have been used:

1) Production functions, acting as positive amplifying feedbacks. These include (a) the channeling of biological productivity and natural renewable resources into economic products that are transferable into market goods and therefore expressable in monetary terms as plant and livestock products and water yields; (b) production functions that can be achieved by "optimization of natural landscape values" of purely bio-ecological nature (e.g. biotic diversity or wildlife) or which may constitute a complex and closely interwoven mixture of ecological, geomorphological, and human-perceptual processes of scenic diversity, attractiveness, and accessibility, lumped together as "recreation amenity."

2) Regulation and protection functions, acting as negative feedbacks and constraints, such as fire hazard resistance, and all other vital ecological processes, and for the conservation of their "living sponge" qualities (Ricklefs *et al.* 1984), expressed as environmental watershed protection.

In Fig. 2, a flow chart of the different management options and their multiple-use benefits is presented for converting degraded maquis (A1), shrubland (A2), and grasslands (A3), either into nature reserves (B), or pine plantations (C), or into multiple benefit natural recreation forests (D1), multi-purpose woodlands (D2) and pastures (D3), or into the new type of planted, semi-natural recreation forest (E1), multi-purpose parkland (E2), and fodder tree-shrub-grass pasture (E3).

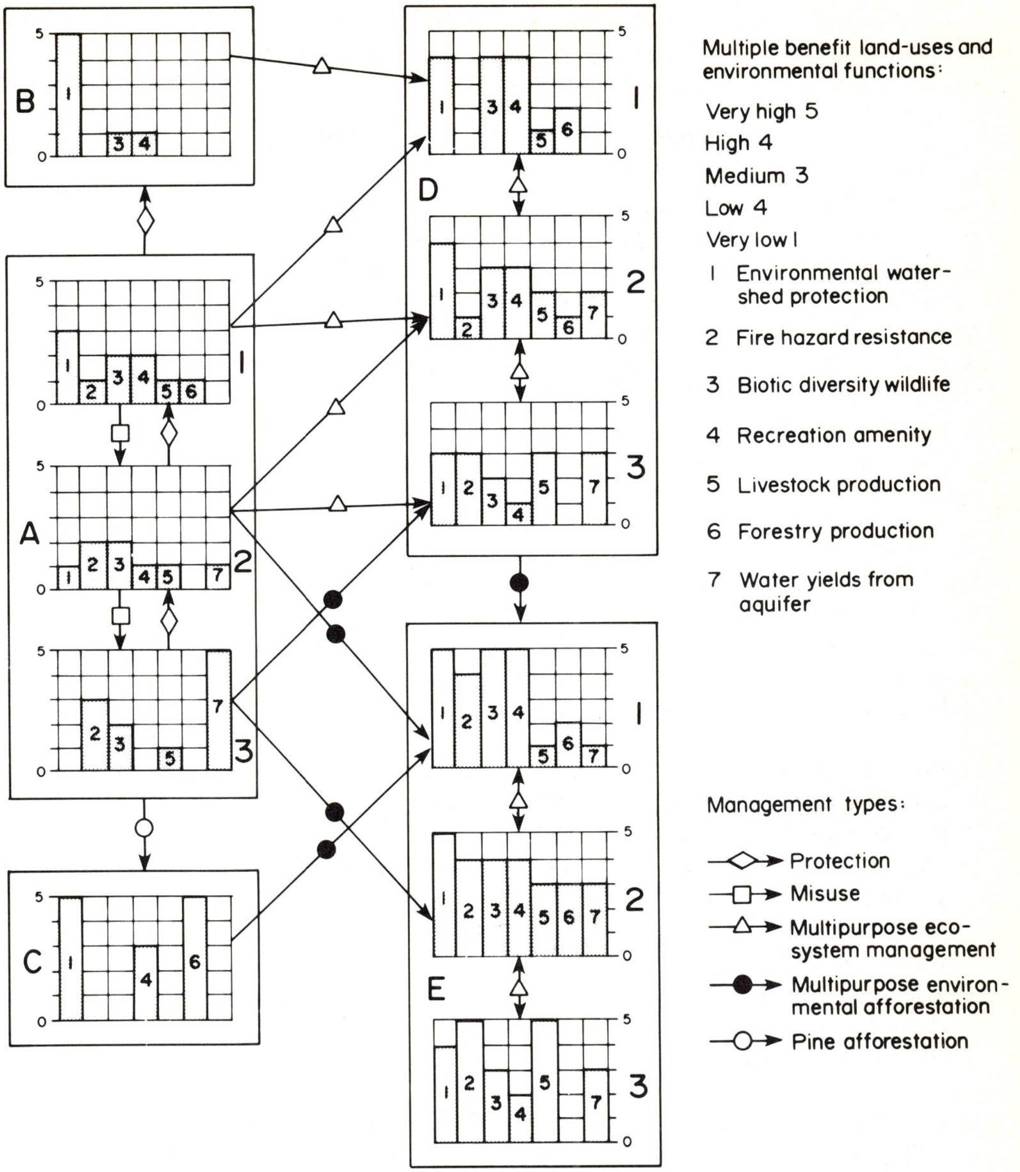

Fig. 2. Upland management options and their relative multiple-use benefits (Naveh and Lieberman 1984). For explanation, see text

From the preliminary balance sheet of multiple benefits (indicated as a relative scaling from very low (0) to very high (5) for each function), it became apparent that highest overall multiple use benefits can be expected from semi-natural, multi-layered park-forests.

CONCLUSIONS

From our discussion, the need for a holistic policy for the conservation and restoration of biological diversity and productivity of Mediterranean open upland landscapes in European countries as a whole emerges. Its major goal should be the prevention of accelerated and uncontrolled replacement and degradation of natural and semi-natural bioecotopes by technoecotopes. This policy should be based on well-coordinated efforts for better public and professional conservation education, more comprehensive landscape conservation planning and management, and more interdisciplinary, integrated ecosystem and conservation research.

The effort to effect such a policy should be guided by system-oriented landscape ecological determinism aimed at reconciling the conflicting demands in resource utilization within these landscapes. Long-term and overall benefits must be optimized. Potential benefits include species refuges; potential as environmental buffering systems; socioecological functions and cultural functions related to scenic beauty, solitude, and wilderness; preservation of unique natural and historical features for future generations, endangered even in protected areas by mass recreation and tourism pressures; socioeconomic functions as sources of food and revenues from diverse plant and animal products such a fodder, fiber, wool, milk, meat, honey, medical, balsam, spice and other plant material; and tourism which can be ensured only by rational management on a long-term sustained basis.

There is an urgent need for interdisciplinary ecosystem research on the ecological and economic implications of different kinds of landscape management models. Foresters should play an important role in these evaluations, applying their rich silvicultural experience in the context of these new approaches. An effort should be made to introduce and establish fodder and forest trees and shrubs of low flamability and to develop fuel management schemes, including prescribed burning.

Ultimately, however, the success of this holistic landscape ecological policy will depend to a great extent on creating awareness for the need, understanding for the sacrifice required, and motivation to achieve Mediterranean landscape conservation among all those who care for, those who live from and those who deal with these landscapes at all levels of decision making. Above all, it is vital to bridge the communication gap and educate a new breed of interdisciplinary landscape managers and replacing the present one-track-minded agronomists, foresters, noninterference-climax protagonists, graziers, recreationalists, and economists. This can be achieved only with the help of more efficient educational tools and high-level ecological, educational, and communication expertise. Of great importance in this respect are interdisciplinary workshops and intensive training programs using environmental simulation games (Vester 1976) and adaptive assessment methods (Holling 1978), which could help to create a common pool of concepts and approaches. Another important venture could be the issuing of special "Redbooks of Threatened Mediterranean Landscapes." They should provide updated information and serve as a monitoring and prediction tool regarding the fate of open landscapes and as a basis for conservation measures and decision making in each country. They should not only be a scientific document but also an important tool for public conservation education to be prepared in an appealing and convincing way by documenting and illustrating the interdependence of the welfare of the people and that of their upland landscapes.

ACKNOWLEDGEMENT

This study was supported by the Funds for Promotion of Research at the Technion, Israel Institute of Technology.

REFERENCES

Ambio (1977) The Mediterrean (Special Issue) 6

Baruch V (1983) The palynology of Late Holocene cores from Lake Kinnereth. Hebrew University, Jerusalem, Dept Archeology (Hebrew)

Beug HJ (1967) Probleme der Vegetationsgeschichte in Südeuropa. Ber Dtsch Bot Ges 80:682-689

Bottema S, van Zeist W (1981) Palynological history of the Near East. In: Prehistoire du Levant, pp 111-123. CNRS, Paris

Brandt J, Agger P (eds) (1984) Methodology in Landscape Ecological Research and Planning, Proc First Intern Sem. The International Association of Landscape Ecology (IALE) Roskilde University Center, Denmark

Di Castri F (1981) Mediterranean-type shrublands of the world. In: Di Castri F, Goodall FW, Specht R (eds) Ecosystems of the World. II. Mediterranean Type Shrublands. Elsevier Scientific Publishing Company, Amsterdam, Oxford, New York, p 1

Ehrlich P, Ehrlich A (1983) Extinction The causes and consequences of the disappearance of species. Ballantine Books, New York

French MH (1970) Observations on the goat. FAO Agr Studies No. 80, Rome

Hallam SJ (1979) Fire and hearth. A study of Aboriginal usage and European usurpation in South-Western Australia. Australia Institute of Aboriginal Studies, Canberra

Holling CS (ed.) (1978) Adaptive Environmental Assessment and Management. John Wiley & Sons, Chichester, New York, Brisbane, Toronto

Horvat I, Glavac V, Ellenberg H (1974) Vegetation Südosteuropas. Gustav Fischer Verlag, Stuttgart

Horowitz A (1979) The Quaternary of Israel. Academic Press, New York

Ilijanic LJ, Hecimovic S (1981) Zur Sukzession der Mediterranean Vegetation auf der Insel Lokrum by Dubrovnik. Vegetatio 46:75-81

IUCN (1978) Categories, objectives and criteria for protected areas. IUCN, Gland, Switzerland

Le Houerou HN (1981) Impacts of man and his animals on Mediterranean vegetation. In: Di Castri F, Goodall W, Specht RL (eds) Ecosystems of the World. II. Mediterranean Type Shrublands. Elsevier Scientific Publishing Company, Amsterdam, Oxford, New York, p 4

Lewis HT (1973) Pattern of Indian burning in California: Ecology and ethnohistory. Ballena Press Anthropolgical Papers. I. Ramona, California

Morandini E (1977) Problems of conservation, management and regeneration of Mediterranean forests: research priorities. In: MAB Techn Notes 2, UNESCO, Paris, p 73

Liacos LA (1977) Fire management in pine forest and evergreen brushland ecosystems in Greece. In: Symposium on the environmental consequences of fire and fuel management in mediterranean ecosystems, Palo Alto, California, USDA FS Gen Tech Rep W-3, Washington, DC, p 289

Messerli P, Brugger EA (1984) Theoretical guidelines for the analysis of problems in mountain areas. In: Brugger EA, Furrer G, Messerli B and Messerli P (eds) The Transformation of Swiss Mountain Regions. Verlag Paul Haupt, Bern and Stuttgart, p 43

Naveh Z (1975) The evolutionary significance of fire in the Mediterranean region. Vegetatio 9:199-306

Naveh Z (1978) A model of multi-purpose ecosystem management for degraded Mediterranean uplands. Environ Mgt 2:31-37

Naveh Z (1982a) Mediterranean landscape evolution and degradation as multivariate biofunctions; theoretical and practical implications. Landscape Planning 9:125-146

Naveh Z (1982b) The dependence of the productivity of a semi-arid Mediterranean hill pasture ecosystem on climatic fluctuations. Agriculture & Environment 7:47-61

Naveh Z (1984) The vegetation of the Carmel and Nahal Sefunim and the evolution of the culture landscape. In: A Ronen The Sefunim Cave on Mt. Carmel and its Archeological Findings. B.A.R. Oxford, p 23

Naveh Z, Dan J (1973) The human degradation of Mediterranean landscapes in Israel. In: Di Castri F, Mooney H (eds) Mediterranean Type Ecosystems, Origin and Structure. Springer Verlag, Heidelberg, Berlin, New York, p 370

Naveh Z, Whittaker RH (1979) Structural and floristic diversity of shrublands and woodlands in Northern Israel and other Mediterranean areas. Vegetatio 41:171-190

Naveh Z, Steinberger E, Chaim S (1981) Photochemical air pollutants - a new threat to Mediterranean conifer forests and upland ecosystems. Environ Conserv 7:301-309

Naveh Z, Lieberman AS (1984) Landscape ecology - theory and applications. Springer Verlag, New York, Heidelberg, Berlin

Perles C (1977) Prehistoire du feu. Masson, Paris

Pignatti S (1978) Evolutionary trends in Mediterranean flora and vegetation. Vegetatio 37:175-185

Pons A (1981) The history of the Mediterranean shrublands. In: Di Castri F, Goodall FW, Specht RL (eds) Ecosystems of the World. II. Mediterranean Type Shrublands. Elsevier Scientific Publishing Company, Amsterdam, Oxford, New York, p 131

Ricklefs RE, Naveh Z, Turner RE (1984) Conservation of ecological processes. Commission on Ecology Paper 8, IUCN, Gland, Switzerland

Susmel L, Viola WF, Bassato G (1976) Ecologia della lecceta del Supramente di Orgosolo. Annali del Centro di Economia Montana dello Venezia. Padova, Vol 20 1969/70

Taylor F (1946) The destruction of the soil in Palestine. Bull Soil Conserv Bd Palest 2

Tchernov E (1984) The fauna of Sefunim Cave, Mt Carmel. In: Ronen A The Sefunim Prehistoric Sites, Mt Carmel. B.A.R., Oxford

Thirgood JV (1981) Man and the Mediterranean Forest. Academic Press, London, New York, Toronto, Sydney, San Francisco

Toffler A (1980) The third wave, William Collins Sons, London

Tomaselli R (1977) Degradation of the Mediterranean maquis. In: Mediterranean Forests and Maquis: Ecology, Conservation and Management. MAB Tech Note 2. UNESCO, Paris, p 33

Vernet JL (1973) Etude sur l´histoire de la vegattion du sudest de la France au Quaternaire, d´apres les charbons de bois principalement. Paleobiol Continent 4:1-90

Vester F (1976) Urban systems in crisis. Understanding and planning of human living space; The biocybernetic approach. Deutscher Verlag, Stuttgart

Vita-Finzi C (1969) The Mediterranean valleys. Geological changes in historical times. Cambridge Press, Cambridge

Westman WE (1977) How much are nature's services worth? Science 197:960-964

Zonneveld IS (1972) Textbook of photo-interpretation. Vol 77, ITC, Enschede

8. Future Directions

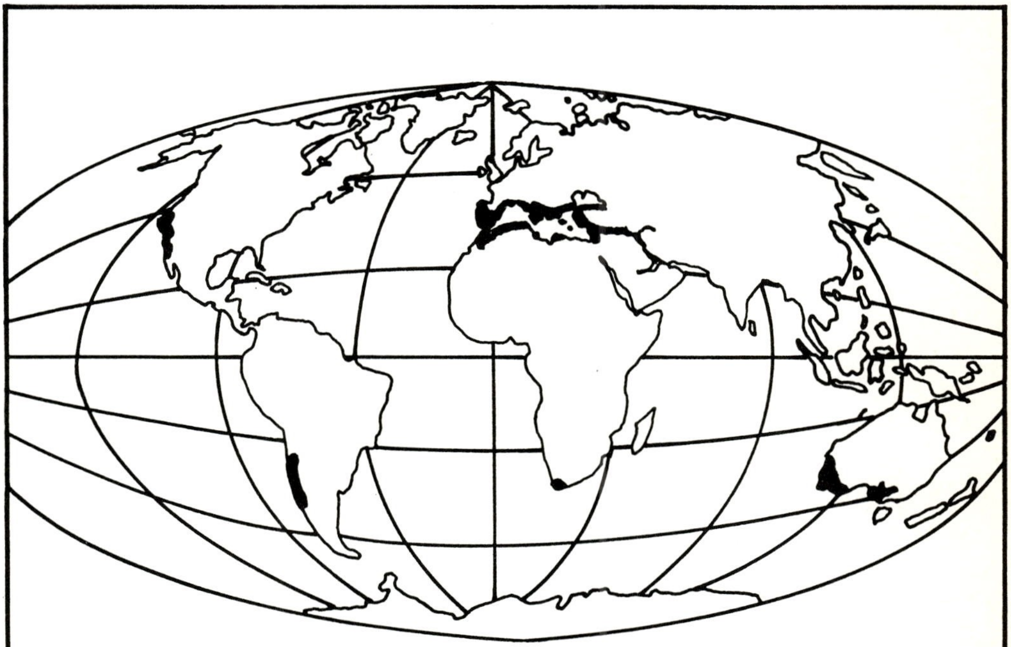

The impact of environmental stress on plant performance in mediterranean-climate ecosystems: Differing levels of analysis

H.A. Mooney

Department of Biological Sciences, Stanford University, Stanford, CA 94305-2493 USA

INTRODUCTION

Study of the effects of naturally-occurring environmental stress on the physiology of mediterranean-climate plants has a long history. Some contemporary approaches, such as assessing responses of photosynthesis of sclerophylls to the seasonal drought period, go back over half a century to the studies of Guttenberg and Buhr (1935), who found reductions in photosynthetic capacity and midday depression of rates. Many recent advances in our understanding of the mechanisms by which mediterranean-climate plants tolerate seasonal drought stress are summarized in this volume.

Progress has resulted from a variety of studies that have examined function at different levels of organization and integration, extending from the leaf or individual organ, to the whole plant, and even to entire ecosystems and landscapes. Here I have tried to present an historical perspective on these recent advances, pointing out certain events which have made this progress possible. I also call attention to those areas, where given our present state of knowledge, more research effort might be particularly rewarding. Since a majority of the studies reported in this volume have investigated problems at the leaf level of organization, i.e. with a fine resolution, I have more extensively critiqued our present understanding in this realm. To balance the emphasis placed in this volume on detailed processes, I propose an approach to considering the effects of stress on plants which unifies information from different levels of integration.

LEAF LEVEL APPROACHES

Enormous progress has been made at the leaf level in understanding the physiological ecology of plants, particularly in understanding controls on gas exchange. This progress has involved the work of scientists,who to a great extent, have centered their work in mediterranean or neighboring desert ecosystems. It is instructive to look at the key elements of this success, in order to identify what has been responsible for it. Both technological and conceptual advances have been important as might be expected, but in addition, an unusually

NATO ASI Series, Vol. G15
Plant Response to Stress. Edited by J. D. Tenhunen et al.

stimulating research climate has prevailed in a number of laboratories in different countries in the last two decades that significantly promoted these advances.

Photosynthesis Research

The technological advances in gas exchange research relate to the development of portable infrared gas analyzers and controlled-climate cuvettes. These advances made it possible to experiment on plants growing under natural conditions with the same precision as utilized in the laboratory. The history of these developments is given in Field and Mooney (1987) and was discussed also by Lange *et al.* (this volume). Conceptual advances relate to the development and linking of photosynthesis, transpiration, and microclimate models which enabled predictions of responses of plants under a wide range of environmental conditions utilizing basic physiological data measured for a given species (reviewed by Mooney 1984; see also Meister *et al.* and Tenhunen *et al.* regarding the integrated response of *Quercus coccifera* in this volume). These advancements made physiological ecology a predictive science.

An exciting and important research collaboration has developed in a number of countries, further fostering rapid progress in the study of stress impacts on plants. This collaboration has involved biochemists, physiologists, and ecologists in centers in Australia (Canberra--I. Cowan, G. Farquhar, B. Osmond, and R. Slatyer), the United States (Stanford--J. Berry, O. Björkman, C. Field, and H. Mooney) and Germany (Würzburg--U. Heber, W. Kaiser, O. Lange, J. Tenhunen, and U. Schreiber). These groups have studied the mechanistic basis of stress response or more generally the adaptive features of plants. All have concentrated on gas exchange with a particular emphasis on photosynthesis. All have worked on mediterranean or desert climate plants. In the quest for satisfying explanations, they have brought a holistic approach to the study of photosynthesis.

The results of these many efforts have permitted us to progress beyond simple measurements of metabolic responses of plants in the field to analysis of the mechanistic bases of plant response. In the past few years, methods have been developed and utilized to distinguish between stomatal versus biochemical limitations on photosynthetic capacity under field conditions (as so elegantly shown by the work of the Würzburg group in this volume). More recently, exciting new fluorescence techniques have been used to determine limitations to photochemical processes under field conditions as described in this book by Schreiber and Bilger. These studies have begun to provide information that help unravel the interactive effects of multiple stress factors (e.g. drought and temperature stress) on photoinhibition. Progress to date in studying the effects of naturally-induced environmental stress on processes other than photosynthesis has not been as spectacular.

Water Relations

The introduction of the pressure bomb (Scholander *et al.* 1964) and the development of simple techniques for looking at water potential components in the field (Ritchie and Hinckley 1975) have led to widespread investigations under natural conditions of such phenomena as osmotic adjustment. However, the significance of various components of water potential as determinants of plant metabolism and physiological regulation has yet to be elaborated (Sinclair and Ludlow 1986). Even the function of turgor in relation to growth is still a matter of some debate. It is thus exciting to see the introduction of promising new approaches for directly relating plant water status to metabolism as described by Kaiser in this volume.

Nutrient Relations

In general, the relationship between nutrient stress and leaf level metabolism has not received much attention in physiological plant ecology. However, recent conceptual advances (Field 1983; Bloom *et al.*1985) may stimulate research in this area. The evergreen sclerophylls of mediterrranean-climates offer particularly interesting material for studies on nutrient stress relationships because of their capacity for nutrient storage in leaves (see also Kruger in this volume).

General Conclusions

The progress that has been made in studying the influence of natural environmental stress on leaf level phenomena such as photosynthesis is quite obvious and similar progress may be possible with respect to other processes. There are, however, some intrinsic problems with leaf-level studies of plant stress relations and of utilizing sclerophylls:

1) Sclerophylls are generally very poor for biochemical studies because of their high concentrations of protein-precipitating tannins; hence the basis for responses measured at the whole leaf physiological level often cannot be verified at the biochemical level.

2) Undue attention has at times been paid to leaf level processes in studies that were intended to examine the overall carbon or water economy of plants. As has been shown a number of times, the effects of dissimilar leaf-level metabolic responses may be minimal in the context of the annual balances for a whole plant, e.g. C_3 versus C_4 (Caldwell *et al.* 1977).

3) The emphasis in leaf level experiments has been on response to single stress factors whereas plants are generally subjected to multiple stresses simultaneously under natural conditions, e.g. high radiation plus drought stress.

4) Rapid progress in studying the effects of environmental stress on leaf level metabolism has been made by compiling a limited number of case histories, i.e. by limiting complexities due to variety in genetic material and adaptive potentials, and by carrying out studies at a limited number of research sites - - e.g. *Nerium oleander* and *Heteromeles arbutifolia* at Stanford; *Arbutus unedo* in Portugal and Germany. This limit on comparative possibilities becomes of even more concern, when one considers that it is extremely difficult to make physiological comparisons between studies separated in time (see Lawrence in this volume) due to continual changes in technology and in the principles guiding experimental design.

5) Finally, leaf-level studies in isolation do not provide strong clues concerning the adaptiveness of a given process. Studies begun at the leaf level rarely extend to demonstrating the impacts of stress on whole-plant reproductive output and success.

WHOLE PLANT APPROACHES

There has been a pronounced lack of studies on stress effects at the whole plant level, especially in the case of mediterranean-climate sclerophyllous shrubs. This is understandable in view of the size and complexity of most of these plants and due to the difficulty of studying their root systems (see Virginia and Jarrell in this volume). Yet we must undertake such studies, since the interpretation of certain leaf level phenomena depends upon knowledge of whole plant responses. Whole plant response data is required if we are to progress in understanding:

1) The adaptive significance of various traits, including metabolic pathways, which can only be assessed in terms of reproductive success.

2) The patterns in seasonal development and phenology of plants (see Montenegro in this volume) which depend on changing resource availability to different organs over time.

3) The basis of community phenomena such as competition or convergent evolution (see also Specht in this volume).

It is difficult to define the degree of stress to which a plant may be subjected. Physical measures of stress such as soil moisture content or soil temperature can only be relative to the species which is being studied as well as the season of the study because of acclimation. Physiological measures (such as plant water potential) may also be difficult to interpret since different tissues on the same plant may have very different physiological states (e.g. young versus mature tissue). I would like to propose a way of examining whole plant response to stress which would link naturally with other organizational levels (finer and coarser) and which clarify whether a plant is tolerating or avoiding stress. In the nonlethal range, stress "consists of phenomena which restrict photosynthetic production" (Grime 1979). One can use the plant itself as the stress indicator, since it is integrating the effects of various stressors and of variable intensities of single stresses.

Field, Roy, and Mooney (unpublished) found when comparing the physiology of watered and unwatered *Eriodictyon* that there was remarkably little difference in plant water potentials, conductances, and photosynthetic capacities. Yet there were large differences in whole plant carbon gain between treatments. This result was obtained due to differences in the leaf areas maintained, in leaf angles and in absorptances. The unwatered plants maintained less leaf area and held their leaves at a steeper angle (absorbing less radiation) exposing a greater fraction of the lower and more reflective leaf surfaces. The unwatered plants thus continued to maintain favorable water relations by manipulation of the whole plant water and energy balance. The stressed plants may have obtained their stress signal to modify leaf charactistics from the roots as described by Davies *et al.* in this volume, since there was little change in plant water potential with increasing drought duration. The point is that the impact of stress could be seen as a reduction of whole plant production although there was little reduction in the carbon gain of individual leaves. Measures of physiological properties of individual leaves, in this case, gave very little indication that the plants were actually experiencing drought stress.

By looking at whole plant production responses one can thus assess the cost of stress adaptation whether the mechanisms which are utilized are "behavioral" or physiological. Stress avoidance (angled leaves, leaf loss, etc.) incurs a cost in terms of loss of potential production, just as does so-called stress adaptation, such as osmotic adjustment or photosynthetic performance at low water potentials.

Another example may be used to demonstrate how information obtained at the leaf level is not always indicative of whole plant processes. As noted above, phenological patterns of plants are often explicable in terms of uneven resource availability to the plant. This uneven resource supply may also be involved in controlling such leaf features as sclerophylly, which is generally taken as an indication of plant stress tolerance. It is somewhat of an anomally that some of the most sclerophytic leaves actually experience less water stress than co-occurring species with less sclerophytic leaves (Davis and Mooney 1986). An example of this is found among various species of *Quercus* , cf. Tenhunen *et al.* in this volume. These deep-rooted plants may exhibit their highest photosynthetic capacity during the summer period when they have water but no nutrients. They produce carbon rich but protein-poor leaves. The differential seasonal supply rate of resources may be very important in explaining both plant morphology and plant function.

In plant physiological ecology, considerable emphasis has been placed on gas exchange, almost to the total exclusion of other important processes. As an example, the movement of water, minerals and assimilates is influenced by internal and external factors. The flux rates and directionality in movement of these substances have important consequences for plant growth and production. The patterns in redistribution of these essential components depend on resource availability and uptake and on plant as well as atmospheric sources and sinks. In addition, they may be dependent simply on the plumbing system of any particular plant species. One need only read the recent paper of Carlquist (1985) on vasicentric tracheids to sense the great importance of plumbing on whole plant water balance. Relationships between plant anatomy and whole plant function have unfortunately not received much attention in recent years.

Physiological experiments done on whole woody plants can be enormously revealing yet are rarely done for obvious reasons. This generally means that predictions of whole plant response are made via modeling of individual plant part responses. There has been little direct experimental validation of these modeling predictions.

ECOSYSTEM APPROACH

Stress effects can also be observed at the level of the community and ecosystem. Study at this particular level is of great interest at the moment because of the realization that interactions between vegetation and the atmosphere are being altered by human-induced perturbations. Some of these atmospheric changes represent stresses for ecosystems which will result in reduced system production --others are actually potential enhancers of production such as the increase in atmospheric CO_2 resulting from burning of fossil fuels.

How do we evaluate these changes when the background year-to-year variation in atmospheric and climatic properties is so high? Ecosystem models have been constructed based on physiological processes of plants and their response to environmental conditions. These models predict system primary productivity, usually with some variance due to randomly generated fluctuations in climate. These models can simulate potential changes that result from a significant shift in atmospheric inputs. Ecosystem scientists working in mediterranean-climate regions have been heavily involved in the development of these sorts of models. The San Diego research group has utilized their production model extensively to summarize information and generate new hypotheses (see Miller 1981) and the South African fynbos group is developing similar models.

Because of the urgent needs for information on altered ecosystem productivity due to human disturbances, an intense argument has developed as to whether models which are built on detailed processes such as individual leaf metabolism, root uptake patterns, etc., and operating over very short time intervals, can reliably predict long-term production trends. Intermediate level models which include very simple physiological responses but also incorporate population characteristics are being utilized most extensively for predictions of the effects of air composition changes (Shugart and McLaughlin 1985). Just as in the case of the whole plant approach, we must devise meaningful system-level experiments that may be used to validate model predictions of stress effects on production.

LANDSCAPE APPROACHES

Naveh in this volume discussed extensively the human impacts on mediterranean landscapes. The effects of stress on landscapes can also be evaluated in production terms. It is widely accepted that mediterranean landscapes have lost considerable productive capacity through land use patterns which have resulted in severe erosion. Ozone and SO_2 are known to reduce production and atmospheric pollution with these gases is common in certain mediterranean-climate regions. There has been a massive restructuring of the biota of mediterranean landscapes, both by the purposeful and inadvertent introduction of foreign species. We know little of the immediate, much less long-term, impacts of these changes on the productive capacity of mediterranean ecosystems. However, these impacts can and should be determined. Mediterranean ecosystems are particularly well suited for evaluating human effects on system properties, since five distinct mediterranean-climate regions offer in a sense matched sites with dissimilar land-use histories. Such comparisons were made to a certain extent during IBP and we learned a great deal. Many more opportunities exist and this resource as well as the experience gained in previous comparative studies may now be drawn on to address new landscape and global questions.

REFERENCES

Bloom A, Chapin FS III, Mooney HA (1985) Resource limitation in plants--an economic analogy. Ann Rev Ecol and Syst 16:363-392

Caldwell M, White R, Moor R, Camp L (1977) Carbon balance, productivity, and water use of cold-winter desert shrub communities dominated by C_3 and C_4 species. Oecologia (Berl) 29:275-300

Davis SD, Mooney HA (1986) Water use patterns of four co-occurring chaparral shrubs. Oecologia (Berl) 70:172-177

Field C (1983) Allocating leaf nitrogen for maximization of carbon gain: leaf age as a control on the allocation program. Oecologia (Berl) 56:341-347

Field C and Mooney HA (1987) Measuring photosynthesis under field conditions--past and present approaches. In: Kramer PJ, Strain BR, Funada S, Hashimoto Y (eds), Scientific Instruments in Physiological Plant Ecology. Academic Press, NY

Guttenberg HV, Buhr H (1935) Studien über die Assimilation und Atmung mediterraner Macchiapflanzen während der Regen- und Trockenzeit. Planta 24:276-287

Grime JP (1979) Plant Strategies and Vegetation Processes. John Wiley and Sons, Chichester, 222 p

Miller PC (ed) (1981) Resource Use by Chaparral and Matorral: A Comparison of Vegetation Function in Two Mediterranean-type Ecosystems. Springer, Berlin-Heidelberg-New York, 455 p

Mooney HA (1984) Progress and promise in plant physiological ecology. In: Cooley HH, Golley FB (eds), Trends in Ecological Research for the 1980's. Plenum Publishing Corp, New York and London, pp 5-17

Ritchie GA, Hinckley TM (1975) The pressure chamber as an instrument for ecological research. Adv Ecolog Res 9:166-254

Scholander PF, Hammel HT, Hemmingsen EA, Bradstreet E (1964) Hydrostatic pressure and osmotic potential in leaves of mangroves and other plants. Proc US Nat Acad Sci 52:119-125

Shugart HH, McLaughlin SB Jr (1985) Modeling SO_2 effects on forest growth and community dynamics. In: Winner WE, Mooney HA, Goldstein RA (eds), Sulfur Dioxide and Vegetation. Stanford University Press, Stanford, pp 478-491

NATO ASI Series G

Vol. 1: **Numerical Taxonomy.** Edited by J. Felsenstein. 644 pages. 1983.

Vol. 2: **Immunotoxicology.** Edited by P.W. Mullen. 161 pages. 1984.

Vol. 3: **In Vitro Effects of Mineral Dusts.** Edited by E.G. Beck and J. Bignon. 548 pages. 1985.

Vol. 4: **Environmental Impact Assessment, Technology Assessment, and Risk Analysis.** Edited by V.T. Covello, J.L. Mumpower, P.J.M. Stallen, and V.R.R. Uppuluri. 1068 pages. 1985.

Vol. 5: **Genetic Differentiation and Dispersal in Plants.** Edited by P. Jacquard, G. Heim, and J. Antonovics. 452 pages. 1985.

Vol. 6: **Chemistry of Multiphase Atmospheric Systems.** Edited by W. Jaeschke. 773 pages. 1986.

Vol. 7: **The Role of Freshwater Outflow in Coastal Marine Ecosystems.** Edited by S. Skreslet. 453 pages. 1986.

Vol. 8: **Stratospheric Ozone Reduction, Solar Ultraviolet Radiation and Plant Life.** Edited by R.C. Worrest and M.M. Caldwell. 374 pages. 1986.

Vol. 9: **Strategies and Advanced Techniques for Marine Pollution Studies: Mediterranean Sea.** Edited by C.S. Giam and H.J.-M. Dou. 475 pages. 1986.

Vol. 10: **Urban Runoff Pollution.** Edited by H.C. Torno, J. Marsalek, and M. Desbordes. 893 pages. 1986.

Vol. 11: **Pest Control: Operations and Systems Analysis in Fruit Fly Management.** Edited by M. Mangel, J.R. Carey, and R.E. Plant. 465 pages. 1986.

Vol. 12: **Mediterranean Marine Avifauna: Population Studies and Conservation.** Edited by MEDMARAVIS and X. Monbailliu. 535 pages. 1986.

Vol. 13: **Taxonomy of Porifera from the N.E. Atlantic and Mediterranean Sea.** Edited by J. Vacelet and N. Boury-Esnault. 332 pages. 1987.

Vol. 14: **Developments in Numerical Ecology.** Edited by P. Legendre and L. Legendre. 585 pages. 1987.

Vol. 15: **Plant Response to Stress. Functional Analysis in Mediterranean Ecosystems.** Edited by J.D. Tenhunen, F.M. Catarino, O.L. Lange, and W.C. Oechel. 668 pages. 1987.